Student Solutions Manual for

APPLIED CALCULUS FOR THE LIFE AND SOCIAL SCIENCES

Larson

Houghton Mifflin Harcourt Publishing Company

Boston New York

Publisher: Richard Stratton
Senior Sponsoring Editor: Cathy Cantin
Senior Marketing Manager: Jennifer Jones
Development Editor: Peter Galuardi
Associate Editor: Jeannine Lawless
Editorial Assistant: Joanna Carter-O'Connell
New Title Project Manager: Susan Peltier

Printed in the U.S.A.

ISBN 13: 978-0-618-96342-3
ISBN 10: 0-618-96342-1

23456789-CRS-12 11 10 09 08

PREFACE

This *Student Solutions Manual* is designed as a supplement to *Applied Calculus for the Life and Social Sciences*, by Ron Larson. All references to chapters, theorems, and exercises relate to the main text. Solutions to every odd-numbered exercise in the text are given with all essential algebraic steps included. Although this supplement is not a substitute for good study habits, it can be valuable when incorporated into a well-planned course of study.

We have made every effort to see that the solutions are correct. However, we would appreciate hearing about any errors or other suggestions for improvement. Good luck with your study of calculus.

Ron Larson
Larson Texts, Inc.

CONTENTS

CHAPTER 0
A Precalculus Review

CHAPTER 0
A Precalculus Review

Section 0.1 The Real Number Line and Order

1. Because $0.25 = \frac{1}{4}$, it is rational.

3. Because π is irrational, $\dfrac{3\pi}{2}$ is irrational.

5. Because $4.3\overline{451}$ has a repeating decimal expansion, it is rational.

7. Because $\sqrt[3]{64} = 4$, it is rational.

9. Because 60 is not the cube of a rational number, $\sqrt[3]{60}$ is irrational.

11. $5x - 12 > 0$

$5x > 12$

$x > \frac{12}{5}$

(a) Yes, if $x = 3$, then $x = 3 = \frac{15}{5}$ is greater than $\frac{12}{5}$.

(b) No, if $x = -3$, then $x = -3 = -\frac{15}{5}$ is not greater than $\frac{12}{5}$.

(c) Yes, if $x = \frac{5}{2}$, then $x = \frac{5}{2} = \frac{25}{10}$ is greater than $\frac{12}{5} = \frac{24}{10}$.

13. $0 < \dfrac{x-2}{4} < 2$

$0 < x - 2 < 8$

$2 < x < 10$

(a) Yes, if $x = 4$, then $2 < x < 10$.

(b) No, if $x = 10$, then x is not less than 10.

(c) No, if $x = 0$, then x is not greater than 2.

15. $x - 5 \geq 7$

$x - 5 + 5 \geq 7 + 5$

$x \geq 12$

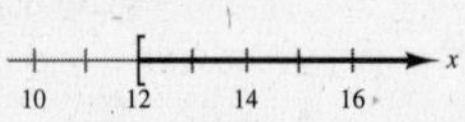

17. $4x + 1 < 2x$

$4x + 1 - 2x - 1 < 2x - 2x - 1$

$2x < -1$

$\frac{1}{2}(2x) < \frac{1}{2}(-1)$

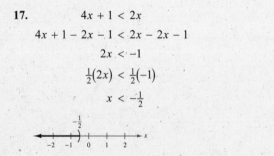

$x < -\frac{1}{2}$

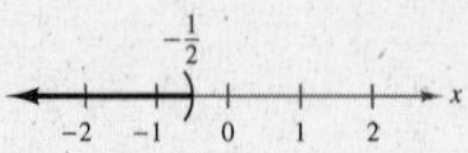

19. $4 - 2x < 3x - 1$

$4 - 2x - 4 - 3x < 3x - 1 - 4 - 3x$

$-5x < -5$

$-\frac{1}{5}(-5x) > -\frac{1}{5}(-5)$

$x > 1$

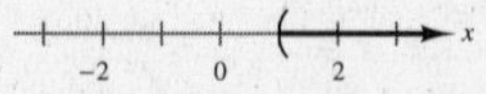

21. $-4 < 2x - 3 < 4$

$-4 + 3 < 2x - 3 + 3 < 4 + 3$

$-1 < 2x < 7$

$\dfrac{-1}{2} < \dfrac{2x}{2} < \dfrac{7}{2}$

$-\dfrac{1}{2} < x < \dfrac{7}{2}$

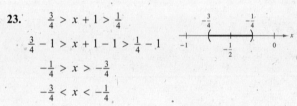

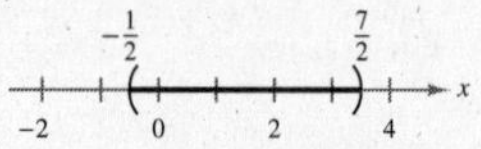

23. $\frac{3}{4} > x + 1 > \frac{1}{4}$

$\frac{3}{4} - 1 > x + 1 - 1 > \frac{1}{4} - 1$

$-\frac{1}{4} > x > -\frac{3}{4}$

$-\frac{3}{4} < x < -\frac{1}{4}$

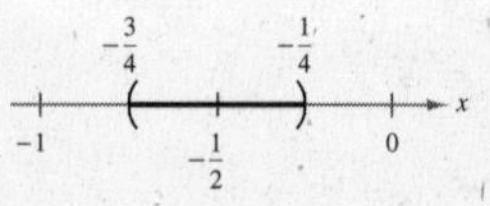

25. $\dfrac{x}{2} + \dfrac{x}{3} > 5$

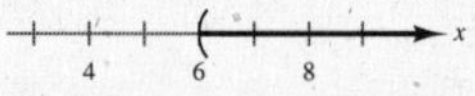

$6\left(\dfrac{x}{2}\right) + 6\left(\dfrac{x}{3}\right) > 6(5)$

$3x + 2x > 30$

$5x > 30$

$\dfrac{1}{5}(5x) > \dfrac{1}{5}(30)$

$x > 6$

27. $2x^2 - x < 6$

$2x^2 - x - 6 < 0$

$(2x + 3)(x - 2) < 0$

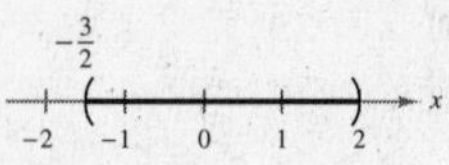

Zeros of the polynomial $(2x + 3)(x - 2)$ are $x = -\frac{3}{2}$ and $x = 2$. By testing the intervals $\left(-\infty, -\frac{3}{2}\right)$, $\left(-\frac{3}{2}, 2\right)$, and $(2, \infty)$, the solution set is $-\frac{3}{2} < x < 2$.

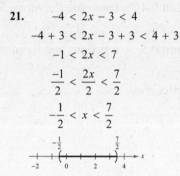

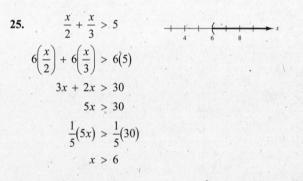

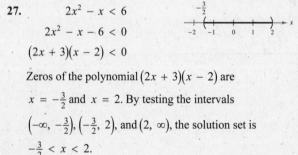

29. Let g represent the cost per gallon, in dollars. Then $2.39 \le g \le 3.25$.

31. Let w represent the person's weight loss for the next three months, in pounds. Then $w \ge 18$.

33. Let h represent hydrochloric acid, l represent lemon juice, o represent oven cleaner, b represent baking soda, p represent pure water, and c represent black coffee.

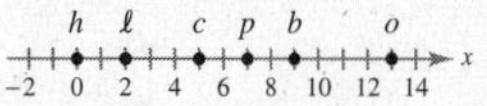

35. Let F represent the temperature of the terrarium, in degrees Fahrenheit. So, you can write $68 \le F \le 75$.

You can use the equation $F = \frac{9}{5}C + 32$ to convert a temperature in degrees Celsius to degrees Fahrenheit.

$F = \frac{9}{5}C + 32 = \frac{9}{5}(18.3) + 32 = 64.94$

Because $64.94 < 68$, the current temperature does not fall within the acceptable range.

37. Given $a < b$,

(a) False. Because $a < b$, $-2a < -2b$.

(b) True. Because $a < b$, $a + 2 < b + 2$.

(c) True. Because $a < b$, $6a < 6b$.

(d) False, if $ab > 0$, then $\dfrac{1}{a} < \dfrac{1}{b}$.

True, if $ab < 0$, then $\dfrac{1}{a} < \dfrac{1}{b}$.

Section 0.2 Absolute Value and Distance on the Real Number Line

1. (a) The directed distance from a to b is $75 - 126 = -51$.

(b) The directed distance from b to a is $126 - 75 = 51$.

(c) The distance between a and b is $|75 - 126| = 51$.

3. (a) The directed distance from a to b is $-5.65 - 9.34 = -14.99$.

(b) The directed distance from b to a is $9.34 - (-5.65) = 14.99$.

(c) The distance between a and b is $|-5.65 - 9.34| = 14.99$.

5. (a) The directed distance from a to b is $\frac{112}{75} - \frac{16}{5} = -\frac{128}{75}$.

(b) The directed distance from b to a is $\frac{16}{5} - \frac{112}{75} = \frac{128}{75}$.

(c) The distance between a and b is $\left|\frac{112}{75} - \frac{16}{5}\right| = \frac{128}{75}$.

7. $|x| \le 2$

9. $|x| > 2$

11. $|x - 5| \le 3$

13. $|x - 2| > 2$

15. $|x - 5| < 3$

17. $|y - a| \le 2$

19. $-4 < x < 4$

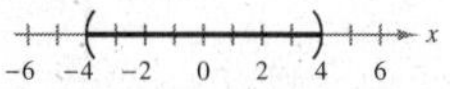

21.

$\dfrac{x}{2} < -3$ or $\dfrac{x}{2} > 3$

$2\left(\dfrac{x}{2}\right) < 2(-3)$ $\quad$ $2\left(\dfrac{x}{2}\right) > 2(3)$

$x < -6$ $\quad$ $x > 6$

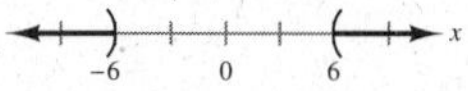

23.

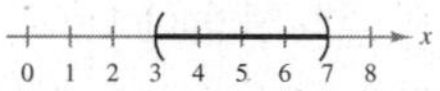

$-2 < x - 5 < 2$

$-2 + 5 < x - 5 + 5 < 2 + 5$

$3 < x < 7$

25.

$\dfrac{x - 3}{2} \le -5$ or $\dfrac{x - 3}{2} \ge 5$

$\dfrac{x - 3}{2}(2) \le -5(2)$ $\quad$ $\dfrac{x - 3}{2}(2) \ge 5(2)$

$x - 3 \le -10$ $\quad$ $x - 3 \ge 10$

$x - 3 + 3 \le -10 + 3$ $\quad$ $x - 3 + 3 \ge 10 + 3$

$x \le -7$ $\quad$ $x \ge 13$

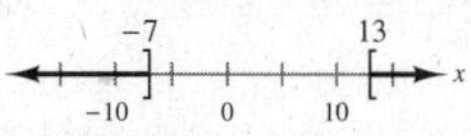

27.

$10 - x < -4$ or $10 - x > 4$

$10 - x - 10 < -4 - 10$ $\quad$ $10 - x - 10 > 4 - 10$

$-x < -14$ $\quad$ $-x > -6$

$x > 14$ $\quad$ $x < 6$

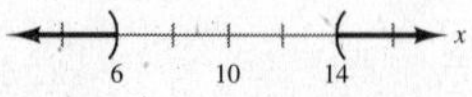

29. $-1 < 9 - 2x < 1$

$$-1 - 9 < 9 - 2x - 9 < 1 - 9$$
$$-10 < -2x < -8$$
$$-\tfrac{1}{2}(-10) > -\tfrac{1}{2}(-2)x > -\tfrac{1}{2}(-8)$$
$$5 > x > 4$$
$$4 < x < 5$$

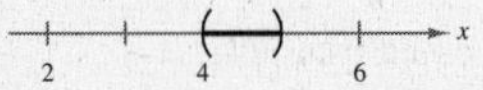

31. $-b \le x - a \le b$

$$-b + a \le x - a + a \le b + a$$
$$a - b \le x \le a + b$$

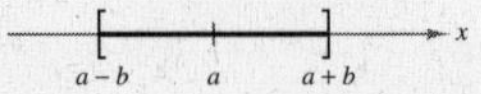

33. $-2b < \dfrac{3x - a}{4} < 2b$

$$4(-2b) < 4\left(\frac{3x - a}{4}\right) < 4(2b)$$
$$-8b < 3x - a < 8b$$
$$-8b + a < 3x - a + a < 8b + a$$
$$\frac{1}{3}(a - 8b) < x < \frac{1}{3}(a + 8b)$$

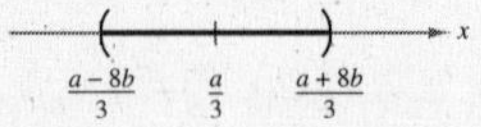

35. Midpoint $= \dfrac{8 + 24}{2} = 16$

37. Midpoint $= \dfrac{-6.85 + 9.35}{2} = 1.25$

39. Midpoint $= \dfrac{-\frac{1}{2} + \frac{3}{4}}{2} = \dfrac{\frac{1}{4}}{2} = \dfrac{1}{8}$

41. $|M - 1083.4| \le 0.2$

43. $\left|\dfrac{h - 68.5}{2.7}\right| \le 1$

$$-1 \le \frac{h - 68.5}{2.7} \le 1$$
$$-2.7 \le h - 68.5 \le 2.7$$
$$65.8 \le h \le 71.2$$

The heights of two-thirds of the members of a population lie between 65.8 inches and 71.2 inches.

45. $|n - 850| \le 100$

$$-100 \le n - 850 \le 100$$
$$750 \le n \le 950$$

The low amount of grass is 750 blades per square foot, and the high amount of grass is 950 blades per square foot.

47. (a) Margin of error: $\pm 3\%$

Candidate X expects to get $45\% \pm 3\%$ of likely voters. Let X represent the percent (in decimal form) of likely voters for Candidate X. You can write the following inequality.

$$0.45 - 0.03 \le X \le 0.45 + 0.03$$
$$0.42 \le X \le 0.48$$

(b) Letting V be the number of people out of 80,000 who vote for Candidate X, it follows that $V = 80{,}000X$.

$$0.42(80{,}000) \le 80{,}000X \le 0.48(80{,}000)$$
$$33{,}600 \le V \le 38{,}400$$

Candidate X can expect at least 33,600 votes and at most 38,400 votes.

Section 0.3 Exponents and Radicals

1. $-2(3)^3 = -2(27) = -54$

3. $4(2)^{-3} = 4\left(\frac{1}{8}\right) = \frac{1}{2}$

5. $\dfrac{1 + 3^{-1}}{3^{-1}} = \dfrac{1 + 1/3}{1/3} = \dfrac{4/3}{1/3} = 4$

7. $3(-2)^2 - 4(-2)^3 = 3(4) - 4(-8) = 12 + 32 = 44$

9. $6(10)^0 - [6(10)]^0 = 6(1) - (60)^0 = 6 - 1 = 5$

11. $\sqrt[3]{27^2} = \sqrt[3]{729} = 9$

13. $4^{-1/2} = \dfrac{1}{\sqrt{4}} = \dfrac{1}{2}$

15. $(-32)^{-2/5} = \dfrac{1}{(-32)^{2/5}} = \dfrac{1}{\left(\sqrt[5]{-32}\right)^2} = \dfrac{1}{(-2)^2} = \dfrac{1}{4}$

17. $500(1.01)^{60} \approx 908.3483$

19. $\sqrt[3]{-154} \approx -5.3601$

21. $6y^{-2}(2y^4)^{-3} = 6y^{-2}(2^{-3}y^{-12}) = 6\left(\frac{1}{2^3}\right)y^{-2-12}$

$= 6\left(\frac{1}{8}\right)y^{-14} = \frac{3}{4y^{14}}$

23. $10(x^2)^2 = 10x^4$

25. $\frac{7x^2}{x^{-3}} = 7x^{2+3} = 7x^5$

27. $\frac{10(x+y)^3}{4(x+y)^{-2}} = \frac{5}{2}(x+y)^{3+2} = \frac{5}{2}(x+y)^5$

29. $\frac{3x\sqrt{x}}{x^{1/2}} = \frac{3x(x^{1/2})}{x^{1/2}} = 3x^{1+1/2-1/2} = 3x$

31. $\sqrt{8} = \sqrt{4\cdot 2} = \sqrt{4}\sqrt{2} = 2\sqrt{2}$

33. $\sqrt[3]{54x^5} = \sqrt[3]{(27x^3)(2x^2)} = \sqrt[3]{27x^3}\sqrt[3]{2x^2} = 3x\sqrt[3]{2x^2}$

35. $\sqrt[3]{144x^9y^{-4}z^5} = \sqrt[3]{(8x^9y^{-3}z^3)(18y^{-1}z^2)}$

$= 2x^3y^{-1}z\sqrt[3]{18y^{-1}z^2}$

$= \frac{2x^3z}{y}\sqrt[3]{\frac{18z^2}{y}}$

37. $3x^3 - 12x = 3x(x^2-4) = 3x(x+2)(x-2)$

39. $2x^{5/2} + x^{-1/2} = x^{-1/2}(2x^3+1) = \frac{2x^3+1}{x^{1/2}}$

41. $3x(x+1)^{3/2} - 6(x+1)^{1/2} = 3(x+1)^{1/2}(x(x+1)-2)$

$= 3(x+1)^{1/2}(x^2+x-2)$

$= 3(x+1)^{1/2}(x-1)(x+2)$

43. $\frac{(x+1)(x-1)^2-(x-1)^3}{(x+1)^2} = \frac{(x-1)^2}{(x+1)^2}((x+1)-(x-1))$

$= \frac{(x-1)^2}{(x+1)^2}(2)$

$= \frac{2(x-1)^2}{(x+1)^2}$

45. $\sqrt{x-4}$ is defined when $x \ge 4$.
Therefore, the domain is $[4, \infty)$.

47. $\sqrt{x^2+3}$ is defined for all real numbers.
Therefore, the domain is $(-\infty, \infty)$.

49. $\frac{1}{\sqrt[3]{x-4}}$ is defined for all real numbers except $x = 4$.
Therefore, the domain is $(-\infty, 4) \cup (4, \infty)$.

51. $\frac{\sqrt{x+2}}{1-x}$

The numerator is defined when $x \ge -2$.
The denominator is defined when $x \ne 1$.
Therefore, the domain is $[-2, 1) \cup (1, \infty)$.

53. $B = \sqrt{\frac{147(48)}{3600}} = 1.4$ square meters

55. $B = \sqrt{\frac{180(80)}{3600}} = 2$ square meters

57. $P = N(1+r)^t = 2{,}000{,}000(1+0.03)^4$

$= 2{,}000{,}000(1.03)^4$

$\approx 2{,}251{,}018$ mollusks

59. (a)

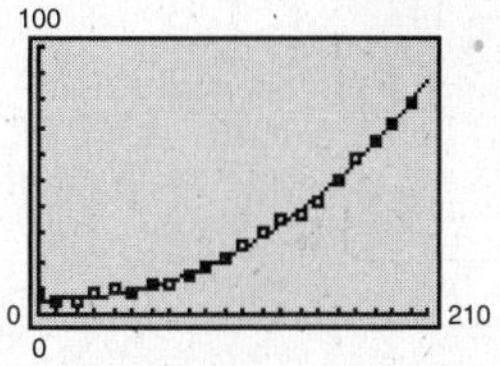

(b)

Year	Population per square mile, P (actual)	Population per square mile, P (model)
1800	6.1	5.5
1810	4.3	5.6
1820	5.5	6.0
1830	7.4	6.6
1840	9.8	7.5
1850	7.9	8.9
1860	10.6	10.5
1870	11.2	12.6
1880	14.2	15.0
1890	17.8	17.9
1900	21.5	21.1
1910	26.0	24.9
1920	29.9	29.0
1930	34.7	33.6
1940	37.2	38.7
1950	42.6	44.2
1960	50.6	50.2
1970	57.5	56.7
1980	64.0	63.7
1990	70.3	71.1
2000	79.6	79.1

(c) The model values are close to the actual values, so the model is a good fit.

(d) Yes, you could use the model to predict the population per square mile for future years. According to the model, the population per square mile in 2015 will be

$$5.54 + 0.000524(215)^{2.237} \approx 92.0.$$

This answer seems reasonable.

(e) Yes, as t gets very large the population per square mile should reach a maximum and begin to decrease, because eventually the capacity of the land area will be reached.

Section 0.4 Factoring Polynomials

1. $a = 6, b = -7$, and $c = 1$

$$x = \frac{7 \pm \sqrt{49 - 24}}{12} = \frac{7 \pm 5}{12}$$

So, $x = \dfrac{7 + 5}{12} = 1$ or $x = \dfrac{7 - 5}{12} = \dfrac{1}{6}$.

3. $a = 4, b = -12$, and $c = 9$

$$x = \frac{12 \pm \sqrt{144 - 144}}{8} = \frac{12}{8} = \frac{3}{2}$$

5. $a = 1, b = 4$, and $c = 1$

$$y = \frac{-4 \pm \sqrt{16 - 4}}{2} = \frac{-4 \pm 2\sqrt{3}}{2} = -2 \pm \sqrt{3}$$

7. $a = 2, b = 3$, and $c = -4$

$$x = \frac{-3 \pm \sqrt{9 + 32}}{4} = \frac{-3 \pm \sqrt{41}}{4}$$

9. $x^2 - 4x + 4 = (x - 2)^2$

11. $4x^2 + 4x + 1 = (2x + 1)^2$

13. $3x^2 - 4x + 1 = (3x - 1)(x - 1)$

15. $3x^2 - 5x + 2 = (3x - 2)(x - 1)$

17. $x^2 - 4xy + 4y^2 = (x - 2y)^2$

19. $81 - y^4 = (9 + y^2)(9 - y^2)$

$$= (9 + y^2)(3 + y)(3 - y)$$

21. $x^3 - 8 = x^3 - 2^3$

$$= (x - 2)(x^2 + 2x + 4)$$

23. $y^3 + 64 = y^3 + 4^3$

$$= (y + 4)(y^2 - 4y + 16)$$

25. $x^3 - y^3 = (x - y)(x^2 + xy + y^2)$

27. $x^3 - 4x^2 - x + 4 = x^2(x - 4) - (x - 4)$

$$= (x - 4)(x^2 - 1)$$

$$= (x - 4)(x + 1)(x - 1)$$

29. $2x^3 - 3x^2 + 4x - 6 = x^2(2x - 3) + 2(2x - 3)$

$$= (2x - 3)(x^2 + 2)$$

31. $2x^3 - 4x^2 - x + 2 = 2x^2(x - 2) - (x - 2)$

$$= (x - 2)(2x^2 - 1)$$

33. $x^4 - 15x^2 - 16 = (x^2 - 16)(x^2 + 1)$

$$= (x - 4)(x + 4)(x^2 + 1)$$

35. $x^2 - 5x = 0$

$$x(x - 5) = 0$$

$$x = 0, 5$$

37. $x^2 - 9 = 0$

$$(x + 3)(x - 3) = 0$$

$$x = -3, 3$$

39. $x^2 - 3 = 0$

$$\left(x + \sqrt{3}\right)\left(x - \sqrt{3}\right) = 0$$

$$x = \pm\sqrt{3}$$

41. $(x - 3)^2 - 9 = 0$

$$x^2 - 6x + 9 - 9 = 0$$

$$x(x - 6) = 0$$

$$x = 0, 6$$

43. $x^2 + x - 2 = 0$

$$(x + 2)(x - 1) = 0$$

$$x = -2, 1$$

45. $x^2 - 5x - 6 = 0$

$(x + 1)(x - 6) = 0$

$x = -1, 6$

47. $3x^2 + 5x + 2 = 0$

$(3x + 2)(x + 1) = 0$

$x = -\frac{2}{3}, -1$

49. $x^3 + 64 = 0$

$x^3 = -64$

$x = \sqrt[3]{-64} = -4$

51. $x^4 - 16 = 0$

$x^4 = 16$

$x = \pm\sqrt[4]{16} = \pm 2$

53. $x^3 - x^2 - 4x + 4 = 0$

$x^2(x - 1) - 4(x - 1) = 0$

$(x - 1)(x^2 - 4) = 0$

$(x - 1)(x - 2)(x + 2) = 0$

$x = 1, \pm 2$

55. Because $\sqrt{x^2 - 4} = \sqrt{(x + 2)(x - 2)}$, the roots are $x = \pm 2$. By testing points inside and outside the interval $[-2, 2]$, we find that the expression is defined when $x \le -2$ or $x \ge 2$. So, the domain is $(-\infty, -2] \cup [2, \infty)$.

57. Because $\sqrt{x^2 - 7x + 12} = \sqrt{(x - 3)(x - 4)}$, the roots are $x = 3$ and $x = 4$. By testing points inside and outside the interval $[3, 4]$, we find that the expression is defined when $x \le 3$ or $x \ge 4$. So, the domain is $(-\infty, 3] \cup [4, \infty)$.

59. Because $\sqrt{5x^2 + 6x + 1} = \sqrt{(5x + 1)(x + 1)}$, the roots are $x = -\frac{1}{5}$ and $x = -1$. By testing the intervals $(-\infty, -1)$, $\left(-1, -\frac{1}{5}\right)$, and $\left(-\frac{1}{5}, \infty\right)$, we find that the expression is defined when $x \le -1$ or $x \ge -\frac{1}{5}$. So, the domain is $(-\infty, -1] \cup \left[-\frac{1}{5}, \infty\right)$.

61.

−1	1	−3	−6	−2
		−1	4	2
	1	−4	−2	0

So, the factorization is

$x^3 - 3x^2 - 6x - 2 = (x + 1)(x^2 - 4x - 2)$.

63.

−1	2	−1	−2	1
		−2	3	−1
	2	−3	1	0

So, the factorization is

$2x^3 - x^2 - 2x + 1 = (x + 1)(2x^2 - 3x + 1)$.

65. Possible rational zeros: $\pm 8, \pm 4, \pm 2, \pm 1$

Using synthetic division for $x = -1$, we have

−1	1	−1	−10	−8
		−1	2	8
	1	−2	−8	0

So,

$x^3 - x^2 - 10x - 8 = 0$

$(x + 1)(x^2 - 2x - 8) = 0$

$(x + 1)(x + 2)(x - 4) = 0$

$x = -1, -2, 4$

67. Possible rational roots: $\pm 1, \pm 2, \pm 3, \pm 6$

Using synthetic division for $x = 1$, we have

1	1	−6	11	−6
		1	−5	6
	1	−5	6	0

So,

$x^3 - 6x^2 + 11x - 6 = 0$

$(x - 1)(x^2 - 5x + 6) = 0$

$(x - 1)(x - 2)(x - 3) = 0$

$x = 1, 2, 3.$

69. Possible rational zeros:

$\pm 6, \pm 3, \pm 2, \pm 1, \pm\frac{3}{2}, \pm\frac{1}{2}, \pm\frac{2}{3}, \pm\frac{1}{3}, \pm\frac{1}{6}$

Using synthetic division for $x = 3$, we have

3	6	−11	−19	−6
		18	21	6
	6	7	2	0

So,

$6x^3 - 11x^2 - 19x - 6 = 0$

$(x - 3)(6x^2 + 7x + 2) = 0$

$(x - 3)(3x + 2)(2x + 1) = 0$

$x = 3, -\frac{2}{3}, -\frac{1}{2}.$

71. Possible rational roots: $\pm 1, \pm 2, \pm 4$

Using synthetic division for $x = 4$, we have

$$\begin{array}{r|rrrr} 4 & 1 & -3 & -3 & -4 \\ & & 4 & 4 & 4 \\ \hline & 1 & 1 & 1 & 0 \end{array}$$

So,

$$x^3 - 3x^2 - 3x - 4 = 0$$
$$(x - 4)(x^2 + x + 1) = 0.$$

Since $x^2 + x + 1$ has no real solutions, $x = 4$ is the only real solution.

73. Possible rational zeros: $\pm\frac{1}{4}, \pm\frac{1}{2}, \pm 1, \pm 2$

Using synthetic division for $x = -1$, we have

$$\begin{array}{r|rrrr} -1 & 4 & 11 & 5 & -2 \\ & & -4 & -7 & 2 \\ \hline & 4 & 7 & -2 & 0 \end{array}$$

So,

$$4x^3 + 11x^2 + 5x - 2 = 0$$
$$(x + 1)(4x^2 + 7x - 2) = 0$$
$$(x + 1)(4x - 1)(x + 2) = 0$$
$$x = -1, \tfrac{1}{4}, -2.$$

75. Volume of pyramid $= \frac{1}{3}s^2h$

$$4 = \tfrac{1}{3}s^2(s + 1)$$
$$12 = s^3 + s^2$$
$$0 = s^3 + s^2 - 12$$

Possible rational roots: $\pm 1, \pm 2, \pm 3, \pm 4, \pm 6, \pm 12$

Using synthetic division for $s = 2$, you have the following.

$$\begin{array}{r|rrrr} 2 & 1 & 1 & 0 & -12 \\ & & 2 & 6 & 12 \\ \hline & 1 & 3 & 6 & 0 \end{array}$$

So,

$$s^3 + s^2 - 12 = 0$$
$$(s - 2)(s^2 + 3s + 6) = 0.$$

Because $s^2 + 3s + 6$ has no real solutions, $s = 2$ is the only real solution.

The length of each side of the base of the mold is 2 feet. The height of the mold is $2 + 1 = 3$ feet.

77.

$$1.8 \times 10^{-5} = \frac{x^2}{1.0 \times 10^{-4} - x}$$
$$1.8 \times 10^{-9} - 1.8 \times 10^{-5}x = x^2$$
$$x^2 + 1.8 \times 10^{-5}x - 1.8 \times 10^{-9} = 0$$

By the Quadratic Formula:

$$x = \frac{-1.8 \times 10^{-5} \pm \sqrt{\left(1.8 \times 10^{-5}\right)^2 + 4 \times 1.8 \times 10^{-9}}}{2}$$
$$\approx \frac{-1.8 \times 10^{-5} \pm \sqrt{7.524 \times 10^{-9}}}{2}$$
$$\approx 3.437 \times 10^{-5}\left[H^+\right]$$

Section 0.5 Fractions and Rationalization

1. $\dfrac{x}{x-2} + \dfrac{3}{x-2} = \dfrac{x+3}{x-2}$

3. $\dfrac{2x}{x^2+2} - \dfrac{1-3x}{x^2+2} = \dfrac{2x-(1-3x)}{x^2+2} = \dfrac{5x-1}{x^2+2}$

5. $\dfrac{2}{x^2-4} - \dfrac{1}{x-2} = \dfrac{2}{(x-2)(x+2)} - \dfrac{1}{x-2}\cdot\dfrac{(x+2)}{(x+2)}$

$$= \frac{2-(x+2)}{(x-2)(x+2)} = -\frac{x}{x^2-4}$$

7. $\dfrac{5}{x-3} + \dfrac{3}{3-x} = \dfrac{5}{x-3} + \dfrac{-3}{x-3} = \dfrac{2}{x-3}$

9. $\dfrac{A}{x+1} + \dfrac{B}{(x+1)^2} + \dfrac{C}{x-2}$

$$= \frac{A(x+1)(x-2)}{(x+1)^2(x-2)} + \frac{B(x-2)}{(x+1)^2(x-2)} + \frac{C(x+1)^2}{(x+1)^2(x-2)}$$
$$= \frac{A(x^2-x-2) + B(x-2) + C(x^2+2x+1)}{(x+1)^2(x-2)}$$
$$= \frac{Ax^2 - Ax - 2A + Bx - 2B + Cx^2 + 2Cx + C}{(x+1)^2(x-2)}$$
$$= \frac{(A+C)x^2 - (A-B-2C)x - (2A+2B-C)}{(x+1)^2(x-2)}$$

11. $\dfrac{A}{x-6}+\dfrac{Bx+C}{x^2+3}=\dfrac{A(x^2+3)}{(x-6)(x^2+3)}+\dfrac{(Bx+C)(x-6)}{(x-6)(x^2+3)}=\dfrac{(A+B)x^2+(C-6B)x+3(A-2C)}{(x-6)(x^2+3)}$

13. $-\dfrac{2}{x}+\dfrac{1}{x^2+2}=\dfrac{-2(x^2+2)}{x(x^2+2)}+\dfrac{1(x)}{x(x^2+2)}=\dfrac{-2x^2-4+x}{x(x^2+2)}=\dfrac{-2x^2+x-4}{x(x^2+2)}$

15. $\dfrac{1}{x^2-x-2}-\dfrac{x}{x^2-5x+6}=\dfrac{1}{(x+1)(x-2)}-\dfrac{x}{(x-2)(x-3)}$

$=\dfrac{1(x-3)}{(x+1)(x-2)(x-3)}-\dfrac{x(x+1)}{(x+1)(x-2)(x-3)}$

$=\dfrac{-x^2-3}{(x+1)(x-2)(x-3)}$

$=-\dfrac{x^2+3}{(x+1)(x-2)(x-3)}$

17. $\dfrac{-x}{(x+1)^{3/2}}+\dfrac{2}{(x+1)^{1/2}}=\dfrac{-x}{(x+1)^{3/2}}+\dfrac{2(x+1)^1}{(x+1)^{3/2}}$ *$1-\frac{3}{2}=\frac{1}{2}$

$=\dfrac{x+2}{(x+1)^{3/2}}$

19. $\dfrac{2-t}{2\sqrt{1+t}}-\sqrt{1+t}$

$=\dfrac{2-t}{2\sqrt{1+t}}-\dfrac{\sqrt{1+t}}{1}\cdot\dfrac{2\sqrt{1+t}}{2\sqrt{1+t}}$

$=\dfrac{(2-t)-2(1+t)}{2\sqrt{1+t}}$

$=\dfrac{-3t}{2\sqrt{1+t}}$

21. $\left(2x\sqrt{x^2+1}-\dfrac{x^3}{\sqrt{x^2+1}}\right)\div(x^2+1)$

$=\left(\dfrac{2x(x^2+1)}{\sqrt{x^2+1}}-\dfrac{x^3}{\sqrt{x^2+1}}\right)\dfrac{1}{x^2+1}$

$=\dfrac{2x^3+2x-x^3}{\sqrt{x^2+1}}\cdot\dfrac{1}{x^2+1}$

$=\dfrac{x^3+2x}{\sqrt{x^2+1}(x^2+1)}$

$=\dfrac{x(x^2+2)}{(x^2+1)^{3/2}}$

23. $\dfrac{(x^2+2)^{1/2}-x^2(x^2+2)^{-1/2}}{x^2}$

$=\dfrac{(x^2+2)^{-1/2}\left[(x^2+2)^1-x^2\right]}{x^2}$

$=\dfrac{2}{x^2\sqrt{x^2+2}}$

25. $\dfrac{-x^2}{(2x+3)^{3/2}}+\dfrac{2x}{(2x+3)^{1/2}}$

$=\dfrac{-x^2}{(2x+3)^{3/2}}+\dfrac{2x(2x+3)}{(2x+3)^{3/2}}$

$=\dfrac{3x^2+6x}{(2x+3)^{3/2}}$

$=\dfrac{3x(x+2)}{(2x+3)^{3/2}}$

27. $\dfrac{2}{\sqrt{10}}=\dfrac{2}{\sqrt{10}}\cdot\dfrac{\sqrt{10}}{\sqrt{10}}=\dfrac{2\sqrt{10}}{10}=\dfrac{\sqrt{10}}{5}$

29. $\dfrac{4x}{\sqrt{x-1}}=\dfrac{4x}{\sqrt{x-1}}\cdot\dfrac{\sqrt{x-1}}{\sqrt{x-1}}=\dfrac{4x\sqrt{x-1}}{x-1}$

31. $\dfrac{49(x-3)}{\sqrt{x^2-9}}=\dfrac{49(x-3)}{\sqrt{x^2-9}}\cdot\dfrac{\sqrt{x^2-9}}{\sqrt{x^2-9}}$

$=\dfrac{49(x-3)\sqrt{x^2-9}}{(x+3)(x-3)}$

$=\dfrac{49\sqrt{x^2-9}}{x+3},\ x\neq 3$

33. $\dfrac{5}{\sqrt{14}-2}=\dfrac{5}{\sqrt{14}-2}\cdot\dfrac{\sqrt{14}+2}{\sqrt{14}+2}$

$=\dfrac{5(\sqrt{14}+2)}{14-4}$

$=\dfrac{\sqrt{14}+2}{2}$

35. $\dfrac{2x}{5-\sqrt{3}} = \dfrac{2x}{5-\sqrt{3}} \cdot \dfrac{5+\sqrt{3}}{5+\sqrt{3}} = \dfrac{2x\left(5+\sqrt{3}\right)}{25-3} = \dfrac{x\left(5+\sqrt{3}\right)}{11}$

37. $\dfrac{1}{\sqrt{6}+\sqrt{5}} = \dfrac{1}{\sqrt{6}+\sqrt{5}} \cdot \dfrac{\sqrt{6}-\sqrt{5}}{\sqrt{6}-\sqrt{5}}$

$= \dfrac{\sqrt{6}-\sqrt{5}}{6-5}$

$= \sqrt{6}-\sqrt{5}$

39. $\dfrac{2}{\sqrt{x}+\sqrt{x-2}} = \dfrac{2}{\sqrt{x}+\sqrt{x-2}} \cdot \dfrac{\sqrt{x}-\sqrt{x-2}}{\sqrt{x}-\sqrt{x-2}}$

$= \dfrac{2\left(\sqrt{x}-\sqrt{x-2}\right)}{x-(x-2)}$

$= \sqrt{x}-\sqrt{x-2}$

41. $\dfrac{\sqrt{x+2}-\sqrt{2}}{x} = \dfrac{\sqrt{x+2}-\sqrt{2}}{x} \cdot \dfrac{\sqrt{x+2}+\sqrt{2}}{\sqrt{x+2}+\sqrt{2}}$

$= \dfrac{x+2-2}{x\left(\sqrt{x+2}+\sqrt{2}\right)}$

$= \dfrac{1}{\sqrt{x+2}+\sqrt{2}},\ x \neq 0$

43. $\dfrac{\dfrac{\sqrt{4-x^2}}{x^4} - \dfrac{2}{x^2\sqrt{4-x^2}}}{4-x^2} = \left(\dfrac{\left(4-x^2\right)}{x^4\sqrt{4-x^2}} - \dfrac{2x^2}{x^4\sqrt{4-x^2}}\right)\dfrac{1}{4-x^2}$

$= \dfrac{\left(4-x^2\right)-2x^2}{x^4\left(4-x^2\right)^{3/2}}$

$= \dfrac{4-3x^2}{x^4\left(4-x^2\right)^{3/2}}$

45. $P = \dfrac{0.6+0.85(n-1)}{1+0.85(n-1)},\ n > 0$

n	1	2	3	4	5	6
P	0.6	0.7838	0.8519	0.8873	0.9091	0.9238

n	7	8	9	10
P	0.9344	0.9424	0.9487	0.9538

Practice Test for Chapter 0

1. Determine whether $\sqrt[4]{81}$ is rational or irrational.

2. Determine whether the given value of x satisfies the inequality $3x + 4 \le x/2$.

(a) $x = -2$ (b) $x = 0$ (c) $x = -\frac{8}{5}$ (d) $x = -6$

3. Solve the inequality $3x + 4 \ge 13$.

4. Solve the inequality $x^2 < 6x + 7$.

5. Determine which of the two given real numbers is greater, $\sqrt{19}$ or $\frac{13}{3}$.

6. Given the interval $[-3, 7]$, find (a) the distance between -3 and 7 and (b) the midpoint of the interval.

7. Solve the inequality $|3x + 1| \le 10$.

8. Solve the inequality $|4 - 5x| > 29$.

9. Solve the inequality $\left|3 - \frac{2x}{5}\right| < 8$.

10. Use absolute values to describe the interval $[-3, 5]$.

11. Simplify $\frac{12x^3}{4x^{-2}}$.

12. Simplify $\left(\frac{\sqrt{3}\sqrt{x^3}}{x}\right)^0$, $x \ne 0$.

13. Remove all possible factors from the radical $\sqrt[3]{32x^4y^3}$.

14. Complete the factorization: $\frac{3}{2}(x + 1)^{-1/3} + \frac{1}{4}(x + 1)^{2/3} = \frac{1}{4}(x + 1)^{-1/3}(\underline{\quad})$

15. Find the domain: $\frac{1}{\sqrt{5 - x}}$

16. Factor completely: $3x^2 - 19x - 14$

17. Factor completely: $25x^2 - 81$

18. Factor completely: $x^3 + 8$

19. Use the Quadratic Formula to find all real roots of $x^2 + 6x - 2 = 0$.

20. Use the Rational Zero Theorem to find all real roots of $x^3 - 4x^2 + x + 6 = 0$.

21. Combine terms and simplify: $\frac{x}{x^2 + 2x - 3} - \frac{1}{x - 1}$

22. Combine terms and simplify: $\dfrac{3 - x}{2\sqrt{x + 5}} + \sqrt{x + 5}$

23. Combine terms and simplify: $\dfrac{\dfrac{\sqrt{x + 2}}{\sqrt{x}} - \dfrac{\sqrt{x}}{\sqrt{x + 2}}}{2(x + 2)}$

24. Rationalize the denominator: $\dfrac{3y}{\sqrt{y^2 + 9}}$

25. Rationalize the numerator: $\dfrac{\sqrt{x} + \sqrt{x + 7}}{14}$

Graphing Calculator Required

26. Use a graphing calculator to find the real solutions of $x^3 - 5x^2 + 2x + 8 = 0$ by graphing $y = x^3 - 5x^2 + 2x + 8$ and finding the x-intercepts.

CHAPTER 1
Functions, Graphs, and Limits

CHAPTER 1
Functions, Graphs, and Limits

Section 1.1 The Cartesian Plane and the Distance Formula

Skills Review

1. $\sqrt{(3-6)^2+[1-(-5)]^2} = \sqrt{(-3)^2+6^2}$
$= \sqrt{9+36}$
$= \sqrt{45}$
$= 3\sqrt{5}$

2. $\sqrt{(-2-0)^2+[-7-(-3)]^2} = \sqrt{(-2)^2+(-4)^2}$
$= \sqrt{4+16}$
$= \sqrt{20}$
$= 2\sqrt{5}$

3. $\dfrac{5+(-4)}{2} = \dfrac{1}{2}$

4. $\dfrac{-3+(-1)}{2} = \dfrac{-4}{2} = -2$

5. $\sqrt{27}+\sqrt{12} = 3\sqrt{3}+2\sqrt{3} = 5\sqrt{3}$

6. $\sqrt{8}-\sqrt{18} = 2\sqrt{2}-3\sqrt{2} = -\sqrt{2}$

7. $\sqrt{(3-x)^2+(7-4)^2} = \sqrt{45}$
$\left(\sqrt{(3-x)^2+(7-4)^2}\right)^2 = \left(\sqrt{45}\right)^2$
$(3-x)^2+(7-4)^2 = 45$
$(3-x)^2+3^2 = 45$
$(3-x)^2+9 = 45$
$(3-x)^2 = 36$
$3-x = \pm 6$
$-x = -3 \pm 6$
$x = 3 \mp 6$
$x = -3, 9$

8. $\sqrt{(6-2)^2+(-2-y)^2} = \sqrt{52}$
$\left(\sqrt{(6-2)^2+(-2-y)^2}\right)^2 = \left(\sqrt{52}\right)^2$
$(6-2)^2+(-2-y)^2 = 52$
$4^2+(-2-y)^2 = 52$
$16+(-2-y)^2 = 52$
$(-2-y)^2 = 36$
$-2-y = \pm 6$
$-y = \pm 6 + 2$
$y = \mp 6 - 2$
$y = -8, 4$

9. $\dfrac{x+(-5)}{2} = 7$
$x+(-5) = 14$
$x = 19$

10. $\dfrac{-7+y}{2} = -3$
$-7+y = -6$
$y = 1$

1.

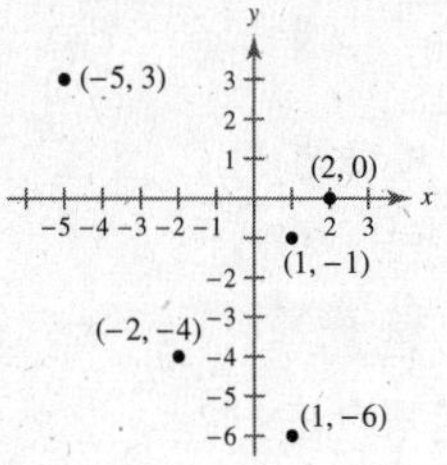

3. (a)

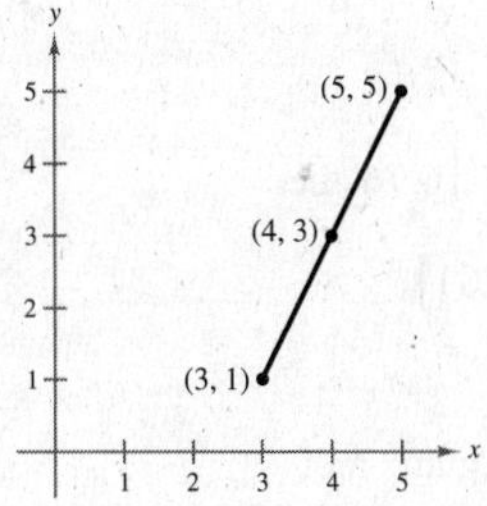

(b) $d = \sqrt{(5-3)^2 + (5-1)^2}$

$= \sqrt{4 + 16} = 2\sqrt{5}$

(c) Midpoint $= \left(\dfrac{3+5}{2}, \dfrac{1+5}{2}\right) = (4, 3)$

5. (a)

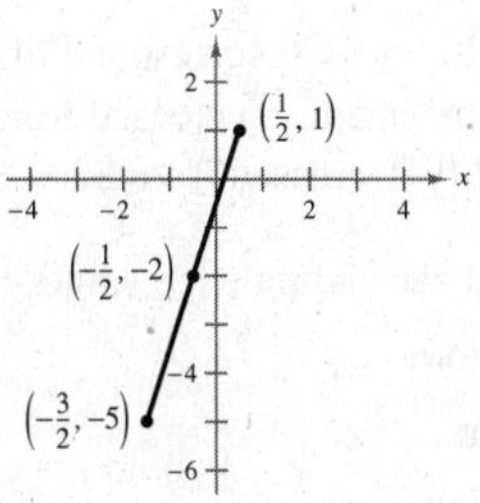

(b) $d = \sqrt{[(3/2) - (1/2)]^2 + (5-1)^2}$

$= \sqrt{4 + 36}$

$= 2\sqrt{10}$

(c) Midpoint $= \left(\dfrac{(1/2) + (-3/2)}{2}, \dfrac{1 + (-5)}{2}\right)$

$= \left(-\dfrac{1}{2}, -2\right)$

7. (a)

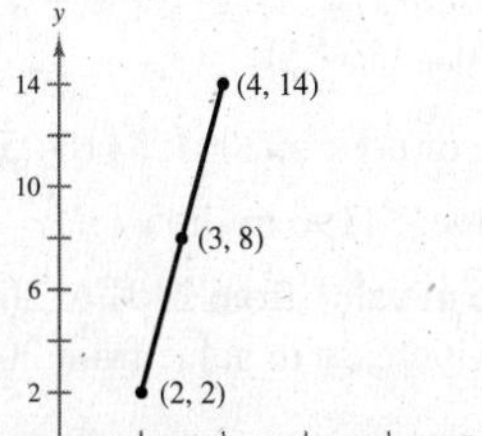

(b) $d = \sqrt{(4-2)^2 + (14-2)^2} = \sqrt{4 + 144} = 2\sqrt{37}$

(c) Midpoint $= \left(\dfrac{2+4}{2}, \dfrac{2+14}{2}\right) = (3, 8)$

9. (a)

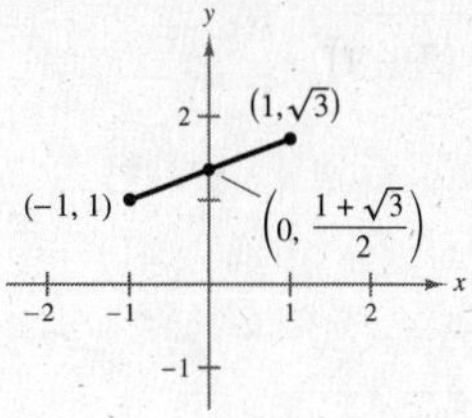

(b) $d = \sqrt{(-1-1)^2 + \left(1 - \sqrt{3}\right)^2}$

$= \sqrt{4 + 1 - 2\sqrt{3} + 3}$

$= \sqrt{8 - 2\sqrt{3}}$

(c) Midpoint $= \left(\dfrac{1 + (-1)}{2}, \dfrac{\sqrt{3} + 1}{2}\right) = \left(0, \dfrac{\sqrt{3} + 1}{2}\right)$

11. (a)

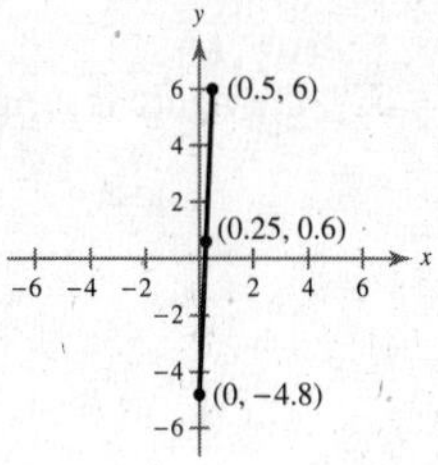

(b) $d = \sqrt{(0.5 - 0)^2 + (6 - (-4.8))^2}$

$= \sqrt{0.25 + 116.64}$

$= \sqrt{116.89}$

(c) Midpoint $= \left(\dfrac{0 + 0.5}{2}, \dfrac{-4.8 + 6}{2}\right) = (0.25, 0.6)$

13. (a) $a = 4$

$b = 3$

$c = \sqrt{(4-0)^2 + (3-0)^2} = \sqrt{16 + 9} = 5$

(b) $a^2 + b^2 = 16 + 9 = 25 = c^2$

15. (a) $a = 10$

$b = 3$

$c = \sqrt{(7+3)^2 + (4-1)^2} = \sqrt{100 + 9} = \sqrt{109}$

(b) $a^2 + b^2 = 100 + 9 = 109 = c^2$

17. $d_1 = \sqrt{(3-0)^2 + (7-1)^2}$
$= \sqrt{9 + 36}$
$= \sqrt{45}$
$= 3\sqrt{5}$

$d_2 = \sqrt{(4-0)^2 + (-1-1)^2}$
$= \sqrt{16 + 4}$
$= \sqrt{20}$
$= 2\sqrt{5}$

$d_3 = \sqrt{(3-4)^2 + [7-(-1)]^2}$
$= \sqrt{1 + 64}$
$= \sqrt{65}$

Because $d_1^2 + d_2^2 = d_3^2$, the figure is a right triangle.

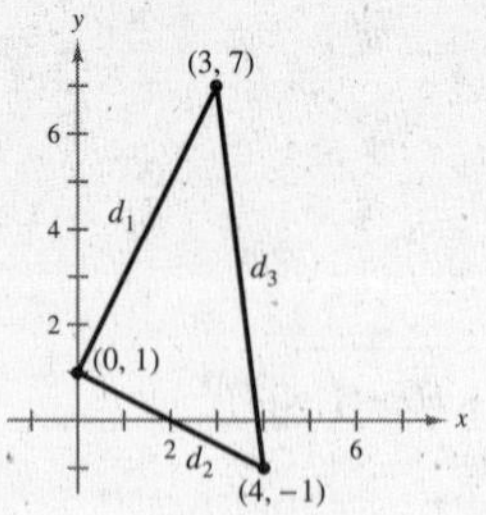

19. $d_1 = \sqrt{(1-0)^2 + (2-0)^2} = \sqrt{1+4} = \sqrt{5}$
$d_2 = \sqrt{(3-1)^2 + (3-2)^2} = \sqrt{4+1} = \sqrt{5}$
$d_3 = \sqrt{(2-3)^2 + (1-3)^2} = \sqrt{1+4} = \sqrt{5}$
$d_4 = \sqrt{(0-2)^2 + (0-1)^2} = \sqrt{4+1} = \sqrt{5}$

Because $d_1 = d_2 = d_3 = d_4$, the figure is a parallelogram.

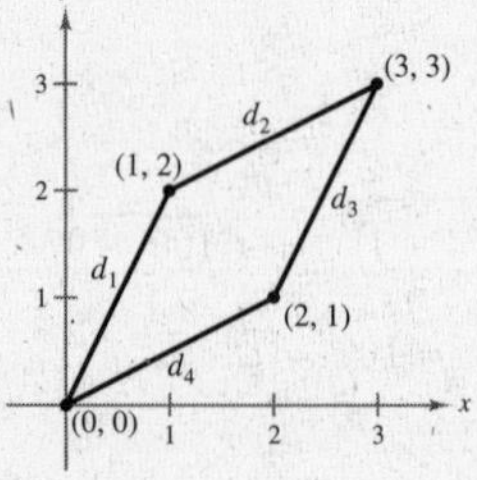

21. $d = \sqrt{(x-1)^2 + (-4-0)^2} = 5$
$\sqrt{x^2 - 2x + 17} = 5$
$x^2 - 2x + 17 = 25$
$x^2 - 2x - 8 = 0$
$(x-4)(x+2) = 0$
$x = 4, -2$

23. $d = \sqrt{(3-0)^2 + (y-0)^2} = 8$
$\sqrt{9 + y^2} = 8$
$9 + y^2 = 64$
$y^2 = 55$
$y = \pm\sqrt{55}$

25. (a) $d^2 = 16^2 + 5^2, d > 0$
$d^2 = 281$
$d = \sqrt{281} \approx 16.76$ feet

(b) $A = 2(40)(\sqrt{281})$
$= 80\sqrt{281}$
≈ 1341.04 square feet

27.

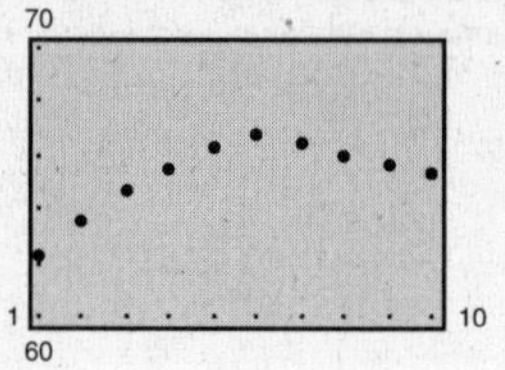

Answers will vary. Let $x = 6$ correspond to 1996. The number of subscribers steadily increased from 1996 to 2001 and then steadily decreased from 2001 to 2005.

29. From the graph, you can estimate the values to be:
(a) 1992: 26.5 million
(b) 1997: 24 million
(c) 2001: 21.9 million
(d) 2004: 19.9 million

31. From the graph, you can estimate the values to be:
(a) 1991: 22.5 million
(b) 1994: 27.8 million
(c) 1997: 23.0 million
(d) 2005: 25.9 million

33. (a) Midpoint $= \left(\frac{2001 + 2005}{2}, \frac{1048 + 1281}{2}\right)$
$= (2003, 1164.5)$

Estimated value of pork in 2003: \$1164.5 million

(b) Actual 2003 value: \$1190 million
(c) No, the increase in value from 2001 to 2003 is greater than the increase in value from 2003 to 2005.

35. (a) Midpoint $= \left(\frac{2001 + 2005}{2}, \frac{2250 + 3763}{2}\right)$

$= (2003, 3006.5)$

Estimated value of wine in 2003: \$3006.5 million

(b) Actual 2003 value: \$3268 million

(c) No, the increase in value from 2001 to 2003 is greater than the increase in value from 2003 to 2005.

37. (a)

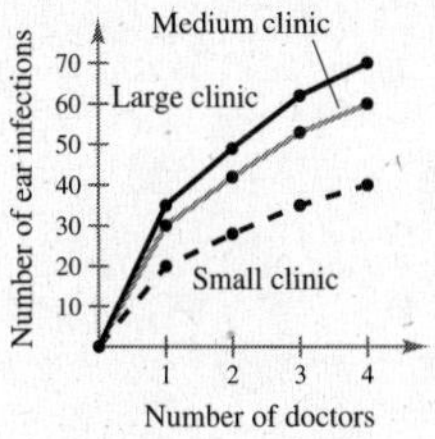

(b) The larger the clinic, the more patients a doctor can treat.

39. (a) $(0, 0)$ is translated to $(0 + 2, 0 + 3) = (2, 3)$.

$(-3, -1)$ is translated to

$(-3 + 2, -1 + 3) = (-1, 2)$.

$(-1, -2)$ is translated to $(-1 + 2, -2 + 3) = (1, 1)$.

(b)

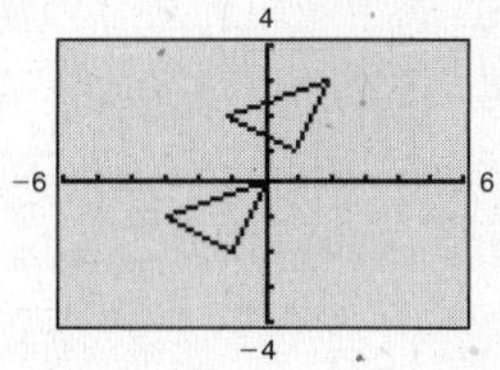

41. Midpoint $= \left(\frac{x_1 + x_2}{2}, \frac{y_1 + y_2}{2}\right)$

The point one-fourth of the way between (x_1, y_1) and (x_2, y_2) is the midpoint of the line segment from (x_1, y_1) to

$\left(\frac{x_1 + x_2}{2}, \frac{y_1 + y_2}{2}\right)$, which is $\left(\frac{x_1 + \frac{x_1 + x_2}{2}}{2}, \frac{y_1 + \frac{y_1 + y_2}{2}}{2}\right) = \left(\frac{3x_1 + x_2}{4}, \frac{3y_1 + y_2}{4}\right)$.

The point three-fourths of the way between (x_1, y_1) and (x_2, y_2) is the midpoint of the line segment from

$\left(\frac{x_1 + x_2}{2}, \frac{y_1 + y_2}{2}\right)$ to (x_2, y_2), which is $\left(\frac{\frac{x_1 + x_2}{2} + x_2}{2}, \frac{\frac{y_1 + y_2}{2} + y_2}{2}\right) = \left(\frac{x_1 + 3x_2}{4}, \frac{y_1 + 3y_2}{4}\right)$.

Thus,

$\left(\frac{3x_1 + x_2}{4}, \frac{3y_1 + y_2}{4}\right), \left(\frac{x_1 + x_2}{2}, \frac{y_1 + y_2}{2}\right)$, and $\left(\frac{x_1 + 3x_2}{4}, \frac{y_1 + 3y_2}{4}\right)$

are the three points that divide the line segment joining (x_1, y_1) and (x_2, y_2) into four equal parts.

43. (a) $\left(\frac{3(1) + 4}{4}, \frac{3(-2) - 1}{4}\right) = \left(\frac{7}{4}, -\frac{7}{4}\right)$

$\left(\frac{1 + 4}{2}, \frac{-2 - 1}{2}\right) = \left(\frac{5}{2}, -\frac{3}{2}\right)$

$\left(\frac{1 + 3(4)}{4}, \frac{-2 + 3(-1)}{4}\right) = \left(\frac{13}{4}, -\frac{5}{4}\right)$

(b) $\left(\frac{3(-2) + 0}{4}, \frac{3(-3) + 0}{4}\right) = \left(-\frac{3}{2}, -\frac{9}{4}\right)$

$\left(\frac{-2 + 0}{2}, \frac{-3 + 0}{2}\right) = \left(-1, -\frac{3}{2}\right)$

$\left(\frac{-2 + 3(0)}{4}, \frac{-3 + 3(0)}{4}\right) = \left(-\frac{1}{2}, -\frac{3}{4}\right)$

Section 1.2 Graphs of Equations

Skills Review

1. $5y - 12 = x$
$5y = x + 12$
$y = \dfrac{x + 12}{5}$

2. $-y = 15 - x$
$y = x - 15$

3. $x^3y + 2y = 1$
$y(x^3 + 2) = 1$
$y = \dfrac{1}{x^3 + 2}$

4. $x^2 + x - y^2 - 6 = 0$
$-y^2 = 6 - x^2 - x$
$y^2 = x^2 + x - 6$
$y = \sqrt{x^2 + x - 6}$

5. $(x - 2)^2 + (y + 1)^2 = 9$
$(y + 1)^2 = 9 - (x - 2)^2$
$y + 1 = \sqrt{9 - (x - 2)^2}$
$y = \left(\sqrt{9 - (x - 2)^2}\right) - 1$
$= \sqrt{9 - (x^2 - 4x + 4)} - 1$
$= \sqrt{5 + 4x - x^2} - 1$

6. $(x + 6)^2 + (y - 5)^2 = 81$
$(y - 5)^2 = 81 - (x + 6)^2$
$y - 5 = \sqrt{81 - (x + 6)^2}$
$y = 5 + \sqrt{81 - (x + 6)^2}$
$= 5 + \sqrt{81 - (x^2 + 12x + 36)}$
$= 5 + \sqrt{45 - 12x - x^2}$

7. $x^2 - 4x + \left(\dfrac{-4}{2}\right)^2$
$x^2 - 4x + (-2)^2$
$x^2 - 4x + 4$

8. $x^2 + 6x + \left(\frac{6}{2}\right)^2$
$x^2 + 6x + 3^2$
$x^2 + 6x + 9$

9. $x^2 - 5x + \left(\dfrac{-5}{2}\right)^2$
$x^2 - 5x + \dfrac{25}{4}$

10. $x^2 + 3x + \left(\frac{3}{2}\right)^2$
$x^2 + 3x + \frac{9}{4}$

11. $x^2 - 3x + 2$
$(x - 1)(x - 2)$

12. $x^2 + 5x + 6$
$(x + 2)(x + 3)$

13. $y^2 - 3y + \frac{9}{4}$
$\left(y - \frac{3}{2}\right)^2$

14. $y^2 - 7y + \frac{49}{4}$
$\left(y - \frac{7}{2}\right)^2$

1. (a) This is not a solution point because
$2(1) - 2 - 3 = -3 \neq 0.$
(b) This is a solution point because
$2(1) - (-1) - 3 = 0.$
(c) This is a solution point because
$2(4) - 5 - 3 = 0.$

3. (a) This is a solution point because
$(1)^2 + (-\sqrt{3})^2 = 4.$
(b) This is not a solution point because
$\left(\frac{1}{2}\right)^2 + (-1)^2 = \frac{5}{4} \neq 4.$
(c) This is not a solution point because
$\left(\frac{3}{2}\right)^2 + \left(\frac{7}{2}\right)^2 = \frac{29}{2} \neq 4.$

5. The graph of $y = x - 2$ is a straight line with y-intercept at $(0, -2)$. So, it matches (e).

7. The graph of $y = x^2 + 2x$ is a parabola opening up with vertex at $(-1, -1)$. So, it matches (c).

9. The graph of $y = |x| - 2$ has a y-intercept at $(0, -2)$ and has x-intercepts at $(-2, 0)$ and $(2, 0)$. So, it matches (a).

11. Let $y = 0$. Then,

$2x - (0) - 3 = 0$

$x = \frac{3}{2}$.

Let $x = 0$. Then,

$2(0) - y - 3 = 0$

$y = -3$.

x-intercept: $\left(\frac{3}{2}, 0\right)$

y-intercept: $(0, -3)$

13. Let $y = 0$. Then,

$0 = x^2 + x - 2$

$0 = (x + 2)(x - 1)$

$x = -2, 1$.

Let $x = 0$. Then,

$y = (0)^2 + (0) - 2$

$y = -2$

x-intercepts: $(-2, 0), (1, 0)$

y-intercept: $(0, -2)$

15. Let $y = 0$. Then,

$0 = \sqrt{4 - x^2}$

$x^2 = 4$

$x = \pm 2$.

Let $x = 0$. Then,

$y = \sqrt{4 - (0)^2}$

$y = 2$.

x-intercepts: $(-2, 0), (2, 0)$

y-intercept: $(0, 2)$

17. Let $y = 0$. Then,

$0 = \dfrac{x^2 - 4}{x - 2}$

$0 = (x - 2)(x + 2)$

$x = \pm 2$.

Let $x = 0$. Then,

$y = \dfrac{(0)^2 - 4}{(0) - 2}$

$y = 2$.

x-intercept: Because the equation is undefined when $x = 2$, the only x-intercept is $(-2, 0)$.

y-intercept: $(0, 2)$

19. Let $y = 0$. Then,

$x^2(0) - x^2 + 4(0) = 0$

$x^2 = 0$

$x = 0$.

Let $x = 0$. Then,

$(0)^2 y - (0)^2 + 4y = 0$

$y = 0$.

x-intercept: $(0, 0)$

y-intercept: $(0, 0)$

21. $y = 2x + 3$

x	-2	$-\frac{3}{2}$	-1	0	1	2
y	-1	0	1	3	5	7

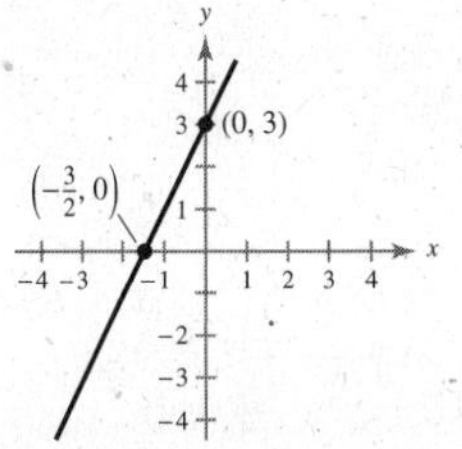

23. $y = x^2 - 3$

x	-2	-1	0	1	2	3
y	1	-2	-3	-2	1	6

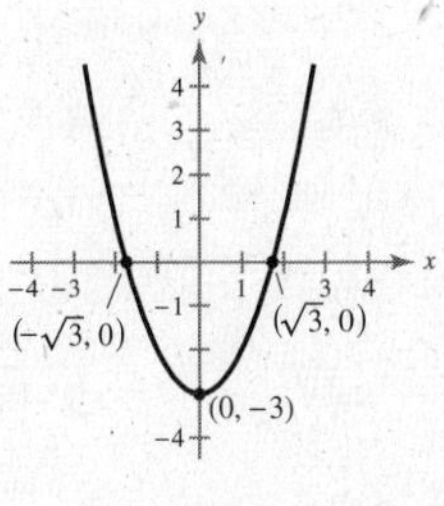

25. $y = (x - 1)^2$

x	-2	-1	0	1	2
y	9	4	1	0	1

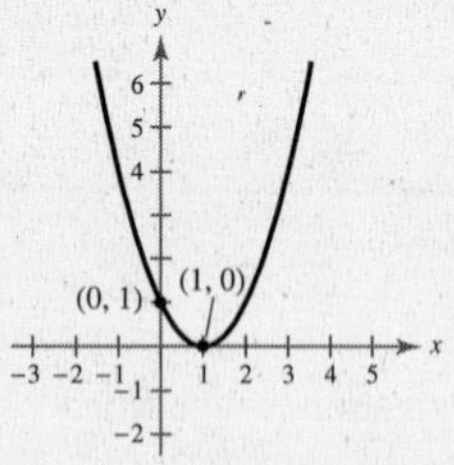

27. $y = x^3 + 2$

x	-2	-1	0	1	2
y	-6	1	2	3	10

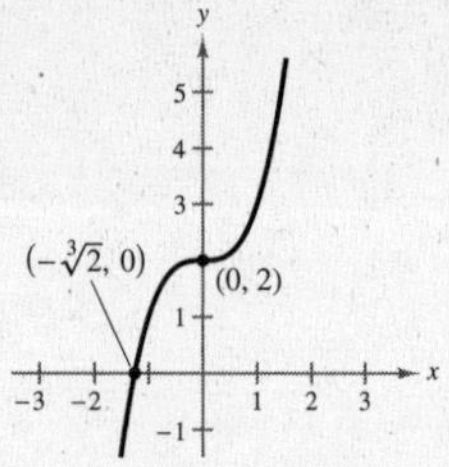

29. $y = -\sqrt{x - 1}$

x	1	2	3	4	5
y	0	-1	-1.41	-1.73	-2

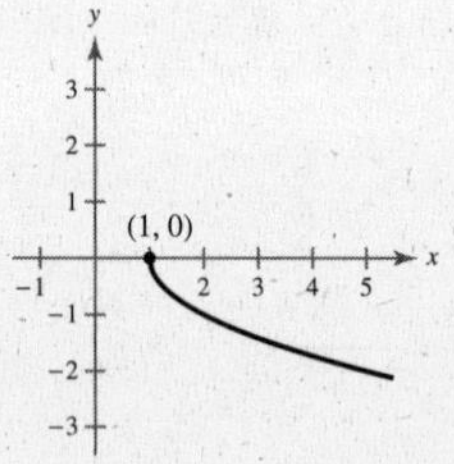

31. $y = |x + 1|$

x	-3	-2	-1	0	1
y	2	1	0	1	2

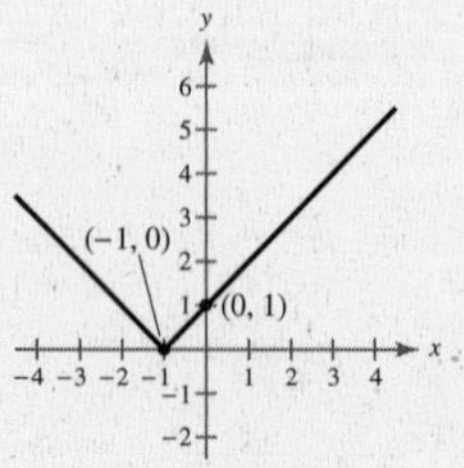

33. $y = \dfrac{1}{x - 3}$

x	-1	0	1	2	2.5	3.5	4	5	6
y	$-\frac{1}{4}$	$-\frac{1}{3}$	$-\frac{1}{2}$	-1	-2	2	1	$\frac{1}{2}$	$\frac{1}{3}$

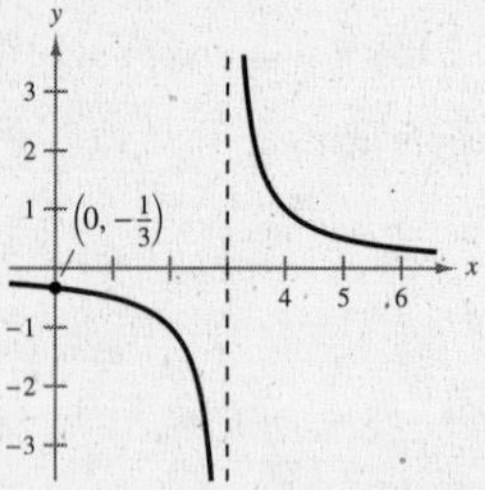

35. $x = y^2 - 4$

x	5	0	-3	-4
y	± 3	± 2	± 1	0

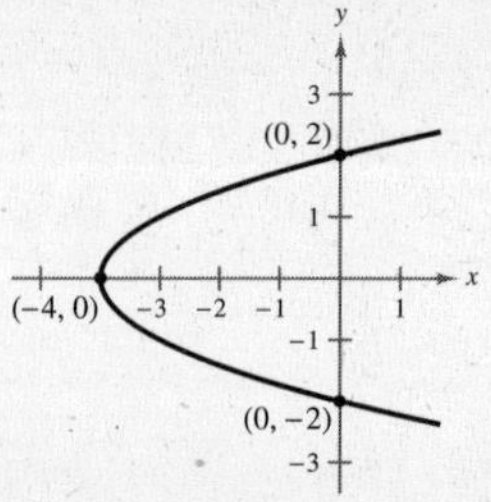

37.
$$(x - 0)^2 + (y - 0)^2 = 4^2$$
$$x^2 + y^2 = 16$$
$$x^2 + y^2 - 16 = 0$$

39.
$$(x - 2)^2 + (y + 1)^2 = 3^2$$
$$x^2 - 4x + 4 + y^2 + 2y + 1 = 9$$
$$x^2 + y^2 - 4x + 2y - 4 = 0$$

41. Since the point $(0, 0)$ lies on the circle, the radius must be the distance between $(0, 0)$ and $(-1, 2)$.

$$r = \sqrt{(0 + 1)^2 + (0 - 2)^2} = \sqrt{5}$$
$$(x + 1)^2 + (y - 2)^2 = 5$$
$$x^2 + y^2 + 2x - 4y = 0$$

43. Center = Midpoint = $\left(\dfrac{0+6}{2}, \dfrac{0+8}{2}\right) = (3, 4)$

r = distance from center to an endpoint

$= \sqrt{(0-3)^2 + (0-4)^2} = 5$

$$(x-3)^2 + (y-4)^2 = 5^2$$

$$x^2 - 6x + 9 + y^2 - 8y + 16 = 25$$

$$x^2 + y^2 - 6x - 8y = 0$$

45. $(x^2 - 2x + 1) + (y^2 + 6y + 9) = -6 + 1 + 9$

$$(x-1)^2 + (y+3)^2 = 4$$

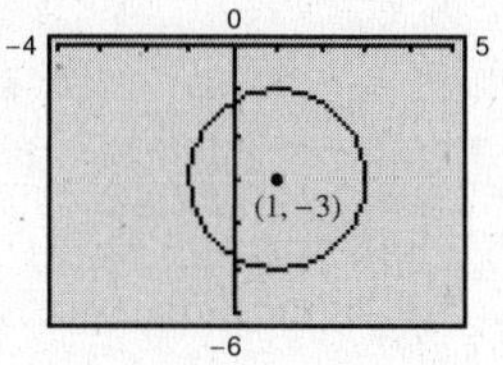

47. $(x^2 - 4x + 4) + (y^2 - 2y + 1) = -1 + 4 + 1$

$$(x-2)^2 + (y-1)^2 = 4$$

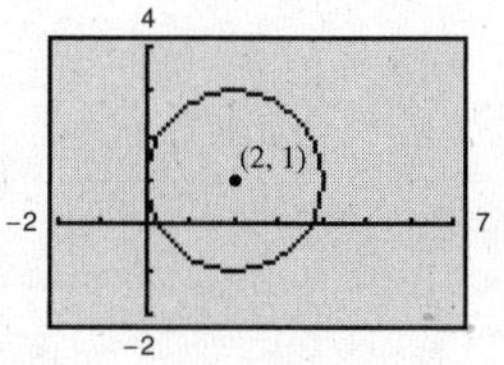

49.
$$x^2 + y^2 - x - y = \frac{3}{2}$$

$$\left(x^2 - x + \tfrac{1}{4}\right) + \left(y^2 - y + \tfrac{1}{4}\right) = \tfrac{3}{2} + \tfrac{1}{4} + \tfrac{1}{4}$$

$$\left(x - \tfrac{1}{2}\right)^2 + \left(y - \tfrac{1}{2}\right)^2 = 2$$

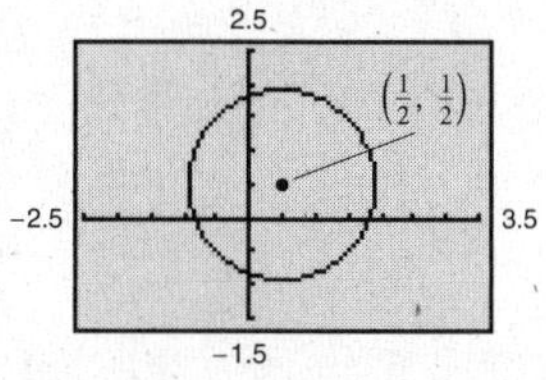

51.
$$x^2 + y^2 + 3x - 6y + \frac{41}{4} = 0$$

$$\left(x^2 + 3x + \tfrac{9}{4}\right) + \left(y^2 - 6y + 9\right) = -\tfrac{41}{4} + \tfrac{9}{4} + 9$$

$$\left(x + \tfrac{3}{2}\right)^2 + (y-3)^2 = 1$$

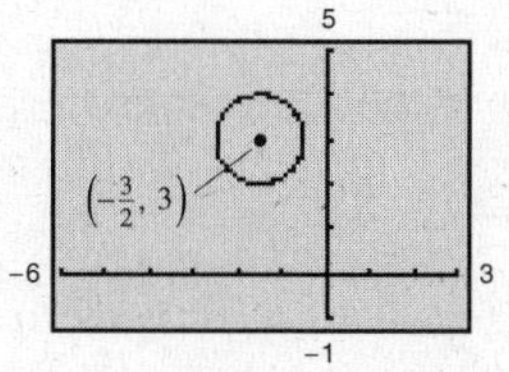

53. Solving for y in the equation $x + y = 2$ yields $y = 2 - x$, and solving for y in the equation $2x - y = 1$ yields $y = 2x - 1$. Then setting these two y-values equal to each other, we have

$$2 - x = 2x - 1$$

$$3 = 3x$$

$$x = 1.$$

The corresponding y-value is $y = 2 - 1 = 1$, so the point of intersection is $(1, 1)$.

55. Solving for y in the second equation yields $y = 10 - 2x$ and substituting this into the first equation gives

$$x^2 + (10 - 2x)^2 = 25$$

$$x^2 + 100 + 4x^2 - 40x = 25$$

$$5x^2 - 40x + 75 = 0$$

$$x^2 - 8x + 15 = 0$$

$$(x-3)(x-5) = 0$$

$$x = 3, 5.$$

The corresponding y-values are $y = 4$ and $y = 0$, so the points of intersection are $(3, 4)$ and $(5, 0)$.

57. By equating the y-values for the two equations, we have

$$x^3 = 2x$$

$$x^3 - 2x = 0$$

$$x(x^2 - 2) = 0$$

$$x = 0, \pm\sqrt{2}.$$

The corresponding y-values are $y = 0$, $y = -2\sqrt{2}$, and $y = 2\sqrt{2}$, so the points of intersection are $(0, 0)$, $(-\sqrt{2}, -2\sqrt{2})$, and $(\sqrt{2}, 2\sqrt{2})$.

59. By equating the y-values for the two equations, we have

$$x^4 - 2x^2 + 1 = 1 - x^2$$

$$x^4 - x^2 = 0$$

$$x^2(x+1)(x-1) = 0$$

$$x = 0, \pm 1.$$

The corresponding y-values are $y = 1$, 0, and 0, so the points of intersection are $(-1, 0)$, $(0, 1)$, and $(1, 0)$.

61. (a) Population of Oregon = Population of Oklahoma

$$3431 + 43.0t = 3448 + 20.1t$$
$$3431 + 22.9t = 3448$$
$$22.9t = 17$$
$$t \approx 0.74$$

So, you would expect that the population of Oregon would have exceeded the population of Oklahoma after $t \approx 0.74$ year, which was sometime during 2000.

(b) Population of Oregon in 2010: $P = 3431 + 43.0(10) = 3861 \Rightarrow 3{,}861{,}000$

Population of Oklahoma in 2010: $P = 3448 + 20.1(10) = 3649 \Rightarrow 3{,}649{,}000$

63. (a) Model: $G = 0.120t^2 + 0.64t + 7.5$

Year	2000	2001	2002	2003	2004	2005
Actual	7.600	8.270	8.753	11.000	12.000	13.600
Model	7.500	8.260	9.260	10.500	11.980	13.700

The model is a good fit. Answers will vary.

(b) When $t = 13$: $G = 36.1$ trillion Btu

65. (a) Model: $y = 0.136t^2 + 3.00t + 20.2$

t	2000	2001	2002	2003	2004	2005
Model	20.2	23.3	26.7	30.4	34.3	38.6
Actual	20.6	22.9	25.9	31.5	34.4	38.3

The model is a good fit. Answers will vary.

(b) When $t = 12$: $y \approx \$75.8$ billion

67. (a) Model: $y = 30.57t^2 + 381.4t + 13{,}852$

Year	2001	2002	2003	2004	2005
Transplants	14,264	14,737	15,271	15,867	16,523

(b)

Year	2001	2002	2003	2004	2005
Transplants (actual)	14,265	14,780	15,136	16,004	16,477

The model seems accurate. Answers will vary.

(c) When $t = 11$: $y \approx 21{,}746$

Yes, this prediction seems reasonable. Answers will vary.

69.

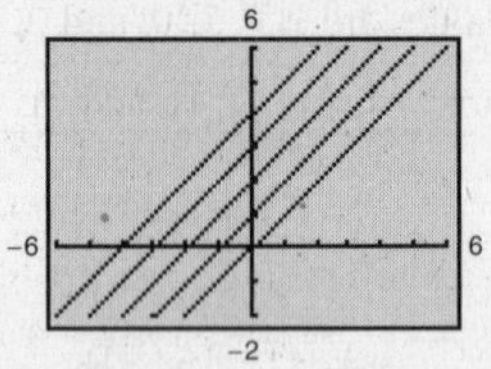

The constant term c is the y-intercept of the graph of the equation.

71.

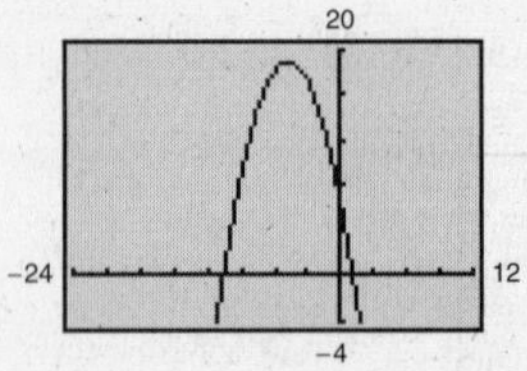

Intercepts: $(0, 6.25)$, $(1.0539, 0)$, $(-10.5896, 0)$

73. 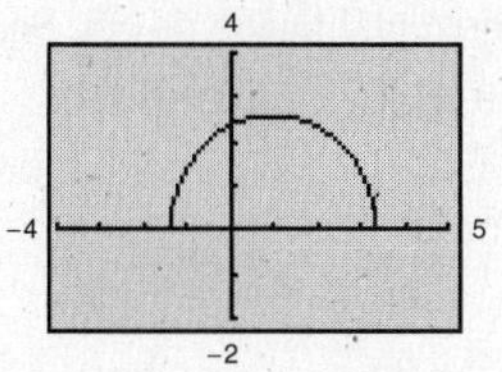

Intercepts: $(3.3256, 0), (-1.3917, 0), (0, 2.3664)$

75. 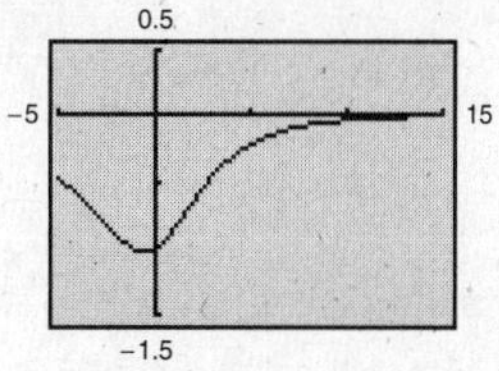

Intercepts: $(0, -1), (13.25, 0)$

Section 1.3 Lines in the Plane and Slope

Skills Review

1. $\frac{5-(-2)}{-3-4} = \frac{7}{-7} = -1$

2. $\frac{-7-(-10)}{4-1} = \frac{3}{3} = 1$

3. $-\frac{1}{m}, m = -3$

$-\frac{1}{-3} = \frac{1}{3}$

4. $-\frac{1}{m}, m = \frac{6}{7}$

$-\frac{1}{\frac{6}{7}} = -\frac{7}{6}$

5. $-4x + y = 7$

$y = 4x + 7$

6. $3x - y = 7$

$-y = 7 - 3x$

$y = 3x - 7$

7. $y - 2 = 3(x - 4)$

$y = 3(x - 4) + 2$

$y = 3x - 12 + 2$

$y = 3x - 10$

8. $y - (-5) = -1[x - (-2)]$

$y + 5 = -x - 2$

$y = -x - 7$

9. $y - (-3) = \frac{4-(-3)}{2-1}(x - 2)$

$y + 3 = \frac{7}{1}(x - 2)$

$y + 3 = 7x - 14$

$y = 7x - 17$

10. $y - 1 = \frac{-3-1}{-7-(-1)}[x - (-1)]$

$y - 1 = \frac{-4}{-6}(x + 1)$

$y - 1 = \frac{2}{3}(x + 1)$

$y - 1 = \frac{2}{3}x + \frac{2}{3}$

$y = \frac{2}{3}x + \frac{5}{3}$

1. The slope is $m = 1$ because the line rises one unit vertically for each unit the line moves to the right.

3. The slope is $m = 0$ because the line is horizontal.

5. 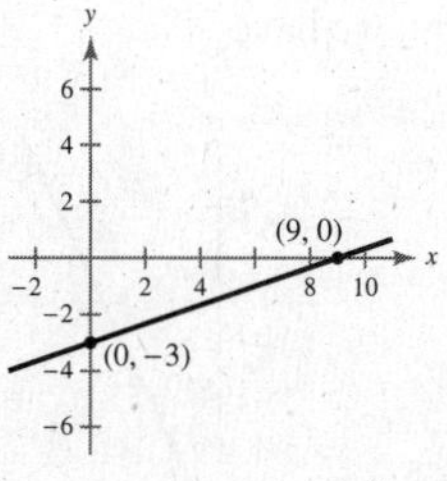

The slope is $m = \frac{0-(-3)}{9-0} = \frac{1}{3}$.

7. 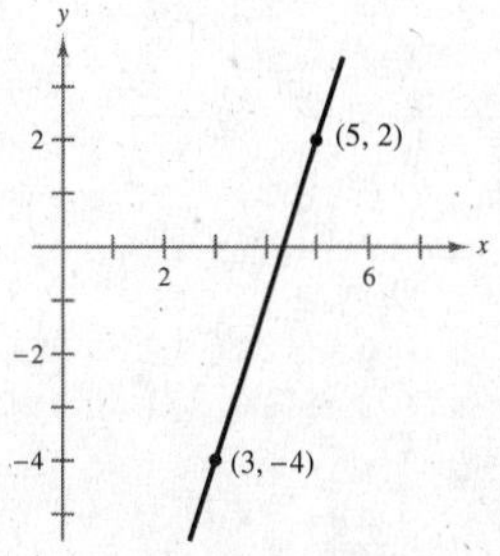

The slope is $m = \frac{2-(-4)}{5-3} = 3$.

9.

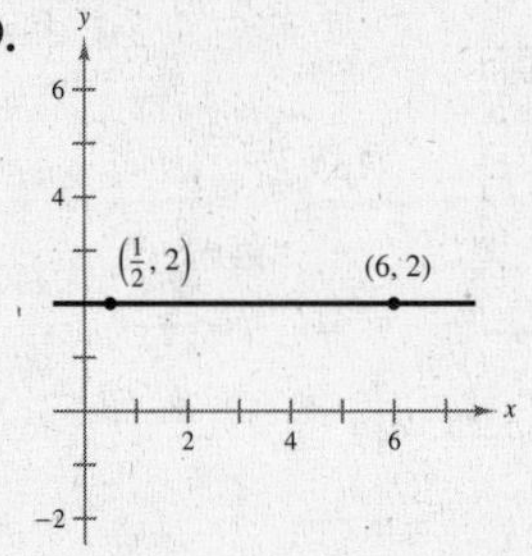

The slope is $m = \dfrac{2 - 2}{6 - (1/2)} = 0.$

So, the line is horizontal.

11.

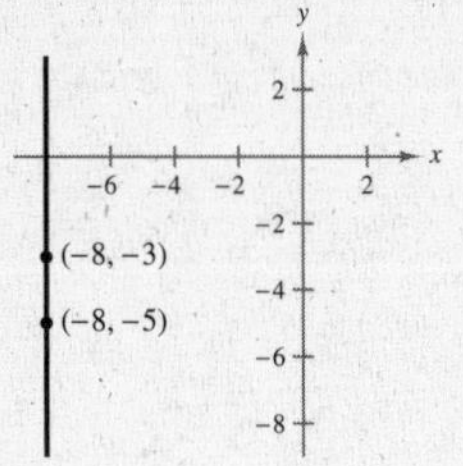

The slope is undefined because $m = \dfrac{-5 - (-3)}{-8 - (-8)}$ and division by zero is undefined. So, the line is vertical.

13.

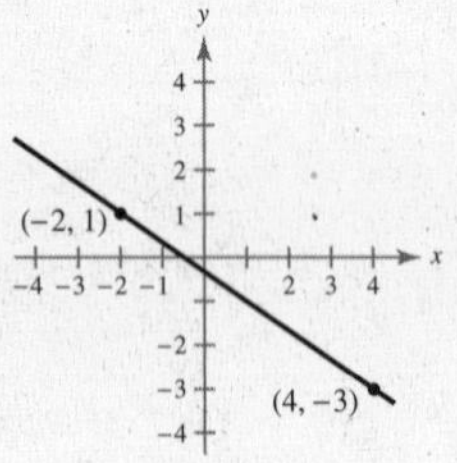

The slope is $m = \dfrac{-3 - 1}{4 - (-2)} = \dfrac{-4}{6} = -\dfrac{2}{3}.$

15.

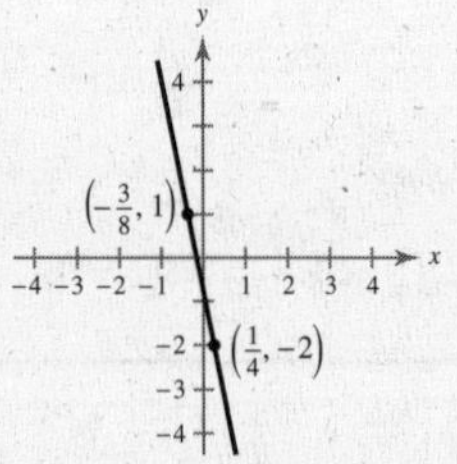

The slope is $m = \dfrac{1 - (-2)}{-\frac{3}{8} - \frac{1}{4}} = \dfrac{3}{-\frac{5}{8}} = -\dfrac{24}{5}.$

17.

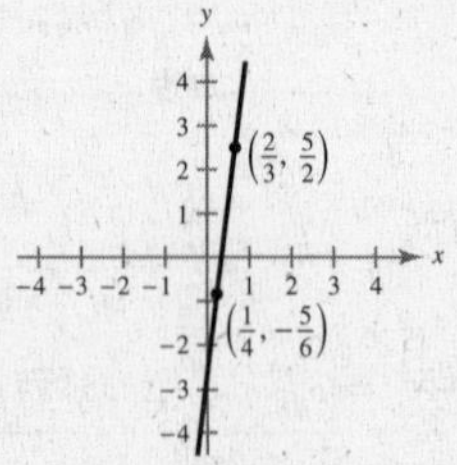

The slope is $m = \dfrac{\frac{5}{2} - \left(-\frac{5}{6}\right)}{\frac{2}{3} - \frac{1}{4}} = \dfrac{\frac{10}{3}}{\frac{5}{12}} = \dfrac{10}{3} \cdot \dfrac{12}{5} = 8,$

19. The equation of this horizontal line is $y = 1$. So, three additional points are $(0, 1)$, $(1, 1)$, and $(3, 1)$.

21. The equation of this line is

$$y + 4 = \frac{2}{3}(x - 6)$$
$$y = \frac{2}{3}x - 8.$$

So, three additional points are $(3, -6)$, $(9, -2)$, and $(12, 0)$.

23. The equation of the line is

$$y - 7 = -3(x - 1)$$
$$y = -3x + 10.$$

So, three additional points are $(0, 10)$, $(2, 4)$, and $(3, 1)$.

25. The equation of this vertical line is $x = -8$. So, three additional points are $(-8, 0)$, $(-8, 2)$, and $(-8, 3)$.

27. $x + 5y = 20$

$$y = -\tfrac{1}{5}x + 4$$

So, the slope is $m = -\frac{1}{5}$, and the y-intercept is $(0, 4)$.

29. $7x + 6y = 30$

$$y = -\tfrac{7}{6}x + 5$$

So, the slope is $m = -\frac{7}{6}$, and the y-intercept is $(0, 5)$.

31. $3x - y = 15$

$$y = 3x - 15$$

So, the slope is $m = 3$, and the y-intercept is $(0, -15)$.

33. $x = 4$

Because the line is vertical, the slope is undefined. There is no y-intercept.

35. $y - 4 = 0$

$$y = 4$$

So, the slope is $m = 0$, and the y-intercept is $(0, 4)$.

37. The slope of the line is $m = \dfrac{3 - (-5)}{4 - 0} = 2.$

Using the point-slope form, we have

$$y + 5 = 2(x - 0)$$
$$y = 2x - 5$$
$$0 = 2x - y - 5.$$

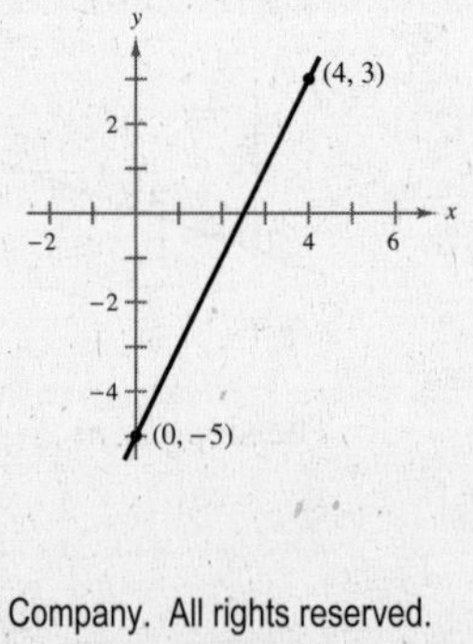

39. The slope of the line is $m = \dfrac{3 - 0}{-1 - 0} = -3$.

Using the point-slope form, we have

$$y = -3x$$
$$3x + y = 0.$$

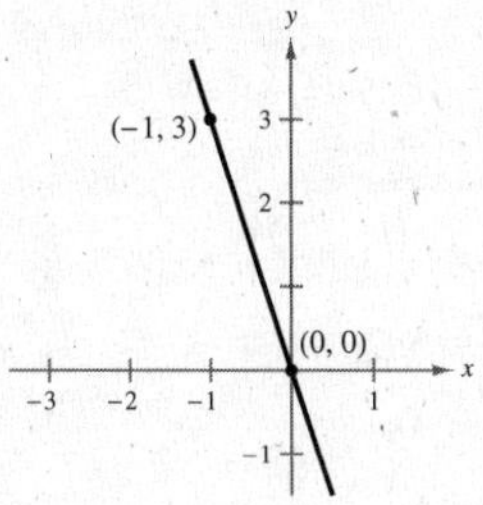

41. The slope of the line is $m = \dfrac{-2 - 3}{2 - 2} = \text{undefined}$.

So, the line is vertical, and its equation is

$$x = 2$$
$$x - 2 = 0.$$

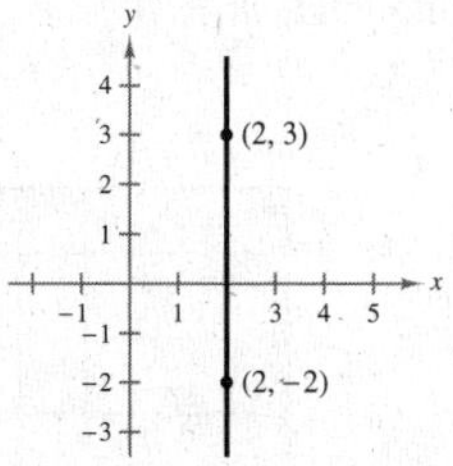

43. The slope of the line is $m = \dfrac{-1 - (-1)}{-2 - 3} = 0$. So, the line is horizontal, and its equation is

$$y = -1$$
$$y + 1 = 0.$$

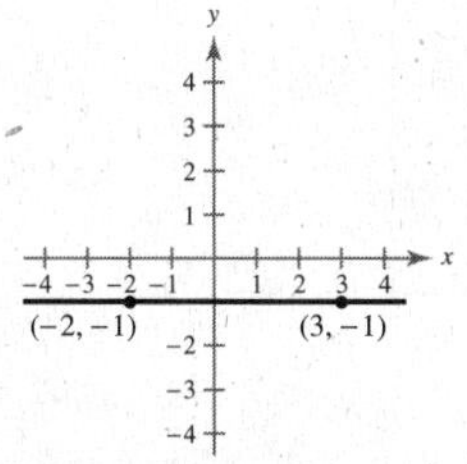

45. The slope of the line is $m = \dfrac{1 - 5/6}{(-1/3) + 2/3} = \dfrac{1}{2}$.

Using the point-slope form, we have

$$y - 1 = \frac{1}{2}\left(x + \frac{1}{3}\right)$$
$$y = \frac{1}{2}x + \frac{7}{6}$$
$$3x - 6y + 7 = 0.$$

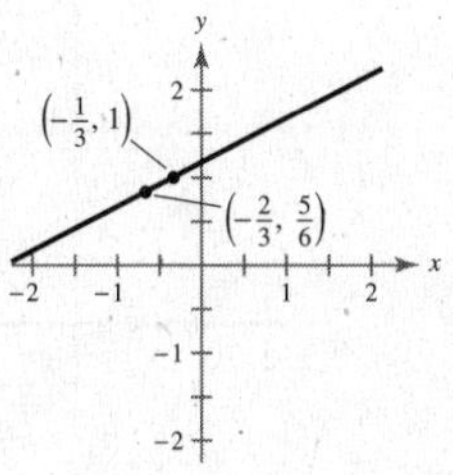

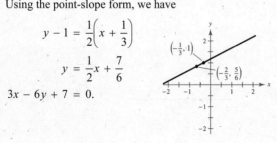

47. The slope of the line is $m = \dfrac{8 - 4}{1/2 + 1/2} = 4$.

Using the point-slope form, we have

$$y - 8 = 4\left(x - \frac{1}{2}\right)$$
$$y = 4x + 6$$
$$0 = 4x - y + 6.$$

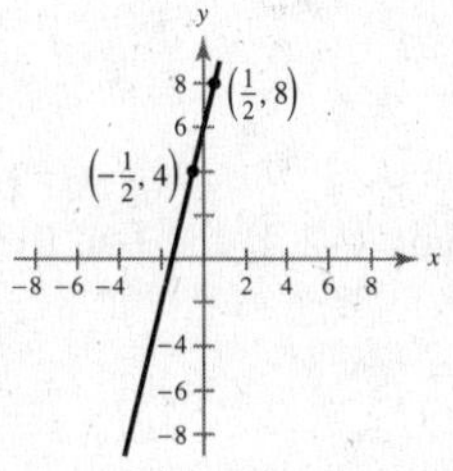

49. Using the slope-intercept form, we have

$$y = \tfrac{3}{4}x + 3$$
$$4y = 3x + 12$$
$$3x - 4y + 12 = 0.$$

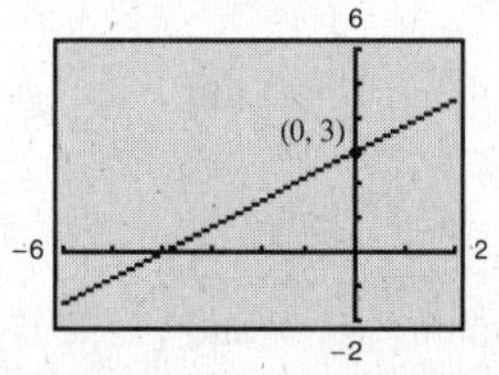

51. Because the slope is undefined, the line is vertical and its equation is $x = -1$.

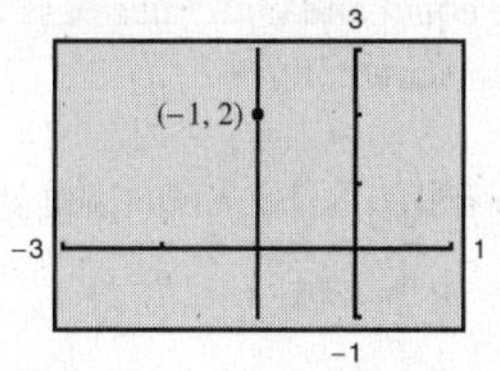

53. Because the slope is 0, the line is horizontal and its equation is $y = 7$.

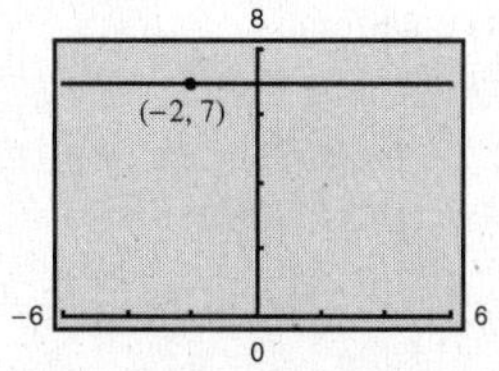

55. Using the point-slope form, we have

$$y + 2 = -4(x - 0)$$
$$y = -4x - 2$$
$$4x + y + 2 = 0.$$

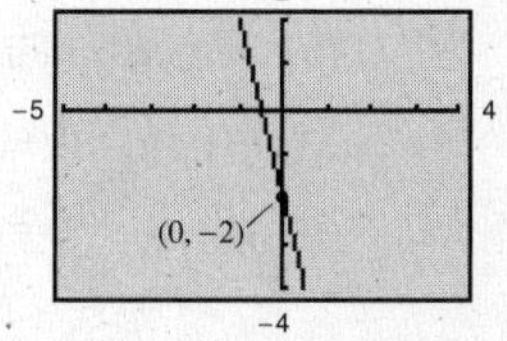

57. Using the slope-intercept form, we have

$$y = \tfrac{3}{4}x + \tfrac{2}{3}$$
$$12y = 9x + 8$$
$$9x - 12y + 8 = 0.$$

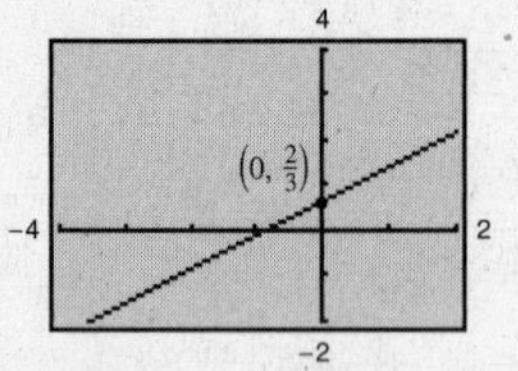

59. The slope of the line joining $(-2, 1)$ and $(-1, 0)$ is

$$\frac{1 - 0}{-2 - (-1)} = \frac{1}{-1} = -1.$$

The slope of the line joining $(-1, 0)$ and $(2, -2)$ is

$$\frac{0 - (-2)}{-1 - 2} = \frac{2}{-3} = -\frac{2}{3}.$$

Because the slopes are different, the points are not collinear.

61. The slope of the line joining $(2, 7)$ and $(-2, -1)$ is

$$\frac{-1 - 7}{-2 - 2} = 2.$$

The slope of the line joining $(0, 3)$ and $(-2, -1)$ is

$$\frac{-1 - 3}{-2 - 0} = 2.$$

Because the slopes are equal and both lines pass through $(-2, -1)$, the three points are collinear.

63. Because the line is vertical, it has an undefined slope, and its equation is

$$x = 3$$
$$x - 3 = 0.$$

65. Because the line is parallel to a horizontal line, it has a slope of $m = 0$, and its equation is

$$y = -10.$$

67. Given line: $y = -x + 7,\ m = -1$

(a) Parallel: $m_1 = -1$

$$y - 2 = -1(x + 3)$$
$$x + y + 1 = 0$$

(b) Perpendicular: $m_2 = 1$

$$y - 2 = 1(x + 3)$$
$$x - y + 5 = 0$$

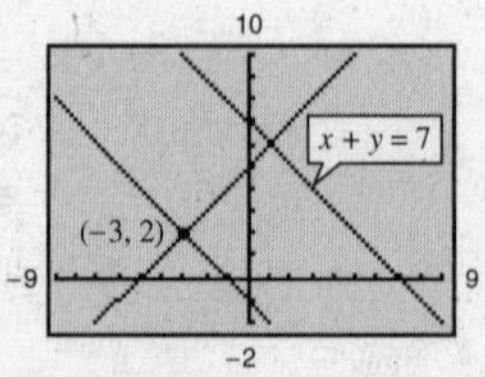

69. Given line: $y = -\tfrac{3}{4}x + \tfrac{7}{4},\ m = -\tfrac{3}{4}$

(a) Parallel: $m_1 = -\tfrac{3}{4}$

$$y - \tfrac{7}{8} = -\tfrac{3}{4}\left(x + \tfrac{2}{3}\right) = -\tfrac{3}{4}x - \tfrac{1}{2}$$
$$8y - 7 = -6x - 4$$
$$6x + 8y - 3 = 0$$

(b) Perpendicular: $m_2 = \tfrac{4}{3}$

$$y - \tfrac{7}{8} = \tfrac{4}{3}\left(x + \tfrac{2}{3}\right) = \tfrac{4}{3}x + \tfrac{8}{9}$$
$$72y - 63 = 96x + 64$$
$$96x - 72y + 127 = 0$$

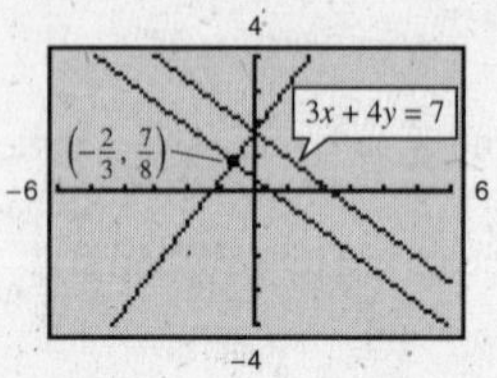

71. Given line: $y = -3$ is horizontal, $m = 0$

(a) Parallel: $m_1 = 0$

$$y - 0 = 0(x + 1)$$
$$y = 0$$

(b) Perpendicular: m_2 is undefined

$$x = -1$$

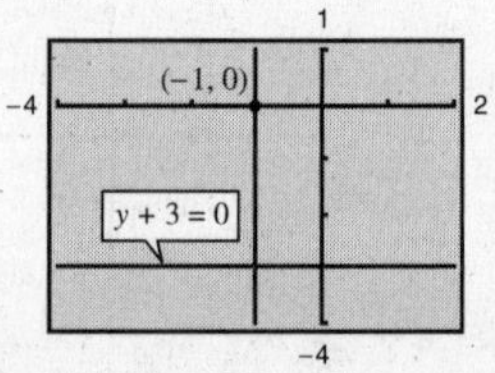

73. Given line: $x - 2 = 0$ is vertical, m is undefined

(a) Parallel: m_1 is undefined, $x = 1$

(b) Perpendicular: $m_2 = 0,\ y - 1 = 0(x - 1),\ y = 1$

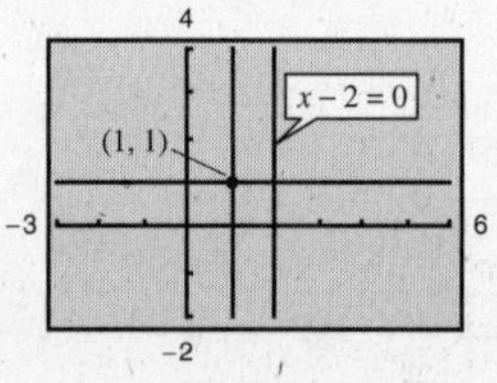

75. $y = -2$

x	-2	-1	0	1
y	-2	-2	-2	-2

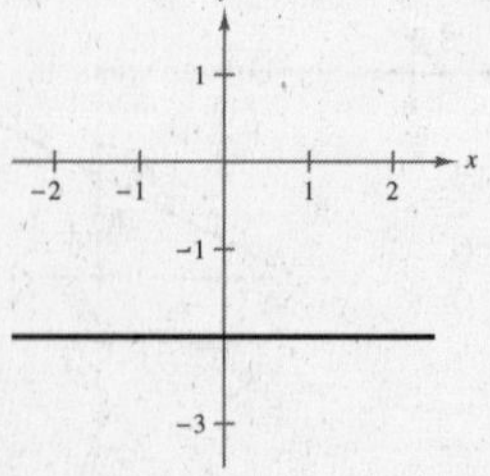

77. $y = 2x - 3$

x	-1	0	1	2
y	-5	-3	-1	1

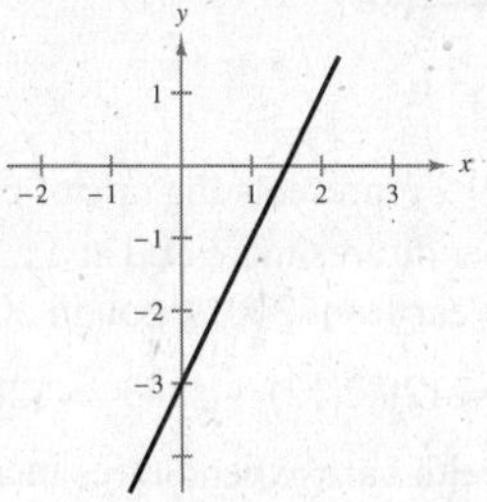

79. $y = -2x + 1$

x	-1	0	1	2
y	3	1	-1	-3

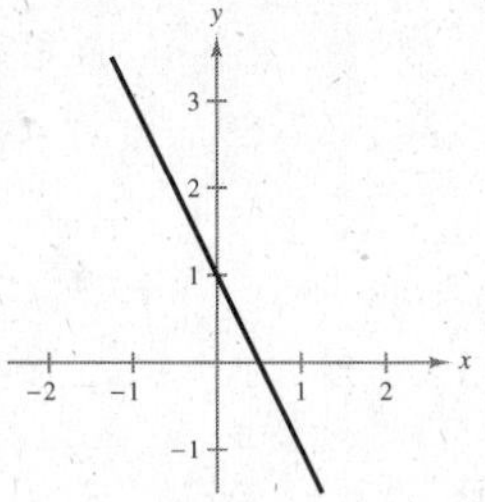

81. $y = -\frac{3}{5}x - 3$

x	0	2	4	5
y	-3	$-\frac{21}{5}$	$-\frac{27}{5}$	-6

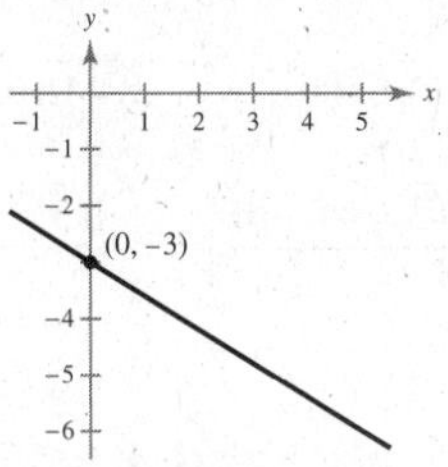

83. $y = -4x - 6$

x	$-\frac{3}{2}$	-1	0	1
y	0	-2	-6	-10

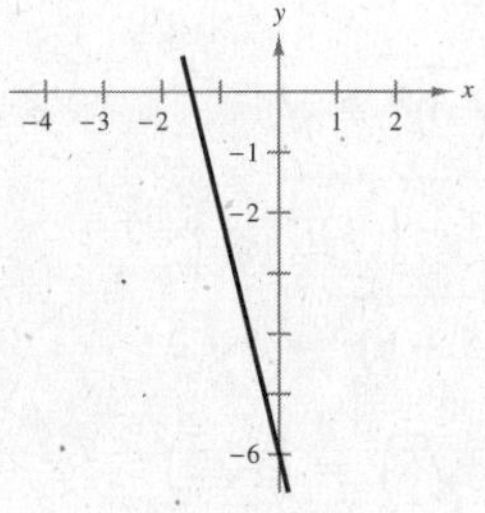

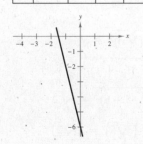

85. $\text{Slope} = \dfrac{\text{vertical change}}{\text{horizontal change}}$

$$= \frac{36 \text{ in.}}{32 \text{ ft}}$$

$$= \frac{3 \text{ ft}}{32 \text{ ft}} = 0.09375$$

Because $\frac{1}{12} \approx 0.083$, the ramp is steeper than recommended.

87. $P = 60t + 1300$ (t represents 2000.)

(a) The slope $m = 60$ tells you that the deer population increases by 60 each year. The y-intercept $(0, 1300)$ tells you that the deer population in 2000 was 1300.

(b) Population in 2005: $P = 60(5) + 1300 = 1600$

(c) Population in 2012: $P = 60(12) + 1300 = 2020$

89. (a) When $t = 8$, $y = 37{,}107$ and when $t = 15$, $y = 39{,}189$.

Slope: $m = \dfrac{39{,}189 - 37{,}107}{15 - 8} = \dfrac{2082}{7} \approx 297.4$

Equation: $y - 37{,}107 = 297.4(t - 8)$

$$y - 37{,}107 = 297.4t - 2379.2$$

$$y = 297.4t + 34{,}727.8$$

(b) When $t = 12$: $y = 297.4(12) + 34{,}727.8 \approx 38{,}297$ fatal crashes

(c) When $t = 18$: $y = 297.4(18) + 34{,}727.8 \approx 40{,}081$ fatal crashes

91. (a) $(0, 4024)$, $(5, 4255)$

$$m = \frac{4255 - 4024}{5 - 0} = \frac{231}{5} = 46.2$$

$$y - 4024 = 46.2(t - 0)$$

$$y = 46.2t + 4024$$

The slope $m = 46.2$ tells you that the population is increasing by 46.2 thousand per year.

(b) When $t = 2$: $y = 4116.4$

In 2002, the population was about 4,116,400.

(c) When $t = 4$: $y = 4208.8$

In 2004, the population was about 4,208,800.

(d) 2002: 4,024,000

2004: 4,198,000

The estimates were close to the exact values.

(e) The model could possibly be used to predict the population in 2009 if the population continues to grow at the same linear rate.

93. $(0, 32), (100, 212)$

$$F - 32 = \frac{212 - 32}{100 - 0}(C - 0)$$

$$F = 1.8C + 32 = \frac{9}{5}C + 32$$

or

$$C = \frac{5}{9}(F - 32)$$

95. (a) The equipment depreciates $\frac{45{,}000}{5} = \$9000$ per year, so the value is $y = -9000t + 45{,}000$, where $0 \le t \le 5$.

(b) 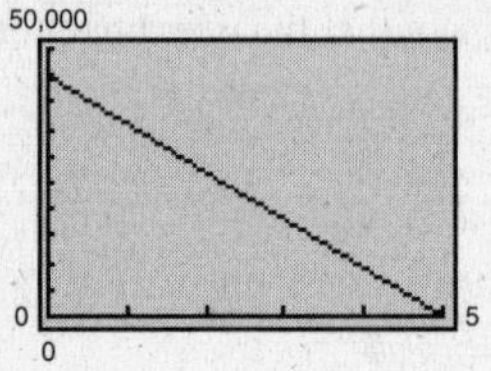

(c) When $t = 3$, the value is \$18,000.

(d) The value is \$28,000 when $t \approx 1.88$ years.

97. (a) 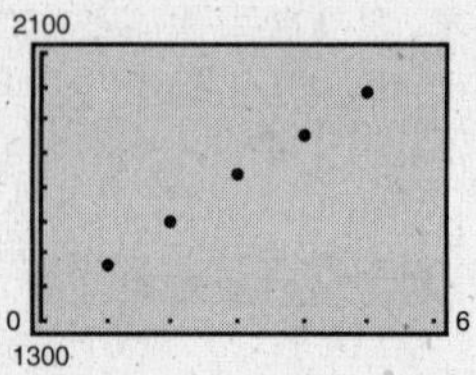

(b) $y = 129.2t + 1343$

(c) The slope $m = 129.2$ represents the rate of change. So, health care expenditures increased at a rate of \$129.2 billion per year from 2001 through 2005.

(d) When $t = 12$: $y = 129.2(12) + 1343 = \$2893.4$ billion. Because health care expenditures increased each year from 2001 to 2005, this prediction seems reasonable.

Mid-Chapter Quiz Solutions

1. (a) 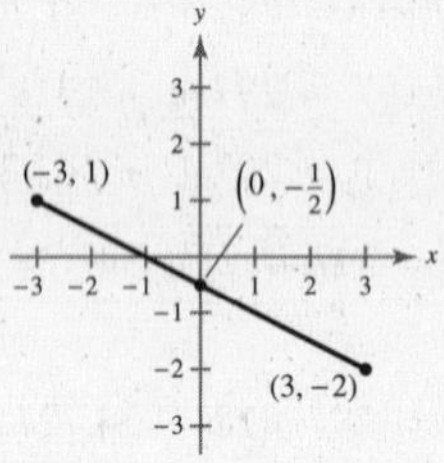

(b) $d = \sqrt{(-3-3)^2 + (1-(-2))^2} = \sqrt{36+9} = 3\sqrt{5}$

(c) $\text{Midpoint} = \left(\frac{-3+3}{2}, \frac{1-2}{2}\right) = \left(0, -\frac{1}{2}\right)$

2. (a) 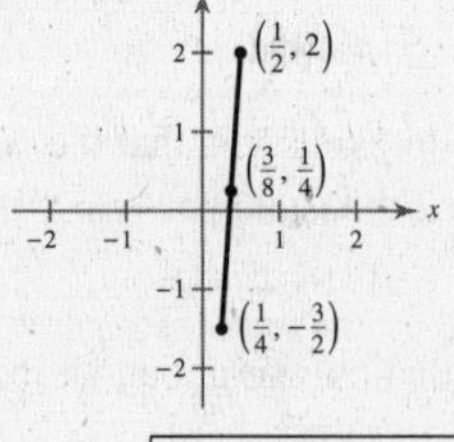

(b) $d = \sqrt{\left(\frac{1}{2} - \frac{1}{4}\right)^2 + \left(2 - \left(-\frac{3}{2}\right)\right)^2}$

$= \sqrt{\frac{1}{16} + \frac{49}{4}} = \sqrt{\frac{197}{16}} = \frac{1}{4}\sqrt{197}$

(c) $\text{Midpoint} = \left(\frac{\frac{1}{2} + \frac{1}{4}}{2}, \frac{2 - \frac{3}{2}}{2}\right) = \left(\frac{3}{8}, \frac{1}{4}\right)$

3. (a) 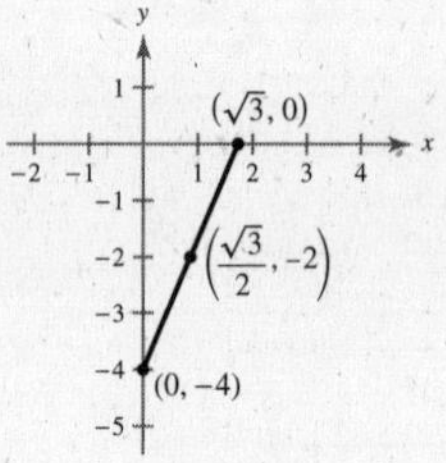

(b) $d = \sqrt{(0 - \sqrt{3})^2 + (-4 - 0)^2} = \sqrt{3 + 16} = \sqrt{19}$

(c) $\text{Midpoint} = \left(\frac{0 + \sqrt{3}}{2}, \frac{-4 - 0}{2}\right) = \left(\frac{\sqrt{3}}{2}, -2\right)$

4. 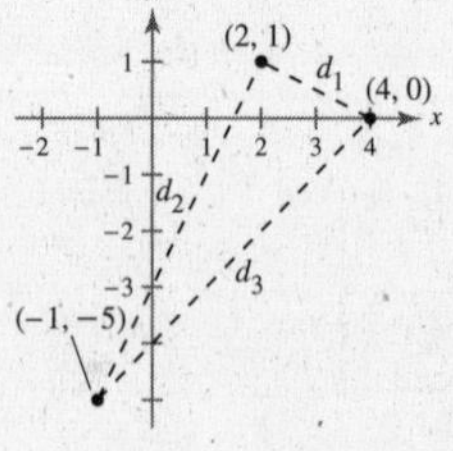

$a = \sqrt{(2-4)^2 + (1-0)^2} = \sqrt{5}$

$b = \sqrt{(2-(-1))^2 + (1-(-5))^2} = 3\sqrt{5}$

$c = \sqrt{(-1-4)^2 + (-5-0)^2} = 5\sqrt{2}$

$a^2 + b^2 = (\sqrt{5})^2 + (3\sqrt{5})^2 = (5\sqrt{2})^2 = c^2$

5. Use the points $(2003,\ 5719)$ and $(2005,\ 5800)$.

$$\text{Midpoint} = \left(\frac{2003 + 2005}{2}, \frac{5719 + 5800}{2}\right)$$
$$= (2004,\ 5759.5)$$

The population in 2004 was about 5,759,500 people.

6. $y = 5x + 2$

x	$-\frac{2}{5}$	0	$\frac{1}{5}$	1
y	0	2	3	7

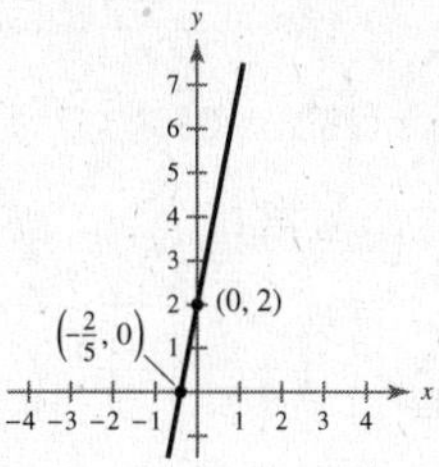

7. $y = x^2 + x - 6$

x	−3	−2	−1	−0.5	0	1	2
y	0	−4	−6	−6.25	−6	−4	0

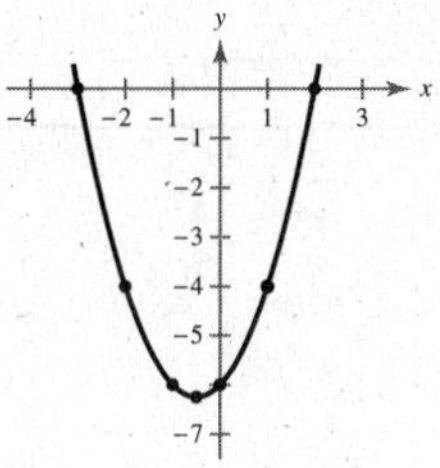

8. $y = |x - 3|$

x	0	1	2	3	4	5	6
y	3	2	1	0	1	2	3

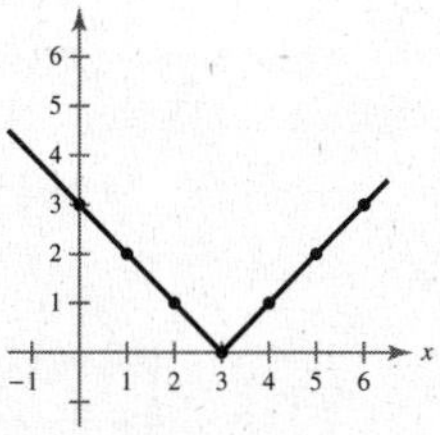

13. Model: $S = 4.42t + 22.3$

Year	2000	2001	2002	2003	2004	2005
Sales	\$22.3	\$26.72	\$31.14	\$35.56	\$39.98	\$44.4

9. $(x + 1)^2 + (y - 0)^2 = 37$

$$x^2 + 2x + 1 + y^2 = 37$$
$$x^2 + y^2 + 2x - 36 = 0$$

10. Because the point $(-1,\ 2)$ lies on the circle, the radius must be the distance between $(-1,\ 2)$ and $(2,\ -2)$.

$$r = \sqrt{(-1 - 2)^2 + (2 + 2)^2} = 5$$
$$(x - 2)^2 + (y + 2)^2 = 25$$
$$x^2 - 4x + 4 + y^2 + 4y + 4 = 25$$
$$x^2 + y^2 - 4x + 4y - 17 = 0$$

11.
$$x^2 + y^2 + 8x - 6y + 16 = 0$$
$$(x^2 + 8x + 16) + (y^2 - 6y + 9) = -16 + 16 + 9$$
$$(x + 4)^2 + (y - 3)^2 = 9$$

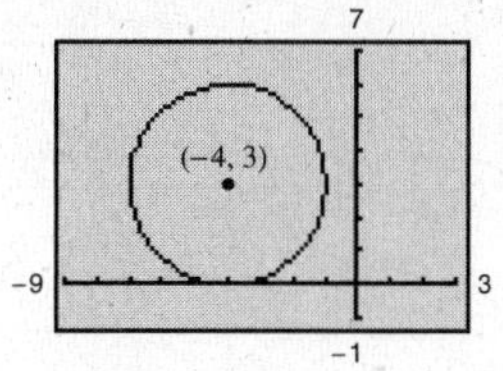

12.
$$4x^2 + 4y^2 - 8x + 4y - 11 = 0$$
$$x^2 + y^2 - 2x + y - \frac{11}{4} = 0$$
$$(x^2 - 2x + 1) + \left(y^2 + y + \frac{1}{4}\right) = \frac{11}{4} + 1 + \frac{1}{4}$$
$$(x - 1)^2 + \left(y + \frac{1}{2}\right)^2 = 4$$

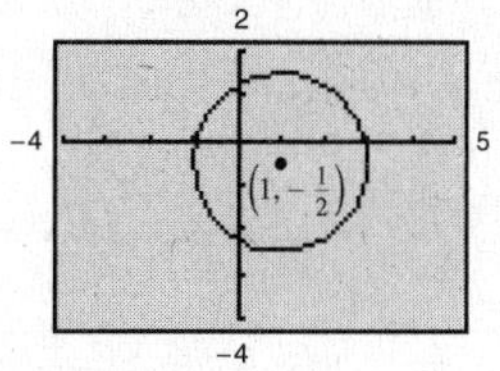

14. $(1, -1), (-4, 5)$

$m = \frac{5+1}{-4-1} = -\frac{6}{5}$

$y + 1 = -\frac{6}{5}(x - 1)$

$y = -\frac{6}{5}x + \frac{1}{5}$

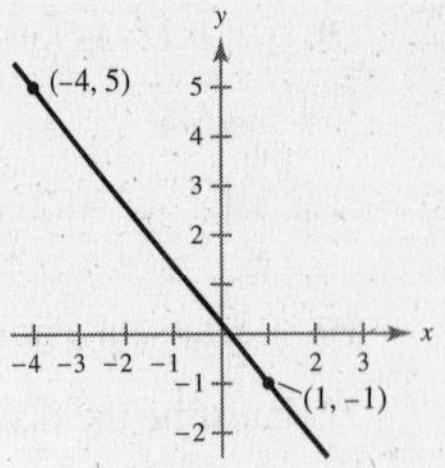

15. $(-2, 3), (-2, 2)$

$m = \frac{2-3}{-2+2} = \text{undefined}$

Because the slope is undefined, the line is vertical and its equation is $x = -2$.

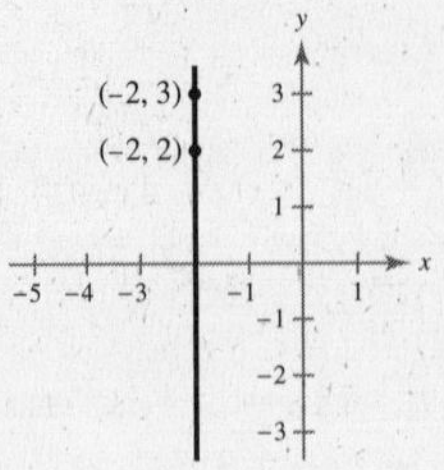

16. $\left(\frac{5}{2}, 2\right), (0, 2)$

$m = \frac{2-2}{0-\frac{5}{2}} = 0$

Because the slope is 0, the line is horizontal and its equation is $y = 2$.

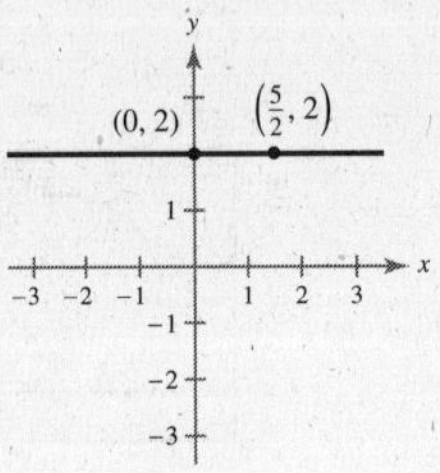

17. Given line: $y = -\frac{1}{4}x - \frac{1}{2}, m = -\frac{1}{4}$

(a) Parallel: $m_1 = -\frac{1}{4}$

$y + 5 = -\frac{1}{4}(x - 3)$

$y = -\frac{1}{4}x - \frac{17}{4}$

(b) Perpendicular: $m_2 = 4$

$y + 5 = 4(x - 3)$

$y = 4x - 17$

18. (a) When $d = 0$, $p = 1$ and when $d = 132$, $p = 5$.

Slope: $m = \frac{5-1}{132-0} = \frac{4}{132} = \frac{1}{33}$

Equation: $p - 1 = \frac{1}{33}(d - 0)$

$p = \frac{1}{33}d + 1$

(b) The rate of change is represented by the slope. So, pressure increases at a rate of $\frac{1}{33} \approx 0.03$ atmosphere per foot.

Section 1.4 Functions

Skills Review

1. $5(-1)^2 - 6(-1) + 9 = 5(1) + 6 + 9 = 20$

2. $(-2)^3 + 7(-2)^2 - 10 = -8 + 7(4) - 10 = -18 + 28 = 10$

3. $(x - 2)^2 + 5x - 10 = x^2 - 4x + 4 + 5x - 10 = x^2 + x - 6$

4. $(3 - x) + (x + 3)^3 = (3 - x) + (x + 3)(x^2 + 6x + 9)$

$= (3 - x) + x^3 + 3x^2 + 6x^2 + 18x + 9x + 27$

$= x^3 + 9x^2 + 26x + 30$

5. $\frac{1}{1 - (1 - x)} = \frac{1}{1 - 1 + x} = \frac{1}{x}$

6. $1 + \frac{x-1}{x} = \frac{x}{x} + \frac{x-1}{x} = \frac{x + x - 1}{x} = \frac{2x-1}{x}$

7. $2x + y - 6 = 11$

$y = -2x + 17$

8. $5y - 6x^2 - 1 = 0$

$5y = 6x^2 + 1$

$y = \frac{6x^2 + 1}{5}$

$= \frac{6}{5}x^2 + \frac{1}{5}$

Skills Review —*continued*—

9. $(y - 3)^2 = 5 + (x + 1)^2$

$y - 3 = \sqrt{5 + (x + 1)^2}$

$y - 3 = \sqrt{5 + x^2 + 2x + 1}$

$y = \sqrt{x^2 + 2x + 6} + 3$

10. $y^2 - 4x^2 = 2$

$y^2 = 2 + 4x^2$

$y = \sqrt{2 + 4x^2}$

11. $x = \dfrac{2y - 1}{4}$

$4x = 2y - 1$

$4x + 1 = 2y$

$\dfrac{4x + 1}{2} = y$

$2x + \dfrac{1}{2} = y$

12. $x = \sqrt[3]{2y - 1}$

$x^3 = 2y - 1$

$-2y = -x^3 - 1$

$y = \frac{1}{2}x^3 + \frac{1}{2}$

1. $y = \pm\sqrt{4 - x^2}$

y is *not* a function of *x* since there are two values of *y* for some *x*.

3. $\frac{1}{2}x - 6y = -3$

$y = \frac{1}{12}x + \frac{1}{2}$

y is a function of *x* since there is only one value of *y* for each *x*.

5. $y = 4 - x^2$

y is a function of *x* since there is only one value of *y* for each *x*.

7. $y = |x + 2|$

y is a function of *x* since there is only one value of *y* for each *x*.

9. 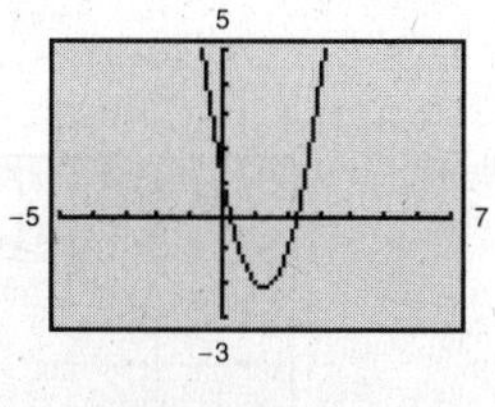

Domain: $(-\infty, \infty)$

Range: $[-2.125, \infty)$

11. 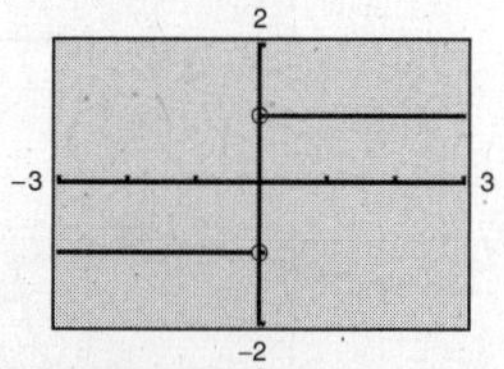

Domain: $(-\infty, 0) \cup (0, \infty)$

Range: $\{-1, 1\}$

13. 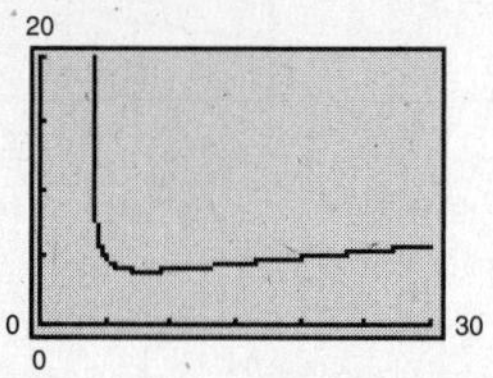

Domain: $(4, \infty)$

Range: $[4, \infty)$

15. 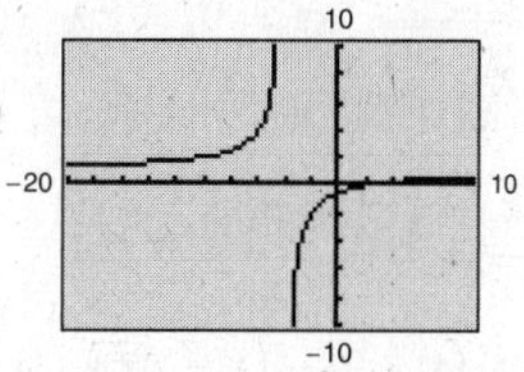

Domain: $(-\infty, -4) \cup (-4, \infty)$

Range: $(-\infty, 1) \cup (1, \infty)$

17. Domain: $(-\infty, \infty)$

Range: $(-\infty, \infty)$

19. Domain: $(-\infty, \infty)$

Range: $(-\infty, 4]$

21. (a) $f(0) = 3(0) - 2 = 0 - 2 = -2$

(b) $f(x - 1) = 3(x - 1) - 2$

$= 3x - 3 - 2 = 3x - 5$

(c) $f(x + \Delta x) = 3(x + \Delta x) - 2$

$= 3x + 3\Delta x - 2$

23. (a) $g\left(\dfrac{1}{4}\right) = \dfrac{1}{1/4} = 4$

(b) $g(x + 4) = \dfrac{1}{x + 4}$

(c) $g(x + \Delta x) - g(x) = \dfrac{1}{x + \Delta x} - \dfrac{1}{x}$

$= \dfrac{x - (x + \Delta x)}{x(x + \Delta x)}$

$= \dfrac{-\Delta x}{x(x + \Delta x)}$

25. $\dfrac{f(x+\Delta x)-f(x)}{\Delta x}$

$$= \frac{(x+\Delta x)^2 - 5(x+\Delta x) + 2 - (x^2 - 5x + 2)}{\Delta x}$$

$$= \frac{[x^2 + 2x\Delta x + (\Delta x)^2 - 5x - 5\Delta x + 2] - [x^2 - 5x + 2]}{\Delta x}$$

$$= \frac{2x\Delta x + (\Delta x)^2 + 5\Delta x}{\Delta x}$$

$$= 2x + \Delta x - 5,\ \Delta x \neq 0$$

27. $\dfrac{g(x+\Delta x)-g(x)}{\Delta x}$

$$= \frac{\sqrt{x+\Delta x+1} - \sqrt{x+1}}{\Delta x} \cdot \frac{\sqrt{x+\Delta x+1} + \sqrt{x+1}}{\sqrt{x+\Delta x+1} + \sqrt{x+1}}$$

$$= \frac{(x+\Delta x+1) - (x+1)}{\Delta x\left[\sqrt{x+\Delta x+1} + \sqrt{x+1}\right]}$$

$$= \frac{1}{\sqrt{x+\Delta x+1} + \sqrt{x+1}},\ \Delta x \neq 0$$

29. $\dfrac{f(x+\Delta x)-f(x)}{\Delta x} = \dfrac{\dfrac{1}{x+\Delta x-2} - \dfrac{1}{x-2}}{\Delta x}$

$$= \frac{(x-2) - (x+\Delta x-2)}{(x+\Delta x-2)(x-2)\Delta x}$$

$$= \frac{-1}{(x+\Delta x-2)(x-2)},\ \Delta x \neq 0$$

31. y is *not* a function of x.

33. y *is* a function of x.

35. (a) $f(x) + g(x) = (2x-5) + 5 = 2x$

(b) $f(x)\cdot g(x) = (2x-5)(5) = 10x - 25$

(c) $\dfrac{f(x)}{g(x)} = \dfrac{2x-5}{5}$

(d) $f(g(x)) = f(5) = 2(5) - 5 = 5$

(e) $g(f(x)) = g(2x-5) = 5$

37. (a) $f(x) + g(x) = (x^2+1) + (x-1) = x^2 + x$

(b) $f(x)\cdot g(x) = (x^2+1)(x-1) = x^3 - x^2 + x - 1$

(c) $\dfrac{f(x)}{g(x)} = \dfrac{x^2+1}{x-1},\ x \neq 1$

(d) $f(g(x)) = f(x-1) = (x-1)^2 + 1 = x^2 - 2x + 2$

(e) $g(f(x)) = g(x^2+1) = (x^2+1) - 1 = x^2$

39. $f(x) = \sqrt{x},\ g(x) = x^2 - 1$

(a) $f(g(1)) = f(1^2 - 1) = f(0) = 0$

(b) $g(f(1)) = g(\sqrt{1}) = g(1) = 0$

(c) $g(f(0)) = g(0) = -1$

(d) $f(g(-4)) = f(15) = \sqrt{15}$

(e) $f(g(x)) = f(x^2 - 1) = \sqrt{x^2 - 1}$

(f) $g(f(x)) = g(\sqrt{x}) = x - 1,\ x \geq 0$

41. $f(g(x)) = f\left(\dfrac{x-1}{5}\right) = 5\left(\dfrac{x-1}{5}\right) + 1 = x$

$g(f(x)) = g(5x+1) = \dfrac{(5x+1)-1}{5} = x$

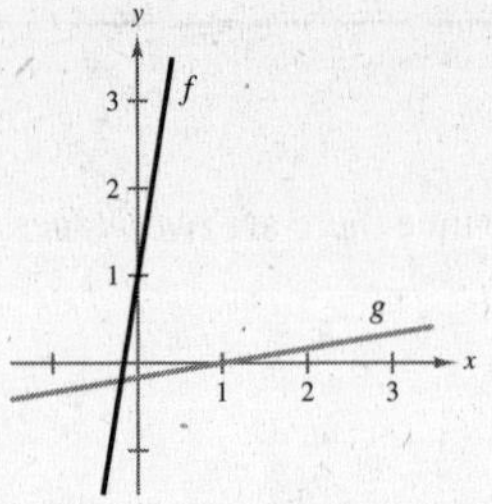

43. $f(g(x)) = f(\sqrt{9-x}) = 9 - (\sqrt{9-x})^2 = x$

$g(f(x)) = g(9 - x^2)$

$= \sqrt{9 - (9 - x^2)} = \sqrt{x^2} = x,\ x \geq 0$

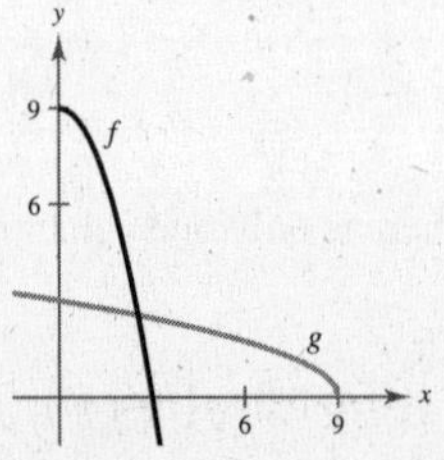

45. $f(x) = 2x - 3 = y$

$2y - 3 = x$

$y = \dfrac{x+3}{2}$

$f^{-1}(x) = \dfrac{x+3}{2}$

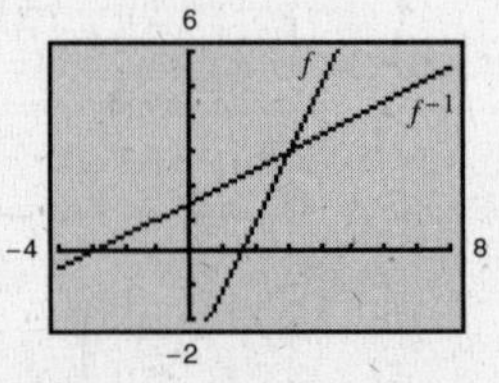

47. $f(x) = x^5 = y$

$y^5 = x$

$y = \sqrt[5]{x}$

$f^{-1}(x) = \sqrt[5]{x}$

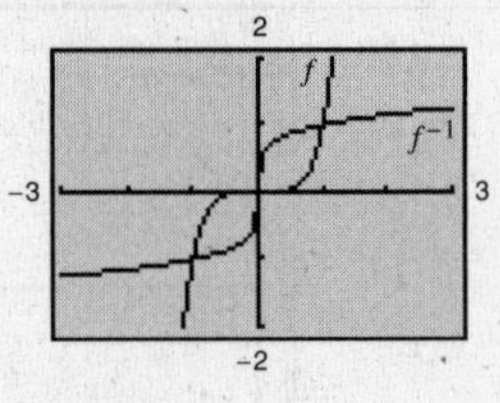

49. $f(x) = \sqrt{9 - x^2} = y,\ 0 \le x \le 3$

$$\sqrt{9 - y^2} = x$$
$$9 - y^2 = x^2$$
$$y^2 = 9 - x^2$$
$$y = \sqrt{9 - x^2}$$
$$f^{-1}(x) = \sqrt{9 - x^2},\ 0 \le x \le 3$$

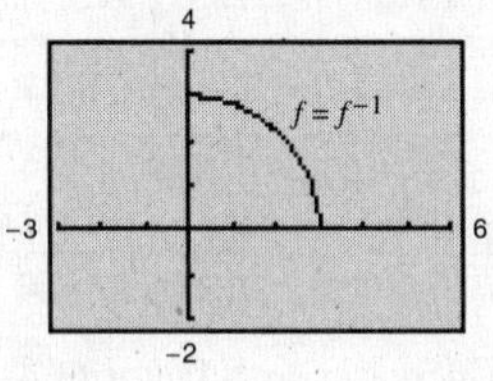

51. $f(x) = x^{2/3} = y,\ x \ge 0$

$$y^{2/3} = x$$
$$y = x^{3/2}$$
$$f^{-1}(x) = x^{3/2}$$

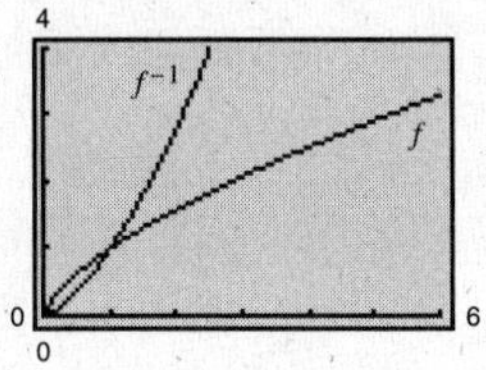

53.

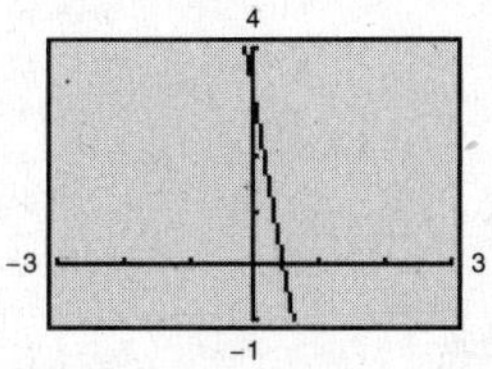

$f(x) = 3 - 7x$ is one-to-one.

$$y = 3 - 7x$$
$$x = 3 - 7y$$
$$y = \frac{3 - x}{7}$$

55.

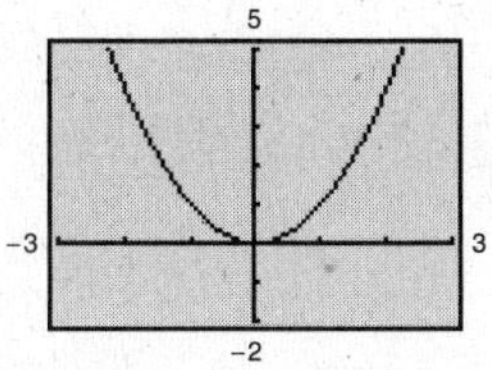

$f(x) = x^2$

f is *not* one-to-one because $f(1) = 1 = f(-1)$.

57. $f(x) = |x + 3|$

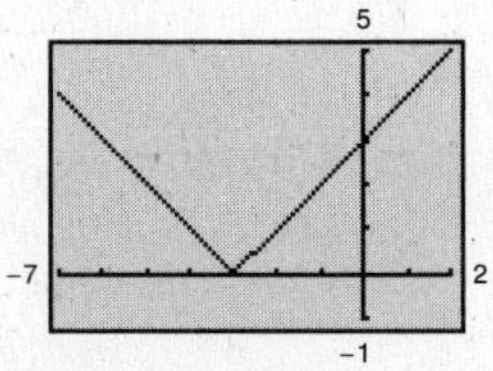

f is *not* one-to-one because $f(-5) = 2 = f(-1)$.

59. (a) $y = \sqrt{x} + 2$

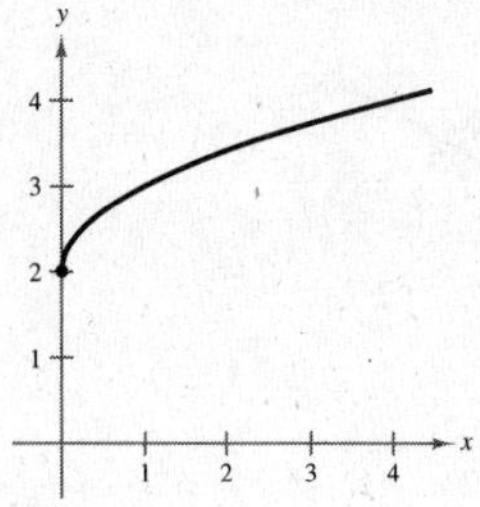

(b) $y = -\sqrt{x}$

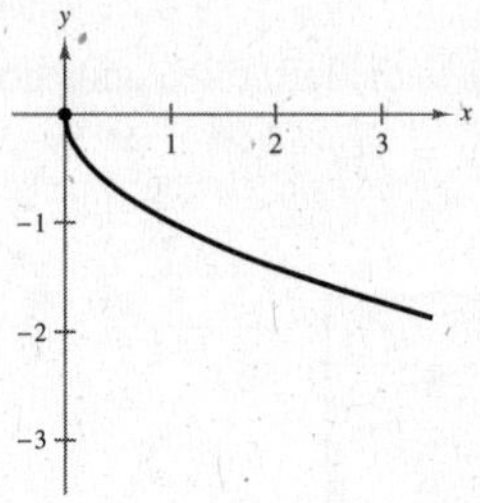

(c) $y = \sqrt{x - 2}$

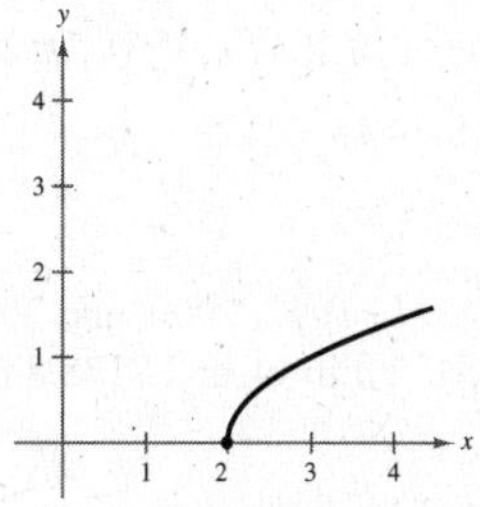

(d) $y = \sqrt{x + 3}$

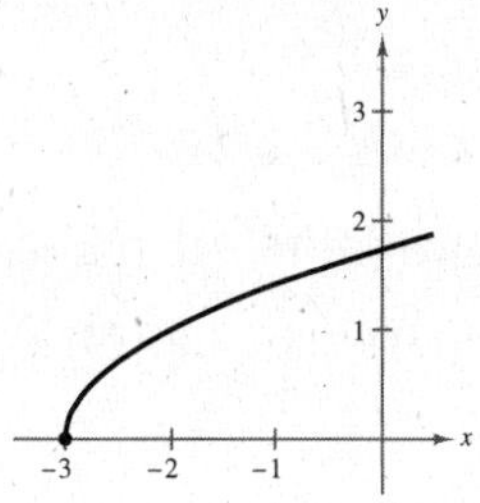

(e) $y = \sqrt{x - 4}$

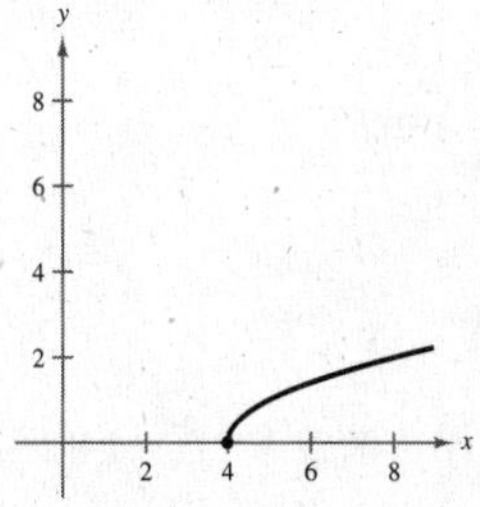

(f) $y = 2\sqrt{x}$

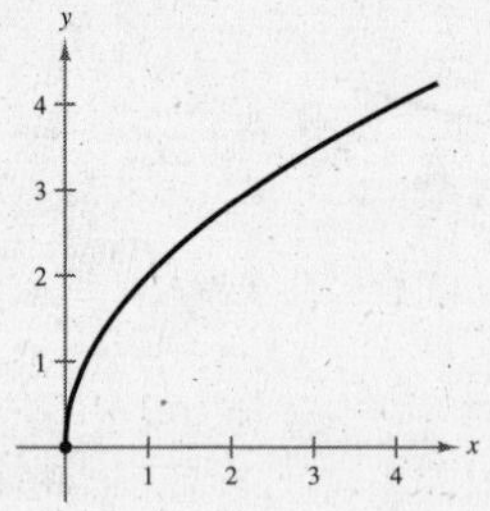

61. (a) Shifted three units to the left: $y = (x + 3)^2$

(b) Shifted six units to the left, three units downward, and reflected: $y = -(x + 6)^2 - 3$

63. (a)

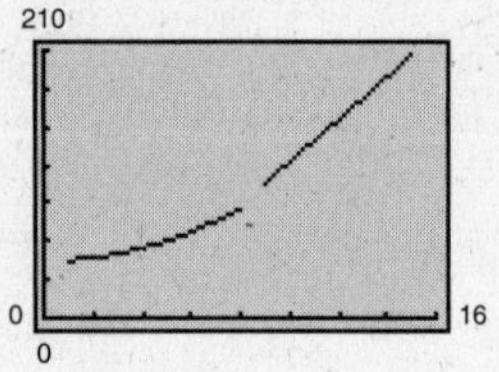

(b) $d(7) = 0.68(7)^2 - 0.3(7) + 45 = 76.22$

$d(10) = 16.7(10) - 45 = 122$

$d(14) = 16.7(14) - 45 = 188.8$

The amounts spent in 1997, 2000, and 2004 were \$76.22 billion, \$122 billion, and \$188.8 billion, respectively.

65. Total distance = Distance ran in first two races + Distance ran in 3rd race

$D = 1100 + 35r$

67. $h(l) = 1.95l + 28.7,\ 15 \le l \le 24$

(a) The domain is given as $[15, 24]$. This corresponds to a range of $[1.95(15) + 28.7, 1.95(24) + 28.7]$, or $[57.95, 75.5]$.

(b) $h(16) = 1.95(16) + 28.7$

$= 59.9$ inches tall ≈ 5 feet tall

(c) $70 = 1.95l + 28.7$

$41.3 = 1.95l$

21.18 inches $\approx l$

69. $N(T) = 8T^2 - 14T + 200,\ T(t) = 2t + 2$

$(N \circ T)(t) = N(T(t)) = 8(2t + 2)^2 - 14(2t + 2) + 200$

$= 8(4t^2 + 8t + 4) - 28t - 28 + 200$

$= 32t^2 + 64t + 32 - 28t + 172$

$= 32t^2 + 36t + 204$

$(N \circ T)(t)$ is the number of bacteria N after t hours.

71. (a) The radius increases at a rate of 0.75 foot per second. So, $r = 0.75t$.

(b) $A(t) = \pi r^2$

$= \pi(0.75t)^2$

$= 0.5625\pi t^2$

Time, t	1	2	3	4	5
Radius, r	0.75	1.5	2.25	3	3.75
Area, A	1.77	7.07	15.90	28.27	44.18

(c) $\dfrac{A(2)}{A(1)} \approx \dfrac{7.07}{1.77} \approx 4$ $\quad \dfrac{A(4)}{A(2)} \approx \dfrac{28.27}{7.07} \approx 4$

When t doubles, A increases by a factor of 4.

When $t = 8$: $A = 4 \cdot A(4) = 4(28.27)$

$= 113.08\ \text{ft}^2$

$A(8) = 0.5625\pi(8)^2 \approx 113.10\ \text{ft}^2$

73. $f(x) = 9x - 4x^2$

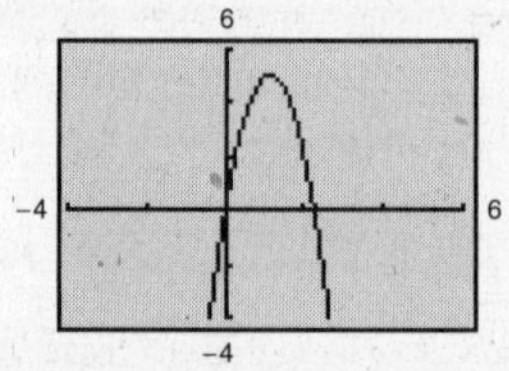

Zeros: $x(9 - 4x) = 0 \Rightarrow x = 0, \frac{9}{4}$

The function is not one-to-one.

75. $g(t) = \dfrac{t + 3}{1 - t}$

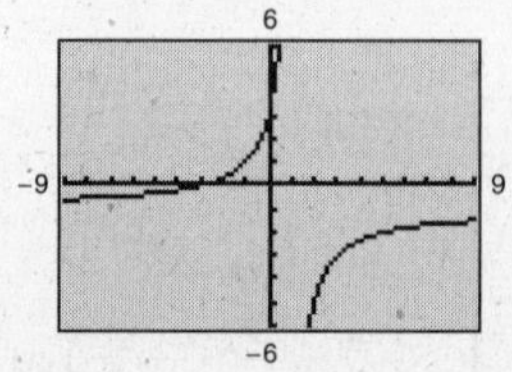

Zero: $t = -3$

The function is one-to-one.

77. $g(x) = x^2\sqrt{x^2 - 4}$

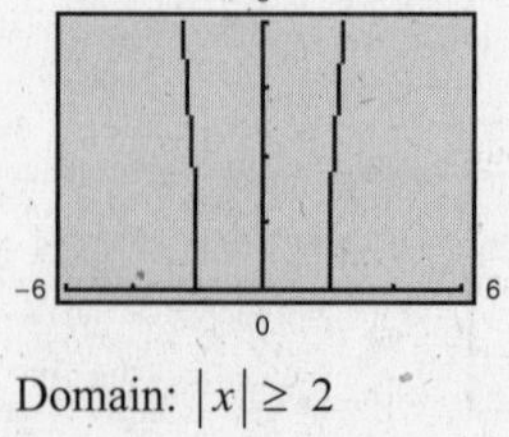

Domain: $|x| \ge 2$

Zeros: $x = \pm 2$

The function is not one-to-one.

79. Answers will vary.

Section 1.5 Limits

Skills Review

1. $f(x) = x^2 - 3x + 3$

(a) $f(-1) = (-1)^2 - 3(-1) + 3$
$= 1 + 3 + 3 = 7$

(b) $f(c) = c^2 - 3c + 3$

(c) $f(x + h) = (x + h)^2 - 3(x + h) + 3$
$= x^2 + 2xh + h^2 - 3x - 3h + 3$

2. $f(x) = \begin{cases} 2x - 2, & x < 1 \\ 3x + 1, & x \geq 1 \end{cases}$

(a) $f(-1) = 2(-1) - 2$
$= -2 - 2$
$= -4$

(b) $f(3) = 3(3) + 1$
$= 9 + 1$
$= 10$

(c) $f(t^2 + 1) = 3(t^2 + 1) + 1$
$= 3t^2 + 3 + 1$
$= 3t^2 + 4$

3. $f(x) = x^2 - 2x + 2$

$$\frac{f(1 + h) - f(1)}{h}$$
$$= \frac{(1 + h)^2 - 2(1 + h) + 2 - (1^2 - 2(1) + 2)}{h}$$
$$= \frac{1 + 2h + h^2 - 2 - 2h + 2 - 1 + 2 - 2}{h}$$
$$= \frac{h^2}{h} = h$$

4. $f(x) = 4x$

$$\frac{f(2 + h) - f(2)}{h} = \frac{4(2 + h) - 4(2)}{h}$$
$$= \frac{8 + 4h - 8}{h}$$
$$= \frac{4h}{h} = 4$$

5. $h(x) = -\dfrac{5}{x}$

Domain: $(-\infty, 0) \cup (0, \infty)$

Range: $(-\infty, 0) \cup (0, \infty)$

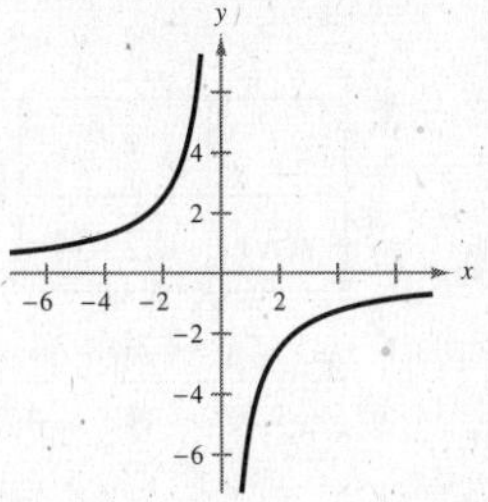

6. $g(x) = \sqrt{25 - x^2}$

Domain: $[-5, 5]$

Range: $[0, 5]$

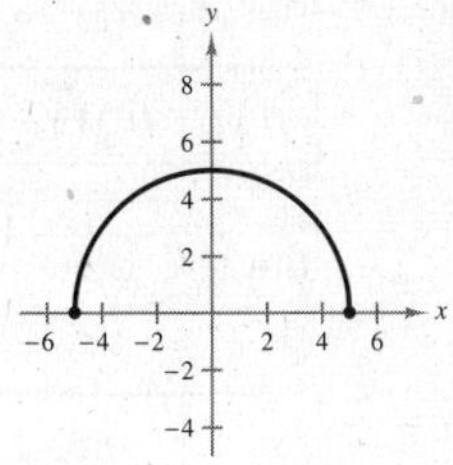

7. $f(x) = |x - 3|$

Domain: $(-\infty, \infty)$

Range: $[0, \infty)$

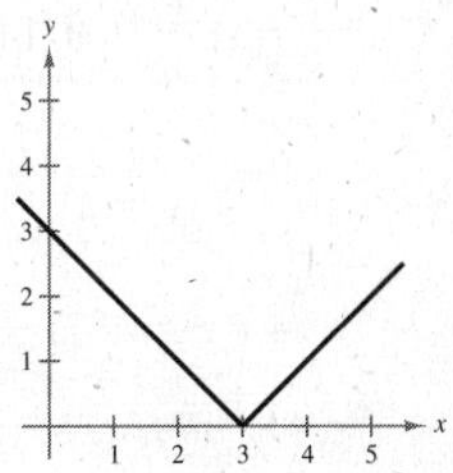

8. $f(x) = \dfrac{|x|}{x}$

Domain: $(-\infty, 0) \cup (0, \infty)$

Range: $y = -1$, $y = 1$

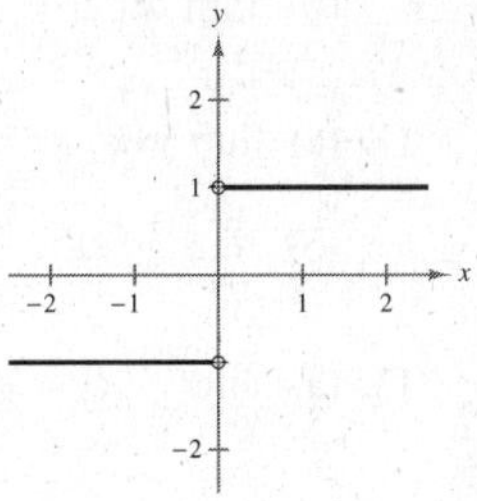

9. $9x^2 + 4y^2 = 49$

$4y^2 = 49 - 9x^2$

$y^2 = \dfrac{49 - 9x^2}{4}$

$y = \dfrac{\pm\sqrt{49 - 9x^2}}{2}$

Not a function of x (fails the vertical line test).

10. $2x^2y + 8x = 7y$

$2x^2y - 7y = -8x$

$y(2x^2 - 7) = -8x$

$y = -\dfrac{8x}{2x^2 - 7}$

Yes, y is a function of x.

1.

x	1.9	1.99	1.999	2	2.001	2.01	2.1
$f(x)$	8.8	8.98	8.998	?	9.002	9.02	9.2

$\lim_{x\to 2}(2x+5) = 9$

3.

x	1.9	1.99	1.999	2	2.001	2.01	2.1
$f(x)$	0.2564	0.2506	0.2501	?	0.2499	0.2494	0.2439

$\lim_{x\to 2}\frac{x-2}{x^2-4} = \frac{1}{4} = 0.25$

5.

x	−0.1	−0.01	−0.001	0	0.001	0.01	0.1
$f(x)$	0.5132	0.5013	0.5001	?	0.4999	0.4988	0.4881

$\lim_{x\to 0}\frac{\sqrt{x+1}-1}{x} = 0.5$

7.

x	−0.5	−0.1	−0.01	−0.001	0
$f(x)$	−0.0714	−0.0641	−0.0627	−0.0625	?

$\lim_{x\to 0^-}\frac{\left(\frac{1}{x+4}\right)-\left(\frac{1}{4}\right)}{x} = -\frac{1}{16} = -0.0625$

9. (a) $\lim_{x\to 0} f(x) = 1$

(b) $\lim_{x\to -1} f(x) = 3$

11. (a) $\lim_{x\to 0} g(x) = 1$

(b) $\lim_{x\to -1} g(x) = 3$

13. (a) $\lim_{x\to c}[f(x)+g(x)] = \lim_{x\to c} f(x) + \lim_{x\to c} g(x)$

$= 3 + 9 = 12$

(b) $\lim_{x\to c}[f(x)g(x)] = \left[\lim_{x\to c} f(x)\right]\left[\lim_{x\to c} g(x)\right]$

$= 3 \cdot 9 = 27$

(c) $\lim_{x\to c}\frac{f(x)}{g(x)} = \frac{\lim_{x\to c} f(x)}{\lim_{x\to c} g(x)} = \frac{3}{9} = \frac{1}{3}$

15. (a) $\lim_{x\to c}\sqrt{f(x)} = \sqrt{16} = 4$

(b) $\lim_{x\to c}[3f(x)] = 3(16) = 48$

(c) $\lim_{x\to c}[f(x)]^2 = 16^2 = 256$

17. (a) $\lim_{x\to 3^+} f(x) = 1$

(b) $\lim_{x\to 3^-} f(x) = 1$

(c) $\lim_{x\to 3} f(x) = 1$

19. (a) $\lim_{x\to 3^+} f(x) = 0$

(b) $\lim_{x\to 3^-} f(x) = 0$

(c) $\lim_{x\to 3} f(x) = 0$

21. (a) $\lim_{x\to 3^+} f(x) = 3$

(b) $\lim_{x\to 3^-} f(x) = -3$

(c) $\lim_{x\to 3} f(x)$ does not exist.

23. $\lim_{x\to 2} x^2 = 2^2 = 4$

25. $\lim_{x\to -3}(2x+5) = \lim_{x\to -3} 2x + \lim_{x\to -3} 5 = 2(-3) + 5 = -1$

27. $\lim_{x\to 1}(1-x^2) = \lim_{x\to 1} 1 - \lim_{x\to 1} x^2 = 1 - 1^2 = 0$

29. $\lim_{x\to 3}\sqrt{x+6} = \sqrt{3+6} = 3$

31. $\lim_{x\to -3}\frac{2}{x+2} = \frac{2}{-3+2} = -2$

33. $\lim_{x\to -2}\frac{x^2-1}{2x} = \frac{(-2)^2-1}{2(-2)} = \frac{3}{-4} = -\frac{3}{4}$

35. $\lim_{x\to 7} \frac{5x}{x+2} = \frac{5(7)}{7+2} = \frac{35}{9}$

37. $\lim_{x\to 3} \frac{\sqrt{x+1}-1}{x} = \frac{\sqrt{3+1}-1}{3} = \frac{1}{3}$

39. $\lim_{x\to 1} \frac{\frac{1}{x+4}-\frac{1}{4}}{x} = \frac{\frac{1}{5}-\frac{1}{4}}{1} = \frac{-1}{20} = -\frac{1}{20}$

41. $\lim_{x\to 1} \frac{x^2-1}{x-1} = \lim_{x\to 1} \frac{(x+1)(x-1)}{x-1}$

$= \lim_{x\to 1}(x+1) = 2$

43. $\lim_{x\to 2} \frac{x-2}{x^2-4x+4} = \lim_{x\to 2} \frac{x-2}{(x-2)(x-2)}$

$= \lim_{x\to 2} \frac{1}{x-2}$ does not exist.

45. $\lim_{t\to 4} \frac{t+4}{t^2-16} = \lim_{t\to 4} \frac{t+4}{(t+4)(t-4)}$

$= \lim_{t\to 4} \frac{1}{t-4}$ does not exist.

47. $\lim_{x\to -2} \frac{x^3+8}{x+2} = \lim_{x\to -2} \frac{(x+2)(x^2-2x+4)}{x+2}$

$= \lim_{x\to -2}(x^2-2x+4) = 12$

49. $\lim_{t\to -2^-} \frac{|x+2|}{x+2} = \frac{-(x+2)}{x+2} = -1$

$\lim_{t\to -2^+} \frac{|x+2|}{x+2} = \frac{(x+2)}{x+2} = 1$

So, $\lim_{t\to -2} \frac{|x+2|}{x+2}$ does not exist.

51. $\lim_{x\to 2^-}(4-x) = 2$

$\lim_{x\to 2^+}(4-x) = 2$

So, $\lim_{x\to 2} f(x) = 2$

53. $\lim_{x\to 3^-} f(x) = \lim_{x\to 3^-}\left(\frac{1}{3}x - 2\right) = -1$

$\lim_{x\to 3^+} f(x) = \lim_{x\to 3^+}(-2x+5) = -1$

So, $\lim_{x\to 3} f(x) = -1$.

55. $\lim_{\Delta x\to 0} \frac{2(x+\Delta x)-2x}{\Delta x} = \lim_{\Delta x\to 0} \frac{2x+2\Delta x-2x}{\Delta x}$

$= \lim_{\Delta x\to 0} 2 = 2$

57. $\lim_{\Delta x\to 0} \frac{\sqrt{x+2+\Delta x}-\sqrt{x+2}}{\Delta x} = \lim_{\Delta x\to 0} \frac{\sqrt{x+2+\Delta x}-\sqrt{x+2}}{\Delta x}\left(\frac{\sqrt{x+2+\Delta x}+\sqrt{x+2}}{\sqrt{x+2+\Delta x}+\sqrt{x+2}}\right)$

$= \lim_{\Delta x\to 0} \frac{(x+2+\Delta x)-(x+2)}{\Delta x\left[\sqrt{x+2+\Delta x}+\sqrt{x+2}\right]}$

$= \lim_{\Delta x\to 0} \frac{1}{\sqrt{x+2+\Delta x}+\sqrt{x+2}} = \frac{1}{2\sqrt{x+2}}$

59. $\lim_{\Delta t\to 0} \frac{(t+\Delta t)^2 - 5(t+\Delta t) - (t^2-5t)}{\Delta t} = \lim_{\Delta t\to 0} \frac{t^2+2t(\Delta t)+(\Delta t)^2-5t-5(\Delta t)-t^2+5t}{\Delta t}$

$= \lim_{\Delta t\to 0} \frac{2t(\Delta t)+(\Delta t)^2-5(\Delta t)}{\Delta t}$

$= \lim_{\Delta t\to 0} 2t + (\Delta t) - 5$

$= 2t - 5$

61. 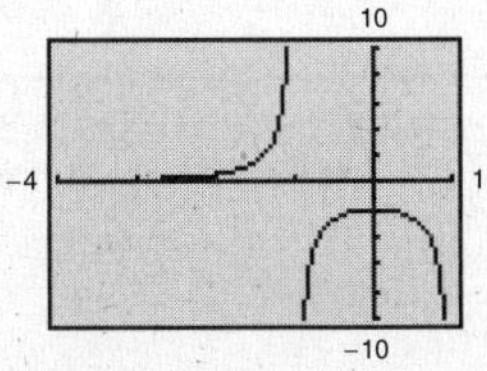

$\lim_{x\to 1^-} \frac{2}{x^2-1} = -\infty$

x	0	0.5	0.9	0.99	0.999	0.9999	1
$f(x)$	−2	−2.67	−10.53	−100.5	−1000.5	−10,000.5	undefined

Because $f(x) = \frac{2}{x^2-1}$ decreases without bound as x tends to 1 from the left, the limit is $-\infty$.

63.

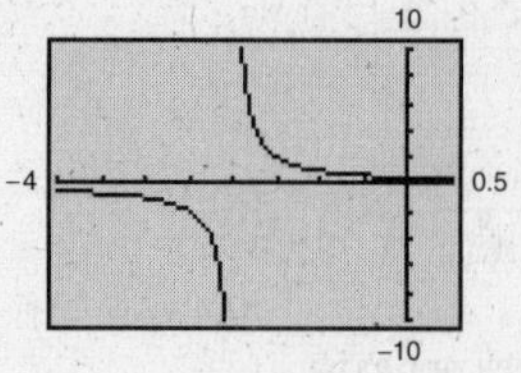

$$\lim_{x \to -2^-} \frac{1}{x + 2} = -\infty$$

x	−3	−2.5	−2.1	−2.01	−2.001	−2.0001	−2
$f(x)$	−1	−2	−10	−100	−1000	−10,000	undefined

Because $f(x) = \dfrac{1}{x + 2}$ decreases without bound as x tends to −2 from the left, the limit is $-\infty$.

65.

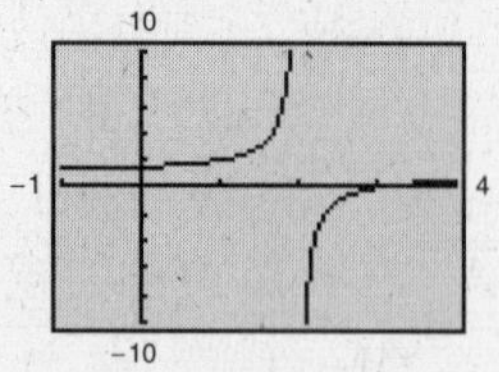

$$\lim_{x \to 2} \frac{x^2 - 5x + 6}{x^2 - 4x + 4} \text{ does not exist.}$$

67.

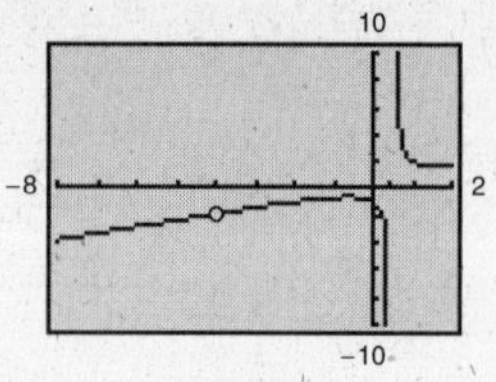

$$\lim_{x \to -4} \frac{x^3 + 4x^2 + x + 4}{2x^2 + 7x - 4} \approx -1.889$$

69. $F(t) = 98 + \dfrac{3}{t + 1}$

$$\lim_{t \to 4} F(t) = 98 + \frac{3}{4 + 1} = 98 + \frac{3}{5} = 98.6$$

It takes 4 hours for the fever-reducing drug to bring the body temperature back to normal.

71. $P = 5000(1 + r)^{10}$

$$\lim_{r \to 0.06} P = 5000(1 + 0.06)^{10}$$

$$\approx \$8954.24$$

73. (a)

x	−0.01	−0.001	−0.0001	0	0.0001	0.001	0.01
$f(x)$	2.732	2.720	2.718	undefined	2.718	2.717	2.705

$$\lim_{x \to 0} (1 + x)^{1/x} \approx 2.718$$

(b)

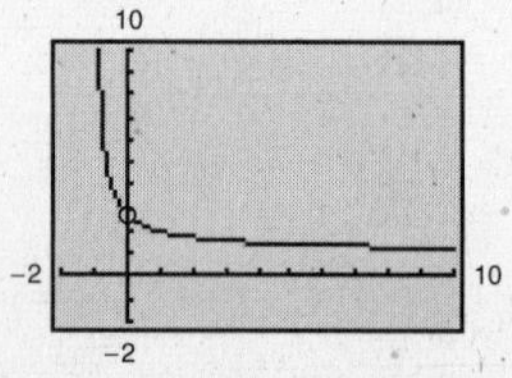

$$\lim_{x \to 0} (1 + x)^{1/x} \approx 2.718$$

(c) Domain: $(-1, 0) \cup (0, \infty)$

Range: $(1, e) \cup (e, \infty)$

Section 1.6 Continuity

Skills Review

1. $\dfrac{x^2 + 6x + 8}{x^2 - 6x - 16} = \dfrac{(x + 4)(x + 2)}{(x - 8)(x + 2)} = \dfrac{x + 4}{x - 8}$

2. $\dfrac{x^2 - 5x - 6}{x^2 - 9x + 18} = \dfrac{(x - 6)(x + 1)}{(x - 6)(x - 3)} = \dfrac{x + 1}{x - 3}$

Skills Review —*continued*—

3. $$\frac{2x^2 - 2x - 12}{4x^2 - 24x + 36} = \frac{2(x^2 - x - 6)}{4(x^2 - 6x + 9)} = \frac{2(x-3)(x+2)}{4(x-3)(x-3)} = \frac{x+2}{2(x-3)}$$

4. $$\frac{x^3 - 16x}{x^3 + 2x^2 - 8x} = \frac{x(x^2 - 16)}{x(x^2 + 2x - 8)} = \frac{x(x^2-16)}{x(x+4)(x-2)} = \frac{x(x+4)(x-4)}{x(x+4)(x-2)} = \frac{x-4}{x-2}$$

5. $x^2 + 7x = 0$
$x(x + 7) = 0$
$x = 0$
$x + 7 = 0 \Rightarrow x = -7$

6. $x^2 + 4x - 5 = 0$
$(x + 5)(x - 1) = 0$
$x + 5 = 0 \Rightarrow x = -5$
$x - 1 = 0 \Rightarrow x = 1$

7. $3x^2 + 8x + 4 = 0$
$(3x + 2)(x + 2) = 0$
$3x + 2 = 0 \Rightarrow x = -\frac{2}{3}$
$x + 2 = 0 \Rightarrow x = -2$

8. $x^3 + 5x^2 - 24x = 0$
$x(x^2 + 5x - 24) = 0$
$x(x - 3)(x + 8) = 0$
$x = 0$
$x - 3 = 0 \Rightarrow x = 3$
$x + 8 = 0 \Rightarrow x = -8$

9. $$\lim_{x \to 3}(2x^2 - 3x + 4) = 2(3^2) - 3(3) + 4 = 2(9) - 9 + 4 = 13$$

10. $$\lim_{x \to -2}(3x^3 - 8x + 7) = 3(-2)^3 - 8(-2) + 7 = 3(-8) + 16 + 7 = -24 + 23 = -1$$

1. Continuous; The function is a polynomial.

3. Not continuous; The rational function is not defined at $x = \pm 2$.

5. Continuous; The rational function's domain is the entire real line.

7. Not continuous; The rational function is not defined at $x = 3$ or $x = 5$.

9. Not continuous; The rational function is not defined at $x = \pm 2$.

11. $f(x) = \dfrac{x^2 - 1}{x}$ is continuous on $(-\infty, 0)$ and $(0, \infty)$ because the domain of f consists of all real numbers except $x = 0$. There is a discontinuity at $x = 0$ because $f(0)$ is not defined and $\lim_{x \to 0} f(x)$ does not exist.

13. $f(x) = \dfrac{x^2 - 1}{x + 1}$ is continuous on $(-\infty, -1)$ and $(-1, \infty)$ because the domain of f consists of all real numbers except $x = -1$. There is a discontinuity at $x = -1$ because $f(-1)$ is not defined and $\lim_{x \to -1} f(x) \neq f(-1)$.

15. $f(x) = x^2 - 2x + 1$ is continuous on $(-\infty, \infty)$ because the domain of f consists of all real numbers.

17. $f(x) = \dfrac{x}{x^2 - 1} = \dfrac{x}{(x + 1)(x - 1)}$ is continuous on $(-\infty, -1)$, $(-1, 1)$, and $(1, \infty)$ because the domain of f consists of all real numbers except $x = \pm 1$. There are discontinuities at $x = \pm 1$ because $f(1)$ and $f(-1)$ are not defined and $\lim_{x \to 1} f(x)$ and $\lim_{x \to -1} f(x)$ do not exist.

19. $f(x) = \dfrac{1}{x^2 + 1}$ is continuous on $(-\infty, \infty)$ because the domain of f consists of all real numbers.

21. $f(x) = \dfrac{x - 5}{x^2 - 9x + 20} = \dfrac{x - 5}{(x - 5)(x - 4)}$ is continuous on $(-\infty, 4)$, $(4, 5)$, and $(5, \infty)$ because the domain of f consists of all real numbers except $x = 4$ and $x = 5$. There is a discontinuity at $x = 4$ and $x = 5$ because $f(4)$ and $f(5)$ are not defined and $\lim_{x \to 4} f(x)$ does not exist and $\lim_{x \to 5} f(x) \neq f(5)$.

23. $f(x) = [\![2x]\!] + 1$ is continuous on all intervals of the form $\left(\frac{1}{2}c, \frac{1}{2}c + \frac{1}{2}\right)$, where c is an integer. That is, f is continuous on $\ldots, \left(-\frac{1}{2}, 0\right), \left(0, \frac{1}{2}\right), \left(\frac{1}{2}, 1\right), \ldots$. f is not continuous at all points $\frac{1}{2}c$, where c is an integer. There are discontinuities at $x = \dfrac{c}{2}$, where c is an integer, because $\lim_{x \to c/2} f(x)$ does not exist.

25. $f(x) = \begin{cases} -2x + 3, & x < 1 \\ x^2, & x \geq 1 \end{cases}$

is continuous on $(-\infty, \infty)$ because the domain of f consists of all real numbers, $f(1)$ is defined, $\lim_{x \to 1} f(x)$ exists, and $\lim_{x \to 1} f(x) = f(1)$.

27. $f(x) = \begin{cases} \frac{1}{2}x + 1, & x \leq 2 \\ 3 - x, & x > 2 \end{cases}$

is continuous on $(-\infty, 2]$ and $(2, \infty)$. There is a discontinuity at $x = 2$ because $\lim_{x \to 2} f(x)$ does not exist.

29. $f(x) = \dfrac{|x + 1|}{x + 1}$ is continuous on $(-\infty, -1)$ and $(-1, \infty)$ because the domain of f consists of all real numbers except $x = -1$. There is a discontinuity at $x = -1$ because $f(-1)$ is not defined, and $\lim_{x \to -1} f(x)$ does not exist.

31. $f(x) = [\![x - 1]\!]$ is continuous on all intervals $(c, c + 1)$. There are discontinuities at $x = c$, where c is an integer, because $\lim_{x \to c} f(x)$ does not exist.

33. $h(x) = f(g(x)) = f(x - 1) = \dfrac{1}{\sqrt{x - 1}}, \; x > 1$

h is continuous on its entire domain $(1, \infty)$.

35. Continuous on $[-1, 5]$ because $f(x) = x^2 - 4x - 5$ is a polynomial.

37. Continuous on $[1, 2)$ and $(2, 4]$ because $f(x) = \dfrac{1}{x - 2}$ has a nonremovable discontinuity at $x = 2$.

39. $f(x) = \dfrac{x^2 - 16}{x - 4} = \dfrac{(x + 4)(x - 4)}{x - 4} = x + 4, \; x \neq 4$

f has a removable discontinuity at $x = 4$;
Continuous on $(-\infty, 4)$ and $(4, \infty)$.

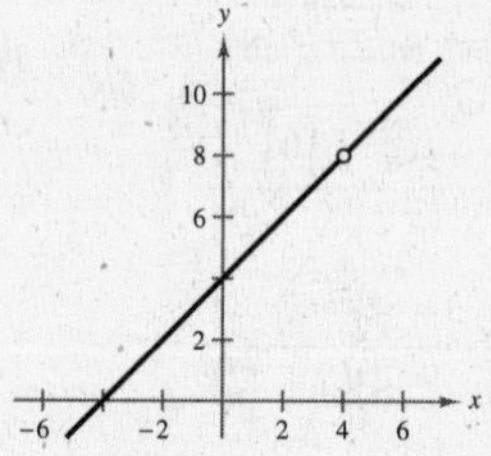

41. $f(x) = \dfrac{x^3 + x}{x} = \dfrac{x(x^2 + 1)}{x} = x^2 + 1, \; x \neq 0$

f has a removable discontinuity at $x = 0$;
Continuous on $(-\infty, 0)$ and $(0, \infty)$.

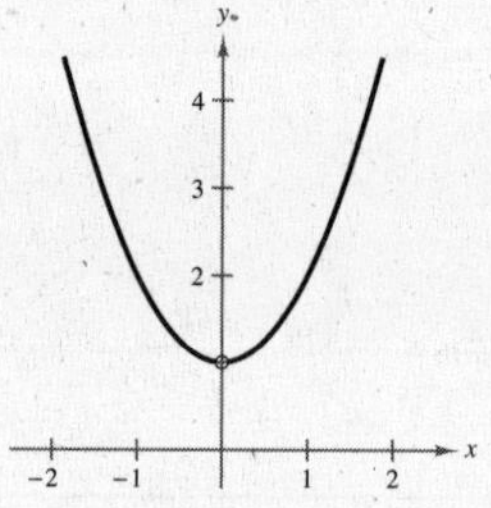

43. $f(x) = \begin{cases} x^2 + 1, & x < 0 \\ x - 1, & x \geq 0 \end{cases}$

f has a nonremovable discontinuity at $x = 0$;
Continuous on $(-\infty, 0)$ and $(0, \infty)$.

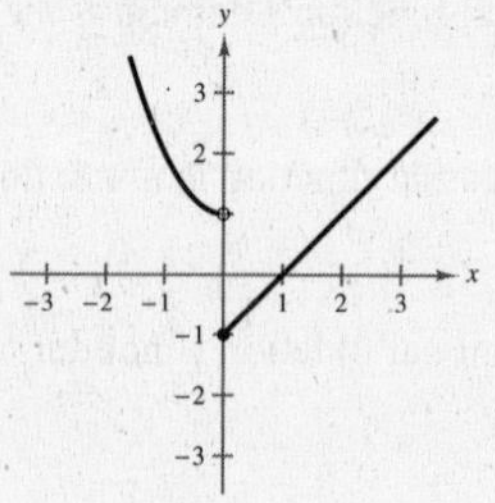

45. $\lim_{x \to 2^-} f(x) = \lim_{x \to 2^-} x^3 = 8$

$\lim_{x \to 2^+} f(x) = \lim_{x \to 2^+} ax^2 = 4a$

So, $8 = 4a$ and $a = 2$.

47.

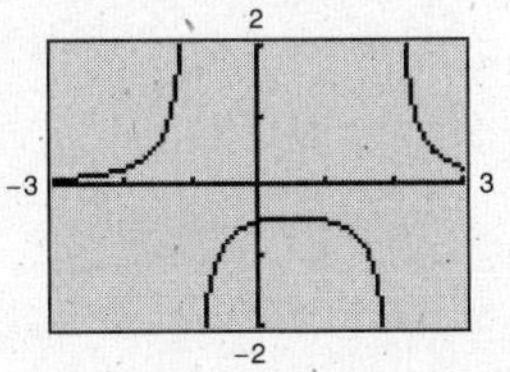

From the graph, you can see that $h(2)$ and $h(-1)$ are not defined, so h is not continuous at $x = 2$ and $x = -1$.

49.

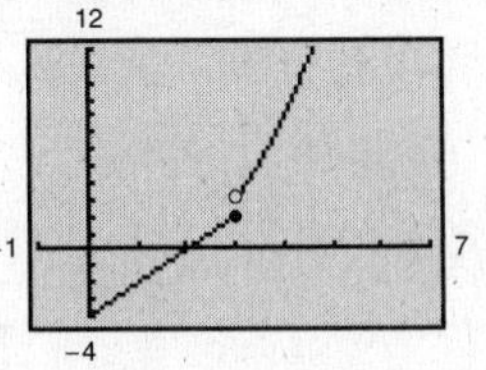

From the graph, you can see that $\lim_{x \to 3} f(x)$ does not exist, so f is not continuous at $x = 3$.

51.

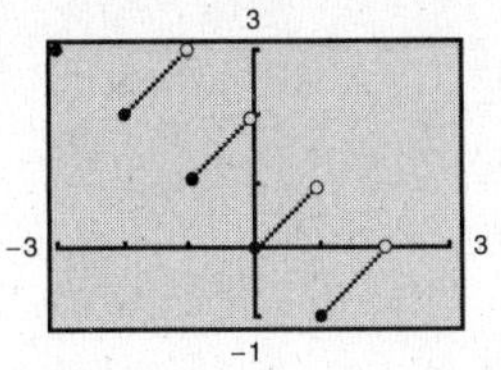

From the graph, you can see that $\lim_{x \to c} \left(x - 2[\![x]\!]\right)$, where c is an integer, does not exist. So f is not continuous at all integers c.

53. $f(x) = \dfrac{x}{x^2 + 1}$ is continuous on $(-\infty, \infty)$.

55. $f(x) = \dfrac{1}{2}[\![2x]\!]$ is continuous on all intervals of the form $\left(\dfrac{c}{2}, \dfrac{c+1}{2}\right)$, where c is an integer.

57.

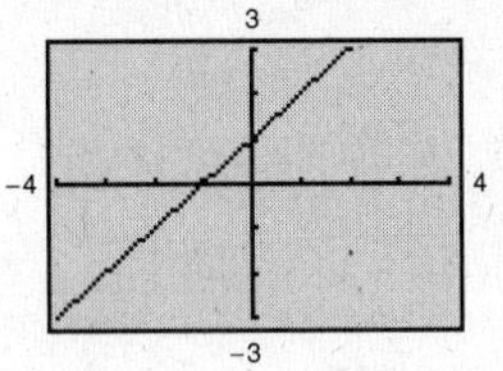

$f(x) = \dfrac{x^2 + x}{x} = \dfrac{x(x+1)}{x}$ appears to be continuous on $[-4, 4]$. But it is not continuous at $x = 0$ (removable discontinuity). Examining a function analytically can reveal removable discontinuities that are difficult to find just from analyzing its graph.

59. (a) $A = 225(1.02)^{[\![t]\!]}, \ t \geq 0$

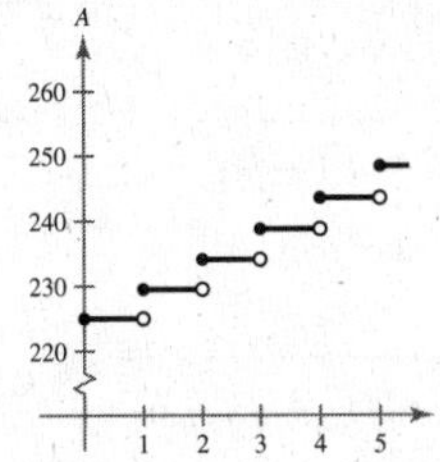

The graph has nonremovable discontinuities at $t = 1, 2, \ldots$.

(b) For $t = 1.5$,

$A = 225(1.02)^{[\![1.5]\!]} = 225(1.02)^1 = \$229.50$.

61. $C = 3.50 - 1.90[\![1 - x]\!] = \begin{cases} 3.50 + 1.90[\![x]\!], & x > 0, x \text{ not an integer} \\ 3.50 + 1.90[\![x - 1]\!], & x > 0, x \text{ an integer} \end{cases}$

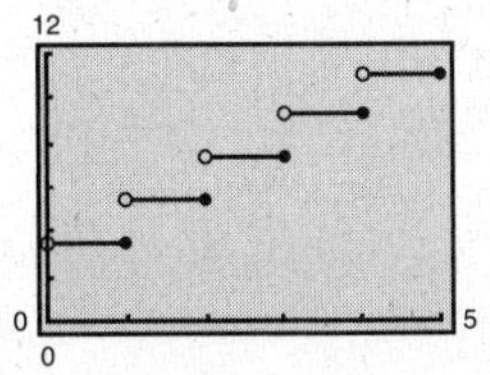

C is not continuous at $x = 1, 2, 3, \ldots$.

63. (a)

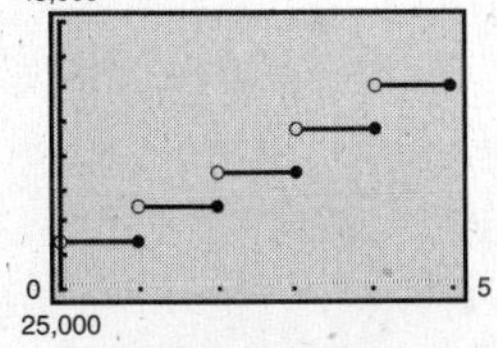

Nonremovable discontinuities at $t = 1, 2, 3, 4, 5$; s is not continuous at $t = 1, 2, 3, 4,$ or 5.

(b) For $t = 5$, $S = \$43,850,78$.

The salary during the fifth year is \$43,850.78.

65. Yes, a linear model is a continuous function. No, actual population would probably not be continuous because population is usually recorded over larger units of time (hourly, daily, or monthly). In these cases, the population may jump between different units of time.

Review Exercises for Chapter 1

1.

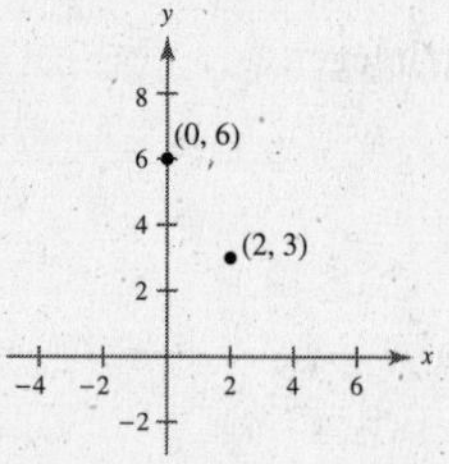

3.

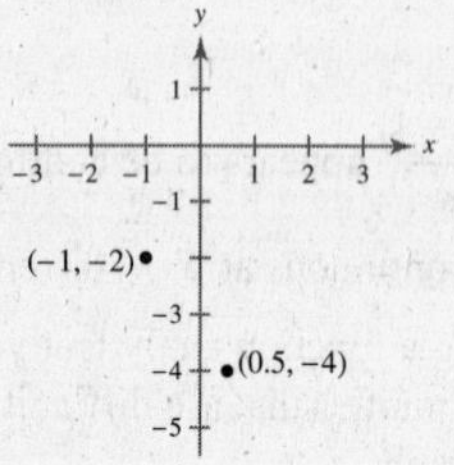

5. Matches (a)

7. Matches (b)

9. $\text{Distance} = \sqrt{(0-5)^2 + (0-2)^2}$
$= \sqrt{25+4} = \sqrt{29}$

11. $\text{Distance} = \sqrt{[-1-(-4)]^2 + (3-6)^2}$
$= \sqrt{9+9} = 3\sqrt{2}$

13. $\text{Midpoint} = \left(\frac{5+9}{2}, \frac{6+2}{2}\right) = (7, 4)$

15. $\text{Midpoint} = \left(\frac{-10-6}{2}, \frac{4+8}{2}\right) = (-8, 6)$

17. From the graph, you can estimate the values to be:
(a) 1996: 752,000
(b) 1999: 826,000
(c) 2002: 912,000
(d) 2005: 907,000

19. The translated vertices are:
$(1+3, 3+4) = (4, 7)$
$(2+3, 4+4) = (5, 8)$
$(5+3, 6+4) = (8, 10)$

21. $y = 4x - 12$

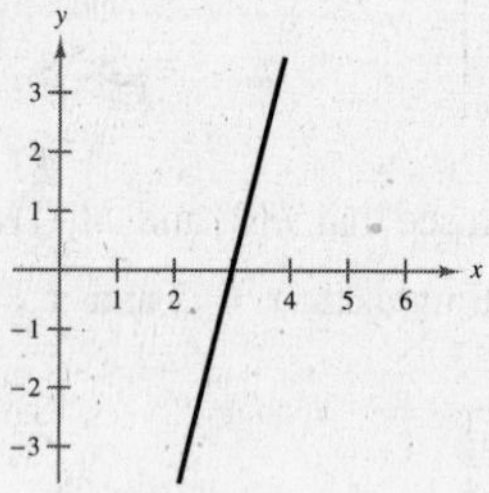

23. $y = x^2 + 5$

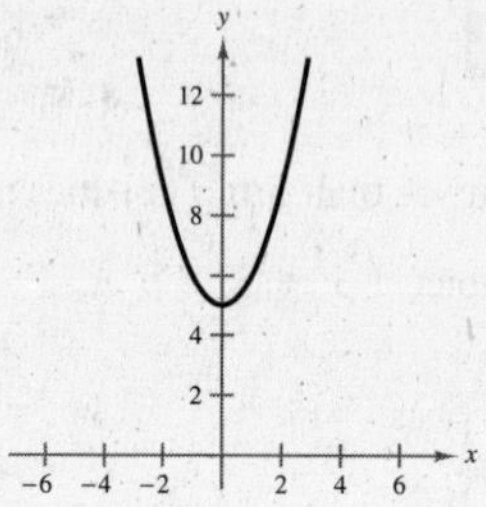

25. $y = |4 - x|$

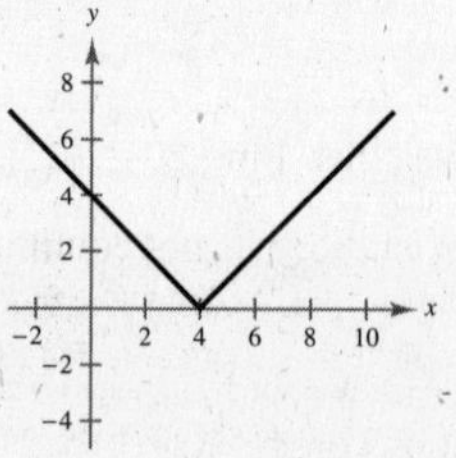

27. $y = x^3 + 4$

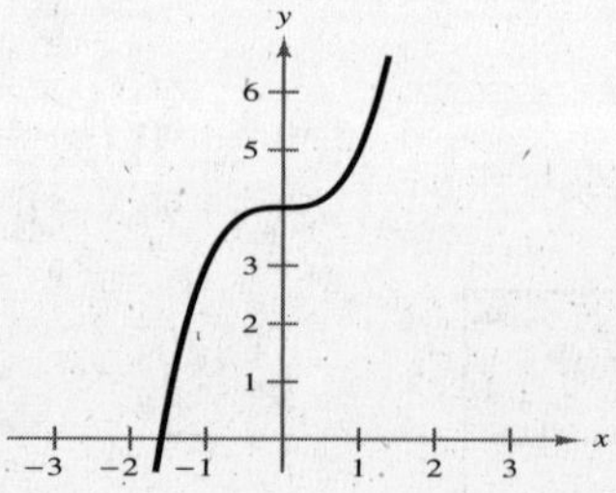

29. $y = \sqrt{4x + 1}$

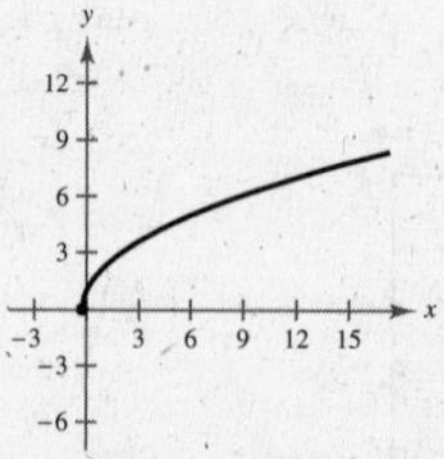

31. Let $y = 0$. Then,

$$4x + 0 + 3 = 0$$
$$x = -\tfrac{3}{4}.$$

Let $x = 0$. Then,

$$4(0) + y + 3 = 0$$
$$y = -3.$$

x-intercept: $\left(-\tfrac{3}{4}, 0\right)$

y-intercept: $(0, -3)$

33.
$$(x - 0)^2 + (y - 0)^2 = r^2$$
$$x^2 + y^2 = r^2$$
$$2^2 + \left(\sqrt{5}\right)^2 = r^2$$
$$9 = r^2$$
$$x^2 + y^2 = 9$$

35.
$$x^2 + y^2 - 6x + 8y = 0$$
$$\left(x^2 - 6x + 9\right) + \left(y^2 + 8y + 16\right) = 9 + 16$$
$$(x - 3)^2 + (y + 4)^2 = 25$$

Center: $(3, -4)$

r: 5

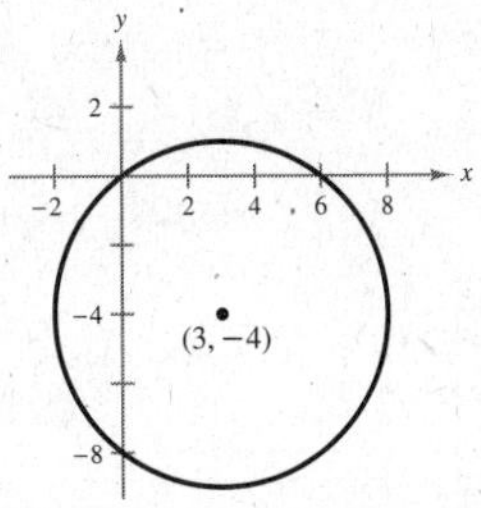

37. Solving both equations for y yields

$$-3y = 13 - 2x \qquad\qquad 3y = 1 - 5x$$
$$y = -\frac{13}{3} + \frac{2}{3}x \qquad\qquad y = \frac{1}{3} - \frac{5}{3}x$$
$$-\frac{13}{3} + \frac{2}{3}x = \frac{1}{3} - \frac{5}{3}x$$
$$\frac{7}{3}x = \frac{14}{3}$$
$$x = 2$$

$$2(2) - 3y = 13$$
$$-3y = 9$$
$$y = -3$$

Point of intersection: $(2, -3)$

39. By equating the y-values for the two equations, we have

$$x^3 = x$$
$$x\left(x^2 - 1\right) = 0$$
$$x = -1, 0, 1$$

The corresponding y-values are $y = -1$, $y = 0$, and $y = 1$, so the points of intersection are $(-1, -1)$, $(0, 0)$, and $(1, 1)$.

41. (a) Population of Kentucky = Population of South Carolina

$$4042 + 26.1t = 4010 + 48.5t$$
$$4042 = 4010 + 22.4t$$
$$32 = 22.4t$$
$$1.43 \approx t$$

So, you would expect that the population of South Carolina would have exceeded the population of Kentucky after $t \approx 1.43$ years, which was sometime during 2001.

(b) Population of Kentucky in 2010:

$$P = 4042 + 26.1(10) = 4303 \Rightarrow 4{,}303{,}000$$

Population of South Carolina in 2010:

$$P = 4010 + 48.5(10) = 4495 \Rightarrow 4{,}495{,}000$$

43.
$$3x + y = -2$$
$$y = -3x - 2$$

Slope: $m = -3$

y-intercept: $(0, -2)$

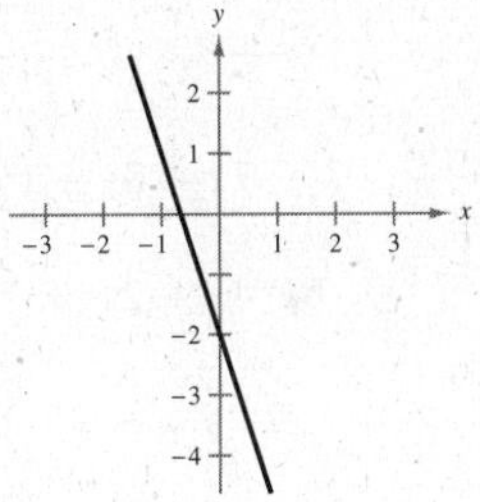

45. $y = -\frac{5}{3}$

Slope: $m = 0$ (horizontal line)

y-intercept: $\left(0, -\frac{5}{3}\right)$

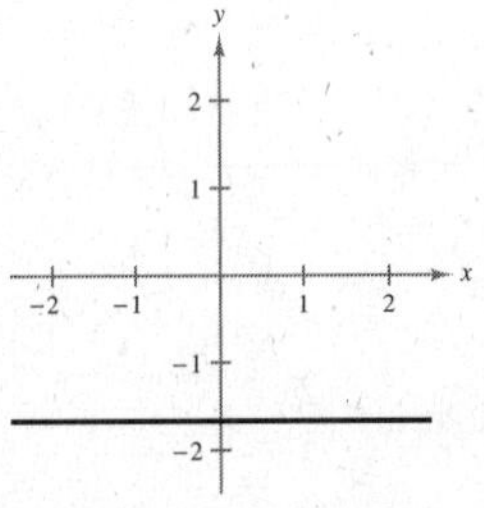

47.
$$-2x - 5y - 5 = 0$$
$$5y = -2x - 5$$
$$y = -\tfrac{2}{5}x - 1$$

Slope: $m = -\frac{2}{5}$

y-intercept: $(0, -1)$

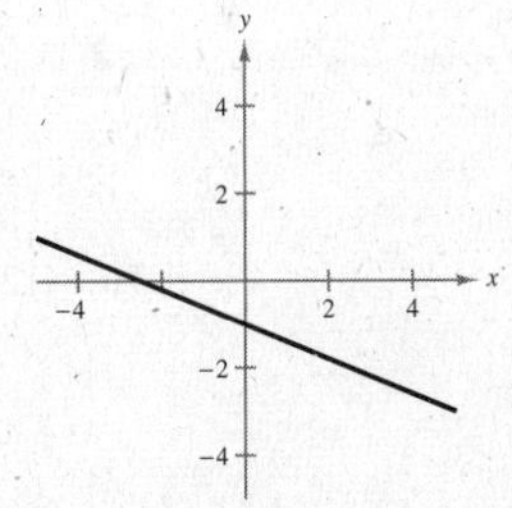

49. Slope $= \dfrac{6-0}{7-0} = \dfrac{6}{7}$

51. Slope $= \dfrac{17-(-3)}{10-(-11)} = \dfrac{20}{21}$

53. $y - (-1) = -2(x - 3)$

$y = -2x + 5$

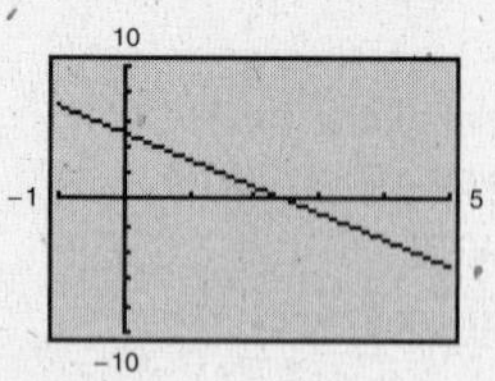

55. $y - (-4) = 0(x - 1.5)$

$y = -4$

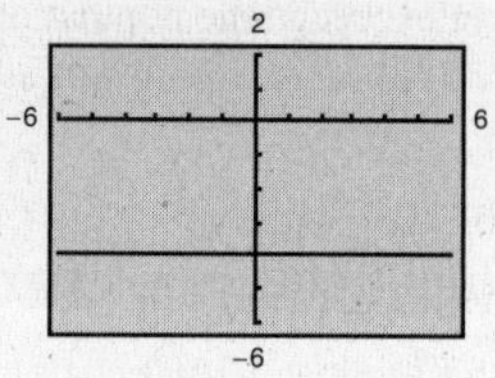

57. (a) $y - 6 = \dfrac{7}{8}[x - (-3)]$

$y = \dfrac{7}{8}x + \dfrac{69}{8}$

$7x - 8y + 69 = 0$

(b) $4x + 2y = 7 \Rightarrow y = -2x + \dfrac{7}{2}$; slope $= -2$

$y - 6 = -2[x - (-3)]$

$y = -2x$

$2x + y = 0$

(c) The line through $(0, 0)$ and $(-3, 6)$ has slope

$\dfrac{6}{-3} = -2$

$y = -2x$

$2x + y = 0$

(d) $3x - 2y = 2 \Rightarrow y = \dfrac{3}{2}x - 1$

Slope of perpendicular is $-\dfrac{2}{3}$.

$y - 6 = -\dfrac{2}{3}[x - (-3)]$

$y = -\dfrac{2}{3}x + 4$

$2x + 3y - 12 = 0$

59. (a) When $t = 5$, $L = 429$ and when $t = 8$, $L = 564$.

Slope: $m = \dfrac{564 - 429}{8 - 5} = \dfrac{135}{3} = 45$

Equation: $L - 429 = 45(t - 5)$

$L - 429 = 45t - 225$

$L = 45t + 204$

(b) When $t = 6$: $L = 45(6) + 204 = 474$ residents

(c) When $t = 13$: $L = 45(13) + 204 = 789$ residents

61. Yes, $y = -x^2 + 2$ is a function of x.

63. No, $y^2 - \frac{1}{4}x^2 = 4$ is not a function of x.

65. (a) $f(1) = 3(1) + 4 = 7$

(b) $f(x + 1) = 3(x + 1) + 4 = 3x + 7$

(c) $f(2 + \Delta x) = 3(2 + \Delta x) + 4 = 10 + 3\Delta x$

67.

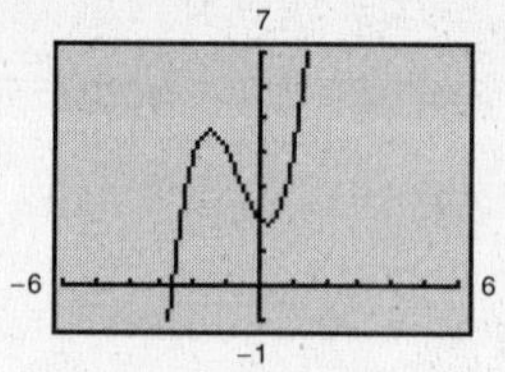

Domain: $(-\infty, \infty)$

Range: $(-\infty, \infty)$

69.

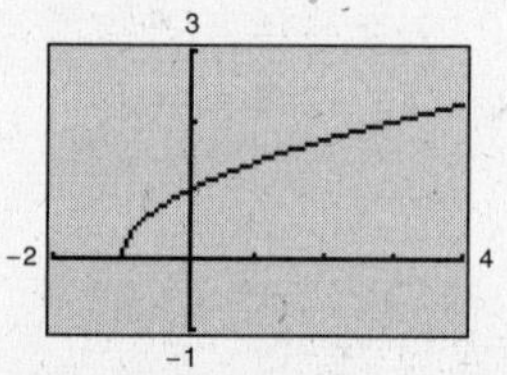

Domain: $[-1, \infty)$

Range: $[0, \infty)$

71.

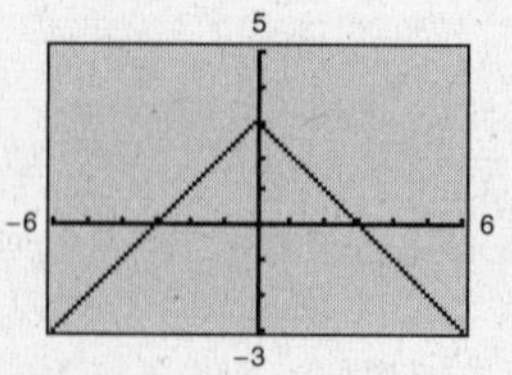

Domain: $(-\infty, \infty)$

Range: $(-\infty, 3]$

73. (a) $f(x) + g(x) = (1 + x^2) + (2x - 1) = x^2 + 2x$

(b) $f(x) - g(x) = (1 + x^2) - (2x - 1) = x^2 - 2x + 2$

(c) $f(x)g(x) = (1 + x^2)(2x - 1) = 2x^3 - x^2 + 2x - 1$

(d) $\dfrac{f(x)}{g(x)} = \dfrac{1 + x^2}{2x - 1}$

(e) $f(g(x)) = f(2x - 1)$

$= 1 + (2x - 1)^2$

$= 4x^2 - 4x + 2$

(f) $g(f(x)) = g(1 + x^2)$

$= 2(1 + x^2) - 1$

$= 2x^2 + 1$

75. $f(x) = \frac{3}{2}x$ has an inverse by the horizontal line test.

$y = \frac{3}{2}x$

$x = \frac{3}{2}y$

$y = \frac{2}{3}x$

$f^{-1}(x) = \frac{2}{3}x$

77. $f(x) = -x^2 + \frac{1}{2}$ does not have an inverse by the horizontal line test.

79. $\lim_{x\to 2} (5x - 3) = 5(2) - 3 = 7$

81. $\lim_{x\to 2} (5x - 3)(2x + 3) = [5(2) - 3][2(2) + 3] = 49$

83. $\lim_{t\to 3} \dfrac{t^2 + 1}{t} = \dfrac{(3)^2 + 1}{3} = \dfrac{10}{3}$

85. $\lim_{t\to 1} \dfrac{t + 1}{t - 2} = \dfrac{1 + 1}{1 - 2} = -2$

87. $\lim_{x\to -2} \dfrac{x + 2}{x^2 - 4} = \lim_{x\to -2} \dfrac{x + 2}{(x + 2)(x - 2)}$

$= \lim_{x\to -2} \dfrac{1}{x - 2}$

$= -\dfrac{1}{4}$

89. $\lim_{x\to 0^+} \left(x - \dfrac{1}{x}\right) = \lim_{x\to 0^+} \dfrac{x^2 - 1}{x} = -\infty$

91. $\lim_{x\to 0} \dfrac{[1/(x - 2)] - 1}{x} = \lim_{x\to 0} \dfrac{1 - (x - 2)}{x(x - 2)}$

$= \lim_{x\to 0} \dfrac{3 - x}{x(x - 2)}$ does not exist.

93. $$\lim_{t\to 0} \frac{\dfrac{1}{\sqrt{t + 4}} - \dfrac{1}{2}}{t} \cdot \left(\frac{\dfrac{1}{\sqrt{t + 4}} + \dfrac{1}{2}}{\dfrac{1}{\sqrt{t + 4}} + \dfrac{1}{2}}\right) = \lim_{t\to 0} \frac{\dfrac{1}{t + 4} - \dfrac{1}{4}}{t\left[\dfrac{1}{\sqrt{t + 4}} + \dfrac{1}{2}\right]}$$

$$= \lim_{t\to 0} \frac{4 - (t + 4)}{t\left[\dfrac{1}{\sqrt{t + 4}} + \dfrac{1}{2}\right][4(t + 4)]}$$

$$= \lim_{t\to 0} \frac{-1}{\left[\dfrac{1}{\sqrt{t + 4}} + \dfrac{1}{2}\right][4t + 16]}$$

$$= -\frac{1}{16}$$

95. $$\lim_{\Delta x\to 0} \frac{(x + \Delta x)^3 - (x + \Delta x) - (x^3 - x)}{\Delta x} = \lim_{\Delta x\to 0} \frac{x^3 + 3x^2\Delta x + 3x(\Delta x)^2 + (\Delta x)^3 - x - \Delta x - x^3 + x}{\Delta x}$$

$$= \lim_{\Delta x\to 0} \frac{3x^2\Delta x + 3x(\Delta x)^2 + (\Delta x)^3 - \Delta x}{\Delta x}$$

$$= \lim_{\Delta x\to 0} \left[3x^2 + 3x\Delta x + (\Delta x)^2 - 1\right]$$

$$= 3x^2 - 1$$

97.

x	1.1	1.01	1.001	1.0001
$f(x)$	0.5680	0.5764	0.5773	0.5773

$$\lim_{x\to 1^+} \frac{\sqrt{2x+1}-\sqrt{3}}{x-1} = \frac{1}{\sqrt{3}} \approx 0.5774$$

99. The statement is false since $\lim_{x\to 0^-} \frac{|x|}{x} = -1.$

101. The statement is false since $\lim_{x\to 0^-} \sqrt{x}$ is undefined.

103. The statement is false since $\lim_{x\to 2^+} f(x) = \lim_{x\to 2^+} 0 = 0.$

105. $f(x) = \frac{1}{(x+4)^2}$ is continuous on the intervals $(-\infty, -4)$ and $(-4, \infty)$ because the domain of f consists of all real numbers except $x = -4$. There is a discontinuity at $x = -4$ because $f(4)$ is not defined.

107. $f(x) = \frac{3}{x+1}$ is continuous on the intervals $(-\infty, -1)$ and $(-1, \infty)$ because the domain of f consists of all real numbers except $x = -1$. There is a discontinuity at $x = -1$ because $f(-1)$ is not defined.

109. $f(x) = [\![x + 3]\!]$ is continuous on all intervals of the form $(c, c + 1)$, where c is an integer. There are discontinuities at all integer values c because $\lim_{x\to c} f(x)$ does not exist.

111. $f(x) = \begin{cases} x, & x \le 0 \\ x + 1, & x > 0 \end{cases}$ is continuous on the intervals $(-\infty, 0)$ and $(0, \infty)$. There is a discontinuity at $x = 0$ because $\lim_{x\to 0} f(x)$ does not exist.

113. $\lim_{x\to 3^-} f(x) = \lim_{x\to 3^-} (-x + 1) = -2$

$\lim_{x\to 3^+} f(x) = \lim_{x\to 3^+} (ax - 8) = 3a - 8$

So, $-2 = 3a - 8$ and $a = 2$.

115. (a)

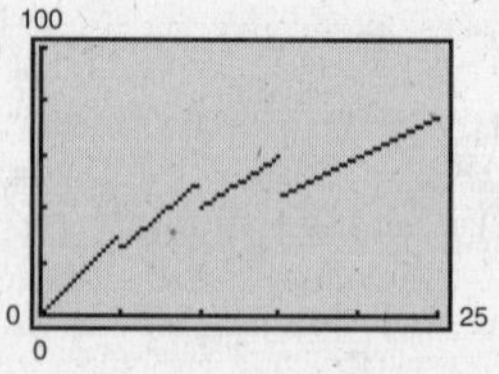

The function is continuous for all $x > 0$ except $x = 5$, $x = 10$, and $x = 15$. There are discontinuities at these values because $\lim_{x\to 5} C(x)$, $\lim_{x\to 10} C(x)$, and $\lim_{x\to 15} C(x)$ do not exist.

(b) $C(10) = 4.99(10) = 49.90$

The cost of purchasing 10 bottles is \$49.90.

117. $C(x) = 5 - 4[\![1 - x]\!]$

$$= \begin{cases} 5 + 4[\![x]\!], & x > 0,\ x \text{ not an integer} \\ 5 + 4[\![x - 1]\!], & x > 0,\ x \text{ an integer} \end{cases}$$

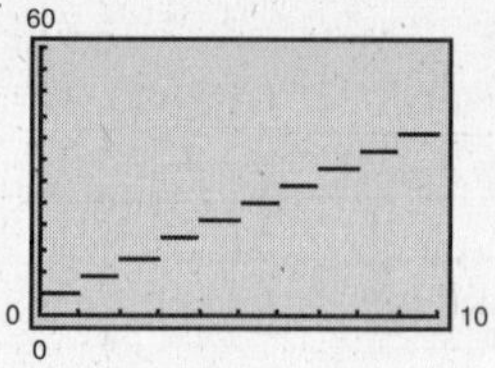

The function is continuous for all noninteger values of $x > 0$.

119. (a)

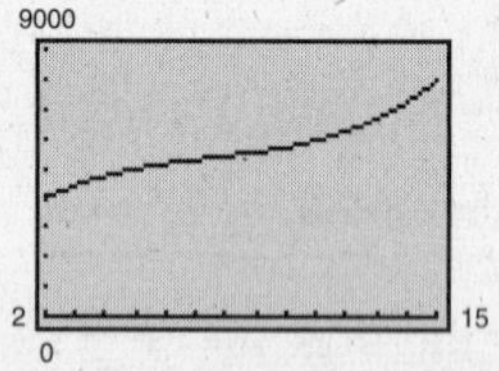

(b)

t	2	3	4	5	6	7	8
D (actual)	4001.8	4351.0	4643.3	4920.6	5181.5	5369.2	5478.2
D (model)	3937.0	4391.3	4727.4	4971.0	5147.8	5283.6	5404.0

t	9	10	11	12	13	14	15
D (actual)	5605.5	5628.7	5769.9	6198.4	6760.0	7354.7	7905.3
D (model)	5534.7	5701.5	5930.1	6246.1	6675.3	7243.3	7976.0

(c) When $t = 20$, $D = 15{,}007.9$. In 2010, the national debt will be about \$15,007.9 billion.

Chapter Test Solutions

1. (a) $d = \sqrt{(1+4)^2 + (-1-4)^2}$

$= \sqrt{25 + 25}$

$= 5\sqrt{2}$

(b) Midpoint $= \left(\dfrac{1-4}{2}, \dfrac{-1+4}{2}\right) = \left(-\dfrac{3}{2}, \dfrac{3}{2}\right)$

(c) $m = \dfrac{4+1}{-4-1} = -1$

2. (a) $d = \sqrt{\left(\dfrac{5}{2} - 0\right)^2 + (2-2)^2}$

$= \sqrt{\dfrac{25}{4} + 0}$

$= \dfrac{5}{2}$

(b) Midpoint $= \left(\dfrac{\frac{5}{2}+0}{2}, \dfrac{2+2}{2}\right) = \left(\dfrac{5}{4}, 2\right)$

(c) $m = \dfrac{2-2}{0-\frac{5}{2}} = 0$

3. (a) $d = \sqrt{\left(3\sqrt{2} - \sqrt{2}\right)^2 + (2-1)^2}$

$= \sqrt{8+1}$

$= 3$

(b) Midpoint $= \left(\dfrac{3\sqrt{2}+\sqrt{2}}{2}, \dfrac{2+1}{2}\right) = \left(2\sqrt{2}, \dfrac{3}{2}\right)$

(c) $m = \dfrac{1-2}{\sqrt{2}-3\sqrt{2}} = \dfrac{-1}{-2\sqrt{2}} \cdot \dfrac{\sqrt{2}}{\sqrt{2}} = \dfrac{\sqrt{2}}{4}$

4. $x^2 + y^2 - 4x - 2y - 4 = 0$

$\left(x^2 - 4x + 4\right) + \left(y^2 - 2y + 1\right) = 4 + 4 + 1$

$(x-2)^2 + (y-1)^2 = 9$

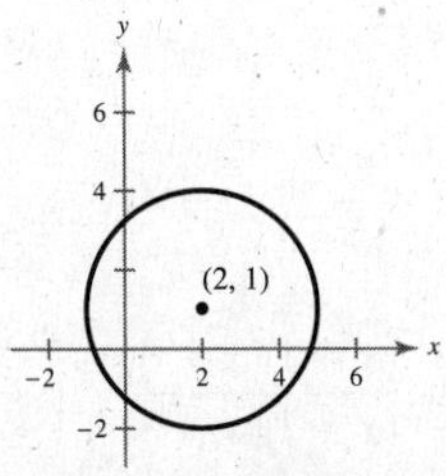

5. Population of Missouri = Population of Maryland

$5603 + 38.9t = 5331 + 51.6t$

$5603 = 5331 + 12.7t$

$272 = 12.7t$

$21.42 \approx t$

So, you would expect that the population of Maryland will exceed the population of Missouri after $t \approx 21.42$ years, which will be sometime during 2021.

6. $m = \frac{1}{5}$.

When $x = 0$:

$y = \frac{1}{5}(0) - 2 = -2$

y-intercept: $(0, -2)$

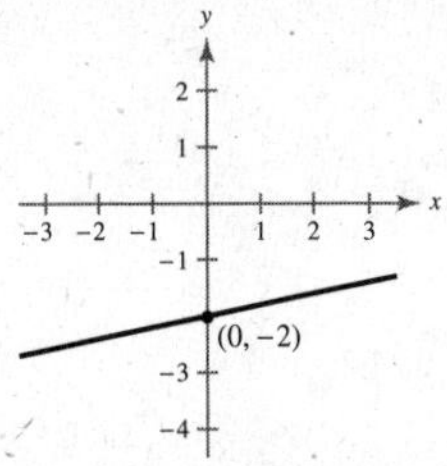

7. The line is vertical, so its slope is undefined, and it has no y-intercept.

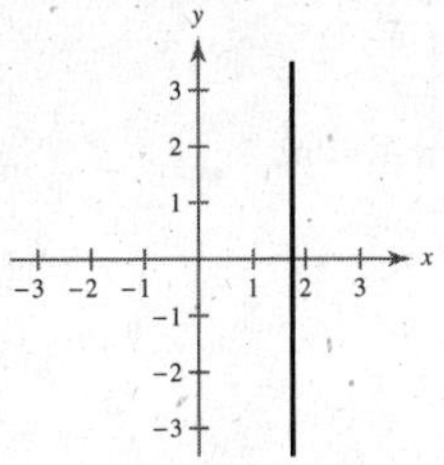

8. $y = -2.5x + 6.25$

$m = -2.5$

When $x = 0$:

$y = -2.5(0) + 6.25 = 6.25$

y-intercept: $(0, 6.25)$

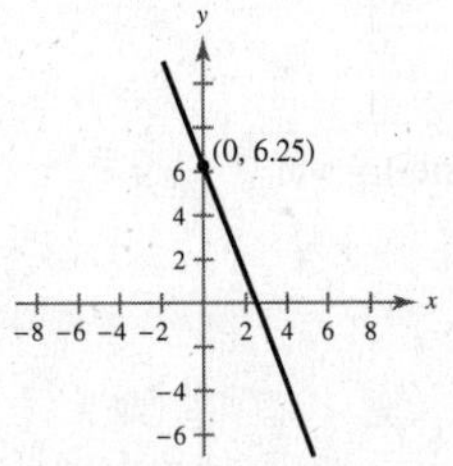

9. (a)

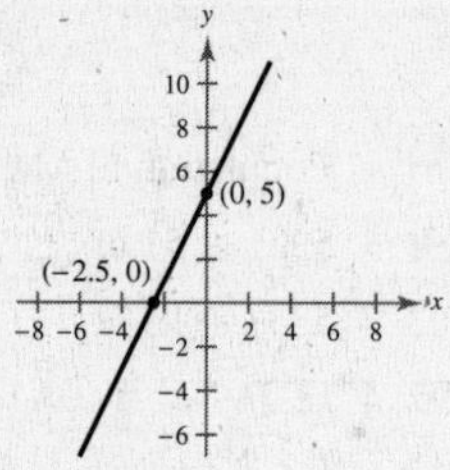

(b) Domain: $(-\infty, \infty)$

Range: $(-\infty, \infty)$

(c)

x	−3	−2	3
$f(x)$	−1	1	11

(d) The function is one-to-one.

10. (a)

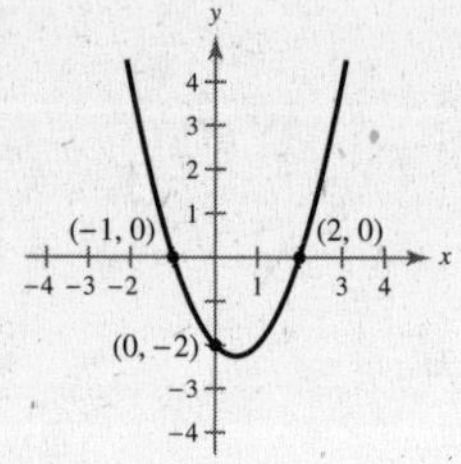

(b) Domain: $(-\infty, \infty)$

Range: $\left[-\frac{9}{4}, \infty\right)$

(c)

x	−3	−2	3
$f(x)$	10	4	4

(d) The function is not one-to-one.

11. (a)

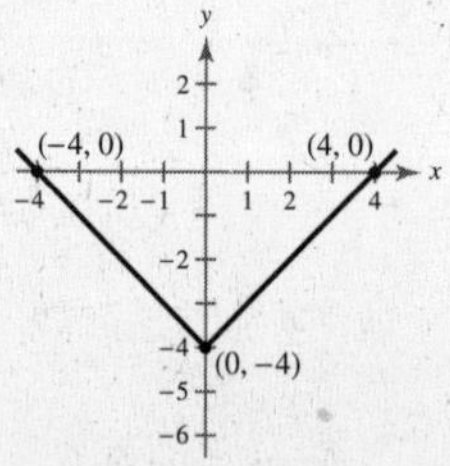

(b) Domain: $(-\infty, \infty)$

Range: $[-4, \infty)$

(c)

x	−3	−2	3
$f(x)$	−1	−2	−1

(d) The function is not one-to-one.

12.
$$\begin{aligned}
y &= 4x + 6\\
x &= 4y + 6\\
y &= \tfrac{1}{4}x - \tfrac{3}{2}\\
f^{-1}(x) &= \tfrac{1}{4}x - \tfrac{3}{2}\\
f\left(f^{-1}(x)\right) &= f\left(\tfrac{1}{4}x - \tfrac{3}{2}\right)\\
&= 4\left(\tfrac{1}{4}x - \tfrac{3}{2}\right) + 6\\
&= x - 6 + 6 = x\\
f^{-1}\left(f(x)\right) &= f^{-1}(4x + 6)\\
&= \tfrac{1}{4}(4x + 6) - \tfrac{3}{2}\\
&= x + \tfrac{3}{2} - \tfrac{3}{2} = x
\end{aligned}$$

13.
$$\begin{aligned}
y &= \sqrt[3]{8 - 3x}\\
x &= \sqrt[3]{8 - 3y}\\
x^3 &= 8 - 3y\\
y &= -\tfrac{1}{3}x^3 + \tfrac{8}{3}\\
f^{-1}(x) &= -\tfrac{1}{3}x^3 + \tfrac{8}{3}\\
f\left(f^{-1}(x)\right) &= f\left(-\tfrac{1}{3}x^3 + \tfrac{8}{3}\right)\\
&= \sqrt[3]{8 - 3\left(-\tfrac{1}{3}x^3 + \tfrac{8}{3}\right)}\\
&= \sqrt[3]{8 + x^3 - 8} = x\\
f^{-1}\left(f(x)\right) &= f^{-1}\left(\sqrt[3]{8 - 3x}\right)\\
&= -\tfrac{1}{3}\left(\sqrt[3]{8 - 3x}\right)^3 + \tfrac{8}{3}\\
&= -\tfrac{1}{3}(8 - 3x) + \tfrac{8}{3}\\
&= -\tfrac{8}{3} + x + \tfrac{8}{3} = x
\end{aligned}$$

14. $\displaystyle \lim_{x\to 0} \frac{x + 5}{x - 5} = \lim_{x\to 0} \frac{0 + 5}{0 - 5} = -1$

15.

x	4.9	4.99	4.999	5.001	5.01	5.1
$f(x)$	−99	−999	−9999	10,001	1001	101

$\displaystyle \lim_{x\to 5^-} \frac{x + 5}{x - 5} = -\infty$

$\displaystyle \lim_{x\to 5^+} \frac{x + 5}{x - 5} = \infty$

$\displaystyle \lim_{x\to 5} f(x)$ does not exist.

16.
$$\begin{aligned}
\lim_{x\to -3} \frac{x^2 + 2x - 3}{x^2 + 4x + 3} &= \lim_{x\to -3} \frac{(x - 1)(x + 3)}{(x + 1)(x + 3)}\\
&= \lim_{x\to -3} \frac{x - 1}{x + 1}\\
&= 2
\end{aligned}$$

17.

x	-0.01	-0.001	-0.0001
$f(x)$	0.16671	0.16667	0.16667

x	0.0001	0.001	0.01
$f(x)$	0.16667	0.16666	0.16662

$$\lim_{x \to 0} \frac{\sqrt{x+9} - 3}{x} \approx 0.16667$$

18. $f(x) = \dfrac{x^2 - 16}{x - 4}$ is continuous on the intervals $(-\infty, 4)$ and $(4, \infty)$ because the domain of f consists of all real numbers except $x = 4$. There is a discontinuity at $x = 4$ because $f(4)$ is not defined.

19. $f(x) = \sqrt{5 - x}$ is continuous on the interval $(5, \infty)$ because the domain of f consists of all $x > 5$.

20. $\lim_{x \to 1^-} f(x) = \lim_{x \to 1^-} (1 - x) = 0$

$\lim_{x \to 1^+} f(x) = \lim_{x \to 1^+} (x - x^2) = 0$

So, $\lim_{x \to 1} f(x) = 0$.

Because $f(1)$ is defined, $\lim_{x \to 1} f(x)$ exists, and $\lim_{x \to 1} f(x) = f(1)$, the function is continuous on the interval $(-\infty, \infty)$.

21. (a)

t	0	1	2
y (actual)	2167	2149	2135
y (model)	2166	2151	2137

t	3	4	5
y (actual)	2127	2113	2101
y (model)	2124	2113	2103

The model fits the actual data very well.

(b) When $t = 9$: $y = 2071.14$

In 2009, the number of farms will be about 2,071,140.

Practice Test for Chapter 1

1. Find the distance between $(3, 7)$ and $(4, -2)$.

2. Find the midpoint of the line segment joining $(0, 5)$ and $(2, 1)$.

3. Determine whether the points $(0, -3)$, $(2, 5)$, and $(-3, -15)$ are collinear.

4. Find x so that the distance between $(0, 3)$ and $(x, 5)$ is 7.

5. Sketch the graph of $y = 4 - x^2$.

6. Sketch the graph of $y = \sqrt{x - 2}$.

7. Sketch the graph of $y = |x - 3|$.

8. Write the equation of the circle in standard form and sketch its graph.

   $x^2 + y^2 - 8x + 2y + 8 = 0$

9. Find the points of intersection of the graphs of $x^2 + y^2 = 25$ and $x - 2y = 10$.

10. Find the general equation of the line passing through the points $(7, 4)$ and $(6, -2)$.

11. Find the general equation of the line passing through the point $(-2, -1)$ with a slope of $m = \frac{2}{3}$.

12. Find the general equation of the line passing through the point $(6, -8)$ with undefined slope.

13. Find the general equation of the line passing through the point $(0, 3)$ and perpendicular to the line given by $2x - 5y = 7$.

14. Given $f(x) = x^2 - 5$, find the following.

    (a) $f(3)$

    (b) $f(-6)$

    (c) $f(x - 5)$

    (d) $f(x + \Delta x)$

15. Find the domain and range of $f(x) = \sqrt{3 - x}$.

16. Given $f(x) = 2x + 3$ and $g(x) = x^2 - 1$, find the following.

    (a) $f(g(x))$

    (b) $g(f(x))$

17. Given $f(x) = x^3 + 6$, find $f^{-1}(x)$.

18. Find $\lim_{x \to -4} (2 - 5x)$.

19. Find $\lim_{x \to 6} \frac{x^2 - 36}{x - 6}$.

20. Find $\lim_{x\to -1} \frac{|x+1|}{x+1}$.

21. Find $\lim_{x\to 0} \frac{\sqrt{x+5}-\sqrt{5}}{x}$.

22. Find $\lim_{x\to 1} f(x)$, where $f(x) = \begin{cases} 2x+3, & x \le 1 \\ x^2+4, & x > 1 \end{cases}$.

23. Find the discontinuities of $f(x) = \frac{x-8}{x^2-64}$. Which are removable?

24. Find the discontinuities of $f(x) = \frac{|x-3|}{x-3}$. Which are removable?

25. Sketch the graph of $f(x) = \frac{x^2-5x+6}{x-3}$.

Graphing Calculator Required

26. Solve the equation for y and graph the resulting two equations on the same set of coordinate axes.

$x^2 + y^2 + 6x + 5 = 0$

27. Use a graphing calculator to graph $f(x) = \frac{x^2-9}{x-3}$ and find $\lim_{x\to 3} f(x)$. Is the graph displayed correctly at $x = 3$?

CHAPTER 2
Differentiation

CHAPTER 2
Differentiation

Section 2.1 The Derivative and the Slope of a Graph

Skills Review

1. $P(2, 1)$, $Q(2, 4)$

   $x = 2$

2. $P(2, 2)$, $Q = (-5, 2)$

   $m = \dfrac{2 - 2}{-5 - 2} = 0$

   $y - 2 = 0(x - 2)$

   $y = 2$

3. $P(2, 0)$, $Q = (3, -1)$

   $m = \dfrac{-1 - 0}{3 - 2} = -1$

   $y - 0 = -1(x - 2)$

   $y = 2 - x$

4. $\displaystyle\lim_{\Delta x \to 0} \frac{2x\Delta x + (\Delta x)^2}{\Delta x} = \lim_{\Delta x \to 0} \frac{\Delta x(2x + \Delta x)}{\Delta x}$

   $= \displaystyle\lim_{\Delta x \to 0} 2x + \Delta x$

   $= 2x$

5. $\displaystyle\lim_{\Delta x \to 0} \frac{3x^2\Delta x + 3x(\Delta x)^2 + (\Delta x)^3}{\Delta x}$

   $= \displaystyle\lim_{\Delta x \to 0} \frac{\Delta x\left[3x^2 + 3x\Delta x + (\Delta x)^2\right]}{\Delta x}$

   $= \displaystyle\lim_{\Delta x \to 0} 3x^2 + 3x\Delta x + (\Delta x)^2$

   $= 3x^2$

6. $\displaystyle\lim_{\Delta x \to 0} \frac{1}{x(x + \Delta x)} = \frac{1}{x^2}$

7. $\displaystyle\lim_{\Delta x \to 0} \frac{(x + \Delta x)^2 - x^2}{\Delta x}$

   $= \displaystyle\lim_{\Delta x \to 0} \frac{x^2 + 2x\Delta x + (\Delta x)^2 - x^2}{\Delta x}$

   $= \displaystyle\lim_{\Delta x \to 0} \frac{2x\Delta x + (\Delta x)^2}{\Delta x}$

   $= \displaystyle\lim_{\Delta x \to 0} \frac{\Delta x(2x + \Delta x)}{\Delta x}$

   $= 2x$

8. $f(x) = \dfrac{1}{x - 1}$

   Domain: $(-\infty, 1) \cup (1, \infty)$

9. $f(x) = \frac{1}{5}x^3 - 2x^2 + \frac{1}{3}x - 1$

   Domain: $(-\infty, \infty)$

10. $f(x) = \dfrac{6x}{x^3 + x}$

    Domain: $(-\infty, 0) \cup (0, \infty)$

1.

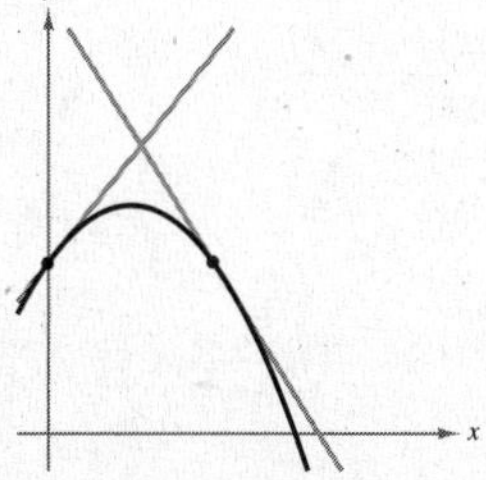

3.

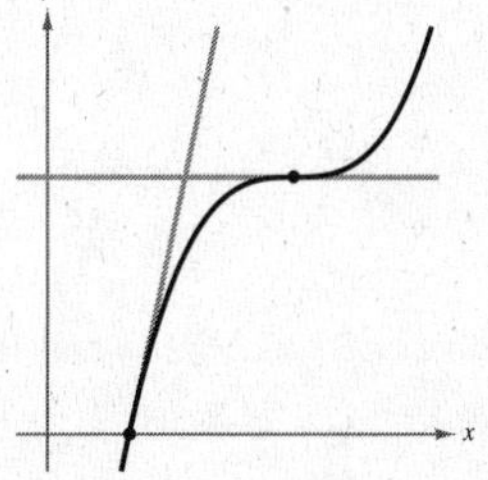

5. The slope is $m = 1$.

7. The slope is $m = 0$.

9. The slope is $m = -\frac{1}{3}$.

11. For $t = 2$: $m \approx 300$

For $t = 8$: $m \approx 0$

For $t = 11$: $m \approx -600$

13. For 2002, $t = 2$ and $m \approx 400$.

For 2004, $t = 4$ and $m \approx 500$.

15.
$$\begin{aligned} m_{\text{sec}} &= \frac{f(2+\Delta x) - f(2)}{\Delta x} \\ &= \frac{6 - 2(2+\Delta x) - [6 - 2(2)]}{\Delta x} \\ &= \frac{6 - 4 - 2\Delta x - 2}{\Delta x} \\ &= \frac{-2\Delta x}{\Delta x} \\ &= -2 \end{aligned}$$

$$m = \lim_{\Delta x \to 0} m_{\text{sec}} = \lim_{\Delta x \to 0} -2 = -2$$

17.
$$\begin{aligned} m_{\text{sec}} &= \frac{f(0+\Delta x) - f(0)}{\Delta x} \\ &= \frac{-1 - (-1)}{\Delta x} \\ &= \frac{0}{\Delta x} \\ &= 0 \end{aligned}$$

$$m = \lim_{\Delta x \to 0} m_{\text{sec}} = \lim_{\Delta x \to 0} 0 = 0$$

19.
$$\begin{aligned} m_{\text{sec}} &= \frac{f(2+\Delta x) - f(2)}{\Delta x} \\ &= \frac{(2+\Delta x)^2 - 1 - [(2)^2 - 1]}{\Delta x} \\ &= \frac{4 + 4\Delta x + (\Delta x)^2 - 1 - 3}{\Delta x} \\ &= \frac{4\Delta x + (\Delta x)^2}{\Delta x} \\ &= \frac{\Delta x(4 + \Delta x)}{\Delta x} \\ &= 4 + \Delta x \end{aligned}$$

$$m = \lim_{\Delta x \to 0} m_{\text{sec}} = \lim_{\Delta x \to 0} (4 + \Delta x) = 4$$

21.
$$\begin{aligned} m_{\text{sec}} &= \frac{f(2+\Delta x) - f(2)}{\Delta x} \\ &= \frac{(2+\Delta x)^3 - (2+\Delta x) - [(2)^3 - (2)]}{\Delta x} \\ &= \frac{8 + 12\Delta x + 6(\Delta x)^2 + (\Delta x)^3 - 2 - \Delta x - 6}{\Delta x} \\ &= \frac{11\Delta x + 6(\Delta x)^2 + (\Delta x)^3}{\Delta x} \\ &= \frac{\Delta x(11 + 6\Delta x + (\Delta x)^2)}{\Delta x} \\ &= 11 + 6\Delta x + (\Delta x)^2 \end{aligned}$$

$$m = \lim_{\Delta x \to 0} m_{\text{sec}} = \lim_{\Delta x \to 0} \left(11 + 6\Delta x + (\Delta x)^2\right) = 11$$

23.
$$\begin{aligned} m_{\text{sec}} &= \frac{f(4+\Delta x) - f(4)}{\Delta x} \\ &= \frac{2\sqrt{4+\Delta x} - 2\sqrt{4}}{\Delta x} \\ &= \frac{2\sqrt{4+\Delta x} - 4}{\Delta x} \cdot \frac{2\sqrt{4+\Delta x} + 4}{2\sqrt{4+\Delta x} + 4} \\ &= \frac{4(4+\Delta x) - 16}{\Delta x\left(2\sqrt{4+\Delta x} + 4\right)} \\ &= \frac{16 + 4\Delta x - 16}{\Delta x\left(2\sqrt{4+\Delta x} + 4\right)} \\ &= \frac{4\Delta x}{\Delta x\left(2\sqrt{4+\Delta x} + 4\right)} \\ &= \frac{4}{2\sqrt{4+\Delta x} + 4} \end{aligned}$$

$$\begin{aligned} m &= \lim_{\Delta x \to 0} m_{\text{sec}} \\ &= \lim_{\Delta x \to 0} \frac{4}{2\sqrt{4+\Delta x} + 4} \\ &= \frac{4}{2\sqrt{4} + 4} \\ &= \frac{1}{2} \end{aligned}$$

25.
$$\begin{aligned} f'(x) &= \lim_{\Delta x \to 0} \frac{f(x+\Delta x) - f(x)}{\Delta x} \\ &= \lim_{\Delta x \to 0} \frac{3 - 3}{\Delta x} \\ &= \lim_{\Delta x \to 0} \frac{0}{\Delta x} \\ &= \lim_{\Delta x \to 0} 0 \\ &= 0 \end{aligned}$$

27. $$\begin{aligned} f'(x) &= \lim_{\Delta x\to 0} \frac{f(x+\Delta x)-f(x)}{\Delta x} \\ &= \lim_{\Delta x\to 0} \frac{-5(x+\Delta x)-(-5x)}{\Delta x} \\ &= \lim_{\Delta x\to 0} \frac{-5x-5\Delta x+5x}{\Delta x} \\ &= \lim_{\Delta x\to 0} \frac{-5\Delta x}{\Delta x} \\ &= \lim_{\Delta x\to 0} -5 \\ &= -5 \end{aligned}$$

29. $$\begin{aligned} g'(s) &= \lim_{\Delta s\to 0} \frac{g(s+\Delta s)-g(s)}{\Delta s} \\ &= \lim_{\Delta s\to 0} \frac{\frac{1}{3}(s+\Delta s)+2-\left(\frac{1}{3}s+2\right)}{\Delta s} \\ &= \lim_{\Delta s\to 0} \frac{\frac{1}{3}s+\frac{1}{3}\Delta s+2-\frac{1}{3}s-2}{\Delta s} \\ &= \lim_{\Delta s\to 0} \frac{\frac{1}{3}\Delta s}{\Delta s} \\ &= \lim_{\Delta s\to 0} \frac{1}{3} \\ &= \frac{1}{3} \end{aligned}$$

31. $$\begin{aligned} f'(x) &= \lim_{\Delta x\to 0} \frac{f(x+\Delta x)-f(x)}{\Delta x} \\ &= \lim_{\Delta x\to 0} \frac{(x+\Delta x)^2-4-(x^2-4)}{\Delta x} \\ &= \lim_{\Delta x\to 0} \frac{x^2+2x\Delta x+(\Delta x)^2-4-x^2+4}{\Delta x} \\ &= \lim_{\Delta x\to 0} \frac{2x\Delta x+(\Delta x)^2}{\Delta x} \\ &= \lim_{\Delta x\to 0} \frac{\Delta x(2x+\Delta x)}{\Delta x} \\ &= \lim_{\Delta x\to 0} 2x+\Delta x \\ &= 2x \end{aligned}$$

33. $$\begin{aligned} h'(t) &= \lim_{\Delta t\to 0} \frac{h(t+\Delta t)-h(t)}{t} \\ &= \lim_{\Delta t\to 0} \frac{\sqrt{t+\Delta t-1}-\sqrt{t-1}}{\Delta t} \\ &= \lim_{\Delta t\to 0} \frac{\sqrt{t+\Delta t-1}-\sqrt{t-1}}{\Delta t}\cdot\frac{\sqrt{t+\Delta t-1}+\sqrt{t-1}}{\sqrt{t+\Delta t-1}+\sqrt{t-1}} \\ &= \lim_{\Delta t\to 0} \frac{t+\Delta t-1-(t-1)}{\Delta t\left(\sqrt{t+\Delta t-1}+\sqrt{t-1}\right)} \\ &= \lim_{\Delta t\to 0} \frac{\Delta t}{\Delta t\left(\sqrt{t+\Delta t-1}+\sqrt{t-1}\right)} \\ &= \lim_{\Delta t\to 0} \frac{1}{\sqrt{t+\Delta t-1}+\sqrt{t-1}} \\ &= \frac{1}{2\sqrt{t-1}} \end{aligned}$$

35. $$f'(t) = \lim_{\Delta t \to 0} \frac{f(t + \Delta t) - f(t)}{\Delta t}$$

$$= \lim_{\Delta t \to 0} \frac{(t + \Delta t)^3 - 12(t + \Delta t) - (t^3 - 12t)}{\Delta t}$$

$$= \lim_{\Delta t \to 0} \frac{t^3 + 3t^2\Delta t + 3t(\Delta t)^2 + (\Delta t)^3 - 12t - 12\Delta t - t^3 + 12t}{\Delta t}$$

$$= \lim_{\Delta t \to 0} \frac{3t^2\Delta t + 3t(\Delta t)^2 + (\Delta t)^3 - 12\Delta t}{\Delta t}$$

$$= \lim_{\Delta t \to 0} \frac{\Delta t\left(3t^2 + 3t\Delta t + (\Delta t)^2 - 12\right)}{\Delta t}$$

$$= \lim_{\Delta t \to 0} \left(3t^2 + 3t\Delta t + (\Delta t)^2 - 12\right)$$

$$= 3t^2 - 12$$

37. $$f'(x) = \lim_{\Delta x \to 0} \frac{f(x + \Delta x) - f(x)}{\Delta x}$$

$$= \lim_{\Delta x \to 0} \frac{\dfrac{1}{x + \Delta x + 2} - \dfrac{1}{x + 2}}{\Delta x}$$

$$= \lim_{\Delta x \to 0} \frac{\dfrac{1}{x + \Delta x + 2} \cdot \dfrac{x + 2}{x + 2} - \dfrac{1}{x + 2} \cdot \dfrac{x + \Delta x + 2}{x + \Delta x + 2}}{\Delta x}$$

$$= \lim_{\Delta x \to 0} \frac{\dfrac{x + 2 - (x + \Delta x + 2)}{(x + \Delta x + 2)(x + 2)}}{\Delta x}$$

$$= \lim_{\Delta x \to 0} \frac{-\Delta x}{\Delta x(x + \Delta x + 2)(x + 2)}$$

$$= \lim_{\Delta x \to 0} \frac{-1}{(x + \Delta x + 2)(x + 2)}$$

$$= -\frac{1}{(x + 2)^2}$$

39. $$f'(x) = \lim_{\Delta x \to 0} \frac{f(x + \Delta x) - f(x)}{\Delta x}$$

$$= \lim_{\Delta x \to 0} \frac{\frac{1}{2}(x + \Delta x)^2 - \frac{1}{2}x^2}{\Delta x}$$

$$= \lim_{\Delta x \to 0} \frac{\frac{1}{2}\left(x^2 + 2x\Delta x + (\Delta x)^2\right) - \frac{1}{2}x^2}{\Delta x}$$

$$= \lim_{\Delta x \to 0} \frac{x\Delta x + (\Delta x)^2}{\Delta x}$$

$$= \lim_{\Delta x \to 0} \frac{\Delta x(x + \Delta x)}{\Delta x}$$

$$= \lim_{\Delta x \to 0} (x + \Delta x)$$

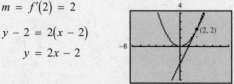

$$= x$$

$$m = f'(2) = 2$$

$$y - 2 = 2(x - 2)$$

$$y = 2x - 2$$

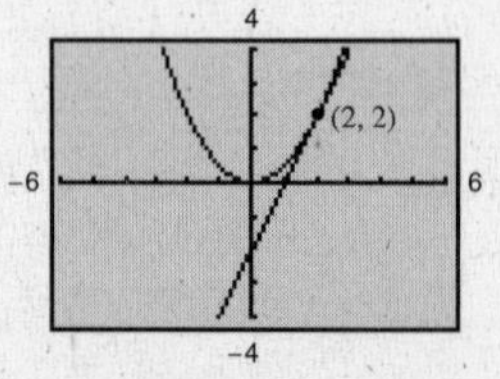

41. $f'(x) = \lim_{\Delta x \to 0} \frac{f(x + \Delta x) - f(x)}{\Delta x}$

$= \lim_{\Delta x \to 0} \frac{(x + \Delta x - 1)^2 - (x - 1)^2}{\Delta x}$

$= \lim_{\Delta x \to 0} \frac{x^2 + 2x\Delta x - 2x + (\Delta x)^2 - 2\Delta x + 1 - x^2 + 2x - 1}{\Delta x}$

$= \lim_{\Delta x \to 0} \frac{2x\Delta x + (\Delta x)^2 - 2\Delta x}{\Delta x}$

$= \lim_{\Delta x \to 0} \frac{\Delta x(2x + \Delta x - 2)}{\Delta x}$

$= \lim_{\Delta x \to 0} (2x + \Delta x - 2)$

$= 2x - 2$

$m = f'(-2) = 2(-2) - 2 = -6$

$y - 9 = -6[x - (-2)]$

$y = -6x - 3$

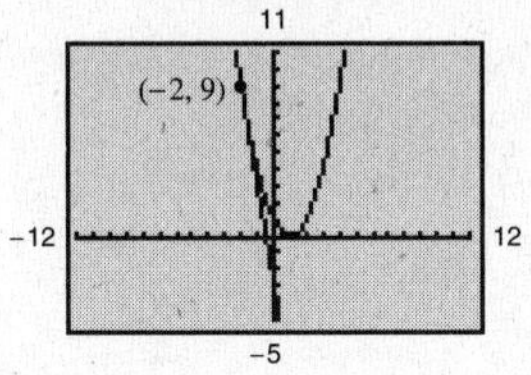

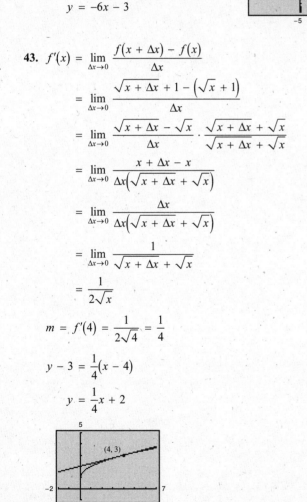

43. $f'(x) = \lim_{\Delta x \to 0} \frac{f(x + \Delta x) - f(x)}{\Delta x}$

$= \lim_{\Delta x \to 0} \frac{\sqrt{x + \Delta x} + 1 - (\sqrt{x} + 1)}{\Delta x}$

$= \lim_{\Delta x \to 0} \frac{\sqrt{x + \Delta x} - \sqrt{x}}{\Delta x} \cdot \frac{\sqrt{x + \Delta x} + \sqrt{x}}{\sqrt{x + \Delta x} + \sqrt{x}}$

$= \lim_{\Delta x \to 0} \frac{x + \Delta x - x}{\Delta x(\sqrt{x + \Delta x} + \sqrt{x})}$

$= \lim_{\Delta x \to 0} \frac{\Delta x}{\Delta x(\sqrt{x + \Delta x} + \sqrt{x})}$

$= \lim_{\Delta x \to 0} \frac{1}{\sqrt{x + \Delta x} + \sqrt{x}}$

$= \frac{1}{2\sqrt{x}}$

$m = f'(4) = \frac{1}{2\sqrt{4}} = \frac{1}{4}$

$y - 3 = \frac{1}{4}(x - 4)$

$y = \frac{1}{4}x + 2$

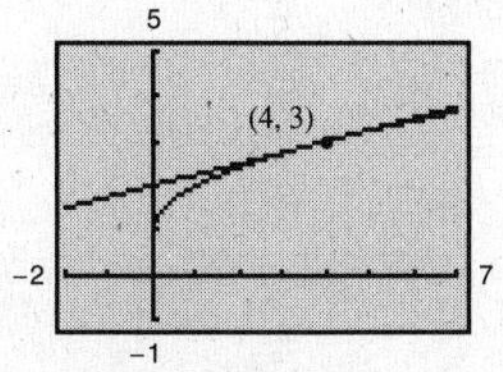

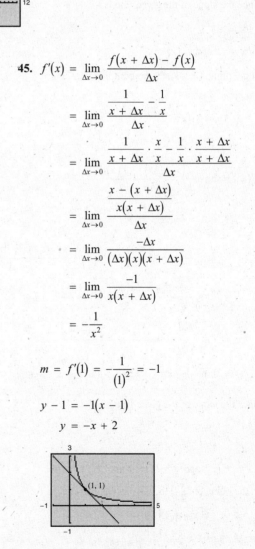

45. $f'(x) = \lim_{\Delta x \to 0} \frac{f(x + \Delta x) - f(x)}{\Delta x}$

$= \lim_{\Delta x \to 0} \frac{\frac{1}{x + \Delta x} - \frac{1}{x}}{\Delta x}$

$= \lim_{\Delta x \to 0} \frac{\frac{1}{x + \Delta x} \cdot \frac{x}{x} - \frac{1}{x} \cdot \frac{x + \Delta x}{x + \Delta x}}{\Delta x}$

$= \lim_{\Delta x \to 0} \frac{\frac{x - (x + \Delta x)}{x(x + \Delta x)}}{\Delta x}$

$= \lim_{\Delta x \to 0} \frac{-\Delta x}{(\Delta x)(x)(x + \Delta x)}$

$= \lim_{\Delta x \to 0} \frac{-1}{x(x + \Delta x)}$

$= -\frac{1}{x^2}$

$m = f'(1) = -\frac{1}{(1)^2} = -1$

$y - 1 = -1(x - 1)$

$y = -x + 2$

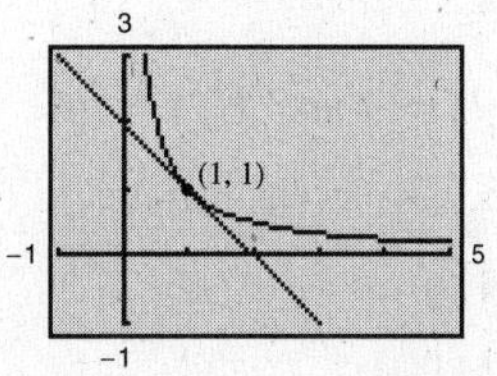

47. $$\begin{aligned} f'(x) &= \lim_{\Delta x \to 0} \frac{f(x+\Delta x) - f(x)}{\Delta x} \\ &= \lim_{\Delta x \to 0} \frac{-\frac{1}{4}(x+\Delta x)^2 - \left(-\frac{1}{4}x^2\right)}{\Delta x} \\ &= \lim_{\Delta x \to 0} \frac{-\frac{1}{4}x^2 - \frac{1}{2}x\Delta x - \frac{1}{4}(\Delta x)^2 + \frac{1}{4}x^2}{\Delta x} \\ &= \lim_{\Delta x \to 0} \frac{-\frac{1}{2}x\Delta x - \frac{1}{4}(\Delta x)^2}{\Delta x} \\ &= \lim_{\Delta x \to 0} \frac{\Delta x\left(-\frac{1}{2}x - \frac{1}{4}\Delta x\right)}{\Delta x} \\ &= \lim_{\Delta x \to 0} \left(-\frac{1}{2}x - \frac{1}{4}\Delta x\right) \\ &= -\frac{1}{2}x \end{aligned}$$

Since the slope of the given line is -1,

$$\begin{aligned} -\tfrac{1}{2}x &= -1 \\ x &= 2 \text{ and } f(2) = -1. \end{aligned}$$

At the point $(2, -1)$, the tangent line parallel to $x + y = 0$ is

$$\begin{aligned} y - (-1) &= -1(x - 2) \\ y &= -x + 1. \end{aligned}$$

49. $$\begin{aligned} f'(x) &= \lim_{\Delta x \to 0} \frac{f(x+\Delta x) - f(x)}{\Delta x} \\ &= \lim_{\Delta x \to 0} \frac{-\frac{1}{2}(x+\Delta x)^3 - \left(-\frac{1}{2}x^3\right)}{\Delta x} \\ &= \lim_{\Delta x \to 0} \frac{-\frac{1}{2}x^3 - \frac{3}{2}x^2\Delta x - \frac{3}{2}x(\Delta x)^2 - \frac{1}{2}(\Delta x)^3 + \frac{1}{2}x^3}{\Delta x} \\ &= \lim_{\Delta x \to 0} \frac{-\frac{3}{2}x^2\Delta x - \frac{3}{2}x(\Delta x)^2 - \frac{1}{2}(\Delta x)^3}{\Delta x} \\ &= \lim_{\Delta x \to 0} \frac{\Delta x\left(-\frac{3}{2}x^2 - \frac{3}{2}x\Delta x - \frac{1}{2}(\Delta x)^2\right)}{\Delta x} \\ &= \lim_{\Delta x \to 0} \left(-\frac{3}{2}x^2 - \frac{3}{2}x\Delta x - \frac{1}{2}(\Delta x)^2\right) = -\frac{3}{2}x^2 \end{aligned}$$

Since the slope of the given line is -6,

$$\begin{aligned} -\tfrac{3}{2}x^2 &= -6 \\ x^2 &= 4 \\ x &= \pm 2 \text{ and } f(-2) = 4 \text{ and } f(2) = -4. \end{aligned}$$

At the point $(-2, 4)$ the tangent line parallel to $6x + y + 4 = 0$ is

$$\begin{aligned} y - 4 &= -6[x - (-2)] \\ y &= -6x - 8. \end{aligned}$$

At the point $(2, -4)$ the tangent line parallel to $6x + y + 4 = 0$ is

$$\begin{aligned} y - (-4) &= -6(x - 2) \\ y &= -6x + 8. \end{aligned}$$

51. y is differentiable for all $x \neq -3$.

At $(-3, 0)$ the graph has a node.

53. y is differentiable for all $x \neq 3$.

At (3, 0) the graph has a cusp.

55. y is differentiable for all $x > 1$.

The derivative does not exist at endpoints.

57. y is differentiable for all $x \neq 0$.

The function is discontinuous at $x = 0$.

59. $f(x)$ is differentiable everywhere except at $x = 1$.

61. Since $f'(x) = -3$ for all x, f is a line of the form

$f(x) = -3x + b$.

$f(0) = 2$, so $2 = (-3)(0) + b$, or $b = 2$.

Thus, $f(x) = -3x + 2$.

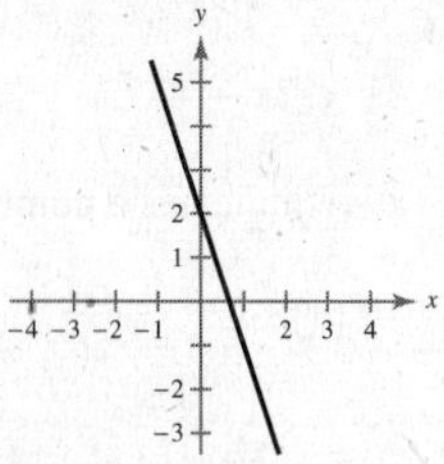

63.

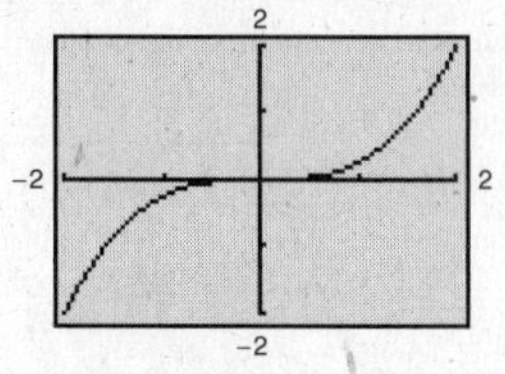

x	-2	$-\frac{3}{2}$	-1	$-\frac{1}{2}$	0	$\frac{1}{2}$	1	$\frac{3}{2}$	2
$f(x)$	−2	−0.8438	−0.25	−0.0313	0	0.0313	0.25	0.8438	2
$f'(x)$	3	1.6875	0.75	0.1875	0	0.1875	0.75	1.6875	3

Analytically, the slope of $f(x) = \frac{1}{4}x^3$ is

$$m = \lim_{\Delta x \to 0} \frac{f(x + \Delta x) - f(x)}{\Delta x}$$

$$= \lim_{\Delta x \to 0} \frac{\frac{1}{4}(x + \Delta x)^3 - \frac{1}{4}x^3}{\Delta x} = \lim_{\Delta x \to 0} \frac{\frac{1}{4}\left[3x^2\Delta x + 3x(\Delta x)^2 + (\Delta x)^3\right]}{\Delta x} = \lim_{\Delta x \to 0} \frac{1}{4}\left[3x^2 + 3x\Delta x + (\Delta x)^2\right] = \frac{3}{4}x^2.$$

65.

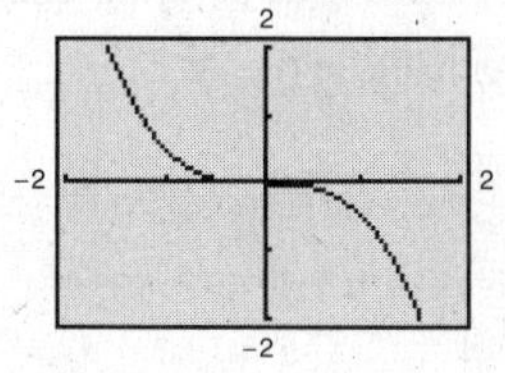

x	-2	$-\frac{3}{2}$	-1	$-\frac{1}{2}$	0	$\frac{1}{2}$	1	$\frac{3}{2}$	2
$f(x)$	4	1.6875	0.5	0.0625	0	−0.0625	−0.5	−1.6875	−4
$f'(x)$	−6	−3.375	−1.5	−0.375	0	−0.375	−1.5	−3.375	−6

Analytically, the slope of $f(x) = -\frac{1}{2}x^3$ is

$$m = \lim_{\Delta x \to 0} \frac{f(x + \Delta x) - f(x)}{\Delta x}$$

$$= \lim_{\Delta x \to 0} \frac{-\frac{1}{2}(x + \Delta x)^3 + \frac{1}{2}x^3}{\Delta x}$$

$$= \lim_{\Delta x \to 0} \frac{-\frac{1}{2}\left[x^3 + 3x^2\Delta x + 3x(\Delta x)^2 + (\Delta x)^3\right] + \frac{1}{2}x^3}{\Delta x}$$

$$= \lim_{\Delta x \to 0} \frac{-\frac{1}{2}\left[3x^2\Delta x + 3x(\Delta x)^2 + (\Delta x)^3\right]}{\Delta x} = \lim_{\Delta x \to 0} -\frac{1}{2}\left[3x^2 + 3x(\Delta x) + (\Delta x)^2\right] = -\frac{3}{2}x^2.$$

67. $f'(x) = \lim_{\Delta x \to 0} \frac{f(x + \Delta x) - f(x)}{\Delta x}$

$= \lim_{\Delta x \to 0} \frac{(x + \Delta x)^2 - 4(x + \Delta x) - (x^2 - 4x)}{\Delta x}$

$= \lim_{\Delta x \to 0} \frac{2x\Delta x + (\Delta x)^2 - 4\Delta x}{\Delta x}$

$= \lim_{\Delta x \to 0} (2x + \Delta x - 4) = 2x - 4$

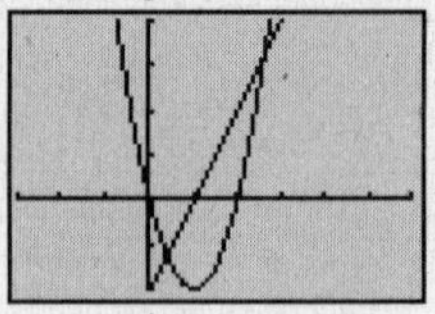

The x-intercept of the derivative indicates a point of horizontal tangency for f.

69. $f'(x) = \lim_{\Delta x \to 0} \frac{f(x + \Delta x) - f(x)}{\Delta x}$

$= \lim_{\Delta x \to 0} \frac{(x + \Delta x)^3 - 3(x + \Delta x) - (x^3 - 3x)}{\Delta x}$

$= \lim_{\Delta x \to 0} \frac{x^3 + 3x^2\Delta x + 3x(\Delta x)^2 + (\Delta x)^3 - 3x - 3\Delta x - x^3 + 3x}{\Delta x}$

$= \lim_{\Delta x \to 0} \frac{3x^2\Delta x + 3x(\Delta x)^2 + (\Delta x)^3 - 3\Delta x}{\Delta x}$

$= \lim_{\Delta x \to 0} \left(3x^2 + 3x\Delta x + (\Delta x)^2 - 3\right) = 3x^2 - 3$

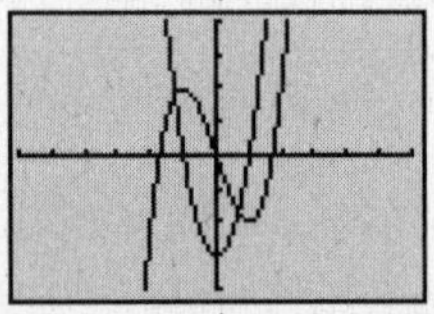

The x-intercepts of the derivative indicate points of horizontal tangency for f.

71. True. The slope of the graph is given by $f'(x) = 2x$, which is different for each different x value.

73. True. See page 122.

75. The graph of $f(x) = x^2 + 1$ is smooth at (0, 1), but the graph of $g(x) = |x| + 1$ has a node at (0, 1). The function g is not differentiable at (0, 1).

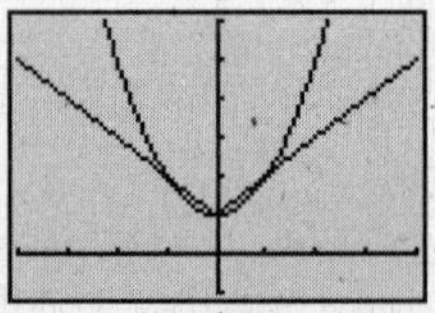

Section 2.2 Some Rules for Differentiation

Skills Review

1. (a) $2x^2,\ x = 2$

$2(2^2) = 2(4) = 8$

(b) $(2x)^2,\ x = 2$

$[2(2)]^2 = 4^2 = 16$

(c) $2x^{-2},\ x = 2$

$2(2)^{-2} = 2\left(\frac{1}{4}\right) = \frac{1}{2}$

Skills Review —*continued*—

2. (a) $\dfrac{1}{(3x)^2}$, $x = 2$

$\dfrac{1}{[3(2)]^2} = \dfrac{1}{6^2} = \dfrac{1}{36}$

(b) $\dfrac{1}{4x^3}$, $x = 2$

$\dfrac{1}{4(2^3)} = \dfrac{1}{4(8)} = \dfrac{1}{32}$

(c) $\dfrac{(2x)^{-3}}{4x^{-2}}$, $x = 2$

$\dfrac{[2(2)]^{-3}}{4(2)^{-2}} = \dfrac{4^{-3}}{4(2)^{-2}} = \dfrac{2^2}{4(4^3)} = \dfrac{1}{64}$

3. $4(3)x^3 + 2(2)x = 12x^3 + 4x = 4x(3x^2 + 1)$

4. $\frac{1}{2}(3)x^2 - \frac{3}{2}x^{1/2} = \frac{3}{2}x^2 - \frac{3}{2}\sqrt{x} = \frac{3}{2}x^{1/2}(x^{3/2} - 1)$

5. $\left(\dfrac{1}{4}\right)x^{-3/4} = \dfrac{1}{4x^{3/4}}$

6. $\dfrac{1}{3}(3)x^2 - 2\left(\dfrac{1}{2}\right)x^{-1/2} + \dfrac{1}{3}x^{-2/3} = x^2 - x^{-1/2} + \dfrac{1}{3}x^{-2/3}$

$= x^2 - \dfrac{1}{\sqrt{x}} + \dfrac{1}{3\sqrt[3]{x^2}}$

7. $3x^2 + 2x = 0$

$x(3x + 2) = 0$

$x = 0$

$3x + 2 = 0 \rightarrow x = -\frac{2}{3}$

8. $x^3 - x = 0$

$x(x^2 - 1) = 0$

$x(x + 1)(x - 1) = 0$

$x = 0$

$x + 1 = 0 \rightarrow x = -1$

$x - 1 = 0 \rightarrow x = 1$

9. $x^2 + 8x - 20 = 0$

$(x + 10)(x - 2) = 0$

$x + 10 = 0 \rightarrow x = -10$

$x - 2 = 0 \rightarrow x = 2$

10. $x^2 - 10x - 24 = 0$

$(x - 12)(x + 2) = 0$

$x - 12 = 0 \rightarrow x = 12$

$x + 2 = 0 \rightarrow x = -2$

1. (a) $y = x^2$

$y' = 2x$

At $(1, 1)$, $y' = 2$.

(b) $y = x^{1/2}$

$y' = \dfrac{1}{2}x^{-1/2} = \dfrac{1}{2\sqrt{x}}$

At $(1, 1)$, $y' = \frac{1}{2}$.

3. (a) $y = x^{-1}$

$y' = -x^{-2} = -\dfrac{1}{x^2}$

At $(1, 1)$, $y' = -1$.

(b) $y = x^{-1/3}$

$y' = -\dfrac{1}{3}x^{-4/3} = -\dfrac{1}{3x^{4/3}}$

At $(1, 1)$, $y' = -\frac{1}{3}$.

5. $y' = 0$

7. $y' = 4x^3$

9. $f'(x) = 4$

11. $g'(x) = 2x + 5$

13. $f'(t) = -6t + 2$

15. $s'(t) = 3t^2 - 2$

17. $y' = 4\left(\frac{4}{3}\right)t^{1/3} = \frac{16}{3}t^{1/3}$

19. $f(x) = 4\sqrt{x} = 4x^{1/2}$

$f'(x) = 4\left(\dfrac{1}{2}x^{-1/2}\right) = \dfrac{2}{\sqrt{x}}$

21. $y' = -8x^{-3} + 4x^1 = -\dfrac{8}{x^3} + 4x$

	Function	*Rewrite*	*Differentiate*	*Simplify*
23.	$y = \dfrac{1}{x^3}$	$y = x^{-3}$	$y' = -3x^{-4}$	$y' = -\dfrac{3}{x^4}$
25.	$y = \dfrac{1}{(4x)^3}$	$y = \dfrac{1}{64}x^{-3}$	$y' = -\dfrac{3}{64}x^{-4}$	$y' = -\dfrac{3}{64x^4}$
27.	$y = \dfrac{\sqrt{x}}{x}$	$y = x^{-1/2}$	$y' = -\dfrac{1}{2}x^{-3/2}$	$y' = -\dfrac{1}{2x^{3/2}}$

29. $f(x) = \dfrac{1}{x} = x^{-1}$

$f'(x) = -x^{-2} = -\dfrac{1}{x^2}$

$f'(1) = -1$

31. $f(x) = -\frac{1}{2}x(1 + x^2) = -\frac{1}{2}x - \frac{1}{2}x^3$

$f'(x) = -\frac{1}{2} - \frac{3}{2}x^2$

$f'(1) = -\frac{1}{2} - \frac{3}{2} = -2$

33. $y = (2x + 1)^2 = 4x^2 + 4x + 1$

$y' = 8x + 4$

$y'(0) = (8)(0) + 4 = 4$

35. $f(x) = x^2 - 4x^{-1} - 3x^{-2}$

$f'(x) = 2x + 4x^{-2} + 6x^{-3} = 2x + \dfrac{4}{x^2} + \dfrac{6}{x^3}$

37. $f(x) = x^2 - 2x - \dfrac{2}{x^4} = x^2 - 2x - 2x^{-4}$

$f'(x) = 2x - 2 + 8x^{-5} = 2x - 2 + \dfrac{8}{x^5}$

39. $f(x) = x(x^2 + 1) = x^3 + x$

$f'(x) = 3x^2 + 1$

41. $f(x) = (x + 4)(2x^2 - 1) = 2x^3 + 8x^2 - x - 4$

$f'(x) = 6x^2 + 16x - 1$

43. $f(x) = \dfrac{2x^3 - 4x^2 + 3}{x^2} = 2x - 4 + 3x^{-2}$

$f'(x) = 2 - 6x^{-3} = 2 - \dfrac{6}{x^3} = \dfrac{2x^3 - 6}{x^3} = \dfrac{2(x^3 - 3)}{x^3}$

45. $f(x) = \dfrac{4x^3 - 3x^2 + 2x + 5}{x^2} = 4x - 3 + 2x^{-1} + 5x^{-2}$

$f'(x) = 4 - 2x^{-2} - 10x^{-3}$

$= 4 - \dfrac{2}{x^2} - \dfrac{10}{x^3} = \dfrac{4x^3 - 2x - 10}{x^3}$

47. $f(x) = x^{4/5} + x$

$f'(x) = \dfrac{4}{5}x^{-1/5} + 1 = \dfrac{4}{5x^{1/5}} + 1$

49. (a) $y = -2x^4 + 5x^2 - 3$

$y' = -8x^3 + 10x$

$m = y'(1) = -8 + 10 = 2$

The equation of the tangent line is

$y - 0 = 2(x - 1)$

$y = 2x - 2.$

(b)

3.1

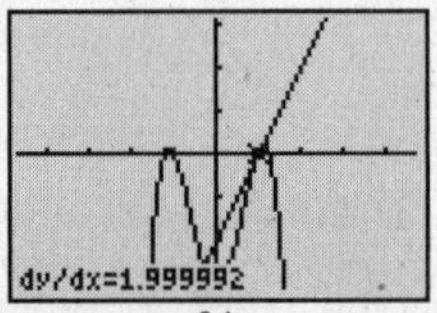

−4.7 4.7

−3.1

51. (a) $f(x) = \sqrt[3]{x} + \sqrt[5]{x} = x^{1/3} + x^{1/5}$

$f'(x) = \dfrac{1}{3}x^{-2/3} + \dfrac{1}{5}x^{-4/5} = \dfrac{1}{3x^{2/3}} + \dfrac{1}{5x^{4/5}}$

$m = f'(1) = \dfrac{1}{3} + \dfrac{1}{5} = \dfrac{8}{15}$

The equation of the tangent line is

$y - 2 = \dfrac{8}{15}(x - 1)$

$y = \dfrac{8}{15}x + \dfrac{22}{15}.$

(b)

3.1

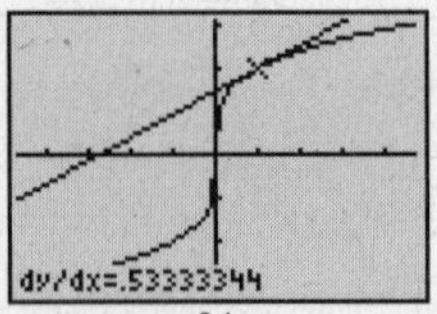

−4.7 4.7

−3.1

53. $y' = -4x^3 + 6x = 2x(3 - 2x^2) = 0$ when

$x = 0, \pm\dfrac{\sqrt{6}}{2}$

If $x = \pm\dfrac{\sqrt{6}}{2}$, then

$$y = -\left(\pm\frac{\sqrt{6}}{2}\right)^4 + 3\left(\pm\frac{\sqrt{6}}{2}\right)^2 - 1$$
$$= -\frac{9}{4} + 3\left(\frac{3}{2}\right) - 1$$
$$= \frac{5}{4}.$$

The function has horizontal tangent lines at the points

$(0, -1), \left(-\dfrac{\sqrt{6}}{2}, \dfrac{5}{4}\right)$, and $\left(\dfrac{\sqrt{6}}{2}, \dfrac{5}{4}\right)$.

55. $y' = x + 5 = 0$ when $x = -5$.

The function has a horizontal tangent line at the point $\left(-5, -\frac{25}{2}\right)$.

57. (a)

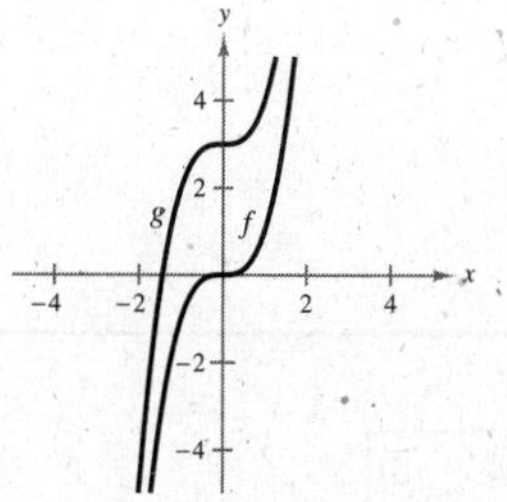

(b) $f'(x) = g'(x) = 3x^2$

$f'(1) = g'(1) = 3$

(c) Tangent line to f at $x = 1$:

$f(1) = 1$

$y - 1 = 3(x - 1)$

$y = 3x - 2$

Tangent line to g at $x = 1$:

$g(1) = 4$

$y - 4 = 3(x - 1)$

$y = 3x + 1$

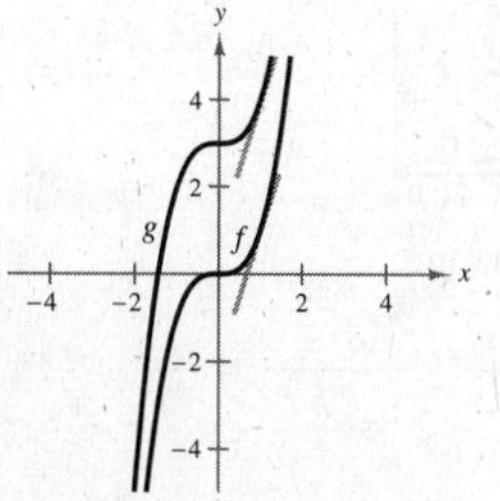

(d) f' and g' are the same.

59. (a) $h(x) = f(x) - 2$

$h'(x) = f'(x) - 0 = f'(x)$

$h'(1) = f'(1) = 3$

(b) $h(x) = 2f(x)$

$h'(x) = 2f'(x)$

$h'(1) = 2f'(1) = 2(3) = 6$

(c) $h(x) = -f(x)$

$h'(x) = -f'(x)$

$h'(1) = -f'(1) = -3$

(d) $h(x) = -1 + 2f(x)$

$h'(x) = 0 + 2f'(x) = 2f'(x)$

$h'(1) = 2f'(1) = 2(3) = 6$

61. $U = 1.6875t^4 - 0.255t^3 + 57.91t^2 - 507.9t + 3118$

($t = 0$ corresponds to 2000; $0 \le t \le 5$)

(a) $U'(t) = 6.75t^3 - 0.765t^2 + 115.82t - 507.9$

2001: $U'(1) \approx -386.10$

2004: $U'(4) = 375.14$

(b) These results are close to the estimates.

(c) The units of U' are thousands of pounds per year per year. So, the slope of the graph at time t is the rate at which production is increasing or decreasing in thousands of pounds per year per year.

63. $T = -0.0418t^3 + 0.605t^2 - 1.16t + 73.4$

($t = 1$ corresponds to January; $1 \le t \le 12$)

(a) $T'(t) = -0.1254t^2 + 1.21t - 1.16$

$T'(1) = -0.0754$

$T'(8) = 0.4944$

$T'(12) = -4.6976$

(b) The units of T' are degrees Fahrenheit per month. So, the slope of the graph at time t is the rate at which temperature is increasing or decreasing in degrees Fahrenheit per month.

65. $f(x) = 4.1x^3 - 12x^2 + 2.5x$

$f'(x) = 12.3x^2 - 24x + 2.5$

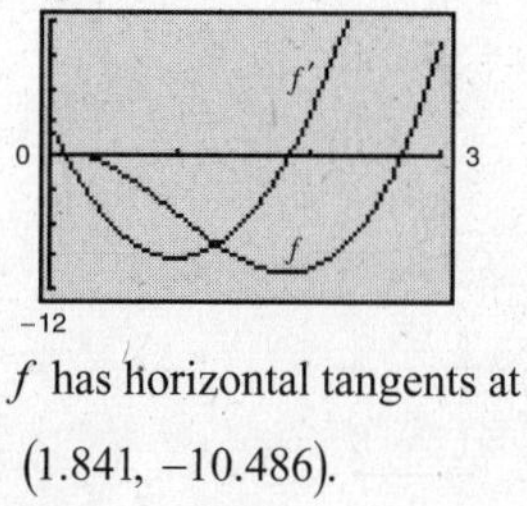

f has horizontal tangents at $(0.110, 0.135)$ and $(1.841, -10.486)$.

67. False. Let $f(x) = x$ and $g(x) = x + 1$.

Then $f'(x) = g'(x) = 1$, but $f(x) \neq g(x)$.

Section 2.3 Rates of Change

Skills Review

1. $\dfrac{-63-(-105)}{21-7} = \dfrac{42}{14} = 3$

2. $\dfrac{-37-54}{16-3} = \dfrac{-91}{13} = -7$

3. $y = 4x^2 - 2x + 7$
   $y' = 8x - 2$

4. $y = -3t^3 + 2t^2 - 8$
   $y' = -9t^2 + 4t$

5. $s = -16t^2 + 24t + 30$
   $s' = -32t + 24$

6. $y = -16x^2 + 54x + 70$
   $y' = -32x + 54$

7. $A = \frac{1}{10}(-2r^3 + 3r^2 + 5r)$
   $A' = \frac{1}{10}(-6r^2 + 6r + 5)$
   $A' = -\frac{3}{5}r^2 + \frac{3}{5}r + \frac{1}{2}$

8. $y = \frac{1}{9}(6x^3 - 18x^2 + 63x - 15)$
   $y' = \frac{1}{9}(18x^2 - 36x + 63)$
   $y' = 2x^2 - 4x + 7$

9. $y = 12x - \dfrac{x^2}{5000}$
   $y' = 12 - \dfrac{2x}{5000}$
   $y' = 12 - \dfrac{x}{2500}$

10. $y = 138 + 74x - \dfrac{x^3}{10,000}$
    $y' = 74 - \dfrac{3x^2}{10,000}$

1. (a) 1980-1985: $\dfrac{115-63}{5} = \$10.4$ billion per year

   (b) 1985-1990: $\dfrac{152-115}{5} = \$7.4$ billion per year

   (c) 1990-1995: $\dfrac{184-152}{5} = \$6.4$ billion per year

   (d) 1995-2000: $\dfrac{267-184}{5} = \$16.6$ billion per year

   (e) 1980-2004: $\dfrac{312-63}{24} = \$10.375$ billion per year

   (f) 1990-2004: $\dfrac{312-152}{14} \approx \$11.429$ billion per year

3.

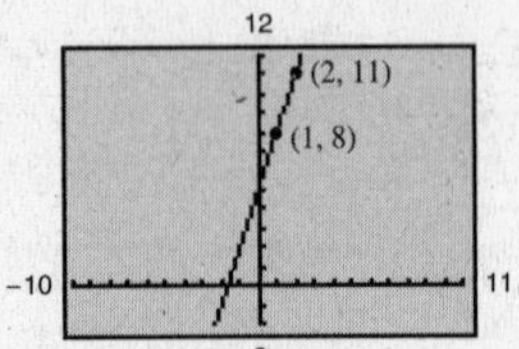

Average rate of change:

$\dfrac{\Delta y}{\Delta t} = \dfrac{f(2) - f(1)}{2-1} = \dfrac{11-8}{1} = 3$

$f'(t) = 3$

Instantaneous rates of change: $f'(1) = 3,\ f'(2) = 3$

5.

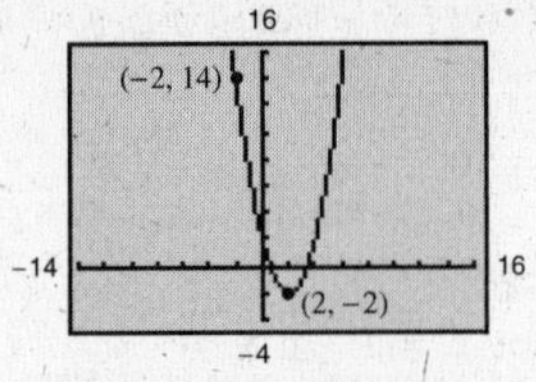

Average rate of change:

$\dfrac{\Delta y}{\Delta x} = \dfrac{h(2) - h(-2)}{2-(-2)} = \dfrac{-2-14}{4} = -4$

$h'(x) = 2x - 4$

Instantaneous rates of change: $h'(-2) = -8,\ h'(2) = 0$

7.

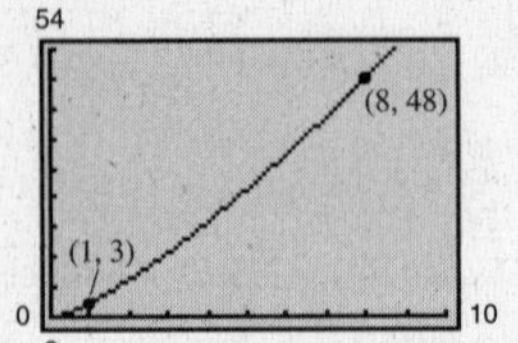

Average rate of change:

$\dfrac{\Delta y}{\Delta x} = \dfrac{f(8) - f(1)}{8-1} = \dfrac{48-3}{7} = \dfrac{45}{7}$

$f'(x) = 4x^{1/3}$

Instantaneous rates of change: $f'(1) = 4,\ f'(8) = 8$

9.

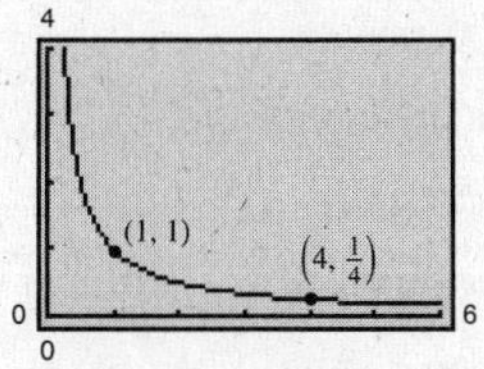

Average rate of change:

$$\frac{\Delta y}{\Delta x} = \frac{f(4) - f(1)}{4 - 1} = \frac{(1/4) - 1}{3} = -\frac{1}{4}$$

$$f'(x) = -\frac{1}{x^2}$$

Instantaneous rates of change:

$$f'(1) = -1, \ f'(4) = -\frac{1}{16}$$

11.

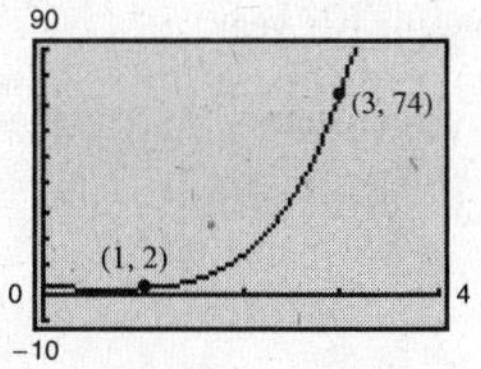

Average rate of change:

$$\frac{\Delta y}{\Delta x} = \frac{g(3) - g(1)}{3 - 1} = \frac{74 - 2}{2} = 36$$

$$g'(x) = 4x^3 - 2x$$

Instantaneous rates of change:

$$g'(1) = 2, \ g'(3) = 102$$

13. (a) The average rate of change is the greatest over $[5, 6]$.

(b) Answers will vary. Sample answer: $[2, 5]$

Both the instantaneous rate of change at $t = 4$ and the average rate of change on $[2, 5]$ is about zero.

15. (a)

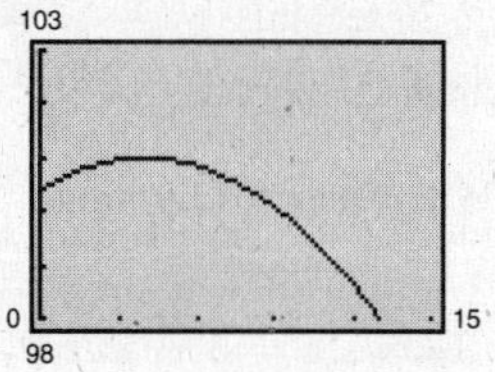

(b) For $t < 4$, the slopes are positive, and the fever is increasing. For $t > 4$, the slopes are negative, and the fever is decreasing.

(c) $T(0) = 100.4°\text{F}$

$T'(4) = 101°\text{F}$

$T(8) = 100.4°\text{F}$

$T(12) = 98.6°\text{F}$

(d) $\dfrac{dT}{dt} = -0.075t + 0.3$; the rate of change of temperature with respect to time

(e) $T'(0) = 0.3°\text{F}$ per hour

$T'(4) = 0°\text{F}$ per hour

$T'(8) = -0.3°\text{F}$ per hour

$T(12) = -0.6°\text{F}$ per hour

17. (a) $H'(v) = 33\left[10\left(\frac{1}{2}v^{-1/2}\right) - 1\right] = 33\left[\frac{5}{\sqrt{v}} - 1\right]$

Rate of change of heat loss with respect to velocity.

(b)
$$\begin{aligned} H'(2) &= 33\left[\frac{5}{\sqrt{2}} - 1\right] \\ &\approx 83.673 \frac{\text{kcal/m}^2\text{/hr}}{\text{m/sec}} \\ &= 83.673 \frac{\text{kcal}}{\text{m}^3} \cdot \frac{\text{sec}}{\text{hr}} \\ &= 83.673 \frac{\text{kcal}}{\text{m}^3} \cdot \frac{1}{3600} \\ &= 0.023 \ \text{kcal/m}^3 \end{aligned}$$

$$\begin{aligned} H'(5) &= 33\left[\frac{5}{\sqrt{5}} - 1\right] \\ &\approx 40.790 \frac{\text{kcal/m}^2\text{/hr}}{\text{m/sec}} \\ &= 40.790 \frac{\text{kcal}}{\text{m}^3} \cdot \frac{\text{sec}}{\text{hr}} \\ &= 40.790 \frac{\text{kcal}}{\text{m}^3} \cdot \frac{1}{3600} \\ &= 0.11 \ \text{kcal/m}^3 \end{aligned}$$

19. First leg: 0.75 km in 20 seconds

Second leg: 0.75 km in 25 seconds

(a) $\dfrac{0.75}{20} = 0.0375 \text{ km/sec} = 37.5 \text{ m/sec}$

(b) $\dfrac{1.50}{20 + 25} = 0.\bar{3} \text{ km/sec} = 33.\bar{3} \text{ m/sec}$

21.

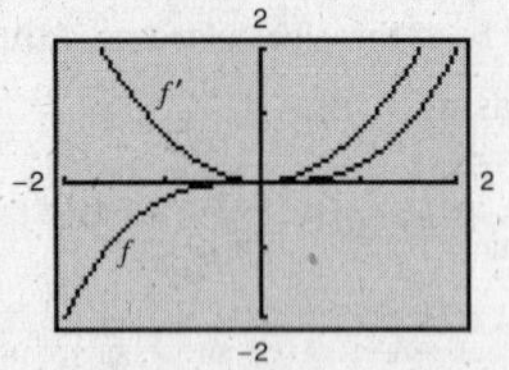

f has a horizontal tangent at (0, 0).

23. (a) $P(0) = 117{,}216{,}000$ people

$P(10) = 123{,}600{,}000$ people

$P(15) \approx 125{,}689{,}000$ people

$P(20) = 127{,}042{,}000$ people

$P(25) \approx 127{,}660{,}000$ people

The population is growing.

(b) $\dfrac{dP}{dt} = -29.42t + 785.5$

(c) $P'(0) \approx 786{,}000$ people per year

$P'(10) \approx 491{,}000$ people per year

$P'(15) \approx 344{,}000$ people per year

$P'(20) \approx 197{,}000$ people per year

$P'(25) = 50{,}000$ people per year

The rate of growth is decreasing.

25. $C(x) = \dfrac{15{,}000 \text{ miles/year}}{x \text{ miles/gallon}}(2.95 \text{ dollars/gallon}) = \dfrac{44{,}250}{x} \text{ dollars/year}$

$\dfrac{dC}{dx} = -\dfrac{44{,}250}{x^2}$

x	10	15	20	25	30	35	40
C	4425	2950	2212.5	1770	1475	1264.3	1106.3
dC/dx	−442.5	−196.7	−110.6	−70.8	−49.2	−36.1	−27.7

The marginal cost can be used to approximate the savings from a 1 mile per gallon increase in fuel efficiency. So, the driver of the car that gets 15 miles per gallon would benefit more.

27. $f(x) = \dfrac{4}{x} = 4x^{-1}$

$f'(x) = -4x^{-2} = -\dfrac{4}{x^2}$

(a)

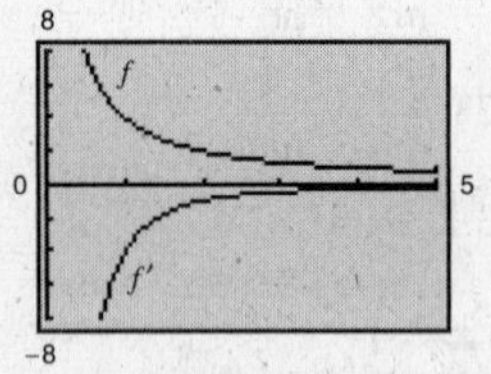

(b)

x	$\frac{1}{8}$	$\frac{1}{4}$	$\frac{1}{2}$	1	2	3	4	5
$f(x)$	32	16	8	4	2	$\frac{4}{3}$	1	$\frac{4}{5}$
$f'(x)$	−256	−64	−16	−4	−1	$-\frac{4}{9}$	$-\frac{1}{4}$	$-\frac{4}{25}$

(c) For $\left[\frac{1}{8}, \frac{1}{4}\right]$: $\dfrac{\Delta f(x)}{\Delta x} = \dfrac{16 - 32}{\frac{1}{4} - \frac{1}{8}} = \dfrac{-16}{\frac{1}{8}} = -128$

For $\left[\frac{1}{4}, \frac{1}{2}\right]$: $\dfrac{\Delta f(x)}{\Delta x} = \dfrac{8 - 16}{\frac{1}{2} - \frac{1}{4}} = \dfrac{-8}{\frac{1}{4}} = -32$

For $\left[\frac{1}{2}, 1\right]$: $\dfrac{\Delta f(x)}{\Delta x} = \dfrac{4 - 8}{1 - \frac{1}{2}} = \dfrac{-4}{\frac{1}{2}} = -8$

For $[1, 2]$: $\dfrac{\Delta f(x)}{\Delta x} = \dfrac{2 - 4}{2 - 1} = \dfrac{-2}{1} = -2$

For $[2, 3]$: $\dfrac{\Delta f(x)}{\Delta x} = \dfrac{\frac{4}{3} - 2}{3 - 2} = \dfrac{-\frac{2}{3}}{1} = -\dfrac{2}{3}$

For $[3, 4]$: $\dfrac{\Delta f(x)}{\Delta x} = \dfrac{1 - \frac{4}{3}}{4 - 3} = \dfrac{-\frac{1}{3}}{1} = -\dfrac{1}{3}$

For $[4, 5]$: $\dfrac{\Delta f(x)}{\Delta x} = \dfrac{\frac{4}{5} - 1}{5 - 4} = \dfrac{-\frac{1}{5}}{1} = -\dfrac{1}{5}$

29. (a) $\dfrac{\Delta A}{\Delta t} = \dfrac{1.5 - 0}{1 - 0} = 1.5$ millimeters per month

(b) $\dfrac{\Delta A}{\Delta t} = \dfrac{1.7 - 1.5}{6 - 3} = \dfrac{0.2}{3} = \dfrac{1}{15}$ millimeter per month

(c) $\dfrac{\Delta A}{\Delta t} = \dfrac{1.9 - 1.7}{9 - 6} = \dfrac{0.2}{3} = \dfrac{1}{15}$ millimeter per month

(d) $\dfrac{\Delta A}{\Delta t} = \dfrac{1.7 - 1.9}{12 - 9} = \dfrac{-0.2}{3} = -\dfrac{1}{15}$ millimeter per month

Mid-Chapter Quiz Solutions

1.
$$\begin{aligned} f'(x) &= \lim_{\Delta x \to 0} \frac{f(x + \Delta x) - f(x)}{\Delta x} \\ &= \lim_{\Delta x \to 0} \frac{-(x + \Delta x) + 2 - (-x + 2)}{\Delta x} \\ &= \lim_{\Delta x \to 0} -\frac{\Delta x}{\Delta x} \\ &= \lim_{\Delta x \to 0} -1 = -1 \end{aligned}$$

At $(2, 0)$: $m = -1$

2.
$$\begin{aligned} f'(x) &= \lim_{\Delta x \to 0} \frac{f(x + \Delta x) - f(x)}{\Delta x} \\ &= \lim_{\Delta x \to 0} \frac{\sqrt{x + \Delta x + 3} - \sqrt{x + 3}}{\Delta x} \\ &= \lim_{\Delta x \to 0} \frac{x + \Delta x + 3 - (x + 3)}{\Delta x\left(\sqrt{x + \Delta x + 3} + \sqrt{x + 3}\right)} \\ &= \lim_{\Delta x \to 0} \frac{1}{\sqrt{x + \Delta x + 3} + \sqrt{x + 3}} \\ &= \frac{1}{2\sqrt{x + 3}} \end{aligned}$$

At $(1, 2)$: $m = \dfrac{1}{2\sqrt{1 + 3}} = \dfrac{1}{4}$

3.
$$\begin{aligned} f'(x) &= \lim_{\Delta x \to 0} \frac{f(x + \Delta x) - f(x)}{\Delta x} \\ &= \lim_{\Delta x \to 0} \frac{\frac{4}{x + \Delta x} - \frac{4}{x}}{\Delta x} \\ &= \lim_{\Delta x \to 0} \frac{\frac{4x - 4(x + \Delta x)}{x(x + \Delta x)}}{\Delta x} \\ &= \lim_{\Delta x \to 0} \frac{\frac{-4\Delta x}{x(x + \Delta x)}}{\Delta x} \\ &= \lim_{\Delta x \to 0} -\frac{4}{x(x + \Delta x)} \\ &= -\frac{4}{x^2} \end{aligned}$$

At $(1, 4)$: $m = -\dfrac{4}{1^2} = -4$

4. $f'(x) = 0$

5. $f'(x) = 19$

6. $f'(x) = -6x$

7. $f'(x) = 3x^{-3/4} = \dfrac{3}{x^{3/4}}$

8. $f'(x) = -8x^{-3} = -\dfrac{8}{x^3}$

9. $f(x) = 2\sqrt{x} = 2x^{1/2}$

$f'(x) = x^{-1/2} = \dfrac{1}{\sqrt{x}}$

10. 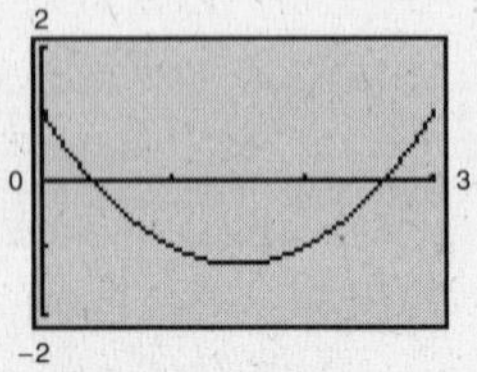

Average rate of change:

$$\frac{\Delta y}{\Delta x} = \frac{f(3) - f(0)}{3 - 0} = \frac{1 - 1}{3} = 0$$

$f'(x) = 2x - 3$

Instantaneous rates of change: $f'(0) = -3,\ f'(3) = 3$

11. 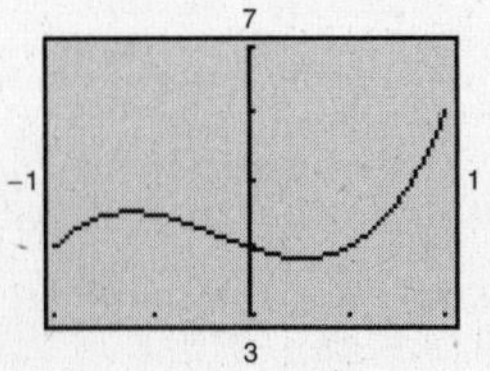

Average rate of change:

$$\frac{\Delta y}{\Delta x} = \frac{f(1) - f(-1)}{1 - (-1)} = \frac{6 - 4}{2} = 1$$

$f'(x) = 6x^2 + 2x - 1$

Instantaneous rates of change: $f'(-1) = 3,\ f'(1) = 7$

12. 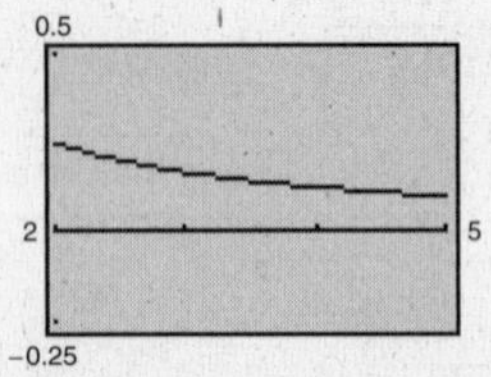

Average rate of change:

$$\frac{\Delta y}{\Delta x} = \frac{f(5) - f(2)}{5 - 2} = \frac{\frac{1}{10} - \frac{1}{4}}{3} = -\frac{1}{20}$$

$f(x) = \dfrac{1}{2x} = \dfrac{1}{2}x^{-1}$

$f'(x) = -\dfrac{1}{2x^2}$

Instantaneous rates of change:

$f'(2) = -\dfrac{1}{8},\ f'(5) = -\dfrac{1}{50}$

13. 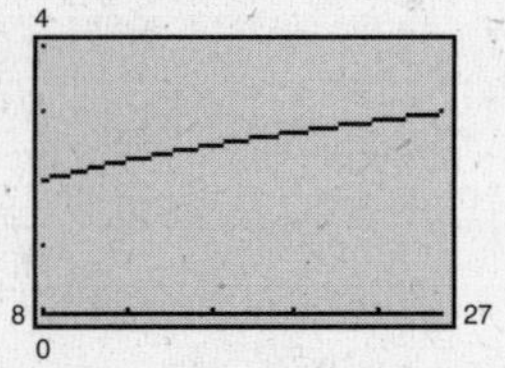

Average rate of change:

$$\frac{\Delta y}{\Delta x} = \frac{f(27) - f(8)}{27 - 8} = \frac{3 - 2}{19} = \frac{1}{19}$$

$f(x) = \sqrt[3]{x} = x^{1/3}$

$f'(x) = \dfrac{1}{3}x^{-2/3} = \dfrac{1}{3x^{2/3}}$

Instantaneous rates of change:

$f'(8) = \dfrac{1}{12},\ f'(27) = \dfrac{1}{27}$

14. For 2002, $t = 2$ and $m \approx 900$.

For 2004, $t = 4$ and $m \approx 0$.

15. $f'(x) = 10x + 6$

At $(-1, -2)$, $m = -4$.

$y + 2 = -4(x + 1)$

$y = -4x - 6$

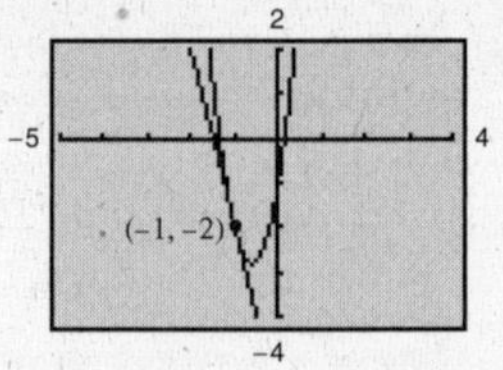

16. $f(x) = x^2 - 1$

$f'(x) = 2x$

At $(0, -1)$, $m = 0$.

$y + 1 = 0(x - 0)$

$y = -1$

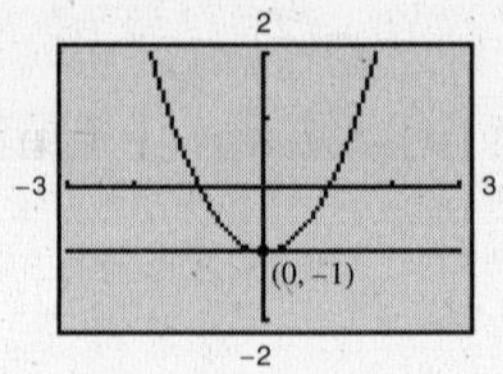

17. $h = -16t^2 + 15t + 25$

(a) $-16t^2 + 15t + 25 = 0$

$$t = \frac{-15 \pm \sqrt{15^2 - 4(-16)(25)}}{2(-16)} = \frac{-15 \pm \sqrt{1825}}{-32} = \frac{-15 \pm 5\sqrt{73}}{-32} \approx 1.80 \text{ or } -0.87$$

The diver will hit the water after about 1.80 seconds ($t \approx -0.87$ does not make sense in this situation).

(b) $h' = -32t + 15$

When $t = 1.80$, $h' = -32(1.80) + 15 = -42.6$. The diver's velocity at impact is −42.6 feet per second.

Section 2.4 The Product and Quotient Rules

Skills Review

1. $\begin{aligned}(x^2 + 1)(2) + (2x + 7)(2x) &= 2x^2 + 2 + 4x^2 + 14x \\ &= 6x^2 + 14x + 2 \\ &= 2(3x^2 + 7x + 1)\end{aligned}$

2. $\begin{aligned}(2x - x^3)(8x) + (4x^2)(2 - 3x^2) &= 16x^2 - 8x^4 + 8x^2 - 12x^4 \\ &= 24x^2 - 20x^4 \\ &= 4x^2(6 - 5x^2)\end{aligned}$

3. $x(4)(x^2 + 2)^3(2x) + (x^2 + 4)(1) = 8x^2(x^2 + 2)^3(x^2 + 4)$

4. $\begin{aligned}x^2(2)(2x + 1)(2) + (2x + 1)^4(2x) &= 4x^2(2x + 1) + 2x(2x + 1)^4 \\ &= 2x(2x + 1)\left[2x + (2x + 1)^3\right]\end{aligned}$

5. $\begin{aligned}\frac{(2x + 7)(5) - (5x + 6)(2)}{(2x + 7)^2} &= \frac{10x + 35 - 10x - 12}{(2x + 7)^2} \\ &= \frac{23}{(2x + 7)^2}\end{aligned}$

6. $\begin{aligned}\frac{(x^2 - 4)(2x + 1) - (x^2 + x)(2x)}{(x^2 - 4)^2} &= \frac{2x^3 + x^2 - 8x - 4 - 2x^3 - 2x^2}{(x^2 - 4)^2} \\ &= \frac{-x^2 - 8x - 4}{(x^2 - 4)^2}\end{aligned}$

7. $\begin{aligned}\frac{(x^2 + 1)(2) - (2x + 1)(2x)}{(x^2 + 1)^2} &= \frac{2x^2 + 2 - 4x^2 - 2x}{(x^2 + 1)^2} \\ &= \frac{-2x^2 - 2x + 2}{(x^2 + 1)^2} \\ &= \frac{-2(x^2 + x - 1)}{(x^2 + 1)^2}\end{aligned}$

8. $\begin{aligned}\frac{(1 - x^4)(4) - (4x - 1)(-4x^3)}{(1 - x^4)^2} &= \frac{4 - 4x^4 + 16x^4 - 4x^3}{(1 - x^4)^2} \\ &= \frac{12x^4 - 4x^3 + 4}{(1 - x^4)^2} \\ &= \frac{4(3x^4 - x^3 + 1)}{(1 - x^4)^2}\end{aligned}$

Skills Review —*continued*—

9. $\left(x^{-1} + x\right)(2) + (2x - 3)\left(-x^{-2} + 1\right) = 2x^{-1} + 2x + \left(-2x^{-1} + 2x + 3x^{-2} - 3\right)$

$= 4x + 3x^{-2} - 3$

$= 4x + \dfrac{3}{x^2} - 3$

$= \dfrac{4x^3 - 3x^2 + 3}{x^2}$

10. $\dfrac{\left(1 - x^{-1}\right)(1) - (x - 4)\left(x^{-2}\right)}{\left(1 - x^{-1}\right)^2} = \left(\dfrac{1 - x^{-1} - x^{-1} + 4x^{-2}}{1 - 2x^{-1} + x^{-2}}\right)\left(\dfrac{x^2}{x^2}\right)$

$= \dfrac{x^2 - 2x + 4}{x^2 - 2x + 1}$

$= \dfrac{x^2 - 2x + 4}{(x - 1)^2}$

11. $f(x) = 3x^2 - x + 4$

$f'(x) = 6x - 1$

$f'(2) = 6(2) - 1$

$= 12 - 1$

$= 11$

12. $f(x) = -x^3 + x^2 + 8x$

$f'(x) = -3x^2 + 2x + 8$

$f'(2) = -3\left(2^2\right) + 2(2) + 8$

$= -3(4) + 4 + 8$

$= 0$

13. $f(x) = \dfrac{1}{x}$

$f'(x) = -\dfrac{1}{x^2}$

$f'(2) = -\dfrac{1}{2^2}$

$= -\dfrac{1}{4}$

14. $f(x) = x^2 - \dfrac{1}{x^2}$

$f'(x) = 2x + \dfrac{2}{x^3}$

$f'(2) = 2(2) + \dfrac{2}{2^3}$

$= 4 + \dfrac{2}{8}$

$= 4 + \dfrac{1}{4}$

$= \dfrac{17}{4}$

In Exercises 1–15, the differentiation rule(s) may vary. A sample answer is provided.

1. $f'(x) = x(2x) + 1\left(x^2 + 3\right) = 3x^2 + 3$

$f'(2) = 15$

Product Rule

3. $f'(x) = x^2\left(9x^2\right) + 2x\left(3x^3 - 1\right) = 15x^4 - 2x$

$f'(1) = 13$

Product Rule

5. $f'(x) = \frac{1}{3}\left(6x^2\right) = 2x^2$

$f'(0) = 0$

Power Rule

7. $g'(x) = \left(x^2 - 4x + 3\right)(1) + (2x - 4)(x - 2)$

$= 3x^2 - 12x + 11$

$g'(4) = 11$

Product Rule

9. $h'(x) = \dfrac{(x-5)(1)-(x)(1)}{(x-5)^2} = \dfrac{-5}{(x-5)^2}$

$h'(6) = -5$

Quotient Rule

11. $f'(t) = \dfrac{(3t+1)(4t)-(2t^2-3)3}{(3t+1)^2}$

$= \dfrac{6t^2+4t+9}{(3t+1)^2}$

$f'(3) = \dfrac{3}{4}$

Quotient Rule

13. $g'(x) = \dfrac{(x-5)(2)-(2x+1)(1)}{(x-5)^2} = \dfrac{-11}{(x-5)^2}$

$g'(6) = -11$

Quotient Rule

15. $f'(t) = \dfrac{(t+4)(2t)-(t^2-1)(1)}{(t+4)^2} = \dfrac{t^2+8t+1}{(t+4)^2}$

$f'(1) = \dfrac{2}{5}$

Quotient Rule

	Function	*Rewrite*	*Differentiate*	*Simplify*
17.	$y = \dfrac{x^2+2x}{x}$	$y = x+2,\ x \neq 0$	$y' = 1,\ x \neq 0$	$y' = 1,\ x \neq 0$
19.	$y = \dfrac{7}{3x^3}$	$y = \dfrac{7}{3}x^{-3}$	$y' = -7x^{-4}$	$y' = -\dfrac{7}{x^4}$
21.	$y = \dfrac{4x^2-3x}{8\sqrt{x}}$	$y = \dfrac{1}{2}x^{3/2} - \dfrac{3}{8}x^{1/2},\ x \neq 0$	$y' = \dfrac{3}{4}x^{1/2} - \dfrac{3}{16}x^{-1/2}$	$y' = \dfrac{3}{4}\sqrt{x} - \dfrac{3}{16\sqrt{x}}$
23.	$y = \dfrac{x^2-4x+3}{x-1}$	$y = x-3,\ x \neq 1$	$y' = 1,\ x \neq 1$	$y' = 1,\ x \neq 1$

In Exercises 25–39, the differentiation rule(s) may vary. A sample answer is provided.

25. $f'(x) = (x^3-3x)(4x+3) + (3x^2-3)(2x^2+3x+5)$

$= 4x^4 + 3x^3 - 12x^2 - 9x + 6x^4 + 9x^3 + 9x^2 - 9x - 15$

$= 10x^4 + 12x^3 - 3x^2 - 18x - 15$

Product Rule

27. $g(t) = (2t^3-1)^2 = (2t^3-1)(2t^3-1)$

$g'(t) = (2t^3-1)(6t^2) + (2t^3-1)(6t^2)$

$= 12t^2(2t^3-1)$

Product Rule

29. $f(x) = \sqrt[3]{x}(\sqrt{x}+3) = x^{1/3}(x^{1/2}+3)$

$f'(x) = (x^{1/2}+3)\left(\dfrac{1}{3}x^{-2/3}\right) + x^{1/3}\left(\dfrac{1}{2}x^{-1/2}\right)$

$= \dfrac{1}{3}x^{-1/6} + x^{-2/3} + \dfrac{1}{2}x^{-1/6}$

$= \dfrac{5}{6x^{1/6}} + \dfrac{1}{x^{2/3}}$

Product Rule

31. $f'(x) = \dfrac{(2x-3)(3)-(3x-2)(2)}{(2x-3)^2} = -\dfrac{5}{(2x-3)^2}$

Quotient Rule

33. $f(x) = \dfrac{3-2x-x^2}{x^2-1}$

$= \dfrac{(3+x)(1-x)}{(x+1)(x-1)} = \dfrac{-(3+x)}{x+1},\ x \neq 1$

$f'(x) = \dfrac{(x+1)(-1)+(3+x)(1)}{(x+1)^2}$

$= \dfrac{2}{(x+1)^2},\ x \neq 1$

Quotient Rule

35. $f(x) = x\left(1 - \dfrac{2}{x+1}\right) = x - \dfrac{2x}{x+1}$

$f'(x) = 1 - \dfrac{(x+1)(2)-(2x)(1)}{(x+1)^2}$

$= 1 - \dfrac{2}{(x+1)^2} = \dfrac{x^2+2x-1}{(x+1)^2}$

Quotient Rule

37. $g(s) = \dfrac{s^2 - 2s + 5}{\sqrt{s}} = \dfrac{s^2 - 2s + 5}{s^{1/2}}$

$$g'(s) = \frac{s^{1/2}(2s - 2) - (s^2 - 2s + 5)\left(\frac{1}{2}s^{-1/2}\right)}{s}$$

$$= \frac{2s^{3/2} - 2s^{1/2} - \frac{1}{2}s^{3/2} + s^{1/2} - \frac{5}{2}s^{-1/2}}{s}$$

$$= \frac{3}{2}s^{1/2} - s^{-1/2} - \frac{5}{2}s^{-3/2} = \frac{3s^2 - 2s - 5}{2s^{3/2}}$$

Quotient Rule

39. $g(x) = \dfrac{x - 3}{x + 4}(x^2 + 2x + 1) = \dfrac{x^3 - x^2 - 5x - 3}{x + 4}$

$$g'(x) = \frac{(x + 4)(3x^2 - 2x - 5) - (x^3 - x^2 - 5x - 3)(1)}{(x + 4)^2}$$

$$= \frac{3x^3 + 10x^2 - 13x - 20 - x^3 + x^2 + 5x + 3}{(x + 4)^2}$$

$$= \frac{2x^3 + 11x^2 - 8x - 17}{(x + 4)^2}$$

Quotient Rule

41. $f(x) = (x - 1)^2(x - 2) = (x^2 - 2x + 1)(x - 2)$

$f'(x) = (x^2 - 2x + 1)(1) + (x - 2)(2x - 2)$

$f'(0) = 1 + 4 = 5$

$y + 2 = 5(x - 0)$

$y = 5x - 2$

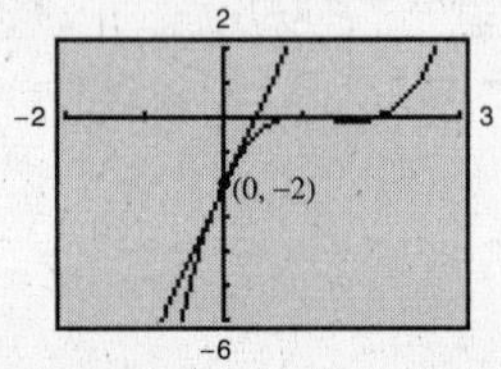

43. $f'(x) = \dfrac{(x + 1) - (x - 2)}{(x + 1)^2} = \dfrac{3}{(x + 1)^2}$

$f'(1) = \dfrac{3}{4}$

$y + \dfrac{1}{2} = \dfrac{3}{4}(x - 1)$

$y = \dfrac{3}{4}x - \dfrac{5}{4}$

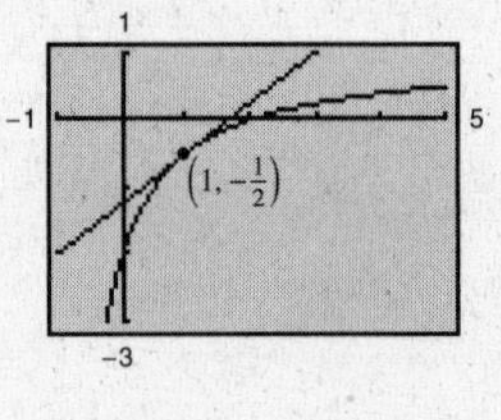

45. $f(x) = \dfrac{x + 5}{x - 1}(2x + 1) = \dfrac{2x^2 + 11x + 5}{(x - 1)}$

$$f'(x) = \frac{(x - 1)(4x + 11) - (2x^2 + 11x + 5)}{(x - 1)^2}$$

$$= \frac{2x^2 - 4x - 16}{(x - 1)^2}$$

$f'(0) = -16$

$y + 5 = -16(x - 0)$

$y = -16x - 5$

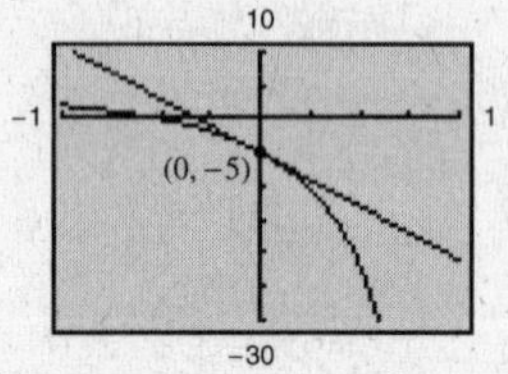

47. $f'(x) = \dfrac{(x - 1)(2x) - x^2(1)}{(x - 1)^2} = \dfrac{x^2 - 2x}{(x - 1)^2}$

$f'(x) = 0$ when $x^2 - 2x = x(x - 2) = 0$, which implies that $x = 0$ or $x = 2$. Thus, the horizontal tangent lines occur at $(0, 0)$ and $(2, 4)$.

49. $f'(x) = \dfrac{(x^3 + 1)(4x^3) - x^4(3x^2)}{(x^3 + 1)^2} = \dfrac{x^6 + 4x^3}{(x^3 + 1)^2}$

$f'(x) = 0$ when $x^6 + 4x^3 = x^3(x^3 + 4) = 0$, which implies that $x = 0$ or $x = \sqrt[3]{-4}$. Thus, the horizontal tangent lines occur at $(0, 0)$ and $\left(\sqrt[3]{-4}, -2.117\right)$.

51. $f(x) = x(x + 1) = x^2 + x$

$f'(x) = 2x + 1$

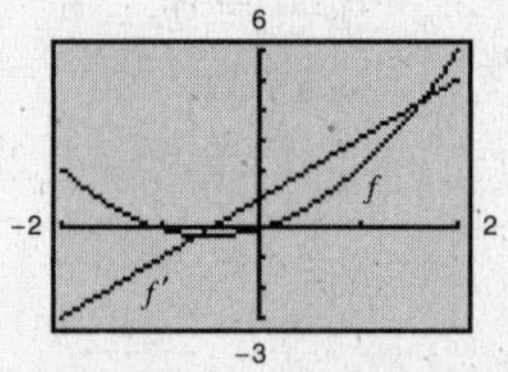

53. $f(x) = x(x + 1)(x - 1)$

$= x^3 - x$

$f'(x) = 3x^2 - 1$

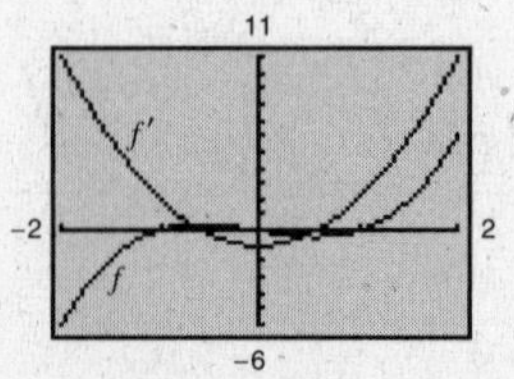

55. $f'(t) = \dfrac{(t^2+1)(2t-1)-(t^2-t+1)(2t)}{(t^2+1)^2} = \dfrac{t^2-1}{(t^2+1)^2}$

(a) $f'(0.5) = -0.48$ per week

(b) $f'(2) = 0.12$ per week

(c) $f'(8) \approx 0.015$ per week

57. $P' = 500\left[\dfrac{(50+t^2)(4)-(4t)(2t)}{(50+t^2)^2}\right] = 500\left[\dfrac{200-4t^2}{(50+t^2)^2}\right]$

When $t = 2$, $P' = 500\left[\dfrac{184}{(54)^2}\right] \approx 31.55$ bacteria/hour.

59. $\dfrac{dP}{dt} = 0.0177t^2 + 0.0083[(t-133.13)(1)+(t-5.10)(1)]$

$= 0.0177t^2 + 0.0083(2t - 138.23)$

$= 0.0177t^2 + 0.0166t - 1.147309$

When $t = 6$, $\dfrac{dP}{dt} = 0.0177(6)^2 + 0.0166(6) - 1.147309 = -0.410509$. The average precipitation for June is decreasing at a rate of about 0.41 inch per month.

61. $T = \dfrac{12t}{t^2+8} + 98.6$

$T' = \dfrac{(t^2+8)(12)-12t(2t)}{(t^2+8)^2}$

$= \dfrac{12t^2+96-24t^2}{(t^2+8)^2}$

$= \dfrac{-12t^2+96}{(t^2+8)^2}$

(a) $T'(0.5) = \dfrac{93}{68.0625} \approx 1.37$ degrees per hour

$T'(4) = \dfrac{-96}{576} \approx -\dfrac{1}{6}$ degrees per hour

(b)

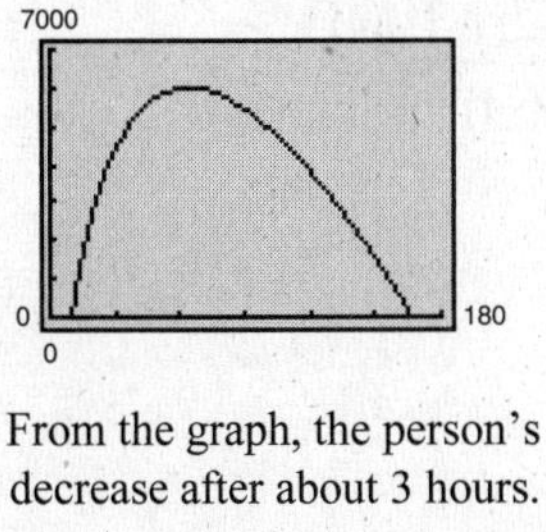

(c) From the graph, the person's temperature starts to decrease after about 3 hours.

(d) $\dfrac{12t}{t^2+8} + 98.6 = 99.6$

$\dfrac{12t}{t^2+8} = 1$

$12t = t^2 + 8$

$0 = t^2 - 12t + 8$

$t = \dfrac{12 \pm \sqrt{(-12)^2 - 4(1)(8)}}{2(1)}$

$= \dfrac{12 \pm \sqrt{112}}{2}$

$= \dfrac{12 \pm 4\sqrt{7}}{2}$

$= 6 \pm 2\sqrt{7}$

≈ 11.29 or 0.71

Choose the larger value of t. After about 11.29 hours, the person has a body temperature that is considered normal.

63. $C = \dfrac{124p}{(10 + p)(100 - p)}, \quad 0 \le p < 100$

$$C' = \frac{(10 + p)(100 - p)(124) - 124p[(10 + p)(-1) + (100 - p)(1)]}{[(10 + p)(100 - p)]^2}$$

$$= \frac{124(1000 + 90p - p^2) - 124p(90 - 2p)}{[(10 + p)(100 - p)]^2}$$

$$= \frac{124{,}000 + 11{,}160p - 124p^2 - 11{,}160p + 248p^2}{[(10 + p)(100 - p)]^2}$$

$$= \frac{124p^2 + 124{,}000}{[(10 + p)(100 - p)]^2}$$

(a) $C'(25) = \dfrac{201{,}500}{6{,}890{,}625} \approx 0.029 \Rightarrow \$2900$ per percent

$C'(75) = \dfrac{821{,}500}{4{,}515{,}625} \approx 0.182 \Rightarrow \$18{,}200$ per percent

(b)

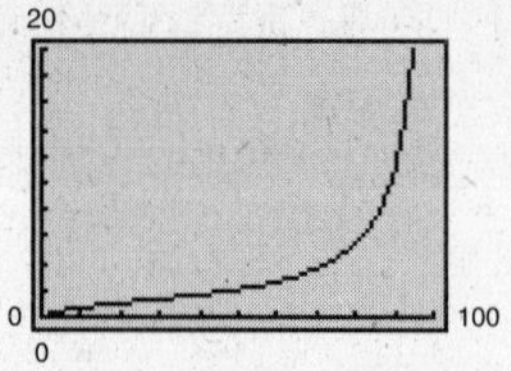

(c) As p approaches 100, C increases without bound and C' increases without bound.

(d) Answers will vary.

65. $M(t) = \dfrac{300t}{t^2 + 1} + 8$

(a) $M'(t) = \dfrac{(t^2 + 1)(300) - 300t(2t)}{(t^2 + 1)^2}$

$= \dfrac{300(1 - t^2)}{(t^2 + 1)^2}$

(b) $M(3) = 98$ represents the sales of memberships 3 months after opening.

$M'(3) = -24$ represents the rate sales are changing per month 3 months after opening.

(c) $M(24) \approx 20.48$ represents the sales of memberships 24 months after opening.

$M'(24) = -0.52$ represents the ate sales are changing per month 24 months after opening.

67. $f(x) = 3 - g(x)$

$f'(x) = -g'(x)$

$f'(2) = -(-2)$

$= 2$

69. $f(x) = \dfrac{g(x)}{h(x)}$

$f'(x) = \dfrac{h(x)g'(x) - g(x)h'(x)}{[h(x)]^2}$

$f'(2) = \dfrac{(-1)(-2) - (3)(4)}{(-1)^2}$

$= -10$

Section 2.5 The Chain Rule

Skills Review

1. $\sqrt[5]{(1-5x)^2} = (1-5x)^{2/5}$

2. $\sqrt[4]{(2x-1)^3} = (2x-1)^{3/4}$

3. $\dfrac{1}{\sqrt{4x^2+1}} = (4x^2+1)^{-1/2}$

4. $\dfrac{1}{\sqrt[3]{x-6}} = (x-6)^{-1/3}$

5. $\dfrac{\sqrt{x}}{\sqrt[3]{1-2x}} = x^{1/2}(1-2x)^{-1/3}$

6. $\dfrac{\sqrt{(3-7x)^3}}{2x} = \dfrac{(3-7x)^{3/2}}{2x}$
$= (2x)^{-1}(3-7x)^{3/2}$

7. $3x^3 - 6x^2 + 5x - 10 = 3x^2(x-2) + 5(x-2)$
$= (3x^2+5)(x-2)$

8. $5x\sqrt{x} - x - 5\sqrt{x} + 1 = x(5\sqrt{x}-1) - 1(5\sqrt{x}-1)$
$= (x-1)(5\sqrt{x}-1)$

9. $4(x^2+1)^2 - x(x^2+1)^3 = (x^2+1)^2[4 - x(x^2+1)]$
$= (x^2+1)^2(4 - x^3 - x)$

10. $-x^5 + 3x^3 + x^2 - 3 = x^3(-x^2+3) - 1(-x^2+3)$
$= (x^3-1)(-x^2+3)$
$= (x-1)(x^2+x+1)(3-x^2)$

$y = f(g(x))$	$u = g(x)$	$y = f(u)$
1. $y = (6x-5)^4$	$u = 6x-5$	$y = u^4$
3. $y = (4-x^2)^{-1}$	$u = 4-x^2$	$y = u^{-1}$
5. $y = \sqrt{5x-2}$	$u = 5x-2$	$y = \sqrt{u}$
7. $y = \dfrac{1}{3x+1}$	$u = 3x+1$	$y = u^{-1}$

9. $\dfrac{dy}{du} = 2u$

$\dfrac{du}{dx} = 4$

$\dfrac{dy}{dx} = (2u)(4) = 8(4x+7) = 32x + 56$

11. $\dfrac{dy}{du} = \dfrac{1}{2}u^{-1/2}$

$\dfrac{du}{dx} = -2x$

$\dfrac{dy}{dx} = \left(\dfrac{1}{2}u^{-1/2}\right)(-2x) = -(3-x^2)^{-1/2}(x) = -\dfrac{x}{\sqrt{3-x^2}}$

13. $\dfrac{dy}{du} = \dfrac{2}{3}u^{-1/3}$

$\dfrac{du}{dx} = 20x^3 - 2$

$\dfrac{dy}{dx} = \dfrac{2}{3}u^{-1/3}(20x^3-2)$

$= \dfrac{2}{3}(5x^4-2x)^{-1/3}(20x^3-2)$

$= \dfrac{4(10x^3-1)}{3\sqrt[3]{5x^4-2x}}$

15. $f(x) = \dfrac{2}{1-x^3} = 2(1-x^3)^{-1}$; (c) General Power Rule

17. $f(x) = \sqrt[3]{8^2}$; (b) Constant Rule

19. $f(x) = \dfrac{x^2+2}{x} = x + 2x^{-1}$; (a) Simple Power Rule

21. $f(x) = \dfrac{2}{x-2} = 2(x-2)^{-1}$;
(c) General Power Rule

23. $y' = 3(2x-7)^2(2) = 6(2x-7)^2$

25. $g'(x) = 3(4-2x)^2(-2) = -6(4-2x)^2$

27. $h'(x) = 2(6x-x^3)(6-3x^2) = 6x(6-x^2)(2-x^2)$

29. $f'(x) = \frac{2}{3}(x^2 - 9)^{-1/3}(2x) = \frac{4x}{3(x^2 - 9)^{1/3}}$

31. $f(t) = \sqrt{t + 1} = (t + 1)^{1/2}$

$f'(t) = \frac{1}{2}(t + 1)^{-1/2}(1) = \frac{1}{2\sqrt{t + 1}}$

33. $s(t) = \sqrt{2t^2 + 5t + 2} = (2t^2 + 5t + 2)^{1/2}$

$s'(t) = \frac{1}{2}(2t^2 + 5t + 2)^{-1/2}(4t + 5) = \frac{4t + 5}{2\sqrt{2t^2 + 5t + 2}}$

35. $y = \sqrt[3]{9x^2 + 4} = (9x^2 + 4)^{1/3}$

$y' = \frac{1}{3}(9x^2 + 4)^{-2/3}(18x) = \frac{6x}{(9x^2 + 4)^{2/3}}$

37. $f(x) = -3\sqrt[4]{2 - 9x} = -3(2 - 9x)^{1/4}$

$f'(x) = -3\left(\frac{1}{4}\right)(2 - 9x)^{-3/4}(-9) = \frac{27}{4(2 - 9x)^{3/4}}$

39. $h'(x) = -\frac{4}{3}(4 - x^3)^{-7/3}(-3x^2) = \frac{4x^2}{(4 - x^3)^{7/3}}$

41. $f'(x) = 2(3)(x^2 - 1)^2(2x) = 12x(x^2 - 1)^2$

$f'(2) = 24(3^2) = 216$

$f(2) = 54$

$y - 54 = 216(x - 2)$

$y = 216x - 378$

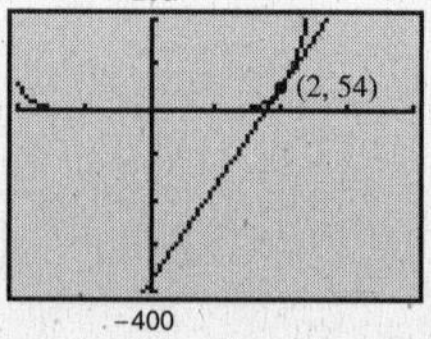

43. $f(x) = \sqrt{4x^2 - 7} = (4x^2 - 7)^{1/2}$

$f'(x) = \frac{1}{2}(4x^2 - 7)^{-1/2}(8x) = \frac{4x}{\sqrt{4x^2 - 7}}$

$f'(2) = \frac{8}{3}$

$f(2) = 3$

$y - 3 = \frac{8}{3}(x - 2)$

$y = \frac{8}{3}x - \frac{7}{3}$

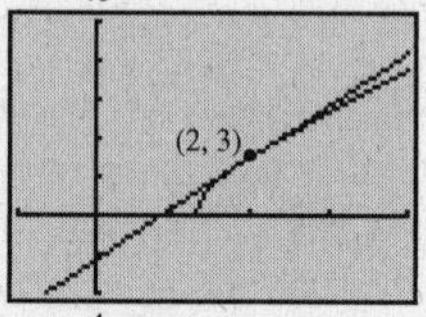

45. $f(x) = \sqrt{x^2 - 2x + 1} = (x^2 - 2x + 1)^{1/2}$

$f'(x) = \frac{1}{2}(x^2 - 2x + 1)^{-1/2}(2x - 2)$

$= \frac{x - 1}{\sqrt{x^2 - 2x + 1}}$

$= \frac{x - 1}{|x - 1|}$

$f'(2) = 1$

$f(2) = 1$

$y - 1 = 1(x - 2)$

$y = x - 1$

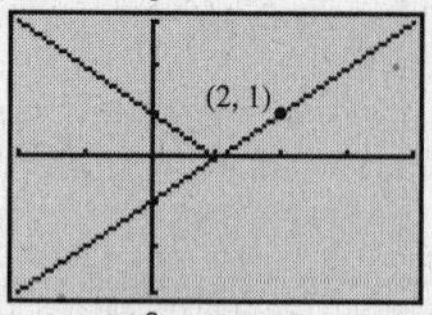

47. $f'(x) = \frac{1 - 3x^2 - 4x^{3/2}}{2\sqrt{x}(x^2 + 1)^2}$

f has a horizontal tangent when $f' = 0$.

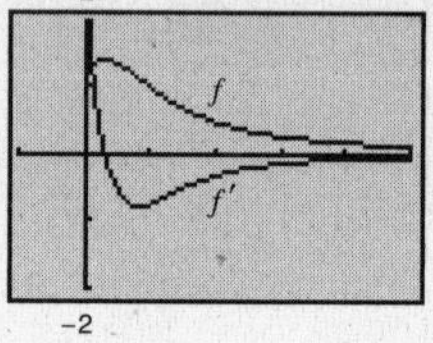

49. $f'(x) = -\frac{1}{2x^{3/2}\sqrt{x + 1}}$

f' is never 0.

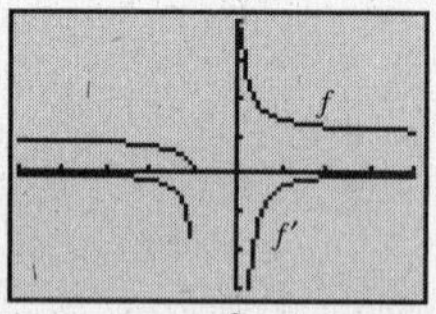

In Exercises 51–65, the differentiation rule(s) may vary. A sample answer is provided.

51. $y = \frac{1}{x - 2} = (x - 2)^{-1}$

$y' = (-1)(x - 2)^{-2} = -\frac{1}{(x - 2)^2}$

General Power Rule

53. $y = -\frac{4}{(t + 2)^2} = -4(t + 2)^{-2}$

$y' = 8(t + 2)^{-3} = \frac{8}{(t + 2)^3}$

General Power Rule

55. $f(x) = \frac{1}{(x^2 - 3x)^2} = (x^2 - 3x)^{-2}$

$f'(x) = -2(x^2 - 3x)^{-3}(2x - 3)$

$= \frac{6 - 4x}{(x^2 - 3x)^3} = -\frac{2(2x - 3)}{x^3(x - 3)^3}$

General Power Rule

57. $g(t) = \dfrac{1}{t^2 - 2} = (t^2 - 2)^{-1}$

$g'(t) = -(t^2 - 2)^{-2}(2t) = -\dfrac{2t}{(t^2 - 2)^2}$

General Power Rule

59. $f'(x) = x(3)(3x - 9)^2(3) + (3x - 9)^3(1)$

$= (3x - 9)^2[9x + (3x - 9)]$

$= 9(x - 3)^2(12x - 9)$

$= 27(x - 3)^2(4x - 3)$

Product and General Power Rules

61. $y = x\sqrt{2x + 3} = x(2x + 3)^{1/2}$

$y' = x\left[\dfrac{1}{2}(2x + 3)^{-1/2}(2)\right] + (2x + 3)^{1/2}$

$= (2x + 3)^{-1/2}[x + (2x + 3)]$

$= \dfrac{3(x + 1)}{\sqrt{2x + 3}}$

Product and General Power Rules

63. $y = t^2\sqrt{t - 2} = t^2(t - 2)^{1/2}$

$y' = t^2\left[\dfrac{1}{2}(t - 2)^{-1/2}(1)\right] + 2t(t - 2)^{1/2}$

$= \dfrac{1}{2}(t - 2)^{-1/2}[t^2 + 4t(t - 2)]$

$= \dfrac{t^2 + 4t(t - 2)}{2\sqrt{t - 2}}$

$= \dfrac{t(5t - 8)}{2\sqrt{t - 2}}$

Product and General Power Rules

65. $y = \left(\dfrac{6 - 5x}{x^2 - 1}\right)^2$

$y' = 2\left(\dfrac{6 - 5x}{x^2 - 1}\right)\left[\dfrac{(x^2 - 1)(-5) - (6 - 5x)(2x)}{(x^2 - 1)^2}\right]$

$= \dfrac{2(6 - 5x)(5x^2 - 12x + 5)}{(x^2 - 1)^3}$

Quotient and General Power Rules

67. $f(t) = \dfrac{36}{(3 - t)^2} = 36(3 - t)^{-2}$

$f'(t) = -72(3 - t)^{-3}(-1) = \dfrac{72}{(3 - t)^3}$

$f'(0) = \dfrac{72}{27} = \dfrac{8}{3}$

$y - 4 = \dfrac{8}{3}(t - 0)$

$y = \dfrac{8}{3}t + 4$

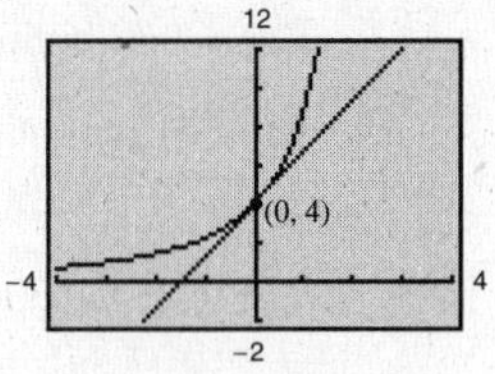

69. $f(t) = (t^2 - 9)\sqrt{t + 2} = (t^2 - 9)(t + 2)^{1/2}$

$f'(t) = (t^2 - 9)\left[\dfrac{1}{2}(t + 2)^{-1/2}\right] + (t + 2)^{1/2}(2t)$

$= \dfrac{1}{2}(t^2 - 9)(t + 2)^{-1/2} + 2t(t + 2)^{1/2}$

$= (t + 2)^{-1/2}\left[\dfrac{1}{2}(t^2 - 9) + 2t(t + 2)\right]$

$= (t + 2)^{-1/2}\left[\dfrac{1}{2}t^2 - \dfrac{9}{2} + 2t^2 + 4t\right]$

$= (t + 2)^{-1/2}\left(\dfrac{5}{2}t^2 + 4t - \dfrac{9}{2}\right)$

$= \dfrac{\frac{5}{2}t^2 + 4t - \frac{9}{2}}{\sqrt{t + 2}}$

$f'(-1) = -6$

$y - (-8) = -6[t - (-1)]$

$y = -6t - 14$

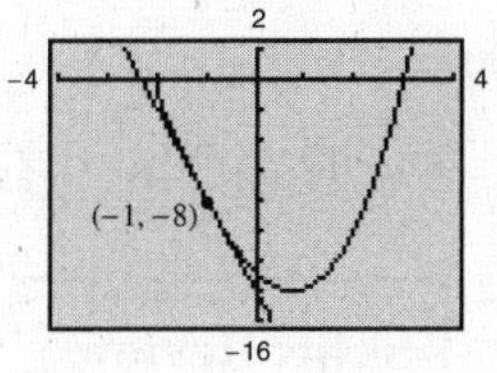

71. $f(x) = \dfrac{x + 1}{\sqrt{2x - 3}} = \dfrac{x + 1}{(2x - 3)^{1/2}}$

$f'(x) = \dfrac{(2x - 3)^{1/2}(1) - (x + 1)\left(\frac{1}{2}\right)(2x - 3)^{-1/2}(2)}{(2x - 3)}$

$= \dfrac{(2x - 3) - (x + 1)}{(2x - 3)^{3/2}}$

$= \dfrac{x - 4}{(2x - 3)^{3/2}}$

$f'(2) = \dfrac{1 - 3}{1} = -2$

$y - 3 = -2(x - 2)$

$y = -2x + 7$

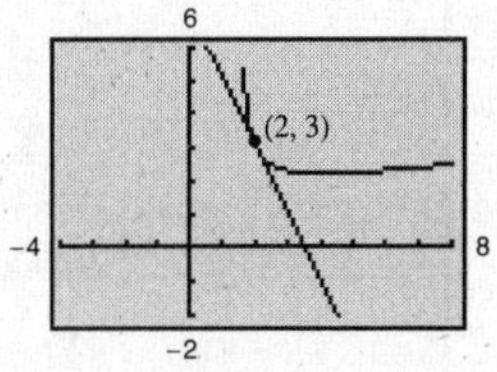

73. $A = \left(7.54t^2 + 29.8t + 2348\right)^{1/2}$

$A' = \frac{1}{2}\left(7.54t^2 + 29.8t + 2348\right)^{-1/2}(15.08t + 29.8)$

$= \dfrac{15.08t + 29.8}{2\sqrt{7.54t^2 + 29.8t + 2348}}$

$A'(2) \approx 0.61$ billions of pounds per year

$A'(4) \approx 0.89$ billions of pounds per year

75. $N = 400\left[1 - 3\left(t^2 + 2\right)^{-2}\right] = 400 - 1200\left(t^2 + 2\right)^{-2}$

$\dfrac{dN}{dt} = 2400\left(t^2 + 2\right)^{-3}(2t) = \dfrac{4800t}{\left(t^2 + 2\right)^3}$

t	0	1	2	3	4
$\dfrac{dN}{dt}$	0	177.78	44.44	10.82	3.29

The rate of growth of N is decreasing.

77. (a) $V = \dfrac{k}{\sqrt[3]{t + 1}}$

When $t = 0$, $V = 10{,}000 \Rightarrow k = 10{,}000$.

Therefore, $V = \dfrac{10{,}000}{\sqrt[3]{t + 1}}$.

(b) $V' = -\dfrac{10{,}000}{3(t + 1)^{4/3}}$

When $t = 1$, $\dfrac{dV}{dt} = -\dfrac{10{,}000}{3(2)^{4/3}}$

$\approx -\$1322.83$ per year.

(c) When $t = 3$, $\dfrac{dV}{dt} = -\dfrac{10{,}000}{3(4)^{4/3}}$

$\approx -\$524.97$ per year.

79. False. By the chain rule,

$y' = \frac{1}{2}(1 - x)^{-1/2}(-1) = -\frac{1}{2}(1 - x)^{-1/2}$

81. $f(x) = h(g(x))$

$f'(x) = h'(g(x))g'(x)$

(a) $f'(2) = h'(g(2))g'(2) = h'(-6)(5) = (3)(5) = 15$

(b) $f'(2) = h'(-1)(-2) = (5)(-2) = -10$

Section 2.6 Higher-Order Derivatives

Skills Review

1. $-16t^2 + 24t = 0$

$t(-16t + 24) = 0$

$t = 0$

$-16t + 24 = 0 \rightarrow t = 1.5$

2. $-16t^2 + 80t + 224 = 0$

$-16\left(t^2 - 5t - 14\right) = 0$

$-16(t - 7)(t + 2) = 0$

$t - 7 = 0 \rightarrow t = 7$

$t + 2 = 0 \rightarrow t = -2$

3. $-16t^2 + 128t + 320 = 0$

$-16\left(t^2 - 8t - 20\right) = 0$

$-16(t - 10)(t + 2) = 0$

$t - 10 = 0 \rightarrow t = 10$

$t + 2 = 0 \rightarrow t = -2$

4. $-16t^2 + 9t + 1440 = 0$

$t = \dfrac{-9 \pm \sqrt{9^2 - 4(-16)(1440)}}{2(-16)}$

$= \dfrac{-9 \pm \sqrt{92{,}241}}{-32} = \dfrac{9 \pm 3\sqrt{10{,}249}}{32}$

$t \approx -9.21$ and $t \approx 9.77$

5. $y = x^2(2x + 7)$

$\dfrac{dy}{dx} = x^2(2) + 2x(2x + 7)$

$= 2x^2 + 4x^2 + 14x = 6x^2 + 14x$

6. $y = \left(x^2 + 3x\right)\left(2x^2 - 5\right)$

$\dfrac{dy}{dx} = \left(x^2 + 3x\right)(4x) + (2x + 3)\left(2x^2 - 5\right)$

$= 4x^3 + 12x^2 + 4x^3 - 10x + 6x^2 - 15$

$= 8x^3 + 18x^2 - 10x - 15$

Skills Review —*continued*—

7. $y = \dfrac{x^2}{2x + 7}$

$\dfrac{dy}{dx} = \dfrac{(2x + 7)(2x) - (x^2)(2)}{(2x + 7)^2}$

$= \dfrac{4x^2 + 14x - 2x^2}{(2x + 7)^2} = \dfrac{2x^2 + 14x}{(2x + 7)^2} = \dfrac{2x(x + 7)}{(2x + 7)^2}$

8. $y = \dfrac{x^2 + 3x}{2x^2 - 5}$

$\dfrac{dy}{dx} = \dfrac{(2x^2 - 5)(2x + 3) - (x^2 + 3x)(4x)}{(2x^2 - 5)^2}$

$= \dfrac{4x^3 + 6x^2 - 10x - 15 - 4x^3 - 12x^2}{(2x^2 - 5)^2}$

$= \dfrac{-6x^2 - 10x - 15}{(2x^2 - 5)^2}$

9. $f(x) = x^2 - 4$

Domain: $(-\infty, \infty)$

Range: $[-4, \infty)$

10. $f(x) = \sqrt{x - 7}$

Domain: $[7, \infty)$

Range: $[0, \infty)$

1. $f'(x) = -2$

$f''(x) = 0$

3. $f'(x) = 2x + 7$

$f''(x) = 2$

5. $g'(t) = t^2 - 8t + 2$

$g''(t) = 2t - 8$

7. $f(t) = \dfrac{3}{4t^2} = \dfrac{3}{4}t^{-2}$

$f'(t) = -\dfrac{3}{2}t^{-3}$

$f''(t) = \dfrac{9}{2}t^{-4} = \dfrac{9}{2t^4}$

9. $f'(x) = 9(2 - x^2)^2(-2x) = -18x(2 - x^2)^2$

$f''(x) = (-18x)2(2 - x^2)(-2x) + (2 - x^2)^2(-18)$

$= 18(2 - x^2)[4x^2 - (2 - x^2)]$

$= 18(2 - x^2)(5x^2 - 2)$

11. $y' = 4(x^3 - 2x)^3(3x^2 - 2) = (x^3 - 2x)^3(12x^2 - 8)$

$y'' = (3x^2 - 2)(4)(3)(x^3 - 2x)^2(3x^2 - 2)$

$+ 4(x^3 - 2x)^3(6x)$

$= (x^3 - 2x)^2\left[12(3x^2 - 2)^2 + 24x(x^3 - 2x)\right]$

$= 12(x^3 - 2x)^2(9x^4 - 12x^2 + 4 + 2x^4 - 4x^2)$

$= 12(x^3 - 2x)^2(11x^4 - 16x^2 + 4)$

13. $f'(x) = \dfrac{(x - 1)(1) - (x + 1)(1)}{(x - 1)^2}$

$= -\dfrac{2}{(x - 1)^2} = -2(x - 1)^{-2}$

$f''(x) = 4(x - 1)^{-3}(1) = \dfrac{4}{(x - 1)^3}$

15. $y = x^2(x^2 + 4x + 8) = x^4 + 4x^3 + 8x^2$

$y' = 4x^3 + 12x^2 + 16x$

$y'' = 12x^2 + 24x + 16$

17. $f'(x) = 5x^4 - 12x^3$

$f''(x) = 20x^3 - 36x^2$

$f'''(x) = 60x^2 - 72x$

19. $f(x) = 5x(x+4)^3$

$= 5x(x^3 + 12x^2 + 48x + 64)$

$= 5x^4 + 60x^3 + 240x^2 + 320x$

$f'(x) = 20x^3 + 180x^2 + 480x + 320$

$f''(x) = 60x^2 + 360x + 480$

$f'''(x) = 120x + 360$

21. $f(x) = \dfrac{3}{16x^2} = \dfrac{3}{16}x^{-2}$

$f'(x) = -\dfrac{3}{8}x^{-3}$

$f''(x) = \dfrac{9}{8}x^{-4}$

$f'''(x) = -\dfrac{9}{2}x^{-5} = -\dfrac{9}{2x^5}$

23. $g'(t) = 20t^3 + 20t$

$g''(t) = 60t^2 + 20$

$g''(2) = 60(4) + 20 = 260$

25. $f(x) = \sqrt{4-x} = (4-x)^{1/2}$

$f'(x) = -\dfrac{1}{2}(4-x)^{-1/2}$

$f''(x) = -\dfrac{1}{4}(4-x)^{-3/2}$

$f'''(x) = -\dfrac{3}{8}(4-x)^{-5/2} = \dfrac{-3}{8(4-x)^{5/2}}$

$f'''(-5) = \dfrac{-3}{8(9)^{5/2}} = -\dfrac{1}{648}$

27. $f(x) = x^2(3x^2 + 3x - 4) = 3x^4 + 3x^3 - 4x^2$

$f'(x) = 12x^3 + 9x^2 - 8x$

$f''(x) = 36x^2 + 18x - 8$

$f'''(x) = 72x + 18$

$f'''(-2) = 72(-2) + 18 = -126$

29. $f''(x) = 4x$

31. $f'''(x) = \dfrac{3x-1}{x} = 3 - \dfrac{1}{x}$

$f^{(4)}(x) = 0 - \left(-\dfrac{1}{x^2}\right) = \dfrac{1}{x^2}$

33. $f^{(5)}(x) = 2(x^2+1)(2x) = 4x^3 + 4x$

$f^{(6)}(x) = 12x^2 + 4$

35. $f'(x) = 3x^2 - 18x + 27$

$f''(x) = 6x - 18$

$f''(x) = 0 \Rightarrow 6x = 18$

$x = 3$

37. $f(x) = (x+3)(x-4)(x+5) = x^3 + 4x^2 - 17x - 60$

$f'(x) = 3x^2 + 8x - 17$

$f''(x) = 6x + 8$

$f''(x) = 0 \Rightarrow 6x = -8$

$x = -\dfrac{4}{3}$

39. $f(x) = x\sqrt{x^2-1} = x(x^2-1)^{1/2}$

$f'(x) = x\dfrac{1}{2}(x^2-1)^{-1/2}(2x) + (x^2-1)^{1/2} = \dfrac{x^2}{(x^2-1)^{1/2}} + (x^2-1)^{1/2}$

$$f''(x) = \frac{(x^2-1)^{1/2}(2x) - x^2\left(\frac{1}{2}\right)(x^2-1)^{-1/2}(2x)}{x^2-1} + \frac{1}{2}(x^2-1)^{-1/2}(2x) = \frac{(x^2-1)(2x) - x^3}{(x^2-1)^{3/2}} + \frac{x}{(x^2-1)^{1/2}} \cdot \frac{x^2-1}{x^2-1} = \frac{2x^3 - 3x}{(x^2-1)^{3/2}}$$

$f''(x) = 0 \Rightarrow 2x^3 - 3x = x(2x^2 - 3) = 0$

$x = \pm\sqrt{\dfrac{3}{2}} = \pm\dfrac{\sqrt{6}}{2}$

$x = 0$ is not in the domain of f.

41. $f'(x) = \dfrac{(x^2+3)(1)-(x)(2x)}{(x^2+3)^2}$

$= \dfrac{3-x^2}{(x^2+3)^2} = (3-x^2)(x^2+3)^{-2}$

$f''(x) = (3-x^2)\left[-2(x^2+3)^{-3}(2x)\right] + (x^2+3)^{-2}(-2x)$

$= -2x(x^2+3)^{-3}\left[2(3-x^2)+(x^2+3)\right]$

$= \dfrac{-2x(9-x^2)}{(x^2+3)^3} = \dfrac{2x(x^2-9)}{(x^2+3)^3}$

$f''(x) = 0 \Rightarrow 2x(x^2-9) = 0$

$x = 0, \pm 3$

43. (a) $s(t) = -16t^2 + 144t$

$v(t) = s'(t) = -32t + 144$

$a(t) = v'(t) = -32$

(b) $v(t) = 0$ when $32t = 144$, or $t = \frac{144}{32} = 4.5$ sec.

$s(4.5) = 324$ feet

(c) $s(t) = 0$ when $16t^2 = 144t$, or $t = 0, 9$ sec.

$v(9) = -144$ ft/sec, which is the same speed as the initial velocity.

45. $\dfrac{d^2s}{dt^2} = \dfrac{(t+10)(90)-(90t)(1)}{(t+10)^2} = \dfrac{900}{(t+10)^2}$

t	0	10	20	30	40	50	60
$\dfrac{ds}{dt}$	0	45	60	67.5	72	75	77.14
$\dfrac{d^2s}{dt^2}$	9	2.25	1	0.56	0.36	0.25	0.18

As time increases, the acceleration decreases. After 1 minute, the automobile is traveling at about 77.14 feet per second.

47. $f(x) = x^2 - 6x + 6$

$f'(x) = 2x - 6$

$f''(x) = 2$

The degrees of the successive derivatives decrease by 1.

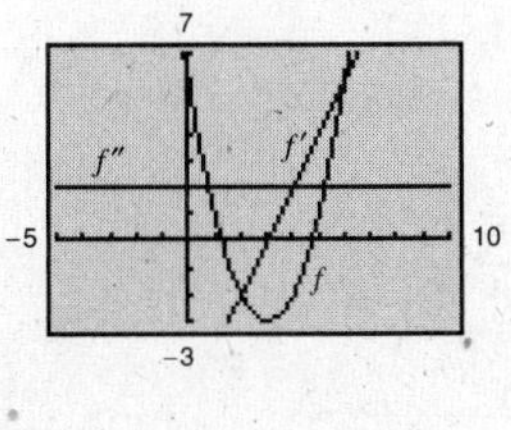

49. The degree of f is 3, and the degrees of the successive derivatives decrease by 1.

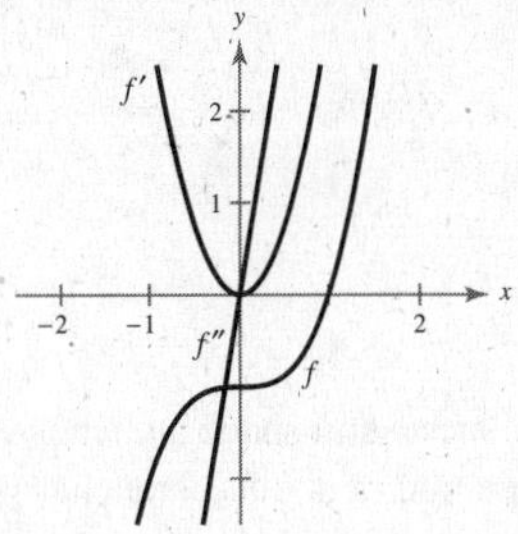

51. (a) $y(t) = -41.333t^3 + 226.54t^2 - 299.9t + 8374$

(b)

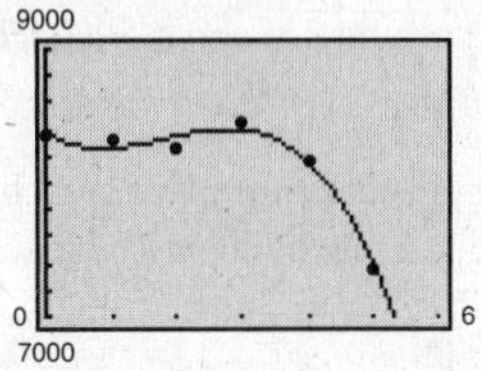

The model fits the data well.

(c) $y'(t) = -123.999t^2 + 453.08t - 299.9$

$y''(t) = -247.998t + 453.08$

(d)

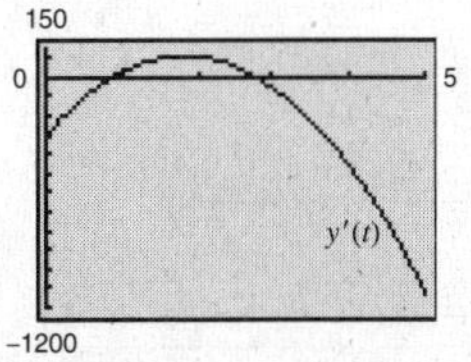

The function $y'(t)$ is negative on the interval $[3, 5]$, so the function y representing utilized production was decreasing from 2003 to 2005.

(e) $y''(t) = 0$

$-247.998t + 453.08 = 0$

$-247.998t = -453.08$

$t \approx 1.83$

So, utilized production was increasing at the greatest rate in 2001.

(f) Sample answer: A function is decreasing when its derivative is negative. A function is increasing at the greatest rate when its derivative is at a maximum.

53. False. The Product Rule is

$[f(x)g(x)]' = f'(x)g(x) + f(x)g'(x)$.

55. True. $h'(c) = f'(c)g(c) + f(c)g'(c) = 0$

57. Let $y = xf(x)$.

Then, $y' = xf'(x) + f(x)$

$y'' = xf''(x) + f'(x) + f'(x) = xf''(x) + 2f'(x)$

$y''' = xf'''(x) + f''(x) + 2f''(x) = xf'''(x) + 3f''(x)$.

In general

$y^{(n)} = [xf(x)]^{(n)} = xf^{(n)}(x) + nf^{(n-1)}(x)$.

Review Exercises for Chapter 2

1. Slope $\approx \frac{-4}{2} = -2$

3. Slope ≈ 0

5. At $t = 1$, slope ≈ -3.

In 2001, the number of successful space launches worldwide was decreasing at a rate of about 3 launches per year.

At $t = 2$, slope ≈ 5.

In 2002, the number of successful space launches worldwide was increasing at a rate of about 5 launches per year.

At $t = 4$, slope ≈ -10.

In 2004, the number of successful space launches worldwide was decreasing at a rate of about 10 launches per year.

7. At $t = 1$, slope ≈ 300.

At $t = 4$, slope ≈ -70.

At $t = 5$, slope ≈ -350.

9.
$$\begin{aligned} f'(x) &= \lim_{\Delta x \to 0} \frac{f(x + \Delta x) - f(x)}{\Delta x} \\ &= \lim_{\Delta x \to 0} \frac{-3(x + \Delta x) - 5 - (-3x - 5)}{\Delta x} \\ &= \lim_{\Delta x \to 0} \frac{-3\Delta x}{\Delta x} = -3 \end{aligned}$$

$$f'(-2) = -3$$

11.
$$\begin{aligned} f'(x) &= \lim_{\Delta x \to 0} \frac{f(x + \Delta x) - f(x)}{\Delta x} \\ &= \lim_{\Delta x \to 0} \frac{(x + \Delta x)^2 - 4(x + \Delta x) - (x^2 - 4x)}{\Delta x} \\ &= \lim_{\Delta x \to 0} \frac{x^2 + 2x\Delta x + (\Delta x)^2 - 4\Delta x - x^2}{\Delta x} \\ &= \lim_{\Delta x \to 0} \frac{2x\Delta x + (\Delta x)^2 - 4\Delta x}{\Delta x} \\ &= \lim_{\Delta x \to 0} (2x + \Delta x - 4) = 2x - 4 \end{aligned}$$

$$f'(1) = 2(1) - 4 = -2$$

13.
$$\begin{aligned} f'(x) &= \lim_{\Delta x \to 0} \frac{f(x + \Delta x) - f(x)}{\Delta x} \\ &= \lim_{\Delta x \to 0} \frac{\sqrt{x + \Delta x + 9} - \sqrt{x + 9}}{\Delta x} \cdot \frac{\sqrt{x + \Delta x + 9} + \sqrt{x + 9}}{\sqrt{x + \Delta x + 9} + \sqrt{x + 9}} \\ &= \lim_{\Delta x \to 0} \frac{(x + \Delta x + 9) - (x + 9)}{\Delta x\left[\sqrt{x + \Delta x + 9} + \sqrt{x + 9}\right]} \\ &= \lim_{\Delta x \to 0} \frac{1}{\sqrt{x + \Delta x + 9} + \sqrt{x + 9}} = \frac{1}{2\sqrt{x + 9}} \end{aligned}$$

$$f'(-5) = \frac{1}{4}$$

15.
$$\begin{aligned} f'(x) &= \lim_{\Delta x \to 0} \frac{f(x + \Delta x) - f(x)}{\Delta x} \\ &= \lim_{\Delta x \to 0} \frac{\frac{1}{x + \Delta x - 5} - \frac{1}{x - 5}}{\Delta x} \\ &= \lim_{\Delta x \to 0} \frac{(x - 5) - (x + \Delta x - 5)}{\Delta x(x + \Delta x - 5)(x - 5)} \\ &= \lim_{\Delta x \to 0} \frac{-1}{(x + \Delta x - 5)(x - 5)} = -\frac{1}{(x - 5)^2} \end{aligned}$$

$$f'(6) = -1$$

17. $f'(x) = -3$

$f'(1) = -3$

19. $f'(x) = -x + 2$

$f'(2) = -2 + 2 = 0$

21. $f(x) = \sqrt{x} + 2 = x^{1/2} + 2$

$f'(x) = \frac{1}{2}x^{-1/2} = \frac{1}{2\sqrt{x}}$

$f'(9) = \frac{1}{2\sqrt{9}} = \frac{1}{6}$

23. $f(x) = \frac{5}{x} = 5x^{-1}$

$f'(x) = -5x^{-2} = -\frac{5}{x^2}$

$f'(1) = -\frac{5}{1^2} = -5$

25. y is not differentiable at $x = 1$, a discontinuity.

27. y is not differentiable at $x = 0$, a discontinuity.

29. $g(t) = \dfrac{2}{3t^2} = \dfrac{2}{3}t^{-2}$

$g'(t) = -\dfrac{4}{3}t^{-3} = -\dfrac{4}{3t^3}$

$g'(1) = -\dfrac{4}{3}$

$y - \dfrac{2}{3} = -\dfrac{4}{3}(t - 1)$

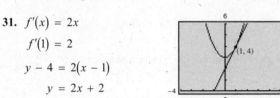

$y = -\dfrac{4}{3}t + 2$

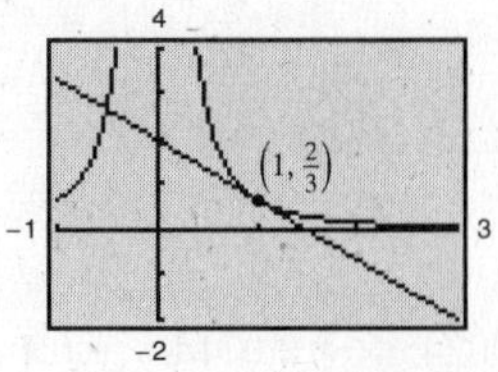

31. $f'(x) = 2x$

$f'(1) = 2$

$y - 4 = 2(x - 1)$

$y = 2x + 2$

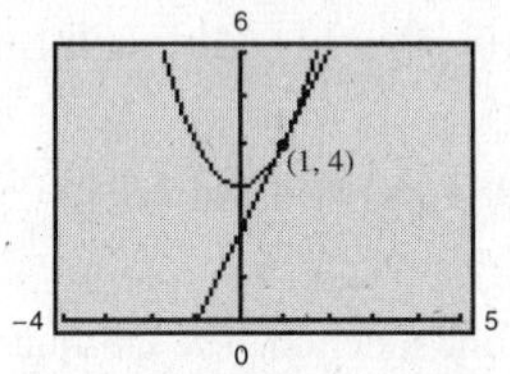

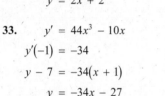

33. $y' = 44x^3 - 10x$

$y'(-1) = -34$

$y - 7 = -34(x + 1)$

$y = -34x - 27$

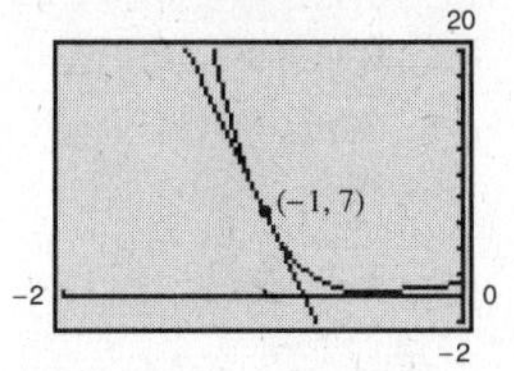

35. $f(x) = \sqrt{x} - \dfrac{1}{\sqrt{x}} = x^{1/2} - x^{-1/2}$

$f'(x) = \dfrac{1}{2}x^{-1/2} + \dfrac{1}{2}x^{-3/2}$

$= \dfrac{1}{2\sqrt{x}} + \dfrac{1}{2x^{3/2}}$

$f'(1) = 1$

$y - 0 = 1(x - 1)$

$y = x - 1$

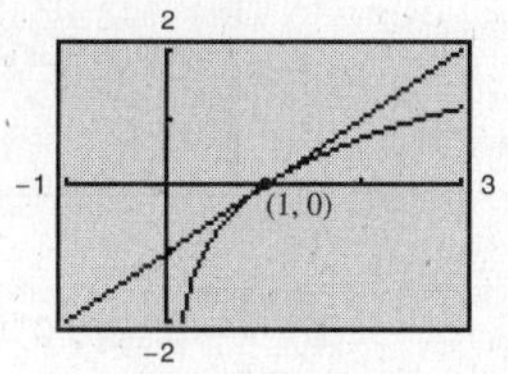

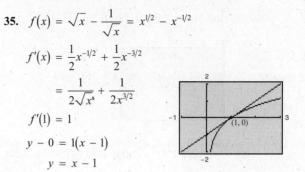

37. $f(x) = \dfrac{x^2 + 3}{x} = x + 3x^{-1}$

$f'(x) = 1 - 3x^{-2}$

$f'(1) = -2$

$y - 4 = -2(x - 1)$

$y = -2x + 6$

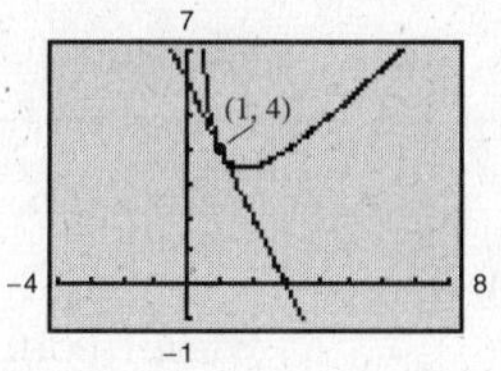

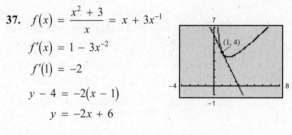

39. Average rate of change $= \dfrac{f(1) - f(0)}{1 - 0} = \dfrac{0 - (-4)}{1} = 4$

$f'(x) = 2x + 3$

$f'(0) = 3$

$f'(1) = 5$

41. $S = 0.9792t^4 - 10.819t^3 + 38.98t^2 - 52.8t + 82$

(a) Average rate of change $= \dfrac{S(5) - S(0)}{5 - 0} = \dfrac{52.1 - 82}{5} \approx -6$ launches per year

(b) $S'(t) = 3.9168t^3 - 32.457t^2 + 77.96t - 52.8$

$S'(0) \approx -53$ launches per year

$S'(5) \approx 15$ launches per year

(c) From 2000 to 2005, the average rate of change in world-wide successful space launches was about −6 launches per year. In 2000, the number of world-wide successful space launches was decreasing at a rate of about 53 launches per year. In 2005, the number of world-wide successful space launches was increasing at a rate of about 15 launches per year.

43. (a) $N'(t) = \dfrac{1}{\sqrt{t}} + 1$

(b)

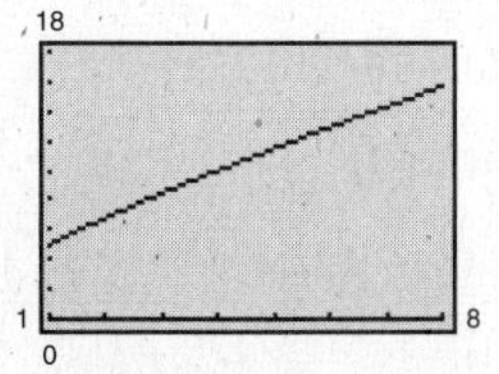

The rate of change appears to be greatest when $t = 1$, or in 2001.

(c)

t	1	2	3	4	5	6	7	8
$N'(t)$	2	1.7071	1.5774	1.5	1.4472	1.4082	1.3780	1.3536

The rate of change is the greatest when $t = 1$, or in 2001.

(d) The rate of change for a function is greatest when the derivative of the function is a maximum.

45. $P = \left(0.035t^2 + 2.04t + 70.7\right)^2$

(a)

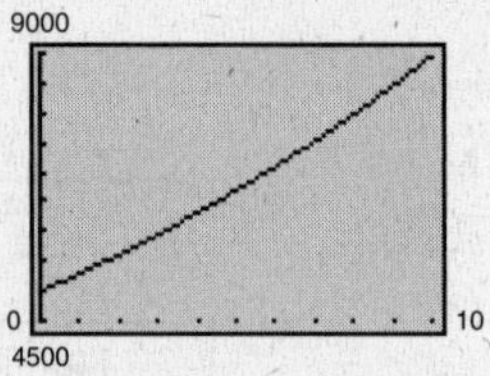

(b) $\dfrac{dP}{dt} = 2\left(0.035t^2 + 2.04t + 70.7\right)\left(0.07t + 2.04\right) = 2\left(0.00245t^3 + 0.0714t^2 + 0.1428t^2 + 4.1616t + 4.949t + 144.228\right)$

$= 2\left(0.00245t^3 + 0.2142t^2 + 9.1106t + 144.228\right)$

When $t = 0$, $\dfrac{dP}{dt} = 2\left[0.00245(0)^3 + 0.2142(0)^2 + 9.1106(0) + 144.228\right] \approx 288$ mosquitos per day

When $t = 5$, $\dfrac{dP}{dt} = 2\left[0.00245(5)^3 + 0.2142(5)^2 + 9.1106(5) + 144.228\right] \approx 391$ mosquitos per day

When $t = 10$, $\dfrac{dP}{dt} = 2\left[0.00245(10)^3 + 0.2142(10)^2 + 9.1106(10) + 144.228\right] \approx 518$ mosquitos per day

(c) $\dfrac{dP}{dt}$ is positive for $t \ge 0$, which agrees with the graph of the function. The mosquito population is increasing over the 10-day period.

47. (a) $s(t) = -16t^2 + 276$

(b) Average velocity $= \dfrac{s(2) - s(0)}{2 - 0} = \dfrac{-64}{2} = -32$ ft/sec

(c) $v(t) = s'(t) = -32t$

$v(2) = -64$ ft/sec

$v(3) = -96$ ft/sec

(d) $s(t) = -16t^2 + 276 = 0 \Rightarrow t^2 = \dfrac{276}{16} = 17.25$

$t \approx 4.15$ sec

(e) $v(4.15) \approx -32(4.15) = -132.8$ ft/sec

The speed of the rock is 132.8 ft/sec at impact.

In Exercises 49–67, the differentiation rule(s) may vary. A sample answer is provided.

49. $f(x) = x^3\left(5 - 3x^2\right) = 5x^3 - 3x^5$

$f'(x) = 15x^2 - 15x^4 = 15x^2\left(1 - x^2\right)$

Simple Power Rule

51. $y = \left(4x - 3\right)\left(x^3 - 2x^2\right)$

$y' = \left(4x - 3\right)\left(3x^2 - 4x\right) + 4\left(x^3 - 2x^2\right)$

$= 12x^3 - 25x^2 + 12x + 4x^3 - 8x^2$

$= 16x^3 - 33x^2 + 12x$

Product Rule

53. $f(x) = \dfrac{6x - 5}{x^2 + 1}$

$f'(x) = \dfrac{\left(x^2 + 1\right)(6) - (6x - 5)(2x)}{\left(x^2 + 1\right)^2}$

$= \dfrac{6 + 10x - 6x^2}{\left(x^2 + 1\right)^2} = \dfrac{2\left(3 + 5x - 3x^2\right)}{\left(x^2 + 1\right)^2}$

Quotient Rule

55. $f(x) = \left(5x^2 + 2\right)^3$

$f'(x) = 3\left(5x^2 + 2\right)^2(10x) = 30x\left(5x^2 + 2\right)^2$

General Power Rule

57. $h(x) = \dfrac{2}{\sqrt{x+1}} = 2(x+1)^{-1/2}$

$h'(x) = 2\left(-\dfrac{1}{2}\right)(x+1)^{-3/2} = -\dfrac{1}{(x+1)^{3/2}}$

General Power Rule

59. $g(x) = x\sqrt{x^2+1} = x(x^2+1)^{1/2}$

$g'(x) = x\left[\dfrac{1}{2}(x^2+1)^{-1/2}(2x)\right] + (1)(x^2+1)^{1/2}$

$= (x^2+1)^{-1/2}\left[x^2 + (x^2+1)\right] = \dfrac{2x^2+1}{\sqrt{x^2+1}}$

Product and General Power Rules

61. $f(x) = x(1-4x^2)^2$

$f'(x) = x(2)(1-4x^2)(-8x) + (1-4x^2)^2$

$= -16x^2(1-4x^2) + (1-4x^2)^2$

$= (1-4x^2)\left[-16x^2 + (1-4x^2)\right]$

$= (1-4x^2)(1-20x^2)$

Product and General Power Rules

63. $h(x) = \left[x^2(2x+3)\right]^3 = x^6(2x+3)^3$

$h'(x) = x^6\left[3(2x+3)^2(2)\right] + 6x^5(2x+3)^3$

$= 6x^5(2x+3)^2\left[x + (2x+3)\right]$

$= 18x^5(2x+3)^2(x+1)$

Product and General Power Rules

65. $f(x) = x^2(x-1)^5$

$f'(x) = 5x^2(x-1)^4 + 2x(x-1)^5$

$= x(x-1)^4\left[5x + 2(x-1)\right] = x(x-1)^4(7x-2)$

Product and General Power Rules

67. $h(t) = \dfrac{\sqrt{3t+1}}{(1-3t)^2} = \dfrac{(3t+1)^{1/2}}{(1-3t)^2}$

$h'(t) = \dfrac{(1-3t)^2(1/2)(3t+1)^{-1/2}(3) - (3t+1)^{1/2}(2)(1-3t)(-3)}{(1-3t)^4} = \dfrac{(3t+1)^{-1/2}\left[(1-3t)(3/2) + (3t+1)6\right]}{(1-3t)^3} = \dfrac{3(9t+5)}{2\sqrt{3t+1}(1-3t)^3}$

Quotient and General Power Rules

69. $P(t) = 400\left(1 + \dfrac{3t}{40+t^2}\right)$

$P'(t) = 400\left[\dfrac{3(40+t^2) - 3t(2t)}{(40+t^2)^2}\right] = 1200\left[\dfrac{40-t^2}{(40+t^2)^2}\right]$

(a) $P'(2) = 1200\left(\dfrac{36}{1936}\right) \approx 22$ bacteria per hr

$P'(4) = 1200\left(\dfrac{24}{3136}\right) \approx 9$ bacteria per hr

$P'(6) = 1200\left(\dfrac{4}{5776}\right) \approx 0.83$ bacteria per hr

$P'(20) = 1200\left(-\dfrac{360}{193{,}600}\right) \approx -2$ bacteria per hr

(b) 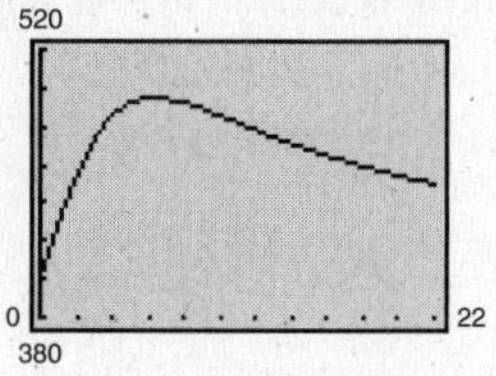

The rate of decrease is approaching zero.

71. When $L = 12$,

$$V = \frac{L}{16}(D-4)^2 = \frac{12}{16}(D-4)^2 = \frac{3}{4}(D-4)^2$$

$$\frac{dV}{dD} = \frac{3}{2}(D-4).$$

(a) When $D = 8$, $\frac{dV}{dD} = \left(\frac{3}{2}\right)(8-4) = 6$ board ft per in.

(b) When $D = 16$, $\frac{dV}{dD} = \left(\frac{3}{2}\right)(16-4) = 18$ board ft per in.

(c) When $D = 24$, $\frac{dV}{dD} = \left(\frac{3}{2}\right)(24-4) = 30$ board ft per in.

(d) When $D = 36$, $\frac{dV}{dD} = \left(\frac{3}{2}\right)(36-4) = 48$ board ft per in.

73. $f(x) = 3x^2 + 7x + 1$

$f'(x) = 6x + 7$

$f''(x) = 6$

75. $f'''(x) = -6x^{-4}$

$f^{(4)}(x) = 24x^{-5}$

$f^{(5)}(x) = -120x^{-6} = -\frac{120}{x^6}$

77. $f'(x) = 7x^{5/2}$

$f''(x) = \frac{35}{2}x^{3/2}$

79. $f''(x) = 6\sqrt[3]{x} = 6x^{1/3}$

$f'''(x) = 2x^{-2/3} = \frac{2}{x^{2/3}}$

81. (a) $s(t) = -16t^2 + 5t + 30$

(b) $s(t) = 0 = -16t^2 + 5t + 30$

Using the Quadratic Formula, $t \approx 1.534$ seconds.

(c) $v(t) = s'(t) = -32t + 5$

$v(1.534) \approx -44.09$ ft/sec

(d) $a(t) = v'(t) = -32$ ft/sec^2

83. (a) $n(t) = 0.2602t^3 - 2.914t^2 + 0.75t + 612.4$

(b)

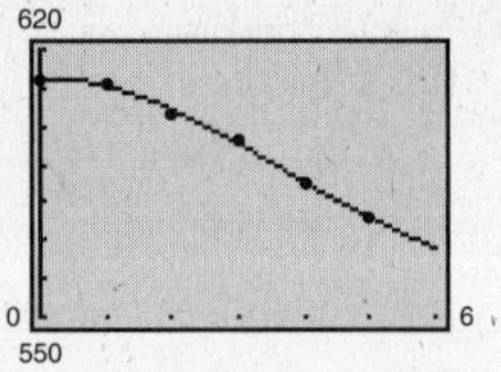

The model fits the data well.

(c) $n'(t) = 0.7806t^2 - 5.828t + 0.75$

$n''(t) = 1.5612t - 5.828$

(d)

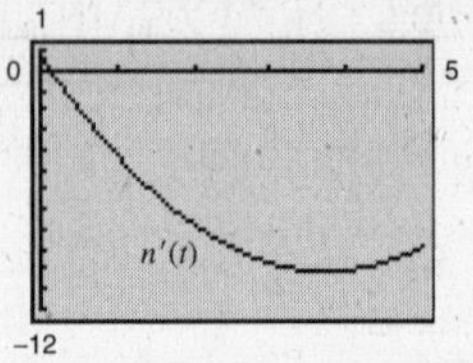

Graphing $n'(t)$ shows that it is negative on the interval $[3, 5]$, so the function n representing students enrolled in public schools in Puerto Rico was decreasing from 2003 to 2005.

(e) $n''(t) = 0$

$1.5612t - 5.828 = 0$

$1.5612t = 5.828$

$t \approx 3.73$

So, enrollment was decreasing at the greatest rate in 2003.

(f) Sample answer: A function is decreasing when its derivative is negative. A function is decreasing at the greatest rate when its derivative is at a minimum.

Chapter Test Solutions

1. $f'(x) = \lim_{\Delta x \to 0} \frac{f(x + \Delta x) - f(x)}{\Delta x}$

$= \lim_{\Delta x \to 0} \frac{(x + \Delta x)^2 + 1 - (x^2 + 1)}{\Delta x}$

$= \lim_{\Delta x \to 0} \frac{2x\Delta x + (\Delta x)^2}{\Delta x} = \lim_{\Delta x \to 0} 2x + \Delta x = 2x$

At $(2, 5)$: $m = 2(2) = 4$

2. $f'(x) = \lim_{\Delta x \to 0} \frac{f(x + \Delta x) - f(x)}{\Delta x}$

$= \lim_{\Delta x \to 0} \frac{\sqrt{x + \Delta x} - 2 - (\sqrt{x} - 2)}{\Delta x}$

$= \lim_{\Delta x \to 0} \frac{\sqrt{x + \Delta x} - \sqrt{x}}{\Delta x}$

$= \lim_{\Delta x \to 0} \frac{1}{\sqrt{x + \Delta x} + \sqrt{x}} = \frac{1}{2\sqrt{x}}$

At $(4, 0)$: $m = \frac{1}{2\sqrt{4}} = \frac{1}{4}$

3. $f'(t) = 3t^2 + 2$

4. $f'(x) = 8x - 8$

5. $f'(x) = \frac{3}{2}x^{1/2} = \frac{3}{2\sqrt{x}}$

6. $f'(x) = (x + 3) + (x - 3)$

$f'(x) = 2x$

7. $f'(x) = 9x^{-4} = \frac{9}{x^4}$

8. $f(x) = \sqrt{x}(5 + x) = 5x^{1/2} + x^{3/2}$

$f'(x) = \frac{5}{2}x^{-1/2} + \frac{3}{2}x^{1/2} = \frac{5}{2\sqrt{x}} + \frac{3\sqrt{x}}{2}$

9. $f'(x) = 2(3x^2 + 4)(6x) = 36x^3 + 48x$

10. $f(x) = \sqrt{1 - 2x} = (1 - 2x)^{1/2}$

$f'(x) = \frac{1}{2}(1 - 2x)^{-1/2}(-2) = -\frac{1}{\sqrt{1 - 2x}}$

11. $f'(x) = \frac{x(3)(5x - 1)^2(5) - (5x - 1)^3}{x^2}$

$= \frac{(5x - 1)^2[15x - (5x - 1)]}{x^2}$

$= \frac{(5x - 1)^2(10x + 1)}{x^2}$

12. $f(x) = x - \frac{1}{x}$

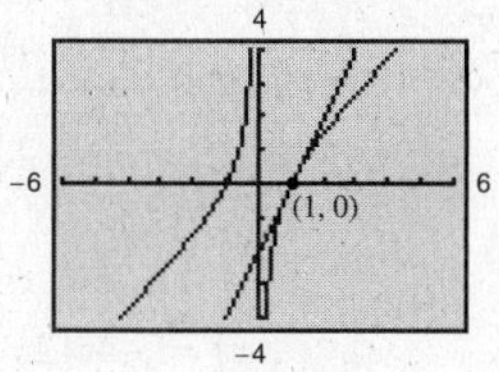

$f'(x) = 1 + \frac{1}{x^2}$

$f'(1) = 1 + \frac{1}{1^2} = 2$

$y - 0 = 2(x - 1)$

$y = 2x - 2$

13. $N = 12.95t^2 + 110.3t + 780$

(a) Average rate of change

$= \frac{N(4) - N(1)}{4 - 1}$

$= \frac{1428.4 - 903.25}{3}$

$= 175.05$ thousand procedures per year

(b) $N'(t) = 25.9t + 110.3$

$N'(1) = 136.2$ thousand procedures per year

$N'(4) = 213.9$ thousand procedures per year

(c) From 2001 to 2004, the average rate of change in the number of bone graft procedures was 175,050 procedures per year. In 2001, the number of bone graft procedures was increasing at a rate of 136,200 procedures per year. In 2004, the number of bone graft procedures was increasing at a rate of 213,900 procedures per year.

14. $f(x) = 2x^2 + 3x + 1$

$f'(x) = 4x + 3$

$f''(x) = 4$

$f'''(x) = 0$

15. $f(x) = \sqrt{3 - x} = (3 - x)^{1/2}$

$f'(x) = \frac{1}{2}(3 - x)^{-1/2}(-1) = -\frac{1}{2}(3 - x)^{-1/2}$

$f''(x) = -\frac{1}{2}\left(-\frac{1}{2}\right)(3 - x)^{-3/2}(-1) = -\frac{1}{4}(3 - x)^{-3/2}$

$f'''(x) = -\frac{1}{4}\left(-\frac{3}{2}\right)(3 - x)^{-5/2}(-1)$

$= -\frac{3}{8}(3 - x)^{-5/2} = -\frac{3}{8(3 - x)^{5/2}}$

16. $f(x) = \dfrac{2x+1}{2x-1}$

$$f'(x) = \frac{(2x-1)(2) - (2x+1)(2)}{(2x-1)^2} = \frac{4}{(2x-1)^2} = -4(2x-1)^{-2}$$

$$f''(x) = 8(2x-1)^{-3}(2) = 16(2x-1)^{-3}$$

$$f'''(x) = -48(2x-1)^{-4}(2) = -\frac{96}{(2x-1)^4}$$

17. (a)

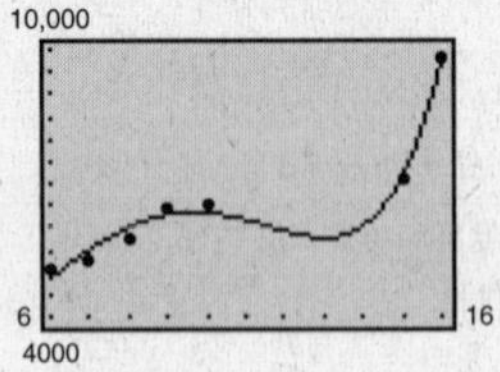

(b) $N'(t) = 0.01578t^5 - 4.14t^3 + 249t$

$N''(t) = 0.0789t^4 - 12.42t^2 + 249$

(c)

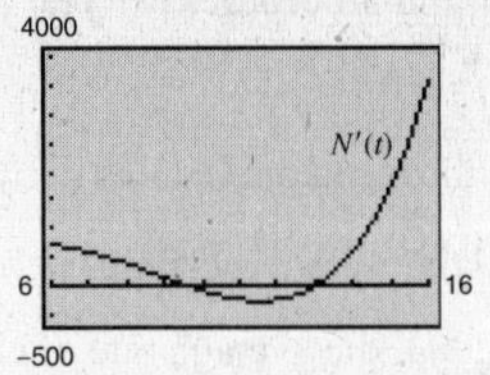

N' is negative on the interval $[10, 12]$, so the function N representing bald eagle breeding pairs was decreasing from 2000-2002.

(d) $N''(t) = 0$

$$0.0789t^4 - 12.42t^2 + 249 = 0$$

$$t^2 = \frac{12.42 \pm \sqrt{(-12.42)^2 - 4(0.0789)(249)}}{2(0.0789)}$$

$$t^2 \approx 133.83, 23.58$$

$$t \approx 11.57, 4.86$$

So, the number of breeding pairs was decreasing at the greatest rate in 2001 ($t = 4.86$ is not in the domain of the original function).

Practice Test for Chapter 2

1. Use the definition of the derivative to find the derivative of $f(x) = 2x^2 + 3x - 5$.

2. Use the definition of the derivative to find the derivative of $f(x) = \dfrac{1}{x - 4}$.

3. Use the definition of the derivative to find equation of the tangent line to the graph of $f(x) = \sqrt{x - 2}$ at the point $(6, 2)$.

4. Find $f'(x)$ for $f(x) = 5x^3 - 6x^2 + 15x - 9$.

5. Find $f'(x)$ for $f(x) = \dfrac{6x^2 - 4x + 1}{x^2}$.

6. Find $f'(x)$ for $f(x) = \sqrt[3]{x^2} + \sqrt[5]{x^3}$.

7. Find the average rate of change of $f(x) = x^3 - 11$ over the interval $[0, 2]$. Compare this to the instantaneous rate of change at the endpoints of the interval.

8. Given the cost function $C = 6200 + 4.31x - 0.0001x^2$, find the marginal cost of producing x units.

9. Find $f'(x)$ for $f(x) = (x^3 - 4x)(x^2 + 7x - 9)$.

10. Find $f'(x)$ for $f(x) = \dfrac{x + 7}{x^2 - 8}$.

11. Find $f'(x)$ for $f(x) = x^3\left(\dfrac{x - 3}{x + 5}\right)$.

12. Find $f'(x)$ for $f(x) = \dfrac{\sqrt{x}}{x^2 + 4x - 1}$.

13. Find $f'(x)$ for $f(x) = (6x - 5)^{12}$.

14. Find $f'(x)$ for $f(x) = 8\sqrt{4 - 3x}$.

15. Find $f'(x)$ for $f(x) = -\dfrac{3}{(x^2 + 1)^3}$.

16. Find $f'(x)$ for $f(x) = \sqrt{\dfrac{10x}{x + 2}}$.

17. Find $f'''(x)$ for $f(x) = x^4 - 9x^3 + 17x^2 - 4x + 121$.

18. Find $f^{(4)}(x)$ for $f(x) = \sqrt{3 - x}$.

Graphing Calculator Required

19. Graph $f(x) = \dfrac{x^2}{x - 2}$ and its derivative on the same set of coordinate axes. From the graph of $f(x)$, determine any points at which the graph has horizontal tangent lines. What is the value of $f'(x)$ at these points?

20. Use a graphing utility to graph $\sqrt[3]{x} + \sqrt[3]{y} = 3$. Then find and sketch the tangent line at the point $(8, 1)$.

CHAPTER 3
Applications of the Derivative

CHAPTER 3
Applications of the Derivative

Section 3.1 Increasing and Decreasing Functions

Skills Review

1.
$$\begin{aligned} x^2 &= 8x \\ x^2 - 8x &= 0 \\ x(x-8) &= 0 \\ x &= 0 \\ x - 8 = 0 &\Rightarrow x = 8 \end{aligned}$$

2.
$$\begin{aligned} 15x &= \tfrac{5}{8}x^2 \\ 15x - \tfrac{5}{8}x^2 &= 0 \\ x\left(15 - \tfrac{5}{8}x\right) &= 0 \\ x &= 0 \\ 15 - \tfrac{5}{8}x = 0 &\Rightarrow x = 24 \end{aligned}$$

3.
$$\begin{aligned} \frac{x^2 - 25}{x^3} &= 0 \\ \frac{1}{x} - \frac{25}{x^3} &= 0 \\ \frac{1}{x} &= \frac{25}{x^3} \\ x^3 &= 25x \\ x^3 - 25x &= 0 \\ x\left(x^2 - 25\right) &= 0 \\ x(x+5)(x-5) &= 0 \\ x &= 0 \text{ Extraneous} \\ x + 5 = 0 &\Rightarrow x = -5 \\ x - 5 = 0 &\Rightarrow x = 5 \end{aligned}$$

4.
$$\begin{aligned} \frac{2x}{\sqrt{1 - x^2}} &= 0 \\ 2x &= 0 \\ x &= 0 \end{aligned}$$

5. The domain of $\dfrac{x+3}{x-3}$ is $(-\infty, 3) \cup (3, \infty)$.

6. The domain of $\dfrac{2}{\sqrt{1-x}}$ is $(-\infty, 1)$.

7. The domain of $\dfrac{2x+1}{x^2 - 3x - 10}$ is $(-\infty, -2) \cup (-2, 5) \cup (5, \infty)$.

8. The domain of $\dfrac{3x}{\sqrt{9 - 3x^2}}$ is $\left(-\sqrt{3}, \sqrt{3}\right)$.

9. When $x = -2$: $-2(-2+1)(-2-1) = -6$

When $x = 0$: $-2(0+1)(0-1) = 2$

When $x = 2$: $-2(2+1)(2-1) = -6$

10. When $x = -2$: $4\left[2(-2)+1\right]\left[2(-2)-1\right] = 60$

When $x = 0$: $4(2 \cdot 0 + 1)(2 \cdot 0 - 1) = -4$

When $x = 2$: $4(2 \cdot 2 + 1)(2 \cdot 2 - 1) = 60$

11. When $x = -2$: $\dfrac{2(-2)+1}{(-2-1)^2} = -\dfrac{1}{3}$

When $x = 0$: $\dfrac{2 \cdot 0 + 1}{(0-1)^2} = 1$

When $x = 2$: $\dfrac{2 \cdot 2 + 1}{(2-1)^2} = 5$

12. When $x = -2$: $\dfrac{-2(-2+1)}{(-2-4)^2} = \dfrac{1}{18}$

When $x = 0$: $\dfrac{-2(0+1)}{(0-4)^2} = -\dfrac{1}{8}$

When $x = 2$: $\dfrac{-2(2+1)}{(2-4)^2} = -\dfrac{3}{2}$

1. $f(x) = \dfrac{x^2}{x^2 + 4}$

$$f'(x) = \frac{(x^2 + 4)(2x) - (x^2)(2x)}{(x^2 + 4)^2} = \frac{8x}{(x^2 + 4)^2}$$

$f'(-1) = -\dfrac{8}{25}$

$f'(0) = 0$

$f'(1) = \dfrac{8}{25}$

3. $f(x) = (x + 2)^{2/3}$

$$f'(x) = \frac{2}{3}(x + 2)^{-1/3} = \frac{2}{3\sqrt[3]{x + 2}}$$

$f'(-3) = -\dfrac{2}{3}$

$f'(-2)$ is undefined.

$f'(-1) = \dfrac{2}{3}$

5. $f(x) = -(x + 1)^2$

$f'(x) = -2(x + 1)$

f has a critical number at $x = -1$. Moreover, f is increasing on $(-\infty, -1)$ and decreasing on $(-1, \infty)$.

7. $f(x) = x^4 - 2x^2$

$f'(x) = 4x^3 - 4x = 4x(x^2 - 1)$

f has critical numbers at $x = 0, \pm1$. Moreover, f is increasing on $(-1, 0)$ and $(1, \infty)$ and decreasing on $(-\infty, -1)$, $(0, 1)$.

9. $f(x) = 2x - 3$

$f'(x) = 2$

There are no critical numbers. Because the derivative is positive for all x, the function is increasing on $(-\infty, \infty)$.

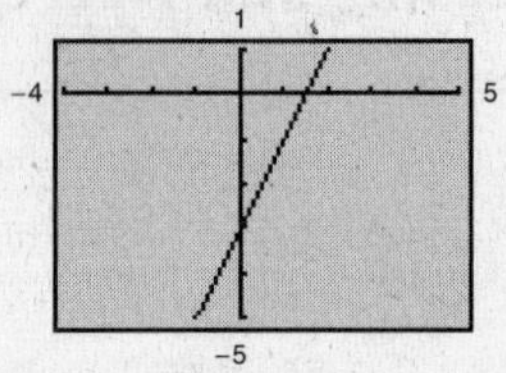

11. $g(x) = (x - 1)^2$

$g'(x) = -2(x - 1)$

Critical number: $x = 1$

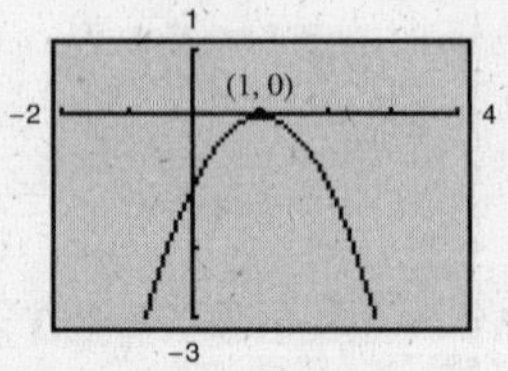

Interval	$-\infty < x < 1$	$1 < x < \infty$
Sign of g'	$g' > 0$	$g' < 0$
Conclusion	Increasing	Decreasing

13. $y = x^2 - 6x$

$y' = 2x - 6$

Critical number: $x = 3$

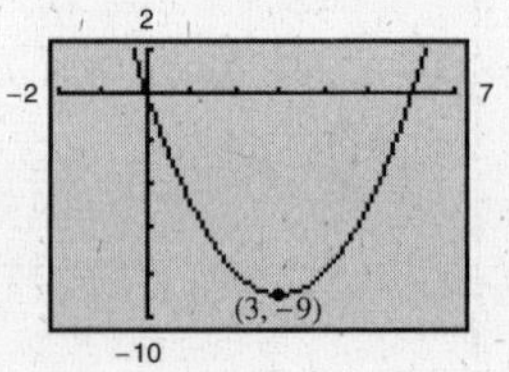

Interval	$-\infty < x < 3$	$3 < x < \infty$
Sign of y'	$y' < 0$	$y' > 0$
Conclusion	Decreasing	Increasing

15. $y = x^3 - 6x^2$

$y' = 3x^2 - 12x = 3x(x - 4)$

Critical numbers: $x = 0$, $x = 4$

Interval	$-\infty < x < 0$	$0 < x < 4$	$4 < x < \infty$
Sign of y'	$y' > 0$	$y' < 0$	$y' > 0$
Conclusion	Increasing	Decreasing	Increasing

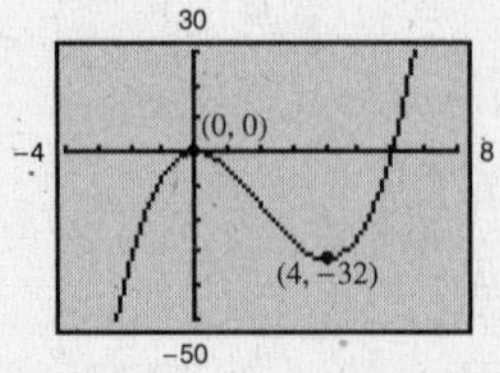

17. $f(x) = \sqrt{x^2 - 1} = (x^2 - 1)^{1/2}$

Domain: $(-\infty, -1] \cup [1, \infty)$

$f'(x) = \frac{1}{2}(x^2 - 1)^{-1/2}(2x) = \frac{x}{\sqrt{x^2 - 1}}$

Critical numbers: $x = \pm 1$ ($x = 0$ not in domain)

Interval	$-\infty < x < -1$	$1 < x < \infty$
Sign of f'	$f' < 0$	$f' > 0$
Conclusion	Decreasing	Increasing

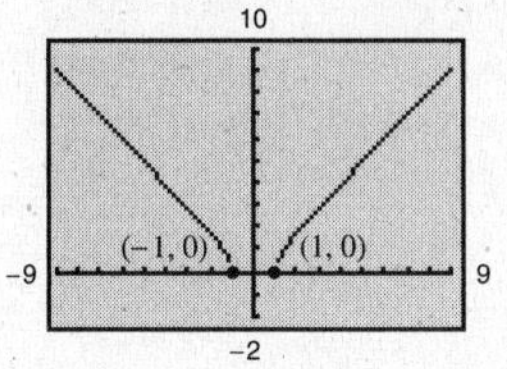

19. $y = x^{1/3} + 1$

$y' = \frac{1}{3}x^{-2/3} = \frac{1}{3x^{2/3}}$

Critical number: $x = 0$

Interval	$-\infty < x < 0$	$0 < x < \infty$
Sign of y'	$y' > 0$	$y' > 0$
Conclusion	Increasing	Increasing

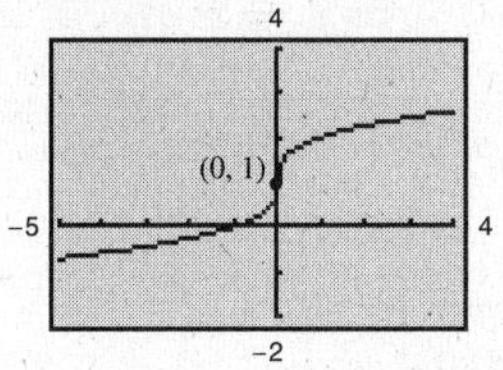

y is increasing on $(-\infty, \infty)$.

21. $g(x) = (x - 1)^{1/3}$

$g'(x) = \frac{1}{3}(x - 1)^{-2/3}$

$= \frac{1}{3(x - 1)^{2/3}}$

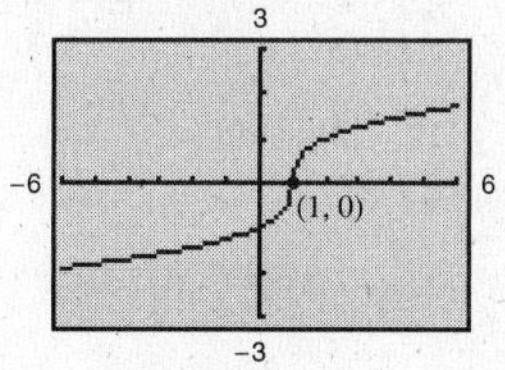

Critical number: $x = 1$

Interval	$-\infty < x < 1$	$1 < x < \infty$
Sign of g'	$g' > 0$	$g' > 0$
Conclusion	Increasing	Increasing

g is increasing on $(-\infty, \infty)$.

23. $f(x) = -2x^2 + 4x + 3$

$f'(x) = -4x + 4$

Critical number: $x = 1$

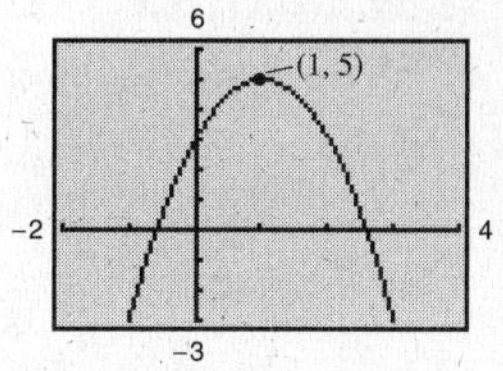

Interval	$-\infty < x < 1$	$1 < x < \infty$
Sign of f'	$f' > 0$	$f' < 0$
Conclusion	Increasing	Decreasing

25. $y = 3x^3 + 12x^2 + 15x$

$y' = 9x^2 + 24x + 15 = 3(x + 1)(3x + 5)$

Critical numbers: $x = -1, -\frac{5}{3}$

Interval	$-\infty < x < -\frac{5}{3}$	$-\frac{5}{3} < x < -1$	$-1 < x < \infty$
Sign of y'	$y' > 0$	$y' < 0$	$y' > 0$
Conclusion	Increasing	Decreasing	Increasing

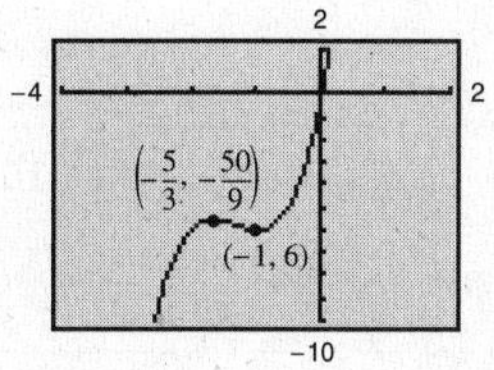

27. $f(x) = x\sqrt{x+1} = x(x+1)^{1/2}$

Domain: $[-1, \infty)$

$$f'(x) = x\left[\frac{1}{2}(x+1)^{-1/2}\right] + (x+1)^{1/2} = \frac{1}{2}(x+1)^{-1/2}\left[x + 2(x+1)\right] = \frac{3x+2}{2\sqrt{x+1}}$$

Critical numbers: $x = -1, -\frac{2}{3}$

Interval	$-1 < x < -\frac{2}{3}$	$-\frac{2}{3} < x < \infty$
Sign of f'	$f' < 0$	$f' > 0$
Conclusion	Decreasing	Increasing

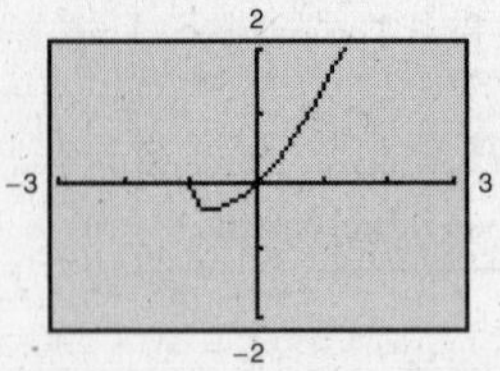

29. $f(x) = x^4 - 2x^3$

$f'(x) = 4x^3 - 6x^2 = 2x^2(2x - 3)$

Critical numbers: $x = 0, x = \frac{3}{2}$

Interval	$-\infty < x < 0$	$0 < x < \frac{3}{2}$	$\frac{3}{2} < x < \infty$
Sign of f'	$f' < 0$	$f' < 0$	$f' > 0$
Conclusion	Decreasing	Decreasing	Increasing

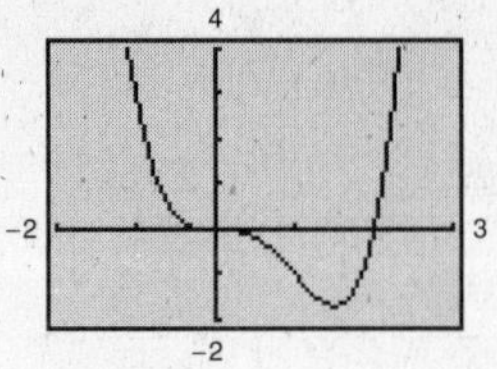

f is decreasing on $\left(-\infty, \frac{3}{2}\right)$ and increasing on $\left(\frac{3}{2}, \infty\right)$.

31. $f(x) = \dfrac{x}{x^2+4}$

$$f'(x) = \frac{(x^2+4)(1) - (x)(2x)}{(x^2+4)^2} = \frac{4 - x^2}{(x^2+4)^2} = \frac{(2-x)(2+x)}{(x^2+4)^2}$$

Critical numbers: $x = \pm 2$

Interval	$-\infty < x < -2$	$-2 < x < 2$	$2 < x < \infty$
Sign of f'	$f' < 0$	$f' > 0$	$f' < 0$
Conclusion	Decreasing	Increasing	Decreasing

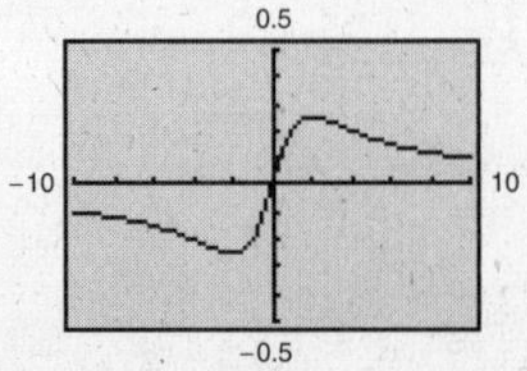

33. $f(x) = \dfrac{2x}{16 - x^2}$

$$f'(x) = \frac{(16 - x^2)2 - 2x(-2x)}{(16 - x^2)^2} = \frac{2x^2 + 32}{(16 - x^2)^2}$$

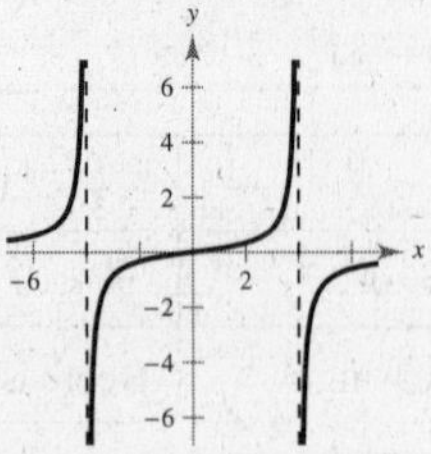

Discontinuities: $x = \pm 4$

Interval	$-\infty < x < -4$	$-4 < x < 4$	$4 < x < \infty$
Sign of f'	$f' > 0$	$f' > 0$	$f' > 0$
Conclusion	Increasing	Increasing	Increasing

35. $y = \begin{cases} 4 - x^2, & x \le 0 \\ -2x, & x > 0 \end{cases}$

$y' = \begin{cases} -2x, & x < 0 \\ -2, & x > 0 \end{cases}$

$y'(0)$ is undefined.

Critical number: $x = 0$

Interval	$-\infty < x < 0$	$0 < x < \infty$
Sign of y'	$y' > 0$	$y' < 0$
Conclusion	Increasing	Decreasing

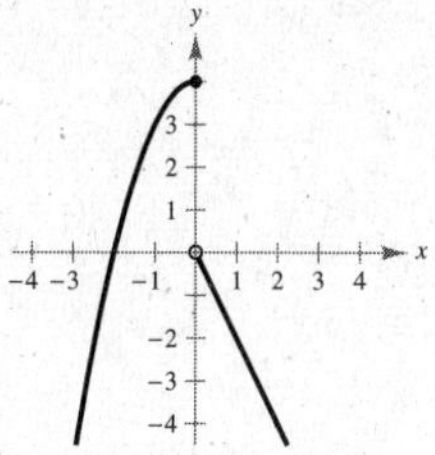

37. $y = \begin{cases} 3x + 1, & x \le 1 \\ 5 - x^2, & x > 1 \end{cases}$

$y' = \begin{cases} 3, & x < 1 \\ -2x, & x > 1 \end{cases}$

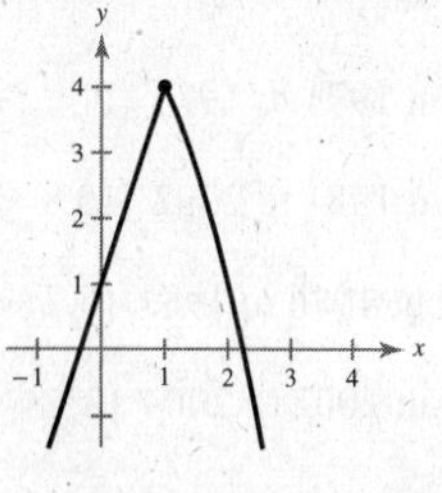

$y'(1)$ is undefined.

Critical number: $x = 1$

($x = 0$ is not a critical number.)

Interval	$-\infty < x < 1$	$1 < x < \infty$
Sign of y'	$y' > 0$	$y' < 0$
Conclusion	Increasing	Decreasing

39. $N = 100\left(\frac{1}{t} + \frac{t}{t+3}\right), 1 \le t \le 31$

(a) $\frac{dN}{dt} = 100\left[-t^{-2} + \frac{(t+3)-(t)}{(t+3)^2}\right] = 100\left[-\frac{1}{t^2} + \frac{3}{(t+3)^2}\right] = 100\left[\frac{-(t+3)^2 + 3t^2}{t^2(t+3)^2}\right] = 100\left[\frac{2t^2 - 6t - 9}{t^2(t+3)^2}\right]$

By the Quadratic Formula, $2t^2 - 6t - 9 = 0$ when

$t = \frac{6 \pm \sqrt{108}}{4} = \frac{3 \pm 3\sqrt{3}}{2}$.

The only critical number in the domain is $t \approx 4.10$. N is decreasing on $[1, 4.10)$ and increasing on $(4.10, 31]$.

(b)

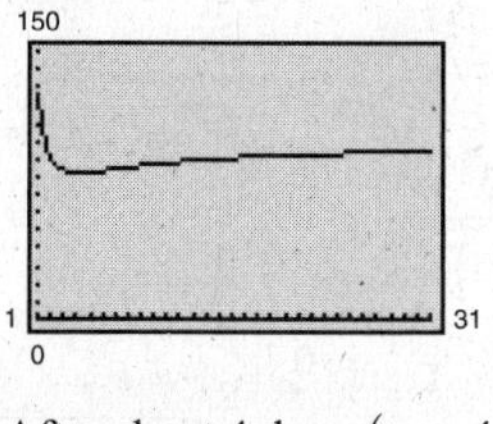

(c) After about 4 days $(t \approx 4.10)$, the number of patients is the lowest.

41. $N = 0.75t^2 - 2.1t + 342, 1 \le t \le 6$

$\frac{dN}{dt} = 1.5t - 2.1 = 0$

$1.5t = 2.1$

Critical number: $t = 1.4$

Interval	$1 < t < 1.4$	$1.4 < t < 6$
Sign of $\frac{dN}{dt}$	$\frac{dN}{dt} < 0$	$\frac{dN}{dt} > 0$
Conclusion	Decreasing	Increasing

43. (a)

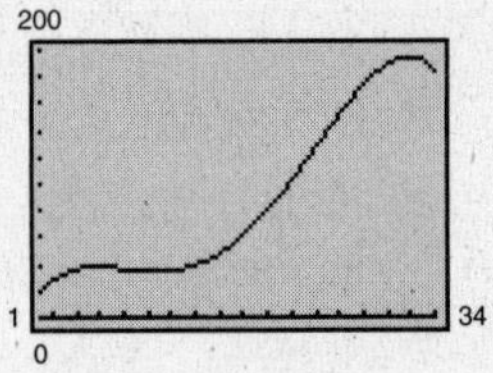

Increasing: from about 1971 to 1976 $(1 < t < 5.7)$

from about 1981 to 2002 $(10.8 < t < 31.8)$

Decreasing: from about 1976 to 1981 $(5.7 < t < 10.8)$

from about 2002 to 2004 $(31.8 < t < 34)$

(b) $y' = -0.0066t^3 + 0.3192t^2 - 3.882t + 13.02$

Domain: $1 \le t \le 34$

Critical numbers: $t \approx 5.7$, $t \approx 10.8$, $t \approx 31.8$

Interval	$1 < t < 5.7$	$5.7 < t < 10.8$	$10.8 < t < 31.8$	$31.8 < t < 34$
Sign of y'	$y' > 0$	$y' < 0$	$y' > 0$	$y' < 0$
Conclusion	Increasing	Decreasing	Increasing	Decreasing

Section 3.2 Extrema and the First-Derivative Test

Skills Review

1. $f(x) = 4x^4 - 2x^2 + 1$

$f'(x) = 16x^3 - 4x = 0$

$4x(4x^2 - 1) = 0$

$4x(2x + 1)(2x - 1) = 0$

$x = 0,\ x = \pm\dfrac{1}{2}$

2. $f(x) = \frac{1}{3}x^3 - \frac{3}{2}x^2 - 10x$

$f'(x) = x^2 - 3x - 10 = 0$

$(x - 5)(x + 2) = 0$

$x = -2,\ x = 5$

3. $f(x) = 5x^{4/5} - 4x$

$f'(x) = 4x^{-1/5} - 4$

$= \dfrac{4}{x^{1/5}} - 4 = 0$

$x^{1/5} = 1$

$x = 1$

4. $f(x) = \frac{1}{2}x^2 - 3x^{5/3}$

$f'(x) = x - 5x^{2/3} = 0$

$x^{2/3}(x^{1/3} - 5) = 0$

$x = 0,\ x = 125$

5. $f(x) = \dfrac{x + 4}{x^2 + 1}$

$f'(x) = \dfrac{-x^2 - 8x + 1}{(x^2 + 1)^2} = 0$

$-x^2 - 8x + 1 = 0$

$x = \dfrac{8 \pm \sqrt{(-8)^2 - 4(-1)(1)}}{2(-1)}$

$= \dfrac{8 \pm \sqrt{68}}{-2}$

$= \dfrac{8 \pm 2\sqrt{17}}{-2}$

$= -4 \pm \sqrt{17}$

Skills Review —*continued*—

6. $f(x) = \dfrac{x - 1}{x^2 + 4}$

$f'(x) = \dfrac{-x^2 + 2x + 4}{(x^2 + 4)^2} = 0$

$-x^2 + 2x + 4 = 0$

$$x = \frac{-2 \pm \sqrt{2^2 - 4(-1)(4)}}{2(-1)} = \frac{-2 \pm \sqrt{20}}{-2} = \frac{-2 \pm 2\sqrt{5}}{-2} = 1 \pm \sqrt{5}$$

7. $g'(x) = -5x^4 - 8x^3 + 12x^2 + 2$

$g'(-4) = -574 < 0$

8. $g'(x) = -5x^4 - 8x^3 + 12x^2 + 2$

$g'(0) = 2 > 0$

9. $g'(x) = -5x^4 - 8x^3 + 12x^2 + 2$

$g'(1) = 1 > 0$

10. $g'(x) = -5x^4 - 8x^3 + 12x^2 + 2$

$g'(3) = -511 < 0$

11. $f(x) = 2x^2 - 11x - 6, (3, 6)$

$f'(x) = 4x - 11$

$f'(4) = 5 > 0$

So, f is increasing on (3, 6).

12. $f(x) = x^3 + 2x^2 - 4x - 8, (-2, 0)$

$f'(x) = 3x^2 + 4x - 4$

$f'(-1) = -5 < 0$

So, f is decreasing on $(-2, 0)$.

1. $f(x) = 2x^2 + 4x + 3$

$f'(x) = 4 - 4x = 4(1 - x)$

Critical number: $x = 1$

Interval	$-\infty < x < 1$	$1 < x < \infty$
Sign of f'	$f' > 0$	$f' < 0$
Conclusion	Increasing	Decreasing

Relative maximum: (1, 5)

3. $f(x) = x^2 - 6x$

$f'(x) = 2x - 6 = 2(x - 3)$

Critical number: $x = 3$

Interval	$-\infty < x < 3$	$3 < x < \infty$
Sign of f'	$f' < 0$	$f' > 0$
Conclusion	Decreasing	Increasing

Relative minimum: $(3, -9)$

5. $g(x) = 6x^3 - 15x^2 + 12x$

$g'(x) = 18x^2 - 30x + 12 = 6(3x^2 - 5x + 2) = 6(x - 1)(3x - 2)$

Critical numbers: $x = 1, x = \frac{2}{3}$

Interval	$-\infty < x < \frac{2}{3}$	$\frac{2}{3} < x < 1$	$1 < x < \infty$
Sign of g'	$g' > 0$	$g' < 0$	$g' > 0$
Conclusion	Increasing	Decreasing	Increasing

Relative maximum: $\left(\frac{2}{3}, \frac{28}{9}\right)$

Relative minimum: $(1, 3)$

7. $h(x) = -(x + 4)^3$

$h'(x) = -3(x + 4)^2$

Critical number: $x = -4$

Interval	$-\infty < x < -4$	$-4 < x < \infty$
Sign of h'	$h' < 0$	$h' < 0$
Conclusion	Decreasing	Decreasing

No relative extrema

9. $f(x) = x^3 - 6x^2 + 15$

$f'(x) = 3x^2 - 12x = 3x(x - 4)$

Critical numbers: $x = 0,\ x = 4$

Interval	$-\infty < x < 0$	$0 < x < 4$	$4 < x < \infty$
Sign of f'	$f' > 0$	$f' < 0$	$f' > 0$
Conclusion	Increasing	Decreasing	Increasing

Relative maximum: $(0, 15)$

Relative minimum: $(4, -17)$

11. $f(x) = x^4 - 2x^3 + x + 1$

$f'(x) = 4x^3 - 6x^2 + 1$

Critical numbers: $x = -0.366, 0.5, 1.366$

Interval	$-\infty < x < -0.366$	$-0.366 < x < 0.5$	$0.5 < x < 1.366$	$1.366 < x < \infty$
Sign of f'	$f' < 0$	$f' > 0$	$f' < 0$	$f' > 0$
Conclusion	Decreasing	Increasing	Decreasing	Increasing

Relative minima: $(-0.366, 0.75)$, $(1.366, 0.75)$

Relative maximum: $(0.5, 1.3125)$

13. $f(x) = (x - 1)^{2/3}$

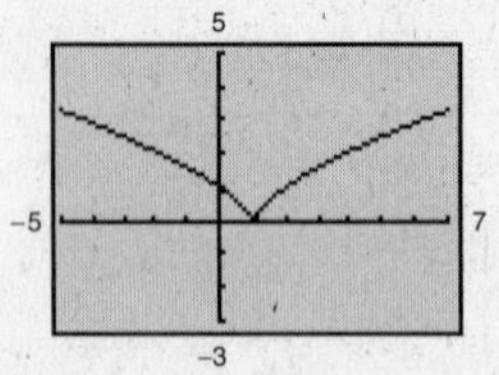

$f'(x) = \frac{2}{3}(x - 1)^{-1/3}$

$= \dfrac{2}{3(x - 1)^{1/3}}$

Critical number: $x = 1$

Interval	$-\infty < x < 1$	$1 < x < \infty$
Sign of f'	$f' < 0$	$f' > 0$
Conclusion	Decreasing	Increasing

Relative minimum: $(1, 0)$

15. $g(t) = t - \dfrac{1}{2t^2} = t - \dfrac{1}{2}t^{-2}$

$g'(t) = 1 + t^{-3}$

$= \dfrac{t^3 + 1}{t^3}$

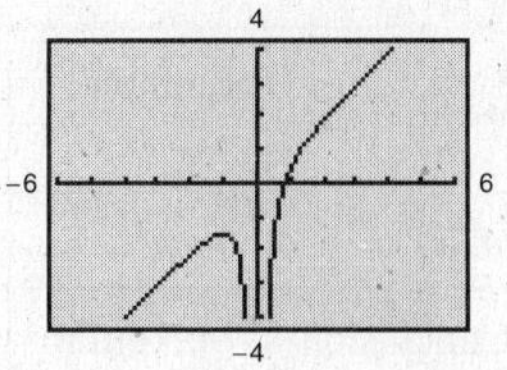

Critical number: $t = -1$

Discontinuity: $t = 0$

Interval	$-\infty < t < -1$	$-1 < t < 0$	$0 < t < \infty$
Sign of g'	$g' > 0$	$g' < 0$	$g' > 0$
Conclusion	Increasing	Decreasing	Increasing

Relative maximum: $\left(-1, -\frac{3}{2}\right)$

17. $f(x) = \dfrac{x}{x + 1}$

$f'(x) = \dfrac{(x + 1) - x}{(x + 1)^2} = \dfrac{1}{(x + 1)^2}$

Discontinuity: $x = -1$

Interval	$-\infty < x < -1$	$-1 < x < \infty$
Sign of f'	$f' > 0$	$f' > 0$
Conclusion	Increasing	Increasing

No relative extrema

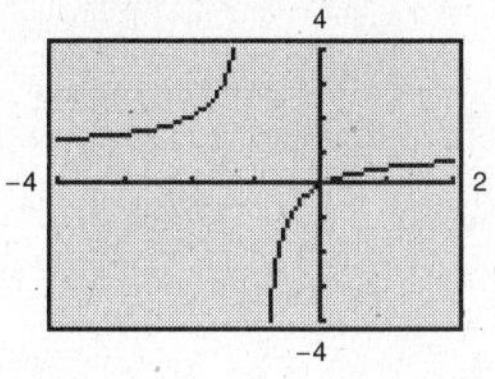

19. $f(x) = 2(3 - x), [-1, 2]$

$f'(x) = -2$

No critical numbers

x-value	Endpoint $x = -1$	Endpoint $x = 2$
$f(x)$	8	2
Conclusion	Maximum	Minimum

21. $f(x) = 5 - 2x^2, [0, 3]$

$f'(x) = -4x$

Critical number: $x = 0$ (endpoint)

x-value	Endpoint $x = 0$	Endpoint $x = 3$
$f(x)$	5	−13
Conclusion	Maximum	Minimum

23. $f(x) = x^3 - 3x^2, [-1, 3]$

$f'(x) = 3x^2 - 6x = 3x(x - 2)$

Critical numbers: $x = 0, x = 2$

x-value	Endpoint $x = -1$	Critical $x = 0$	Critical $x = 2$	Endpoint $x = 3$
$f(x)$	−4	0	−4	0
Conclusion	Minimum	Maximum	Minimum	Maximum

25. $h(s) = \dfrac{1}{3 - s} = (3 - s)^{-1}, [0, 2]$

$h'(s) = -(3 - s)^{-2}(-1) = \dfrac{1}{(3 - s)^2}$

No critical numbers

s-value	Endpoint $s = 0$	Endpoint $s = 2$
$h(s)$	$\frac{1}{3}$	1
Conclusion	Minimum	Maximum

27. $f(x) = 3x^{2/3} - 2x, \ [-1, 2]$

$$f'(x) = 2x^{-1/3} - 2 = \frac{2}{x^{1/3}} - 2 = \frac{2(1 - x^{1/3})}{x^{1/3}}$$

Critical numbers: $x = 0, x = 1$

x-value	Endpoint $x = -1$	Critical $x = 0$	Critical $x = 1$	Endpoint $x = 2$
$f(x)$	5	0	1	0.762
Conclusion	Maximum	Minimum		

29. $h(t) = (t - 1)^{2/3}, \ [-7, 2]$

$$h'(t) = \frac{2}{3}(t - 1)^{-1/3} = \frac{2}{3(t - 1)^{1/3}}$$

Critical number: $t = 1$

t-value	Endpoint $t = -7$	Critical $t = 1$	Endpoint $t = 2$
$h(t)$	4	0	1
Conclusion	Maximum	Minimum	

31. Critical number: $x = 2$

The function has an absolute maximum at the critical number.

33. $f(x) = 0.4x^3 - 1.8x^2 + x - 3, [0, 5]$

Maximum: $(5, 7)$

Minimum: $(2.69, 5.55)$

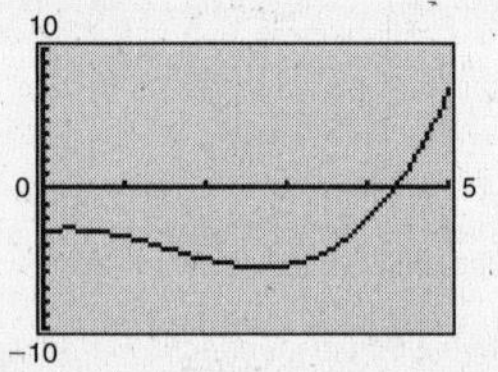

35. $f(x) = \frac{4}{3}x\sqrt{3 - x}, \ [0, 3]$

Maximum: $(2, 2.67)$

Minimum: $(0, 0), (3, 0)$

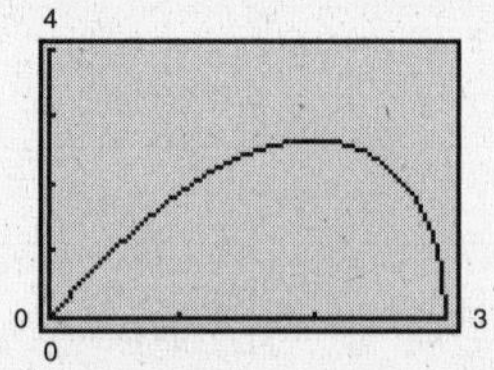

37. $f(x) = \frac{4x}{x^2 + 1}, \ [0, \infty)$

$$f'(x) = \frac{(x^2 + 1)(4) - 4x(2x)}{(x^2 + 1)^2} = \frac{4(1 - x^2)}{(x^2 + 1)^2}$$

Critical number: $x = 1$

x-value	Endpoint $x = 0$	Critical $x = 1$
$f(x)$	0	2
Conclusion	Minimum	Maximum

39. $f(x) = \frac{2x}{x^2 + 4}, \ [0, \infty)$

$$f'(x) = \frac{(x^2 + 4)(2) - 2x(2x)}{(x^2 + 4)^2}$$

$$= \frac{8 - 2x^2}{(x^2 + 4)^2}$$

$$= \frac{2(2 - x)(2 + x)}{(x^2 + 4)^2}$$

Critical number: $x = 2$

x-value	Endpoint $x = 0$	Critical $x = 2$
$f(x)$	0	$\frac{1}{2}$
Conclusion	Minimum	Maximum

41. $f(x) = \sqrt{1 + x^3} = (1 + x^3)^{1/2}$, $[0, 2]$

$$f'(x) = \frac{1}{2}(1 + x^3)^{-1/2}(3x^2) = \frac{3x^2}{2(1 + x^3)^{1/2}}$$

$$f''(x) = \frac{2(1 + x^3)^{1/2}(6x) - 3x^2(1 + x^3)^{-1/2}(3x^2)}{4(1 + x^3)}$$

$$= \frac{3x^4 + 12x}{4(1 + x^3)^{3/2}}$$

$$f'''(x) = \frac{4(1 + x^3)^{3/2}(12x^3 + 12) - 4(3x^4 + 12x)\left[\frac{3}{2}(1 + x^3)^{1/2}(3x^2)\right]}{16(1 + x^3)^3}$$

$$= \frac{4(1 + x^3)(12x^3 + 12) - 18x^2(3x^4 + 12x)}{16(1 + x^3)^{5/2}}$$

$$= -\frac{3(x^6 + 20x^3 - 8)}{8(1 + x^3)^{5/2}}$$

Critical number for f'' in $[0, 2]$: $x \approx 0.73$

x-value	Endpoint $x = 0$	Critical $x = 0.73$	Endpoint $x = 2$
$\lvert f''(x) \rvert$	0	1.47	$\frac{2}{3}$
Conclusion		Maximum	

43. $f(x) = (x + 1)^{2/3}$, $[0, 2]$

$$f'(x) = \frac{2}{3}(x + 1)^{-1/3} = \frac{2}{3(x + 1)^{1/3}}$$

$$f''(x) = \left(\frac{2}{3}\right)\left(-\frac{1}{3}\right)(x + 1)^{-4/3} = -\frac{2}{9(x + 1)^{4/3}}$$

$$f'''(x) = \left(-\frac{2}{9}\right)\left(-\frac{4}{3}\right)(x + 1)^{-7/3} = \frac{8}{27(x + 1)^{7/3}}$$

$$f^{(4)}(x) = \left(\frac{8}{27}\right)\left(-\frac{7}{3}\right)(x + 1)^{10/3} = -\frac{56}{81(x + 1)^{10/3}}$$

$$f^{(5)}(x) = \left(-\frac{56}{81}\right)\left(\frac{10}{3}\right)(x + 1)^{-13/3} = \frac{560}{243(x + 1)^{13/3}}$$

No critical numbers of $f^{(4)}$

x-value	Endpoint $x = 0$	Endpoint $x = 2$
$\lvert f^{(4)}(x) \rvert$	$\frac{56}{81}$	0.018
Conclusion	Maximum	

45. Answers will vary. Sample answer:

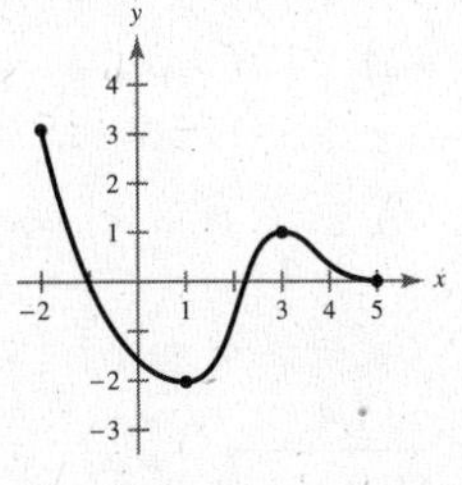

47. $T = 0.0143t^2 - 0.074t + 18$, $1 \le t \le 34$

$T' = 0.0286t - 0.074 = 0$ when $t \approx 2.6$

Critical number: $t \approx 2.6$

Because T is decreasing on $(1, 2.6)$ and increasing on $(2.6, 34)$, the fewest pairs of twins were born mid 1972.

49. $C = \dfrac{3t}{27 + t^3},\ t \geq 0$

(a)

t	0	0.5	1	1.5	2	2.5	3
$C(t)$	0	0.0553	0.1071	0.1481	0.1714	0.1760	0.1667

The concentration reached a maximum at about $t = 2.5$.

(b)

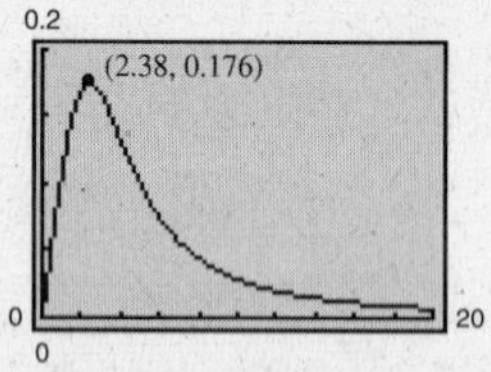

The concentration reached a maximum at $t \approx 2.4$.

(c) $C' = \dfrac{(27 + t^3)(3) - 3t(3t^2)}{(27 + t^3)^2} = \dfrac{81 - 6t^3}{(27 + t^3)^2}$

$C' = 0$ when $t = 3/\sqrt[3]{2} \approx 2.4$.

Because C is increasing on $(0, 2.4)$ and decreasing on $(2.4, \infty)$, the concentration reached a maximum about 2.4 hours.

51. (a) 1970: 2500 per 1000 women

(b) 1985–1990 most rapidly

1975–1980 most slowly

(c) 1970–1975 most rapidly

1980–1985 most slowly

(d) Answers will vary.

Section 3.3 Concavity and the Second-Derivative Test

Skills Review

1. $f(x) = 4x^4 - 9x^3 + 5x - 1$

$f'(x) = 16x^3 - 27x^2 + 5$

$f''(x) = 48x^2 - 54x$

2. $g(s) = (s^2 - 1)(s^2 - 3s + 2)$

$= s^4 - 3s^3 + s^2 + 3s - 2$

$g'(s) = 4s^3 - 9s^2 + 2s + 3$

$g''(s) = 12s^2 - 18s + 2$

3. $g(x) = (x^2 + 1)^4$

$g'(x) = 4(x^2 + 1)^3(2x) = 8x(x^2 + 1)^3$

$g''(x) = 8x\left[3(x^2 + 1)^2(2x)\right] + (x^2 + 1)^3(8)$

$= 8(x^2 + 1)^2(7x^2 + 1)$

4. $f(x) = (x - 3)^{4/3}$

$f'(x) = \dfrac{4}{3}(x - 3)^{1/3}$

$f''(x) = \dfrac{4}{9}(x - 3)^{-2/3} = \dfrac{4}{9(x - 3)^{2/3}}$

Skills Review —*continued*—

5. $h(x) = \dfrac{4x + 1}{5x - 1}$

$h'(x) = \dfrac{(5x - 1)(4) - (4x + 1)(5)}{(5x - 1)^2}$

$= \dfrac{20x - 4 - 20x - 5}{(5x - 1)^2}$

$= -9(5x - 1)^{-2}$

$h''(x) = 18(5x - 1)^{-3}(5) = \dfrac{90}{(5x - 1)^3}$

6. $f(x) = \dfrac{2x - 1}{3x + 2}$

$f'(x) = \dfrac{(3x + 2)(2) - (2x - 1)(3)}{(3x + 2)^2}$

$= \dfrac{6x + 4 - 6x + 3}{(3x + 2)^2}$

$= 7(3x + 2)^{-2}$

$f''(x) = -14(3x + 2)^{-3}(3) = -\dfrac{42}{(3x + 2)^3}$

7. $f(x) = 5x^3 - 5x + 11$

$f'(x) = 15x^2 - 5 = 0$

$15x^2 = 5$

$x^2 = \dfrac{1}{3}$

$x = \pm\dfrac{1}{\sqrt{3}}$

Critical numbers: $x = \pm\dfrac{1}{\sqrt{3}}$

8. $f(x) = x^4 - 4x^3 - 10$

$f'(x) = 4x^3 - 12x^2 = 0$

$4x^2(x - 3) = 0$

$4x^2 = 0 \Rightarrow x = 0$

$x - 3 = 0 \Rightarrow x = 3$

Critical numbers: $x = 0,\ x = 3$

9. $g(t) = \dfrac{16 + t^2}{t} = \dfrac{16}{t} - t$

$g'(t) = -\dfrac{16}{t^2} - 1 = 0$

$t^2 = -16$

No critical numbers

10. $h(x) = \dfrac{x^4 - 50x^2}{8}$

$h'(x) = \dfrac{1}{2}x^3 - \dfrac{25}{2}x = 0$

$\dfrac{1}{2}x(x^2 - 25) = 0$

$\dfrac{1}{2}x(x + 5)(x - 5) = 0$

$\dfrac{1}{2}x = 0 \Rightarrow x = 0$

$x + 5 = 0 \Rightarrow x = -5$

$x - 5 = 0 \Rightarrow x = 5$

Critical numbers: $x = 0,\ x = \pm 5$

1. $y = x^2 - x - 2$

$y' = 2x - 1$

$y'' = 2$

Interval	$-\infty < x < \infty$
Sign of y''	$y'' > 0$
Conclusion	Concave upward

3. $f(x) = \dfrac{x^2 - 1}{2x + 1}$

$$f'(x) = \frac{(2x + 1)(2x) - (x^2 - 1)(2)}{(2x + 1)^2} = \frac{2x^2 + 2x + 2}{(2x + 1)^2} = (2x^2 + 2x + 2)(2x + 1)^{-2}$$

$$\begin{aligned} f''(x) &= (2x^2 + 2x + 2)\left[-2(2x + 1)^{-3}(2)\right] + (2x + 1)^{-2}(4x + 2) \\ &= -8(x^2 + x + 1)(2x + 1)^{-3} + 2(2x + 1)^{-2}(2x + 1) \\ &= 2(2x + 1)^{-3}\left[-4(x^2 + x + 1) + (2x + 1)(2x + 1)\right] \\ &= 2(2x + 1)^{-3}\left[-4x^2 - 4x - 4 + 4x^2 + 4x + 1\right] \\ &= \frac{-6}{(2x + 1)^3} \end{aligned}$$

$f''(x) \neq 0$ for any value of x.

$x = -\frac{1}{2}$ is a discontinuity.

Interval	$-\infty < x < -\frac{1}{2}$	$-\frac{1}{2} < x < \infty$
Sign of f''	$f'' > 0$	$f'' < 0$
Conclusion	Concave upward	Concave downward

5. $f(x) = \dfrac{24}{x^2 + 12} = 24(x^2 + 12)^{-1}$

$$f'(x) = -24(2x)(x^2 + 12)^{-2} = -48x(x^2 + 12)^{-2}$$

$$f''(x) = -48x\left[-2(x^2 + 12)^{-3}(2x)\right] + (x^2 + 12)^{-2}(-48) = \frac{-48(-4x^2 + x^2 + 12)}{(x^2 + 12)^3} = \frac{144(x^2 - 4)}{(x^2 + 12)^3}$$

$f''(x) = 0$ when $x = \pm 2$.

Interval	$-\infty < x < -2$	$-2 < x < 2$	$2 < x < \infty$
Sign of f''	$f'' > 0$	$f'' < 0$	$f'' > 0$
Conclusion	Concave upward	Concave downward	Concave upward

7. $f(x) = -x^3 + 6x^2 - 9x - 1$

$f'(x) = -3x^2 + 12x - 9$

$f''(x) = -6x + 12$

$f''(x) = 0$ when $x = 2$.

Interval	$-\infty < x < 2$	$2 < x < \infty$
Sign of f''	$f'' > 0$	$f'' < 0$
Conclusion	Concave upward	Concave downward

9. $f(x) = 6x - x^2$

$f'(x) = 6 - 2x$

Critical number: $x = 3$

$f''(x) = -2$

$f''(3) = -2 < 0$

$(3, 9)$ is a relative maximum.

11. $f(x) = x^3 - 5x^2 + 7x$

$f'(x) = 3x^2 - 10x + 7 = (3x - 7)(x - 1)$

Critical numbers: $x = 1$, $x = \frac{7}{3}$

$f''(x) = 6x - 10$

$f''(1) = -4 < 0$

$f''\left(\frac{7}{3}\right) = 4 > 0$

So, $(1, 3)$ is a relative maximum and $\left(\frac{7}{3}, \frac{49}{27}\right)$ is a relative minimum.

13. $f(x) = x^{2/3} - 3$

$$f'(x) = \frac{2}{3}x^{-1/3} = \frac{2}{3x^{1/3}}$$

Critical number: $x = 0$

The Second-Derivative Test does not apply, so use the First-Derivative Test to conclude that $(0, -3)$ is a relative minimum.

15. $f(x) = \sqrt{x^2 + 1} = (x + 1)^{1/2}$

$$f'(x) = \frac{1}{2}(x^2 + 1)^{-1/2}(2x) = \frac{x}{(x^2 + 1)^{1/2}}$$

Critical number: $x = 0$

$$f''(x) = \frac{(x^2 + 1)^{1/2} - x\left[\frac{1}{2}(x^2 + 1)^{-1/2}(2x)\right]}{x^2 + 1}$$

$$= \frac{1}{(x^2 + 1)^{3/2}}$$

$f''(0) = 1 > 0$

So, (0, 1) is a relative minimum.

17. $f(x) = \sqrt{9 - x^2} = (9 - x^2)^{1/2}$

$$f'(x) = \frac{1}{2}(9 - x^2)^{-1/2}(-2x) = -\frac{x}{(9 - x^2)^{1/2}}$$

Critical number: $x = 0, x = \pm 3$

$$f''(x) = \frac{(9 - x^2)^{1/2}(-1) - (-x)\left[\frac{1}{2}(9 - x^2)^{-1/2}(-2x)\right]}{9 - x^2}$$

$$= -\frac{9}{(9 - x^2)^{3/2}}$$

$f''(0) = -\frac{1}{3} < 0$

So, (0, 3) is a relative maximum. There are absolute minima at $(\pm 3, 0)$.

19. $f(x) = \dfrac{8}{x^2 + 2} = 8(x^2 + 2)^{-1}$

$$f'(x) = -8(x^2 + 2)^{-2}(2x) = -\frac{16x}{(x^2 + 2)^2}$$

Critical number: $x = 0$

$$f''(x) = \frac{(x^2 + 2)^2(-16) - (-16x)\left[(2)(x^2 + 2)(2x)\right]}{(x^2 + 2)^4}$$

$$= \frac{(x^2 + 2)\left[-16(x^2 + 2) + 64x^2\right]}{(x^2 + 2)^4}$$

$$= \frac{48x^2 - 32}{(x^2 + 2)^3}$$

$f''(0) = -4 < 0$

So, $(0, 4)$ is a relative maximum.

21. $f(x) = \dfrac{x}{x - 1}$

$$f'(x) = \frac{(x - 1) - (x)}{(x - 1)^2} = -\frac{1}{(x - 1)^2}$$

No critical numbers

No relative extrema

23. $f(x) = \frac{1}{2}x^4 - \frac{1}{3}x^3 - \frac{1}{2}x^2$

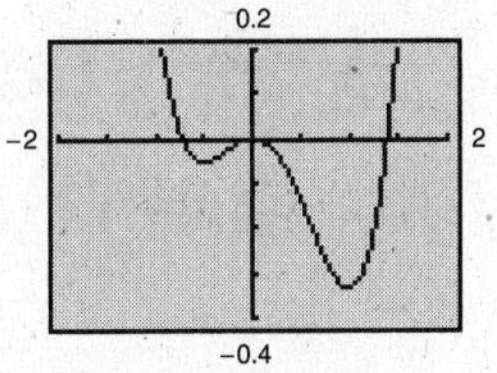

Relative maximum: (0, 0)

Relative minima: $(-0.5, -0.052)$, $(1, -0.333)$

25. $f(x) = 5 + 3x^2 - x^3$

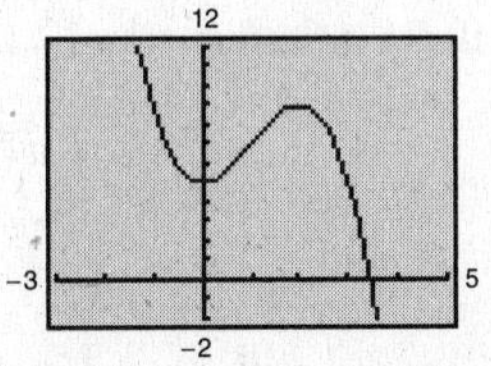

Relative minimum: (0, 5)

Relative maximum: (2, 9)

27. f is increasing so $f' > 0$.

f is concave upward so $f'' > 0$.

29. f is decreasing so $f' < 0$.

f is concave downward so $f'' < 0$.

31. $f(x) = x^3 - 9x^2 + 24x - 18$

$f'(x) = 3x^2 - 18x + 24$

$f''(x) = 6x - 18 = 0$ when $x = 3$.

$f''(x) < 0$ on $(-\infty, 3)$.

$f''(x) > 0$ on $(3, \infty)$.

So, $(3, 0)$ is an inflection point.

33. $f(x) = (x - 1)^3(x - 5)$

$$\begin{aligned} f'(x) &= (x-1)^3 + (x-5)(3)(x-1)^2 \\ &= (x-1)^2[(x-1) + 3(x-5)] \\ &= (x-1)^2(4x-16) \\ &= 4(x-1)^2(x-4) \end{aligned}$$

$$\begin{aligned} f''(x) &= 4(x-1)^2 + (x-4)(8)(x-1) \\ &= 4(x-1)[(x-1) + 2(x-4)] \\ &= 4(x-1)(3x-9) \\ &= 12(x-1)(x-3) = 0 \text{ when } x = 1 \text{ or } x = 3. \end{aligned}$$

$f''(x) > 0$ on $(-\infty, 1)$.

$f''(x) < 0$ on $(1, 3)$.

$f''(x) > 0$ on $(3, \infty)$.

So, $(1, 0)$ and $(3, -16)$ are inflection points.

35. $g(x) = 2x^4 - 8x^3 + 12x^2 + 12x$

$g'(x) = 8x^3 - 24x^2 + 24x + 12$

$g''(x) = 24x^2 - 48x + 24 = 24(x - 1)^2 = 0$ when $x = 1$.

$g'' > 0$ on $(-\infty, 1)$ and $(1, \infty)$.

No inflection points

37. $h(x) = (x - 2)^3(x - 1)$

$h'(x) = (x - 2)^3 + (x - 1)\left[3(x - 2)^2\right] = (4x - 5)(x - 2)^2$

$h''(x) = (x - 2)^2(4) + (4x - 5)(2)(x - 2) = 6(2x - 3)(x - 2) = 0$ when $x = \dfrac{3}{2}$ or $x = 2$.

$h''(x) > 0$ on $\left(-\infty, \dfrac{3}{2}\right)$.

$h''(x) < 0$ on $\left(\dfrac{3}{2}, 2\right)$.

$h''(x) > 0$ on $(2, \infty)$.

So, $\left(\dfrac{3}{2}, -\dfrac{1}{16}\right)$ and $(2, 0)$ are inflection points.

39. $f(x) = x^3 - 12x$

$f'(x) = 3x^2 - 12 = 3(x^2 - 4)$

Critical numbers: $x = \pm 2$

$f''(x) = 6x$

$f''(2) = 12 > 0$

$f''(-2) = -12 < 0$

Relative maximum: $(-2, 16)$

Relative minimum: $(2, -16)$

$f''(x) = 0$ when $x = 0$.

$f''(x) < 0$ on $(-\infty, 0)$.

$f''(x) > 0$ on $(0, \infty)$.

Point of inflection: $(0, 0)$

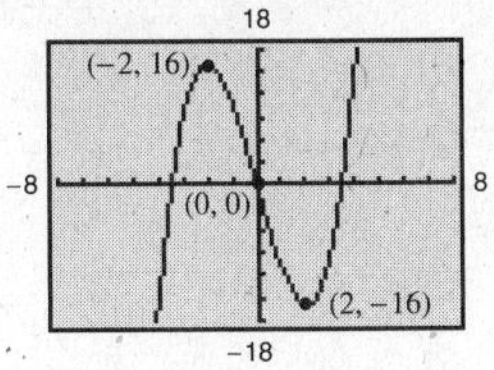

41. $f(x) = x^3 - 6x^2 + 12x$

$f'(x) = 3x^2 - 12x + 12 = 3(x - 2)^2$

Critical number: $x = 2$

$f''(x) = 6(x - 2)$

$f''(2) = 0$

No relative extrema

Because $f'(x) > 0$ when $x \neq 2$ and the concavity changes at $x = 2$, $(2, 8)$ is an inflection point.

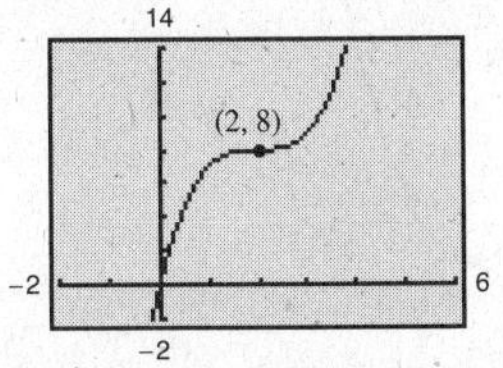

43. $f(x) = \frac{1}{4}x^4 - 2x^2$

$f'(x) = x^3 - 4x = x(x^2 - 4)$

Critical numbers: $x = \pm 2$, $x = 0$

$f''(x) = 3x^2 - 4$

$f''(-2) = 8 > 0$

$f''(0) = -4 < 0$

$f''(2) = 8 > 0$

Relative maximum: $(0, 0)$

Relative minima: $(\pm 2, -4)$

$f''(x) = 3x^2 - 4 = 0$ when $x = \pm\frac{2\sqrt{3}}{3}$.

$f''(x) > 0$ on $\left(-\infty, -\frac{2\sqrt{3}}{3}\right)$.

$f''(x) < 0$ on $\left(-\frac{2\sqrt{3}}{3}, \frac{2\sqrt{3}}{3}\right)$.

$f''(x) > 0$ on $\left(\frac{2\sqrt{3}}{3}, \infty\right)$.

Points of inflection: $\left(-\frac{2\sqrt{3}}{3}, -\frac{20}{9}\right), \left(\frac{2\sqrt{3}}{3}, -\frac{20}{9}\right)$

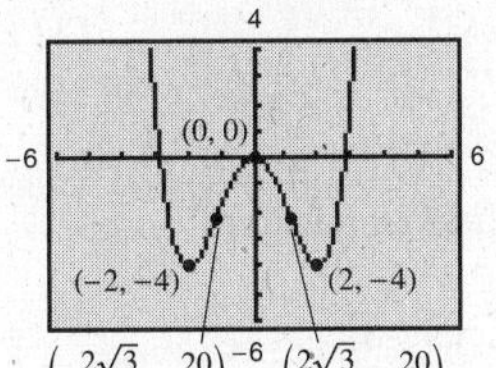

$\left(-\frac{2\sqrt{3}}{3}, -\frac{20}{9}\right)$ $\left(\frac{2\sqrt{3}}{3}, -\frac{20}{9}\right)$

45. $g(x) = (x - 2)(x + 1)^2 = x^3 - 3x - 2$

$g'(x) = (x - 2)[2(x + 1)] + (x + 1)^2 = 3(x^2 - 1)$

Critical numbers: $x = \pm 1$

$g''(x) = 6x$

$g''(-1) = -6 < 0$

$g''(1) = 6 > 0$

Relative maximum: $(-1, 0)$

Relative minimum: $(1, -4)$

$g''(x) = 6x = 0$ when $x = 0$.

$g''(x) < 0$ on $(-\infty, 0)$.

$g''(x) > 0$ on $(0, \infty)$.

Point of inflection: $(0, -2)$

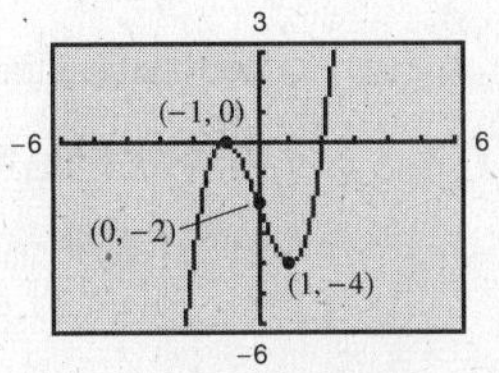

47. $g(x) = x\sqrt{x+3} = x(x+3)^{1/2}$

The domain of g is $[-3, \infty)$.

$$g'(x) = x\left[\frac{1}{2}(x+3)^{-1/2}\right] + (x+3)^{1/2} = \frac{3x+6}{2(x+3)^{1/2}}$$

Critical numbers: $x = -3$, $x = -2$

$$g''(x) = \frac{2(x+3)^{1/2}(3) - (3x+6)\left[2\left(\frac{1}{2}\right)(x+3)^{-1/2}\right]}{4(x+3)}$$

$$= \frac{3(x+4)}{4(x+3)^{3/2}}$$

$g''(-3)$ is undefined.

$g''(-2) = \dfrac{3}{2} > 0$

Relative minimum: $(-2, -2)$

$g''(x) > 0$ for all x in the domain, so there are no points of inflection.

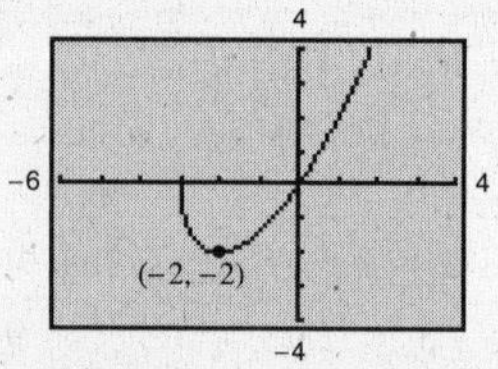

49. $f(x) = \dfrac{4}{1+x^2} = 4(1+x^2)^{-1}$

$$f'(x) = -4(1+x^2)^{-2}(2x) = -\frac{8x}{(1+x^2)^2}$$

Critical number: $x = 0$

$$f''(x) = \frac{(1+x^2)^2(-8) - (-8x)\left[2(1+x^2)(2x)\right]}{(1+x^2)^4}$$

$$= -\frac{8(1-3x^2)}{(1+x^2)^3}$$

$f''(0) = -8 < 0$

Relative maximum: $(0, 4)$

$f''(x) = 1 - 3x^2 = 0$ when $x = \pm\dfrac{\sqrt{3}}{3}$.

$f''(x) > 0$ on $\left(-\infty, -\dfrac{\sqrt{3}}{3}\right)$.

$f''(x) < 0$ on $\left(-\dfrac{\sqrt{3}}{3}, \dfrac{\sqrt{3}}{3}\right)$.

$f''(x) > 0$ on $\left(\dfrac{\sqrt{3}}{3}, \infty\right)$.

Points of inflection: $\left(\dfrac{\sqrt{3}}{3}, 3\right), \left(-\dfrac{\sqrt{3}}{3}, 3\right)$

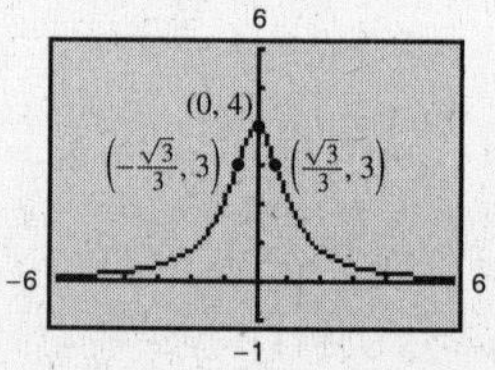

51.

Function	*First Derivative*	*Second Derivative*
$f(2) = 0$	$f'(x) < 0, x < 3$	$f''(x) > 0$
$f(4) = 0$	$f'(3) = 0$	
	$f'(x) > 0, x > 3$	

Answers will vary. Sample answer:

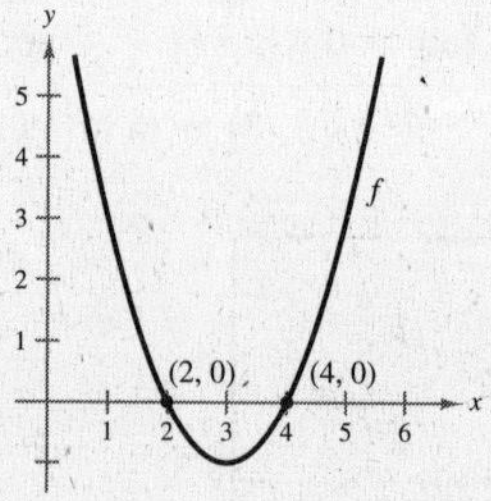

The function has x-intercepts at (2, 0) and (4, 0). On $(-\infty, 3)$, f is decreasing, and on $(3, \infty)$, f is increasing. A relative minimum occurs when $x = 3$. The graph of f is concave upward.

53.

Function	*First Derivative*	*Second Derivative*
$f(0) = 0$	$f'(x) > 0, x < 1$	$f''(x) < 0$
$f(2) = 0$	$f'(1) = 0$	
	$f'(x) < 0, x > 1$	

Answers will vary. Sample answer:

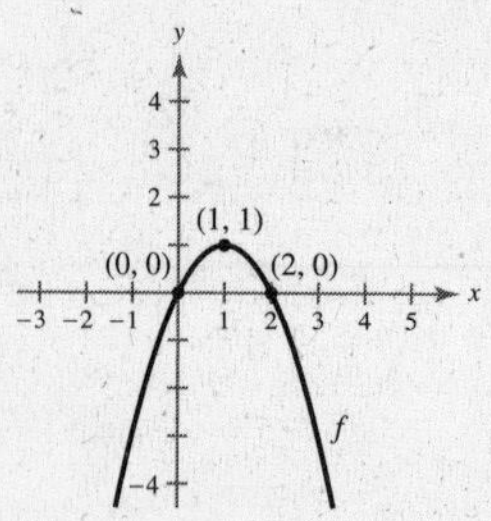

The function has x-intercepts at (0, 0) and (2, 0). On $(-\infty, 1)$, f is increasing and on $(1, \infty)$, f is decreasing. A relative maximum occurs when $x = 1$. The graph of f is concave downward.

55.

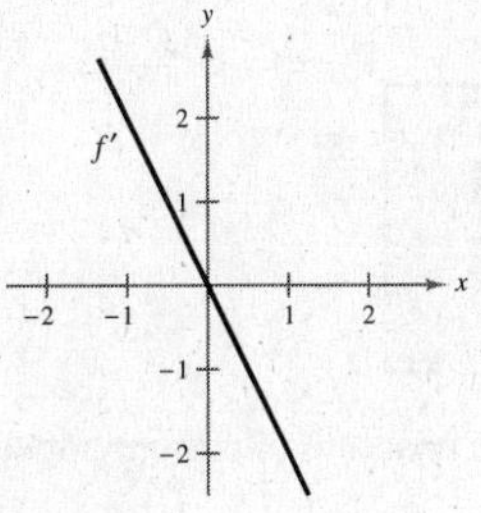

(a) $f'(x) > 0$ on $(-\infty, 0)$ where f is increasing.

(b) $f'(x) < 0$ on $(0, \infty)$ where f is decreasing.

(c) f' is not increasing. f is not concave upward.

(d) f' is decreasing on $(-\infty, \infty)$ where f is concave downward.

57. $f'(x) = 2x + 5$

$f''(x) = 2$

(a) Because the second derivative is positive for all x, f' is increasing on $(-\infty, \infty)$.

(b) Because the second derivative is positive for all x, f is concave upward.

(c) Critical number: $x = -\frac{5}{2}$

$f''\left(-\frac{5}{2}\right) = 2 > 0$

A relative minimum occurs when $x = -\frac{5}{2}$.

$f''(x) > 0$ on $(-\infty, \infty)$.

No points of inflection

(d)

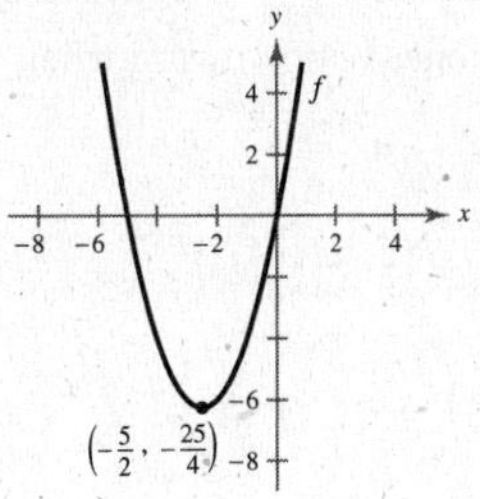

59. $f'(x) = -x^2 + 2x - 1$; Critical number: $x = 1$

$f''(x) = -2x + 2$; $f''(x) = 0$ when $x = 1$

(a) The value of $f''(x)$ is positive on $(-\infty, 1)$ and negative on $(1, \infty)$. So, $f'(x)$ is increasing on $(-\infty, 1)$ and decreasing on $(1, \infty)$.

(b) $f''(x) > 0$ on $(-\infty, 1)$.

$f''(x) < 0$ on $(1, \infty)$.

So, f is concave upward on $(-\infty, 1)$ and concave downward on $(1, \infty)$.

(c) $f''(1) = 0$

The Second-Derivative Test fails. Because $f'(x) \le 0$, f is never increasing and there are no relative extrema.

Concavity changes at $x = 1$, so a point of inflection occurs when $x = 1$.

(d)

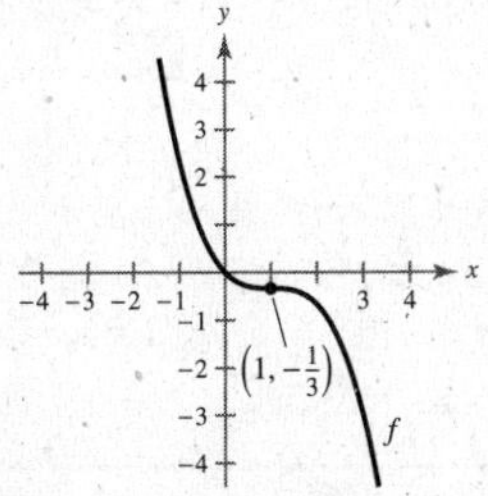

61. $N(t) = -0.12t^3 + 0.54t^2 + 8.22t$, $0 \le t \le 4$

$N'(t) = -0.36t^2 + 1.08t + 8.22$

$N''(t) = -0.72t + 1.08 = 0$ when $t = 1.5$.

$N'''(t) = -0.72$

$N'''(1.5) = -0.72 < 0$

The student is assembling components at the greatest rate when $t = 1.5$, or 8:30 P.M.

63. $f(x) = \frac{1}{2}x^3 - x^2 + 3x - 5, [0, 3]$

$f'(x) = \frac{3}{2}x^2 - 2x + 3$

$f''(x) = 3x - 2$

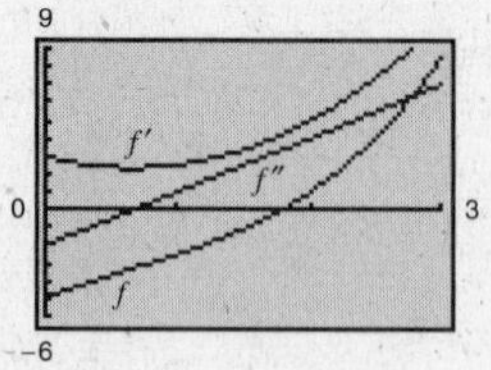

No relative extrema.

Point of inflection: $\left(\frac{2}{3}, -3.30\right)$

f is increasing when f' is positive. f is concave upward when f'' is positive and concave downward when f'' is negative.

65. $f(x) = \dfrac{2}{x^2 + 1} = 2\left(x^2 + 1\right)^{-1}, [-3, 3]$

$f'(x) = -2\left(x^2 + 1\right)^{-2}(2x) = -\dfrac{4x}{\left(x^2 + 1\right)^2}$

$f''(x) = \dfrac{\left(x^2 + 1\right)^2(-4) - (-4x)\left[2\left(x^2 + 1\right)(2x)\right]}{\left(x^2 + 1\right)^4}$

$= \dfrac{12x^2 - 4}{\left(x^2 + 1\right)^3}$

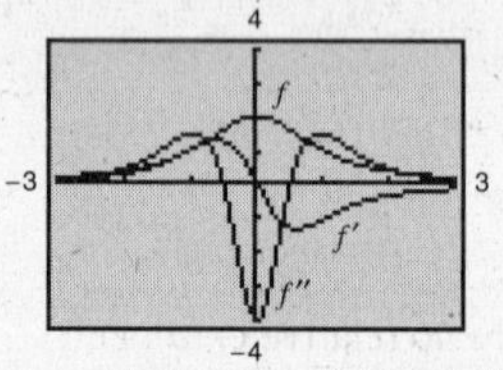

Relative maximum: (0, 2)

Points of inflection: $(0.58, 1.5), (-0.58, 1.5)$

f is increasing when f' is positive and decreasing when f' is negative. f is concave upward when f'' is positive and concave downward when f'' is negative.

67. (a)

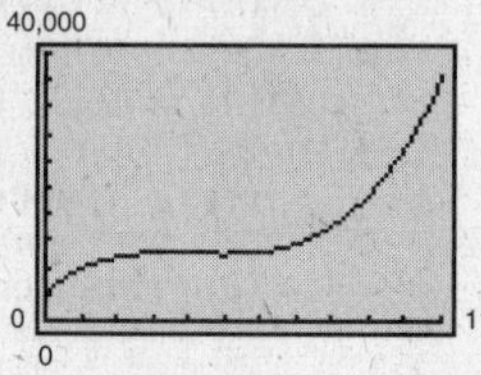

(b) From the graph, the absolute minimum appears to be in November.

(c) From the graph, the absolute maximum appears to be in October.

(d) From the graph, the rate of increase was the greatest in October and the least in April.

69. $d = -20.444t^3 + 152.33t^2 - 266.6t + 1162, 0 \le t \le 5$

(a)

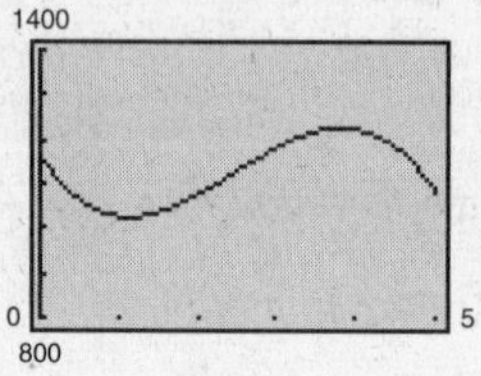

(b) $d' = -61.332t^2 + 304.66t - 266.6$

$d'' = -122.664t + 304.66$

$d'' = 0$ when $t \approx 2.48$

$d'' > 0$ on $(0, 2.48)$, so d'' is concave upward on $(0, 2.48)$.

$d'' < 0$ on $(2.48, 5)$, so d'' is concave downward on $(2.48, 5)$.

(c) From part (b), the point $(2.48, 1125.89)$ is an inflection point.

(d) The number of high school dropouts not in the labor force was increasing at the greatest rate when $t \approx 2.48$, or in 2002.

71. Answers will vary.

Section 3.4 Optimization Problems

Skills Review

1. Let x be the first number and y be the second number.

   $x + \frac{1}{2}y = 12$

2. Let x be the first number and y be the second number.

   $2xy = 24$

3. Let x be the length of the rectangle and y be the width of the rectangle.

   $xy = 24$

Skills Review —continued—

4. Let (x_1, y_1) be the first point and (x_2, y_2) be the second point.

$$\sqrt{(x_1 - x_2)^2 + (y_1 - y_2)^2} = 10$$

5. $y = x^2 + 6x - 9$

$y' = 2x + 6 = 0$

Critical number: $x = -3$

6. $y = 2x^3 - x^2 - 4x$

$y' = 6x^2 - 2x - 4 = 0$

$2(3x^2 - x - 2) = 0$

$2(3x + 2)(x - 1) = 0$

Critical numbers $x = -\frac{2}{3}$, $x = 1$

7. $y = 5x + \dfrac{125}{x}$

$y' = 5 - \dfrac{125}{x^2} = 0$

$\dfrac{5(x^2 - 25)}{x^2} = 0$

$5(x^2 - 25) = 0$

Critical numbers: $x = \pm 5$

8. $y = 3x + \dfrac{96}{x^2}$

$y' = 3 - \dfrac{192}{x^3} = 0$

$\dfrac{3(x^3 - 64)}{x^3} = 0$

$3(x^3 - 64) = 0$

Critical number: $x = 4$

9. $y = \dfrac{x^2 + 1}{x} = x + \dfrac{1}{x}$

$y' = 1 - \dfrac{1}{x^2} = 0$

$\dfrac{x^2 - 1}{x^2} = 0$

$x^2 - 1 = 0$

Critical numbers: $x = \pm 1$

10. $y = \dfrac{x}{x^2 + 9}$

$y' = \dfrac{x^2 + 9 - x(2x)}{(x^2 + 9)^2} = 0$

$\dfrac{9 - x^2}{(x^2 + 9)^2} = 0$

$9 - x^2 = 0$

Critical numbers: $x = \pm 3$

1. Let x be the first number and y be the second number. Then $x + y = 120$ and $y = 120 - x$. The product of x and y is:

$P = xy = x(120 - x)$

$P' = 120 - 2x$

$P'' = -2$

$P' = 0$ when $x = 60$. Because $P''(60) = -2 < 0$, P is a maximum when $x = 60$ and $y = 120 - 60 = 60$.

3. Let x be the first number and y be the second number. Then $x + 2y = 36$ and $x = 36 - 2y$. The product of x and y is:

$P = xy = (36 - 2y)y$

$P' = 36 - 4y$

$P'' = -4$

$P' = 0$ when $y = 9$. Because $P''(9) = -4 < 0$, P is a maximum when $y = 9$ and $x = 36 - 2(9) = 18$.

5. Let x be the first number and y be the second number. Then $xy = 192$ and $y = 192/x$. The sum of x and y is:

$S = x + y = x + \dfrac{192}{x}$

$S' = 1 - \dfrac{192}{x^2} = \dfrac{x^2 - 192}{x^2}$

$S'' = \dfrac{384}{x^3}$

$S' = 0$ when $x = \sqrt{192} = 8\sqrt{3}$. Because $S''(8\sqrt{3}) > 0$, S is minimum when $x = 8\sqrt{3}$ and $y = 192/\sqrt{192} = 8\sqrt{3}$.

7. Let l be the length and w be the width of the rectangle. Then $2l + 2w = 100$ and $w = 50 - l$. The area is:

$A = lw = l(50 - l)$

$A' = 50 - 2l$

$A'' = -2$

$A' = 0$ when $x = 25$. Because $A''(25) = -2 < 0$, A is maximum when $l = 25$ meters and $w = 50 - 25 = 25$ meters.

9. Let l and w be the length and width of the rectangle. Then the area is $lw = 64$ and $w = 64/l$. The perimeter is:

$P = 2l + 2w = 2l + 2\left(\frac{64}{l}\right)$

$P' = 2 - \frac{128}{l^2} = \frac{2(l^2 - 64)}{l^2}$

$P'' = \frac{256}{l^3}$

$P' = 0$ when $l = 8$. Because $P''(8) = \frac{1}{2} > 0$, P is a minimum when $l = 8$ and $w = 64/8 = 8$ feet.

11. The length of fencing is given by $4x + 3y = 200$ and $y = (200 - 4x)/3$. The area of the corrals is:

$A = 2xy = 2x\left(\frac{200 - 4x}{3}\right) = \frac{8}{3}(50x - x^2)$

$A' = \frac{8}{3}(50 - 2x)$

$A'' = -\frac{16}{3}$

$A' = 0$ when $x = 25$. Because $A''(25) = -\frac{16}{3} < 0$, A is maximum when $x = 25$ feet and $y = \frac{100}{3}$ feet.

13. (a) $9 + 9 + 4(3)(11) = 150 \text{ in.}^2$

$6(25) = 150 \text{ in.}^2$

$36 + 36 + 4(6)(3.25) = 150 \text{ in.}^2$

(b) $V = 3(3)(11) = 99 \text{ in.}^3$

$V = 5(5)(5) = 125 \text{ in.}^3$

$V = 6(6)(3.25) = 117 \text{ in.}^3$

(c) Let the base measure x by x, and the height measure y. Then the surface area is $2x^2 + 4xy = 150$ and $y = \frac{1}{2x}(75 - x^2)$. The volume of the solid is:

$V = x^2y = x^2\left[\frac{1}{2x}(75 - x^2)\right] = \frac{75}{2}x - \frac{x^3}{2}$

$V' = \frac{75}{2} - \frac{3x^2}{2}$

$V'' = -3x$

$V' = 0$ when $x = 5$. Because $V''(5) = -15 < 0$, V is maximum when $x = 5$ inches and $y = 5$ inches.

15. (a) Let the base measure x by x and the height measure y. Then the volume is $x^2y = 8000$ and $y = \frac{8000}{x^2}$.

The surface area of the solid is

$S = 2x^2 + 4xy = 2x^2 + 4x\left(\frac{8000}{x^2}\right) = 2x^2 + \frac{32{,}000}{x}$

$S' = 4x - \frac{32{,}000}{x^2}$

$S'' = 4 + \frac{64{,}000}{x^3}$

$S' = 0$ when $x = 20$. Because $S''(20) > 0$, S is minimum when $x = 20$ inches and $y = 20$ inches.

(b) When $x = 20$ and $y = 20$, $S = 2400$ square inches.

17. The volume of the enclosure is $x^2y = 83\frac{1}{3} = \frac{250}{3}$ and $y = \frac{250}{3x^2}$. The surface area of the enclosure is:

$A = 3xy + x^2 = 3x\left(\frac{250}{3x^2}\right) + x^2 = \frac{250}{x} + x^2$

$A' = -\frac{250}{x^2} + 2x = \frac{2x^3 - 250}{x^2}$

$A'' = \frac{500}{x^3} + 2$

$A' = 0$ when $x = 5$. Because $A''(5) > 0$, the surface area is minimum when $x = 5$ meters and

$y = \frac{250}{3(5)^2} = \frac{10}{3}$ meters.

19. Let x be the length of the cut (the height of the box). Then the length of the box is $3 - 2x$ and its width is $2 - 2x$. The volume of the box is:

$V = x(3 - 2x)(2 - 2x) = 4x^3 - 10x^2 + 6x, 0 < x < 1$

$V' = 12x^2 - 20x + 6$

By the Quadratic Formula, $V' = 0$ when $x = (5 \pm \sqrt{7})/6$, but $(5 + \sqrt{7})/6$ is not in the domain.

So the volume is maximum when $x = \frac{5 - \sqrt{7}}{6}$, and the corresponding volume is $\frac{10 + 7\sqrt{7}}{27} \text{ft}^3 \approx 1.056 \text{ ft}^3$.

21. (a) Let x be the number of trees and y be the number of oranges. Use the points $(90, 700)$ and $(91, 675)$ to determine the linear equation relating the number of trees to the yield.

$$y - 700 = \frac{700 - 675}{90 - 91}(x - 90)$$

$$y - 700 = -25(x - 90)$$

$$y = -25x + 2950$$

The number of oranges is

$$N(x) = xy = x(-25x + 2950).$$

$$N(x) = -25x^2 + 2950x$$

$$N'(x) = -50x + 2950$$

$$N''(x) = -50$$

$N'(x) = 0$ when $x = 59$. Because $N''(x) = -50 < 0$, $N(x)$ is maximum when $x = 59$. So, 59 trees should be planted.

(b) When $x = 59$, the maximum yield is 87,025 oranges.

23. The triangles $(0, y), (1, 2), (0, 2)$ and $(1, 2), (x, 0), (1, 0)$ are similar.

(a) $\dfrac{y-2}{0-1} = \dfrac{0-2}{x-1}$

$$y - 2 = \frac{2}{x-1}$$

$$y = 2 + \frac{2}{x-1}$$

$$L = \sqrt{x^2 + y^2}$$

$$= \sqrt{x^2 + \left(2 + \frac{2}{x-1}\right)^2}$$

$$= \sqrt{x^2 + 4 + \frac{8}{x-1} + \frac{4}{(x-1)^2}}, \quad x > 1$$

(b)

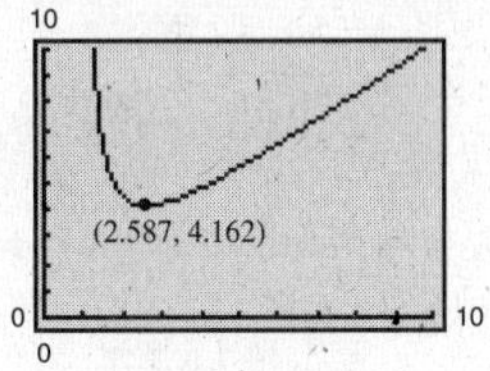

L is a minimum when $x \approx 2.587$ units and $L \approx 4.162$ units.

(c) The area is:

$$A(x) = \frac{1}{2}xy = \frac{1}{2}x\left(2 + \frac{2}{x-1}\right) = x + \frac{x}{x-1}$$

$$A'(x) = 1 + \frac{(x-1) - x}{(x-1)^2} = \frac{(x-1)^2 - 1}{(x-1)^2} = \frac{x(x-2)}{(x-1)^2}$$

$$A''(x) = \frac{2}{(x-1)^3}$$

$A'(x) = 0$ when $x = 2$. Because $A''(2) > 0$, $A(x)$ is minimum when $x = 2$ and $y = 4$.

Vertices: (0, 0), (2, 0), (0, 4)

25. The area is:

$$A = 2xy = 2x\left(r^2 - x^2\right)^{1/2}$$

$$A' = 2\left[x\left(\frac{1}{2}\right)\left(r^2 - x^2\right)^{-1/2}(-2x) + \left(r^2 - x^2\right)^{1/2}\right] = 2\left[\frac{r^2 - 2x^2}{\left(r^2 - x^2\right)^{1/2}}\right]$$

$$A'' = 2\left[\frac{\left(r^2 - x^2\right)^{1/2}(-4x) - \left(r^2 - 2x^2\right)\left(\frac{1}{2}\right)\left(r^2 - x^2\right)^{-1/2}(-2x)}{r^2 - x^2}\right] = 2x\left[\frac{2x^2 - 3r^2}{\left(r^2 - x^2\right)^{3/2}}\right]$$

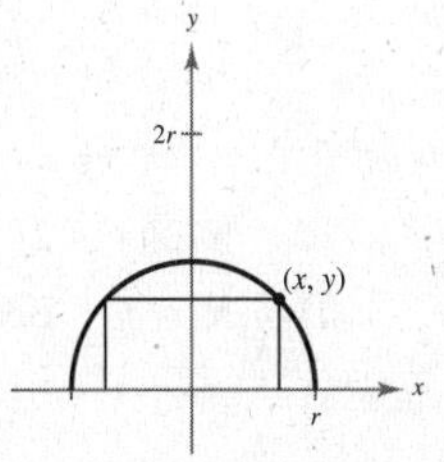

$A' = 0$ when $x = \dfrac{r}{\sqrt{2}} = \dfrac{\sqrt{2}r}{2}$. Because $A''\left(\dfrac{\sqrt{2}r}{2}\right) < 0$, A is maximum when the length is $2x = \sqrt{2}r$ units and the width is $y = \left[r^2 - \left(\dfrac{r}{\sqrt{2}}\right)^2\right]^{1/2} = \dfrac{\sqrt{2}r}{2}$ units.

27. (a) The volume of the cylinder is

$V = \pi r^2 h = (16 \text{ oz})(1.80469 \text{ in.}^3 \text{ per oz}) \approx 28.88 \text{ in.}^3$

which implies that $h = 28.88/(\pi r^2)$. The surface area of the container is:

$$S = 2\pi r^2 + 2\pi rh$$
$$= 2\pi r^2 + 2\pi r\left(\frac{28.88}{\pi r^2}\right)$$
$$= 2\left(\pi r^2 + \frac{28.88}{r}\right)$$
$$S' = 2\left(2\pi r - \frac{28.88}{r^2}\right) = 2\left(\frac{2\pi r^3 - 28.88}{r^2}\right)$$
$$S'' = 2\left(2\pi + \frac{57.76}{r^3}\right)$$

$S' = 0$ when $r = \sqrt[3]{\frac{28.88}{2\pi}} \approx 1.66$ inches. Because $S''(1.66) > 0$, S is minimum when $r \approx 1.66$ inches and $h = \dfrac{28.88}{\pi(1.66)^2} \approx 3.34$ inches.

(b) When $r = 1.66$ and $h = 3.34$, $S \approx 52.15$ square inches.

For each container in Exercises 26 and 27, the height is about twice the length of the radius of the container.

29. The distance between a point (x, y) on the graph and the point $(5, 3)$ is

$$d = \sqrt{(x-5)^2 + (y-3)^2}$$
$$= \sqrt{(x-5)^2 + \left[(x+1)^2 - 3\right]^2}$$

and d can be minimized by minimizing its square $L = d^2$.

$$L = (x-5)^2 + \left[(x+1)^2 - 3\right]^2$$
$$= x^4 + 4x^3 + x^2 - 18x + 29$$
$$L' = 4x^3 + 12x^2 + 2x - 18$$
$$L'' = 12x^2 + 24x + 2$$

$L' = 0$ when $x = 1$. Because $L''(1) > 0$, L is minimum when $x = 1$ and $y = (1+1)^2 = 4$. The point nearest $(5, 3)$ is $(1, 4)$.

31. The distance between a point (x, y) on the graph and the point $(2, 0)$ is

$$d = \sqrt{(x-2)^2 + y^2} = \sqrt{(x-2)^2 + (x-8)}$$

and d can be minimized by minimizing its square $L = d^2$.

$$L = (x-2)^2 + (x-8) = x^2 - 3x - 4$$
$$L' = 2x - 3$$
$$L'' = 2$$

$L' = 0$ when $x = \frac{3}{2}$, however the domain of f is $[8, \infty)$ so consider the end point of the graph $(8, 0)$. Because $L'' > 0$, L' is increasing and L is a minimum when $x = 8$ and $y = 0$. The point nearest $(2, 0)$ is $(8, 0)$.

33. The volume is $\frac{2}{3}\pi r^3 + \pi r^2 h = 12$ and $h = \dfrac{12}{\pi r^2} - \dfrac{2}{3}r$.

The surface area is:

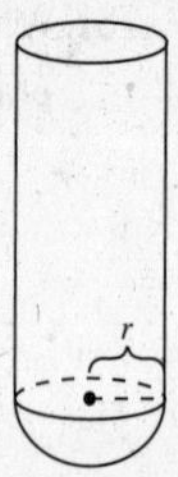

$$A = 2\pi r^2 + 2\pi rh$$
$$= 2\pi r^2 + 2\pi r\left(\frac{12}{\pi r^2} - \frac{2}{3}r\right)$$
$$= 2\pi r^2 + \frac{24}{r} - \frac{4}{3}\pi r^2$$
$$= \frac{2}{3}\pi r^2 + \frac{24}{r}$$
$$A' = \frac{4}{3}\pi r - \frac{24}{r^2} = \frac{4(\pi r^3 - 18)}{3r^2}$$
$$A'' = \frac{4}{3}\pi + \frac{48}{r^3}$$

$A' = 0$ when $r = \sqrt[3]{18/\pi} \approx 1.79$ inches. Because $A''(1.79) > 0$, A is minimum when $r \approx 1.79$ inches.

35. Let x be the length of a side of the square and r be the radius of the circle. Then the combined perimeter is $4x + 2\pi r = 16$ and

$$x = \frac{16 - 2\pi r}{4} = 4 - \frac{\pi r}{2}.$$

The total area is:

$$A = x^2 + \pi r^2 = \left(4 - \frac{\pi r}{2}\right)^2 + \pi r^2$$
$$A' = 2\left(4 - \frac{\pi r}{2}\right)\left(-\frac{\pi}{2}\right) + 2\pi r = \frac{1}{2}\left(\pi^2 r + 4\pi r - 8\pi\right)$$
$$A'' = \frac{1}{2}\left(\pi^2 + 4\pi\right)$$

$A' = 0$ when $r = \dfrac{8\pi}{\pi^2 + 4\pi} = \dfrac{8}{\pi + 4}$. Because $A'' > 0$, A is minimum when $r = \dfrac{8}{\pi + 4}$ units and

$$x = 4 - \frac{\pi\left[8/(\pi+4)\right]}{2} = \frac{16}{\pi + 4}.$$

37. Using the formula Distance = (Rate)(Time), we have $T = D/R$.

$$T = T_{\text{rowed}} + T_{\text{walked}}$$
$$= \frac{D_{\text{rowed}}}{R_{\text{rowed}}} + \frac{D_{\text{walked}}}{R_{\text{walked}}} = \frac{\sqrt{x^2+4}}{2} + \frac{\sqrt{1+(3-x)^2}}{4}$$
$$T' = \frac{x}{2\sqrt{x^2+4}} - \frac{3-x}{4\sqrt{1+(3-x)^2}}$$

By setting $T' = 0$, we have:

$$\frac{x^2}{4(x^2+4)} = \frac{(3-x)^2}{16\left[1+(3-x)^2\right]}$$
$$\frac{x^2}{x^2+4} = \frac{9-6x+x^2}{4(10-6x+x^2)}$$
$$4(x^4-6x^3+10x^2) = (x^2+4)(9-6x+x^2)$$
$$x^4 - 6x^3 + 9x^2 + 8x - 12 = 0$$

Possible rational roots: $\pm1, \pm2, \pm3, \pm4, \pm6, \pm12$

Using a graphing utility, the solution on $[0, 3]$ is $x = 1$ mile.

By testing, we find that $x = 1$ mile.

39. Let w be the number of weeks and p be the price per bushel. Use the points $(1, 30)$ and $(2, 29.20)$ to determine the linear equation relating the price per bushel to the number of weeks that pass.

$$p - 30 = \frac{29.20 - 30}{2-1}(w-1)$$
$$p - 30 = -0.80w + 0.80$$
$$p = -0.80w + 30.80$$

Let b be the number of bushels in the field.

Use the points $(1, 120)$ and $(2, 124)$ to determine the linear equation relating the number of bushels in the field to the number of weeks that pass.

$$b - 120 = \frac{124-120}{2-1}(w-1)$$
$$b - 120 = 4w - 4$$
$$b = 4w + 116$$

The total value of the crop is:

$$R = pb = (-0.80w + 30.80)(4w + 116)$$
$$= -3.2w^2 + 30.4w + 3572.8$$
$$R' = -6.4w + 30.4$$
$$R'' = -6.4$$

$R' = 0$ when $w = 4.75$. Because $R'' < 0$, R is maximum when $w = 4.75$. So the farmer should harvest the strawberries after 4 weeks.

The total bushels harvested is $b(4.75) = 135$ bushels.

The maximum value of the strawberries is $R(4.75) = \$3645$.

41. (a) The perimeter is $4x + 2\pi r = 4$ and $r = \dfrac{2-2x}{\pi}$.

The area is:

$$A(x) = x^2 + \pi r^2$$
$$= x^2 + \pi\left(\frac{2-2x}{\pi}\right)^2$$
$$= x^2 + \frac{(2-2x)^2}{\pi}$$
$$= \left(1 + \frac{4}{\pi}\right)x^2 - \frac{8}{\pi}x + \frac{4}{\pi}$$

(b) Because $x \ge 0$ and $r = \dfrac{2-2x}{\pi} \ge 0$, the domain is $0 \le x \le 1$

(c)

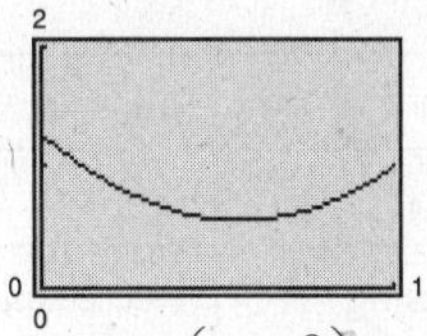

(d) $A'(x) = \left(2 + \dfrac{8}{\pi}\right)x - \dfrac{8}{\pi}$

$A''(x) = 2 + \dfrac{8}{\pi}$

$A'(x) = 0$ when $x = \dfrac{8/\pi}{2 + 8/\pi} = \dfrac{4}{4+\pi}$.

Because $A''(x) > 0$, $A(x)$ is minimum when

$x = \dfrac{4}{4+\pi} \approx 0.56$ feet and

$r = \dfrac{2}{4+\pi} \approx 0.28$ feet. So the total area is minimum when 2.24 feet is used for the square and 1.76 feet is used for the circle.

From the graph, $A(x)$ is maximum when $x = 0$ feet and $r = \dfrac{2}{\pi} \approx 0.64$ feet. So the total area is maximum when all 4 feet is used for the circle.

Mid-Chapter Quiz Solutions

1. $f(x) = x^2 - 6x + 1$

$f'(x) = 2x - 6$

Critical number: $x = 3$

Interval	$-\infty < x < 3$	$3 < x < \infty$
Sign of f'	$f' < 0$	$f' > 0$
Conclusion	Decreasing	Increasing

2. $f(x) = 2x^3 + 12x^2$

$f'(x) = 6x^2 + 24x = 6x(x + 4)$

Critical numbers: $x = 0,\ x = -4$

Interval	$-\infty < x < -4$	$-4 < x < 0$	$0 < x < \infty$
Sign of f'	$f' > 0$	$f' < 0$	$f' > 0$
Conclusion	Increasing	Decreasing	Increasing

3. $f(x) = \dfrac{1}{x^2 + 2} = \left(x^2 + 2\right)^{-1}$

$f'(x) = -\left(x^2 + 2\right)^{-2}(2x) = -\dfrac{2x}{\left(x^2 + 2\right)^2}$

Critical number: $x = 0$

Interval	$-\infty < x < 0$	$0 < x < \infty$
Sign of f'	$f' > 0$	$f' < 0$
Conclusion	Increasing	Decreasing

4. $f(x) = x^3 + 3x^2 - 5$

$f'(x) = 3x^2 + 6x = 3x(x + 2)$

Critical numbers: $x = -2,\ x = 0$

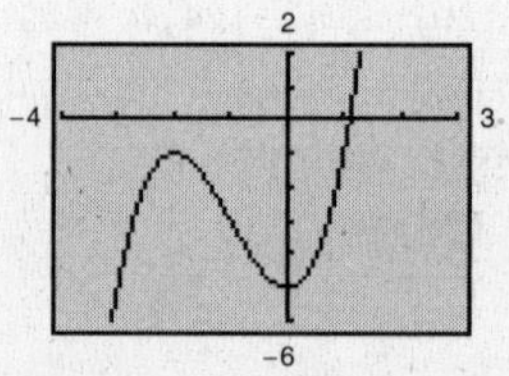

Interval	$-\infty < x < -2$	$-2 < x < 0$	$0 < x < \infty$
Sign of f'	$f' > 0$	$f' < 0$	$f' > 0$
Conclusion	Increasing	Decreasing	Increasing

Relative maximum: $(-2, -1)$

Relative minimum: $(0, -5)$

5. $f(x) = x^4 - 8x^2 + 3$

$f'(x) = 4x^3 - 16x = 4x(x^2 - 4)$

Critical numbers: $x = \pm 2,\ x = 0$

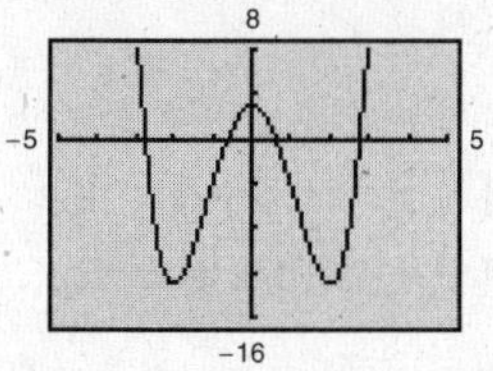

Interval	$-\infty < x < -2$	$-2 < x < 0$	$0 < x < 2$	$2 < x < \infty$
Sign of f'	$f' < 0$	$f' > 0$	$f' < 0$	$f' > 0$
Conclusion	Decreasing	Increasing	Decreasing	Increasing

Relative maximum: $(0, 3)$

Relative minima: $(-2, -13), (2, -13)$

6. $f(x) = 2x^{2/3}$

$f'(x) = \frac{4}{3}x^{-1/3} = \frac{4}{3x^{1/3}}$

Critical number: $x = 0$

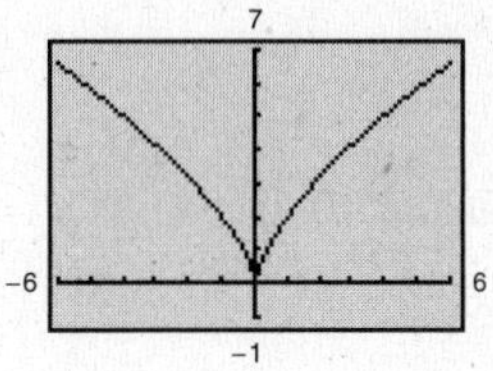

Interval	$-\infty < x < 0$	$0 < x < \infty$
Sign of f'	$f' < 0$	$f' > 0$
Conclusion	Decreasing	Increasing

Relative minimum: $(0, 0)$

7. $f(x) = x^2 + 2x - 8,\ [-2, 1]$

$f'(x) = 2x + 2$

Critical number: $x = -1$

x-value	Endpoint $x = -2$	Critical $x = -1$	Endpoint $x = 1$
$f(x)$	-8	-9	-5
Conclusion		Minimum	Maximum

8. $f(x) = x^3 - 27x,\ [4, 4]$

$f'(x) = 3x^2 - 27$

Critical numbers: $x = \pm 3$

x-value	Endpoint $x = -4$	Critical $x = -3$	Critical $x = 3$	Endpoint $x = 4$
$f(x)$	44	54	-54	44
Conclusion		Maximum	Minimum	

9. $f(x) = \frac{x}{x^2 + 1},\ [0, 2]$

$f'(x) = \frac{(x^2 - 1) - x(2x)}{(x^2 + 1)^2} = \frac{1 - x^2}{(x^2 + 1)^2}$

Critical number: $x = 1$

x-value	Endpoint $x = 0$	Critical $x = 1$	Endpoint $x = 2$
$f(x)$	0	$\frac{1}{2}$	$\frac{2}{5}$
Conclusion	Minimum	Maximum	

10. $f(x) = x^3 - 6x^2 + 7x$

$f'(x) = 3x^2 - 12x + 7$

$f''(x) = 6x - 12$

$f''(x) = 0$ when $x = 2$.

Point of inflection: $(2, -2)$

Interval	$-\infty < x < 2$	$2 < x < \infty$
Sign of $f''(x)$	$f''(x) < 0$	$f''(x) > 0$
Conclusion	Concave downward	Concave upward

11. $f(x) = x^4 - 24x^2$

$f'(x) = 4x^3 - 48x$

$f''(x) = 12x^2 - 48$

$f''(x) = 0$ when $x = \pm 2$.

Interval	$-\infty < x < -2$	$-2 < x < 2$	$2 < x < \infty$
Sign of $f''(x)$	$f''(x) > 0$	$f''(x) < 0$	$f''(x) > 0$
Conclusion	Concave upward	Concave downward	Concave upward

Points of inflection: $(-2, -80), (2, -80)$

12. $f(x) = 2x^3 + 3x^2 - 12x + 16$

$f'(x) = 6x^2 + 6x - 12 = 6(x + 2)(x - 1)$

Critical numbers: $x = -2, x = 1$

$f''(x) = 12x + 6$

$f''(-2) = -18 < 0$

$f''(1) = 18 > 0$

So, $(-2, 36)$ is a relative maximum and $(1, 9)$ is a relative minimum.

13. $f(x) = \dfrac{x^2 + 1}{x} = x + \dfrac{1}{x}$

$f'(x) = 1 - \dfrac{1}{x^2} = \dfrac{x^2 - 1}{x^2}$

Critical numbers: $x = \pm 1$

$f''(x) = \dfrac{2}{x^3}$

$f''(-1) = -2 < 0$

$f''(1) = 2 > 0$

So, $(-1, -2)$ is a relative maximum and $(1, 2)$ is a relative minimum.

14. $C = -166.472t^3 + 1013.92t^2 - 1642.9t + 44{,}316,$
$0 \le t \le 5$

(a) 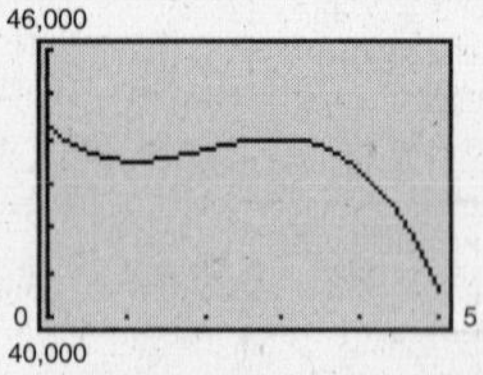

(b) $C' = -499.416t^2 + 2027.84t - 1642.9$

$C'' = 998.832t + 2027.84$

$C'' = 0$ when $t \approx 2.03$

$C'' > 0$ on $(0, 2.03)$, so C is concave upward on $(0, 2.03)$.

$C'' < 0$ on $(2.03, 5)$, so C is concave downward on $(2.03, 5)$.

(c) From part (b), the point (2.03, 43,766.57) is an inflection point. The number of civil rights cases pending was increasing at the greatest rate when $t \approx 2.03$, or in 2002.

15. The perimeter is $x + 2y = 200$ and $y = 100 - \frac{1}{2}x$.

The area is:

$A = xy = x\left(100 - \frac{1}{2}x\right) = 100x - \frac{1}{2}x^2$

$A' = 100 - x$

$A'' = -1$

$A' = 0$ when $x = 100$. Because $A'' = -1 < 0$, A is maximum when $x = 100$ feet and

$y = 100 - \frac{1}{2}(100) = 50$ feet.

16. (a) $P' = 0.8082t^2 - 4.096t - 0.73$

Critical number: $t \approx -0.17$

t-value	Endpoint $t = -1$	Critical $t = -0.17$	Endpoint $t = 5$
$f(t)$	570.31	571.96	550.73
Conclusion		Maximum	Minimum

During 1999 the population was the greatest and during 2005 the population was the least.

(b)

Interval	$-1 < t < -0.17$	$-0.17 < t < 5$
Sign of f'	$f' > 0$	$f' < 0$
Conclusion	Increasing	Decreasing

The population was increasing from 1999 to 2000. The population was decreasing from 2000 to 2005.

Section 3.5 Asymptotes

Skills Review

1. $f(x) = \dfrac{1}{x-5}$

$f(x)$ is undefined when $x = 5$, so the domain is $(-\infty, 5) \cup (5, \infty)$. The range is all real numbers.

2. $f(x) = \dfrac{1}{x^2 - 4}$

$f(x)$ is undefined when $x^2 - 4 = 0$. So,

$x^2 - 4 = 0 \Rightarrow (x-2)(x+2) = 0$

$\Rightarrow x = 2, -2$

So, the domain is $(-\infty, -2) \cup (-2, 2) \cup (2, \infty)$.

The range is all real numbers.

3. $f(x) = \sqrt{x+3}$.

$f(x)$ is undefined when $x + 3 < 0$.

So, $x + 3 < 0 \Rightarrow x < -3$.

So, the domain is all $x \ge -3$. The range is all $y \ge 0$.

4. $f(x) = \dfrac{1}{\sqrt{x-3}}$

$f(x)$ is undefined when $x - 3 \le 0$.

So, $x - 3 \le 0 \Rightarrow x \le 3$.

The domain is $x > 3$. The range is all $y > 0$.

5. $\lim_{x\to 2}(x+1) = 2 + 1 = 3$

6. $\lim_{x\to -1}(3x+4) = 3(-1) + 4 = 1$

7. $\lim_{x\to -3} \dfrac{2x^2 + x - 15}{x+3} = \lim_{x\to -3} \dfrac{(2x-5)(x+3)}{x+3}$
$= \lim_{x\to -3}(2x-5) = 2(-3) - 5 = -11$

8. $\lim_{x\to 2} \dfrac{3x^2 - 8x + 4}{x-2} = \lim_{x\to 2} \dfrac{(3x-2)(x-2)}{x-2}$
$= \lim_{x\to 2}(3x-2) = 3(2) - 2 = 4$

9. $\lim_{x\to 2^+} \dfrac{x^2 - 5x + 6}{x^2 - 4} = \lim_{x\to 2^+} \dfrac{(x-3)(x-2)}{(x+2)(x-2)}$
$= \lim_{x\to 2^+} \dfrac{x-3}{x+2} = \dfrac{2-3}{2+2} = -\dfrac{1}{4}$

10. $\lim_{x\to 1^-} \dfrac{x^2 - 6x + 5}{x^2 - 1} = \lim_{x\to 1^-} \dfrac{(x-5)(x-1)}{(x+1)(x-1)}$
$= \lim_{x\to 1^-} \dfrac{x-5}{x+1} = \dfrac{1-5}{1+1} = -2$

11. $\lim_{x\to 0^+} \sqrt{x} = \sqrt{0} = 0$

12. $\lim_{x\to 1^+} \left(x + \sqrt{x-1}\right) = 1 + \sqrt{1-1} = 1$

1. A horizontal asymptote occurs at $y = 1$ because

$\lim_{x\to\infty} \dfrac{x^2+1}{x^2} = 1, \lim_{x\to -\infty} \dfrac{x^2+1}{x^2} = 1.$

A vertical asymptote occurs at $x = 0$ because

$\lim_{x\to 0^-} \dfrac{x^2+1}{x^2} = \infty, \lim_{x\to 0^+} \dfrac{x^2+1}{x^2} = \infty.$

3. A horizontal asymptote occurs at $y = 1$ because

$\lim_{x\to\infty} \dfrac{x^2-2}{x^2-x-2} = 1, \lim_{x\to -\infty} \dfrac{x^2-2}{x^2-x-2} = 1.$

Vertical asymptotes occur at $x = -1$ and $x = 2$ because

$\lim_{x\to -1^-} \dfrac{x^2-2}{x^2-x-2} = -\infty, \lim_{x\to -1^+} \dfrac{x^2-2}{x^2-x-2} = \infty,$

$\lim_{x\to -2^-} \dfrac{x^2-2}{x^2-x-2} = -\infty, \lim_{x\to -2^+} \dfrac{x^2-2}{x^2-x-2} = \infty.$

5. A horizontal asymptote occurs at $y = \dfrac{3}{2}$ because

$\lim_{x\to\infty} \dfrac{3x^2}{2(x^2+1)} = \dfrac{3}{2}$ and $\lim_{x\to -\infty} \dfrac{3x^2}{2(x^2+1)} = \dfrac{3}{2}.$

The graph has no vertical asymptotes because the denominator is never zero.

7. A horizontal asymptote occurs at $y = \frac{1}{2}$ because

$\lim_{x\to\infty} \dfrac{x^2-1}{2x^2-8} = \dfrac{1}{2}$ and $\lim_{x\to -\infty} \dfrac{x^2-1}{2x^2-8} = \dfrac{1}{2}.$

Vertical asymptotes occur at $x = \pm 2$ because

$\lim_{x\to 2^-} \dfrac{x^2-1}{2x^2-8} = -\infty, \lim_{x\to 2^+} \dfrac{x^2-1}{2x^2-8} = \infty,$

$\lim_{x\to -2^-} \dfrac{x^2-1}{2x^2-8} = \infty, \lim_{x\to -2^+} \dfrac{x^2-1}{2x^2-8} = -\infty.$

9. The graph of f has a horizontal asymptote at $y = 3$. It matches graph (d).

11. The graph of f has a horizontal asymptote at $y = 2$. It matches graph (a).

13. $\lim_{x\to-2^-} \dfrac{1}{(x+2)^2} = \infty$

15. $\lim_{x\to3^+} \dfrac{x-4}{x-3} = -\infty$

17. $\lim_{x\to4^-} \dfrac{x^2}{x^2-16} = -\infty$

19. $\lim_{x\to0^-} \left(1 + \dfrac{1}{x}\right) = 1 + (-\infty) = -\infty$

21.

x	10^0	10^1	10^2	10^3	10^4	10^5	10^6
$f(x)$	2.000	0.348	0.101	0.032	0.010	0.003	0.001

$\lim_{x\to\infty} f(x) = 0$

23.

x	10^0	10^1	10^2	10^3	10^4	10^5	10^6
$f(x)$	0	49.5	49.995	49.99995	50	50	50

$\lim_{x\to\infty} f(x) = 50$

25.

x	-10^6	-10^4	-10^2	10^0	10^2	10^4	10^6
$f(x)$	−2	−2	−1.9996	0.8944	1.9996	2	2

$\lim_{x\to\infty} f(x) = 2$, $\lim_{x\to-\infty} f(x) = -2$

27. (a) $h(x) = \dfrac{5x^3 - 3}{x^2}$

$\lim_{x\to\infty} h(x) = \infty$

(b) $h(x) = \dfrac{5x^3 - 3}{x^3}$

$\lim_{x\to\infty} h(x) = 5$

(c) $h(x) = \dfrac{5x^3 - 4}{x^4}$

$\lim_{x\to\infty} h(x) = 0$

29. (a) $\lim_{x\to\infty} \dfrac{x^2+2}{x^3-1} = 0$

(b) $\lim_{x\to\infty} \dfrac{x^2+2}{x^2-1} = 1$

(c) $\lim_{x\to\infty} \dfrac{x^2+2}{x-1} = \infty$

31. $\lim_{x\to\infty} \dfrac{4x-3}{2x+1} = \dfrac{4}{2} = 2$

33. $\lim_{x\to\infty} \dfrac{3x}{4x^2-1} = 0$

35. $\lim_{x\to-\infty} \dfrac{5x^2}{x+3} = -\infty$

37. $\lim_{x\to\infty} \left(2x - \dfrac{1}{x^2}\right) = \infty - 0 = \infty$

39. $\lim_{x\to-\infty} \left(\dfrac{2x}{x-1} + \dfrac{3x}{x+1}\right) = 2 + 3 = 5$

41. $y = \dfrac{3x}{1-x}$

Intercept: $(0, 0)$

Horizontal asymptote: $y = -3$

Vertical asymptote: $x = 1$

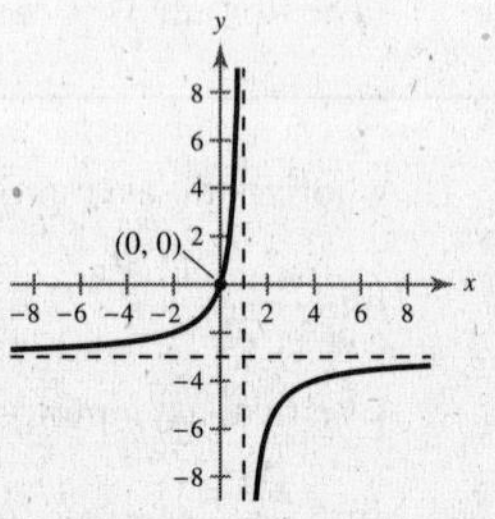

$y' = \dfrac{(1-x)3 - 3x(-1)}{(1-x)^2} = \dfrac{1}{(1-x)^2}$

$y' \neq 0$ so there are no relative extrema.

43. $f(x) = \dfrac{x^2}{x^2 + 9}$

Intercept: $(0, 0)$

Horizontal asymptote: $y = 1$

$$f'(x) = \frac{(x^2 + 9)(2x) - x^2(2x)}{(x^2 + 9)^2} = \frac{18x}{(x^2 + 9)^2}$$

The critical number is $x = 0$ and by the First-Derivative Test $(0, 0)$ is a relative minimum.

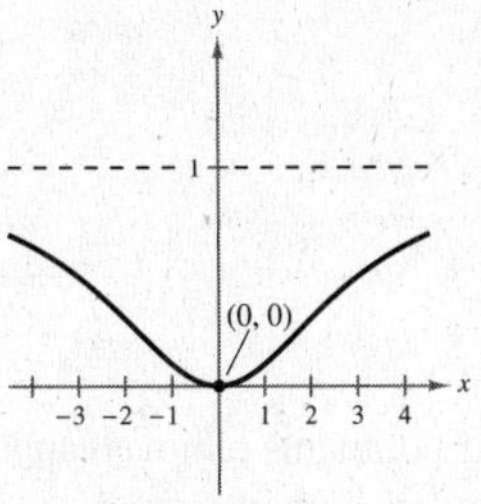

45. $g(x) = \dfrac{x^2}{x^2 - 16}$

Intercept: $(0, 0)$

Horizontal asymptote: $y = 1$

Vertical asymptotes: $x = \pm 4$

$$g'(x) = \frac{(x^2 - 16)(2x) - x^2(2x)}{(x^2 - 16)^2} = \frac{-32x}{(x^2 - 16)^2}$$

The critical number is $x = 0$ and by the First-Derivative Test $(0, 0)$ is a relative maximum.

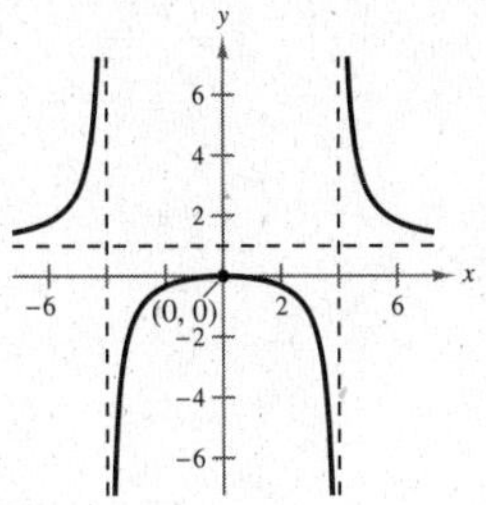

47. $x = \dfrac{4}{y^2}$

No intercepts

Horizontal asymptote: $y = 0$

Vertical asymptote: $x = 0$

$$1 = -\frac{8}{y^3}\frac{dy}{dx}$$

$$\frac{dy}{dx} = -\frac{y^3}{8}$$

$$\frac{dy}{dx} = 0 \text{ when } y = 0.$$

Because $y \neq 0$, there are no relative extrema.

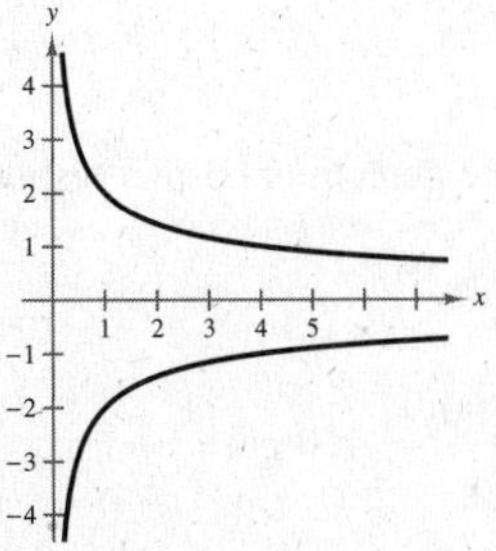

49. $y = \dfrac{2x}{1 - x}$

Intercept: $(0, 0)$

Horizontal asymptote: $y = -2$

Vertical asymptote: $x = 1$

$$y' = \frac{(1 - x)(2) - (2x)(-1)}{(1 - x)^2} = \frac{2}{(1 - x)^2}$$

$y' \neq 0$ so there are no relative extrema.

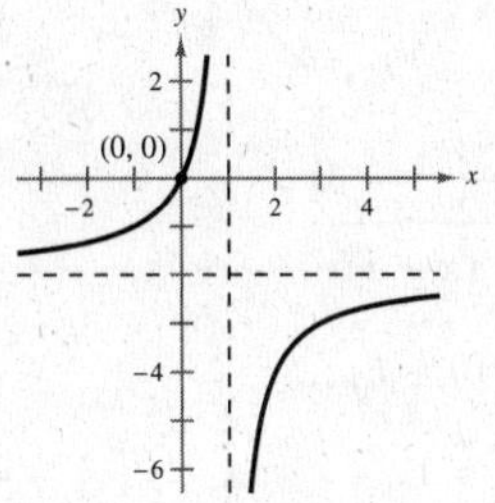

51. $y = 1 - 3x^{-2}$

x-intercepts: $(\pm\sqrt{3}, 0)$

Horizontal asymptote: $y = 1$

Vertical asymptote: $x = 0$

$$y' = \frac{6}{x^3}$$

$y' \neq 0$ so there are no relative extrema.

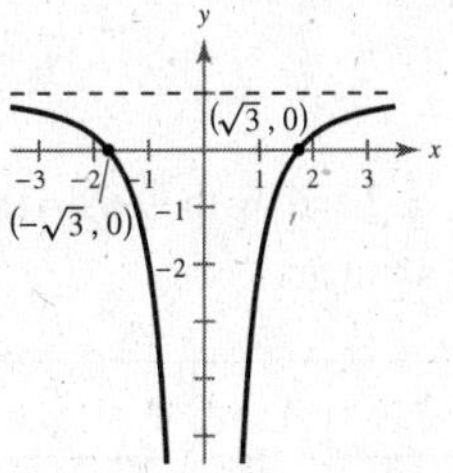

53. $f(x) = \dfrac{1}{x^2 - x - 2} = \dfrac{1}{(x + 1)(x - 2)}$

Intercept: $\left(0, -\dfrac{1}{2}\right)$

Horizontal asymptote: $y = 0$

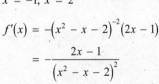

Vertical asymptotes: $x = -1, x = 2$

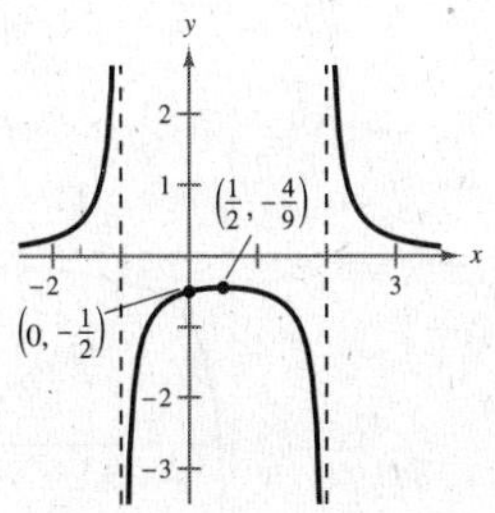

$$f'(x) = -(x^2 - x - 2)^{-2}(2x - 1)$$

$$= -\frac{2x - 1}{(x^2 - x - 2)^2}$$

The critical number is $x = \dfrac{1}{2}$ and by the First-Derivative Test $\left(\dfrac{1}{2}, -\dfrac{4}{9}\right)$ is a relative maximum.

55. $g(x) = \dfrac{x^2 - x - 2}{x - 2} = \dfrac{(x - 2)(x + 1)}{x - 2}$

$= x + 1 \text{ for } x \neq 2$

Intercepts: $(-1, 0), (0, 1)$

No asymptotes

$g'(x) = 1$ for $x \neq 2$

$g'(x) \neq 0$ so there are no relative extrema.

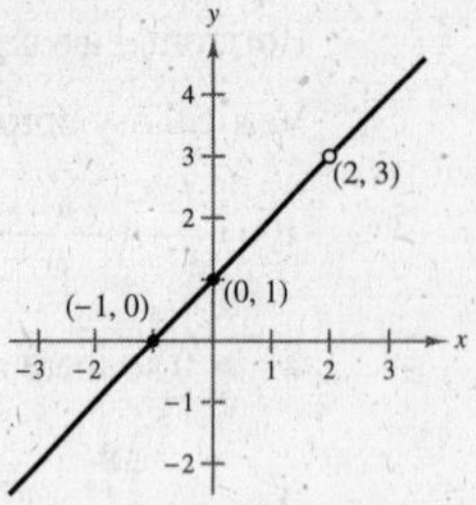

57. $y = \dfrac{2x^2 - 6}{x^2 - 2x + 1} = \dfrac{2(x^2 - 3)}{(x - 1)^2}$

Intercepts: $(\pm\sqrt{3}, 0), (0, -6)$

Vertical asymptote: $x = 1$

Horizontal asymptote: $y = 2$

$$y' = \frac{(x^2 - 2x + 1)(4x) - (2x^2 - 6)(2x - 2)}{(x - 1)^4} = \frac{4(3 - x)}{(x - 1)^3}$$

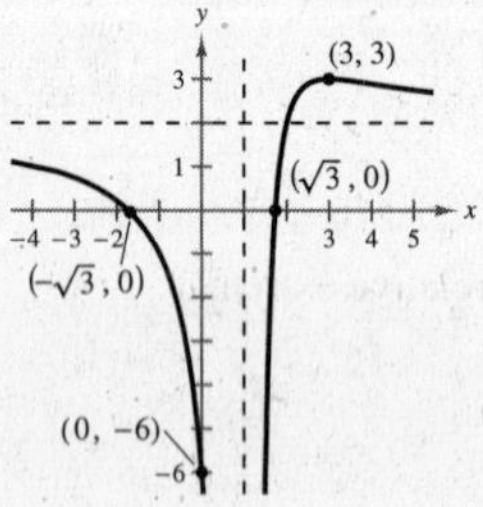

The critical number is $x = 3$ and by the First-Derivative Test $(3, 3)$ is a relative maximum.

59. (a) $\lim\limits_{t \to 0^+} T = 425$

As the time t approaches 0 from the right, the temperature of the apple pie approaches 425° F.

(b) $\lim\limits_{t \to \infty} T = 72$

As the time t increases, the temperature of the apple pie approaches 72°F.

61. $C = \dfrac{80{,}000p}{(100 - p)}, \; 0 \le p < 100$

(a) $C(15) \approx \$14{,}117.65$

$C(50) = \$80{,}000$

$C(90) = \$720{,}000$

(b) $\lim\limits_{x \to 100^-} \dfrac{80{,}000p}{100 - p} = \infty$

As the percent of air pollutants removed approaches 100%, the cost increases without bound.

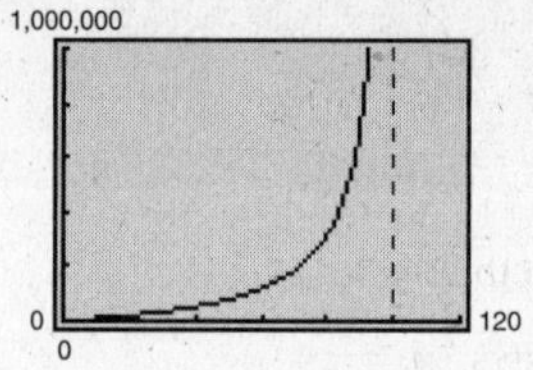

63. (a)

n	1	2	3	4	5	6	7	8	9	10
P	0.50	0.74	0.82	0.86	0.89	0.91	0.92	0.93	0.94	0.95

(b) $\lim\limits_{n \to \infty} \dfrac{0.5 + 0.9(n - 1)}{1 + 0.9(n - 1)} = 1$

(c)

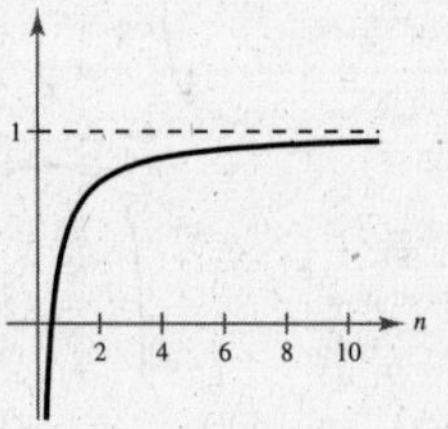

As the number of times the task is performed increases, the percent of correct responses approaches 100%.

65. $N = \dfrac{10(3 + 4t)}{1 + 0.1t} = \dfrac{400t + 300}{t + 10}$

(a) $N(5) \approx 153$ elk

$N(10) = 215$ elk

$N(25) \approx 294$ elk

(b) $\lim\limits_{t \to \infty} \dfrac{400t + 300}{t + 10} = 400$ elk

Section 3.6 Curve Sketching: A Summary

Skills Review

1. A vertical asymptote occurs at $x = 0$ because

$$\lim_{x\to 0^-} \frac{1}{x^2} = \infty \text{ and } \lim_{x\to 0^+} \frac{1}{x^2} = \infty.$$

No horizontal asymptotes.

2. A vertical asymptote occurs at $x = 2$ because

$$\lim_{x\to 2^-} \frac{8}{(x-2)^2} = \infty \text{ and } \lim_{x\to 2^+} \frac{8}{(x-2)^2} = \infty.$$

No horizontal asymptotes.

3. A vertical asymptote occurs at $x = -3$ because

$$\lim_{x\to -3^-} \frac{40x}{x+3} = \infty \text{ and } \lim_{x\to -3^+} \frac{40x}{x+3} = -\infty.$$

A horizontal asymptote occurs at $y = 40$ because

$$\lim_{x\to \infty} \frac{40x}{x+3} = 40 \text{ and } \lim_{x\to -\infty} \frac{40x}{x+3} = 40.$$

4. Vertical asymptotes occur at $x = 1$ and $x = 3$ because

$$\lim_{x\to 1^-} \frac{x^2-3}{x^2-4x+3} = -\infty, \ \lim_{x\to 1^+} \frac{x^2-3}{x^2-4x+3} = \infty,$$

$$\lim_{x\to 3^-} \frac{x^2-3}{x^2-4x+3} = -\infty, \text{ and}$$

$$\lim_{x\to 3^+} \frac{x^2-3}{x^2-4x+3} = \infty.$$ A horizontal asymptote

occurs at $y = 1$ because $\lim_{x\to \infty} \frac{x^2-3}{x^2-4x+3} = 1$ and

$$\lim_{x\to -\infty} \frac{x^2-3}{x^2-4x+3} = 1.$$

5. $f(x) = x^2 + 4x + 2$

$f'(x) = 2x + 4$

Critical number: $x = -2$

Interval	$-\infty < x < -2$	$-2 < x < \infty$
Sign of f'	$f' < 0$	$f' > 0$
Conclusion	Decreasing	Increasing

6. $f(x) = -x^2 - 8x + 1$

$f'(x) = -2x - 8$

Critical number: $x = -4$

Interval	$-\infty < x < -4$	$-4 < x < \infty$
Sign of f'	$f' > 0$	$f' < 0$
Conclusion	Increasing	Decreasing

7. $f(x) = x^3 - 3x + 1$

$f'(x) = 3x^2 - 3$

Critical numbers: $x = \pm 1$

Interval	$-\infty < x < -1$	$-1 < x < 1$	$1 < x < \infty$
Sign of f'	$f' > 0$	$f' < 0$	$f' > 0$
Conclusion	Increasing	Decreasing	Increasing

8. $f(x) = \dfrac{-x^3 + x^2 - 1}{x^2} = -x + 1 - \dfrac{1}{x^2}$

$f'(x) = -1 + \dfrac{2}{x^3} = \dfrac{-x^3 + 2}{x^3}$

Critical number: $x = \sqrt[3]{2}$

Discontinuity: $x = 0$

Interval	$-\infty < x < 0$	$0 < x < \sqrt[3]{2}$	$\sqrt[3]{2} < x < \infty$
Sign of f'	$f' < 0$	$f' > 0$	$f' < 0$
Conclusion	Decreasing	Increasing	Decreasing

Skills Review —*continued*—

9. $f(x) = \dfrac{x-2}{x-1}$

$f'(x) = \dfrac{(x-1)-(x-2)}{(x-1)^2} = \dfrac{1}{(x-1)^2}$

No critical numbers

Discontinuity: $x = 1$

Interval	$-\infty < x < 1$	$1 < x < \infty$
Sign of f'	$f' > 0$	$f' > 0$
Conclusion	Increasing	Increasing

10. $f(x) = -x^3 - 4x^2 + 3x + 2$

$f'(x) = -3x^2 - 8x + 3$

Critical numbers: $x = -3,\ x = \dfrac{1}{3}$

Interval	$-\infty < x < -3$	$-3 < x < \frac{1}{3}$	$\frac{1}{3} < x < \infty$
Sign of f'	$f' < 0$	$f' > 0$	$f' < 0$
Conclusion	Decreasing	Increasing	Decreasing

1. $y = -x^2 - 2x + 3$

$= -(x+3)(x-1)$

$y' = -2x - 2 = -2(x+1)$

$y'' = -2$

Intercepts:

$(0, 3), (1, 0), (-3, 0)$

Relative maximum:

$(-1, 4)$

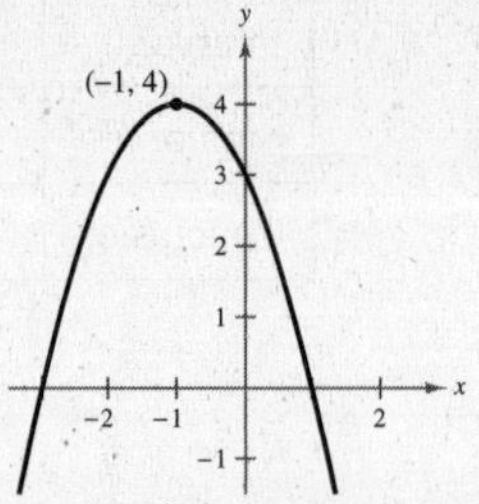

3. $y = x^3 - 4x^2 + 6$

$y' = 3x^2 - 8x = x(3x-8)$

$y'' = 6x - 8 = 2(3x-4)$

Relative maximum: $(0, 6)$

Relative minimum: $\left(\dfrac{8}{3}, -\dfrac{94}{27}\right)$

Point of inflection: $\left(\dfrac{4}{3}, \dfrac{34}{27}\right)$

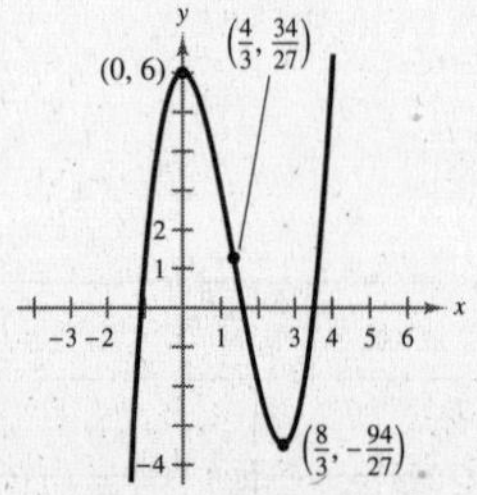

5. $y = 2 - x - x^3$

$y' = -1 - 3x^2$

$y'' = -6x$

No relative extrema

Point of inflection: $(0, 2)$

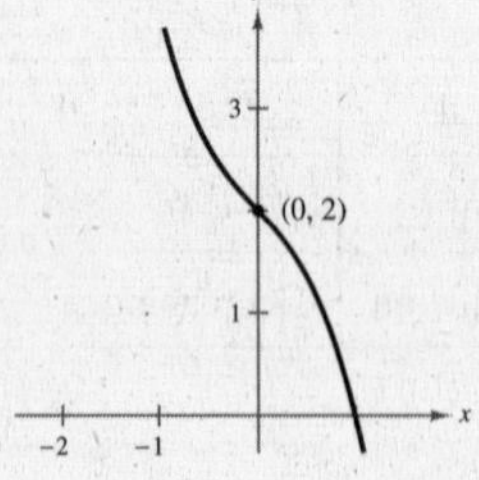

7. $y = 3x^3 - 9x + 1$

$y' = 9x^2 - 9$

$= 9(x-1)(x+1)$

$y'' = 18x$

Relative maximum: $(-1, 7)$

Relative minimum: $(1, -5)$

Point of inflection: $(0, 1)$

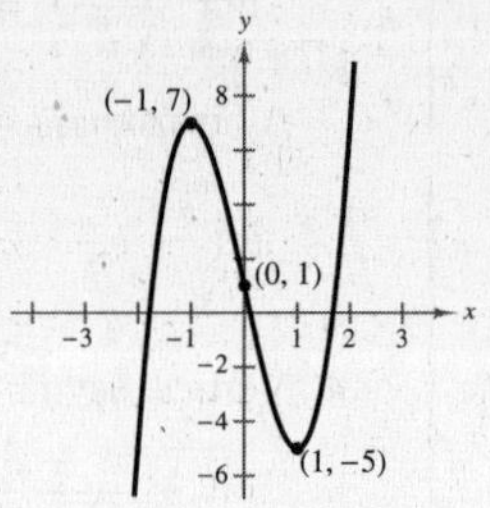

9. $y = 3x^4 + 4x^3 = x^3(3x+4)$

$y' = 12x^3 + 12x^2 = 12x^2(x+1)$

$y'' = 36x^2 + 24x = 12x(3x+2)$

Intercepts: $(0, 0), \left(-\dfrac{4}{3}, 0\right)$

Relative minimum: $(-1, -1)$

Points of inflection: $(0, 0), \left(-\dfrac{2}{3}, -\dfrac{16}{27}\right)$

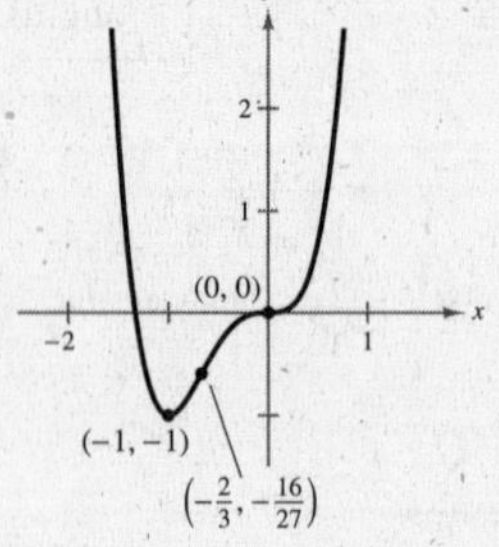

11. $y = x^3 - 6x^2 + 3x + 10 = (x + 1)(x - 2)(x - 5)$

$y' = 3x^2 - 12x + 3 = 3(x^2 - 4x + 1)$

$y'' = 6x - 12 = 6(x - 2)$

Intercepts: $(0, 10), (-1, 0), (2, 0), (5, 0)$

Relative maximum: $\left(2 - \sqrt{3}, 6\sqrt{3}\right)$

Relative minimum: $\left(2 + \sqrt{3}, -6\sqrt{3}\right)$

Point of inflection: $(2, 0)$

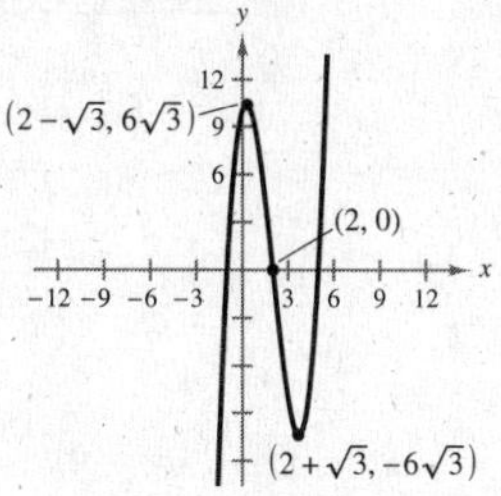

13. $y = x^4 - 8x^3 + 18x^2 - 16x + 5 = (x - 5)(x - 1)^3$

$y' = 4x^3 - 24x^2 + 36x - 16 = 4(x - 4)(x - 1)^2$

$y'' = 12x^2 - 48x + 36 = 12(x - 1)(x - 3)$

Intercepts: $(0, 5), (1, 0), (5, 0)$

Relative minimum: $(4, -27)$

Points of inflection: $(1, 0), (3, -16)$

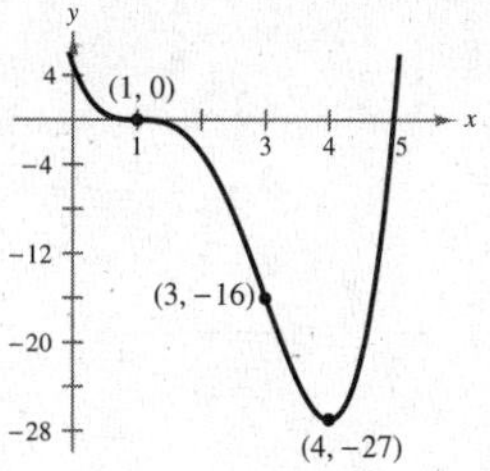

15. $y = x^4 - 4x^3 + 16x = x(x^3 - 4x^2 + 16)$

$y' = 4x^3 - 12x^2 + 16 = 4(x + 1)(x - 2)^2$

$y'' = 12x^2 - 24x = 12x(x - 2)$

Relative minimum: $(-1, -11)$

Points of inflection: $(0, 0), (2, 16)$

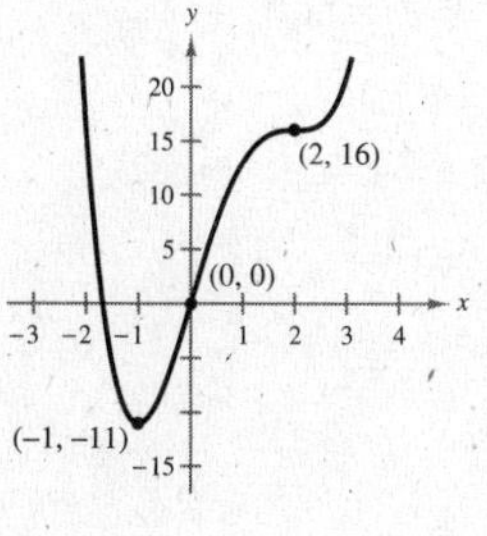

17. $y = x^5 - 5x = x(x^4 - 5)$

$y' = 5x^4 - 5 = 5(x + 1)(x - 1)(x^2 + 1)$

$y'' = 20x^3$

Intercepts: $(0, 0), \left(\pm\sqrt[4]{5}, 0\right)$

Relative maximum: $(-1, 4)$

Relative minimum: $(1, -4)$

Point of inflection: $(0, 0)$

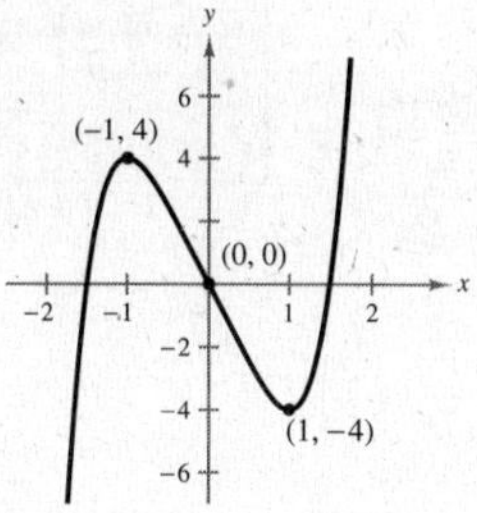

19. $y = \dfrac{x^2 + 1}{x} = x + \dfrac{1}{x}$

$y' = 1 - \dfrac{1}{x^2} = \dfrac{x^2 - 1}{x^2}$

$y'' = \dfrac{2}{x^3}$

No intercepts

Relative maximum: $(-1, -2)$

Relative minimum: $(1, 2)$

No points of inflection

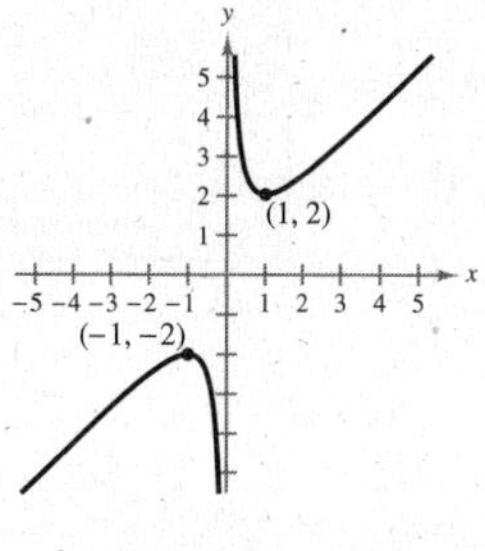

21. $y = \begin{cases} x^2 + 1, & x \le 0 \\ 1 - 2x, & x > 0 \end{cases}$

$y' = \begin{cases} 2x, & x < 0 \\ -2 & x > 0 \end{cases}$

$y'' = \begin{cases} 2, & x < 0 \\ 0 & x > 0 \end{cases}$

Intercept: $\left(\frac{1}{2}, 0\right)$

No relative extrema

No points of inflection

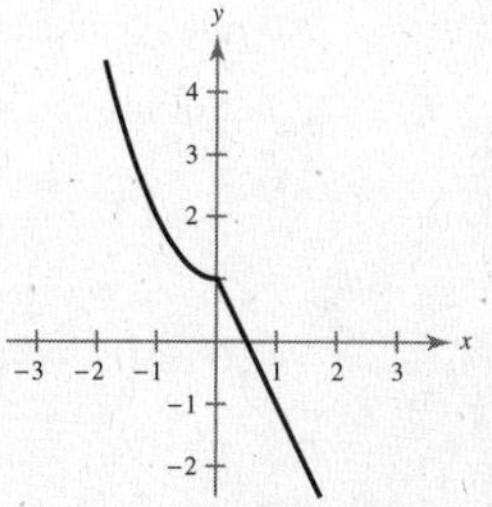

23. $y = \dfrac{x^2}{x^2 + 3}$

$$y' = \frac{(x^2 + 3)(2x) - x^2(2x)}{(x^2 + 3)^2} = \frac{6x}{(x^2 + 3)^2}$$

$$y'' = \frac{(x^2 + 3)^2(6) - 6x[2(x^2 + 3)(2x)]}{(x^2 + 3)^4} = \frac{6(3 - x^2)}{(x^2 + 3)^3}$$

Intercept: $(0, 0)$

Relative minimum: $(0, 0)$

Points of inflection: $\left(\pm\sqrt{3}, \dfrac{1}{2}\right)$

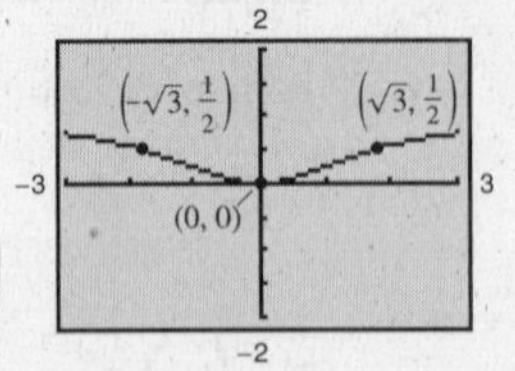

25. $y = 3x^{2/3} - 2x$

$$y' = 2x^{-1/3} - 2 = 2\left(\frac{1}{x^{1/3}} - 1\right)$$

$$y'' = -\frac{2}{3x^{4/3}}$$

Intercepts: $(0, 0), \left(\dfrac{27}{8}, 0\right)$

Relative maximum: $(1, 1)$

Relative minimum: $(0, 0)$

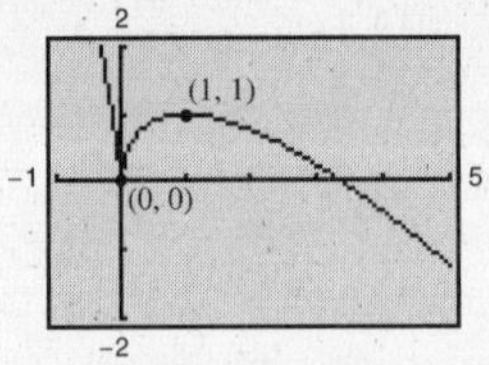

27. $y = 1 - x^{2/3}$

$$y' = -\frac{2}{3}x^{-1/3} = -\frac{2}{3x^{1/3}}$$

$$y'' = \frac{2}{9}x^{-4/3} = \frac{2}{9x^{4/3}}$$

Intercepts: $(0, 1), (\pm 1, 0)$

Relative maximum: $(0, 1)$

No points of inflection

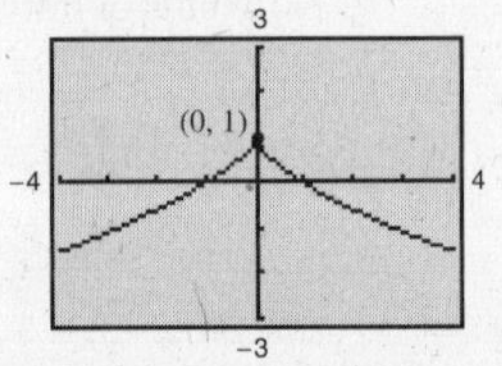

29. $y = x^{1/3} + 1$

$$y' = \frac{1}{3}x^{-2/3} = \frac{1}{3x^{2/3}}$$

$$y'' = -\frac{2}{9}x^{-5/3} = -\frac{2}{9x^{5/3}}$$

Intercepts: $(0, 1), (-1, 0)$

No relative extrema

Point of inflection: $(0, 1)$

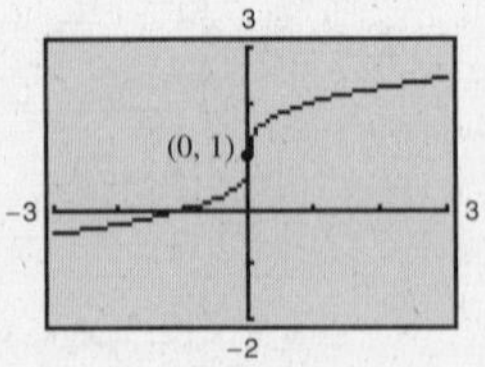

31. $y = x^{5/3} - 5x^{2/3} = x^{2/3}(x - 5)$

$$y' = \frac{5}{3}x^{2/3} - \frac{10}{3}x^{-1/3} = \frac{5(x - 2)}{3x^{1/3}}$$

$$y'' = \frac{10}{9}x^{-1/3} + \frac{10}{9}x^{-4/3} = \frac{10(x + 1)}{9x^{4/3}}$$

Intercepts: $(0, 0), (5, 0)$

Relative maximum: $(0, 0)$

Relative minimum: $\left(2, -3\sqrt[3]{4}\right)$

Point of inflection: $(-1, -6)$

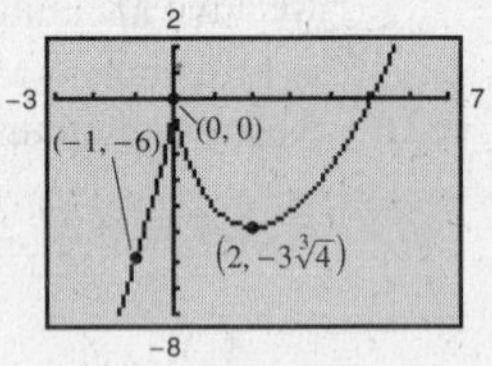

33. $y = x\sqrt{x^2 - 9}, \ |x| \geq 3$

$$y' = (x^2 - 9)^{1/2} + x\left[\frac{1}{2}(x^2 - 9)^{-1/2}(2x)\right] = \frac{2x^2 - 9}{(x^2 - 9)^{1/2}}$$

$$y'' = \frac{(x^2 - 9)^{1/2}(4x) - (2x^2 - 9)\left[\frac{1}{2}(x^2 - 9)^{-1/2}(2x)\right]}{x^2 - 9}$$

$$= \frac{x(2x^2 - 27)}{(x^2 - 9)^{3/2}}$$

Intercepts: $(\pm 3, 0)$

No relative extrema

Points of inflection:

$$\left(-\frac{3\sqrt{6}}{2}, -\frac{9\sqrt{3}}{2}\right), \left(\frac{3\sqrt{6}}{2}, \frac{9\sqrt{3}}{2}\right)$$

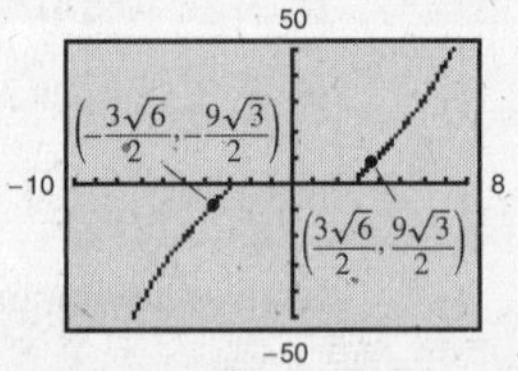

35. $y = \dfrac{5 - 3x}{x - 2}$

$$y' = \frac{(x - 2)(-3) - (5 - 3x)}{(x - 2)^2} = \frac{1}{(x - 2)^2}$$

$$y'' = -2(x - 2)^{-3} = -\frac{2}{(x - 2)^3}$$

Intercepts: $\left(0, -\dfrac{5}{2}\right), \left(\dfrac{5}{3}, 0\right)$

No relative extrema

No points of inflection

Vertical asymptote: $x = 2$

Horizontal asymptote: $y = -3$

Domain: $(-\infty, 2) \cup (2, \infty)$

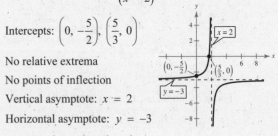

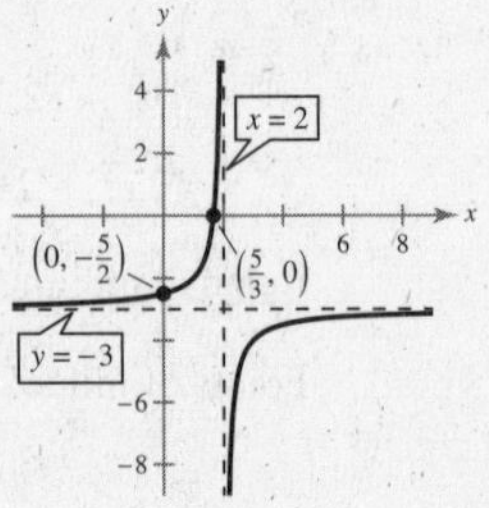

37. $y = \dfrac{2x}{x^2 - 1}$

$$y' = \frac{(x^2 - 1)(2) - 2x(2x)}{(x^2 - 1)^2} = -\frac{-2(x^2 + 1)}{(x^2 - 1)^2}$$

$$y'' = \frac{(x^2 - 1)^2(-4x) - (-2x^2 - 2)[2(x^2 - 1)(2x)]}{(x^2 - 1)^4}$$

$$= \frac{4x(x^2 + 3)}{(x^2 - 1)^3}$$

Intercept: $(0, 0)$

No relative extrema

Point of inflection: $(0, 0)$

Horizontal asymptote: $y = 0$

Vertical asymptotes: $x = \pm 1$

Domain: $(-\infty, -1) \cup (-1, 1) \cup (1, \infty)$

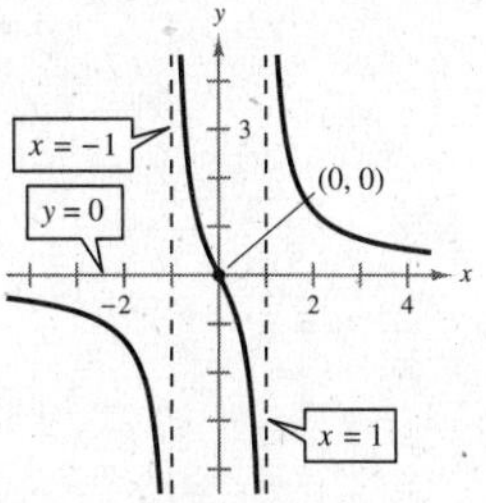

39. $y = x\sqrt{4 - x}$

$$y' = \frac{8 - 3x}{2\sqrt{4 - x}}$$

$$y'' = \frac{3x - 16}{4(4 - x)^{3/2}}$$

Intercepts: $(0, 0)$, $(4, 0)$

Relative maximum: $\left(\dfrac{8}{3}, \dfrac{16}{3\sqrt{3}}\right)$

No points of inflection

No asymptotes

Domain: $(-\infty, 4]$

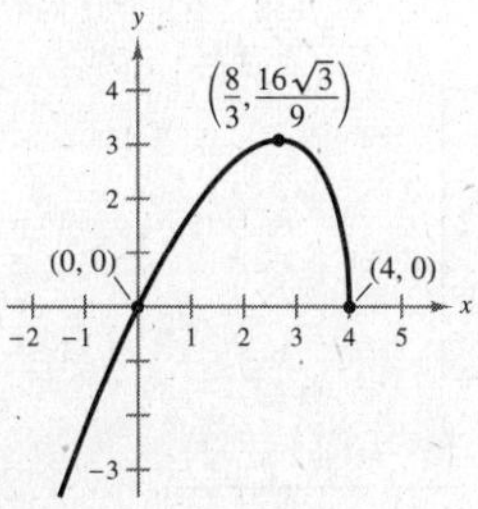

41. $y = \dfrac{x - 3}{x} = 1 - \dfrac{3}{x}$

$$y' = \frac{3}{x^2}$$

$$y'' = -\frac{6}{x^3}$$

Intercept: $(3, 0)$

No relative extrema

Horizontal asymptote: $y = 1$

Vertical asymptote: $x = 0$

No points of inflection

Domain: $(-\infty, 0) \cup (0, \infty)$

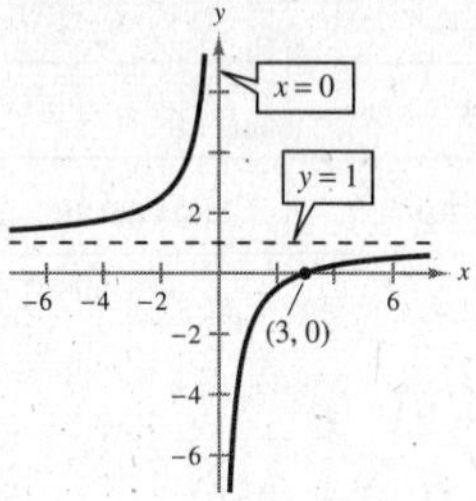

43. $y = \dfrac{x^3}{x^3 - 1}$

$$y' = \frac{(x^3 - 1)(3x^2) - x^3(3x^2)}{(x^3 - 1)^2} = -\frac{3x^2}{(x^3 - 1)^2}$$

$$y'' = \frac{(x^3 - 1)^2(-6x) - (-3x^2)[2(x^3 - 1)(3x^2)]}{(x^3 - 1)^4}$$

$$= \frac{6x(2x^3 + 1)}{(x^3 - 1)^3}$$

Intercept: $(0, 0)$

No relative extrema

Points of inflection: $(0, 0)$, $\left(-\dfrac{1}{\sqrt[3]{2}}, \dfrac{1}{3}\right)$

Horizontal asymptote: $y = 1$

Vertical asymptote: $x = 1$

Domain: $(-\infty, 1) \cup (1, \infty)$

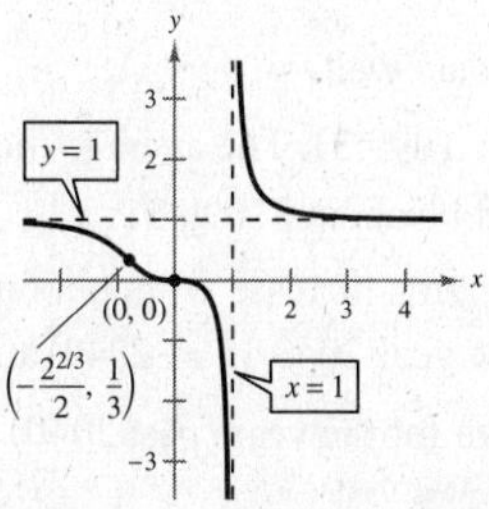

45. Answers will vary. Sample answer:

Let $a = -1$, $b = 1$, $c = 1$, and $d = 1$ so

$f(x) = -x^3 + x^2 + x + 1$.

47. Because $f'(x) = 2$, the graph of f is a line with a slope of 2.

Answers will vary.

Sample answer:

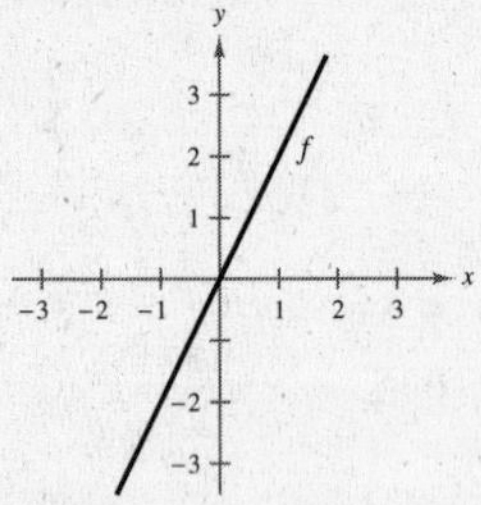

49. $f''(x) = 2 > 0$ so $f(x)$ is a concave upward parabola.

Answers will vary. Sample answer:

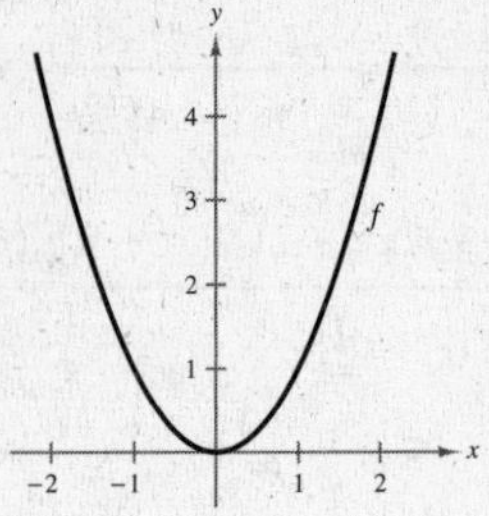

51.

Interval	$-\infty < x < -1$	$-1 < x < 0$	$0 < x < \infty$
Sign of f'	$f' > 0$	$f' < 0$	$f' > 0$
Conclusion	Increasing	Decreasing	Increasing

Intercepts: $(-2, 0), (0, 0)$

Relative maximum: $(-1, f(-1))$

Relative minimum: $(0, f(0)) = (0, 0)$

Answers will vary.

Sample answer:

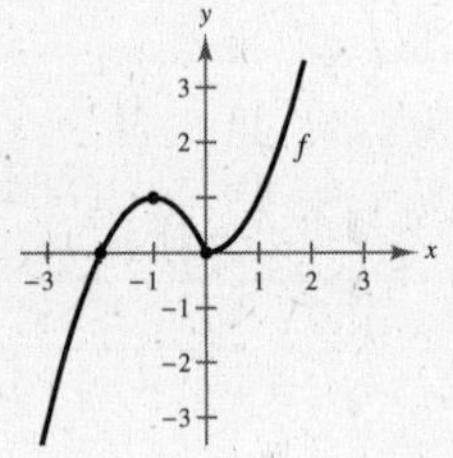

53. Answers will vary. Sample answer:

$$f(x) = \frac{1}{x - 5}$$

55. (a)

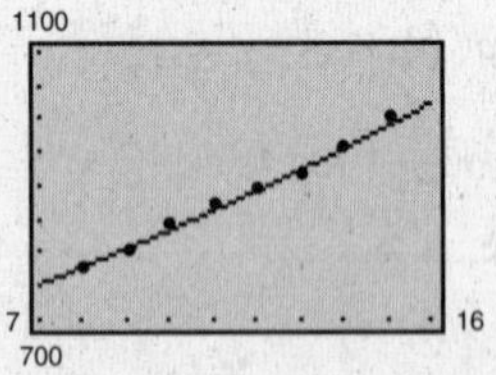

The model fits the data well.

(b) When $t = 18$, $B \approx 1099.31$. The average monthly benefit in 2008 will be about \$1099.31.

(c) No, because the benefits increase without bound as time approaches the year 2040 $(t = 2040)$ and the benefits are negative for the years past 2040.

57.

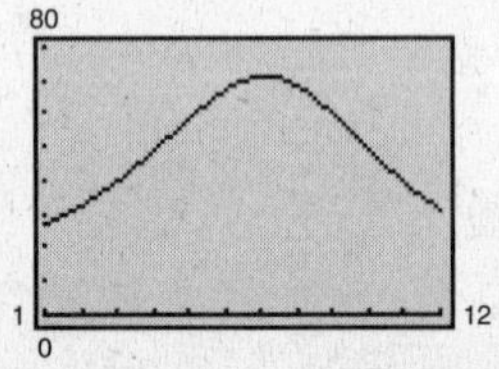

Absolute maximum: $(7.1, 71.2)$

Absolute minimum: $(1, 26.7)$

The model predicts the average monthly temperature to be lowest in January $(26.7°\text{F})$ and highest in July $(71.2°\text{F})$.

59.

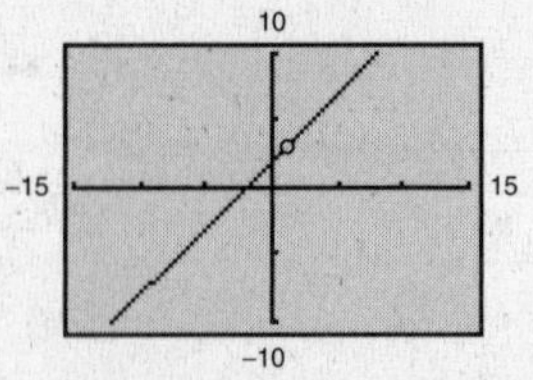

$$g(x) = \frac{x^2 + x - 2}{x - 1} = \frac{(x - 1)(x + 2)}{x - 1} = x + 2, \; x \neq 1$$

The rational function simplifies to a linear function that is undefined at $x = 1$.

Section 3.7 Differentials: Linear Approximation

Skills Review

1. $C = 44 + 0.09x^2$

   $\dfrac{dC}{dx} = 0.18x$

2. $C = 250 + 0.15x$

   $\dfrac{dC}{dx} = 0.15$

3. $R = x\left(1.25 + 0.02\sqrt{x}\right) = 1.25x + 0.02x^{3/2}$

   $\dfrac{dR}{dx} = 1.25 + 0.03\sqrt{x}$

4. $R = x(15.5 - 1.55x) = 15.5x - 1.55x^2$

   $\dfrac{dR}{dx} = 15.5 - 3.1x$

5. $P = -0.03x^{1/3} + 1.4x - 2250$

   $\dfrac{dP}{dx} = \dfrac{-0.01}{x^{2/3}} + 1.4$

6. $P = -0.02x^2 + 25x - 1000$

   $\dfrac{dP}{dx} = -0.04x + 25$

7. $A = \dfrac{1}{4}\sqrt{3}x^2$

   $\dfrac{dA}{dx} = \dfrac{1}{2}\sqrt{3}x$

8. $A = 6x^2$

   $\dfrac{dA}{dx} = 12x$

9. $C = 2\pi r$

   $\dfrac{dC}{dr} = 2\pi$

10. $P = 4w$

    $\dfrac{dP}{dw} = 4$

11. $S = 4\pi r^2$

    $\dfrac{dS}{dr} = 8\pi r$

12. $P = 2x + \sqrt{2}x$

    $\dfrac{dP}{dx} = 2 + \sqrt{2}$

13. $A = \pi r^2$

14. $A = x^2$

15. $V = x^3$

16. $V = \frac{4}{3}\pi r^3$

1. $y = 3x^2 - 4$

$$\frac{dy}{dx} = 6x$$

$$dy = 6x\,dx$$

3. $y = (4x - 1)^3$

$$\frac{dy}{dx} = 3(4x - 1)^2(4)$$

$$dy = 12(4x - 1)^2\,dx$$

5. $y = \sqrt{9 - x^2}$

$$\frac{dy}{dx} = \frac{1}{2}\left(9 - x^2\right)^{-1/2}(-2x) = -\frac{x}{\sqrt{9 - x^2}}$$

$$dy = -\frac{x}{\sqrt{9 - x^2}}\,dx$$

7. $f(x) = 5x^2 - 1,\ x = 1,\ \Delta x = 0.01$

$$\Delta y = f(x + \Delta x) - f(x)$$

$$= \left[5(1.01)^2 - 1\right] - \left[5(1)^2 - 1\right] = 0.1005$$

9. $f(x) = \dfrac{4}{x^{1/3}},\ x = 1,\ \Delta x = 0.01$

$$\Delta y = f(x + \Delta x) - f(x)$$

$$= \frac{4}{(1.01)^{1/3}} - 4 \approx -0.013245$$

11. $y = 0.5x^3,\ x = 2,\ \Delta x = dx = 0.1$

$$dy = 1.5x^2\,dx = 1.5(2)^2(0.1) = 0.6$$

$$\Delta y = 0.5(2 + 0.1)^3 - 0.5(2)^3 = 0.6305$$

$$dy \approx \Delta y$$

13. $y = x^4 + 1,\ x = -1,\ \Delta x = dx = 0.01$

$dy = 4x^3\,dx = 4(-1)^3(0.01) = -0.04$

$\Delta y = (-1 + 0.01)^4 + 1 - \left[(-1)^4 + 1\right] \approx -0.0394$

$dy \approx \Delta y$

15. $dy = 2x\,dx,\ x = 2,\ \Delta y = (x + \Delta x)^2 - x^2$

$dx = \Delta x$	dy	Δy	$\Delta y - dy$	$dy/\Delta y$
1.000	4	5	1	0.8
0.500	2	2.25	0.25	0.889
0.100	0.4	0.41	0.010	0.976
0.010	0.04	0.040	0.000	0.998
0.001	0.004	0.004	0.000	1.000

17. $dy = -\dfrac{2}{x^3}\,dx,\ x = 2,\ \Delta y = \dfrac{1}{(x + \Delta x)^2} - \dfrac{1}{x^2}$

$dx = \Delta x$	dy	Δy	$\Delta y - dy$	$dy/\Delta y$
1.000	−0.25	−0.139	0.111	1.8
0.500	−0.125	−0.09	0.035	1.389
0.100	−0.025	−0.023	0.002	1.076
0.010	−0.003	−0.002	0.000	1.008
0.001	0.000	0.000	0.000	1.001

19. $dy = \dfrac{1}{4}x^{-3/4}\,dx = \dfrac{1}{4x^{3/4}}\,dx,\ x = 2,$

$\Delta y = (x + \Delta x)^{1/4} - x^{1/4}$

$dx = \Delta x$	dy	Δy	$\Delta y - dy$	$dy/\Delta y$
1.000	0.149	0.127	−0.022	1.172
0.500	0.074	0.068	−0.006	1.089
0.100	0.015	0.015	0.000	1.019
0.010	0.001	0.001	0.000	1.002
0.001	0.000	0.000	0.000	1.000

21. $f(x) = 2x^3 - x^2 + 1,\ (-2, -19)$

$f'(x) = 6x^2 - 2x$

$f'(-2) = 24 + 4 = 28$

$y + 19 = 28(x + 2)$

$y = 28x + 37$ Tangent line

$f(-2 + 0.01) \approx -18.72$

$y(-2 + 0.01) = -18.72$

$f(-2 - 0.01) \approx -19.28$

$y(-2 - 0.01) = -19.28$

23. $f(x) = \dfrac{x}{x^2 + 1},\ (0, 0)$

$f'(x) = \dfrac{(x^2 + 1) - x(2x)}{(x^2 + 1)^2} = \dfrac{1 - x^2}{(x^2 + 1)^2}$

$f'(0) = 1$

$y - 0 = 1(x - 0)$

$y = x$ Tangent line

$f(0 + 0.01) \approx 0.009999$

$y(0 + 0.01) = 0.01$

$f(0 - 0.01) \approx -0.009999$

$y(0 - 0.01) = -0.01$

25. $N = 32t^2 + 36t + 204$

$\dfrac{dN}{dt} = 64t + 36$

$\Delta N \approx dN = (64t + 36)dt$

When $t = 12$ and $dt = 4$: $dN = (64 \cdot 12 + 36)(4)$

$= 3216.$

The number of bacteria increases by 3216.

27. $p = 0.537t^3 - 4.46t^2 + 9.4t + 27,\ 0 \le t \le 5$

$\dfrac{dP}{dt} = 1.611t^2 - 8.92t + 9.4$

$\Delta p \approx dp = (1.611t^2 - 8.92t + 9.4)dt$

(a) When $t = 1$ and

$dt = 1$: $dp = \left[1.611(1)^2 - 8.92(1) + 9.4\right](1)$

$= 2.091.$

The percent increases by 2.091%. When $t = 1$, $p = 32.477\%$. So, when $t = 2$, $p = 32.477\% + 2.091\% = 34.568\%$.

The percent change is $\dfrac{34.568 - 32.477}{32.477} \approx 6.44\%$.

(b) When $t = 4$ and

$$dt = 1: dp = \left[1.611(4)^2 - 8.92(4) + 9.4\right](1)$$
$$= -0.504.$$

The percent decreases by 0.504%. When $t = 4$, $p = 27.608\%$. So, when $t = 5$,

$$p = 27.608\% - 0.504\% = 27.104\%$$

The percent change is $\dfrac{27.104 - 27.608}{27.608} \approx -1.83\%$.

29. $C = \dfrac{3t}{27 + t^3}$

$$dC = \frac{(27 + t^3)(3) - (3t)(3t^2)}{(27 + t^3)^2} = \frac{3(27 - 2t^3)}{(27 + t^3)^2}\,dt$$

When $t = 1$ and $dt = \dfrac{1}{2}$, you have

$$dC = \frac{3(25)}{(28)^2}\left(\frac{1}{2}\right) \approx 0.0478.$$

31. (a) $P = \dfrac{25t^2 + 125}{t^2 + 1}$

$$\frac{dP}{dt} = \frac{(t^2 + 1)(50t) - (25t^2 + 125)(2t)}{(t^2 + 1)^2}$$
$$= \frac{50t^3 + 50t - 50t^3 - 250t}{(t^2 + 1)^2}$$
$$= \frac{-200t}{(t^2 + 1)^2}$$

$$\Delta P \approx dP = \frac{-200t}{(t^2 + 1)^2}dt$$

When $t = 8$ and $dt = 1$:

$$dP = \frac{-200(8)}{(8^2 + 1)^2}(1) \approx -0.38$$

The change in pressure is approximately -0.38 millimeter of mercury.

(b)

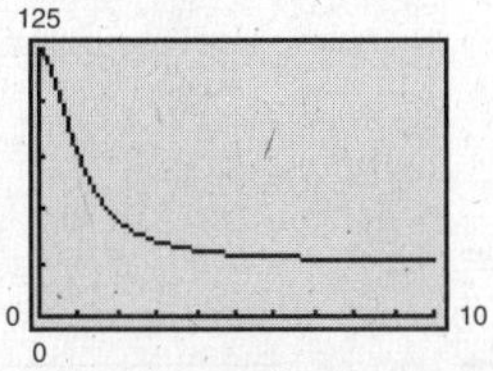

The change in pressure is approximately -0.32 millimeter of mercury.

(c) The results are approximately the same.

33. (a) $A = x^2$

$$\frac{dA}{dx} = 2x$$
$$dA = 2x\,dx \Rightarrow \Delta A = 2x\Delta x \Rightarrow dA \approx \Delta A$$

(b) The two gray rectangles have an area of $2x\Delta x$, or $2xdx = dA$.

(c) $\Delta A - dA = 2x\Delta x - 2x\,dx \approx 0$

35. $A = \pi r^2$

$$dA = 2\pi r\,dr$$

When $r = 10$ inches and $dr = \pm\frac{1}{8}$ inch,

$$dA = 2\pi(10)\left(\pm\tfrac{1}{8}\right) = \pm\tfrac{5}{2}\pi \text{ in.}^2 \approx \pm 7.85 \text{ in.}^2$$

When $A = 100\pi$, the relative error is

$$\frac{dA}{A} = \frac{\pm\frac{5}{2}\pi}{100\pi} = \pm 0.025.$$

37. $V = \dfrac{4}{3}\pi r^3$

$$dV = 4\pi r^2\,dr$$

When $r = 6$ inches and $dr = \pm 0.02$ inch,

$$dV = 4\pi(6)^2(\pm 0.02) = \pm 2.88\pi \text{ in.}^3$$

When $r = 6$, the relative error is

$$\frac{dV}{V} = \frac{4\pi r^2 dr}{(4/3)\pi r^3} = \frac{3\,dr}{r} = \frac{3(\pm 0.02)}{6} = \pm 0.01.$$

39. True; $\Delta y = y(x + \Delta x) - y(x)$

$$= a(x + \Delta x) + b - (ax + b)$$
$$= a\Delta x \text{ and } \frac{\Delta y}{\Delta x} = \frac{a\Delta x}{\Delta x} = a = \frac{dy}{dx}.$$

Review Exercises for Chapter 3

1. $f(x) = -x^2 + 2x + 4$

$f'(x) = -2x + 2$

Critical number: $x = 1$

3. $f(x) = x^{3/2} - 3x^{1/2}$

$f'(x) = \frac{3}{2}x^{1/2} - \frac{3}{2}x^{-1/2} = \frac{3}{2}x^{-1/2}(x - 1) = \frac{3(x - 1)}{2\sqrt{x}}$

Critical numbers: $x = 0,\ x = 1$

5. $f(x) = x^2 + x - 2$

$f'(x) = 2x + 1$

Critical number: $x = -\frac{1}{2}$

Increasing on $\left(-\frac{1}{2}, \infty\right)$

Decreasing on $\left(-\infty, -\frac{1}{2}\right)$

7. $h(x) = \dfrac{x^2 - 3x - 4}{(x - 3)}$

$h'(x) = \dfrac{(x - 3)(2x - 3) - (x^2 - 3x - 4)}{(x - 3)^2}$

$= \dfrac{x^2 - 6x + 13}{(x - 3)^2} > 0$ for all $x \neq 3$.

Increasing on $(-\infty, 3)$ and $(3, \infty)$

9. $T = 0.0380t^4 - 1.092t^3 + 9.23t^2 - 19.6t + 44,$

$1 \leq t \leq 12$

$T' = 0.152t^3 - 3.276t^2 + 18.46t - 19.6$

Using a graphing utility, $T^2 = 0$ when $t \approx 1.38$ and $t \approx 7.24$.

(a) Increasing on $(1.38, 7.24)$

(b) Decreasing on $(1, 1.38)$ and $(7.24, 12)$

(c) Normal monthly temperature increasing from January to July, and decreasing from July to January.

(d)

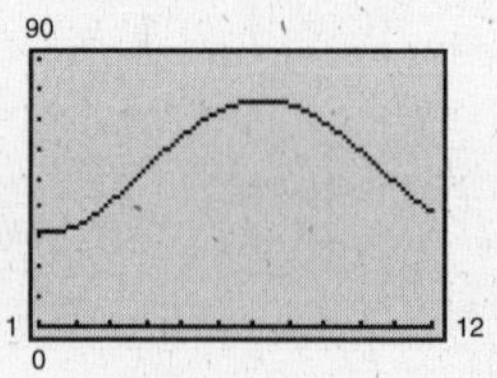

11. $f(x) = 4x^3 - 6x^2 - 2$

$f'(x) = 12x^2 - 12x = 12x(x - 1)$

Critical numbers: $x = 0,\ x = 1$

Relative maximum: $(0, -2)$

Relative minimum: $(1, -4)$

Interval	$-\infty < x < 0$	$0 < x < 1$	$1 < x < \infty$
Sign of f'	$f' > 0$	$f' < 0$	$f' > 0$
Conclusion	Increasing	Decreasing	Increasing

13. $g(x) = x^2 - 16x + 12$

$g'(x) = 2x - 16 = 2(x - 8)$

Critical number: $x = 8$

Relative minimum: $(8, -52)$

Interval	$-\infty < x < 8$	$8 < x < \infty$
Sign of g'	$g' < 0$	$g' > 0$
Conclusion	Decreasing	Increasing

15. $h(x) = 2x^2 - x^4$

$h'(x) = 4x - 4x^3 = 4x(1 - x)(1 + x)$

Critical numbers: $x = 0,\ x = \pm 1$

Relative maxima: $(-1, 1), (1, 1)$

Relative minimum: $(0, 0)$

Interval	$-\infty < x < -1$	$-1 < x < 0$	$0 < x < 1$	$1 < x < \infty$
Sign of h'	$h' > 0$	$h' < 0$	$h' > 0$	$h' < 0$
Conclusion	Increasing	Decreasing	Increasing	Decreasing

17. $f(x) = \dfrac{6}{x^2 + 1}$

$f'(x) = -6(x^2 + 1)^{-2}(2x) = -\dfrac{12x}{(x^2 + 1)^2}$

Critical number: $x = 0$

Relative maximum: $(0, 6)$

Interval	$-\infty < x < 0$	$0 < x < \infty$
Sign of f'	$f' > 0$	$f' < 0$
Conclusion	Increasing	Decreasing

19. $h(x) = \dfrac{x^2}{x - 2}$

$h'(x) = \dfrac{(x - 2)(2x) - x^2}{(x - 2)^2} = \dfrac{x(x - 4)}{(x - 2)^2}$

Critical numbers: $x = 0, x = 4$

Discontinuity: $x = 2$

Relative maximum: $(0, 0)$

Relative minimum: $(4, 8)$

Interval	$-\infty < x < 0$	$0 < x < 2$	$2 < x < 4$	$4 < x < \infty$
Sign of h'	$h' > 0$	$h' < 0$	$h' < 0$	$h' > 0$
Conclusion	Increasing	Decreasing	Decreasing	Increasing

21. $f(x) = x^2 + 5x + 6, [-3, 0]$

$f'(x) = 2x + 5$

Critical number: $x = -\frac{5}{2}$

x-value	Endpoint $x = -3$	Critical $x = -\frac{5}{2}$	Endpoint $x = 0$
$f(x)$	0	$-\frac{1}{4}$	6
Conclusion		Minimum	Maximum

23. $f(x) = x^3 - 12x + 1, [-4, 4]$

$f'(x) = 3x^2 - 12 = 3(x - 2)(x + 2)$

Critical numbers: $x = \pm 2$

x-value	Endpoint $x = -4$	Critical $x = -2$	Critical $x = 2$	Endpoint $x = 4$
$f(x)$	-15	17	-15	17
Conclusion	Minimum	Maximum	Minimum	Maximum

25. $f(x) = 4\sqrt{x} - x^2, [0, 3]$

$f'(x) = 2x^{-1/2} - 2x = \dfrac{2 - 2x^{3/2}}{\sqrt{x}}$

Critical numbers: $x = 0$ (endpoint), $x = 1$

x-value	Endpoint $x = 0$	Critical $x = 1$	Endpoint $x = 3$
$f(x)$	0	3	$4\sqrt{3} - 9$
Conclusion		Maximum	Minimum

27. $f(x) = \dfrac{x}{\sqrt{x^2+1}}, \; [0, 2]$

$$f'(x) = \frac{(x^2+1)^{1/2} - x\left[\frac{1}{2}(x^2+1)^{-1/2}(2x)\right]}{x^2+1} = \frac{1}{(x^2+1)^{3/2}}$$

No critical numbers

x-value	Endpoint $x = 0$	Endpoint $x = 2$
$f(x)$	0	$\frac{2\sqrt{5}}{5}$
Conclusion	Minimum	Maximum

29. $f(x) = \dfrac{2x}{x^2+1}, \; [-1, 2]$

$$f'(x) = \frac{(x^2+1)(2) - 2x(2x)}{(x^2+1)^2} = -\frac{2(x^2-1)}{(x^2+1)^2}$$

Critical numbers: $x = 1$, $x = -1$ (endpoint)

x-value	Endpoint $x = -1$	Critical $x = 1$	Endpoint $x = 2$
$f(x)$	-1	1	$\frac{4}{5}$
Conclusion	Minimum	Maximum	

31. $S = \pi r^2 + \dfrac{\pi}{r}$

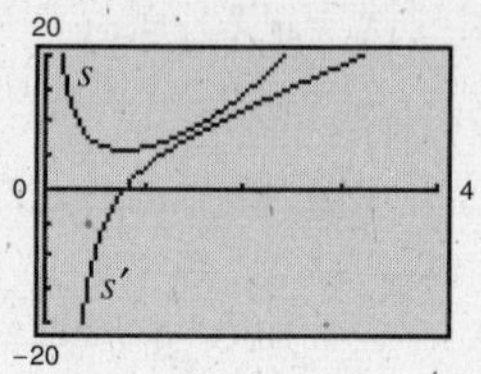

$S' = 2\pi r - \dfrac{\pi}{r^2}$

When $S' = 0$, $2\pi r = \dfrac{\pi}{r^2}$

$2\pi r^3 = \pi$

$r^3 = \dfrac{1}{2}$

$r \approx 0.79$ inch

Minimum: $(0.79, 5.94)$

33. $f(x) = (x-2)^3$

$f'(x) = 3(x-2)^2$

$f''(x) = 6(x-2)$

$f''(x) = 0$ when $x = 2$.

Interval	$-\infty < x < 2$	$2 < x < \infty$
Sign of f''	$f'' < 0$	$f'' > 0$
Conclusion	Concave downward	Concave upward

35. $g(x) = \dfrac{1}{4}(-x^4 + 8x^2)$

$g'(x) = -x^3 + 4x$

$g''(x) = -3x^2 + 4$

$g''(x) = 0$ when $x = \pm\dfrac{2}{\sqrt{3}}$.

Interval	$-\infty < x < -\frac{2}{\sqrt{3}}$	$-\frac{2}{\sqrt{3}} < x < \frac{2}{\sqrt{3}}$	$\frac{2}{\sqrt{3}} < x < \infty$
Sign of g''	$g'' < 0$	$g'' > 0$	$g'' < 0$
Conclusion	Concave downward	Concave upward	Concave downward

37. $f(x) = \dfrac{1}{2}x^4 - 4x^3$

$f'(x) = 2x^3 - 12x^2$

$f''(x) = 6x^2 - 24x = 6x(x-4) = 0$

when $x = 0$ or $x = 4$.

$f''(x) > 0$ on $(-\infty, 0)$.

$f''(x) < 0$ on $(0, 4)$.

$f''(x) > 0$ on $(4, \infty)$.

$(0, 0)$ and $(4, -128)$ are points of inflection.

39. $f(x) = x^3(x-3)^2$

$f'(x) = (x-3)^2(3x^2) + x^3[2(x-3)]$

$= 5x^4 - 24x^3 + 27x^2$

$f''(x) = 20x^3 - 72x^2 + 54x$

$= 2x(10x^2 - 36x + 27) = 0$

when $x = 0$, $x \approx 1.0652$, or $x \approx 2.5348$.

$f''(x) < 0$ on $(-\infty, 0)$.

$f''(x) > 0$ on $(0, 1.0652)$.

$f''(x) < 0$ on $(1.0652, 2.5348)$.

$f''(x) > 0$ on $(2.5348, \infty)$.

$(0, 0)$, $(1.0652, 4.5244)$, and $(2.5348, 3.5246)$ are points of inflection.

41. $f(x) = x^5 - 5x^3$

$f'(x) = 5x^4 - 15x^2 = 5x^2(x^2 - 3)$

$f''(x) = 20x^3 - 30x$

Critical numbers: $x = 0,\ x = \pm\sqrt{3}$

$f''(\sqrt{3}) = 30\sqrt{3} > 0$

Relative minimum: $(\sqrt{3}, -6\sqrt{3})$

$f''(-\sqrt{3}) = -30\sqrt{3} < 0$

Relative maximum: $(-\sqrt{3}, 6\sqrt{3})$

$f''(0) = 0$

By the First-Derivative Test, $(0, 0)$ is not a relative extremum.

43. $f(x) = 2x^2(1 - x^2)$

$f'(x) = (1 - x^2)(4x) + 2x^2(-2x)$

$= 4x - 8x^3 = 4x(1 - 2x^2)$

$f''(x) = 4 - 24x^2$

Critical numbers: $x = 0,\ x = \pm\dfrac{1}{\sqrt{2}}$

$f''(0) = 4 > 0$

Relative minimum: $(0, 0)$

$f''\left(-\dfrac{1}{\sqrt{2}}\right) = -8 < 0$

$f''\left(\dfrac{1}{\sqrt{2}}\right) = -8 < 0$

Relative maxima: $\left(\pm\dfrac{1}{\sqrt{2}}, \dfrac{1}{2}\right)$

45. $d = 0.0793t^4 - 3.463t^3 + 54.74t^2 - 370.8t + 953,$

$5 \le t \le 16$

(a)

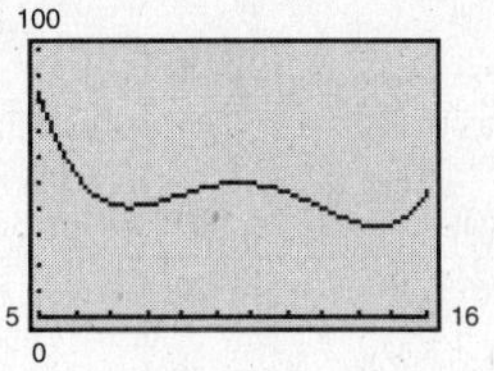

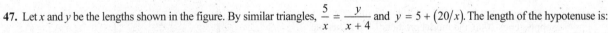

(b) $d' = 0.3172t^3 - 10.389t^2 + 109.48t - 370.8$

$d'' = 0.9516t^2 - 20.778t + 109.48$

$d'' = 0$ when $t \approx 8.88$ and $t \approx 12.95$.

$d'' > 0$ on (0, 8.88) and (12.95, 16), so d'' is concave upward on (0, 8.88) and (12.95, 16).

$d'' < 0$ on (8.88, 12.95), so d'' is concave downward on (8.88, 12.95).

(c) From part (b), the points (8.88, 44.99) and (12.95, 40.66) are inflection points.

(d) The number of fatalities due to lightning was increasing at the greatest rate when $t \approx 8.88$, or in 1998. The number of fatalities due to lightning was decreasing at the greatest rate when $t \approx 12.95$, or in 2002.

47. Let x and y be the lengths shown in the figure. By similar triangles, $\dfrac{5}{x} = \dfrac{y}{x + 4}$ and $y = 5 + (20/x)$. The length of the hypotenuse is:

$$L = \sqrt{(x + 4)^2 + y^2} = \sqrt{(x + 4)^2 + \left(5 + \frac{20}{x}\right)^2} = \sqrt{x^2 + 8x + 41 + \frac{200}{x} + \frac{400}{x^2}}$$

$$L' = \frac{1}{2}\left[x^2 + 8x + 41 + \frac{200}{x} + \frac{400}{x^2}\right]^{-1/2}\left(2x + 8 - \frac{200}{x^2} - \frac{800}{x^3}\right)$$

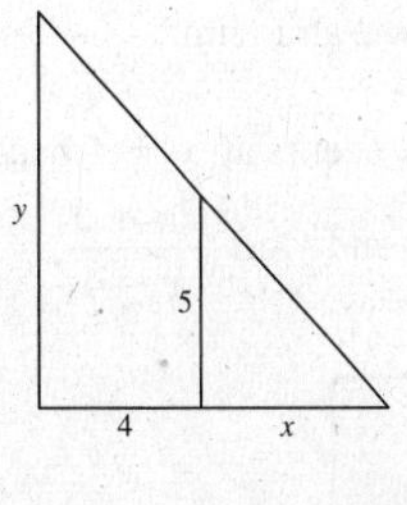

$L' = 0$ when

$$x + 4 - \frac{100}{x^2} - \frac{400}{x^3} = 0$$

$$x^4 + 4x^3 - 100x - 400 = 0$$

$$(x + 4)(x^3 - 100) = 0 \text{ or } x = \sqrt[3]{100}.$$

By the First-Derivative Test, when $x = \sqrt[3]{100}$, L is minimum. When $x = \sqrt[3]{100}$, $z \approx 12.7$ feet.

49. (a) Let l be the length and w be the width of the rectangular region. Then the area is $lw = 4900$ and $w = 4900/l$. The perimeter is:

$$P = 2l + 2w = 2l + 2\left(\frac{4900}{l}\right)$$

$$P' = 2 - \frac{9800}{l^2} = \frac{2(l^2 - 4900)}{l^2}$$

$$P'' = \frac{19,600}{l^3}$$

$P' = 0$ when $l = 70$. Because $P''(70) > 0$, P is a minimum when $l = 70$ feet and $w = 70$ feet.

(b) Let l be the length and w be the width of the rectangular region. Then the area is $lw = 10,000$ and $w = 10,000/l$. The perimeter is:

$$P = 2l + 2w = 2l + 2\left(\frac{10,000}{l}\right)$$

$$P' = 2 - \frac{20,000}{l^2} = \frac{2(l^2 - 10,000)}{l^2}$$

$$P'' = \frac{40,000}{l^3}$$

$P' = 0$ when $l = 100$. Because $P''(100) > 0$, P is a minimum when $l = 100$ feet and $w = 100$ feet.

51. (a) $N = -0.382t^3 - 0.97t^2 + 30.5t + 1466,$ $0 \le t \le 5$

$$\frac{dN}{dt} = -1.146t^2 - 1.94t + 30.5$$

$$\frac{d^2N}{dt^2} = -2.292t - 1.94$$

Because $\frac{d^2N}{dt^2} < 0$ for all $t > 0$, $\frac{dN}{dt}$ is decreasing on $(0, 5)$.

(b) $\lim_{t\to\infty} -0.382t^3 - 0.97t^2 + 30.5t + 1466 = -\infty$

(c) The rate of TV usage in the United States per year was decreasing from 2000 to 2005. As the year approaches 2000, the model predicts 1466 hours per person per year of TV usage.

53. A horizontal asymptote occurs at $y = 2$ because

$$\lim_{x\to\infty} \frac{2x + 3}{x - 4} = \frac{2}{1} = 2 \text{ and } \lim_{x\to-\infty} \frac{2x + 3}{x - 4} = \frac{2}{1} = 2.$$

A vertical asymptote occurs at $x = 4$ because

$$\lim_{x\to 4^-} \frac{2x + 3}{x - 4} = -\infty \text{ and } \lim_{x\to 4^+} \frac{2x + 3}{x - 4} = \infty.$$

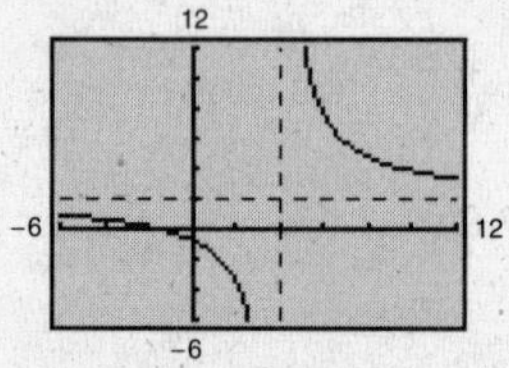

55. Horizontal asymptotes occur at $y = \pm 3$ because

$$\lim_{x\to\infty} \frac{\sqrt{9x^2 + 1}}{x} = 3 \text{ and } \lim_{x\to-\infty} \frac{\sqrt{9x^2 + 1}}{x} = -3.$$

A vertical asymptote occurs at $x = 0$ because

$$\lim_{x\to 0^-} \frac{\sqrt{9x^2 + 1}}{x} = -\infty \text{ and } \lim_{x\to 0^+} \frac{\sqrt{9x^2 + 1}}{x} = \infty.$$

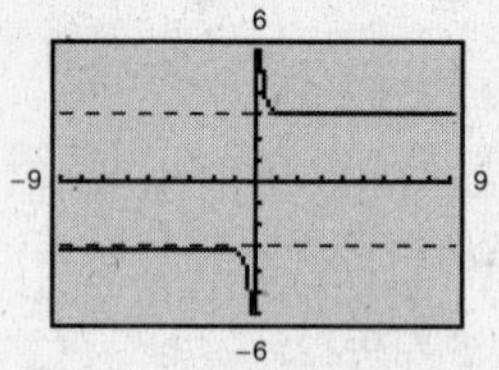

57. A horizontal asymptote occurs at $y = 0$ because

$$\lim_{x\to\infty} \frac{4}{x^2 + 1} = 0 \text{ and } \lim_{x\to-\infty} \frac{4}{x^2 + 1} = 0.$$

The graph has no vertical asymptotes because the denominator is never zero.

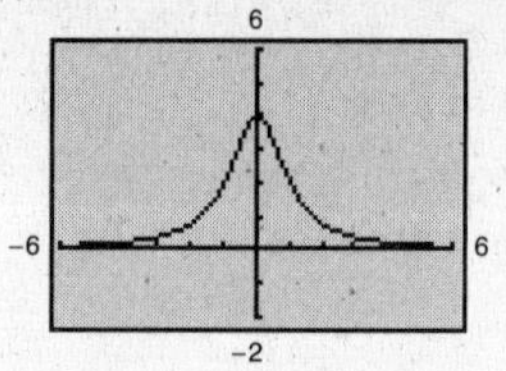

59. $\lim_{x\to 0^+} \left(x - \frac{1}{x^3}\right) = 0 - \infty = -\infty$

61. $\lim_{x\to -1^+} \frac{x^2 - 2x + 1}{x + 1} = \infty$

63. $\lim_{x\to\infty} \frac{2x^2}{3x^2 + 5} = \frac{2}{3}$

65. $\lim_{x\to-\infty} \frac{3x}{x^2 + 1} = 0$

67. (a)

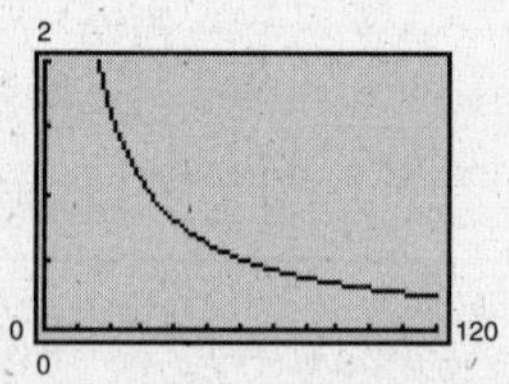

(b) $\lim_{s\to\infty} T = \lim_{s\to\infty} \frac{-0.03s + 33.6}{s} = -0.03$

As s increases, T approaches -0.03.

69.

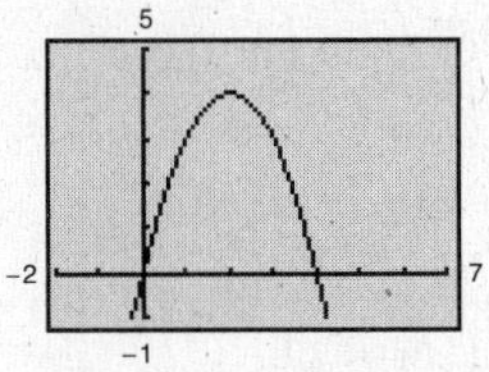

Intercepts: $(0, 0), (4, 0)$

Relative maximum: $(2, 4)$

No points of inflection

No asymptotes

Domain: $(-\infty, \infty)$

71.

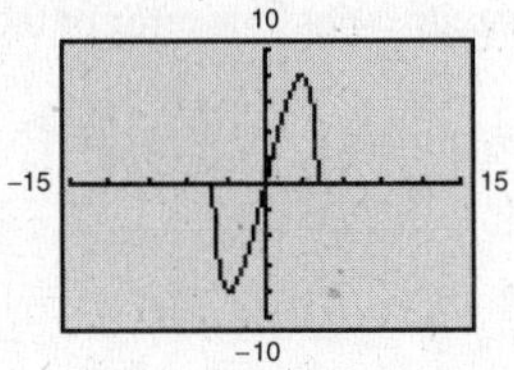

Intercepts: $(0, 0), (4, 0), (-4, 0)$

Relative maximum: $(2.83, 8)$

Relative minimum: $(-2.83, -8)$

Point of inflection: $(0, 0)$

No asymptotes

Domain: $[-4, 4]$

73.

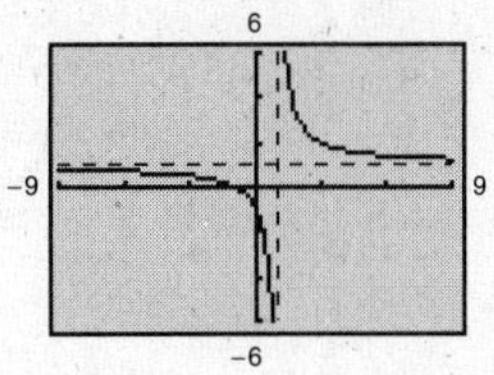

Intercepts: $(-1, 0), (0, -1)$

No relative extrema

No points of inflection

Horizontal asymptote: $y = 1$

Vertical asymptote: $x = 1$

Domain: $(-\infty, 1) \cup (1, \infty)$

75.

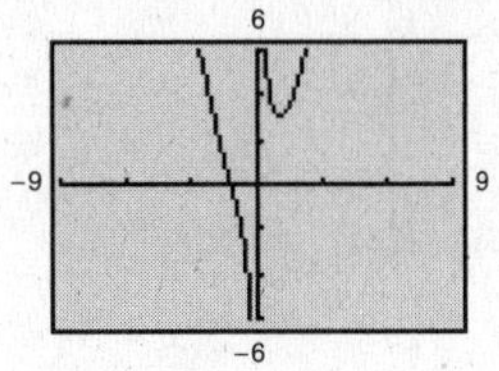

Intercept: $(-1.26, 0)$

Relative minimum: $(1, 3)$

Point of inflection: $(-1.26, 0)$

Vertical asymptote: $x = 0$

Domain: $(-\infty, 0) \cup (0, \infty)$

77. (a)

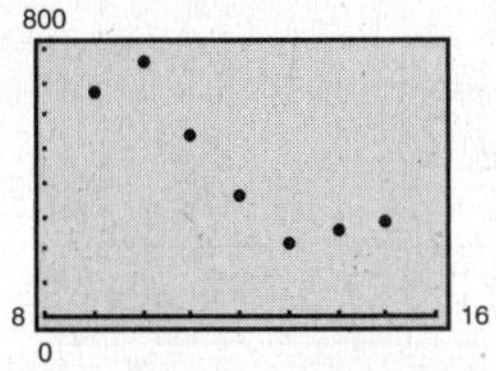

(b) The numbers of flight overbooking complaints increased from 1999 to 2000, decreased from 2000 to 2003, and then increased again from 2003 to 2005.

(c)

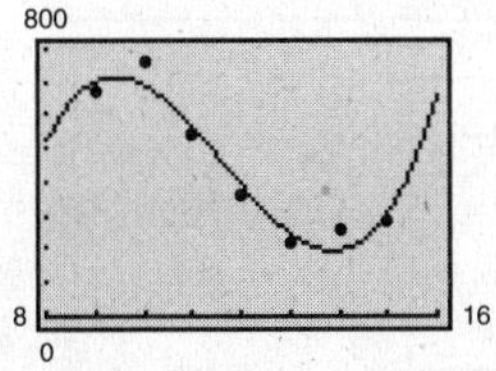

(d) $A' = 35.166t^2 - 819.28t + 4596.9, \ 9 \le t \le 15$

Critical numbers: $t \approx 9.4, \ t \approx 13.9$

Interval	$9 < t < 9.4$	$9.4 < t < 13.9$	$13.9 < t < 15$
Sign of A'	$A' > 0$	$A' < 0$	$A' > 0$
Conclusion	Increasing	Decreasing	Increasing

(e) $A'' = 70.332t - 819.28, \ 9 \le t \le 15$

Critical number $t \approx 11.6$

Interval	$9 < t < 11.6$	$11.6 < t < 15$
Sign of A''	$A'' < 0$	$A'' > 0$
Conclusion	Decreasing	Increasing

79. $y = x(1-x) = x - x^2$

$\frac{dy}{dx} = 1 - 2x$

$dy = (1 - 2x)dx$

81. $y = \sqrt{36 - x^2}$

$\frac{dy}{dx} = \frac{1}{2}(36 - x^2)^{-1/2}(-2x)$

$dy = -\frac{x}{\sqrt{36 - x^2}}dx$

83. $E = 22.5t + 7.5t^2 - 2.5t^3,\ 0 \le t \le 4.5$

(a)

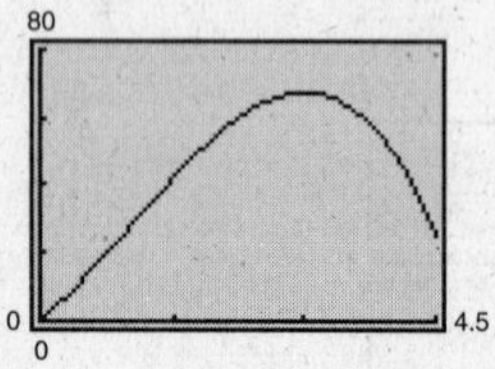

(b) $E' = 22.5 + 15t - 7.5t^2 = -7.5(x-3)(x+1)$

$E'' = 15 - 15t$

$E' = 0$ when $x = 3$. Because $E''(3) = -30 < 0$, E is maximum when $x = 3$, and $E(3) = 67.5$.

85. $p = 0.000235w^2 - 0.054w + 7.1$

$\frac{dp}{dw} = 0.00047w - 0.054$

$\Delta p \approx dp = (0.00047w - 0.054)\,dw$

(a) When $w = 10$ and $dw = 10$:

$dp = (0.00047 \cdot 10 - 0.054)(10) = -0.493$

The percent of biomass decreases by 0.493 percent.

(b) When $w = 40$ and $dw = 20$:

$dp = (0.00047 \cdot 40 - 0.054)(20) = -0.704$

The percent of biomass decreases by 0.704 percent.

87. The radius is 9 inches with a possible error of 0.025 inch.

$S = 4\pi r^2$

$dS = 8\pi r\ dr$

$V = \frac{4}{3}\pi r^3$

$dV = 4\pi r^2 dr$

When $r = 9$ and $dr = \pm 0.025$:

$dS = 8\pi(9)(\pm 0.025) = \pm 1.8\pi$ in.^{2}

$dV = 4\pi(9)^2(\pm 0.025) = \pm 8.1\pi$ in.^{3}

Chapter Test Solutions

1. $f(x) = 3x^2 - 4$

$f'(x) = 6x$

Critical number: $x = 0$

Interval	$-\infty < x < 0$	$0 < x < \infty$
Sign of f'	$f' < 0$	$f' > 0$
Conclusion	Decreasing	Increasing

2. $f(x) = x^3 - 12x$

$f'(x) = 3x^2 - 12 = 3(x-2)(x+2)$

Critical numbers: $x = \pm 2$

Interval	$-\infty < x < -2$	$-2 < x < 2$	$2 < x < \infty$
Sign of f'	$f' > 0$	$f' < 0$	$f' > 0$
Conclusion	Increasing	Decreasing	Increasing

3. $f(x) = (x-5)^4$

$f'(x) = 4(x-5)^3$

Critical number: $x = 5$

Interval	$-\infty < x < 5$	$5 < x < \infty$
Sign of f'	$f' < 0$	$f' > 0$
Conclusion	Decreasing	Increasing

4. $f(x) = \frac{1}{3}x^3 - 9x + 4$

$f'(x) = x^2 - 9 = (x+3)(x-3)$

Critical numbers: $x = \pm 3$

Relative minimum: $(3, -14)$

Relative maximum: $(-3, 22)$

Interval	$-\infty < x < -3$	$-3 < x < 3$	$3 < x < \infty$
Sign of f'	$f' > 0$	$f' < 0$	$f' > 0$
Conclusion	Increasing	Decreasing	Increasing

5. $f(x) = 2x^4 - 4x^2 - 5$

$f'(x) = 8x^3 - 8x = 8x(x + 1)(x - 1)$

Critical numbers: $x = 0$, $x = \pm 1$

Relative minima: $(-1, -7)$, $(1, -7)$

Relative maximum: $(0, -5)$

Interval	$-\infty < x < -1$	$-1 < x < 0$	$0 < x < 1$	$1 < x < \infty$
Sign of f'	$f' < 0$	$f' > 0$	$f' < 0$	$f' > 0$
Conclusion	Decreasing	Increasing	Decreasing	Increasing

6. $f(x) = \dfrac{5}{x^2 + 2}$

$f'(x) = -5(x^2 + 2)^{-2}(2x) = -\dfrac{10x}{(x^2 + 2)^2}$

Critical number: $x = 0$

Relative maximum: $\left(0, \dfrac{5}{2}\right)$

Interval	$-\infty < x < 0$	$0 < x < \infty$
Sign of f'	$f' > 0$	$f' < 0$
Conclusion	Increasing	Decreasing

7. $f(x) = x^2 + 6x + 8$, $[-4, 0]$

$f'(x) = 2x + 6$

Critical number: $x = -3$

x-value	Endpoint $x = -4$	Critical $x = -3$	Endpoint $x = 0$
$f(x)$	0	-1	8
Conclusion		Minimum	Maximum

8. $f(x) = 12\sqrt{x} - 4x$, $[0, 5]$

$f'(x) = 12\left(\dfrac{1}{2}x^{-1/2}\right) - 4 = \dfrac{6}{x^{1/2}} - 4$

Critical number: $x = \dfrac{9}{4}$

x-value	Endpoint $x = 0$	Critical $x = \frac{9}{4}$	Endpoint $x = 5$
$f(x)$	0	9	$12\sqrt{5} - 20$
Conclusion	Minimum	Maximum	

9. $f(x) = \dfrac{6}{x} + \dfrac{x}{2}$, $[1, 6]$

$f'(x) = -\dfrac{6}{x^2} + \dfrac{1}{2}$

Critical number: $x = 2\sqrt{3}$

x-value	Endpoint $x = 1$	Critical $x = 2\sqrt{3}$	Endpoint $x = 6$
$f(x)$	$\frac{13}{2}$	$2\sqrt{3}$	4
Conclusion	Maximum	Minimum	

10. $f(x) = x^5 - 4x^2$

$f'(x) = 5x^4 - 8x$

$f''(x) = 20x^3 - 8$

$f''(x) = 0$ when $x = \dfrac{\sqrt[3]{50}}{5}$.

Interval	$-\infty < x < \frac{\sqrt[3]{50}}{5}$	$\frac{\sqrt[3]{50}}{5} < x < \infty$
Sign of f''	$f'' < 0$	$f'' > 0$
Conclusion	Concave downward	Concave upward

11. $f(x) = \dfrac{20}{3x^2 + 8}$

$f'(x) = -20(3x^2 + 8)^{-2}(6x) = -\dfrac{120x}{(3x^2 + 8)^2}$

$f''(x) = \dfrac{(3x^2 + 8)^2(-120) - (-120x)[2(3x^2 + 8)(6x)]}{(3x^2 + 8)^4} = \dfrac{120(9x^2 - 8)}{(3x^2 + 8)^3}$

$f''(x) = 0$ when $x = \pm\dfrac{2\sqrt{2}}{3}$

Interval	$-\infty < x < -\dfrac{2\sqrt{2}}{3}$	$-\dfrac{2\sqrt{2}}{3} < x < \dfrac{2\sqrt{2}}{3}$	$\dfrac{2\sqrt{2}}{3} < x < \infty$
Sign of $f''(x)$	$f''(x) > 0$	$f''(x) < 0$	$f''(x) > 0$
Conclusion	Concave upward	Concave downward	Concave upward

12. $f(x) = x^4 + 6$

$f'(x) = 4x^3$

$f''(x) = 12x^2 = 0$ when $x = 0$.

$f''(x) > 0$ on $(-\infty, 0)$.

$f''(x) > 0$ on $(0, \infty)$.

There are no points of inflection.

13. $f(x) = \dfrac{1}{5}x^5 - 4x^2$

$f'(x) = x^4 - 8x$

$f''(x) = 4x^3 - 8 = 0$ when $x = \sqrt[3]{2}$.

$f''(x) < 0$ on $(-\infty, \sqrt[3]{2})$.

$f''(x) > 0$ on $(\sqrt[3]{2}, \infty)$.

$\left(\sqrt[3]{2}, -\dfrac{18\sqrt[3]{4}}{5}\right)$ is a point of inflection.

14. $f(x) = x^3 - 6x^2 - 24x + 12$

$f'(x) = 3x^2 - 12x - 24$

Critical numbers: $x \approx -1.46$ and $x \approx 5.46$

$f''(x) = 6x - 12$

$f''(-1.46) < 0$

$f''(5.46) > 0$

Relative maximum: $(-1.46, 31.14)$

Relative minimum: $(5.46, -135.14)$

15. $f(x) = \dfrac{3}{5}x^5 - 9x^3$

$f'(x) = 3x^4 - 27x^2 = 3x^2(x + 3)(x - 3)$

Critical numbers: $x = 0$, $x = \pm 3$

$f''(x) = 12x^3 - 54x$

$f''(0) = 0$

$f''(-3) = -162 < 0$

$f''(3) = 162 > 0$

Relative maximum: $\left(-3, \dfrac{486}{5}\right)$

Relative minimum: $\left(3, -\dfrac{486}{5}\right)$

By the First-Derivative Test, $(0, 0)$ is not a relative extremum.

16. A vertical asymptote occurs at $x = 5$ because $\lim\limits_{x \to 5^-} \dfrac{3x + 2}{x - 5} = -\infty$ and $\lim\limits_{x \to 5^+} \dfrac{3x + 2}{x - 5} = \infty$.

A horizontal asymptote occurs at $y = 3$ because $\lim\limits_{x \to \infty} \dfrac{3x + 2}{x - 5} = 3$ and $\lim\limits_{x \to -\infty} \dfrac{3x + 2}{x - 5} = 3$.

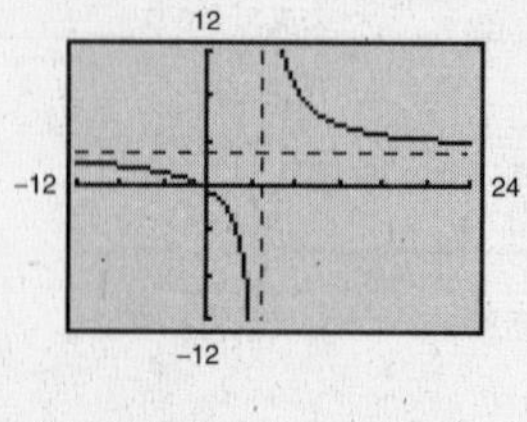

17. There are no vertical asymptotes because the denominator is never zero. A horizontal asymptote occurs at $y = 2$ because $\lim_{x\to\infty} \frac{2x^2}{x^2 + 3} = 2$ and

$\lim_{x\to-\infty} \frac{2x^2}{x^2 + 3} = 2.$

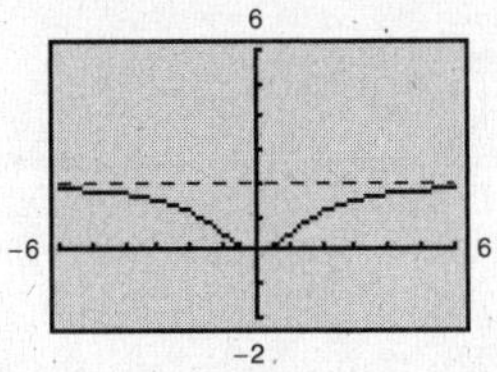

18. A vertical asymptote occurs at $x = 1$ because

$\lim_{x\to1^-} \frac{2x^2 - 5}{x - 1} = \infty$ and $\lim_{x\to1^+} \frac{2x^2 - 5}{x - 1} = -\infty.$

There are no horizontal asymptotes because

$\lim_{x\to\infty} \frac{2x^2 - 5}{x - 1} = \infty$ and $\lim_{x\to-\infty} \frac{2x^2 - 5}{x - 1} = -\infty.$

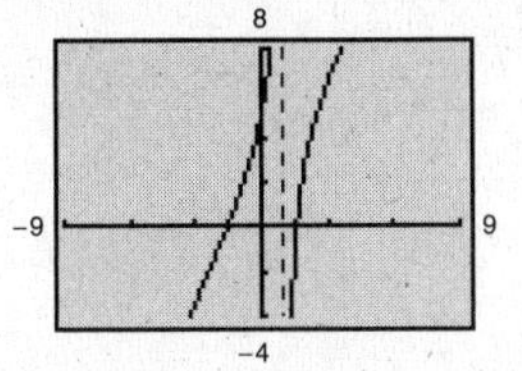

19. $\lim_{x\to\infty} \left(\frac{3}{x} + 1\right) = 0 + 1 = 1$

20. $\lim_{x\to\infty} \frac{3x^2 - 4x + 1}{x - 7} = \infty$

21. $\lim_{x\to-\infty} \frac{6x^2 + x - 5}{2x^2 - 5x} = \frac{6}{2} = 3$

22. $y = 5x^2 - 3$

$\frac{dy}{dx} = 10x$

$dy = 10x\,dx$

23. $y = \frac{1 - x}{x + 3}$

$\frac{dy}{dx} = \frac{(x + 3)(-1) - (1 - x)}{(x + 3)^2} = -\frac{4}{(x + 3)^2}$

$dy = -\frac{4}{(x + 3)^2}dx$

24. $y = (x + 4)^3$

$\frac{dy}{dx} = 3(x + 4)^2$

$dy = 3(x + 4)^2\,dx$

25. $p = 0.500t^3 - 3.71t^2 + 5.8t + 43, 1 \le t \le 5$

$p' = 1.5t^2 - 7.42t + 5.8$

$p' = 0$ when $t \approx 3.97.$

Because p is decreasing on (1, 3.97) and increasing on (3.97, 5), the percent of households that have indoor house plants was the lowest in late 2003.

Practice Test for Chapter 3

1. Find the critical numbers and the intervals on which f is increasing or decreasing for $f(x) = x^3 - 6x^2 + 5$.

2. Find the critical numbers and the intervals on which f is increasing or decreasing for $f(x) = 2x\sqrt{1 - x}$.

3. Find the relative extrema of $f(x) = x^4 - 32x + 3$.

4. Find the relative extrema of $f(x) = (x + 3)^{4/3}$.

5. Find the extrema of $f(x) = x^2 - 4x - 5$ on $[0, 5]$.

6. Find the points of inflection of $f(x) = 3x^4 - 24x + 2$.

7. Find the points of inflection of $f(x) = \dfrac{x^2}{1 + x^2}$.

8. Find two positive numbers whose product is 200 such that the sum of the first plus three times the second is a minimum.

9. Three rectangular fields are to be enclosed by 3000 feet of fencing, as shown in the accompanying figure. What dimensions should be used so that the enclosed area will be a maximum?

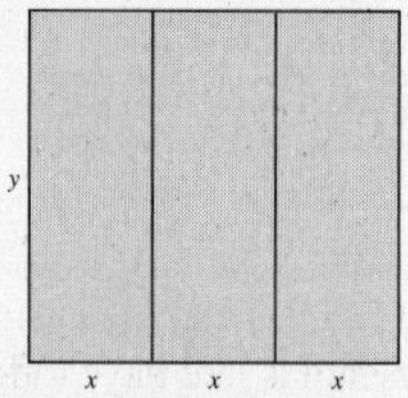

10. Find $\displaystyle\lim_{x \to 3^-} \frac{x + 4}{x - 3}$.

11. Find $\displaystyle\lim_{x \to \infty} \frac{4x^3 - 9x^2 + 1}{1 - 2x^3}$.

12. Sketch the graph of $f(x) = \dfrac{x^2}{x^2 - 9}$.

13. Sketch the graph of $f(x) = \dfrac{x + 2}{x^2 + 5}$.

14. Sketch the graph of $f(x) = x^3 + 3x^2 + 3x - 1$.

15. Sketch the graph of $f(x) = |4 - 2x|$.

16. Sketch the graph of $f(x) = (2 - x)^{2/3}$.

17. Use differentials to approximate $\sqrt[3]{65}$.

18. Find and interpret the point(s) of inflection of the graph of $p = -0.783t^3 + 18.30t^2 - 66.4t + 63$, where p is the number of pancreas transplants over a 17-year period, t is the number of years, and $t = 0$ corresponds to the beginning of the first year.

Graphing Calculator Required

19. Graph $y = \dfrac{5x}{\sqrt{x^2 + 4}}$ on a graphing calculator and find any asymptotes that exist. Are there any relative extrema?

20. Graph $y = \dfrac{2x^4 - 5x + 1}{x^4 + 1}$ on a graphing calculator. Is it possible for a graph to cross its horizontal asymptote?

CHAPTER 4
Exponential and Logarithmic Functions

CHAPTER 4
Exponential and Logarithmic Functions

Section 4.1 Exponential Functions

Skills Review

1. The graph of g is the graph of f translated to the left two units.

2. The graph of g is the graph of f reflected in the x-axis.

3. The graph of g is the graph of f translated downward one unit.

4. The graph of g is the graph of f reflected in the y-axis.

5. The graph of g is the graph of f translated to the right one unit.

6. The graph of g is the graph of f translated upward two units.

7. The function is continuous on $(-\infty, -4)$ and $(-4, \infty)$.

8. The function is continuous on the entire real number line.

9. The function is continuous on $(-\infty, -1)$, $(-1, 1)$, and $(1, \infty)$.

10. The function is continuous on the entire real number line.

11. $2x - 6 = 4$
$2x = 10$
$x = 5$

12. $3x + 1 = 5$
$3x = 4$
$x = \frac{4}{3}$

13. $(x + 4)^2 = 25$
$x + 4 = \pm 5$
$x = 1$ or $x = -9$

14. $(x - 2)^2 = 8$
$x - 2 = \pm 2\sqrt{2}$
$x = 2 \pm 2\sqrt{2}$

15. $x^2 + 4x - 5 = 0$
$(x - 1)(x + 5) = 0$
$x - 1 = 0 \Rightarrow x = 1$
$x + 5 = 0 \Rightarrow x = -5$

16. $2x^2 - 3x + 1 = 0$
$(2x - 1)(x - 1) = 0$
$2x - 1 = 0 \Rightarrow x = \frac{1}{2}$
$x - 1 = 0 \Rightarrow x = 1$

---

1. (a) $5(5^3) = 5^4 = 625$

(b) $27^{2/3} = \left(\sqrt[3]{27}\right)^2 = 3^2 = 9$

(c) $64^{3/4} = \left(2^6\right)^{3/4} = 2^{9/2} = \sqrt{2^9} = 2^4\sqrt{2} = 16\sqrt{2}$

(d) $81^{1/2} = \sqrt{81} = 9$

(e) $25^{3/2} = \left(\sqrt{25}\right)^3 = 5^3 = 125$

(f) $32^{2/5} = \left(\sqrt[5]{32}\right)^2 = 2^2 = 4$

3. (a) $(5^2)(5^3) = 5^5 = 3125$

(b) $(5^2)(5^{-3}) = 5^{-1} = \dfrac{1}{5}$

(c) $(5^2)^2 = 5^4 = 625$

(d) $5^{-3} = \dfrac{1}{5^3} = \dfrac{1}{125}$

5. (a) $\dfrac{5^3}{25^2} = \dfrac{5^3}{(5^2)^2} = \dfrac{5^3}{5^4} = \dfrac{1}{5}$

(b) $(9^{2/3})(3)(3^{2/3}) = (3^2)^{2/3}(3)(3^{2/3})$
$= (3^{4/3})(3^{5/3}) = 3^{9/3} = 3^3 = 27$

(c) $\left[(25^{1/2})5^2\right]^{1/3} = \left[5 \cdot 5^2\right]^{1/3} = \left[5^3\right]^{1/3} = 5$

(d) $(8^2)(4^3) = (64)(64) = 4096$

7. $f(x) = 2^{x-1}$

(a) $f(3) = 2^{3-1} = 2^2 = 4$

(b) $f\left(\frac{1}{2}\right) = 2^{1/2-1} = 2^{-1/2} = \frac{1}{\sqrt{2}} = \frac{\sqrt{2}}{2}$

(c) $f(-2) = 2^{-2-1} = 2^{-3} = \frac{1}{8}$

(d) $f\left(-\frac{3}{2}\right) = 2^{-3/2-1} = 2^{-5/2} = \frac{1}{2^{5/2}} = \frac{1}{4\sqrt{2}} = \frac{\sqrt{2}}{8}$

9. $g(x) = 1.05^x$

(a) $g(-2) = 1.05^{-2} = \frac{1}{1.05^2} = \frac{1}{1.1025} \approx 0.91$

(b) $g(120) = 1.05^{120} \approx 348.91$

(c) $g(12) = 1.05^{12} \approx 1.80$

(d) $g(5.5) = 1.05^{5.5} \approx 1.31$

11. $P(t) = 2184(0.956)^t$ ($t = 0$ corresponds to 2000.)

(a) 2009: $P(9) = 2184(0.956)^9 \approx 1457$

(b) 2013: $P(13) = 2184(0.956)^{13} \approx 1217$

13. The graph of $f(x) = 3^x$ is an exponential curve with the following characteristics.

Passes through $(0, 1), (1, 3), \left(-1, \frac{1}{3}\right)$

Horizontal asymptote: $y = 0$

It matches graph (e).

15. The graph of $f(x) = -3^x = (-1)(3^x)$ is an exponential curve with the following characteristics.

Passes through $(0, -1), (1, -3), \left(-1, -\frac{1}{3}\right)$

Horizontal asymptote: $y = 0$

It matches graph (a).

17. The graph of $f(x) = 3^{-x} - 1 = \left(\frac{1}{3}\right)^x - 1$ is an exponential curve with the following characteristics.

Passes through $(0, 0), \left(1, -\frac{2}{3}\right), (-1, 2)$

Horizontal asymptote: $y = -1$

It matches graph (d).

19. $f(x) = 6^x$

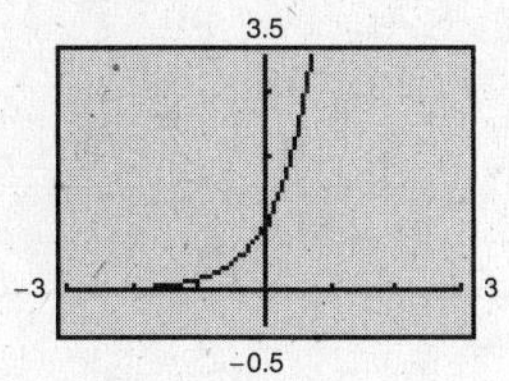

21. $f(x) = 5^{-x}$

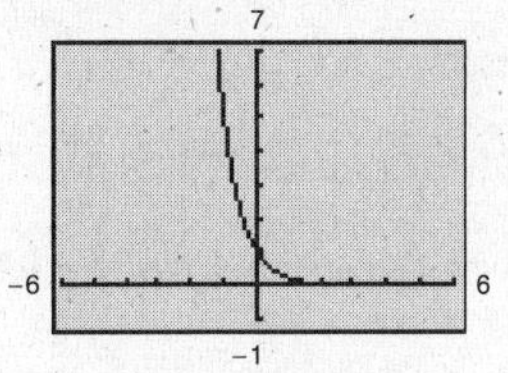

23. $y = 2^{x-1}$

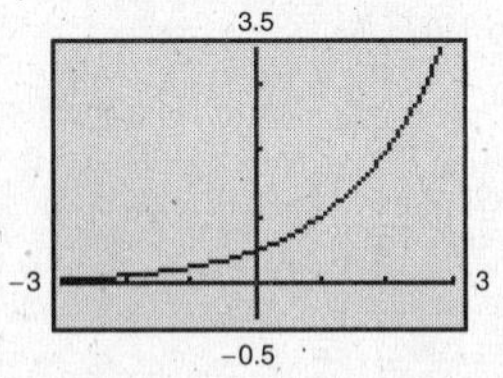

25. $y = -2^x$

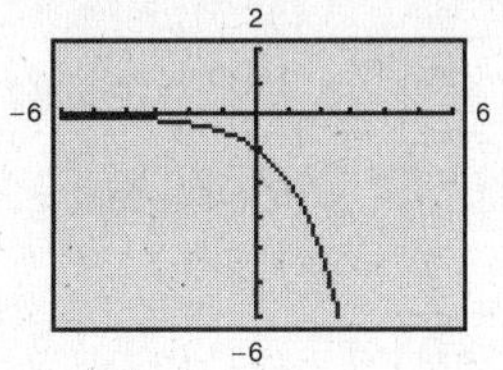

27. $y = 3^{-x^2}$

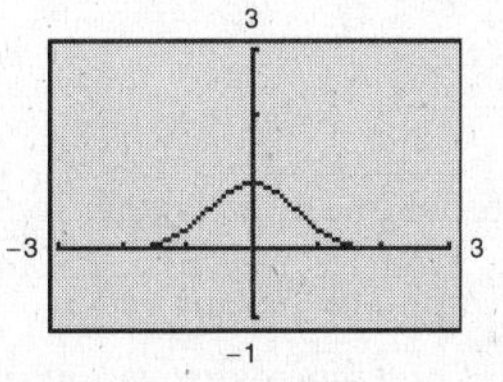

29. $s(t) = \frac{1}{4}(3^{-t}) = \frac{3^{-t}}{4} = \frac{1}{4(3^t)}$

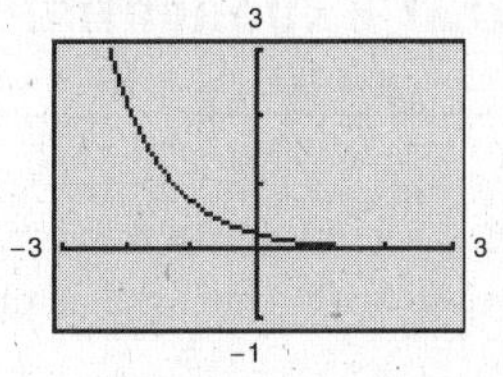

31. (a)

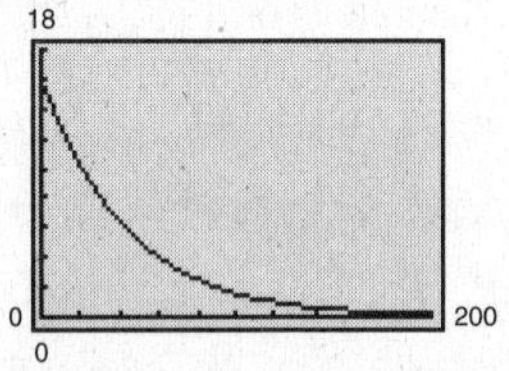

(b) When $t = 150$ years, $y = 16\left(\frac{1}{2}\right)^{150/30} = 16\left(\frac{1}{2}\right)^5 = \frac{1}{2}$.

After 150 years, $\frac{1}{2}$ gram of the initial mass remains.

(c)

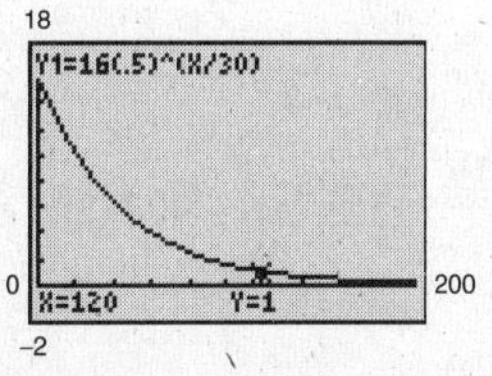

The mass will decay to an amount of 1 gram after 120 years.

33. (a) $V(t) = 30{,}500\left(\frac{7}{8}\right)^t$

(b) When $t = 4$ years,

$$V(t) = V(4) = 30{,}500\left(\tfrac{7}{8}\right)^4 \approx 17{,}878.54.$$

The value of the van 4 years after it was puchased is \$17,878.54.

35. $N(t) = 315.5(1.078)^t,\ 3 \le t \le 46$

$(t = 3$ corresponds to 1963.$)$

(a)

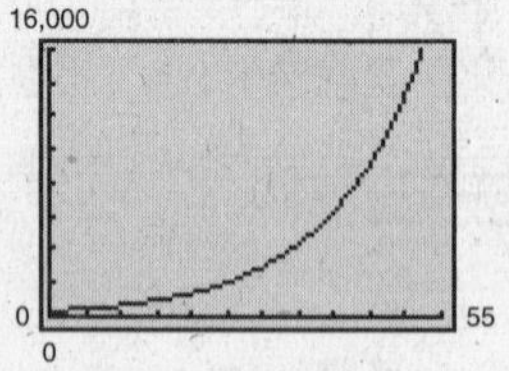

(b) $N(30) = 315.5(1.078)^{30} \approx 3003$ breeding pairs

(c) $N(40) = 315.5(1.078)^{40} \approx 6364$ breeding pairs

(d)

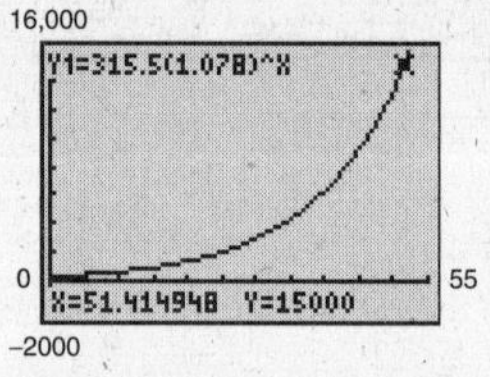

The number of breeding pairs will reach 15,000 in 2011 $(t \approx 51.4)$.

(e) Explanations will vary. Sample answer:

Yes. The number of breeding pairs was increasing exponentially from 1963 to 2006, so 15,000 breeding pairs in 2011 seems reasonable.

Section 4.2 Natural Exponential Functions

Skills Review

1. The function is continuous on the entire real number line.

2. The function is continuous on $(-\infty, -2)$, $(-2, 2)$, and $(2, \infty)$.

3. The function is continuous on $\left(-\infty, -\sqrt{3}\right)$, $\left(-\sqrt{3}, \sqrt{3}\right)$, and $\left(\sqrt{3}, \infty\right)$.

4. The function is continuous on $(-\infty, 4)$ and $(4, \infty)$.

5. $\displaystyle\lim_{x\to\infty} \frac{25}{1 + 4x} = 0$

6. $\displaystyle\lim_{x\to\infty} \frac{16x}{3 + x^2} = 0$

7. $\displaystyle\lim_{x\to\infty} \frac{8x^3 + 2}{2x^3 + x} = 4$

8. $\displaystyle\lim_{x\to\infty} \frac{x}{2x} = \frac{1}{2}$

9. $\displaystyle\lim_{x\to\infty} \frac{3}{2 + (1/x)} = \frac{3}{2 + 0} = \frac{3}{2}$

10. $\displaystyle\lim_{x\to\infty} \frac{6}{1 + x^{-2}} = \frac{6}{1 + \left(1/x^2\right)} = \frac{6}{1 + 0} = 6$

11. $\displaystyle\lim_{x\to\infty} 2^{-x} = \frac{1}{2^x} = 0$

12. $\displaystyle\lim_{x\to\infty} \frac{7}{1 + 5x} = 0$

1. (a) $(e^3)(e^4) = e^{3+4} = e^7$

(b) $(e^3)^4 = e^{3(4)} = e^{12}$

(c) $(e^3)^{-2} = e^{3(-2)} = e^{-6} = \dfrac{1}{e^6}$

(d) $e^0 = 1$

3. (a) $(e^2)^{5/2} = e^{2(5/2)} = e^5$

(b) $(e^2)(e^{1/2}) = e^{2+1/2} = e^{5/2}$

(c) $(e^{-2})^{-3} = e^{-2(-3)} = e^6$

(d) $\dfrac{e^5}{e^{-2}} = e^{5-(-2)} = e^{5+2} = e^7$

5. $f(x) = e^{2x+1}$

Increasing exponential passing through $(0, e)$.

It matches graph (f).

7. $f(x) = e^{x^2}$

Symmetric with respect to y-axis.
It matches graph (d).

9. $f(x) = e^{\sqrt{x}}$

Domain: $x \geq 0$ passing through $(1, e)$.

It matches graph (c).

11. $h(x) = e^{x-3}$

x	-1	0	1	2	3
$h(x)$	0.02	0.05	0.14	0.37	1

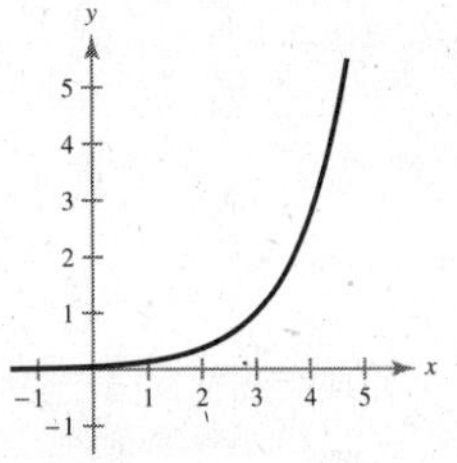

13. $g(x) = e^{1-x}$

x	-1	0	1	2	3
$g(x)$	7.39	2.72	1	0.37	0.14

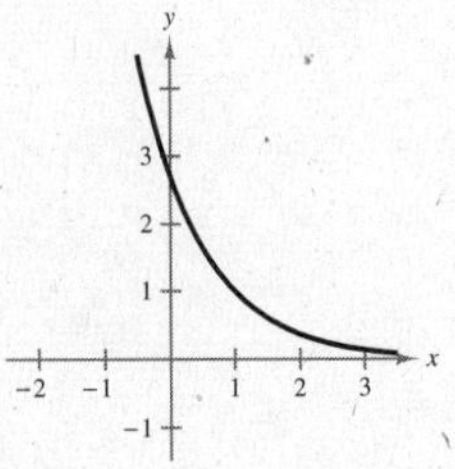

15. $N(t) = 500e^{-0.2t}$

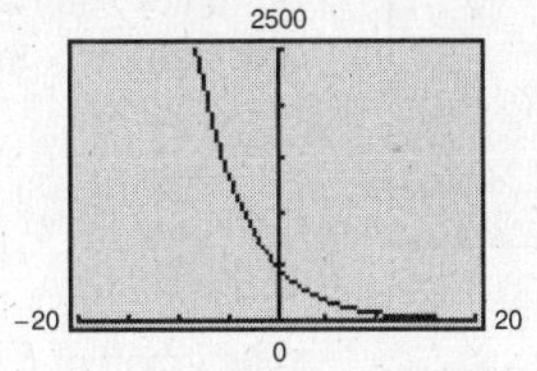

17. $g(x) = \dfrac{2}{1 + e^{x^2}}$

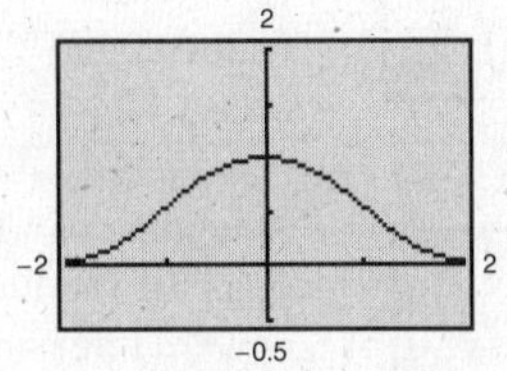

19. $f(x) = \dfrac{e^x + e^{-x}}{2}$

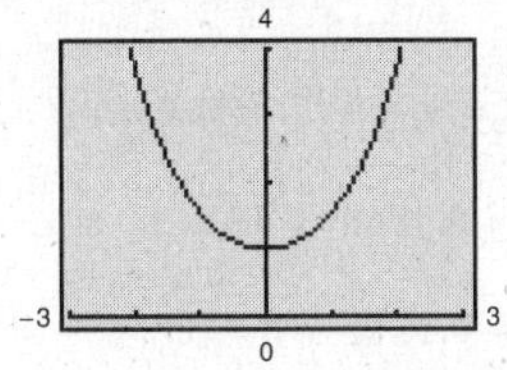

$$\lim_{x\to\infty} \frac{e^x + e^{-x}}{2} = \frac{\infty + 0}{2} = \infty$$

$$\lim_{x\to-\infty} \frac{e^x + e^{-x}}{2} = \frac{0 + \infty}{2} = \infty$$

No horizontal asymptotes

Continuous on the entire real number line

21. $f(x) = \dfrac{2}{1 + e^{1/x}}$

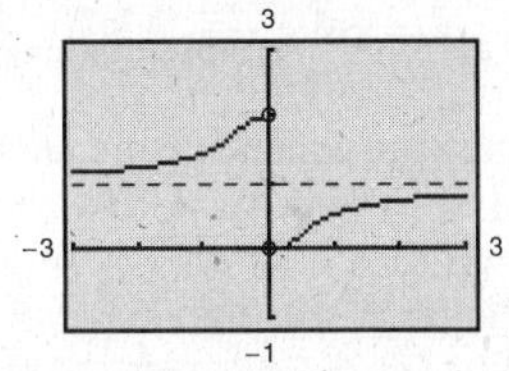

$$\lim_{x\to\infty} \frac{2}{1 + e^{1/x}} = \frac{2}{1 + e^0} = \frac{2}{1+1} = 1$$

$$\lim_{x\to-\infty} \frac{2}{1 + e^{1/x}} = \frac{2}{1 + e^0} = \frac{2}{1+1} = 1$$

Horizontal asymptote: $y = 1$

Discontinuous at $x = 0$

23. (a)

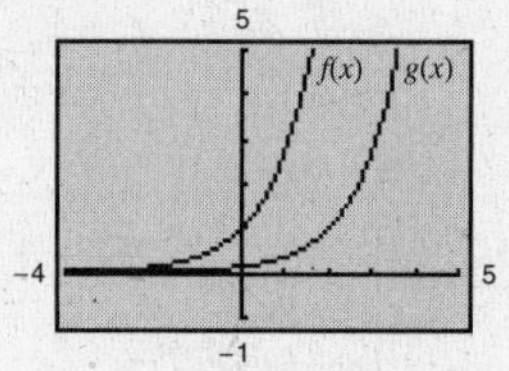

The graph of g is the graph of f shifted to the right two units.

(b)

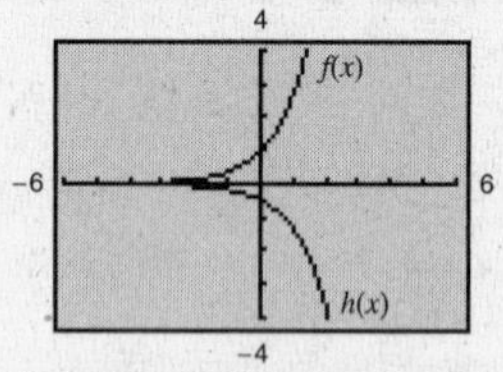

The graph of h is the graph of f reflected in the x-axis and vertically shrunk by a factor of $\frac{1}{2}$.

(c)

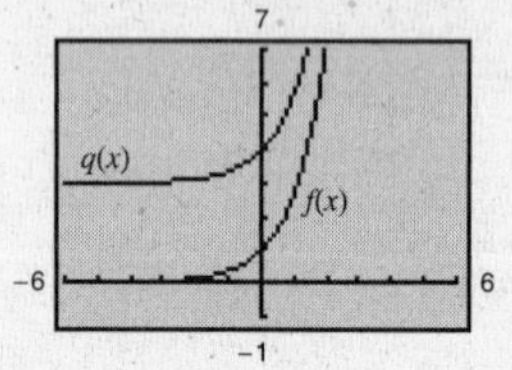

The graph of q is the graph of f shifted upward three units.

25. (a) $P\left(\frac{1}{2}\right) = 1 - e^{-1/2/3} = 1 - e^{-1/6} \approx 0.1535 = 15.35\%$

(b) $P(2) = 1 - e^{-2/3} \approx 0.4866 = 48.66\%$

(c) $P(5) = 1 - e^{-5/3} \approx 0.8111 = 81.11\%$

27. $P = 68.4e^{0.0467t}$ ($t = 0$ corresponds to 1960.)

(a) 1960: $P(0) = 68.4e^{0.0467(0)} = 68.4$ thousand

1970: $P(10) = 68.4e^{0.0467(10)} \approx 109.1$ thousand

1980: $P(20) = 68.4e^{0.0467(20)} \approx 174.1$ thousand

1990: $P(30) = 68.4e^{0.0467(30)} \approx 277.7$ thousand

2000: $P(40) = 68.4e^{0.0467(40)} \approx 442.9$ thousand

2005: $P(45) = 68.4e^{0.0467(45)} \approx 559.4$ thousand

(b) The population is growing exponentially, not by the same rate each year.

(c)

$$\begin{aligned} 900 &= 68.4e^{0.0467t} \\ 13.158 &\approx e^{0.0467t} \\ \ln 13.158 &\approx 0.0467t \\ \frac{\ln 13.158}{0.0467} &\approx t \\ 55.2 &\approx t \end{aligned}$$

By 2015, the population will exceed 900,000.

29.

Interval	1	2	3	4	5	6
Number of cells	1	2	4	8	16	32

Interval	7	8	9	10
Number of cells	64	128	256	512

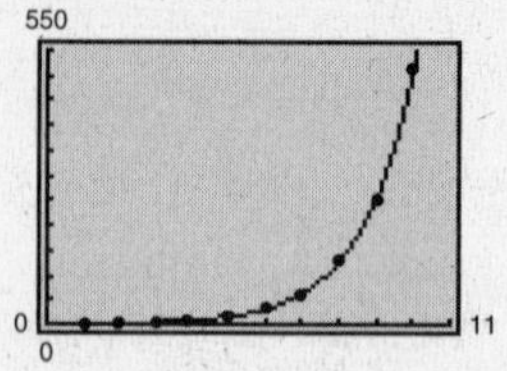

Model: $y = 2^{t-1}$ or $y = e^{(t-1)\ln 2}$

31. $y = \dfrac{300}{3 + 17e^{-0.0625x}}$

(a) When $x = 2$ (2000 egg masses)

$$y = \frac{300}{3 + 17e^{-0.0625(2)}} \approx 16.66.$$

If 2000 egg masses are counted, the percent of defoliation is about 16.66%.

(b)

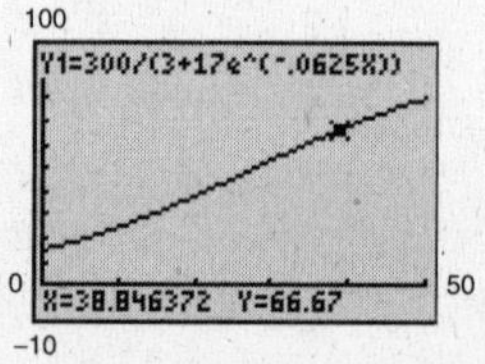

If approximately 66.67% of a forest was defoliated, then there were about 38,846 egg masses.

Section 4.3 Derivatives of Exponential Functions

Skills Review

1. $x^2e^x - \frac{1}{2}e^x = e^x\left(x^2 - \frac{1}{2}\right)$

2. $\left(xe^{-x}\right)^{-1} + e^x = x^{-1}e^x + e^x = e^x\left(x^{-1} + 1\right)$

3. $xe^x - e^{2x} = e^x\left(x - e^x\right)$

4. $e^x - xe^{-x} = e^{-x}\left(e^{2x} - x\right)$

5. $f(x) = \dfrac{3}{7x^2} = \dfrac{3}{7}\left(x^{-2}\right)$

$f'(x) = -2\left(\dfrac{3}{7}\right)x^{-3} = -\dfrac{6}{7x^3}$

6. $g(x) = 3x^2 - \dfrac{x}{6}$

$g'(x) = 6x - \dfrac{1}{6}$

7. $f(x) = (4x - 3)(x^2 + 9)$

$f'(x) = (4x - 3)(2x) + (x^2 + 9)(4)$

$= 8x^2 - 6x + 4x^2 + 36$

$= 12x^2 - 6x + 36$

8. $f(t) = \dfrac{t-2}{\sqrt{t}} = \dfrac{t-2}{t^{1/2}} = t^{1-1/2} - 2t^{-1/2} = t^{1/2} - 2t^{-1/2}$

$f'(t) = \dfrac{1}{2}t^{-1/2} - 2\left(-\dfrac{1}{2}\right)t^{-3/2}$

$= \dfrac{1}{2t^{1/2}} + \dfrac{1}{t^{3/2}}$

$= \dfrac{1}{2\sqrt{t}} + \dfrac{1}{t\sqrt{t}}$

$= \dfrac{t}{2t\sqrt{t}} + \dfrac{2}{2t\sqrt{t}}$

$= \dfrac{t+2}{2t\sqrt{t}}$

9. $f(x) = \dfrac{1}{8}x^3 - 2x$

$f'(x) = \dfrac{3}{8}x^2 - 2 = 0$

$\dfrac{3}{8}x^2 = 2$

$x^2 = \dfrac{16}{3}$

Critical numbers: $x = \pm\dfrac{4}{\sqrt{3}} = \pm\dfrac{4\sqrt{3}}{3}$

Interval	$-\infty < x < -\frac{4\sqrt{3}}{3}$	$-\frac{4\sqrt{3}}{3} < x < \frac{4\sqrt{3}}{3}$	$\frac{4\sqrt{3}}{3} < x < \infty$
Sign of f'	$f' > 0$	$f' < 0$	$f' > 0$
Conclusion	Increasing	Decreasing	Increasing

Relative maximum: $\left(-\dfrac{4\sqrt{3}}{3}, \dfrac{16\sqrt{3}}{9}\right)$

Relative minimum: $\left(\dfrac{4\sqrt{3}}{3}, -\dfrac{16\sqrt{3}}{9}\right)$

Skills Review —*continued*—

10. $f(x) = x^4 - 2x^2 + 5$

$$f'(x) = 4x^3 - 4x = 0$$

$$4x(x^2 - 1) = 0$$

Critical numbers: $x = 0, x = \pm 1$

Interval	$-\infty < x < -1$	$-1 < x < 0$	$0 < x < 1$	$1 < x < \infty$
Sign of f'	$f' < 0$	$f' > 0$	$f' < 0$	$f' > 0$
Conclusion	Decreasing	Increasing	Decreasing	Increasing

Relative maximum: $(0, 5)$

Relative minima: $(-1, 4), (1, 4)$

1. $y = e^{3x}$

$y' = 3e^{3x}$

$y'(0) = 3$

3. $y = e^{-x}$

$y' = -e^{-x}$

$y'(0) = -1$

5. $y = e^{5x}$

$y' = 5e^{5x}$

7. $y = e^{-x^2}$

$y' = (-2x)e^{-x^2} = -2xe^{-x^2}$

9. $f(x) = e^{-1/x^2}$

$$f'(x) = e^{-1/x^2} \cdot \frac{d}{dx}(-x^{-2}) = e^{-1/x^2} \cdot 2x^{-3} = \frac{2}{x^3}e^{-1/x^2}$$

11. $f(x) = (x^2 + 1)e^{4x}$

$f'(x) = (x^2 + 1)4e^{4x} + (2x)e^{4x} = e^{4x}(4x^2 + 2x + 4)$

13. $$f(x) = \frac{2}{(e^x + e^{-x})^3} = 2(e^x + e^{-x})^{-3}$$

$$f'(x) = -6(e^x + e^{-x})^{-4}(e^x - e^{-x}) = -\frac{6(e^x - e^{-x})}{(e^x + e^{-x})^4}$$

15. $y = xe^x - 4e^{-x}$

$y' = xe^x + e^x + 4e^{-x}$

17. $y = e^{-2x+x^2}$

$y' = e^{-2x+x^2}(-2 + 2x)$

$y'(2) = 1(2) = 2$

$y - 1 = 2(x - 2)$

$y = 2x - 3$

19. $y = x^2e^{-x}$

$y' = x^2(-e^{-x}) + 2xe^{-x} = xe^{-x}(-x + 2)$

$y'(2) = 0$

$$y - \frac{4}{e^2} = 0(x - 2)$$

$$y = \frac{4}{e^2} = 4e^{-2} \quad \text{(horizontal line)}$$

21. $y = (e^{2x} + 1)^3$

$y' = 3(e^{2x} + 1)^2(2e^{2x})$

$y'(0) = 3(4)(2) = 24$

$y - 8 = 24(x - 0)$

$y = 24x + 8$

23. $f(x) = 2e^{3x} + 3e^{-2x}$

$f'(x) = 6e^{3x} - 6e^{-2x}$

$f''(x) = 18e^{3x} + 12e^{-2x} = 6(3e^{3x} + 2e^{-2x})$

25. $f(x) = 5e^{-x} - 2e^{-5x}$

$f'(x) = -5e^{-x} + 10e^{-5x}$

$f''(x) = 5e^{-x} - 50e^{-5x} = 5e^{-x}(1 - 10e^{-4x})$

27. $f(x) = \dfrac{1}{2 - e^{-x}}$

$f'(x) = \dfrac{-e^{-x}}{\left(2 - e^{-x}\right)^2}$

$f''(x) = \dfrac{e^{-x}\left(2 + e^{-x}\right)}{\left(2 - e^{-x}\right)^3}$

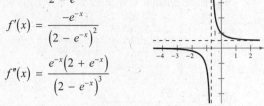

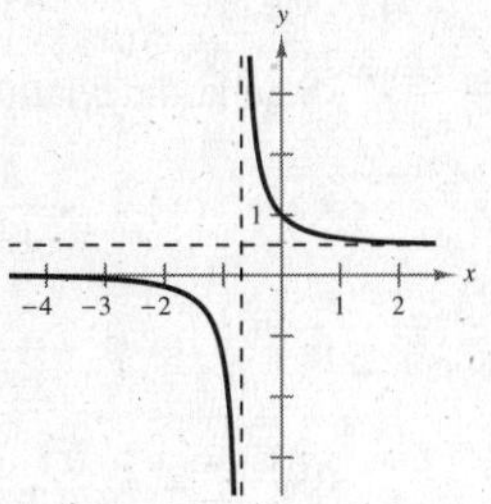

Horizontal asymptote to the right: $y = \dfrac{1}{2}$

Horizontal asymptote to the left: $y = 0$

Vertical asymptote when $2 = e^{-x} \Rightarrow x \approx -0.693$

No relative extrema or inflection points

x	-3	-2	-1	0	1
$f(x)$	-0.055	-0.186	-1.392	1	0.613

29. $f(x) = x^2e^{-x}$

$f'(x) = -x^2e^{-x} + 2xe^{-x} = xe^{-x}(2 - x)$

$f''(x) = x^2e^{-x} - 4xe^{-x} + 2e^{-x} = e^{-x}\left(x^2 - 4x + 2\right)$

$f'(x) = 0$ when $x = 0$ and $x = 2$. Because $f''(0) > 0$ and $f''(2) < 0$, you have the following.

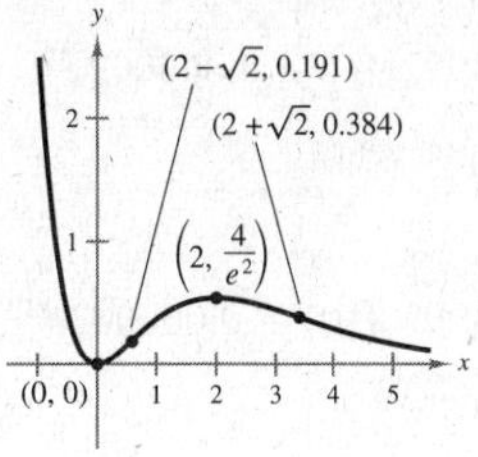

Relative minimum: $(0, 0)$

Relative maximum: $\left(2, \dfrac{4}{e^2}\right)$

Because $f''(x) = 0$ when $x = 2 \pm \sqrt{2}$, the inflection points occur at $\left(2 - \sqrt{2}, 0.191\right)$ and $\left(2 + \sqrt{2}, 0.384\right)$.

Horizontal asymptote: $y = 0$

x	-2	-1	0	1	2	3
$f(x)$	29.56	2.72	0	0.37	0.54	0.45

31. $f(x) = \dfrac{8}{1 + e^{-0.5x}}$

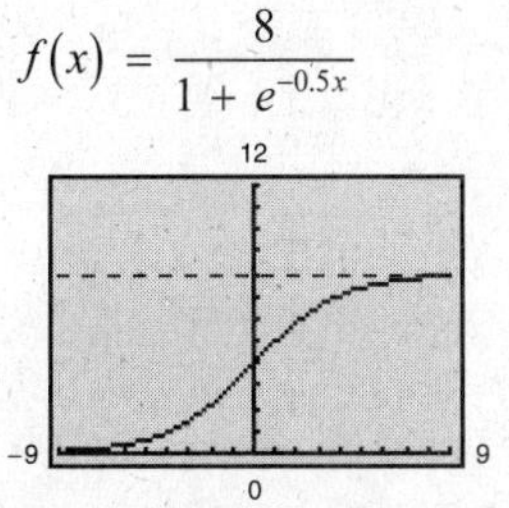

Horizontal asymptotes: $y = 8$ (as $x \to \infty$)

$y = 0$ (as $x \to -\infty$)

No vertical asymptotes

33. $e^{-3x} = e$

$e^{-3x} = e^1$

$-3x = 1$

$x = -\frac{1}{3}$

35. $e^{\sqrt{x}} = e^3$

$\sqrt{x} = 3$

$x = 9$

37. $l = 337 - 276e^{-0.178t}$

$l' = 49.128e^{-0.178t}$

$l'(3) = 49.128e^{-0.178(3)} \approx 28.8$ centimeters per year

39. $p = (100 - 20)e^{-(0.5)t} + 20 = 80e^{-0.5t} + 20$

$p' = -40e^{-0.5t}$

When $t = 1$, $p' = -40e^{-0.5(1)} \approx -24.3\%$ per week.

When $t = 3$, $p' = -40e^{-0.5(3)} \approx -8.9\%$ per week.

41. $y = 98.020 + 6.2472t - 0.24964t^2 + 0.000002e^t$

($t = 6$ corresponds to 1996.)

(a)

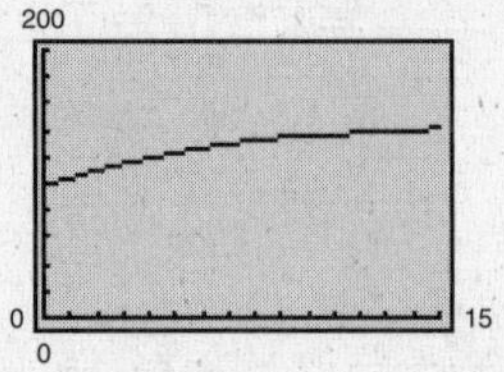

(b) 1996: $y'(6) \approx 3.25$ million per year

2000: $y'(10) \approx 1.30$ million per year

2005: $y'(15) \approx 5.30$ million per year

(c) $y' = 6.2472 - 0.49928t + 0.000002e^t$

When $t = 6$,

$y' = 6.2472 - 0.49928(6) + 0.000002e^{(6)}$

≈ 3.25 million per year

When $t = 10$,

$y' = 6.2472 - 0.49928(10) + 0.000002e^{(10)}$

≈ 1.30 million per year

When $t = 15$,

$y' = 6.2472 - 0.49928(15) + 0.000002e^{(15)}$

≈ 5.30 million per year

43. Mean $= \mu = 650$

Standard deviation $= \sigma = 12.5$

(a) $f(x) = \dfrac{1}{\sigma\sqrt{2\pi}}e^{-(x-\sigma)^2/2\sigma^2}$

$= \dfrac{1}{12.5\sqrt{2\pi}}e^{-(x-650)^2/312.5}$

(b)

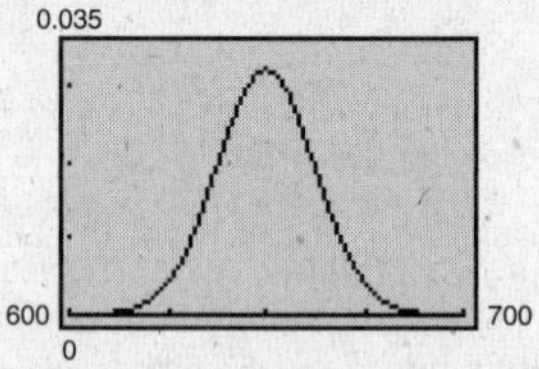

(c) $f'(x) = \dfrac{1}{12.5\sqrt{2\pi}}e^{-(x-650)^2/312.5}\left(\dfrac{-2(x-650)}{312.5}\right)$

$= \dfrac{-1}{1953.125\sqrt{2\pi}}(x-650)e^{-(x-650)^2/312.5}$

(d) Answers will vary.

45. $f(x) = \dfrac{1}{\sigma\sqrt{\pi}}e^{-(x-\mu)^2/2\sigma^2} = \dfrac{1}{\sigma\sqrt{2\pi}}e^{-x^2/2\sigma^2}$

For larger σ, the graph becomes flatter.

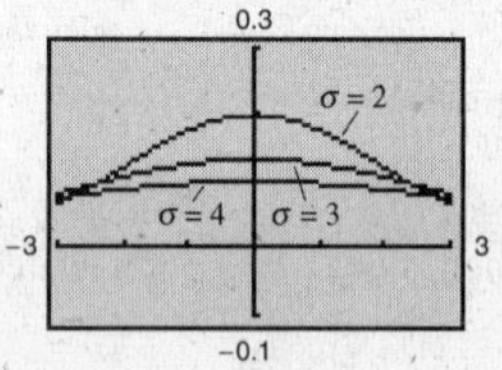

47. $f(x) = \dfrac{1}{\sigma\sqrt{2\pi}}e^{-x^2/2\sigma^2}$

$f'(x) = \dfrac{1}{\sigma\sqrt{2\pi}}\left(-\dfrac{x}{\sigma^2}\right)e^{-x^2/2\sigma^2} = -\dfrac{1}{\sigma^3\sqrt{2\pi}}(x)e^{-x^2/2\sigma^2} \Rightarrow$ Maximum at $x = 0$

$f''(x) = \dfrac{1}{\sigma\sqrt{2\pi}}\left[\left(-\dfrac{x}{\sigma^2}\right)\left(-\dfrac{x}{\sigma^2}\right)e^{-x^2/2\sigma^2} + \left(-\dfrac{1}{\sigma^2}\right)e^{-x^2/2\sigma^2}\right]$

$f''(x) = \dfrac{1}{\sigma\sqrt{2\pi}}\left(e^{-x^2/2\sigma^2}\right)\left(\dfrac{x^2}{\sigma^4} - \dfrac{1}{\sigma^2}\right)$

$0 = \dfrac{1}{\sigma\sqrt{2\pi}}\left(e^{-x^2/2\sigma^2}\right)\left(\dfrac{x^2}{\sigma^4} - \dfrac{1}{\sigma^2}\right)$

$0 = \left(\dfrac{x^2}{\sigma^4} - \dfrac{1}{\sigma^2}\right)$

$\dfrac{1}{\sigma^2} = \dfrac{x^2}{\sigma^4}$

$\sigma^4 = \sigma^2x^2$

$\sigma^2 = x^2$

$\pm\sigma = x$

Interval	$-\infty < x < -\sigma$	$-\sigma < x < \sigma$	$\sigma < x < \infty$
Test value	-2σ	$\frac{1}{2}\sigma$	2σ
Sign of f''	$f'' > 0$	$f'' < 0$	$f'' > 0$
Conclusion	Concave upward	Concave downward	Concave upward

Because concavity changes at $x = \pm\sigma$, you know the graph has points of inflection at $x = \pm\sigma$.

Mid-Chapter Quiz Solutions

1. $4(4)^2 = 4^3 = 64$

2. $\left(\frac{2}{3}\right)^3 = \frac{2^3}{3^3} = \frac{8}{27}$

3. $81^{1/3} = \sqrt[3]{81} = \sqrt[3]{27 \cdot 3} = 3\sqrt[3]{3}$

4. $\left(\frac{4}{9}\right)^2 = \frac{4^2}{9^2} = \frac{16}{81}$

5. $4^3\left(4^2\right) = 4^5 = 1024$

6. $\left(\frac{1}{6}\right)^{-3} = 6^3 = 216$

7. $\frac{3^8}{3^5} = 3^3 = 27$

8. $\left(5^{1/2}\right)\left(3^{1/2}\right) = \sqrt{5 \cdot 3} = \sqrt{15}$

9. $\left(e^2\right)\left(e^5\right) = e^7$

10. $\left(e^{2/3}\right)\left(e^3\right) = e^{11/3}$

11. $\frac{e^2}{e^{-4}} = e^2\left(e^4\right) = e^6$

12. $\left(e^{-1}\right)^{-3} = e^3$

13. $f(x) = 3^x - 2$

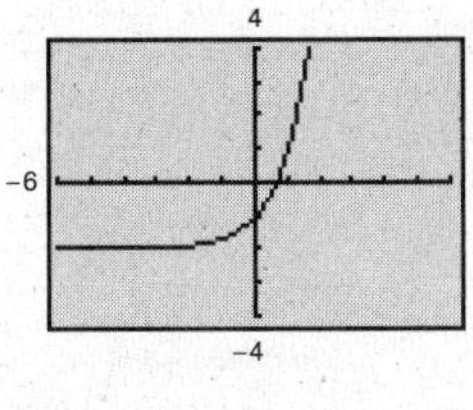

14. $f(x) = 5^{-x} + 2$

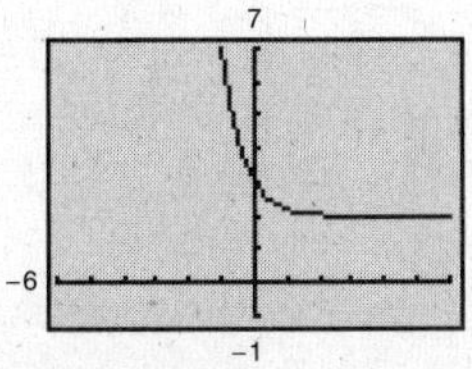

15. $f(x) = 6^{x-3}$

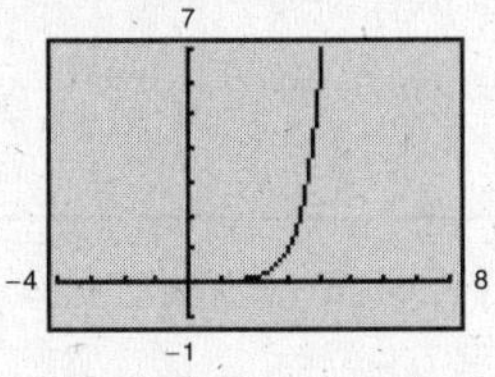

16. $f(x) = e^{x+2}$

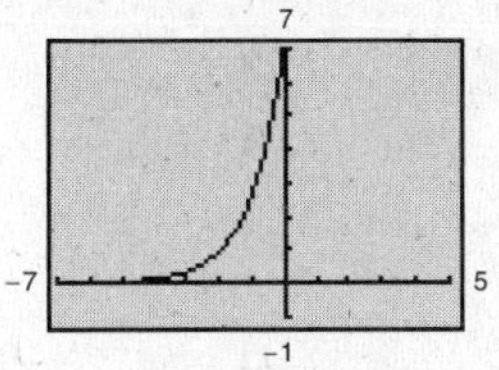

17. $f(x) = 250e^{0.15x}$

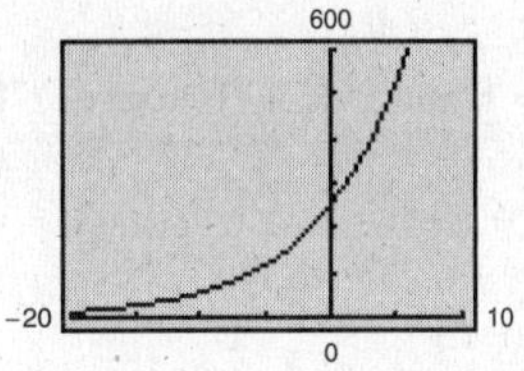

18. $f(x) = \frac{5}{1 + e^x}$

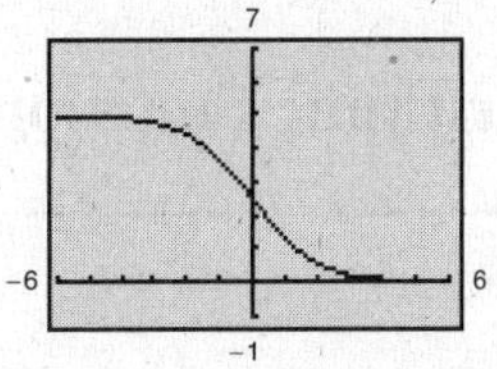

19. (a) $\mu = 5.4, \sigma = 0.5$

$$f(x) = \frac{1}{\sigma\sqrt{2\pi}}e^{-(x-\mu)^2/2\sigma^2} = \frac{1}{0.5\sqrt{2\pi}}e^{-(x-5.4)^2/0.5}$$

(b)

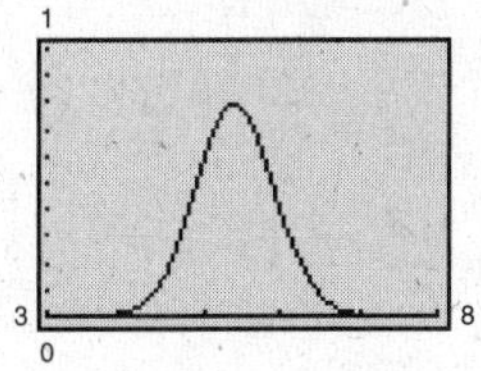

(c) $f'(x) = \frac{1}{0.5\sqrt{2\pi}}\left[-4(x - 5.4)e^{-(x-5.4)^2/0.5}\right]$

(d) Choose a value of x that is less than 5.4, such as $x = 5$.

$f'(5) \approx 0.927 > 0$

Choose a value of x that is greater than 5.4, such as $x = 6$.

$f'(6) \approx -0.932 < 0$

So, $f' > 0$ for $x < \mu$ and $f' < 0$ for $x > \mu$.

20. $y = e^{5x}$

$y' = 5e^{5x}$

21. $y = e^{x-4}$

$y' = e^{x-4}$

22. $y = 5e^{x+2}$

$y' = 5e^{x+2}$

23. $y = 3e^x - xe^x$

$y' = 3e^x - xe^x - e^x = e^x(2 - x)$

24. $y = e^{-2x}$

$y' = -2e^{-2x}$

$y'(0) = -2$

$y - 1 = -2(x - 0)$

$y = -2x + 1$

25. $f(x) = 0.5x^2e^{-0.5x}$

$f'(x) = 0.5x^2(-0.5)e^{-0.5x} + e^{-0.5x}(x) = xe^{-0.5x}(-0.25x + 1)$

$f''(x) = -0.25x^2(-0.5)e^{-0.5x} + e^{-0.5x}(-0.5x) + x(-0.5)e^{-0.5x} + e^{-0.5x} = 0.125x^2e^{-0.5x} - xe^{-0.5x} + e^{-0.5x} = e^{-0.5x}(0.125x^2 - x + 1)$

Horizontal asymptote: $y = 0$

$f'(x) = 0$ when $x = 0$ and $x = 4$. Because $f''(0) > 0$ and $f''(4) < 0$, you have the following.

Relative maximum: $\left(4, \dfrac{8}{e^2}\right)$

Relative minimum: $(0, 0)$

x	−2	−1	0	1	2
$f(x)$	5.44	0.82	0	0.30	0.74

Because $f''(x) = 0$ when $x = 4 \pm 2\sqrt{2}$, the inflection points occur at $(4 + 2\sqrt{2}, 0.77)$ and $(4 - 2\sqrt{2}, 0.38)$.

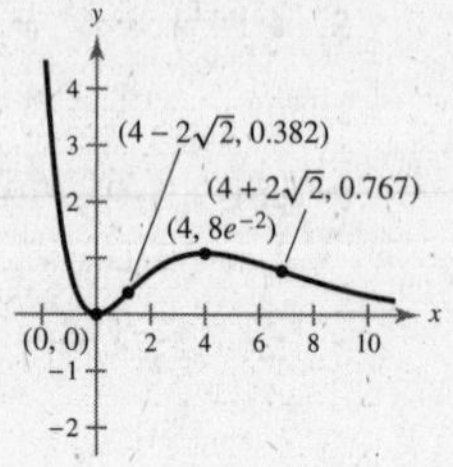

Section 4.4 Logarithmic Functions

Skills Review

1. $(4^2)(4^{-3}) = 4^{-1} = \frac{1}{4}$

2. $(2^3)^2 = 2^6 = 64$

3. $\dfrac{3^4}{3^{-2}} = 3^6 = 729$

4. $\left(\dfrac{3}{2}\right)^{-3} = \left(\dfrac{2}{3}\right)^3 = \dfrac{2^3}{3^3} = \dfrac{8}{27}$

5. $e^0 = 1$

6. $(3e)^4 = 3^4e^4 = 81e^4$

7. $\left(\dfrac{2}{e^3}\right)^{-1} = \dfrac{e^3}{2}$

8. $\left(\dfrac{4e^2}{25}\right)^{-3/2} = \left(\dfrac{25}{4e^2}\right)^{3/2} = \dfrac{125}{8e^3}$

9. $0 < x + 4$

$-4 < x$

10. $0 < x^2 + 1$

x is all real numbers.

11. $0 < \sqrt{x^2 - 1}$

$0 < x^2 - 1$

$1 < x^2 \Rightarrow 1 < x, -1 > x$

$(-\infty, -1)$ and $(1, \infty)$

12. $0 < x - 5$

$5 < x$

13. $f(x) = \dfrac{x}{2} + 3$

$g(x) = 2x - 6$

$f(g(x)) = f(2x - 6) = \dfrac{2x - 6}{2} + 3 = x$

$g(f(x)) = g\left(\dfrac{x}{2} + 3\right) = 2\left(\dfrac{x}{2} + 3\right) - 6 = x$

14. $f(x) = \sqrt[3]{x - 8}$

$g(x) = x^3 + 8$

$f(g(x)) = f(x^3 + 8) = \sqrt[3]{x^3 + 8 - 8} = x$

$g(f(x)) = g(\sqrt[3]{x - 8}) = (\sqrt[3]{x - 8})^3 + 8 = x$

1. $e^{0.6931\ldots} = 2$

3. $e^{-1.6094\ldots} = 0.2$

5. $\ln 1 = 0$

7. $\ln 0.0498 = -3$

9. $f(x) = 2 + \ln x$

The graph is a logarithmic curve that passes through the point $(1, 2)$ with a vertical asymptote at $x = 0$.

It matches graph (c).

11. $f(x) = \ln(x + 2)$

The graph is a logarithmic curve that passes through the point $(-1, 0)$ with a vertical asymptote at $x = -2$. It matches graph (b).

13. $y = \ln(x - 1)$

x	1.5	2	3	4	5
y	−0.69	0	0.69	1.10	1.39

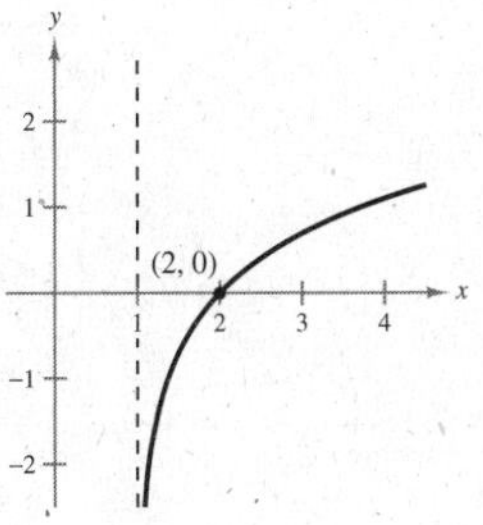

15. $y = \ln 2x$

x	0.25	0.5	1	3	5
y	−0.69	0	0.69	1.79	2.30

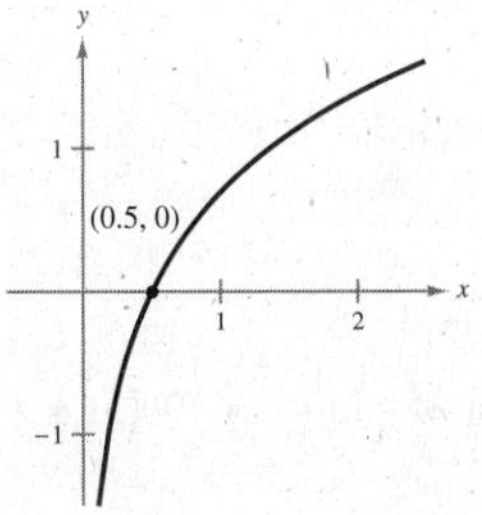

17. $y = 3 \ln x$

x	0.5	1	2	3	4
y	−2.08	0	2.08	3.30	4.16

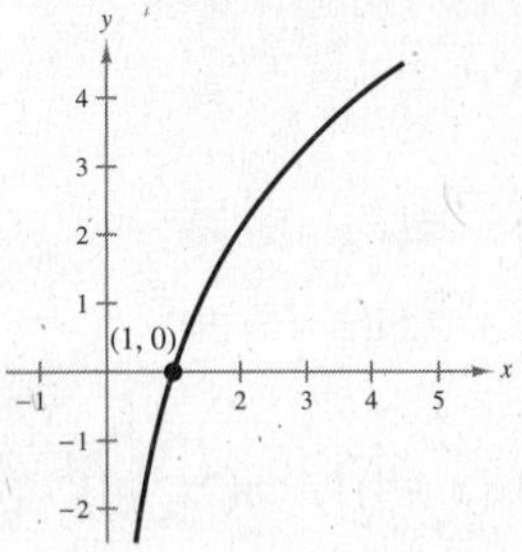

19. $g(x) = \ln \sqrt{x} = \frac{1}{2} \ln x$

$$f(g(x)) = f\left(\tfrac{1}{2} \ln x\right) = e^{2(1/2 \ln x)} = e^{\ln x} = x$$

$$g(f(x)) = g(e^{2x}) = \tfrac{1}{2} \ln e^{2x} = \tfrac{1}{2}(2x) \ln e = x$$

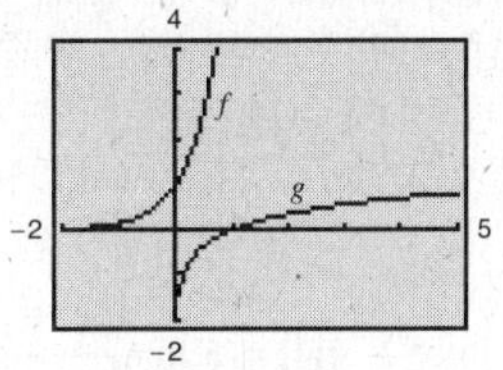

21.
$$\begin{aligned} f(g(x)) &= f\left(\tfrac{1}{2} + \ln \sqrt{x}\right) \\ &= e^{2\left[(1/2) + \ln \sqrt{x}\right] - 1} \\ &= e^{2 \ln x^{1/2}} \\ &= e^{\ln x} \\ &= x \end{aligned}$$

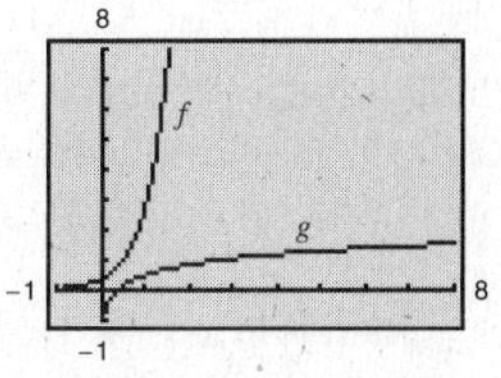

$$\begin{aligned} g(f(x)) &= g(e^{2x-1}) \\ &= \tfrac{1}{2} + \ln \sqrt{e^{2x-1}} \\ &= \tfrac{1}{2} + \tfrac{1}{2} \ln e^{2x-1} \\ &= \tfrac{1}{2} + \tfrac{1}{2}(2x - 1) \\ &= x \end{aligned}$$

23. $\ln e^{x^2} = x^2$

25. $e^{\ln(5x+2)} = 5x + 2$

27. $-1 + \ln e^{2x} = -1 + 2x = 2x - 1$

29. (a) $\ln 6 = \ln(2 \cdot 3) = \ln 2 + \ln 3$
$\approx 0.6931 + 1.0986 = 1.7917$

(b) $\ln \frac{3}{2} = \ln 3 - \ln 2 \approx 1.0986 - 0.6931 = 0.4055$

(c) $\ln 81 = \ln 3^4 = 4 \ln 3 \approx 4(1.0986) = 4.3944$

(d) $\ln \sqrt{3} = \ln 3^{(1/2)} = \left(\frac{1}{2}\right) \ln 3$
$\approx \left(\frac{1}{2}\right)(1.0986) = 0.5493$

31. $\ln \frac{2}{3} = \ln 2 - \ln 3$

33. $\ln xyz = \ln x + \ln y + \ln z$

35. $\ln\sqrt{x^2 + 1} = \ln(x^2 + 1)^{1/2}$

$= \frac{1}{2}\ln(x^2 + 1)$

37. $\ln\left[z(z - 1)^2\right] = \ln z + \ln(z - 1)^2$

$= \ln z + 2\ln(z - 1)$

39. $\ln \dfrac{3x(x + 1)}{(2x + 1)^2} = \ln\left[3x(x + 1)\right] - \ln(2x + 1)^2$

$= \ln 3 + \ln x + \ln(x + 1) - 2\ln(2x + 1)$

41. $\ln(x - 2) - \ln(x + 2) = \ln \dfrac{x - 2}{x + 2}$

43. $3\ln x + 2\ln y - 4\ln z = \ln x^3 + \ln y^2 - \ln z^4$

$= \ln\left(\dfrac{x^3y^2}{z^4}\right)$

45. $3\left[\ln x + \ln(x + 3) - \ln(x + 4)\right] = 3\ln \dfrac{x(x + 3)}{x + 4}$

$= \ln\left[\dfrac{x(x + 3)}{x + 4}\right]^3$

47. $\frac{3}{2}\left[\ln x(x^2 + 1) - \ln(x + 1)\right] = \frac{3}{2}\ln \dfrac{x(x^2 + 1)}{x + 1}$

$= \ln\left[\dfrac{x(x^2 + 1)}{x + 1}\right]^{3/2}$

49. $\frac{1}{3}\ln(x + 1) - \frac{2}{3}\ln(x - 1) = \ln(x + 1)^{1/3} - \ln(x - 1)^{2/3}$

$= \ln\left[\dfrac{(x + 1)^{1/3}}{(x - 1)^{2/3}}\right]$

$= \ln\left[\dfrac{x + 1}{(x - 1)^2}\right]^{1/3}$

$= \ln\sqrt[3]{\dfrac{x + 1}{(x - 1)^2}}$

51. $e^{\ln x} = 4$

$x = 4$

53. $\ln x = 0$

$x = e^0$

$x = 1$

55. $\ln 2x = 2.4$

$2x = e^{2.4}$

$x = \dfrac{e^{2.4}}{2} \approx 5.51$

57. $3\ln 5x = 10$

$\ln 5x = \dfrac{10}{3}$

$5x = e^{10/3}$

$x = \dfrac{e^{10/3}}{5} \approx 5.61$

59. $e^{x+1} = 4$

$x + 1 = \ln 4$

$x = \ln 4 - 1 \approx 0.39$

61. $300e^{-0.2t} = 700$

$e^{-0.2t} = \dfrac{7}{3}$

$-0.2t = \ln 7 - \ln 3$

$t = \dfrac{\ln 7 - \ln 3}{-0.2} \approx -4.24$

63. $4e^{2x-1} - 1 = 5$

$4e^{2x-1} = 6$

$e^{2x-1} = \dfrac{3}{2}$

$2x - 1 = \ln 3 - \ln 2$

$x = \dfrac{\ln 3 - \ln 2 + 1}{2}$

$x = \dfrac{\ln 3 - \ln 2 + 1}{2} \approx 0.70$

65. $\dfrac{10}{1 + 4e^{-0.01x}} = 2.5$

$1 + 4e^{-0.01x} = \dfrac{10}{2.5} = 4$

$4e^{-0.01x} = 3$

$e^{-0.01x} = \dfrac{3}{4}$

$-0.01x = \ln 3 - \ln 4$

$x = -100(\ln 3 - \ln 4) = 100(\ln 4 - \ln 3)$

≈ 28.77

67. $5^{2x} = 15$

$\ln 5^{2x} = \ln 15$

$2x\ln 5 = \ln 15$

$x = \dfrac{\ln 15}{2\ln 5} \approx 0.84$

69. $500(1.07)^t = 1000$

$1.07^t = 2$

$t \ln 1.07 = \ln 2$

$t = \dfrac{\ln 2}{\ln 1.07} \approx 10.24$

71. $\left(1 + \dfrac{0.07}{12}\right)^{12t} = 3$

$12t \ln\left(1 + \dfrac{0.07}{12}\right) = \ln 3$

$t = \dfrac{\ln 3}{12 \ln\left[1 + (0.07/12)\right]} \approx 15.74$

73. $\left(16 - \dfrac{0.878}{26}\right)^{3t} = 30$

$3t \ln\left(16 - \dfrac{0.878}{26}\right) = \ln 30$

$t = \dfrac{\ln 30}{3 \ln\left[16 - (0.878/26)\right]} \approx 0.41$

75. $P = 131e^{0.019t}$ ($t = 0$ corresponds to 1980.)

(a) When $t = 2005$, $P(25) = 131e^{0.019(25)} \approx 210{,}650$.

The population in 2005 was about 210,650.

(b) $131e^{0.019t} = 300$

$e^{0.019t} = \dfrac{300}{131}$

$0.019t = \ln \dfrac{300}{131}$

$t = \dfrac{\ln 300 - \ln 131}{0.019}$

$t \approx 43.6$

The population will be 300,000 in 2023.

77. $0.32 \times 10^{-12} = 10^{-12}\left(\dfrac{1}{2}\right)^{t/5715}$

$0.32 = \left(\dfrac{1}{2}\right)^{t/5715}$

$\ln 0.32 = \dfrac{t}{5715} \ln \dfrac{1}{2}$

$t = \dfrac{5715 \ln 0.32}{\ln 1/2} \approx 9394.6$ years

79. $0.22 \times 10^{-12} = 10^{-12}\left(\dfrac{1}{2}\right)^{t/5715}$

$0.22 = \left(\dfrac{1}{2}\right)^{t/5715}$

$\ln 0.22 = \dfrac{t}{5715} \ln\left(\dfrac{1}{2}\right)$

$t = \dfrac{5715 \ln 0.22}{\ln 1/2} \approx 12{,}484.0$ years

81. $N = 19.257 + 10.64 \ln(t - 8)$

($t = 9$ corresponds to 1999.)

(a) When $t = 13$,

$N = 19.257 + 10.64 \ln(13 - 8)$

$= 19.257 + 10.64 \ln(5)$

≈ 36.38 million Internet surfers.

(b) $19.257 + 10.64 \ln(t - 8) = 40$

$10.64 \ln(t - 8) = 20.743$

$\ln(t - 8) \approx 1.9495$

$e^{\ln(t-8)} \approx e^{1.9495}$

$t - 8 \approx e^{1.9495}$

$t \approx e^{1.9495} + 8 \approx 15.03$

The number of Internet surfers reached 40 million in 2005.

83. $P = \dfrac{0.83}{1 + e^{-0.2n}}$

(a)

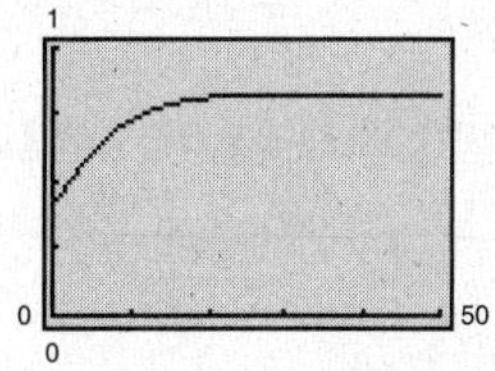

(b) The graph has a horizontal asymptote at $P = 0.83$. As the number of trials increases, the population of correct responses approaches 0.83 or 83%.

(c) $0.60 = \dfrac{0.83}{1 + e^{-0.2n}}$

$0.60\left(1 + e^{-0.2n}\right) = 0.83$

$1 + e^{-0.2n} = \dfrac{83}{60}$

$e^{-0.2n} = \dfrac{23}{60}$

$-0.2n = \ln \dfrac{23}{60}$

$n = \dfrac{\ln 23 - \ln 60}{-0.2}$

$n \approx 5$ trials

85. (a) You cannot find a model of the form $p = a + b \ln n$ for the data because $\ln 0$ is undefined.

(b) $h = 0.86 - 6.447 \ln p$

(c)

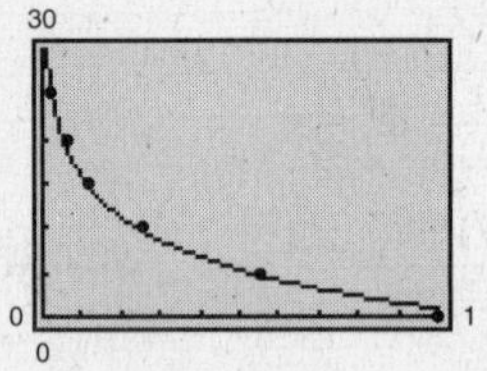

(d) When $p = 0.75$,

$h = 0.86 - 6.447 \ln(0.75) \approx 2.715$ kilometers.

(e) When $h = 13$,

$$13 = 0.86 - 6.447 \ln p$$
$$12.14 = -6.447 \ln p$$
$$-1.883 \approx \ln p$$
$$e^{-1.883} \approx e^{\ln p}$$
$$0.152 \text{ atmosphere} \approx p.$$

87. $f(x) = \dfrac{\ln x}{x}$

x	1	5	10	10^2	10^4	10^6
$f(x)$	0	0.3219	0.2303	0.0461	0.0009	0.00001

(a) $\lim_{x \to \infty} \dfrac{\ln x}{x} = 0$

(b)

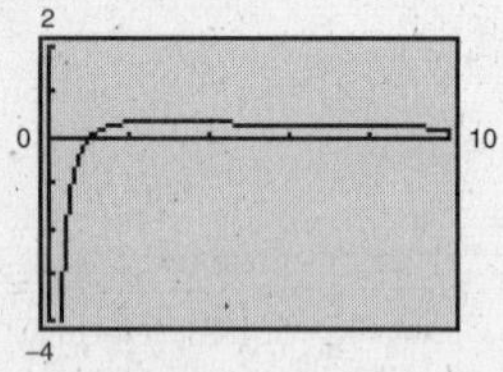

Relative maximum at $(2.7183, 0.3679)$

No relative minima

89.

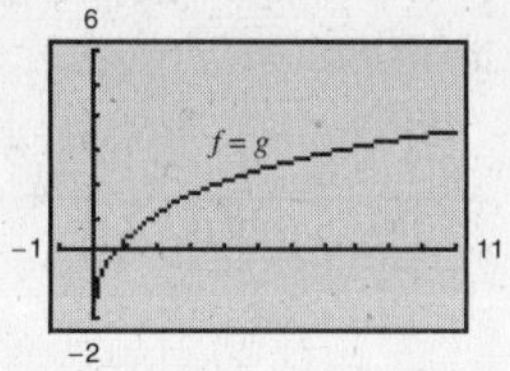

The graphs appear to be identical.

91. True. $\ln(ax) = \ln a + \ln x$

93. False. $\frac{1}{2}f(x) = \frac{1}{2}\ln x = \ln x^{1/2} = \ln\left(\sqrt{x}\right)$

95. True. $\ln x < 0$ for $0 < x < 1$

Section 4.5 Derivatives of Logarithmic Functions

Skills Review

1. $\ln(x+1)^2 = 2\ln(x+1)$

2. $\ln x(x+1) = \ln x + \ln(x+1)$

3. $\ln \dfrac{x}{x+1} = \ln x - \ln(x+1)$

4. $\ln\left(\dfrac{x}{x-3}\right)^3 = 3 \ln \dfrac{x}{x-3} = 3\left[\ln x - \ln(x-3)\right]$

5. $\ln \dfrac{4x(x-7)}{x^2} = \ln 4x + \ln(x-7) - \ln x^2$
$= \ln 4x + \ln(x-7) - 2\ln x$
$= \ln 4 + \ln x + \ln(x-7) - 2\ln x$

6. $\ln x^3(x+1) = \ln x^3 + \ln(x+1) = 3\ln x + \ln(x+1)$

7. $f(x) = x^2 + 2x + 3$; $(0, 3)$
$f'(x) = 2x + 2$
$f'(0) = 2$

8. $f(x) = \dfrac{1}{\sqrt{x+1}}$; $(0, 1)$
$f(x) = (x+1)^{-1/2}$
$f'(x) = -\dfrac{1}{2}(x+1)^{-3/2} = -\dfrac{1}{2(x+1)^{3/2}}$
$f'(0) = -\dfrac{1}{2}$

Skills Review —continued—

9. $f(x) = x^2(x+1) - 3x^3$

$f'(x) = x^2 + 2x(x+1) - 9x^2$

$= x^2 + 2x^2 + 2x - 9x^2 = -6x^2 + 2x$

$f''(x) = -12x + 2$

10. $f(x) = -\dfrac{1}{x^2}$

$f'(x) = \dfrac{2}{x^3}$

$f''(x) = -\dfrac{6}{x^4}$

1. $y = \ln x^3 = 3\ln x$

$y' = \dfrac{3}{x}$

$y'(1) = 3$

3. $y = \ln x^2 = 2\ln x$

$y' = \dfrac{2}{x}$

$y'(1) = 2$

5. $y = \ln x^2 = 2\ln x$

$y' = \dfrac{2}{x}$

7. $y = \ln(x^2 + 3)$

$y' = \dfrac{2x}{x^2 + 3}$

9. $y = \ln\sqrt{x-4}$

$y = \dfrac{1}{2}\ln(x-4) = \dfrac{1}{2}\cdot\dfrac{1}{(x-4)} = \dfrac{1}{2(x-4)}$

11. $y = (\ln x)^4$

$y' = \left[4(\ln x)^3\dfrac{1}{x}\right] = \dfrac{4(\ln x)^3}{x}$

13. $f(x) = 2x\ln x$

$f'(x) = 2x\left(\dfrac{1}{x}\right) + 2\ln x = 2 + 2\ln x$

15. $y = \ln\left(x\sqrt{x^2-1}\right)$

$= \ln x + \ln(x^2+1)^{1/2} = \ln x + \dfrac{1}{2}\ln(x^2-1)$

$y' = \dfrac{1}{x} + \left(\dfrac{1}{2}\right)\dfrac{1}{x^2-1}(2x)$

$= \dfrac{1}{x} + \dfrac{x}{x^2-1} = \dfrac{2x^2-1}{x(x^2-1)}$

17. $y = \ln x - \ln(x+1)$

$y' = \dfrac{1}{x} - \dfrac{1}{x+1} = \dfrac{1}{x(x+1)}$

19. $y = \ln\left(\dfrac{x-1}{x+1}\right)^{1/3} = \dfrac{1}{3}\left[\ln(x-1) - \ln(x+1)\right]$

$y' = \dfrac{1}{3}\left[\dfrac{1}{x-1} - \dfrac{1}{x+1}\right] = \dfrac{1}{3}\left[\dfrac{2}{x^2-1}\right] = \dfrac{2}{3(x^2-1)}$

21. $y = \ln\dfrac{\sqrt{4+x^2}}{x}$

$= \ln\dfrac{(4+x^2)^{1/2}}{x}$

$= \dfrac{1}{2}\ln(4+x^2) - \ln x$

$y' = \dfrac{1}{2}\left(\dfrac{2x}{4+x^2}\right) - \dfrac{1}{x}$

$= \dfrac{x^2 - (4+x^2)}{x(4+x^2)}$

$= -\dfrac{4}{x(4+x^2)}$

23. $g(x) = e^{-x}\ln x$

$g'(x) = e^{-x}\left(\dfrac{1}{x}\right) + (-e^{-x})\ln x = e^{-x}\left(\dfrac{1}{x} - \ln x\right)$

25. $g(x) = \ln\dfrac{e^x + e^{-x}}{2} = \ln(e^x + e^{-x}) - \ln 2$

$g'(x) = \dfrac{e^x - e^{-x}}{e^x + e^{-x}}$

27. $2^x = e^{x(\ln 2)}$

29. $\log_4 x = \dfrac{1}{\ln 4}\ln x$

31. $\log_4 7 = \dfrac{\ln 7}{\ln 4} \approx 1.404$

33. $\log_2 48 = \dfrac{\ln 48}{\ln 2} \approx 5.585$

35. $\log_3 \dfrac{1}{2} = \dfrac{\ln(1/2)}{\ln 3} \approx -0.631$

37. $\log_{1/5}(31) = \dfrac{\ln 31}{\ln(1/5)} \approx -2.134$

39. $y = 3^x$

$y' = (\ln 3)3^x$

41. $f(x) = \log_2 x$

$f'(x) = \dfrac{1}{\ln 2} \cdot \dfrac{1}{x} = \dfrac{1}{x \ln 2}$

43. $h(x) = 4^{2x-3}$

$h'(x) = (\ln 4)4^{2x-3}(2) = (2 \ln 4)4^{2x-3}$

45. $y = \log_{10}(x^2 + 6x)$

$y' = \dfrac{1}{\ln 10}\dfrac{1}{x^2 + 6x}(2x + 6) = \dfrac{2x + 6}{(x^2 + 6x)\ln 10}$

47. $y = x2^x$

$y' = x(\ln 2)2^x + 2^x = 2^x(1 + x \ln 2)$

49. $y = x \ln x$

$y' = x\left(\dfrac{1}{x}\right) + (1)\ln x = 1 + \ln x$

$y'(1) = 1$

$y - 0 = 1(x - 1)$

$y = x - 1$ Tangent Line

51. $y = \log_3 x$

$y' = \dfrac{1}{\ln 3}\dfrac{1}{x}$

$y'(27) = \dfrac{1}{27 \ln 3}$

$y - 3 = \dfrac{1}{27 \ln 3}(x - 27)$

$y = \dfrac{1}{27 \ln 3}x - \dfrac{1}{\ln 3} + 3$ Tangent Line

53. $f(x) = x \ln \sqrt{x} + 2x = \dfrac{1}{2}x \ln x + 2x$

$f'(x) = \dfrac{1}{2}x\left(\dfrac{1}{x}\right) + \dfrac{1}{2}\ln x + 2 = \dfrac{1}{2}\ln x + \dfrac{5}{2}$

$f''(x) = \dfrac{1}{2x}$

55. $f(x) = 2 + x \ln x$

$f'(x) = 1 + \ln x$

$f''(x) = \dfrac{1}{x}$

57. $f(x) = 5^x$

$f'(x) = (\ln 5)5^x$

$f''(x) = (\ln 5)(\ln 5)5^x = (\ln 5)^2 5^x$

59. $\beta = 10 \log_{10} I - 10 \log_{10}(10^{-16})$

$\dfrac{d\beta}{dI} = \dfrac{10}{(\ln 10)I}$

For $I = 10^{-4}$,

$\dfrac{d\beta}{dI} = \dfrac{10}{(\ln 10)10^{-4}} = \dfrac{10^5}{\ln 10}$

$\approx 43{,}429.4$ decibels per watt per cm^2.

61. $f(x) = 1 + 2x \ln x$

$f'(x) = 2x\left(\dfrac{1}{x}\right) + 2 \ln x = 2 + 2 \ln x$

At $(1, 1)$, the slope of the tangent line is $f'(1) = 2$.

Tangent line: $y - 1 = 2(x - 1)$

$y = 2x - 1$

63. $f(x) = \ln\dfrac{5(x + 2)}{x} = \ln 5 + \ln(x + 2) - \ln x$

$f'(x) = \dfrac{1}{x + 2} - \dfrac{1}{x}$

At $(-2.5, 0)$, the slope of the tangent line is

$f'(-2.5) = \dfrac{1}{-2.5 + 2} - \dfrac{1}{-2.5} = -2 + \dfrac{2}{5} = -\dfrac{8}{5}.$

Tangent line: $y - 0 = -\dfrac{8}{5}\left(x + \dfrac{5}{2}\right)$

$y = -\dfrac{8}{5}x - 4$

65. $f(x) = x \log_2 x$

$f'(x) = \log_2 x + x\left(\dfrac{1}{\ln 2} \cdot \dfrac{1}{x}\right) = \log_2 x + \dfrac{1}{\ln 2}$

$f'(1) = \dfrac{1}{\ln 2}$

$y - 0 = \dfrac{1}{\ln 2}(x - 1)$

$y = \dfrac{1}{\ln 2}x - \dfrac{1}{\ln 2}$

67. $y = x - \ln x$

$y' = 1 - \dfrac{1}{x} = \dfrac{x-1}{x}$

$y' = 0$ when $x = 1$.

$y'' = \dfrac{1}{x^2}$

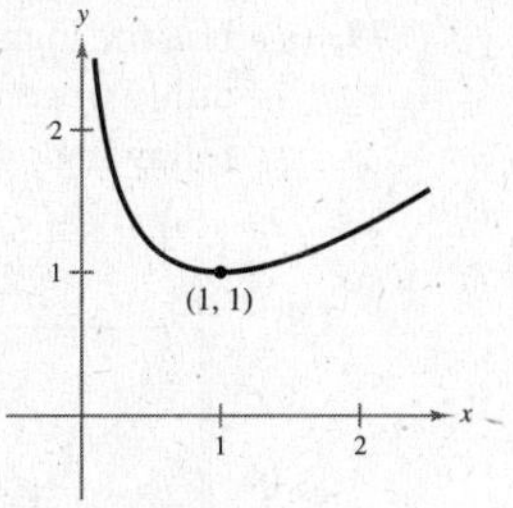

Because $y''(1) = 1 > 0$, there is a relative minimum at $(1, 1)$. Moreover, because $y'' > 0$ on $(0, \infty)$, it follows that the graph is concave upward on its domain and there are no inflection points.

69. The domain of the function

$y = \dfrac{\ln x}{x}$ is $(0, \infty)$.

$y' = \dfrac{1 - \ln x}{x^2}$

$y' = 0$ when $x = e$.

$y'' = \dfrac{2\ln x - 3}{x^3}$

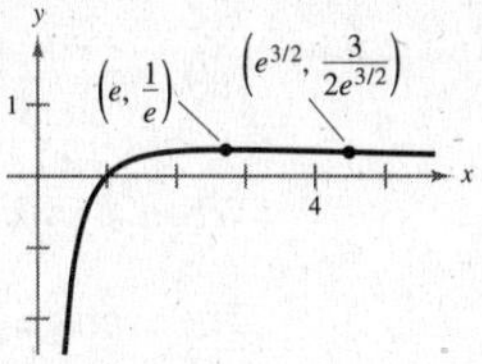

Because $y''(e) < 0$, it follows that $(e, 1/e)$ is a relative maximum. Because $y'' = 0$ when $2\ln x - 3 = 0$ and $x = e^{3/2}$, there is an inflection point at $\left(e^{3/2}, 3/\left(2e^{3/2}\right)\right)$.

71. $y = x^2 \ln \dfrac{x}{4}$

$y' = x^2\left(\dfrac{1}{(1/4)x}\right)\left(\dfrac{1}{4}\right) + 2x\ln\dfrac{x}{4} = x + 2x\ln\dfrac{x}{4}$

$y'' = 1 + 2x\left(\dfrac{1}{(1/4)x}\right)\left(\dfrac{1}{4}\right) + 2\ln\dfrac{x}{4}$

$= 1 + 2 + 2\ln\dfrac{x}{4} = 3 + 2\ln\dfrac{x}{4}$

Because $y''\left(4e^{-1/2}\right) > 0$, there is a relative minimum at $\left(4e^{-1/2}, -8e^{-1}\right)$. Because $y'' = 0$ when $3 + 2\ln\dfrac{x}{4} = 0$ and $x = 4e^{-3/2}$, there is an inflection point at $\left(4e^{-3/2}, -\dfrac{24}{e^3}\right)$.

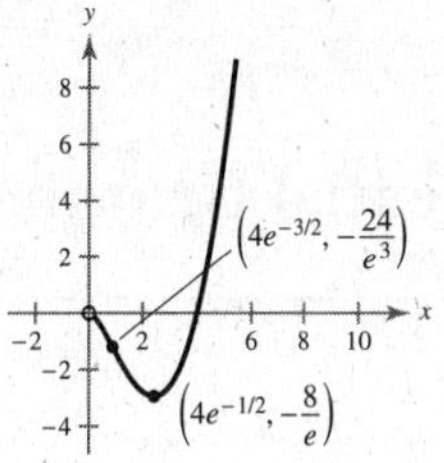

73. $C = 1.42 + 13.33\ln x,\ x \ge 1$

(a)

x	1	10	10^2	10^3	10^4	10^5	10^6
y	1.42	32.11	62.81	93.5	124.19	154.89	185.58

As the depth increases, the concentration increases slowly.

(b) $\dfrac{dC}{dx} = \dfrac{13.33}{x}$

When $x = 10$, $\dfrac{dC}{dx} = \dfrac{13.33}{10} = 1.333$.

The concentration of contamination in the sample is increasing at the rate of 1.333 milligrams per kilogram per centimeter.

75. $C = 500 - 220.3\ln(t+1),\ 0 \le t \le 8$

(a)

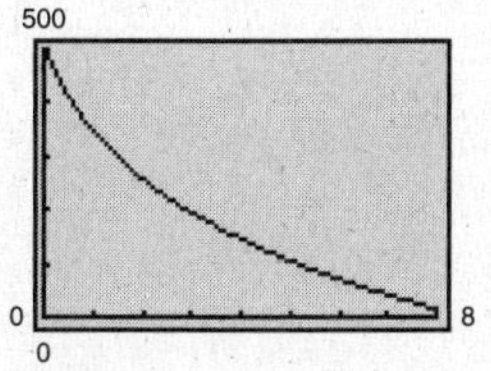

(b) When $t = 4$,

$C = 500 - 220.3\ln(4+1)$

≈ 145.44 milligrams per milliliter.

(c) When $t = 8$,

$C = 500 - 220.3\ln(8+1)$

≈ 15.95 milligrams per milliliter.

(d) $\dfrac{dC}{dt} = -\dfrac{220.3}{t+1}$

When $t = 4$,

$\dfrac{dC}{dt} = -\dfrac{220.3}{4+1}$

$= -44.06$ milligrams per milliliter per hour.

When $t = 8$,

$\dfrac{dC}{dt} = -\dfrac{220.3}{8+1}$

≈ -24.48 milligrams per milliliter per hour.

(e) The model is not valid for determining the concentration 9 or more hours after the injection because for $t \ge 9$, $C < 0$, which does not make sense, in this situation.

77. $T = 20\left[1 + 7\left(2^{-h}\right)\right]$

(a) The temperature is changing most rapidly when $h = 0$.

(b) $\dfrac{dT}{dh} = -20\left[7 \ln 2\left(2^{-h}\right)\right]$

When $h = 0$,

$\dfrac{dT}{dh} = -20\left[7 \ln 2\left(2^{0}\right)\right] \approx -97.04°C$ per hour.

When $h = 1$,

$\dfrac{dT}{dh} = -20\left[7 \ln 2\left(2^{-1}\right)\right] \approx -48.52°C$ per hour.

When $h = 3$,

$\dfrac{dT}{dh} = -20\left[7 \ln 2\left(2^{-3}\right)\right] \approx -12.13°C$ per hour.

79. (a) Use the logarithmic regression feature of a graphing utility or computer algebra system to obtain the following model.

$s(t) = 84.66 - 11.00 \ln x$

(b)

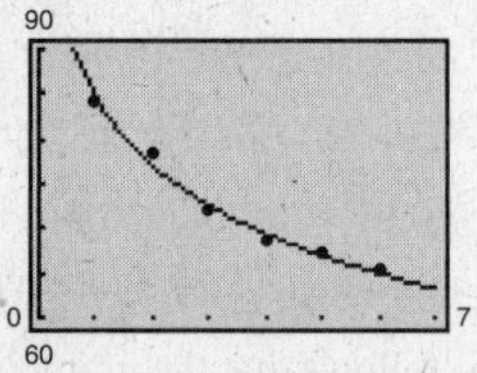

The model fits the data well.

(c) $s(t) = 84.66 - 11.00 \ln x$

$s'(t) = \dfrac{-11.00}{x}$

When $t = 2$, $s'(2) = \dfrac{-11.00}{2} = -5.5$

The average score is decreasing at a rate of 5.5 points per month after 2 months.

Section 4.6 Exponential Growth and Decay

Skills Review

1.
$$\begin{aligned} 12 &= 24e^{4k} \\ \tfrac{1}{2} &= e^{4k} \\ \ln \tfrac{1}{2} &= 4k \\ \tfrac{1}{4}(\ln 1 - \ln 2) &= k \\ \tfrac{1}{4}(\ln 2) &= k \\ -0.1733 &\approx k \end{aligned}$$

2.
$$\begin{aligned} 10 &= 3e^{5k} \\ \tfrac{10}{3} &= e^{5k} \\ \ln \tfrac{10}{3} &= 5k \\ \tfrac{1}{5}(\ln 10 - \ln 3) &= k \\ 0.2408 &\approx k \end{aligned}$$

3.
$$\begin{aligned} 25 &= 16e^{-0.01k} \\ \tfrac{25}{16} &= e^{-0.01k} \\ \ln \tfrac{25}{16} &= -0.01k \\ -100(\ln 25 - \ln 16) &= k \\ -44.6287 &\approx k \end{aligned}$$

4.
$$\begin{aligned} 22 &= 32e^{-0.02k} \\ \tfrac{11}{16} &= e^{-0.02k} \\ \ln \tfrac{11}{16} &= -0.02k \\ -50(\ln 11 - \ln 16) &= k \\ 18.7347 &\approx k \end{aligned}$$

5.
$$\begin{aligned} y &= 32e^{0.23t} \\ y' &= 32e^{0.23t}(0.23) \\ &= 7.36e^{0.23t} \end{aligned}$$

6.
$$\begin{aligned} y &= 18e^{0.072t} \\ y' &= 18e^{0.072t}(0.072) \\ &= 1.296e^{0.072t} \end{aligned}$$

7.
$$\begin{aligned} y &= 24e^{-1.4t} \\ y' &= 24e^{-1.4t}(-1.4) \\ &= -33.6e^{-1.4t} \end{aligned}$$

8.
$$\begin{aligned} y &= 25e^{-0.001t} \\ y' &= 25e^{-0.001t}(-0.001) \\ &= -0.025e^{-0.001t} \end{aligned}$$

9. $e^{\ln 4} = 4$

10. $4e^{\ln 3} = 4(3) = 12$

11. $e^{\ln(2x+1)} = 2x + 1$

12. $e^{\ln\left(x^2+1\right)} = x^2 + 1$

1. Because $y = 1$ when $t = 0$, it follows that $C = 1$. Moreover, because $y = 10$ when $t = 4$, you have $10 = e^{4k}$ and $k = \frac{1}{4}\ln 10 \approx 0.5756$. So, $y = e^{0.5756t}$.

3. Because $y = 1$ when $t = 0$, it follows that $C = 1$. Moreover, because $y = \frac{1}{4}$ when $t = 4$, you have

$$\frac{1}{4} = e^{4k} \text{ and } k = \frac{\ln(1/4)}{4} \approx -0.3466.$$

So, $y = e^{-0.3466t}$.

5. Because $y = 2$ when $t = 0$, it follows that $C = 2$. Moreover, because $y = 3$ when $t = 4$, you have

$$3 = 2e^{4k} \text{ and } k = \frac{\ln(3/2)}{4} \approx 0.1014.$$

So, $y = 2e^{0.1014t}$.

7. Because $y = 4$ when $t = 0$, it follows that $C = 4$. Moreover, because $y = \frac{1}{2}$ when $t = 5$, you have

$$\frac{1}{2} = 4e^{5k} \text{ and } k = \frac{\ln(1/8)}{5} \approx -0.4159.$$

So, $y = 4e^{-0.4159t}$.

9. Because $y = 1$ when $t = 1$ and $y = 5$ when $t = 5$, you have $1 = Ce^{k}$ and $5 = Ce^{5k}$. From these two equations you have

$$Ce^{k} = \left(\frac{1}{5}\right)Ce^{5k}.$$

So, $5 = e^{4k}$

$$k = \frac{\ln 5}{4} \approx 0.4024,$$

and you have $y = Ce^{0.4024t}$. Because $1 = Ce^{0.4024}$, it follows that $C \approx 0.6687$ and $y = 0.6687e^{0.4024t}$.

11. $\frac{dy}{dt} = 2y$, $y = 10$ when $t = 0$

$$y = 10e^{2t}$$

$$\frac{dy}{dt} = 10(2)e^{2t} = 2(10e^{2t}) = 2y$$

Exponential growth

13. $\frac{dy}{dt} = -4y$, $y = 30$ when $t = 0$

$$y = 30e^{-4t}$$

$$\frac{dy}{dt} = 30(-4)e^{-4t} = -4(30e^{-4t}) = -4y$$

Exponential decay

15. From Example 1 you have

$$y = Ce^{kt} = 10e^{[\ln(1/2)/1599]t}.$$

When $t = 1000$,

$$y = 10e^{[\ln(1/2)/1599]1000} \approx 6.48 \text{ grams.}$$

When $t = 10{,}000$,

$$y = 10e^{[\ln(1/2)/1599]10{,}000} \approx 0.13 \text{ gram.}$$

17. From Example 1 you have

$$y = Ce^{kt} = Ce^{[\ln(1/2)/5715]t}$$

$$2 = Ce^{[\ln(1/2)/5715]10{,}000}$$

$$C \approx 6.73.$$

The initial quantity is 6.73 grams.

When $t = 1000$,

$$y = 6.73e^{[\ln(1/2)/5715]1000} \approx 5.96 \text{ grams.}$$

19. From Example 1 you have

$$y = Ce^{[\ln(1/2)/24{,}100]t}$$

$$2.1 = Ce^{[\ln(1/2)/24{,}100]1000}$$

$$C \approx 2.16.$$

The initial quantity is 2.16 grams.

When $t = 10{,}000$,

$$y = 2.16e^{[\ln(1/2)/24{,}000]10{,}000} \approx 1.62 \text{ grams.}$$

21. From Example 1 you have

$$y = Ce^{[\ln(1/2)/1599]}.$$

When $t = 900$,

$$y = Ce^{[\ln(1/2)/1620]900} \approx 0.68C.$$

After 900 years, approximately 68% of the radioactive radium will remain.

23. $0.15C = Ce^{[\ln(1/2)/5715]t}$

$$\ln 0.15 = \frac{\ln(1/2)}{5715}t$$

$$t = \frac{5715 \ln 0.15}{\ln(1/2)} \approx 15{,}641.8 \text{ years}$$

25. At $(0, 5)$, $y_1 = 5e^{k_1 t}$

$$\begin{aligned} \text{At } (12, 20),\ 20 &= 5e^{k_1(12)} \\ 4 &= e^{12k_1} \\ \ln 4 &= 12k_1 \\ \frac{\ln 4}{12} &= k_1 \\ 0.1155 &\approx k_1 \end{aligned}$$

So, $y_1 = 5e^{0.1155t}$.

$k_1 = k_2 \ln 2$

$$\begin{aligned} \text{At } (0, 5),\ y_2 &= 5(2)^{k_2 t} \\ \text{At } (12, 20),\ 20 &= 5(2)^{k_2(12)} \\ 4 &= 2^{12k_2} \\ \log_2 4 &= 12k_2 \\ \frac{\log_2 4}{12} &= k_2 \\ \frac{1}{6} &= k_2 \end{aligned}$$

So, $y_2 = 5(2)^{t/6}$.

27. The model is $y = Ce^{kt}$. Because $y = 150$ when $t = 0$, you have $C = 150$. Furthermore,

$$\begin{aligned} 450 &= 150e^{k(5)} \\ 3 &= e^{5k} \\ \ln 3 &= 5k \\ k &= \frac{\ln 3}{5}. \end{aligned}$$

So, $y = 150e^{[(\ln 3)/5]t} \approx 150e^{0.2197t}$.

(a) When $t = 10$, $y = 150e^{[(\ln 3)/5]10} = 1350$ bacteria.

(b) To find the time required for the population to double, solve for t.

$$\begin{aligned} 300 &= 150e^{[(\ln 3)/5]t} \\ 2 &= e^{[(\ln 3)/5]t} \\ \ln 2 &= \frac{\ln 3}{5}t \\ t &= \frac{5 \ln 2}{\ln 3} \approx 3.2 \text{ hours} \end{aligned}$$

(c) No, the doubling time is always 3.2 hours.

29. (a) **Australia:**

Because $y = 20.1$ when $t = 5$ and $y = 20.9$ when $t = 10$, you have $20.1 = Ce^{5k}$ and $20.9 = Ce^{10k}$. From these two equations you have

$$\begin{aligned} \frac{20.1}{e^{5k}} &= \frac{20.9}{e^{10k}} \\ 20.1e^{10k} &= 20.9e^{5k} \\ e^{5k} &= \frac{20.9}{20.1} \\ 5k &= \ln \frac{20.9}{20.1} \\ k &= \frac{1}{5} \ln \frac{20.9}{20.1} \approx 0.0078. \end{aligned}$$

Then $y = Ce^{0.0078t}$. Because $20.1 = Ce^{5(0.0078)}$, it follows that $C = \dfrac{20.1}{e^{5(0.0078)}} \approx 19.331$.

So, $y = 19.331e^{0.0078t}$.

In 2030, $y = 19.331e^{0.0078(30)} \approx 24.4$ million

Canada:

Because $y = 32.8$ when $t = 5$ and $y = 34.3$ when $t = 10$, you have $32.8 = Ce^{5k}$ and $34.3 = Ce^{10k}$. From these two equations you have

$$\begin{aligned} \frac{32.8}{e^{5k}} &= \frac{34.3}{e^{10k}} \\ 32.8e^{10k} &= 34.3e^{5k} \\ e^{5k} &= \frac{34.3}{32.8} \\ 5k &= \ln \frac{34.3}{32.8} \\ k &= \frac{1}{5} \ln \frac{34.3}{32.8} \approx 0.0089. \end{aligned}$$

Then $y = Ce^{0.0089t}$. Because $32.8 = Ce^{5(0.0089)}$, it follows that $C = \dfrac{32.8}{e^{5(0.0089)}} \approx 31.372$.

So, $y = 31.372e^{0.0089t}$.

In 2030, $y = 31.372e^{0.0089(30)} \approx 41.0$ million.

Philippines:

Because $y = 87.9$ when $t = 5$ and $y = 95.9$ when $t = 10$, you have $87.9 = Ce^{5k}$ and $95.9 = Ce^{10k}$. From these two equations you have

$$\frac{87.9}{e^{5k}} = \frac{95.9}{e^{10k}}$$
$$87.9e^{10k} = 95.9e^{5k}$$
$$e^{5k} = \frac{95.9}{87.9}$$
$$5k = \ln \frac{95.9}{87.9}$$
$$k = \frac{1}{5} \ln \frac{95.9}{87.9} \approx 0.0174.$$

Then $y = Ce^{0.0174t}$. Because $87.9 = Ce^{5(0.0174)}$, it follows that $C = \dfrac{87.9}{e^{5(0.0174)}} \approx 80.576$.

So, $y = 80.576e^{0.0174t}$.

In 2030, $y = 80.576e^{0.0174(30)} \approx 135.8$ million.

South Africa:

Because $y = 44.3$ when $t = 5$ and $y = 43.3$ when $t = 10$, you have $44.3 = Ce^{5k}$ and $43.3 = Ce^{10k}$. From these two equations you have

$$\frac{44.3}{e^{5k}} = \frac{43.3}{e^{10k}}$$
$$44.3e^{10k} = 43.3e^{5k}$$
$$e^{5k} = \frac{43.3}{44.3}$$
$$5k = \ln \frac{43.3}{44.3}$$
$$k = \frac{1}{5} \ln \frac{43.3}{44.3} = -0.0046$$

Then $y = Ce^{-0.0046t}$. Because $44.3 = Ce^{5(-0.0046)}$, it follows that $C = \dfrac{44.3}{e^{5(-0.0046)}} \approx 45.331$.

So, $y = 45.331e^{-0.0046t}$.

In 2030, $y = 45.331e^{-0.0046(30)} \approx 39.5$ million.

Turkey:

Because $y = 69.7$ when $t = 5$ and $y = 73.3$ when $t = 10$, you have $69.7 = Ce^{5k}$ and $73.3 = Ce^{10k}$. From these two equations you have

$$\frac{69.7}{e^{5k}} = \frac{73.3}{e^{10k}}$$
$$69.7e^{10k} = 73.3e^{5k}$$
$$e^{5k} = \frac{73.3}{69.7}$$
$$5k = \ln \frac{73.3}{69.7}$$
$$k = \frac{1}{5} \ln \frac{73.3}{69.7} \approx 0.0101.$$

Then $y = Ce^{0.0101t}$. Because $69.7 = Ce^{5(0.0101)}$, it follows that $C = \dfrac{69.7}{e^{5(0.0101)}} \approx 66.268$.

So, $y = 66.268e^{0.0101t}$.

In 2030, $y = 66.268e^{0.0101(30)} \approx 89.7$ million.

(b) From the different growth rates, you can determine the constant k in each equation.

(c) You know from part (b) that the constant k represents the growth rate. If the constant is positive, then the population is increasing. If the constant is negative, then the population is decreasing.

31. Because $N = 280$ when $t = 10$, you have

$$280 = 250e^{10k}$$
$$1.12 = e^{10k}$$
$$\ln 1.12 = 10k$$
$$\frac{1}{10} \ln 1.12 = k$$

So, $N = 250e^{[(\ln 1.12)/10]t} \approx 250e^{0.0113t}$.

$$500 = 250e^{[(\ln 1.12)/10]t}$$
$$2 = e^{[(\ln 1.12)/10]t}$$
$$\ln 2 = \frac{\ln 1.12}{10}t$$
$$t = \frac{10 \ln 2}{\ln 1.12} \approx 61.2$$

It will take about 61.2 hours for the population to double in size.

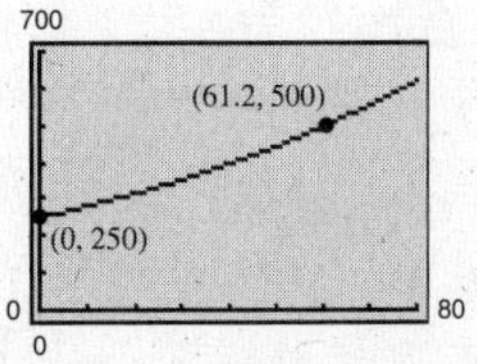

33. Consider the points $(0, 2000)$ and $(2, 500)$.

(a) $V - 2000 = \frac{500 - 2000}{2 - 0}(t - 0)$

$V - 2000 = -750t$

$V = -750t + 2000$

(b) Because $V = 2000$ when $t = 0$, it follows that $C = 2000$. Moreover, because $V = 500$ when $t = 2$, you have $500 = 2000e^{2k}$ and

$k = \frac{\ln\left(\frac{1}{4}\right)}{2} \approx -0.6931$. So, $V = 2000e^{-0.6931t}$.

(c) 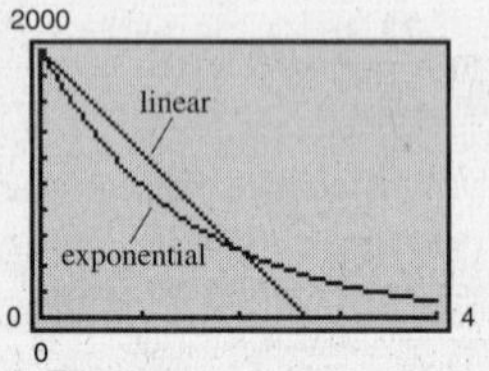

For the first year, the exponential model depreciates faster.

(d) After 1 year, $V = -750(1) + 2000 = \$1250$.

After 3 years, $V = -750(3) + 2000 = -\$250$.

(e) After 1 year, $V = 2000e^{-0.6931(1)} \approx \$1000$.

After 3 years, $V = 2000e^{-0.6931(3)} \approx \$250$.

35. Let y represent the number of substance abuse treatment facilities and let t represent the year, with $t = 5$ corresponding to 1995.

(a) Exponential model: $y = 9635(1.022)^t$

$= 9635e^{(\ln 1.022)t}$

$= 9635e^{0.0218t}$

Linear model: $y = 262.1t + 9436$

(b) In $2011(t = 21)$,

$y = 9635e^{0.0218(21)} \approx 15{,}229$ facilities.

(c) In 2011 $(t = 21)$,

$y = 262.1(21) + 9436 \approx 14{,}940$ facilities.

(d) 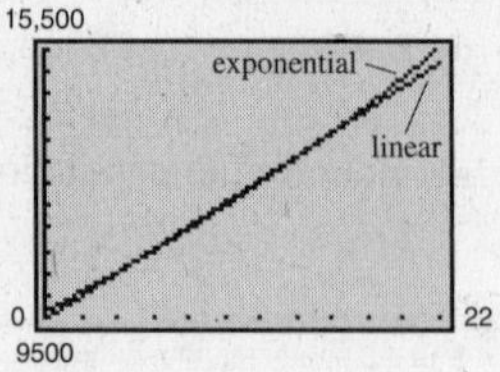

Answers will vary.

37. (a) Because $N = 20$ when $t = 30$, you have

$20 = 30\left(1 - e^{30k}\right)$

$\frac{2}{3} = \left(1 - e^{30k}\right)$

$k = \frac{\ln(1/3)}{30} \approx -0.0366$

$N = 30\left(1 - e^{-0.0366t}\right)$.

(b) $25 = 30\left(1 - e^{-0.0366t}\right)$

$\frac{5}{6} = \left(1 - e^{-0.0366t}\right)$

$t = \frac{\ln(1/6)}{-0.0366} \approx 49.0$ days

39. $A = Ve^{-0.04t} = 100{,}000e^{0.6\sqrt{t}}e^{-0.04t} = 100{,}000e^{\left(0.6\sqrt{t} - 0.04t\right)}$

$A'(t) = 100{,}000\left(\frac{0.6}{2\sqrt{t}} - 0.04\right)e^{\left(0.6\sqrt{t} - 0.04t\right)} = 0$

$\frac{0.6}{2\sqrt{t}} - 0.04 = 0$

$\frac{0.6}{2\sqrt{t}} = 0.04$

$\sqrt{t} = \frac{0.6}{(0.04)2} = 7.5$

$t = 56.25 \approx 56$

The timber should be harvested in 2046.

41. Answers will vary.

Review Exercises for Chapter 4

1. $32^{3/5} = \left[(32)^{1/5}\right]^3 = 2^3 = 8$

3. $\left(\frac{1}{16}\right)^{-3/2} = 16^{3/2} = \left[(16)^{1/2}\right]^3 = 4^3 = 64$

5. $\left(\frac{9}{16}\right)^0 = 1$

7. $\dfrac{6^3}{36^2} = \dfrac{6 \cdot 6^2}{36 \cdot 6^2} = \dfrac{6}{36} = \dfrac{1}{6}$

9. $\left(e^2\right)^5 = e^{2 \cdot 5} = e^{10}$

11. $\left(e^{-1}\right)\left(e^4\right) = e^{-1+4} = e^3$

13. $f(x) = 2^{x+3}$

$f(4) = 2^{4+3} = 2^7 = 128$

15. $f(x) = 1.02^x$

$f(10) = 1.02^{10} \approx 1.22$

17. $V = 258.82(2.317)^B,\ 1.39 \le B \le 1.87$

(a) When $B = 1.4$:

$V = 258.82(2.317)^{1.4} \approx 839.3$ milliliters

When $B = 1.6$:

$V = 258.82(2.317)^{1.6} \approx 992.8$ milliliters

When $B = 1.75$:

$V = 258.82(2.317)^{1.75} \approx 1126.2$ milliliters

(b) Answers will vary.

19. The graph of $f(x) = 6^x$ is an exponential curve with the following characteristics.

Passes through $(0, 1)$, $(1, 6)$, $\left(-1, \frac{1}{6}\right)$

Horizontal asymptote: $y = 0$

It matches graph (c)

21. The graph of $f(x) = 2^{-x^2}$ is an exponential curve with the following characteristics.

Passes through $(0, 1)$, $\left(1, \frac{1}{2}\right)$, $\left(-1, \frac{1}{2}\right)$

Horizontal asymptote: $y = 0$

It matches graph (f).

23. The graph of $f(x) = -2 + e^{x-5}$ is an exponential curve with the following characteristics.

Increasing and passes through $(5, -1)$

Horizontal asymptote: $y = -2$

It matches graph (e).

25. $f(x) = 9^{x/2}$

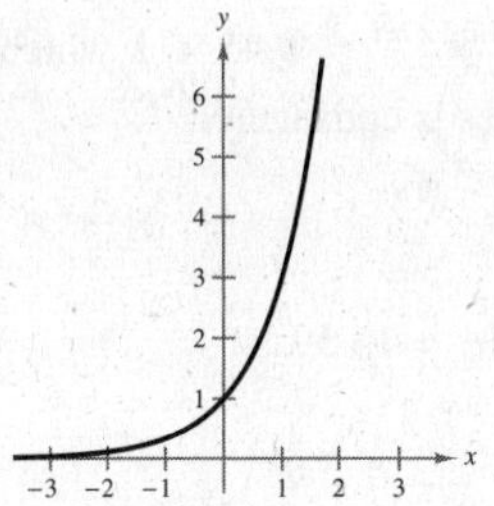

27. $f(t) = \left(\frac{1}{6}\right)^t$

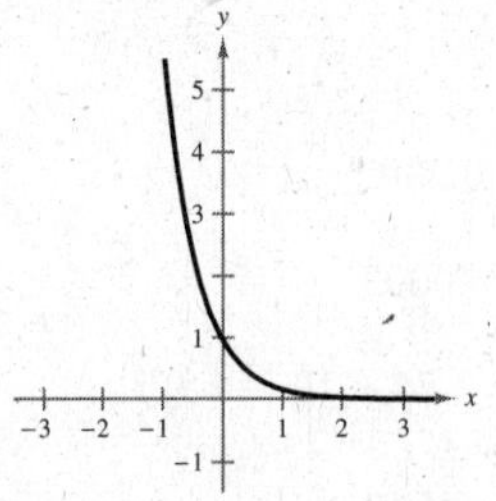

29. $f(x) = \left(\frac{1}{2}\right)^{2x} + 4$

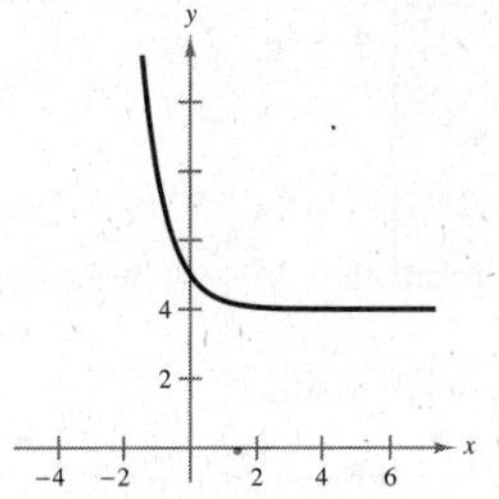

31. $f(x) = e^{-x} + 1$

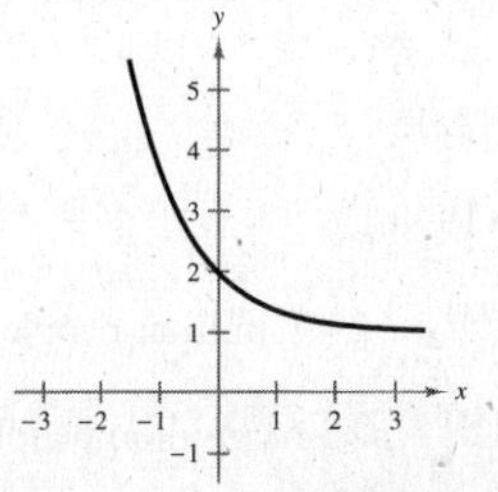

33. $f(x) = 1 - e^x$

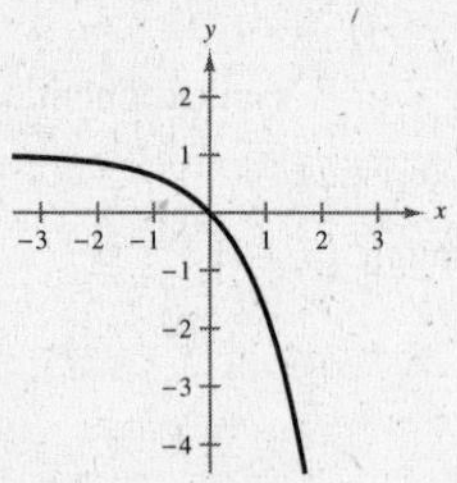

35. $y = 1096e^{-0.39t}$

If $t = 20$, $y = 1096e^{-0.39(20)} = 0.45 < 1$, which indicates that the species is endangered.

37. $f(x) = 5e^{x-1}$

(a) $f(2) = 5e^{2-1} = 5e \approx 13.59$

(b) $f\left(\frac{1}{2}\right) = 5e^{1/2-1} = \dfrac{5}{e^{1/2}} \approx 3.03$

(c) $f(10) = 5e^{10-1} = 5e^9 \approx 40{,}515.42$

39. $g(t) = 6e^{-0.2t}$

(a) $g(17) = 6e^{-3.4} \approx 0.2002$

(b) $g(50) = 6e^{-10} \approx 0.0003$

(c) $g(100) = 6e^{-20} \approx 1.24 \times 10^{-8} \approx 0$

41. (a) $P = \dfrac{10{,}000}{1 + 19e^{-t/5}}$, $t \geq 0$

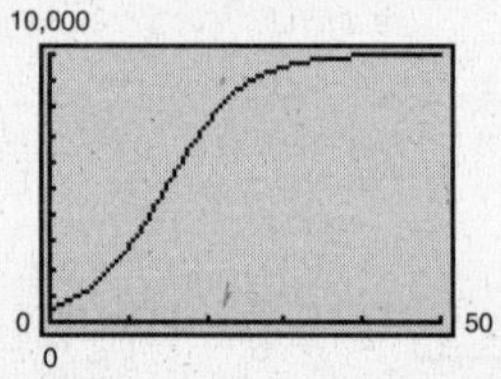

(b) When $t = 4$, $P \approx 1049$ fish.

(c) Yes, P approaches 10,000 fish as $t \to \infty$.

(d) The population is increasing most rapidly at the inflection point, around $t = 15$ months $(P = 5000)$.

43. $P = 29.7e^{0.01t}$, $0 \leq t \leq 15$

$(t = 0$ corresponds to 1990.)

1990: $P(0) = 29.7e^{0.01(0)} = 29.7$ million people

2000: $P(10) = 29.7e^{0.01(10)} \approx 32.8$ million people

2005: $P(15) = 29.7e^{0.01(15)} \approx 34.5$ million people

45. $y = 4e^{x^2}$

$y' = 4e^{x^2}(2x) = 8xe^{x^2}$

47. $y = \dfrac{x}{e^{2x}}$

$y' = \dfrac{e^{2x}(1) - x2e^{2x}}{\left(e^{2x}\right)^2} = \dfrac{1 - 2x}{e^{2x}}$

49. $y = \sqrt{4e^{4x}} = \left(4e^{4x}\right)^{1/2} = 2e^{2x}$

$y' = 4e^{2x}$

51. $y = \dfrac{5}{1 + e^{2x}} = 5\left(1 + e^{2x}\right)^{-1}$

$y' = -5\left(1 + e^{2x}\right)^{-2}\left(2e^{2x}\right) = -\dfrac{10e^{2x}}{\left(1 + e^{2x}\right)^2}$

53. $f(x) = 4e^{-x}$

$f'(x) = -4e^{-x}$

$f''(x) = 4e^{-x}$

No relative extrema

No inflection points

Horizontal asymptote:

$y = 0$

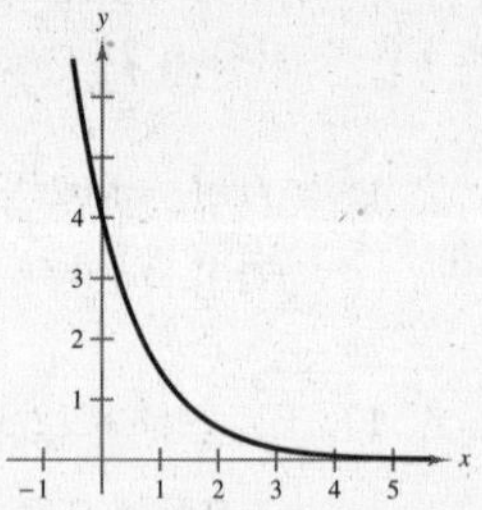

55. $f(x) = x^3e^x$

$f'(x) = x^2e^x(3 + x)$

Critical number: $x = -3$

$f''(x) = xe^x\left(x^2 + 6x + 6\right)$

$f''(-3) = 9e^{-3} \approx 0.45 > 0$

Relative minimum: $(-3, -1.34)$

$f''(x) < 0$ on $\left(-\infty, -3 - \sqrt{3}\right)$

$f''(x) > 0$ on $\left(-3 - \sqrt{3}, -3 + \sqrt{3}\right)$

$f''(x) < 0$ on $\left(-3 + \sqrt{3}, 0\right)$

$f''(x) > 0$ on $(0, \infty)$

Inflections points: $(0, 0)$, $\left(-3 + \sqrt{3}, -0.57\right)$, and $\left(-3 - \sqrt{3}, -0.93\right)$

Horizontal asymptote: $y = 0$

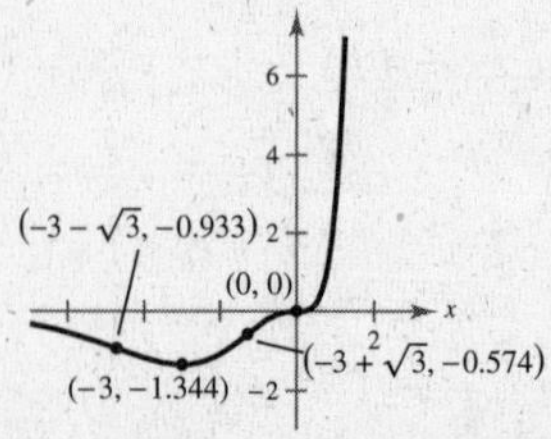

57. $f(x) = \dfrac{1}{xe^x}$

$f'(x) = \dfrac{-x-1}{x^2e^x}$

Critical number: $x = -1$

$f''(x) = \dfrac{x^2+2x+2}{x^3e^x}$

$f''(-1) = -\dfrac{1}{e^{-1}} < 0$

Relative maximum: $(-1, -2.72)$

No inflection points.

Horizontal asymptote: $y = 0$

Vertical asymptote: $x = 0$

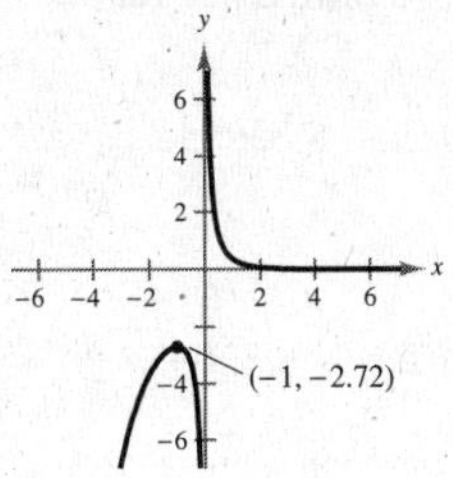

59. $f(x) = xe^{2x}$

$f'(x) = (2x+1)e^{2x}$

Critical number: $x = -\dfrac{1}{2}$

$f''(x) = (4x+4)e^{2x}$

$f''\left(-\dfrac{1}{2}\right) = \dfrac{2}{e} > 0$

$f''(x) < 0$ on $(-\infty, -1)$

$f''(x) > 0$ on $(-1, \infty)$

Relative minimum: $\left(-\dfrac{1}{2}, -\dfrac{1}{2e}\right)$

Inflection point: $\left(-1, -\dfrac{1}{e^2}\right)$

Horizontal asymptote: $y = 0$

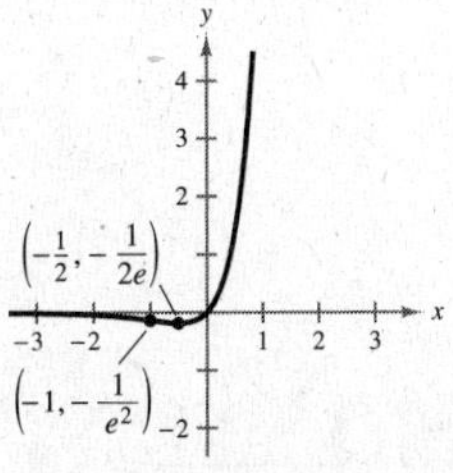

61. $\ln 12 \approx 2.4849$

$e^{2.4849} \approx 12$

63. $e^{1.5} \approx 4.4817$

$\ln 4.4817 \approx 1.5$

65. $y = \ln(4-x)$

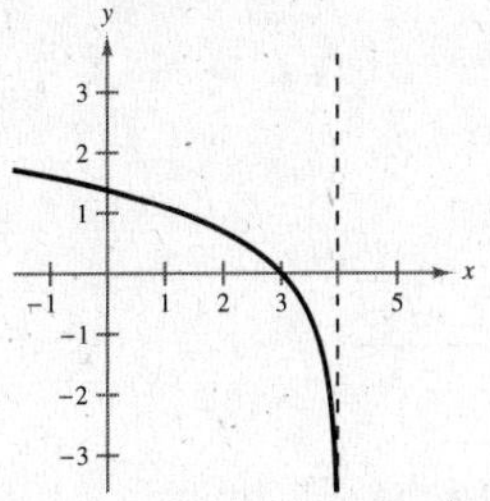

67. $y = \ln\dfrac{x}{3} = \ln x - \ln 3$

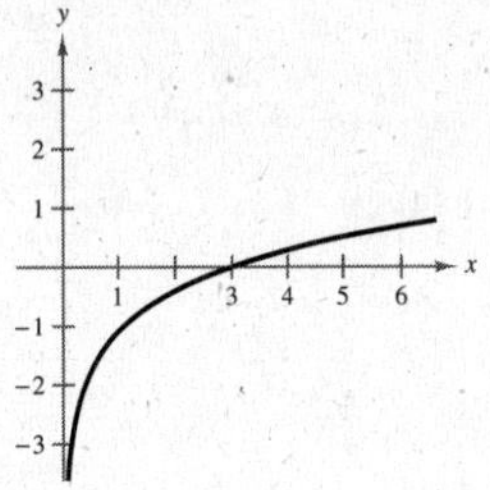

69. $\ln\sqrt{x^2(x-1)} = \frac{1}{2}\ln\left[x^2(x-1)\right]$

$= \frac{1}{2}\left[\ln x^2 + \ln(x-1)\right]$

$= \ln x + \frac{1}{2}\ln(x-1)$

71. $\ln\dfrac{x^2}{(x+1)^3} = \ln x^2 - \ln(x+1)^3 = 2\ln x - 3\ln(x+1)$

73. $\ln\left(\dfrac{1-x}{3x}\right)^3 = 3\ln\left(\dfrac{1-x}{3x}\right)$

$= 3\left[\ln(1-x) - \ln 3x\right]$

$= 3\left[\ln(1-x) - \ln 3 - \ln x\right]$

75. $e^{\ln x} = 3$

$x = 3$

77. $\ln x = 3e^{-1} = \dfrac{3}{e}$

$e^{3/e} = x \approx 3.02$

79. $\ln 2x - \ln(3x-1) = 0$

$\ln 2x = \ln(3x-1)$

$2x = 3x - 1$

$x = 1$

81. $e^{2x-1} - 6 = 0$

$e^{2x-1} = 6$

$2x - 1 = \ln 6$

$x = \dfrac{1+\ln 6}{2} \approx 1.40$

83. $\ln x + \ln(x - 3) = 0$

$$\ln[x(x - 3)] = 0$$
$$x(x - 3) = 1$$
$$x^2 - 3x - 1 = 0$$
$$x = \frac{3 \pm \sqrt{13}}{2}$$

$x = \frac{3 + \sqrt{13}}{2} \approx 3.30$ is the only solution in the domain.

85. $e^{-1.386x} = 0.25$

$$-1.386x = \ln 0.25$$
$$x = \frac{\ln 0.25}{-1.386} \approx 1.00$$

87. $100(1.21)^x = 110$

$$1.21^x = \frac{110}{100} = 1.1$$
$$x \ln 1.21 = \ln 1.1$$
$$x = \frac{\ln 1.1}{\ln 1.21} = 0.5$$

89. $\frac{40}{1 - 5e^{-0.01x}} = 200$

$$1 - 5e^{-0.01x} = \frac{1}{5}$$
$$5e^{-0.01x} = \frac{4}{5}$$
$$e^{-0.01x} = \frac{4}{25}$$
$$-0.01x = \ln\left(\frac{4}{25}\right)$$
$$x = -100 \ln\left(\frac{4}{25}\right) = 100 \ln\left(\frac{25}{4}\right) \approx 183.26$$

91. $V = 30.25 + 5.282t + 0.1131t^2 + 8.82065e^{-t}$

($t = 0$ corresponds to 2000.)

(a)

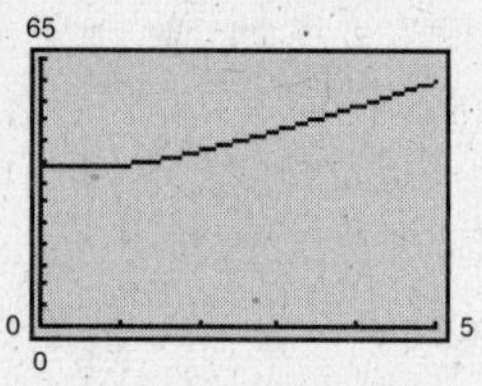

(b) When $t = 1$, $\frac{dV}{dt} \approx$ \$2 billion per year.

When $t = 2$, $\frac{dV}{dt} \approx$ \$5 billion per year.

When $t = 4$, $\frac{dV}{dt} \approx$ \$6 billion per year.

(c) $\frac{dV}{dt} = 5.282 + 0.2262t - 8.82065e^{-t}$

When $t = 1$,

$$\frac{dV}{dt} = 5.282 + 0.2262(1) - 8.82065e^{-1}$$

$\approx$ \$2.26 billion per year.

When $t = 2$,

$$\frac{dV}{dt} = 5.282 + 0.2262(2) - 8.82065e^{-2}$$

$\approx$ \$4.54 billion per year.

When $t = 4$,

$$\frac{dV}{dt} = 5.282 + 0.2262(4) - 8.82065e^{-4}$$

$\approx$ \$6.03 billion per year.

93. $f(x) = \ln 3x^2 = \ln 3 + 2 \ln x$

$$f'(x) = \frac{2}{x}$$

95. $y = \ln \frac{x(x - 1)}{x - 2} = \ln x + \ln(x - 1) - \ln(x - 2)$

$$y' = \frac{1}{x} + \frac{1}{x - 1} - \frac{1}{x - 2}$$

97. $f(x) = \ln e^{2x+1} = 2x + 1$

$$f'(x) = 2$$

99. $y = \frac{\ln x}{x^3}$

$$y' = \frac{x^3(1/x) - 3x^2 \cdot \ln x}{x^6} = \frac{1 - 3 \ln x}{x^4}$$

101. $y = \ln(x^2 - 2)^{2/3} = \frac{2}{3} \ln(x^2 - 2)$

$$y' = \frac{2}{3} \cdot \frac{2x}{x^2 - 2} = \frac{4x}{3(x^2 - 2)}$$

103. $f(x) = \ln\left(x^2\sqrt{x+1}\right) = 2\ln x + \frac{1}{2}\ln(x+1)$

$f'(x) = \frac{2}{x} + \frac{1}{2(x+1)}$

105. $y = \ln\frac{e^x}{1+e^x} = \ln e^x - \ln\left(1+e^x\right) = x - \ln\left(1+e^x\right)$

$y' = 1 - \frac{e^x}{1+e^x} = \frac{1}{1+e^x}$

107. $y = \ln(x+3)$

$y' = \frac{1}{x+3}$

$y'' = -\frac{1}{(x+3)^2}$

No relative extrema

No inflection points

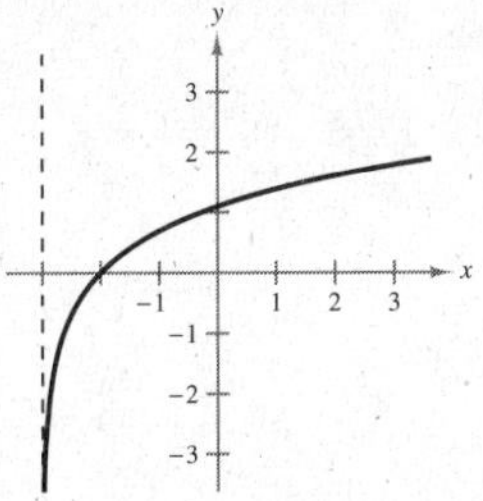

109. $y = \ln\left(\frac{10}{x+2}\right) = \ln 10 - \ln(x+2)$

$y' = -\frac{1}{x+2}$

$y'' = \frac{1}{(x+2)^2}$

No relative extrema

No inflection points

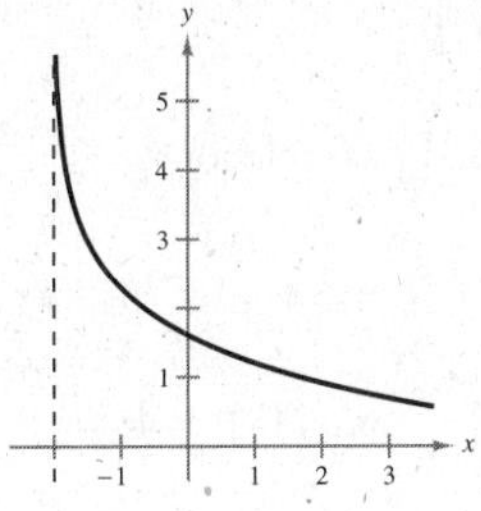

111. $\log_7 49 = \log_7 7^2 = 2\log_7 7 = 2$

113. $\log_{10} 1 = 0$

115. $\log_5 13 = \frac{\ln 13}{\ln 5} \approx 1.594$

117. $\log_{16} 64 = \frac{\ln 64}{\ln 16} = 1.5$

119. $\text{pH} = -\log_{10}\left[\text{H}^+\right]$

$\text{pH} = -\log_{10}\left[11.3 \times 10^{-6}\right] \approx 4.95$

121.
$$\begin{aligned}
\text{pH} &= -\log_{10}\left[\text{H}^+\right]\\
3.2 &= -\log_{10}\left[\text{H}^+\right]\\
-3.2 &= \log_{10}\left[\text{H}^+\right]\\
10^{-3.2} &= \left[\text{H}^+\right]\\
6.3 \times 10^{-4} &\approx \left[\text{H}^+\right]
\end{aligned}$$

123.
$$\begin{aligned}
\text{pH} &= -\log_{10}\left[\text{H}^+\right] & \text{pH} - 1 &= -\log_{10}\left[\text{H}^+\right]\\
-\text{pH} &= \log_{10}\left[\text{H}^+\right] & 1 - \text{pH} &= \log_{10}\left[\text{H}^+\right]\\
10^{-\text{pH}} &= \left[\text{H}^+\right] & 10^{1-\text{pH}} &= \left[\text{H}^+\right]\\
\frac{1}{10^{\text{pH}}} &= \left[\text{H}^+\right] & 10\left(\frac{1}{10^{\text{pH}}}\right) &= \left[\text{H}^+\right]
\end{aligned}$$

So, the hydrogen ion concentration increases by a factor of 10.

125. $y = \log_3(2x-1)$

$y' = \frac{1}{\ln 3}\cdot\frac{2}{2x-1} = \frac{2}{(2x-1)\ln 3}$

127. $y = \log_2\frac{1}{x^2} = \log_2 1 - \log_2 x^2 = -2\log_2 x$

$y' = -2\frac{1}{\ln 2}\cdot\frac{1}{x} = -\frac{2}{x\ln 2}$

129. $V = 695(0.75)^t$

(a)

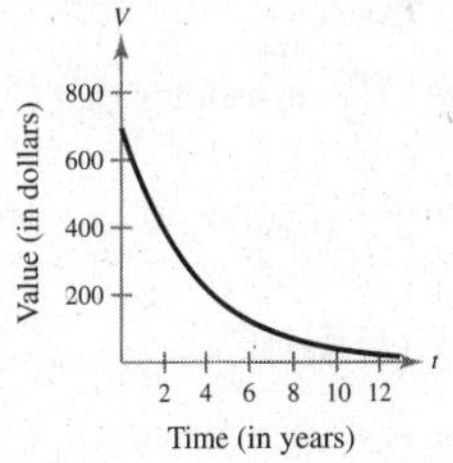

After 2 years, the value is $695(0.75)^2 \approx \$390.94$

(b) $V' = 695(\ln 0.75)(0.75)^t$

$V'(1) = 695(\ln 0.75)(0.75)^1 \approx -\$149.95$ per year

$V'(4) = 695(\ln 0.75)(0.75)^4 \approx -\$63.26$ per year

(c)
$$\begin{aligned}
695(0.75)^t &= 100\\
(0.75)^t &= \frac{20}{139}\\
t &= \log_{0.75}\frac{20}{139}\\
t &= \frac{\ln(20/139)}{\ln 0.75}\\
t &\approx 6.7
\end{aligned}$$

The instrument will be worth \$100 after about 6.7 years.

131. $A = Ce^{kt} = 500e^{kt}$

$A = 500$ when $t = 0$.

$A = 300$ after 40 days.

$$300 = 500e^{40k}$$

$$\frac{3}{5} = e^{40k}$$

$$k = \frac{\ln(3/5)}{40} \approx -0.0128$$

$$A = 500e^{-0.0128t}$$

133. $y = 50e^{kt}$

$$42.031 = 50e^{7k}$$

$$k = \frac{1}{7}\ln\left(\frac{42.031}{50}\right) \approx -0.02480$$

$$25 = 50e^{kt}$$

$$\frac{1}{2} = e^{kt}$$

$$t = \frac{1}{k}\ln\left(\frac{1}{2}\right) \approx 27.95 \text{ years}$$

135. $N = Ce^{kt}$

When $t = 5$, $N = 1885$:

$$1885 = Ce^{5k}$$

$$\frac{1885}{e^{5k}} = C$$

When $t = 16$, $N = 2218$:

$$2218 = Ce^{16k}$$

$$\frac{2218}{e^{16k}} = C$$

$$\frac{1885}{e^{5k}} = \frac{2218}{e^{16k}}$$

$$1885e^{16k} = 2218e^{5k}$$

$$\frac{e^{16k}}{e^{5k}} = \frac{2218}{1885}$$

$$e^{11k} = \frac{2218}{1885}$$

$$11k = \ln\left(\frac{2218}{1885}\right)$$

$$k = \frac{1}{11}\ln\left(\frac{2218}{1885}\right) \approx 0.0148$$

Because $1885 = Ce^{5(0.0148)}$, it follows that $C = \frac{1885}{e^{5(0.0148)}} \approx 1750.55$.

So, $N = 1750.55e^{0.0148t}$. In 2009, $N = 1750.55e^{0.0148(19)} \approx 2319$ associations.

Chapter Test Solutions

1. $3^2(3^{-2}) = \frac{3^2}{3^2} = 3^0 = 1$

2. $\left(\frac{2^3}{2^{-5}}\right)^{-1} = \frac{2^{-3}}{2^5} = \frac{1}{2^3(2^5)} = \frac{1}{2^8} = \frac{1}{256}$

3. $(e^{1/2})(e^4) = e^{9/2}$

4. $(e^3)(e^{-1}) = e^2$

5. $f(x) = 5^{x-2}$

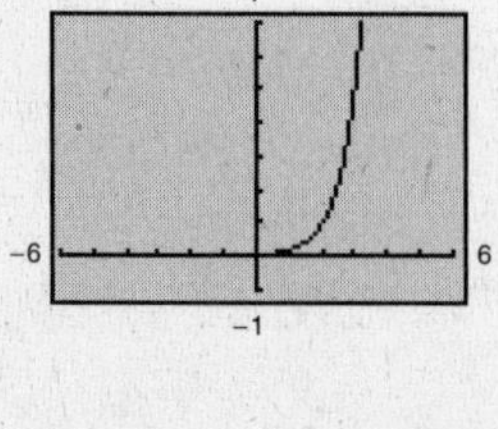

6. $f(x) = 4^{-x}$

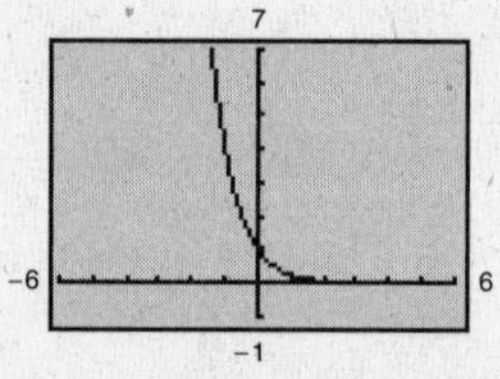

7. $f(x) = 3^{x-3}$

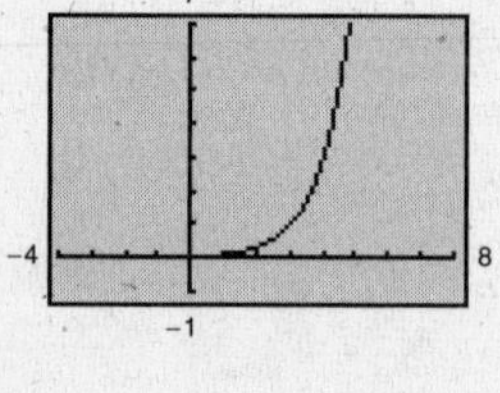

8. $f(x) = 8 + \ln x^2$

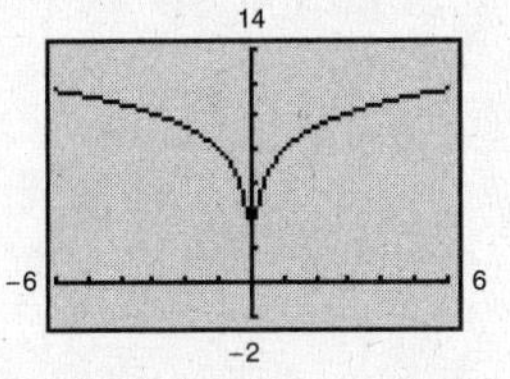

9. $f(x) = \ln(x - 5)$

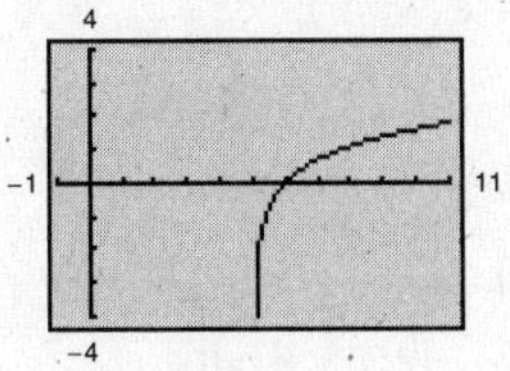

10. $f(x) = 0.5 \ln x$

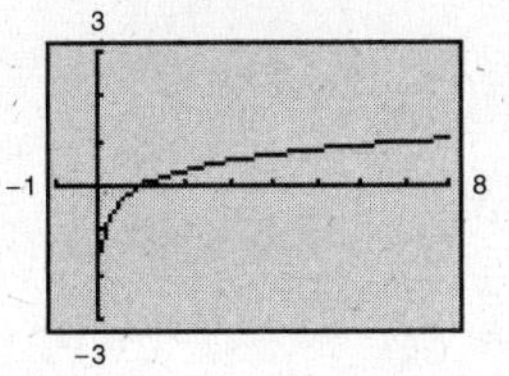

11. $\ln \frac{3}{2} = \ln 3 - \ln 2$

12. $\ln \sqrt{x + y} = \ln(x + y)^{1/2} = \frac{1}{2} \ln(x + y)$

13. $\ln \dfrac{x + 1}{y} = \ln(x + 1) - \ln y$

14. $\ln y + \ln(x + 1) = \ln[y(x + 1)]$

15.
$$\begin{aligned} 3 \ln 2 - 2 \ln(x - 1) &= \ln 2^3 - \ln(x - 1)^2 \\ &= \ln \frac{2^3}{(x - 1)^2} \\ &= \ln \frac{8}{(x - 1)^2} \end{aligned}$$

16.
$$\begin{aligned} 2 \ln x + \ln y - \ln(z + 4) &= \ln x^2 + \ln y - \ln(z + 4) \\ &= \ln(x^2 y) - \ln(z + 4) \\ &= \ln \frac{x^2 y}{z + 4} \end{aligned}$$

17.
$$\begin{aligned} e^{x-1} &= 9 \\ x - 1 &= \ln 9 \\ x &= 1 + \ln 9 \approx 3.20 \end{aligned}$$

18.
$$\begin{aligned} 10e^{2x+1} &= 900 \\ e^{2x+1} &= 90 \\ 2x + 1 &= \ln 90 \\ 2x &= -1 + \ln 90 \\ x &= \frac{-1 + \ln 90}{2} \approx 1.75 \end{aligned}$$

19.
$$\begin{aligned} 50(1.06)^x &= 1500 \\ (1.06)^x &= 30 \\ x &= \log_{1.06} 30 \\ x &= \frac{\ln 30}{\ln 1.06} \approx 58.37 \end{aligned}$$

20. $N = 3767e^{0.02t}$

($t = 0$ corresponds to 2000.)

(a) 2000: $N(0) = 3767e^{0.02(0)}$

$= 3767$ thousand employees

2005: $N(5) = 3767e^{0.02(5)}$

≈ 4163 thousand employees

(b)
$$\begin{aligned} 4000 &= 3767e^{0.02t} \\ \frac{4000}{3767} &= e^{0.02t} \\ \ln\left(\frac{4000}{3767}\right) &= 0.02t \\ 50 \ln\left(\frac{4000}{3767}\right) &= t \\ 3 &\approx t \end{aligned}$$

The number of employees reached 4 million in 2003.

21.
$$\begin{aligned} y &= e^{-3x} + 5 \\ y' &= -3e^{-3x} \end{aligned}$$

22.
$$\begin{aligned} y &= 7e^{x+2} + 2x \\ y' &= 7e^{x+2} + 2 \end{aligned}$$

23.
$$\begin{aligned} y &= \ln(3 + x^2) \\ y' &= \frac{2x}{3 + x^2} \end{aligned}$$

24.
$$\begin{aligned} y &= \ln \frac{5x}{x + 2} = \ln 5x - \ln(x + 2) \\ y' &= \frac{5}{5x} - \frac{1}{x + 2} \\ &= \frac{1}{x} - \frac{1}{x + 2} = \frac{x + 2 - x}{x(x + 2)} = \frac{2}{x(x + 2)} \end{aligned}$$

25. $S = 93 \log_{10} d + 65$

(a)
$$283 = 93 \log_{10} d + 65$$
$$218 = 93 \log_{10} d$$
$$\frac{218}{93} = \log_{10} d$$
$$10^{218/93} = d$$
$$220.8 \approx d$$

The tornado traveled about 220.8 miles.

(b) $\dfrac{dS}{dd} = \left(\dfrac{1}{\ln 10}\right)\left(\dfrac{93}{d}\right)$

When $d = 100$, $\dfrac{dS}{dd} = \left(\dfrac{1}{\ln 10}\right)\left(\dfrac{93}{100}\right) \approx 0.404.$

The rate of change of the wind speed is increasing at about 0.404 mile per hour per mile.

26. $y = Ce^{[\ln(1/2)/1599]t}$

When $t = 1200$, $y = Ce^{[\ln(1/2)/1599]\cdot 1200} \approx 0.59C$. After 1200 years, approximately 59% of the radioactive radium will remain.

27.
$$y = Ce^{0.0175t}$$
$$2C = Ce^{0.0175t}$$
$$2 = e^{0.0175t}$$
$$\ln 2 = 0.0175t$$
$$39.6 \text{ years} \approx t$$

Practice Test for Chapter 4

1. Evaluate each of the following expressions.

(a) $27^{4/3}$

(b) $4^{-5/2}$

(c) $\left(8^{2/3}\right)\left(64^{-1/3}\right)$

2. Solve for x.

(a) $4^{x+1} = 64$

(b) $x^{6/5} = 64$

(c) $(2x + 3)^{10} = 13^{10}$

3. Sketch the graph of (a) $f(x) = 3^x$, and (b) $g(x) = \left(\frac{4}{9}\right)^x$.

4. Differentiate $y = e^{3x^2}$.

5. Differentiate $y = e^{\sqrt[3]{x}}$.

6. Differentiate $y = \sqrt{e^x + e^{-x}}$.

7. Differentiate $y = x^3e^{2x}$.

8. Differentiate $y = \dfrac{e^x + 3}{4x}$.

9. Write $\ln 5 = 1.6094\ldots$ as an exponential equation.

10. Sketch the graph of (a) $y = \ln(x + 2)$, and (b) $y = \ln x + 2$.

11. Write the given expression as a single logarithm.

(a) $\ln(3x + 1) - \ln(2x - 5)$

(b) $4 \ln x - 3 \ln y - \frac{1}{2} \ln z$

12. Solve for x.

(a) $\ln x = 17$

(b) $5^{3x} = 2$

13. Differentiate $y = \ln(6x - 7)$.

14. Differentiate $y = \ln\left(\dfrac{x^3}{4x + 10}\right)$.

15. Differentiate $y = \ln \sqrt[3]{\dfrac{x}{x + 3}}$.

16. Differentiate $y = x^4 \ln x$.

17. Differentiate $y = \sqrt{\ln x + 1}$.

18. Find the exponential function $y = Ce^{kt}$ that passes through the following points.

(a) $(0, 7), \left(4, \frac{1}{3}\right)$

(b) $\left(3, \frac{2}{3}\right), (8, 8)$

19. A population can be modeled by $y = 150e^{0.25t}$, where t is the time in hours. How long will it take for the initial population to double? triple?

Graphing Calculator Required

20. Use a graphing calculator to graph both $y = \ln\left[x^3\sqrt{x+3}\right]$ and $y = 3\ln x + \frac{1}{2}\ln(x+3)$ on the same set of axes. What do you notice about the graphs?

21. Graph the function $f(t) = \dfrac{4200}{7 + e^{-0.9t}}$ and use the graph to find $\lim_{t\to\infty} f(t)$ and $\lim_{t\to-\infty} f(t)$.

CHAPTER 5
Trigonometric Functions

CHAPTER 5
Trigonometric Functions

Section 5.1 Radian Measure of Angles

Skills Review

1. $A = \frac{1}{2}bh$
$= \frac{1}{2}(10)(7)$
$= 35 \text{ cm}^2$

2. $A = \frac{1}{2}bh$
$= \frac{1}{2}(4)(6)$
$= 12 \text{ in.}^2$

3. $a^2 + b^2 = c^2$
$5^2 + 12^2 = c^2$
$169 = c^2$
$13 = c$

4. $a^2 + b^2 = c^2$
$3^2 + b^2 = 5^2$
$9 + b^2 = 25$
$b^2 = 16$
$b = 4$

5. $a^2 + b^2 = c^2$
$8^2 + b^2 = 17^2$
$64 + b^2 = 289$
$b^2 = 225$
$b = 15$

6. $a^2 + b^2 = c^2$
$a^2 + 8^2 = 10^2$
$a^2 + 64 = 100$
$a^2 = 36$
$a = 6$

7. Because all side lengths are 4, the triangle is an equilateral triangle.

8. Because $a = b$, the triangle is an isosceles triangle.

9. Because $a^2 + b^2 = c^2$, the triangle is a right triangle.

10. Because $a = b$ and $a^2 + b^2 = c^2$, the triangle is an isosceles right triangle.

1. (a) Positive: $45° + 360° = 405°$
Negative: $45° - 360° = -315°$
(b) Positive: $-41° + 360° = 319°$
Negative: $-41° - 360° = -401°$

3. (a) Positive: $300° + 360° = 660°$
Negative: $300° - 360° = -60°$
(b) Positive: $740° - 2(360°) = 20°$
Negative: $740° - 3(360°) = -340°$

5. (a) Positive: $\frac{\pi}{9} + 2\pi = \frac{19\pi}{9}$
Negative: $\frac{\pi}{9} - 2\pi = -\frac{17\pi}{9}$
(b) Positive: $\frac{2\pi}{3} + 2\pi = \frac{8\pi}{3}$
Negative: $\frac{2\pi}{3} - 2\pi = -\frac{4\pi}{3}$

7. (a) Positive: $-\frac{9\pi}{4} + 2(2\pi) = \frac{7\pi}{4}$
Negative: $-\frac{9\pi}{4} + 2\pi = -\frac{\pi}{4}$
(b) Positive: $-\frac{2\pi}{15} + 2\pi = \frac{28\pi}{15}$
Negative: $-\frac{2\pi}{15} - 2\pi = -\frac{32\pi}{15}$

9. $30°\left(\frac{\pi \text{ radians}}{180°}\right) = \frac{\pi}{6}$ radian

11. $270°\left(\frac{\pi \text{ radians}}{180°}\right) = \frac{3\pi}{2}$ radians

13. $315°\left(\frac{\pi \text{ radians}}{180°}\right) = \frac{7\pi}{4}$ radians

15. $-20°\left(\frac{\pi \text{ radians}}{180°}\right) = -\frac{\pi}{9}$ radian

17. $-270°\left(\dfrac{\pi \text{ radians}}{180°}\right) = -\dfrac{3\pi}{2}$ radians

19. $330°\left(\dfrac{\pi \text{ radians}}{180°}\right) = \dfrac{11\pi}{6}$ radians

21. $\dfrac{5\pi}{2}\left(\dfrac{180°}{\pi}\right) = 450°$

23. $\dfrac{7\pi}{3}\left(\dfrac{180°}{\pi}\right) = 420°$

25. $-\dfrac{\pi}{12}\left(\dfrac{180°}{\pi}\right) = -15°$

27. $\dfrac{9\pi}{4}\left(\dfrac{180°}{\pi}\right) = 405°$

29. $\dfrac{19\pi}{6}\left(\dfrac{180°}{\pi}\right) = 570°$

31. $-270°\left(\dfrac{\pi \text{ radians}}{180°}\right) = -\dfrac{3\pi}{2}$ radians

33. $144°\left(\dfrac{\pi \text{ radians}}{180°}\right) = \dfrac{4\pi}{5}$ radians

35. The angle θ is $\theta = 90° - 30° = 60°$. By the Pythagorean Theorem, the value of c is

$$c = \sqrt{\left(5\sqrt{3}\right)^2 + 5^2}$$
$$= \sqrt{75 + 25}$$
$$= \sqrt{100}$$
$$= 10.$$

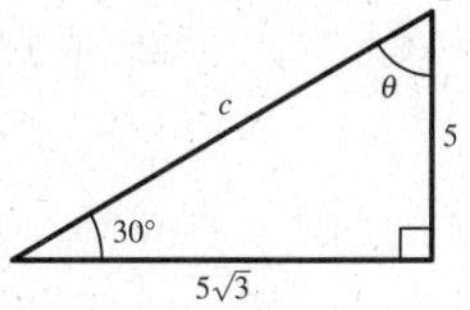

37. The angle θ is $\theta = 90° - 60° = 30°$. By the Pythagorean Theorem, the value of a is

$$a = \sqrt{8^2 - 4^2}$$
$$= \sqrt{64 - 16}$$
$$= \sqrt{48}$$
$$= 4\sqrt{3}.$$

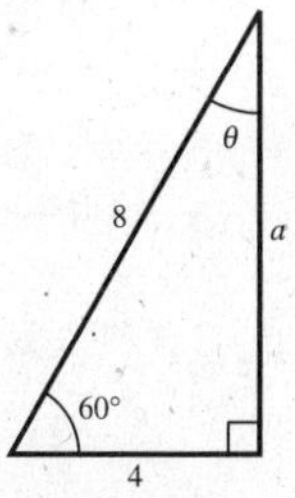

39. Because the triangle is isosceles, you have $\theta = 40°$.

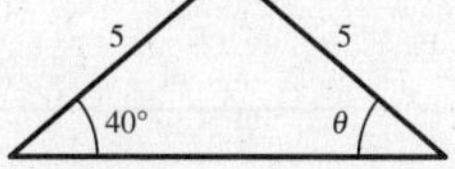

41. Because the largest triangle is similar to the two smaller triangles, $\theta = 60°$. The hypotenuse of the large triangle is

$$\sqrt{2^2 + \left(2\sqrt{3}\right)^2} = \sqrt{16} = 4.$$

By similar triangles you have

$$\frac{s}{2} = \frac{2\sqrt{3}}{4}$$
$$s = \sqrt{3}.$$

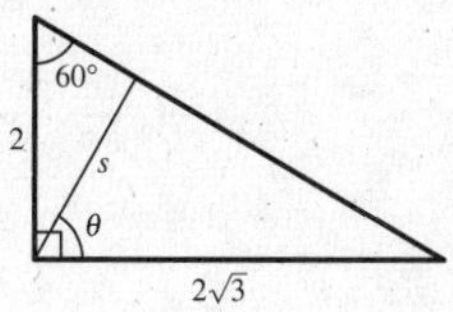

43. $h^2 + 2^2 = 4^2$

$$h^2 = 12$$
$$h = 2\sqrt{3}$$
$$A = \tfrac{1}{2}bh$$
$$= \tfrac{1}{2}(4)\left(2\sqrt{3}\right)$$
$$= 4\sqrt{3} \text{ sq in.}$$

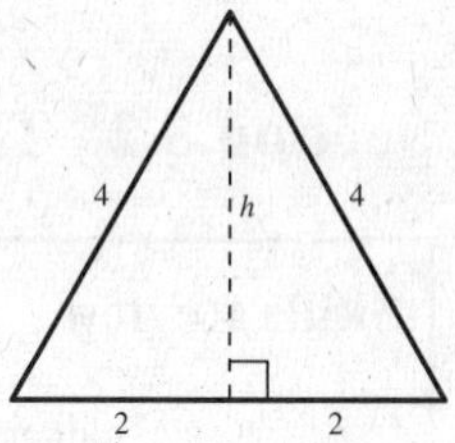

45. $h^2 + \left(\dfrac{5}{2}\right)^2 = 5^2$

$$h^2 = \frac{75}{4}$$
$$h = \frac{5\sqrt{3}}{2}$$
$$A = \frac{1}{2}bh$$
$$= \frac{1}{2}(5)\left(\frac{5\sqrt{3}}{2}\right)$$
$$= \frac{25\sqrt{3}}{4} \text{ sq ft}$$

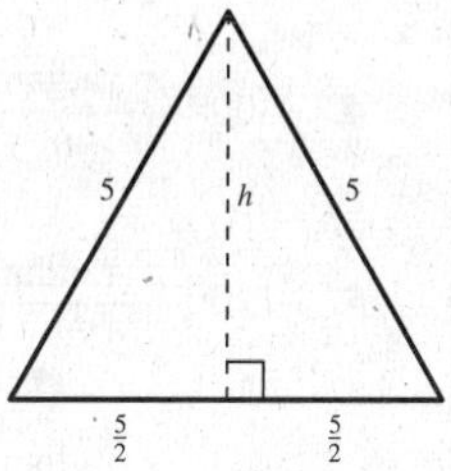

47. Using similar triangles, you have

$$\frac{h}{16 + 8} = \frac{6}{8}$$
$$8h = 144$$

which implies $h = 18$ feet.

49.

r	8 ft	15 in.	85 cm	24 in.	$\dfrac{12{,}963}{\pi}$ mi
s	12 ft	24 in.	$\dfrac{255\pi}{4}$ cm	96 in.	8642 mi
θ	$\dfrac{3}{2}$	$\dfrac{8}{5}$	$\dfrac{3\pi}{4}$	4	$\dfrac{2\pi}{3}$

51. (a) $-80°\left(\dfrac{\pi \text{ radians}}{180°}\right) = -\dfrac{4\pi}{9}$ radians

(b) $S = \dfrac{4\pi}{9}\left(\dfrac{5}{2}\right) = \dfrac{10\pi}{9} \approx 3.49$ ft

53. $120°\left(\dfrac{\pi \text{ radians}}{180°}\right) = \dfrac{2\pi}{3}$ radians

$$\begin{aligned} A &= \frac{1}{2}r^2\theta \\ &= \frac{1}{2}(70^2)\left(\frac{2\pi}{3}\right) \\ &= \frac{4900\pi}{3} \approx 5131.27 \text{ sq ft} \end{aligned}$$

55. False. An obtuse angle is between $90°$ and $180°$.

57. True. The angles would be $90°$, $89°$, and $1°$.

Section 5.2 The Trigonometric Functions

Skills Review

1. $135°\left(\dfrac{\pi \text{ radians}}{180°}\right) = \dfrac{3\pi}{4}$ radians

2. $315°\left(\dfrac{\pi \text{ radians}}{180°}\right) = \dfrac{7\pi}{4}$ radians

3. $-210°\left(\dfrac{\pi \text{ radians}}{180°}\right) = -\dfrac{7\pi}{6}$ radians

4. $-300°\left(\dfrac{\pi \text{ radians}}{180°}\right) = -\dfrac{5\pi}{3}$ radians

5. $-120°\left(\dfrac{\pi \text{ radians}}{180°}\right) = -\dfrac{2\pi}{3}$ radians

6. $-225°\left(\dfrac{\pi \text{ radians}}{180°}\right) = -\dfrac{5\pi}{4}$ radians

7. $540°\left(\dfrac{\pi \text{ radians}}{180°}\right) = 3\pi$ radians

8. $390°\left(\dfrac{\pi \text{ radians}}{180°}\right) = \dfrac{13\pi}{6}$ radians

9.
$$\begin{aligned} x^2 - x &= 0 \\ x(x-1) &= 0 \\ x &= 0 \\ x - 1 = 0 &\Rightarrow x = 1 \\ x = 0 &\text{ or } x = 1 \end{aligned}$$

10.
$$\begin{aligned} 2x^2 + x &= 0 \\ x(2x+1) &= 0 \\ x &= 0 \\ 2x + 1 = 0 &\Rightarrow x = -\tfrac{1}{2} \\ x = 0 &\text{ or } x = -\tfrac{1}{2} \end{aligned}$$

11.
$$\begin{aligned} 2x^2 - x &= 1 \\ 2x^2 - x - 1 &= 0 \\ (2x+1)(x-1) &= 0 \\ 2x + 1 = 0 &\Rightarrow x = -\tfrac{1}{2} \\ x - 1 = 0 &\Rightarrow x = 1 \\ x = -\tfrac{1}{2} &\text{ or } x = 1 \end{aligned}$$

12.
$$\begin{aligned} x^2 - 2x &= 3 \\ x^2 - 2x - 3 &= 0 \\ (x+1)(x-3) &= 0 \\ x + 1 = 0 &\Rightarrow x = -1 \\ x - 3 = 0 &\Rightarrow x = 3 \\ x = -1 &\text{ or } x = 3 \end{aligned}$$

13.
$$\begin{aligned} x^2 - 2x &= -1 \\ x^2 - 2x + 1 &= 0 \\ (x-1)(x-1) &= 0 \\ x - 1 = 0 &\Rightarrow x = 1 \\ x &= 1 \end{aligned}$$

Skills Review —*continued*—

14.
$$\begin{aligned} 2x^2 + x &= 1 \\ 2x^2 + x - 1 &= 0 \\ (2x - 1)(x + 1) &= 0 \\ 2x - 1 = 0 &\Rightarrow x = \tfrac{1}{2} \\ x + 1 = 0 &\Rightarrow x = -1 \end{aligned}$$
$x = \frac{1}{2}$ or $x = -1$

15.
$$\begin{aligned} x^2 - 5x &= -6 \\ x^2 - 5x + 6 &= 0 \\ (x - 3)(x - 2) &= 0 \\ x - 3 = 0 &\Rightarrow x = 3 \\ x - 2 = 0 &\Rightarrow x = 2 \end{aligned}$$
$x = 3$ or $x = 2$

16.
$$\begin{aligned} x^2 + x &= 2 \\ x^2 + x - 2 &= 0 \\ (x + 2)(x - 1) &= 0 \\ x + 2 = 0 &\Rightarrow x = -2 \\ x - 1 = 0 &\Rightarrow x = 1 \end{aligned}$$
$x = -2$ or $x = 1$

17.
$$\begin{aligned} \frac{2\pi}{24}(t - 4) &= \frac{\pi}{2} \\ t - 4 &= 6 \\ t &= 10 \end{aligned}$$

18.
$$\begin{aligned} \frac{2\pi}{12}(t - 2) &= \frac{\pi}{4} \\ t - 2 &= \frac{3}{2} \\ t &= \frac{7}{2} \end{aligned}$$

19.
$$\begin{aligned} \frac{2\pi}{365}(t - 10) &= \frac{\pi}{4} \\ t - 10 &= \frac{365}{8} \\ t &= \frac{445}{8} \end{aligned}$$

20.
$$\begin{aligned} \frac{2\pi}{12}(t - 4) &= \frac{\pi}{2} \\ t - 4 &= 3 \\ t &= 7 \end{aligned}$$

1. Because $x = 3$ and $y = 4$, it follows that $r = \sqrt{3^2 + 4^2} = 5$. So you have

$\sin\theta = \frac{4}{5}$ $\quad\csc\theta = \frac{5}{4}$

$\cos\theta = \frac{3}{5}$ $\quad\sec\theta = \frac{5}{3}$

$\tan\theta = \frac{4}{3}$ $\quad\cot\theta = \frac{3}{4}$.

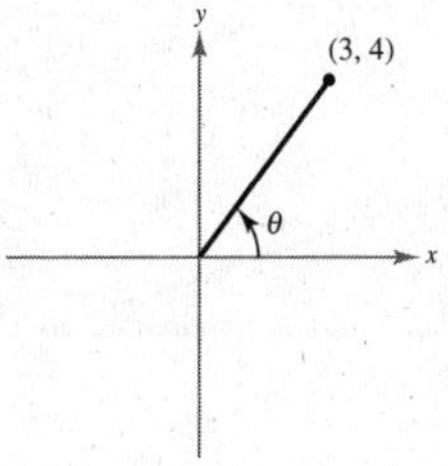

3. Because $x = -12$ and $y = -5$, it follows that $r = \sqrt{(-12)^2 + (-5)^2} = 13$. So you have

$\sin\theta = -\frac{5}{13}$ $\quad\csc\theta = -\frac{13}{5}$

$\cos\theta = -\frac{12}{13}$ $\quad\sec\theta = -\frac{13}{12}$

$\tan\theta = \frac{5}{12}$ $\quad\cot\theta = \frac{12}{5}$.

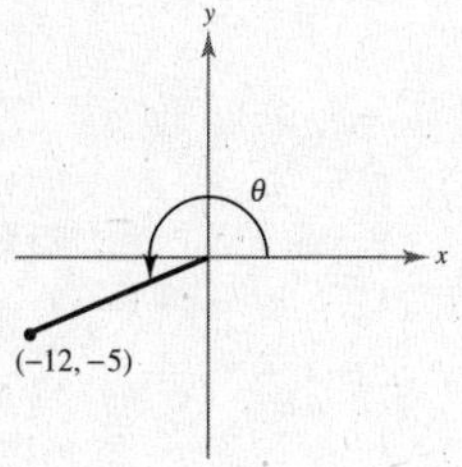

5. Because $x = -\sqrt{3}$ and $y = 1$, it follows that $r = \sqrt{\left(-\sqrt{3}\right)^2 + 1^2} = 2$. So you have

$\sin\theta = \dfrac{1}{2}$ $\quad\csc\theta = 2$

$\cos\theta = -\dfrac{\sqrt{3}}{2}$ $\quad\sec\theta = -\dfrac{2\sqrt{3}}{3}$

$\tan\theta = -\dfrac{\sqrt{3}}{3}$ $\quad\cot\theta = -\sqrt{3}$.

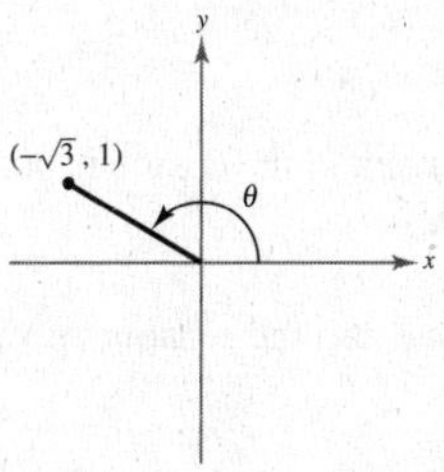

7. $\csc\theta = \dfrac{1}{\sin\theta} = \dfrac{1}{1/2} = 2$

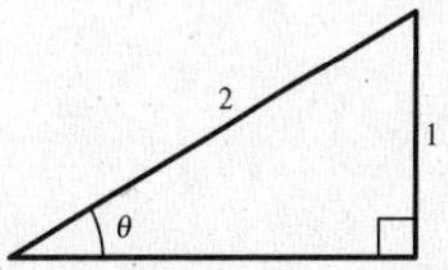

9. Because $x = 4$ and $r = 5$, the length of the opposite side is $y = \sqrt{5^2 - 4^2} = 3$. So you have

$\cot\theta = \dfrac{x}{y} = \dfrac{4}{3}$.

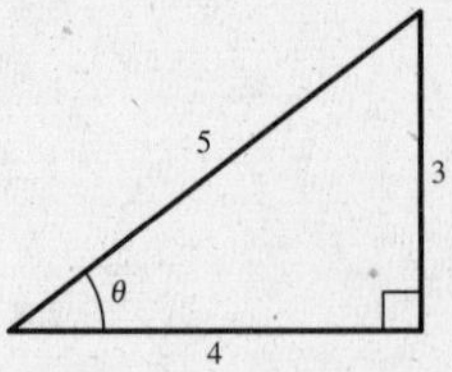

11. Because $x = 15$ and $y = 8$, the length of hypotenuse is $r = \sqrt{15^2 + 8^2} = 17$. So you have

$\sec\theta = \dfrac{r}{x} = \dfrac{17}{15}$.

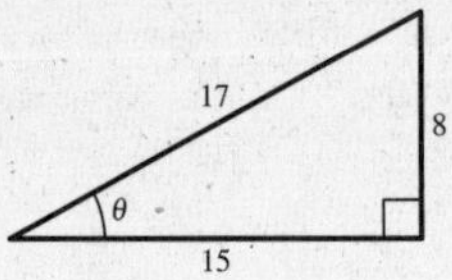

13. Because $y = 1$ and $r = 2$, the length of the third side of the triangle is $x = \sqrt{2^2 - 1^2} = 2\sqrt{2}$. So you have

$\sin\theta = \dfrac{1}{3}$ $\qquad \csc\theta = 3$

$\cos\theta = \dfrac{2\sqrt{2}}{3}$ $\qquad \sec\theta = \dfrac{3}{2\sqrt{2}} = \dfrac{3\sqrt{2}}{4}$

$\tan\theta = \dfrac{1}{2\sqrt{2}} = \dfrac{\sqrt{2}}{4}$ $\qquad \cos\theta = 2\sqrt{2}$.

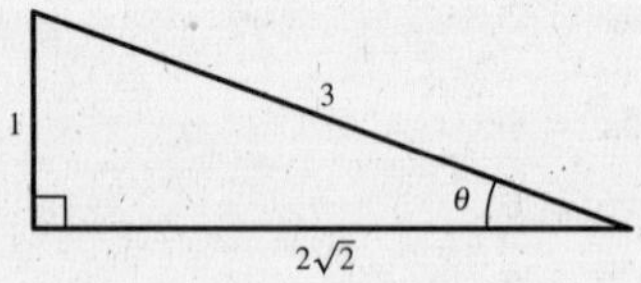

15. Because $x = 1$ and $r = 2$, the length of the opposite side is $y = \sqrt{2^2 - 1^2} = \sqrt{3}$. So you have

$\sin\theta = \dfrac{\sqrt{3}}{2}$ $\qquad \csc\theta = \dfrac{2}{\sqrt{3}} = \dfrac{2\sqrt{3}}{3}$

$\cos\theta = \dfrac{1}{2}$ $\qquad \sec\theta = 2$

$\tan\theta = \sqrt{3}$ $\qquad \cot\theta = \dfrac{1}{\sqrt{3}} = \dfrac{\sqrt{3}}{3}$.

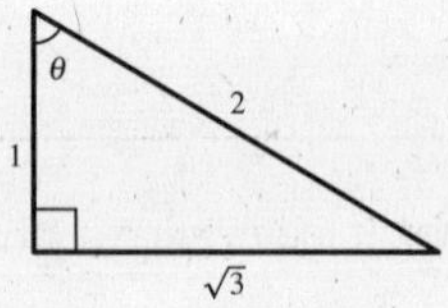

17. Because $x = 1$ and $y = 3$, the length of the hypotenuse is $r = \sqrt{1^2 + 3^2} = \sqrt{10}$.

So you have

$\sin\theta = \dfrac{3}{\sqrt{10}} = \dfrac{3\sqrt{10}}{10}$ $\qquad \csc\theta = \dfrac{\sqrt{10}}{3}$

$\cos\theta = \dfrac{1}{\sqrt{10}} = \dfrac{\sqrt{10}}{10}$ $\qquad \sec\theta = \sqrt{10}$

$\tan\theta = 3$ $\qquad \cot\theta = \dfrac{1}{3}$.

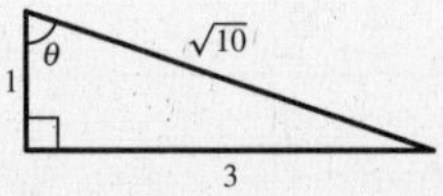

19. Because the sine is negative and the cosine is positive, θ must lie in Quadrant IV.

21. Because the sine is positive and the secant is positive, θ must lie in Quadrant I.

23. Because the cosecant is positive and the tangent is negative, θ must lie in Quadrant II.

25. $\theta = 30°\left(\dfrac{\pi \text{ radians}}{180°}\right) = \dfrac{\pi}{6}$ radians

$\sin 30° = \sin\dfrac{\pi}{6} = \dfrac{1}{2}$

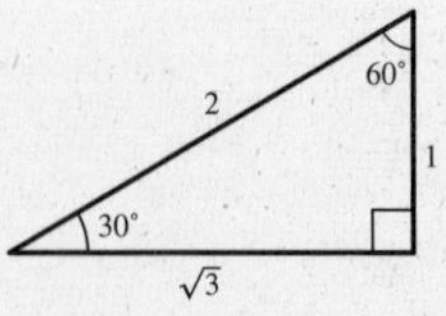

27. $\theta = \frac{\pi}{3}\left(\frac{180°}{\pi}\right) = 60°$

$\tan 60° = \tan\frac{\pi}{3} = \sqrt{3}$

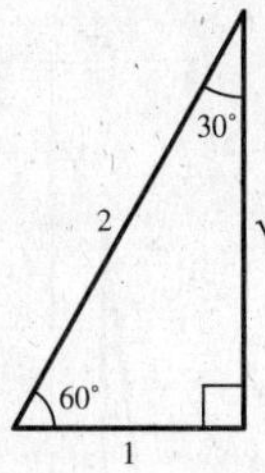

29. $\cot\theta = 1$

$\theta = 45° = 45°\left(\frac{\pi \text{ radians}}{180°}\right) = \frac{\pi}{4}$ radians

$\cot 45° = \cot\frac{\pi}{4} = 1$

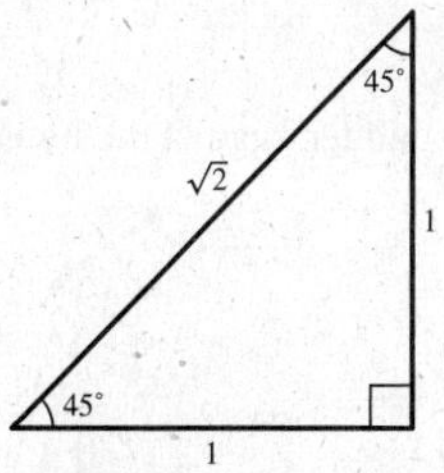

31. (a) $\sin 60° = \frac{\sqrt{3}}{2}$

$\cos 60° = \frac{1}{2}$

$\tan 60° = \sqrt{3}$

(b) $\sin\left(-\frac{2\pi}{3}\right) = -\frac{\sqrt{3}}{2}$

$\cos\left(-\frac{2\pi}{3}\right) = -\frac{1}{2}$

$\tan\left(-\frac{2\pi}{3}\right) = \sqrt{3}$

33. (a) $\sin\left(-\frac{\pi}{6}\right) = -\frac{1}{2}$

$\cos\left(-\frac{\pi}{6}\right) = \frac{\sqrt{3}}{2}$

$\tan\left(-\frac{\pi}{6}\right) = -\frac{\sqrt{3}}{3}$

(b) $\sin 150° = \frac{1}{2}$

$\cos 150° = -\frac{\sqrt{3}}{2}$

$\tan 150° = -\frac{\sqrt{3}}{3}$

35. (a) $\sin 225° = -\frac{\sqrt{2}}{2}$

$\cos 225° = -\frac{\sqrt{2}}{2}$

$\tan 225° = 1$

(b) $\sin(-225°) = \frac{\sqrt{2}}{2}$

$\cos(-225°) = -\frac{\sqrt{2}}{2}$

$\tan(-225°) = -1$

37. (a) $\sin 750° = \frac{1}{2}$

$\cos 750° = \frac{\sqrt{3}}{2}$

$\tan 750° = \frac{\sqrt{3}}{3}$

(b) $\sin 510° = \frac{1}{2}$

$\cos 510° = -\frac{\sqrt{3}}{2}$

$\tan 510° = -\frac{\sqrt{3}}{3}$

39. (a) $\sin 10° \approx 0.1736$

(b) $\csc 10° = \frac{1}{\sin 10°} \approx 5.7588$

41. (a) $\tan\left(\frac{\pi}{9}\right) \approx 0.3640$

(b) $\tan\left(\frac{10\pi}{9}\right) \approx 0.3640$

43. (a) $\cos(-110°) \approx -0.3420$

(b) $\cos 250° \approx -0.3420$

45. (a) $\csc 2.62 = \frac{1}{\sin 2.62} \approx 2.0070$

(b) $\csc 150° = \frac{1}{\sin 150°} = 2.0000$

47. (a) $\theta = \frac{\pi}{6}$ or $\theta = \frac{5\pi}{6}$

(b) $\theta = \frac{7\pi}{6}$ or $\theta = \frac{11\pi}{6}$

49. (a) $\theta = \frac{\pi}{3}$ or $\theta = \frac{2\pi}{3}$

(b) $\theta = \frac{3\pi}{4}$ or $\theta = \frac{7\pi}{4}$

51. (a) $\theta = \dfrac{\pi}{4}$ or $\theta = \dfrac{5\pi}{4}$

(b) $\theta = \dfrac{5\pi}{6}$ or $\theta = \dfrac{11\pi}{6}$

53. $2\sin^2\theta = 1$

$$\sin\theta = \pm\frac{\sqrt{2}}{2}$$

$$\theta = \frac{\pi}{4}, \frac{3\pi}{4}, \frac{5\pi}{4}, \frac{7\pi}{4}$$

55. $\tan^2\theta - \tan\theta = 0$

$$\tan\theta(\tan\theta - 1) = 0$$

$\tan\theta = 0$ or $\tan\theta = 1$

$\theta = 0, \pi, 2\pi$ $\quad\theta = \dfrac{\pi}{4}, \dfrac{5\pi}{4}$

57. $\sin 2\theta - \cos\theta = 0$

$$2\sin\theta\cos\theta - \cos\theta = 0$$

$$\cos\theta\,(2\sin\theta - 1) = 0$$

$\cos\theta = 0$ or $2\sin\theta = 1$

$\theta = \dfrac{\pi}{2}, \dfrac{3\pi}{2}$ $\quad\theta = \dfrac{\pi}{6}, \dfrac{5\pi}{6}$

59. $\sin\theta = \cos\theta$

Dividing both sides by $\cos\theta$ produces

$\tan\theta = 1$

$\theta = \dfrac{\pi}{4}, \dfrac{5\pi}{4}.$

61. $\cos^2\theta + \sin\theta = 1$

Using the identity $\cos^2\theta = 1 - \sin^2\theta$ produces

$$1 - \sin^2\theta + \sin\theta = 1$$

$$\sin^2\theta - \sin\theta = 0$$

$$\sin\theta(\sin\theta - 1) = 0$$

$\sin\theta = 0$ or $\sin\theta = 1$

$\theta = 0, \pi, 2\pi$ $\quad\theta = \dfrac{\pi}{2}.$

63. Because $\tan 30° = \dfrac{1}{\sqrt{3}} = \dfrac{y}{100}$ it follows that

$$y = \frac{100}{\sqrt{3}} = \frac{100\sqrt{3}}{3}.$$

65. Because $\tan 60° = \dfrac{\sqrt{3}}{1} = \dfrac{25}{x}$, it follows that

$$x = \frac{25}{\sqrt{3}} = \frac{25\sqrt{3}}{3}.$$

67. Because $\sin 40° = \dfrac{10}{r}$, it follows that

$$r = \frac{10}{\sin 40°} \approx 15.5572.$$

69. $\tan 82° = \dfrac{x}{45}$

$$45\tan 82° = x$$

$$320.19 \approx x$$

Total height $\approx 123 + 320.19$

$= 443.19$ m

$$\cos 82° = \frac{45}{d}$$

$$d = \frac{45}{\cos 82°}$$

$$d \approx 323.34$$

The distance between you and your friend is about 323.34 m.

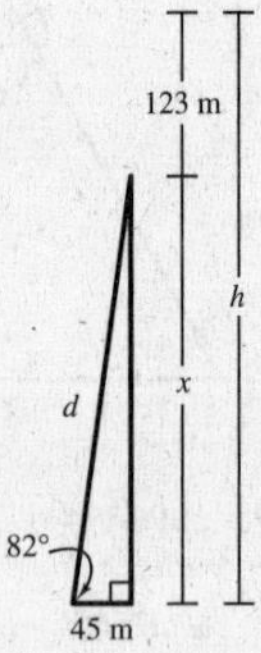

71. Let h be the height of the ladder against the house. Then

$$\sin 75° = \frac{h}{20}$$

and $h = 20\sin 75°$

≈ 19.32 feet.

The ladder reaches about 19.32 feet up the house.

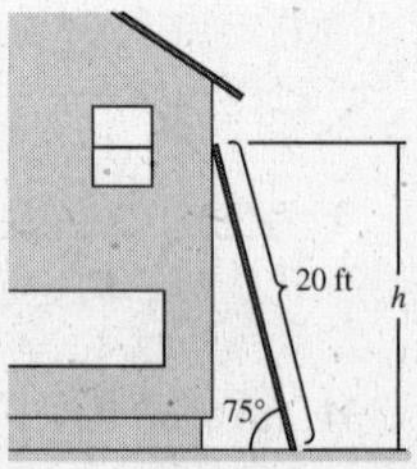

73. Let h be the height of the mountain, and x be the distance from you and the mountain.

$\tan 3.5° = \dfrac{h}{13 + x}$ $\qquad \tan 9° = \dfrac{h}{x}$

$\tan 3.5°(13 + x) = h$ $\qquad x\tan 9° = h$

$$\tan 3.5°\,(13 + x) = x\tan 9°$$

$$13\tan 3.5° + x\tan 3.5° = x\tan 9°$$

$$x\tan 3.5° - x\tan 9° = -13\tan 3.5°$$

$$x(\tan 3.5° - \tan 9°) = -13\tan 3.5°$$

$$x = \frac{-13\tan 3.5°}{\tan 3.5° - \tan 9°}$$

$$\frac{-13\tan 3.5°}{\tan 3.5° - \tan 9°}(\tan 9°) = h$$

$$1.3 \approx h$$

The mountain is about 1.3 miles high.

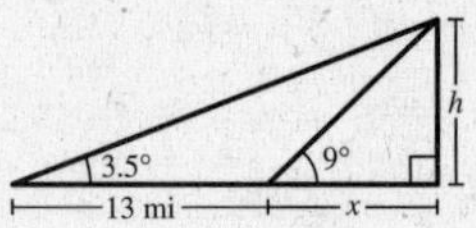

Not drawn to scale

75. (a) 10:00 P.M.:

$$T(0) = 98.6 + 4\cos 0 = 98.6 + 4 = 102.6°$$

(b) 4:00 A.M.:

$$T(6) = 98.6 + 4\cos\left(\frac{6\pi}{36}\right) = 98.6 + 4\left(\frac{\sqrt{3}}{2}\right) \approx 102.1°$$

(c) 10:00 A.M.:

$$T(12) = 98.6 + 4\cos\left(\frac{12\pi}{36}\right) = 98.6 + 4\left(\frac{1}{2}\right) = 100.6°$$

The temperature returns to normal when

$$98.6 = 98.6 + 4\cos\left(\frac{\pi t}{36}\right)$$

$$0 = \cos\left(\frac{\pi t}{36}\right)$$

$$\frac{\pi}{2} = \frac{\pi t}{36}$$

$t = 18$ or 4 P.M.

77.

x	0	2	4	6	8	10
$f(x)$	0	2.7021	2.7756	1.2244	1.2979	4

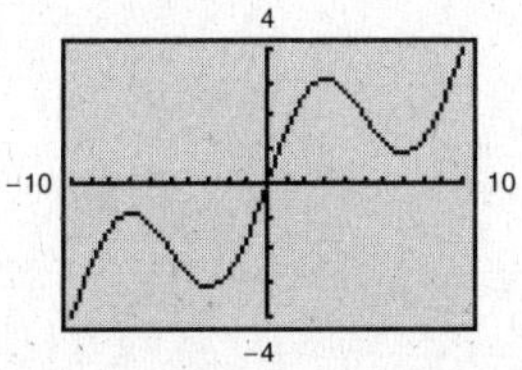

Mid-Chapter Quiz Solutions

1. $15°\left(\frac{\pi \text{ radians}}{180°}\right) = \frac{\pi}{12}$ radian

2. $105°\left(\frac{\pi \text{ radians}}{180°}\right) = \frac{7\pi}{12}$ radians

3. $-80°\left(\frac{\pi \text{ radians}}{180°}\right) = -\frac{4\pi}{9}$ radians

4. $35°\left(\frac{\pi \text{ radians}}{180°}\right) = \frac{7\pi}{36}$ radian

5. $\frac{2\pi}{3}\left(\frac{180°}{\pi}\right) = 120°$

6. $\frac{4\pi}{15}\left(\frac{180°}{\pi}\right) = 48°$

7. $-\frac{4\pi}{3}\left(\frac{180°}{\pi}\right) = -240°$

8. $\frac{11\pi}{12}\left(\frac{180°}{\pi}\right) = 165°$

9. $\sin\left(-\frac{\pi}{4}\right) = -\frac{\sqrt{2}}{2}$

10. $\cos 210° = -\frac{\sqrt{3}}{2}$

11. $\tan\frac{5\pi}{6} = -\frac{\sqrt{3}}{3}$

12. $\cot 45° = 1$

13. $\sec(-60°) = 2$

14. $\csc\frac{3\pi}{2} = -1$

15. $\tan\theta - 1 = 0$

$$\tan\theta = 1$$

$$\theta = \frac{\pi}{4}, \frac{5\pi}{4}$$

16. $\cos^2\theta - 2\cos\theta + 1 = 0$

$$(\cos\theta - 1)(\cos\theta - 1) = 0$$

$$\cos\theta = 1$$

$$\theta = 0, 2\pi$$

17. $\sin^2 \theta = 3 \cos^2 \theta$

$$\frac{\sin^2 \theta}{\cos^2 \theta} = 3$$

Using the identity $\dfrac{\sin \theta}{\cos \theta} = \tan \theta$ produces the following.

$$\tan^2 \theta = 3$$

$$\tan \theta = \pm\sqrt{3}$$

$$\theta = \frac{\pi}{3}, \frac{2\pi}{3}, \frac{4\pi}{3}, \frac{5\pi}{3}$$

18. The angle θ is $90° - 60° = 30°$.

Find a by using the Pythagorean Theorem.

$$a = \sqrt{10^2 - 5^2} = \sqrt{75} = 5\sqrt{3}$$

19. The angle θ is $90° - 50° = 40°$.

$$\cos 50° = \frac{a}{16}$$

$$a = 16 \cos 50°$$

$$a \approx 10.2846$$

20. The angle θ is $90° - 40° = 50°$.

$$\tan 40° = \frac{a}{4}$$

$$a = 4 \tan 40°$$

$$a \approx 3.3564$$

21. $\sin 35° = \dfrac{d}{500}$

$$500 \sin 35° = d$$

$$286.8 \text{ ft} \approx d$$

22. $\sin \theta = \dfrac{3.5}{17.5}$

$$\sin \theta = \frac{1}{5}$$

Use inverse sine function of a calculator.

$$\theta \approx 11.54°$$

Section 5.3 Graphs of Trigonometric Functions

Skills Review

1. $\lim_{x\to 2} \left(x^2 + 4x + 2\right) = 2^2 + 4(2) + 2 = 14$
2. $\lim_{x\to 3} \left(x^3 - 2x^2 + 1\right) = 3^3 - 2(3)^2 + 1 = 10$
3. $\cos \dfrac{\pi}{2} = 0$
4. $\sin \pi = 0$
5. $\tan \dfrac{5\pi}{4} = 1$
6. $\cot \dfrac{2\pi}{3} = -\dfrac{\sqrt{3}}{3}$
7. $\sin \dfrac{11\pi}{6} = -\dfrac{1}{2}$
8. $\cos \dfrac{5\pi}{6} = -\dfrac{\sqrt{3}}{2}$
9. $\cos \dfrac{5\pi}{3} = \dfrac{1}{2}$
10. $\sin \dfrac{4\pi}{3} = -\dfrac{\sqrt{3}}{2}$
11. $\cos 15° \approx 0.9659$
12. $\sin 220° \approx -0.6428$
13. $\sin 275° \approx -0.9962$
14. $\cos 310° \approx 0.6428$
15. $\sin 103° \approx 0.9744$
16. $\cos 72° \approx 0.3090$
17. $\tan 327° \approx -0.6494$
18. $\tan 140° \approx -0.8391$

1. $y = 2 \sin 2x$

Period: $\dfrac{2\pi}{2} = \pi$

Amplitude: 2

3. $y = \dfrac{3}{2} \cos \dfrac{x}{2}$

Period: $\dfrac{2\pi}{1/2} = 4\pi$

Amplitude: $\dfrac{3}{2}$

5. $y = \dfrac{1}{2} \cos \pi x$

Period: $\dfrac{2\pi}{\pi} = 2$

Amplitude: $\dfrac{1}{2}$

7. $y = -2 \sin x$

Period: $\dfrac{2\pi}{1} = 2\pi$

Amplitude: 2

9. $y = -2 \sin 10x$

Period: $\dfrac{2\pi}{10} = \dfrac{\pi}{5}$

Amplitude: 2

11. $y = \dfrac{1}{2} \sin \dfrac{2x}{3}$

Period: $\dfrac{2\pi}{2/3} = 3\pi$

Amplitude: $\dfrac{1}{2}$

13. $y = 3 \sin 4\pi x$

Period: $\dfrac{2\pi}{4\pi} = \dfrac{1}{2}$

Amplitude: 3

15. $y = 3 \tan x$

Period: $\dfrac{\pi}{1} = \pi$

17. $y = 3 \sec 5x$

Period: $\dfrac{2\pi}{5}$

19. $y = \cot \dfrac{\pi x}{6}$

Period: $\dfrac{\pi}{\pi/6} = 6$

21. $y = \sec 2x$

The graph of this function has a period of π and matches graph (c).

23. $y = \cot \dfrac{\pi x}{2}$

The graph of this function has a period of 2 and matches graph (f).

25. $y = 2 \csc \dfrac{x}{2}$

The graph of this function has a period of 4π and matches graph (b).

27. $y = \sin \dfrac{x}{2}$

Period: 4π

Amplitude: 1

x-intercepts:

$(0, 0), (2\pi, 0), (4\pi, 0)$

Maximum: $(\pi, 1)$

Minimum: $(3\pi, -1)$

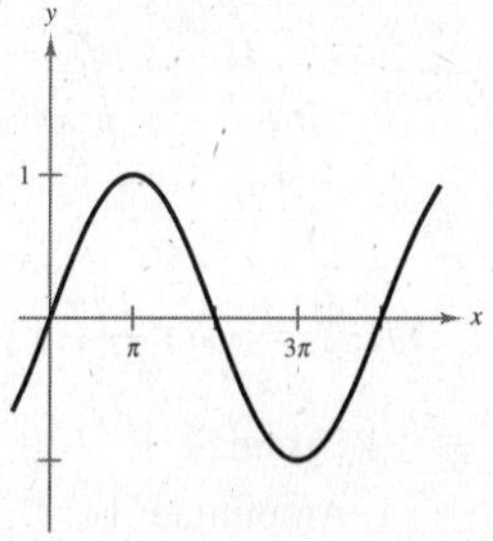

29. $y = 2 \cos \dfrac{\pi x}{3}$

Period: 6

Amplitude: 2

x-intercepts:

$\left(-\frac{3}{2}, 0\right), \left(\frac{3}{2}, 0\right), \left(\frac{9}{2}, 0\right)$

Maxima: $(0, 2), (6, 2)$

Minimum: $(3, -2)$

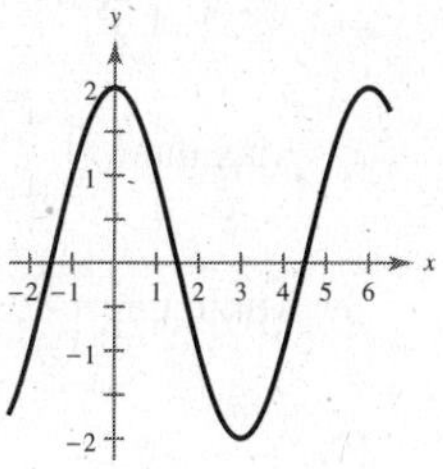

31. $y = -2 \sin 6x$

Period: $\dfrac{\pi}{3}$

Amplitude: 2

x-intercepts:

$(0, 0), \left(\dfrac{\pi}{6}, 0\right), \left(\dfrac{\pi}{3}, 0\right)$

Maximum: $\left(\dfrac{\pi}{4}, 2\right)$

Minimum: $\left(\dfrac{\pi}{12}, -2\right)$

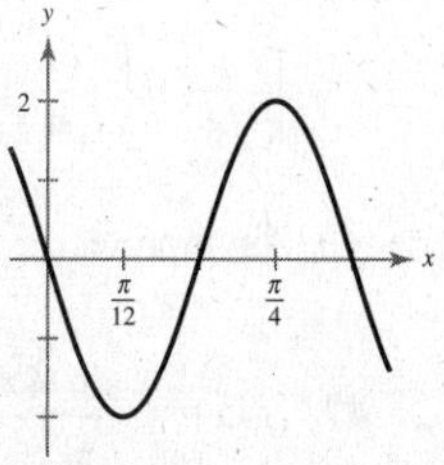

33. $y = \cos 2\pi x$

Period: 1

Amplitude: 1

x-intercepts: $\left(\frac{1}{4}, 0\right), \left(\frac{3}{4}, 0\right)$

Maxima: $(0, 1), (1, 1)$

Minima: $\left(\frac{1}{2}, 1\right)$

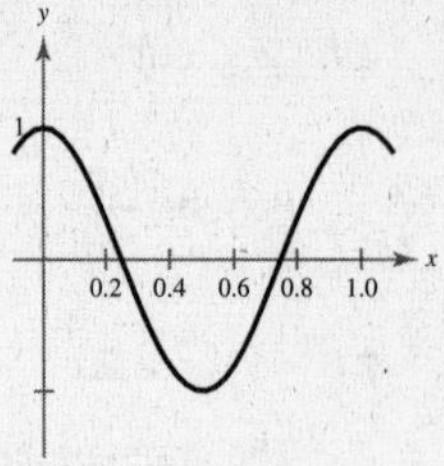

35. $y = 2 \tan x$

Period: π

x-intercepts: $(0, 0), (\pi, 0)$

Asymptotes: $x = -\dfrac{\pi}{2}, x = \dfrac{\pi}{2}, x = \dfrac{3\pi}{2}$

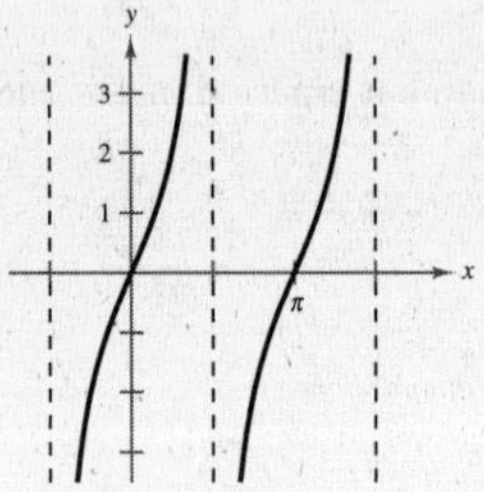

37. $y = -\sin \dfrac{2\pi x}{3}$

Period: 3

Amplitude: 1

x-intercepts:

$(0, 0), \left(\dfrac{3}{2}, 0\right), (3, 0)$

Maximum: $\left(\dfrac{9}{4}, 1\right)$

Minimum: $\left(\dfrac{3}{4}, -1\right)$

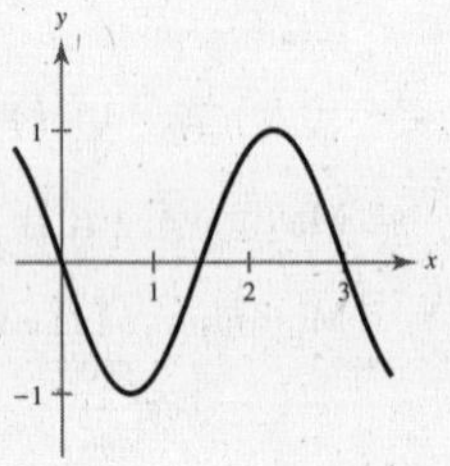

39. $y = \cot 2x$

Period: $\dfrac{\pi}{2}$

x-intercepts:

$\left(-\dfrac{\pi}{4}, 0\right), \left(\dfrac{\pi}{4}, 0\right)$

Asymptotes: $x = -\dfrac{\pi}{2}, x = 0, x = \dfrac{\pi}{2}$

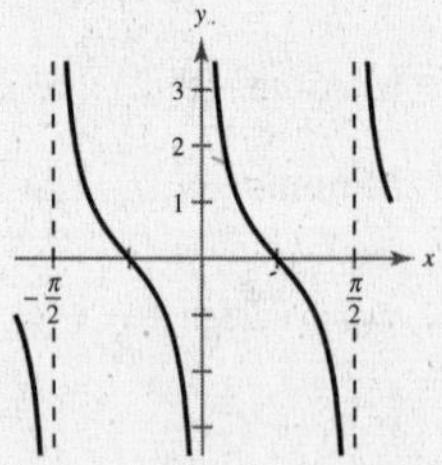

47. $f(x) = \dfrac{\sin 4x}{2x}$

x	−0.1	−0.01	−0.001	0.001	0.01	0.1
$f(x)$	1.9471	1.9995	2.0000	2.0000	1.9995	1.9471

From this table, you can estimate that $\lim\limits_{x \to 0} \dfrac{\sin 4x}{2x} = 2$.

41. $y = \csc \dfrac{2x}{3}$

Period: 3π

Asymptotes:

$x = 0, x = \dfrac{3\pi}{2}, 3\pi$

Relative minimum: $\left(\dfrac{3\pi}{4}, 1\right)$

Relative maximum: $\left(\dfrac{9\pi}{4}, -1\right)$

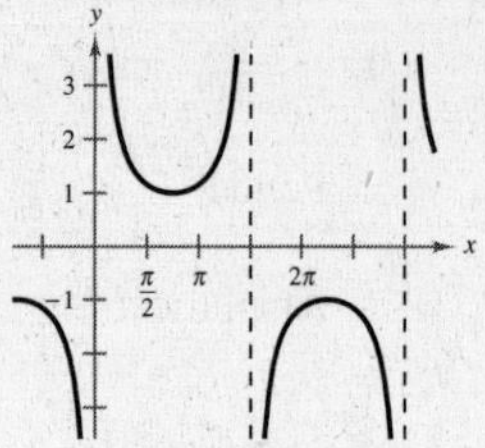

43. $y = 2 \sec 2x$

Period: π

Asymptotes: $x = \dfrac{\pi}{4}, x = \dfrac{3\pi}{4}$

Relative minima: $(0, 2), (\pi, 2)$

Relative maxima: $\left(\dfrac{\pi}{2}, -2\right), \left(\dfrac{3\pi}{2}, -2\right)$

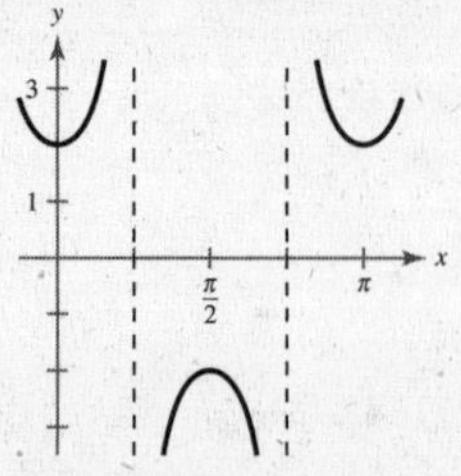

45. $y = \csc 2\pi x$

Period: 1

Asymptotes: $x = 0, x = \frac{1}{2}, x = 1$

Relative minimum: $\left(\frac{1}{4}, 1\right)$

Relative maximum: $\left(\frac{3}{4}, -1\right)$

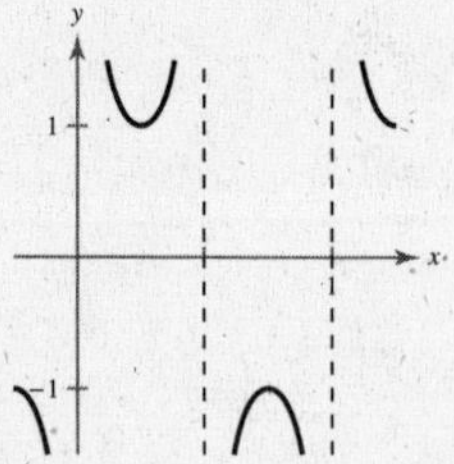

49. $f(x) = \dfrac{\sin x}{5x}$

x	−0.1	−0.01	−0.001	0.001	0.01	0.1
$f(x)$	0.1997	0.2000	0.2000	0.2000	0.2000	0.1997

From this table, you can estimate that $\lim_{x \to 0} \dfrac{\sin x}{5x} = \dfrac{1}{5}$.

51. $f(x) = \dfrac{3(1 - \cos x)}{x}$

x	−0.1	−0.01	−0.001	0.001	0.01	0.1
$f(x)$	−0.1499	−0.015	−0.0015	0.0015	0.015	0.1499

From this table, you can estimate that $\lim_{x \to 0} \dfrac{3(1 - \cos x)}{x} = 0$.

53. $f(x) = \dfrac{\tan 2x}{x}$

x	−0.1	−0.01	−0.001	0.001	0.01	0.1
$f(x)$	2.0271	2.0003	2.0000	2.0000	2.0003	2.0271

From this table, you can estimate that $\lim_{x \to 0} \dfrac{\tan 2x}{x} = 2$.

55. $f(x) = \dfrac{\sin^2 x}{x}$

x	−0.1	−0.01	−0.001	0.001	0.01	0.1
$f(x)$	−0.0997	−0.01	−0.001	0.001	0.01	0.0997

From this table, you can estimate that $\lim_{x \to 0} \dfrac{\sin^2 x}{x} = 0$.

57. $f(x) = \dfrac{\sin x}{2x}$

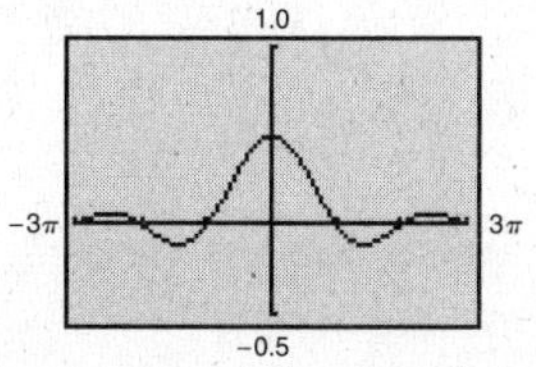

$\lim_{x \to 0} \dfrac{\sin x}{2x} = \dfrac{1}{2}$

59. $f(x) = \dfrac{\sin 5x}{\sin 2x}$

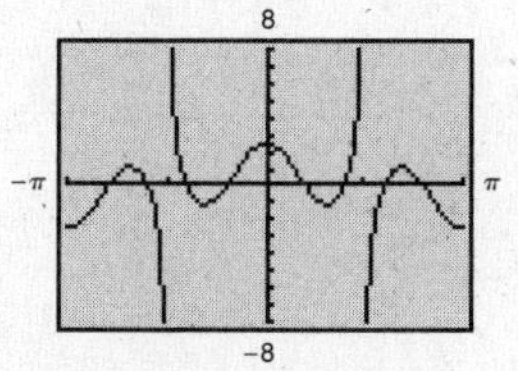

$\lim_{x \to 0} \dfrac{\sin 5x}{\sin 2x} = \dfrac{5}{2}$

61. $f(x) = a \cos x + d$

Amplitude: $\frac{1}{2}[3 - (-1)] = 2 \Rightarrow a = 2$

Vertical shift one unit upward $\Rightarrow d = 1$

So, $f(x) = 2 \cos x + 1$.

63. $f(x) = a \cos x + d$

Amplitude: $\frac{1}{2}[8 - 0] = 4 \Rightarrow a = 4$

Reflection in the x-axis: $-a \Rightarrow a = -4$

Vertical shift four units upward $\Rightarrow d = 4$

So, $f(x) = -4 \cos x + 4$.

65. The graph of this function has a period of 2π and matches graph (b).

67. The graph of this function is the graph of $y = \sin x$ shifted horizontally π units and matches graph (a).

69. $V = 0.9 \sin \dfrac{\pi t}{3}$

(a) Period $= \dfrac{2\pi}{\pi/3} = 6$ sec

(b) Cycles per minute $= \dfrac{60}{6} = 10$

(c)

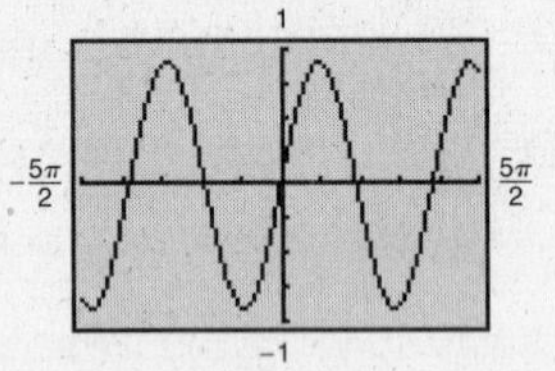

71. $y = 0.001 \sin 880\pi t$

(a) Period $= \dfrac{2\pi}{880\pi} = \dfrac{1}{440}$

(b) Frequency $= \dfrac{1}{\text{period}} = \dfrac{1}{1/440} = 440$

(c)

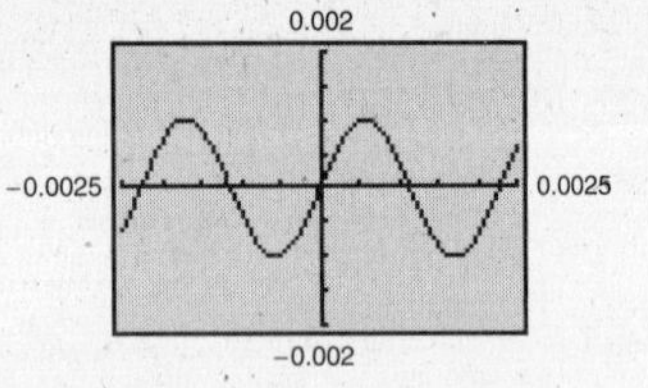

73. (a) $P = 8000 + 2500 \sin \dfrac{2\pi t}{24}$

$p = 12{,}000 + 4000 \cos \dfrac{2\pi t}{24}$

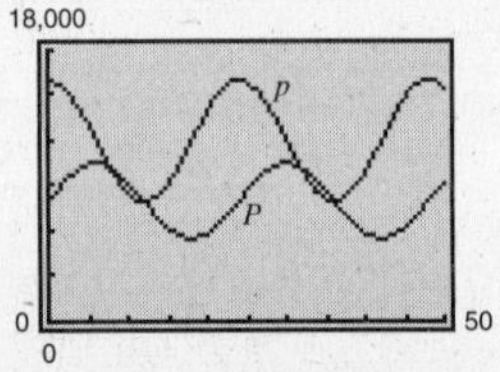

(b) As the population of the prey increases, the population of the predator increases as well. At some point, the predator eliminates the prey faster than the prey can reproduce, and the prey population decreases rapidly. As the prey becomes scarce, the predator population decreases, releasing the prey from predator pressure, and the cycle begins again.

75. (a) and (b)

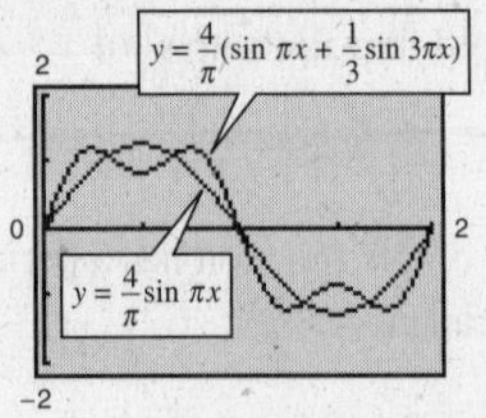

77. $P = \sin \dfrac{2\pi t}{23}$, $E = \sin \dfrac{2\pi t}{28}$, and $I = \sin \dfrac{2\pi t}{33}$ where $t \geq 0$.

$$P(8930) = \sin \frac{2\pi(8930)}{23} \approx 0.9977$$

$$E(8930) = \sin \frac{2\pi(8930)}{28} \approx -0.4339$$

$$I(8930) = \sin \frac{2\pi(8930)}{33} \approx -0.6182$$

The physical cycle is about 0.9977, the emotional cycle is about −0.4339, and the intellectual cycle is about −0.6182.

79. $W = 7594 + 455.2 \sin(0.41t - 1.713)$

($t = 1$ corresponds to January 1, 2006.)

(a)

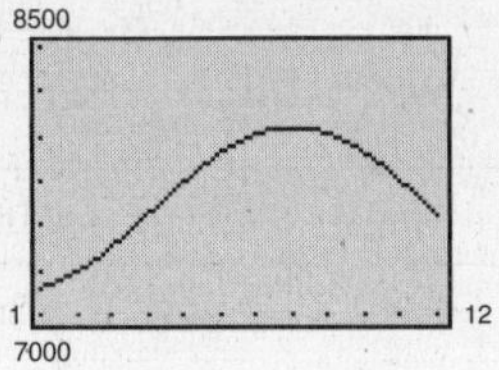

(b) Yes. The graphs of W and $y = 8000$ intersect at $t \approx 6.865$ and $t \approx 9.154$. The number of construction workers exceeded 8 million during June, July, August, and September.

81. (a) A

(b) B

(c) B

(d) They are reciprocals of each other.

83. False, the amplitude is $|-3| = 3$.

85. True

87. Answers will vary.

Section 5.4 Derivatives of Trigonometric Functions

Skills Review

1. $f(x) = 3x^3 - 2x^2 + 4x - 7$

$f'(x) = 9x^2 - 4x + 4$

2. $g(x) = (x^3 + 4)^4$

$g'(x) = 4(x^3 + 4)^3(3x^2)$

$= 12x^2(x^3 + 4)^3$

3. $f(x) = (x - 1)(x^2 + 2x + 3)$

$f'(x) = (x - 1)(2x + 2) + (x^2 + 2x + 3)$

$= 2x^2 - 2 + x^2 + 2x + 3$

$= 3x^2 + 2x + 1$

4. $g(x) = \dfrac{2x}{x^2 + 5}$

$g'(x) = \dfrac{(x^2 + 5)(2) - 2x(2x)}{(x^2 + 5)^2}$

$= \dfrac{2x^2 + 10 - 4x^2}{(x^2 + 5)^2}$

$= \dfrac{-2x^2 + 10}{(x^2 + 5)^2}$

$= \dfrac{2(5 - x^2)}{(x^2 + 5)^2}$

5. $f(x) = x^2 + 4x + 1$

$f'(x) = 2x + 4 = 0$

Critical number: $x = -2$

Relative minimum: $(-2, -3)$

Interval	$-\infty < x < -2$	$-2 < x < \infty$
Sign of f'	$f' < 0$	$f' > 0$
Conclusion	Decreasing	Increasing

6. $f(x) = \frac{1}{3}x^3 - 4x + 2$

$f'(x) = x^2 - 4 = 0$

Critical numbers: $x = \pm 2$

Relative maximum: $\left(-2, \frac{22}{3}\right)$

Relative minimum: $\left(2, -\frac{10}{3}\right)$

Interval	$-\infty < x < -2$	$-2 < x < 2$	$2 < x < \infty$
Sign of f'	$f' > 0$	$f' < 0$	$f' > 0$
Conclusion	Increasing	Decreasing	Increasing

7. $\sin x = \dfrac{\sqrt{3}}{2}$

$x = \dfrac{\pi}{3}, \dfrac{2\pi}{3}$

8. $\cos x = -\dfrac{1}{2}$

$x = \dfrac{2\pi}{3}, \dfrac{4\pi}{3}$

9. $\cos \dfrac{x}{2} = 0$

$\dfrac{x}{2} = \dfrac{\pi}{2}$

$x = \pi$

10. $\sin \dfrac{x}{2} = -\dfrac{\sqrt{2}}{2}$

$\dfrac{x}{2} = \dfrac{5\pi}{4}$

No solution in $0 \le x \le 2\pi$

1. $y = \frac{1}{2} - 3 \sin x$

$y' = -3 \cos x$

3. $y = x^2 - \cos x$

$y' = 2x + \sin x$

5. $f(x) = 4\sqrt{x} + 3 \cos x$

$f'(x) = \dfrac{2}{\sqrt{x}} - 3 \sin x$

7. $f(t) = t^2 \cos t$

$f'(t) = -t^2 \sin t + 2t \cos t$

9. $g(t) = \dfrac{\cos t}{t}$

$g'(t) = \dfrac{t(-\sin t) - (\cos t)(1)}{t^2} = -\dfrac{t \sin t + \cos t}{t^2}$

11. $y = \tan x + x^2$

$y' = \sec^2 x + 2x$

13. $y = e^{x^2} \sec x$

$y' = e^{x^2}(\sec x \tan x) + 2xe^{x^2} \sec x$

$= e^{x^2} \sec x(\tan x + 2x)$

15. $y = \cos 3x + \sin^2 x$

$y' = -\sin 3x(3) + 2 \sin x \cos x$

$= -3 \sin 3x + 2 \sin x \cos x$

17. $y = \sec \pi x$

$y' = \pi \sec \pi x \tan \pi x$

19. $y = x \sin \dfrac{1}{x}$

$y' = x \cos\left(\dfrac{1}{x}\right)\left(-\dfrac{1}{x^2}\right) + \sin \dfrac{1}{x} = \sin \dfrac{1}{x} - \dfrac{1}{x} \cos \dfrac{1}{x}$

21. $y = 3 \tan 4x$

$y' = 3 \sec^2 4x(4)$

$y' = 12 \sec^2 4x$

23. $y = 2 \tan^2 4x$

$y' = 4 \tan(4x) \sec^2 4x(4) = 16 \tan 4x \sec^2 4x$

Equivalently,

$y' = 16 \tan 4x(1 + \tan^2 4x) = 16 \tan 4x + 16 \tan^3 4x.$

25. $y = e^{2x} \sin 2x$

$y' = e^{2x}(\cos 2x(2)) + 2e^{2x} \sin 2x$

$= 2e^{2x}(\cos 2x + \sin 2x)$

27. $y = (\cos x)^2$

$y' = 2(\cos x)(-\sin x) = -2 \cos x \sin x = -\sin 2x$

29. $y = \cos^2 x - \sin^2 x$

$y' = -2 \cos x \sin x - 2 \sin x \cos x$

$= -4 \cos x \sin x = -2 \sin 2x$

31. $y' = \sin x^2 - \cos 2x$

$y' = 2 \sin x \cos x + 2 \sin 2x$

$= \sin 2x + 2 \sin 2x$

$= 3 \sin 2x$

33. $y = \tan x - x$

$y' = \sec^2 x - 1 = \tan^2 x$

35. $y = \frac{1}{3}(\sin x)^3 - \frac{1}{5}(\sin x)^5$

$y' = (\sin x)^2 \cos x - (\sin x)^4 \cos x$

$= \sin^2 x \cos x(1 - \sin^2 x)$

$= \sin^2 x \cos x \cos^2 x$

$= \sin^2 x \cos^3 x$

37. $y = \ln(\sin^2 x)$

$y' = \dfrac{1}{\sin^2 x} 2 \sin x \cos x = \dfrac{2 \cos x}{\sin x} = 2 \cot x$

39. $y = \tan x, \left(-\dfrac{\pi}{4}, -1\right)$

$y' = \sec^2 x, \quad y'\left(-\dfrac{\pi}{4}\right) = 2$

$y - (-1) = 2\left[x - \left(-\dfrac{\pi}{4}\right)\right]$

$y = 2x - 1 + \dfrac{\pi}{2}$

41. $y = \sin 4x, \quad (\pi, 0)$

$y' = 4 \cos 4x, \quad y'(\pi) = 4$

$y - 0 = 4(x - \pi)$

$y = 4x - 4\pi$

43. $y = \cot x, \left(\dfrac{3\pi}{4}, -1\right)$

$y' = -\csc^2 x, \quad y'\left(\dfrac{3\pi}{4}\right) = -2$

$y - (-1) = -2\left(x - \dfrac{3\pi}{4}\right)$

$y = -2x - 1 + \dfrac{3\pi}{2}$

45. $y = \ln(\sin x + 2), \quad \left(\frac{3\pi}{2}, 0\right)$

$y' = \frac{1}{\sin x + 2}(\cos x), \quad y'\left(\frac{3\pi}{2}\right) = 0$

$y - 0 = 0\left(x - \frac{3\pi}{2}\right)$

$y = 0$

47. $y = \sin \frac{5x}{4}$

$y' = \frac{5}{4} \cos \frac{5x}{4}$

The slope of the tangent line at $(0, 0)$ is $\frac{5}{4} = 1.25$.

There is one complete cycle of the graph in the interval $[0, 2\pi]$.

49. $y = \sin 2x$

$y' = 2 \cos 2x$

The slope of the tangent line at $(0, 0)$ is 2. There are two complete cycles of the graph in the interval $[0, 2\pi]$.

51. $y = \sin x$

$y' = \cos x$

The slope of the tangent line at $(0, 0)$ is 1. There is one complete cycle of the graph in the interval $[0, 2\pi]$.

53. $y = 2 \sin x + \sin 2x$

$y' = 2 \cos x + 2 \cos 2x$

$2 \cos x + 2 \cos 2x = 0$

$2\left[\cos x + 2 \cos^2 x - 1\right] = 0$

$2 \cos^2 x + \cos x - 1 = 0$

$(2 \cos x - 1)(\cos x + 1) = 0$

$\cos x = \frac{1}{2}$ or $\cos x = -1$.

Critical numbers: $x = \frac{\pi}{3}, \pi, \frac{5\pi}{3}$

Relative minimum: $\left(\frac{5\pi}{3}, -\frac{3\sqrt{3}}{2}\right)$

Relative maximum: $\left(\frac{\pi}{3}, \frac{3\sqrt{3}}{2}\right)$

55. $y = x - 2 \sin x$

$y' = 1 - 2 \cos x$

$1 - 2 \cos x = 0$

$\cos x = \frac{1}{2}$.

Critical numbers: $x = \frac{\pi}{3}, \frac{5\pi}{3}$

Relative minimum: $\left(\frac{\pi}{3}, \frac{\pi}{3} - \sqrt{3}\right)$

Relative maximum: $\left(\frac{5\pi}{3}, \frac{5\pi}{3} + \sqrt{3}\right)$

57. $y = e^x \cos x$

$y' = -e^x \sin x + e^x \cos x$

$-e^x \sin x + e^x \cos x = 0$

$-e^x(\sin x - \cos x) = 0$

$\sin x = \cos x$

$\tan x = 1$.

Critical numbers: $x = \frac{\pi}{4}, \frac{5\pi}{4}$

Relative minimum: $\left(\frac{5\pi}{4}, -\frac{\sqrt{2}}{2}e^{5\pi/4}\right)$

Relative maximum: $\left(\frac{\pi}{4}, \frac{\sqrt{2}}{2}e^{\pi/4}\right)$

59. $D = 12.13 - 1.87 \cos\frac{\pi(t - 0.07)}{6}, \quad 0 \le t \le 12$

$D' = \frac{1.87\pi}{6} \sin \frac{\pi(t - 0.07)}{6}$

$\frac{1.87\pi}{6} \sin \frac{\pi(t - 0.07)}{6} = 0$

$\frac{\pi(t - 0.07)}{6} = \pi$

$t \approx 6.07$

D has a relative maximum at $(6.07, 14)$. The maximum number of daylight hours occurred in July with 14 hours of daylight.

61. $h(t) = 0.20t + 0.03 \sin 2\pi t$

$h'(t) = 0.20 + 0.06\pi \cos 2\pi t$

$0.20 + 0.06\pi \cos 2\pi t = 0$

$\cos 2\pi t = -\dfrac{10}{3\pi}$

$t = 0, \dfrac{1}{2}$

(a) $h'(t)$ is maximum when $t = 0$. The growth rate is a maximum at midnight.

(b) $h'(t)$ is minimum when $t = \frac{1}{2}$. The growth rate is a minimum at noon.

63. (a) $f(t) = t^2 \sin t, (0, 2\pi)$

$f'(t) = t^2 \cos t + 2t \sin t = t(t \cos t + 2 \sin t)$

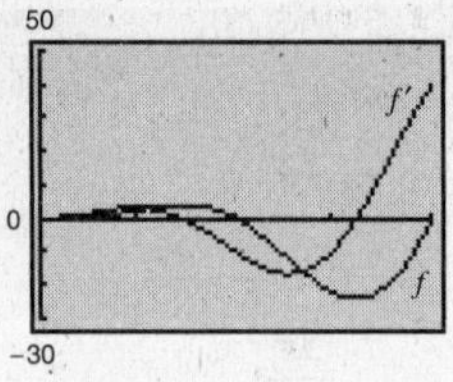

(b) $f'(t) = 0$ when $t = 0$, $t \approx 2.289$, and $t \approx 5.087$.

(c)

Interval	(0, 2.289)	(2.289, 5.087)	(5.087, 2π)
Sign of f'	$f' > 0$	$f' < 0$	$f' > 0$
Conclusion	f is increasing.	f is decreasing.	f is increasing.

65. (a) $f(x) = \sin x - \dfrac{1}{3} \sin 3x + \dfrac{1}{5} \sin 5x, (0, \pi)$

$f'(x) = \cos x - \cos 3x + \cos 5x$

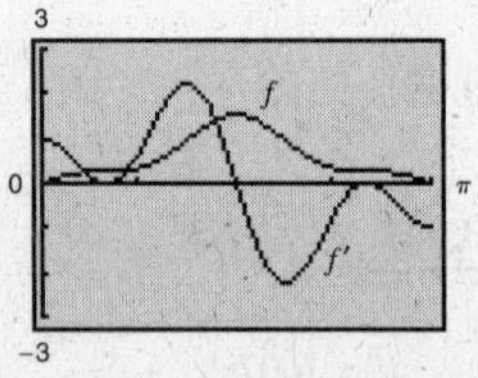

(b) $f'(x) = 0$ when $x \approx 0.524$, $x \approx 1.571$, and $x \approx 2.618$.

(c)

Interval	(0, 0.524)	(0.524, 1.571)	(1.571, 2.618)	(2.618, π)
Sign of f'	$f' > 0$	$f' > 0$	$f' < 0$	$f' < 0$
Conclusion	f is increasing.	f is increasing.	f is decreasing.	f is decreasing.

67. (a) $f(x) = \sqrt{2x}\sin x,\ (0, 2\pi)$

$$f'(x) = \sqrt{2x}\cos x + \frac{\sin x}{\sqrt{2x}}$$

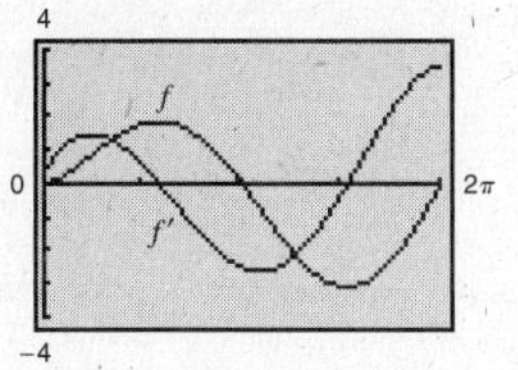

(b) $f'(x) = 0$ when $x \approx 1.837$ and $x \approx 4.816$.

(c)

Interval	$(0, 1.837)$	$(1.837, 4.816)$	$(4.816, 2\pi)$
Sign of f'	$f' > 0$	$f' < 0$	$f' > 0$
Conclusion	f is increasing.	f is decreasing.	f is increasing.

69. $f(x) = \dfrac{x}{\sin x}$

Relative maximum: $(4.49, -4.60)$

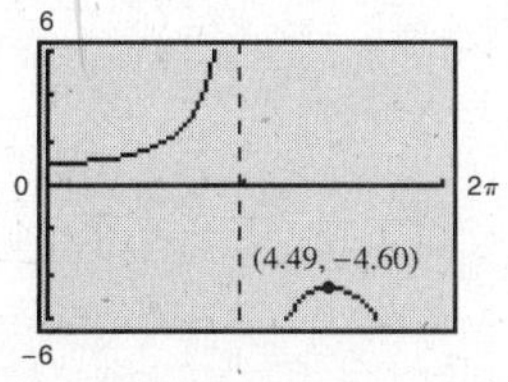

71. $f(x) = \ln x \cos x$

Relative minimum: $(3.38, -1.18)$

Relative maximum: $(1.27, 0.07)$

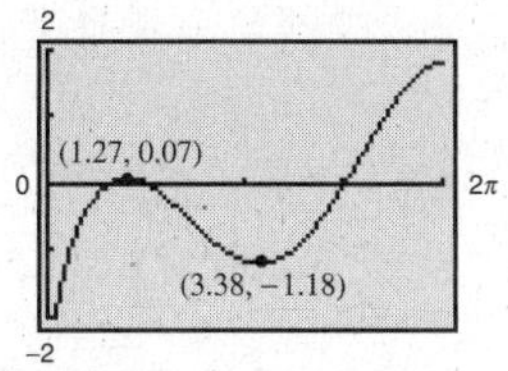

73. $f(x) = \sin(0.1x^2)$

Relative maximum: $(3.96, 1)$

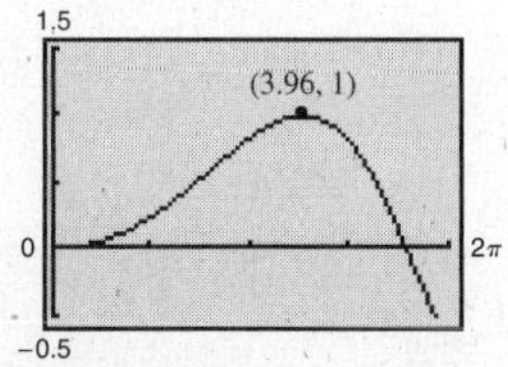

75. False. If $y = (1 - \cos x)^{1/2}$, then

$$y' = \frac{1}{2}(1 - \cos x)^{-1/2}(\sin x).$$

77. False. If $y = x\sin^2 x$, then

$$y' = 2x\sin x\cos x + \sin^2 x = x\sin 2x + \sin^2 x.$$

79. (a) Let l represent the length of the tank.

$$\begin{aligned} V &= \text{Total volume of tank} - \text{Volume of segment} \\ &= \pi r^2 l - \tfrac{1}{2}r^2 l(\theta - \sin\theta) \\ &= \pi(26)^2(100) - \tfrac{1}{2}(26)^2(100)(\theta - \sin\theta) \\ &= 67{,}600\pi - 33{,}800(\theta - \sin\theta) \\ &= 67{,}600\pi - 33{,}800\theta + 33{,}800\sin\theta \end{aligned}$$

(b) Because $h = 40$ and $r = 26$, you know the distance from the chord to the arc is $52 - 40 = 12$. The height of the triangle is $40 - 26 = 14$. Because a radius bisects the angle given by θ, you have the following

$$\cos\left(\frac{\theta}{2}\right) = \frac{14}{26}$$

$$\frac{\theta}{2} \approx 1.002$$

$$\theta \approx 2.004 \text{ radians.}$$

(c) $\dfrac{dv}{d\theta} = -33{,}800 + 33{,}800\cos\theta$

When $\theta \approx 2.004$ rad,

$$\frac{dv}{d\theta} = -33{,}800 + 33{,}800\cos(2.004)$$

$$\approx -47{,}988.6 \text{ cubic inches per radian.}$$

81. Answers will vary.

Review Exercises for Chapter 5

1. Positive coterminal angle: $\frac{7\pi}{4} + 2\pi = \frac{15\pi}{4}$

Negative coterminal angle: $\frac{7\pi}{4} - 2\pi = -\frac{\pi}{4}$

3. Positive coterminal angle: $\frac{3\pi}{2} + 2\pi = \frac{7\pi}{2}$

Negative coterminal angle: $\frac{3\pi}{2} - 2\pi = -\frac{\pi}{2}$

5. Positive coterminal angle: $135° + 360° = 495°$

Negative coterminal angle: $135° - 360° = -225°$

7. Positive coterminal angle: $-405° + 2(360°) = 315°$

Negative coterminal angle: $-405° + 360° = -45°$

9. $210°\left(\frac{\pi \text{ radians}}{180°}\right) = \frac{7\pi}{6}$ radians

11. $-60°\left(\frac{\pi \text{ radians}}{180°}\right) = -\frac{\pi}{3}$ radians

13. $-480°\left(\frac{\pi \text{ radians}}{180°}\right) = -\frac{8\pi}{3}$ radians

15. $110°\left(\frac{\pi \text{ radians}}{180°}\right) = \frac{11\pi}{18}$ radians

17. $\frac{4\pi}{3}\left(\frac{180°}{\pi}\right) = 240°$

19. $-\frac{2\pi}{3}\left(\frac{180°}{\pi}\right) = -120°$

21. $b = \sqrt{8^2 - 4^2} = \sqrt{48} = 4\sqrt{3}$

$\theta = 90° - 30° = 60°$

23. $c = 5$, $\theta = 60°$ (equilateral triangle)

$a = \sqrt{5^2 - \left(\frac{5}{2}\right)^2} = \frac{5\sqrt{3}}{2}$

25. $c = \sqrt{50^2 + 75^2} = \sqrt{8125} \approx 90.14$ feet

27. The reference angle is $\pi - \frac{2\pi}{3} = \frac{\pi}{3}$.

29. The reference angle is $\pi - \frac{5\pi}{6} = \frac{\pi}{6}$.

31. The reference angle is $240° - 180° = 60°$.

33. The reference angle is $420° - 360° = 60°$.

35. $\cos(-45°) = \frac{\sqrt{2}}{2}$

37. $\tan\frac{2\pi}{3} = -\sqrt{3}$

39. $\sin\left(\frac{5\pi}{3}\right) = -\frac{\sqrt{3}}{2}$

41. $\cot\left(-\frac{5\pi}{6}\right) = \sqrt{3}$

43. $\sec(-180°) = -1$

45. $\cos\left(-\frac{4\pi}{3}\right) = -\frac{1}{2}$

47. $\tan 33° \approx 0.6494$

49. $\sec\frac{12\pi}{5} = \frac{1}{\cos(12\pi/5)} = \sqrt{5} + 1 \approx 3.2361$

51. $\sin\left(-\frac{\pi}{9}\right) \approx -0.3420$

53. $\cos 105° \approx -0.2588$

55. $\cos 70° = \frac{50}{r} \Rightarrow r = \frac{50}{\cos 70°} \approx 146.19$

57. $\tan 20° = \frac{25}{x} \Rightarrow x = \frac{25}{\tan 20°} \approx 68.69$

59. $2\cos x + 1 = 0$

$\cos x = -\frac{1}{2}$

$x = \frac{2\pi}{3}, \frac{4\pi}{3}$

61. $2\sin^2 x + 3\sin x + 1 = 0$

$(2\sin x + 1)(\sin x + 1) = 0$

$\sin x = -\frac{1}{2}$ or $\sin x = -1$

$x = \frac{7\pi}{6}, \frac{11\pi}{6}$ or $x = \frac{3\pi}{2}$

63. $\sec^2 x - \sec x - 2 = 0$

$(\sec x - 2)(\sec x + 1) = 0$

$\sec x = 2$ or $\sec x = -1$

$\cos x = \frac{1}{2}$ or $\cos x = -1$

$x = \frac{\pi}{3}, \frac{5\pi}{3}$ or $x = \pi$

65. $\tan 33° = \dfrac{h}{125}$

$h = 125 \tan 33°$

≈ 81.18 feet

67. $y = 2\cos 6x$

Period: $\dfrac{2\pi}{6} = \dfrac{\pi}{3}$

Amplitude: 2

x-intercepts: $\left(\dfrac{\pi}{12}, 0\right), \left(\dfrac{\pi}{4}, 0\right)$

Relative minimum: $\left(\dfrac{\pi}{6}, -2\right)$

Relative maxima: $(0, 2), \left(\dfrac{\pi}{3}, 2\right)$

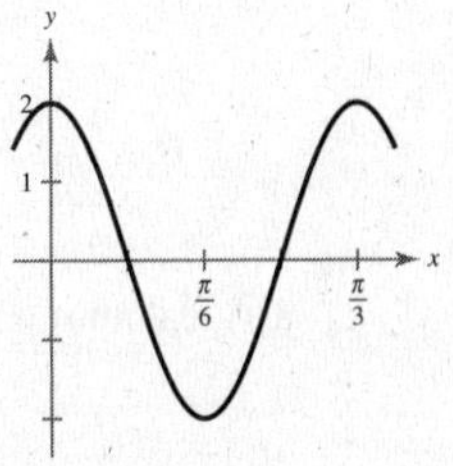

69. $y = \dfrac{1}{3}\tan x$

Period: π

Asymptotes:

$x = -\dfrac{\pi}{2}, x = \dfrac{\pi}{2}, x = \dfrac{3\pi}{2}$

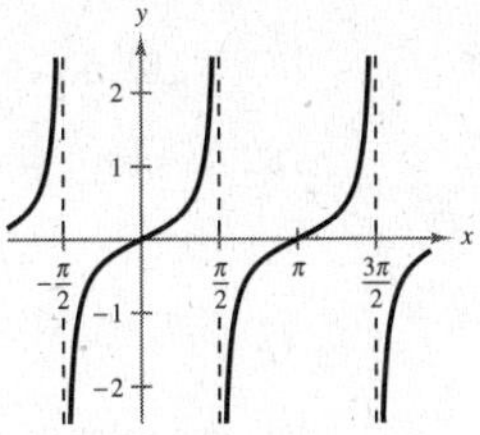

71. $y = 3\sin\dfrac{2x}{5}$

Period: $\dfrac{2\pi}{2/5} = 5\pi$

Amplitude: 3

x-intercepts:

$(0, 0), \left(\dfrac{5\pi}{2}, 0\right), (5\pi, 0)$

Relative minimum: $\left(\dfrac{15\pi}{4}, -3\right)$

Relative maximum: $\left(\dfrac{5\pi}{4}, 3\right)$

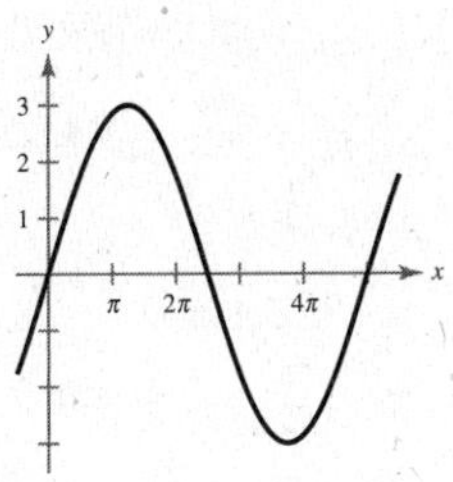

73. $y = \sec 2\pi x$

Period: $\dfrac{2\pi}{2\pi} = 1$

Asymptotes: $x = -\dfrac{1}{4}, x = \dfrac{1}{4}, x = \dfrac{3}{4}, x = \dfrac{5}{4}$

Relative minima: $(0, 1), (1, 1)$

Relative maximum: $\left(\dfrac{1}{2}, -1\right)$

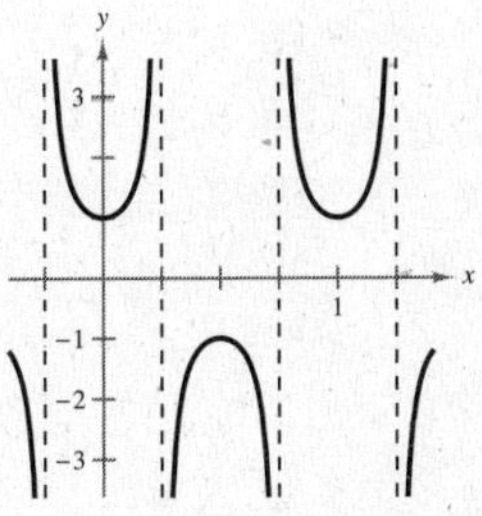

75. (a) $P = 6200 + 1700\sin\dfrac{2\pi t}{24}$

$p = 9650 + 3300\cos\dfrac{2\pi t}{24}$

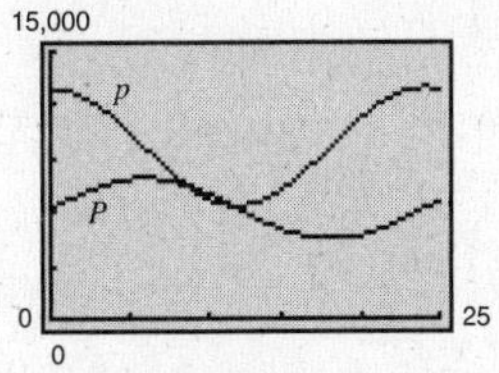

(b) As the population of the prey increases, the population of the predator increases as well until some point when the predator eliminates the prey faster than the prey can reproduce, and the prey population decreases rapidly. As the prey becomes scarce, the predator population decreases, releasing the prey from predator pressure, and the cycle begins again.

77. $y = \sin 5\pi x$

$y' = 5\pi\cos 5\pi x$

79. $y = -x\tan x$

$y' = -x\sec^2 x - \tan x$

81. $y = \dfrac{\cos x}{x^2}$

$y' = \dfrac{x^2(-\sin x) - \cos x(2x)}{\left(x^2\right)^2} = -\dfrac{x\sin x + 2\cos x}{x^3}$

83. $y = \sin^2 x + x$

$y' = 2\sin x\cos x + 1 = \sin 2x + 1$

85. $y = \csc^4 x$

$y' = -4\csc^3 x \csc x \cot x = -4\csc^4 x \cot x$

87. $y = e^x \cot x$

$y' = e^x\left(-\csc^2 x\right) + e^x\cot x = e^x\left(\cot x - \csc^2 x\right)$

89. $y = \cos 2x, \left(\dfrac{\pi}{4}, 0\right)$

$y' = -2\sin 2x, \quad y'\left(\dfrac{\pi}{4}\right) = -2$

$y - 0 = -2\left(x - \dfrac{\pi}{4}\right)$

$y = -2x + \dfrac{\pi}{2}$

91. $y = \dfrac{1}{2}\sin^2 x, \left(\dfrac{\pi}{2}, \dfrac{1}{2}\right)$

$y' = \sin x\cos x, \quad y'\left(\dfrac{\pi}{2}\right) = 0$

$y - \dfrac{1}{2} = 1(x - 0)$

$y = \dfrac{1}{2}$

93. $y = x \tan 2x, \quad (0, 0)$

$y' = 2x \sec^2 2x + \tan 2x, \quad y'(0) = 0$

$y - 0 = 0(x - 0)$

$y = 0$

95. $f(x) = \dfrac{x}{2} + \cos x$

$f'(x) = \dfrac{1}{2} - \sin x$

$\dfrac{1}{2} - \sin x = 0$

$-\sin x = -\dfrac{1}{2}$

$\sin x = \dfrac{1}{2}$

Critical numbers: $x = \dfrac{\pi}{6}, \dfrac{5\pi}{6}$

Relative minimum: $\left(\dfrac{5\pi}{6}, \dfrac{5\pi}{12} - \dfrac{\sqrt{3}}{2}\right)$

Relative maximum: $\left(\dfrac{\pi}{6}, \dfrac{\pi}{12} + \dfrac{\sqrt{3}}{2}\right)$

97. $f(x) = \sin^2 x + \sin x$

$f'(x) = 2 \sin x \cos x + \cos x$

$2 \sin x \cos x + \cos x = 0$

$\cos x(2 \sin x + 1) = 0$

$\cos x = 0$ or $2 \sin x + 1 = 0$

$2 \sin x = -1$

$\sin x = -\dfrac{1}{2}$

Critical numbers: $x = \dfrac{\pi}{2}, \dfrac{7\pi}{6}, \dfrac{3\pi}{2}, \dfrac{11\pi}{6}$

Relative minima: $\left(\dfrac{7\pi}{6}, -\dfrac{1}{4}\right), \left(\dfrac{11\pi}{6}, -\dfrac{1}{4}\right)$

Relative maxima: $\left(\dfrac{\pi}{2}, 2\right), \left(\dfrac{3\pi}{2}, 0\right)$

99. $W = 7594 + 455.2 \sin(0.4t - 1.713)$

$W' = 182.08 \cos(0.4t - 1.713)$

$182.08 \cos(0.4t - 1.713) = 0$

$0.4t - 1.713 = \dfrac{\pi}{2}$

$t \approx 8.2$

W has a relative maximum at $(8.2, 8049.2)$. The number of construction workers was a maximum during August with 8,049,200 workers.

Chapter Test Solutions

	Function	*θ (deg)*	*θ (rad)*	*Function value*
1.	sin	67.5°	$\frac{3\pi}{8}$	0.9239
2.	cos	36°	$\frac{\pi}{5}$	0.8090
3.	tan	15°	$\frac{\pi}{12}$	0.2679
4.	cot	−30°	$-\frac{\pi}{6}$	$-\sqrt{3}$
5.	sec	−40°	$-\frac{2\pi}{9}$	1.3054
6.	csc	−225°	$-\frac{5\pi}{4}$	$\sqrt{2}$

7. $\cos 24° = \dfrac{25}{l}$

$l = \dfrac{25}{\cos 24°}$

$l \approx 27.4$ inches

8. $2 \sin \theta - \sqrt{2} = 0$

$2 \sin \theta = \sqrt{2}$

$\sin \theta = \dfrac{\sqrt{2}}{2}$

$\theta = \dfrac{\pi}{4}, \dfrac{3\pi}{4}$

9. $\cos^2 \theta - \sin^2 \theta = 0$

$\cos^2 \theta = \sin^2 \theta$

$1 = \dfrac{\sin^2 \theta}{\cos^2 \theta}$

$\pm 1 = \tan \theta$

$\theta = \dfrac{\pi}{4}, \dfrac{3\pi}{4}, \dfrac{5\pi}{4}, \dfrac{7\pi}{4}$

10. $\csc \theta = \sqrt{3} \sec \theta$

$\dfrac{\csc \theta}{\sec \theta} = \sqrt{3}$

$\dfrac{\cos \theta}{\sin \theta} = \sqrt{3}$

$\cot \theta = \sqrt{3}$

$\theta = \dfrac{\pi}{6}, \dfrac{7\pi}{6}$

11. $y = 3 \sin 2x$

Period: $\dfrac{2\pi}{2} = \pi$

Amplitude: 3

x-intercepts:

$(0, 0), \left(\dfrac{\pi}{2}, 0\right), (\pi, 0)$

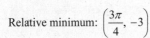

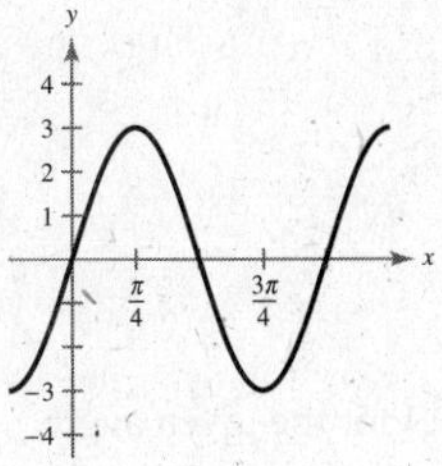

Relative minimum: $\left(\dfrac{3\pi}{4}, -3\right)$

Relative maximum: $\left(\dfrac{\pi}{4}, 3\right)$

12. $y = 4 \cos 3\pi x$

Period: $\dfrac{2\pi}{3\pi} = \dfrac{2}{3}$

Amplitude: 4

x-intercepts:

$\left(\dfrac{1}{6}, 0\right), \left(\dfrac{1}{2}, 0\right), \left(\dfrac{5}{6}, 0\right)$

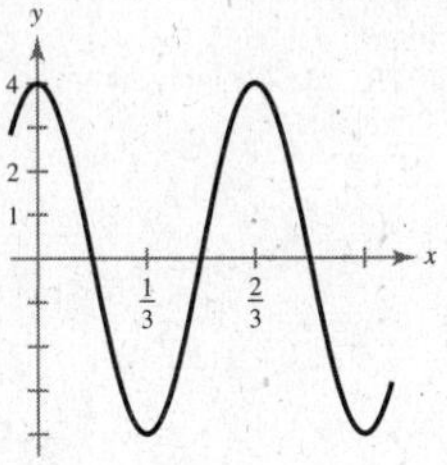

Relative minimum: $\left(\dfrac{1}{3}, -4\right)$

Relative maxima: $(0, 4), \left(\dfrac{2}{3}, 4\right)$

13. $y = \cot \dfrac{\pi x}{5}$

Period: $\dfrac{\pi}{(\pi/5)} = 5$

x-intercepts: $\left(-\dfrac{5}{2}, 0\right), \left(\dfrac{5}{2}, 0\right)$

Asymptotes: $x = -5, x = 5$

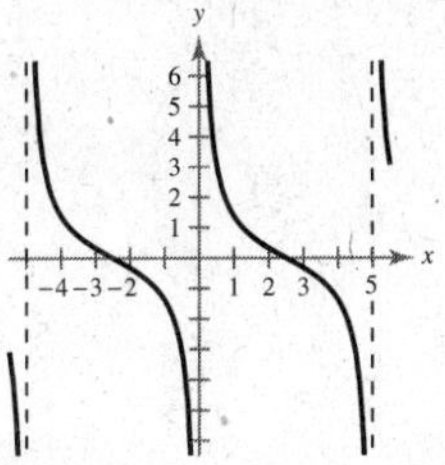

14. (a) $y = \cos x - \cos^2 x$

$y' = -\sin x + 2 \cos x \sin x$

$y' = \sin 2x - \sin x$

(b) $-\sin x + 2 \cos x \sin x = 0$

$-\sin x(1 - 2 \cos x) = 0$

$\sin x = 0$ or $1 - 2 \cos x = 0$

$-2 \cos x = -1$

$\cos x = \dfrac{1}{2}$

Critical numbers: $x = \dfrac{\pi}{3}, \pi, \dfrac{5\pi}{3}$

Relative minimum: $(\pi, -2)$

Relative maxima: $\left(\dfrac{\pi}{3}, \dfrac{1}{4}\right), \left(\dfrac{5\pi}{3}, \dfrac{1}{4}\right)$

15. (a) $y = \sec\left(x - \dfrac{\pi}{4}\right), y' = \sec\left(x - \dfrac{\pi}{4}\right) \tan\left(x - \dfrac{\pi}{4}\right)$

(b) $\sec\left(x - \dfrac{\pi}{4}\right) \tan\left(x - \dfrac{\pi}{4}\right) = 0$

$\sec\left(x - \dfrac{\pi}{4}\right) = 0$ or $\tan\left(x - \dfrac{\pi}{4}\right) = 0$

Critical numbers: $x = \dfrac{\pi}{4}, \dfrac{5\pi}{4}$

Relative minimum: $\left(\dfrac{\pi}{4}, 1\right)$

Relative maximum: $\left(\dfrac{5\pi}{4}, -1\right)$

16. (a) $y = \dfrac{1}{3 - \sin(x + \pi)}, y' = \dfrac{\cos(x + \pi)}{[3 - \sin(x + \pi)]^2}$

(b) $\cos(x + \pi) = 0$

Critical numbers: $x = \dfrac{\pi}{2}, \dfrac{3\pi}{2}$

Relative minimum: $\left(\dfrac{\pi}{2}, \dfrac{1}{4}\right)$

Relative maximum: $\left(\dfrac{3\pi}{2}, \dfrac{1}{2}\right)$

17. (a) $P = 7300 + 2900 \sin \dfrac{2\pi t}{24}$

$p = 10{,}000 + 3700 \cos \dfrac{2\pi t}{24}$

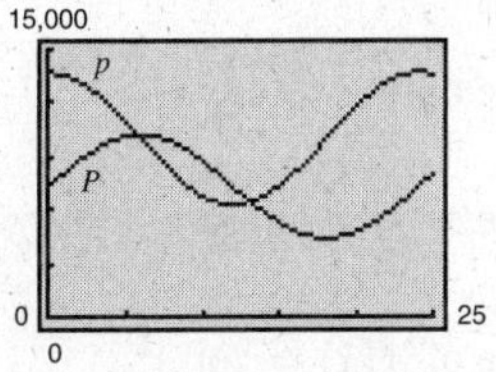

(b) As the population of the prey increases, the population of the predator increases as well until some point when the predator eliminates the prey faster than the prey can reproduce, and the prey population decreases rapidly. As the prey becomes scarce, the predator population decreases, releasing the prey from predator pressure, and the cycle begins again.

Practice Test for Chapter 5

1. (a) Express $\frac{12\pi}{23}$ in degree measure.

(b) Express 105° in radian measure.

2. Determine two coterminal angles (one positive and one negative) for the given angle.

(a) −220°; give your answers in degrees.

(b) $\frac{7\pi}{9}$; give your answers in radians.

3. Find the six trigonometric functions of the angle θ if it is in standard position and the terminal side passes through the point $(12, -5)$.

4. Solve for θ, $(0 \le \theta \le 2\pi)$: $\sin^2 \theta + \cos \theta = 1$

5. Sketch the graph of the given function.

(a) $y = 3 \sin \frac{x}{4}$ (b) $y = \tan 2\pi x$

6. Find the derivative of $y = 3x - 3 \cos x$.

7. Find the derivative of $f(x) = x^2 \tan x$.

8. Find the derivative of $g(x) = \sin^3 x$.

9. Find the derivative of $y = \frac{\sec x}{x^2}$.

10. Find the derivative of $y = \sin 5x \cos 5x$.

11. Find the derivative of $y = \sqrt{\csc x}$.

12. Find the derivative of $y = \ln|\sec x + \tan x|$.

13. Find the derivative of $f(x) = \cot e^{2x}$.

14. Find $\frac{dy}{dx}$: $\sin(x^2 + y) = 3x$

15. Find $\frac{dy}{dx}$: $\tan x - \cot 3y = 4$

Graphing Calculator Required

16. Use a graphing utility to graph $f(x) = \sin x + \cos 2x$. What are the minimum and maximum values that $f(x)$ takes on?

CHAPTER 6
Integration and Its Applications

CHAPTER 6
Integration and Its Applications

Section 6.1 Antiderivatives and Indefinite Integrals

Skills Review

1. $\dfrac{\sqrt{x}}{x} = \dfrac{x^{1/2}}{x} = x^{-1/2}$

2. $\sqrt[3]{2x}(2x) = (2x)^{1/3}(2x) = (2x)^{4/3}$

3. $\sqrt{5x^3} + \sqrt{x^5} = (5x^3)^{1/2} + (x^5)^{1/2} = 5^{1/2}x^{3/2} + x^{5/2}$

4. $\dfrac{1}{\sqrt{x}} + \dfrac{1}{\sqrt[3]{x^2}} = \dfrac{1}{x^{1/2}} + \dfrac{1}{x^{2/3}}$

$= x^{-1/2} + x^{-2/3}$

5. $\dfrac{(x+1)^3}{\sqrt{x+1}} = \dfrac{(x+1)^3}{(x+1)^{1/2}} = (x+1)^{5/2}$

6. $\dfrac{\sqrt{x}}{\sqrt[3]{x}} = \dfrac{x^{1/2}}{x^{1/3}} = x^{1/6}$

7. $y = x^2 + 5x + C$

$2 = 2^2 + 5(2) + C$

$-12 = C$

8. $y = 3x^3 - 6x + C$

$2 = 3(2)^3 - 6(2) + C$

$-10 = C$

9. $y = -16x^2 + 26x + C$

$2 = -16(2)^2 + 26(2) + C$

$14 = C$

10. $y = -\frac{1}{4}x^4 - 2x^2 + C$

$2 = -\frac{1}{4}(2)^4 - 2(2)^2 + C$

$14 = C$

1. $\dfrac{d}{dx}\left[\dfrac{3}{x^3} + C\right] = \dfrac{d}{dx}\left(3x^{-3} + C\right) = -9x^{-4} = -\dfrac{9}{x^4}$

3. $\dfrac{d}{dx}\left[x^4 + \dfrac{1}{x} + C\right] = 4x^3 - \dfrac{1}{x^2}$

5. $\dfrac{d}{dx}\left[\dfrac{4x^{3/2}(x-5)}{5} + C\right] = \dfrac{d}{dx}\left[\dfrac{4}{5}x^{5/2} - 4x^{3/2} + C\right]$

$= 2x^{3/2} - 6x^{1/2}$

$= 2\sqrt{x}(x-3)$

7. $\dfrac{d}{dx}\left[\dfrac{1}{3}x^3 - 4x + C\right] = x^2 - 4 = (x-2)(x+2)$

9. $\int 6\,dx = 6x + C$

$\dfrac{d}{dx}[6x + C] = 6$

11. $\int 5t^2\,dt = \dfrac{5t^3}{3} + C$

$\dfrac{d}{dt}\left[\dfrac{5t^3}{3} + C\right] = 5t^2$

13. $\int 5x^{-3}\,dx = \dfrac{5x^{-2}}{-2} + C = -\dfrac{5}{2x^2} + C$

$\dfrac{d}{dx}\left[-\dfrac{5}{2}x^{-2} + C\right] = 5x^{-3}$

15. $\int du = u + C$

$\dfrac{d}{du}[u + C] = 1$

17. $\int e\,dt = et + C$

$\dfrac{d}{dt}[et + C] = e$

19. $\int y^{3/2}\,dy = \dfrac{y^{5/2}}{5/2} + C = \dfrac{2}{5}y^{5/2} + C$

$\dfrac{d}{dy}\left[\dfrac{2}{5}y^{5/2} + C\right] = y^{3/2}$

Original Integral	*Rewrite*	*Integrate*	*Simplify*
21. $\int \sqrt[3]{x}\,dx$	$\int x^{1/3}\,dx$	$\dfrac{x^{4/3}}{4/3} + C$	$\dfrac{3}{4}x^{4/3} + C$
23. $\int \dfrac{1}{x\sqrt{x}}\,dx$	$\int x^{-3/2}\,dx$	$\dfrac{x^{-1/2}}{-1/2} + C$	$-\dfrac{2}{\sqrt{x}} + C$
25. $\int \dfrac{1}{2x^3}\,dx$	$\dfrac{1}{2}\int x^{-3}\,dx$	$\dfrac{1}{2}\left(\dfrac{x^{-2}}{-2}\right) + C$	$-\dfrac{1}{4x^2} + C$

27. $\int (x + 3)\,dx = \dfrac{x^2}{2} + 3x + C$

$\dfrac{d}{dx}\left[\dfrac{x^2}{2} + 3x + C\right] = x + 3$

29. $\int (x^3 + 2)\,dx = \dfrac{x^4}{4} + 2x + C$

$\dfrac{d}{dx}\left[\dfrac{x^4}{4} + 2x + C\right] = x^3 + 2$

31. $\int \left(\sqrt[3]{x} - \dfrac{1}{2\sqrt[3]{x}}\right) dx = \int \left(x^{1/3} - \dfrac{1}{2}x^{-1/3}\right) dx$

$= \dfrac{3}{4}x^{4/3} - \dfrac{3}{4}x^{2/3} + C$

$\dfrac{d}{dx}\left[\dfrac{3}{4}x^{4/3} - \dfrac{3}{4}x^{2/3} + C\right] = x^{1/3} - \dfrac{1}{2x^{1/3}}$

$= \sqrt[3]{x} - \dfrac{1}{2\sqrt[3]{x}}$

33. $\int \sqrt[3]{x^2}\,dx = \int x^{2/3}\,dx = \dfrac{x^{5/3}}{5/3} + C = \dfrac{3}{5}x^{5/3} + C$

$\dfrac{d}{dx}\left[\dfrac{3}{5}x^{5/3} + C\right] = x^{2/3} = \sqrt[3]{x^2}$

35. $\int \dfrac{1}{x^4}\,dx = \int x^{-4}\,dx = \dfrac{x^{-3}}{-3} + C = -\dfrac{1}{3x^3} + C$

$\dfrac{d}{dx}\left[-\dfrac{1}{3}x^{-3} + C\right] = x^{-4} = \dfrac{1}{x^4}$

37. $\int \dfrac{2x^3 + 1}{x^3}\,dx = \int (2 + x^{-3})\,dx = 2x - \dfrac{1}{2}x^{-2} + C$

$= 2x - \dfrac{1}{2x^2} + C$

$\dfrac{d}{dx}\left[2x - \dfrac{1}{2x^2} + C\right] = 2 + \dfrac{1}{x^3} = \dfrac{2x^3 + 1}{x^3}$

39. $\int u(3u^2 + 1)\,du = \int (3u^3 + u)\,du$

$= \dfrac{3}{4}u^4 + \dfrac{1}{2}u^2 + C$

$\dfrac{d}{du}\left[\dfrac{3}{4}u^4 + \dfrac{1}{2}u^2 + C\right] = 3u^3 + u$

$= u(3u^2 + 1)$

41. $\int (x + 1)(3x - 2)\,dx = \int (3x^2 + x - 2)\,dx$

$= x^3 + \dfrac{1}{2}x^2 - 2x + C$

$\dfrac{d}{dx}\left[x^3 + \dfrac{1}{2}x^2 - 2x + C\right] = 3x^2 + x - 2$

$= (x + 1)(3x - 2)$

43. $\int y^2\sqrt{y}\,dy = \int y^{5/2}\,dy = \dfrac{2}{7}y^{7/2} + C$

$\dfrac{d}{dy}\left[\dfrac{2}{7}y^{7/2} + C\right] = y^{5/2} = y^2\sqrt{y}$

45. If $f'(x) = 2$, then $f(x) = 2x + C$.

Sample answer: $f(x) = 2x$ or $f(x) = 2x + 1$

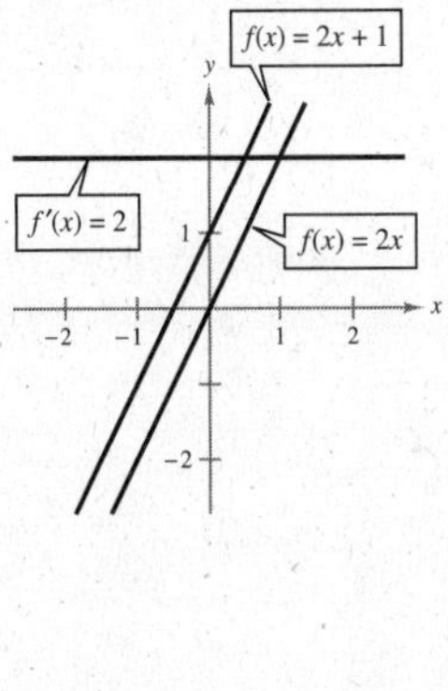

47. If $f'(x) = x$, then $f(x) = \frac{x^2}{2} + C$.

Sample answer: $f(x) = \frac{x^2}{2}$ or $f(x) = \frac{x^2}{2} + 2$

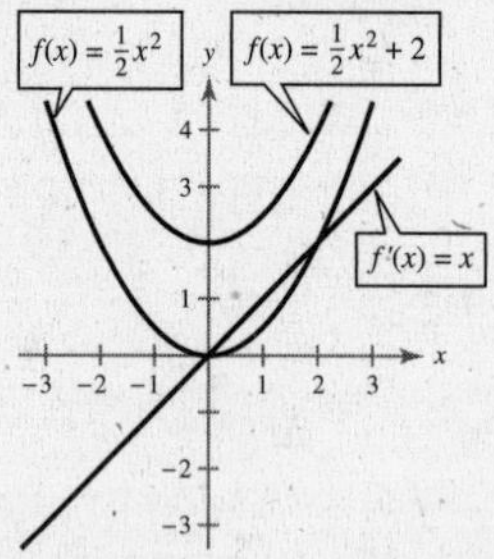

49. $f(x) = \int 4x\, dx = \frac{4}{2}x^2 + C = 2x^2 + C$

$$f(0) = 6$$
$$2(0)^2 + C = 6$$
$$C = 6$$
$$f(x) = 2x^2 + 6$$

51. $f(x) = \int 2(x-1)\, dx = \int (2x-2)\, dx = x^2 - 2x + C$

$$f(3) = 2$$
$$(3)^2 - 2(3) + C = 2$$
$$C = -1$$
$$f(x) = x^2 - 2x - 1$$

53. $f(x) = \int \frac{2-x}{x^3}\, dx$

$$= \int \left(2x^{-3} - x^{-2}\right) dx$$
$$= -x^{-2} + x^{-1} + C$$
$$= -\frac{1}{x^2} + \frac{1}{x} + C$$
$$f(2) = \frac{3}{4}$$
$$-\frac{1}{2^2} + \frac{1}{2} + C = \frac{3}{4}$$
$$C = \frac{1}{2}$$
$$f(x) = -\frac{1}{x^2} + \frac{1}{x} + \frac{1}{2}$$

55. $y = \int (-5x - 2)\, dx$

$$= -\tfrac{5}{2}x^2 - 2x + C$$

At $(0, 2)$: $y(0) = 2$

$$-\tfrac{5}{2}(0)^2 - 2(0) + C = 2$$
$$C = 2$$
$$y = -\tfrac{5}{2}x^2 - 2x + 2$$

57. $f(x) = \int 2x\, dx = x^2 + C$

At $(-2, -2)$: $f(-2) = -2$

$$(-2)^2 + C = -2$$
$$C = -6$$
$$f(x) = x^2 - 6$$

59. $f'(x) = \int 2\, dx = 2x + C_1$

Because $f'(2) = 2(2) + C_1 = 5$, you know that $C_1 = 1$.
So, $f'(x) = 2x + 1$.

$$f(x) = \int (2x + 1)\, dx = x^2 + x + C_2$$

Because $f(2) = 4 + 2 + C_2 = 10$, you know that $C_2 = 4$. So, $f(x) = x^2 + x + 4$.

61. $f'(x) = \int x^{-2/3}\, dx = 3x^{1/3} + C_1$

Because $f'(8) = 3(8)^{1/3} + C_1 = 6$, you know that $C_1 = 0$. So, $f'(x) = 3x^{1/3}$.

$$f(x) = \int 3x^{1/3}\, dx = \tfrac{9}{4}x^{4/3} + C_2$$

Because $f(0) = 0 + C_2 = 0$, you know that $C_2 = 0$. So, $f(x) = \frac{9}{4}x^{4/3}$.

63. $a(t) = -32$

$$v(t) = \int a(t)\, dt = -32t + C_1$$

Letting v_0 be the initial velocity, $v_0 = 60$ and $C_1 = v_0$, so $v(t) = -32t + 60$.

$$s(t) = \int v(t)\, dt = -16t^2 + 60t + C_2.$$

Because $s(0) = 0$, $C_2 = 0$. So, the position function is $s(t) = -16t^2 + 60t$.

$v(t) = 0$ when $t = 1.875$.

The height of the ball is at a maximum when $t = 1.875$ seconds. The maximum height is

$$s(1.875) = -16(1.875)^2 + 60(1.875) = 56.25 \text{ feet.}$$

65. $v(t) = \int -32\, dt = -32t + C_1$

Letting v_0 be the initial velocity, $v(0) = v_0$ and $C_1 = v_0$, so $v(t) = -32t + v_0$.

$s(t) = \int(-32t + v_0)\, dt = -16t^2 + v_0 t + C_2$

Because $s(0) = 0$, $C_2 = 0$. So, the position function is $s(t) = 16t^2 + v_0 t$. At the highest point, the velocity is zero. This occurs when $v(t) = -32t + v_0 = 0$, and $t = v_0/32$ seconds. Substituting this value into the position function, $s\left(\frac{v_0}{32}\right) = -16\left(\frac{v_0}{32}\right)^2 + v_0\left(\frac{v_0}{32}\right) = 550$ which implies that $v_0^2 = 35{,}200$ and the initial velocity should be $v_0 = 40\sqrt{22} \approx 187.62$ ft/sec.

69. $\frac{dP}{dt} = 500t^{1.06}$

$\int dP = \int 500t^{1.06} dt$

$P = \frac{500}{2.06}t^{2.06} + C$

$P \approx 242.72t^{2.06} + C$

When $t = 0$, $P = 50{,}000$.

$50{,}000 = 242.72(0)^{2.06} + C$

$50{,}000 = C$

$P = 242.72t^{2.06} + 50{,}000.$

When $t = 10$:

$P = 242.72(10)^{2.06} + 50{,}000 \approx 77{,}868$

67. (a) $h(t) = \int(1.5t + 5)\, dt = 0.75t^2 + 5t + C$

$h(0) = 0 + 0 + C = 12$ and $C = 12$.

So, $h(t) = 0.75t^2 + 5t + 12$.

(b) $h(6) = 69$

The shrubs are 69 centimeters tall when they are sold.

71. (a) $I(t) = \int(-0.25t^3 + 5.319t^2 - 19.34t + 21.03)\, dt = -0.0625t^4 + 1.773t^3 - 9.67t^2 + 21.03t + C$

$t = 14$ corresponds to the year 2004, so $I(14) = -0.0625(14)^4 + 1.773(14)^3 - 9.67(14)^2 + 21.03(14) + C = 863$ and $C = -0.212$. So, $I(t) = -0.0625t^4 + 1.773t^3 - 9.67t^2 + 21.03t - 0.212$.

(b) $I(22) = 20.072$

According to the model, the number of Internet users in the world will be about 20,072,000 in 2012. This does not seem reasonable because it is drastically less than the 863 million of users in 2004. The number of Internet users should continue to increase.

73. (a) $A(t) = \int(-0.272t + 1.27)\, dt = -0.136t^2 + 1.27t + C$

$t = 5$ corresponds to the year 2005, so $A(5) = -0.136(5)^2 + 1.27(5) + c = 29.9$ and $C = 26.95$.

So, $A(t) = -0.136t^2 + 1.27t + 26.95$.

(b) $A(10) = 26.05$

According to the model, the number of persons living alone will be about 26,050,000 in 2010.

Section 6.2 Integration by Substitution and the General Power Rule

Skills Review

1. $\int(2x^3 + 1)\, dx = \frac{1}{2}x^4 + x + C$

2. $\int(x^{1/2} + 3x - 4)\, dx = \frac{2}{3}x^{3/2} + \frac{3}{2}x^2 - 4x + C$

3. $\int \frac{1}{x^2}\, dx = \int x^{-2}\, dx = -\frac{1}{x} + C$

4. $\int \frac{1}{3t^3}\, dt = \frac{1}{3}\int t^{-3}\, dt = -\frac{1}{6t^2} + C$

Skills Review —continued—

5. $\int (1 + 2t)t^{3/2}\, dt = \int \left(t^{3/2} + 2t^{5/2}\right) dt$
$= \frac{2}{5}t^{5/2} + \frac{4}{7}t^{7/2} + C$

6. $\int \sqrt{x}(2x - 1)\, dx = \int \left(2x^{3/2} - x^{1/2}\right) dx$
$= \frac{4}{5}x^{5/2} - \frac{2}{3}x^{3/2} + C$

7. $\int \frac{5x^3 + 2}{x^2}\, dx = \int \left(5x + 2x^{-2}\right) dx$
$= \frac{5}{2}x^2 - \frac{2}{x} + C = \frac{5x^3 - 4}{2x} + C$

8. $\int \frac{2x^2 - 5}{x^4}\, dx = \int \left(2x^{-2} - 5x^{-4}\right) dx$
$= -\frac{2}{x} + \frac{5}{3x^3} + C$
$= \frac{-6x^2 + 5}{3x^3} + C$

9. $\int \left(x^2 + 1\right)^2 dx = \int \left(x^4 + 2x^2 + 1\right) dx$
$= \frac{1}{5}x^5 + \frac{2}{3}x^3 + x + C$

10. $\int \left(x^3 - 2x + 1\right)^2 dx = \int \left(x^6 - 4x^4 + 2x^3 + 4x^2 - 4x + 1\right) dx$
$= \frac{1}{7}x^7 - \frac{4}{5}x^5 + \frac{1}{2}x^4 + \frac{4}{3}x^3 - 2x^2 + x + C$

11. $\left(-\frac{5}{4}\right)\frac{(x - 2)^4}{4} = -\frac{5(x - 2)^4}{16}$

12. $\left(\frac{1}{6}\right)\frac{(x - 1)^{-2}}{-2} = -\frac{1}{12(x - 1)^2}$

13. $(6)\frac{\left(x^2 + 3\right)^{2/3}}{2/3} = 9\left(x^2 + 3\right)^{2/3}$

14. $\left(\frac{5}{2}\right)\frac{\left(1 - x^3\right)^{-1/2}}{-1/2} = -\frac{5}{\left(1 - x^3\right)^{1/2}}$

$\int u^n \frac{du}{dx}\, dx$	u	$\frac{du}{dx}$
1. $\int \left(5x^2 + 1\right)^2 (10x)\, dx$	$5x^2 + 1$	$10x$
3. $\int \sqrt{1 - x^2}(-2x)\, dx$	$1 - x^2$	$-2x$
5. $\int \left(4 + \frac{1}{x^2}\right)^5 \left(\frac{-2}{x^3}\right) dx$	$4 + \frac{1}{x^2}$	$-\frac{2}{x^3}$
7. $\int \left(1 + \sqrt{x}\right)^3 \frac{1}{2\sqrt{x}}\, dx$	$1 + \sqrt{x}$	$\frac{1}{2\sqrt{x}}$

9. $\int (1 + 2x)^4 (2)\, dx = \frac{(1 + 2x)^5}{5} + C$

$\frac{d}{dx}\left[\frac{(1 + 2x)^5}{5} + C\right] = (1 + 2x)^4 (2)$

11. $\int \sqrt{4x^2 - 5}(8x)\, dx = \frac{\left(4x^2 - 5\right)^{3/2}}{3/2} + C$

$\frac{d}{dx}\left[\frac{\left(4x^2 - 5\right)^{3/2}}{3/2} + C\right] = \left(4x^2 - 5\right)^{1/2}(8x)$
$= \sqrt{4x^2 - 5}(8x)$

13. $\int (x - 1)^4\, dx = \int (x - 1)^4 (1)\, dx = \frac{(x - 1)^5}{5} + C$

$\frac{d}{dx}\left[\frac{(x - 1)^5}{5} + C\right] = (x - 1)^4$

15. $\int 2x\left(x^2 - 1\right)^7 dx = \frac{\left(x^2 - 1\right)^8}{8} + C$

$\frac{d}{dx}\left[\frac{\left(x^2 - 1\right)^8}{8} + C\right] = \left(x^2 - 1\right)^7 (2x)$

17. $\int \frac{x^2}{\left(1 + x^3\right)^2}\, dx = \frac{1}{3}\int \left(1 + x^3\right)^{-2}\left(3x^2\right) dx$
$= \frac{1}{3}\frac{\left(1 + x^3\right)^{-1}}{-1} + C$
$= -\frac{1}{3\left(1 + x^3\right)} + C$

$\frac{d}{dx}\left[-\frac{1}{3\left(1 + x^3\right)} + C\right] = \frac{1}{3}\left(1 + x^3\right)^{-2}\left(3x^2\right) = \frac{x^2}{\left(1 + x^3\right)^2}$

19. $\int \frac{x+1}{(x^2+2x-3)^2}\,dx = \frac{1}{2}\int (x^2+2x-3)^{-2}(2x+2)\,dx = \frac{1}{2}\frac{(x^2+2x-3)^{-1}}{-1} + C = -\frac{1}{2(x^2+2x-3)} + C$

$\frac{d}{dx}\left[-\frac{1}{2(x^2+2x-3)} + C\right] = \frac{1}{2}(x^2+2x-3)^{-2}(2x+2) = \frac{x+1}{(x^2+2x-3)^2}$

21. $\int \frac{x-2}{\sqrt{x^2-4x+3}}\,dx = \frac{1}{2}\int (x^2-4x+3)^{-1/2}(2x-4)\,dx = \frac{1}{2}\left[\frac{(x^2-4x+3)^{1/2}}{1/2}\right] + C = \sqrt{x^2-4x+3} + C$

$\frac{d}{dx}\left[\sqrt{x^2-4x+3} + C\right] = \frac{1}{2}(x^2-4x+3)^{-1/2}(2x-4) = \frac{x-2}{\sqrt{x^2-4x+3}}$

23. $\int 5u\sqrt[3]{1-u^2}\,du = 5\left(-\frac{1}{2}\right)\int (1-u^2)^{1/3}(-2u)\,du$

$= -\frac{5}{2}\left(\frac{3}{4}\right)(1-u^2)^{4/3} + C$

$= -\frac{15(1-u^2)^{4/3}}{8} + C$

$\frac{d}{du}\left[-\frac{15(1-u^2)^{4/3}}{8} + C\right] = -\frac{5}{2}(1-u^2)^{1/3}(-2u)$

$= 5u\sqrt[3]{1-u^2}$

25. $\int \frac{4y}{\sqrt{1+y^2}}\,dy = 4\left(\frac{1}{2}\right)\int (1+y^2)^{-1/2}(2y)\,dy$

$= 2\left[\frac{(1+y^2)^{1/2}}{1/2}\right] + C$

$= 4\sqrt{1+y^2} + C$

$\frac{d}{dy}\left[4\sqrt{1+y^2} + C\right] = 2(1+y^2)^{-1/2}(2y)$

$= \frac{4y}{\sqrt{1+y^2}}$

27. $\int \frac{-3}{\sqrt{2t+3}}\,dt = -\frac{3}{2}\int (2t+3)^{-1/2}(2)\,dt$

$= -\frac{3}{2}\left[\frac{(2t+3)^{1/2}}{1/2}\right] + C$

$= -3\sqrt{2t+3} + C$

$\frac{d}{dt}\left[-3\sqrt{2t+3} + C\right] = -\frac{3}{2}(2t+3)^{-1/2}(2) = \frac{-3}{\sqrt{2t+3}}$

29. $\int \frac{x^3}{\sqrt{1-x^4}}\,dx = -\frac{1}{4}\int (1-x^4)^{-1/2}(-4x^3)\,dx$

$= -\frac{1}{4}\left[\frac{(1-x^4)^{1/2}}{1/2}\right] + C$

$= -\frac{\sqrt{1-x^4}}{2} + C$

31. $\int \left(1+\frac{4}{t^2}\right)^2\left(\frac{1}{t^3}\right)dt = -\frac{1}{8}\int (1+4t^{-2})^2(-8t^{-3})\,dt$

$= -\frac{1}{8}\left[\frac{(1+4t-2)^3}{3}\right] + C$

$= -\frac{1}{24}\left(1+\frac{4}{t^2}\right)^3 + C$

33. $\int (x^3+3x+9)(x^2+1)\,dx = \frac{1}{3}\int (x^3+3x+9)(3x^2+3)\,dx$

$= \frac{1}{3}\left[\frac{(x^3+3x+9)^2}{1/2}\right] + C = \frac{1}{6}(x^3+3x+9)^2 + C$

35. Let $u = 6x^2 - 1$, then $du = 12x\,dx$.

$\int 12x(6x^2-1)^3\,dx = \int (6x^2-1)^3(12x\,dx)$

$= \int u^3\,du = \frac{u^4}{4} + C = \frac{1}{4}(6x^2-1)^4 + C$

37. Let $u = 2 - 3x^3$, then $du = -9x^2\ dx$ which implies that $x^2\ dx = -\frac{1}{9}\ du$.

$$\begin{aligned}\int x^2(2-3x^3)^{3/2}\ dx &= \int (2-3x^3)^{3/2}(x^2)\ dx\\ &= \int u^{3/2}\left(-\frac{1}{9}\right) du\\ &= -\frac{1}{9}\left[\frac{u^{5/2}}{5/2}\right] + C\\ &= -\frac{2}{45}(2-3x^3)^{5/2} + C\end{aligned}$$

39. Let $u = x^2 + 25$, then $du = 2x\ dx$ which implies that $x\ dx = \frac{1}{2}\ du$.

$$\begin{aligned}\int \frac{x}{\sqrt{x^2+25}}\ dx &= \int (x^2+25)^{-1/2}(x)\ dx\\ &= \int u^{-1/2}\left(\frac{1}{2}\right) du\\ &= \frac{1}{2}\left[\frac{u^{1/2}}{1/2}\right] + C\\ &= \sqrt{u} + C\\ &= \sqrt{x^2+25} + C\end{aligned}$$

41. Let $u = x^3 + 3x + 4$, then

$du = (3x^2 + 3)\ dx = 3(x^2 + 1)\ dx$

which implies that $(x^2 + 1)\ dx = \frac{1}{3}\ du$.

$$\begin{aligned}\int \frac{x^2+1}{\sqrt{x^3+3x+4}}\ dx &= \int (x^3+3x+4)^{-1/2}(x^2+1)\ dx\\ &= \int u^{-1/2}\left(\frac{1}{3}\right) du\\ &= \frac{1}{3}\left[\frac{u^{1/2}}{1/2}\right] + C\\ &= \frac{2}{3}\sqrt{u} + C\\ &= \frac{2}{3}\sqrt{x^3+3x+4} + C\end{aligned}$$

43. (a) $$\begin{aligned}\int (x-1)^2\ dx &= \int (x^2 - 2x + 1)\ dx\\ &= \frac{1}{3}x^3 - x^2 + x + C_1\end{aligned}$$

$$\begin{aligned}\int (x-1)^2\ dx &= \frac{(x-1)^3}{3} + C_2\\ &= \frac{1}{3}(x^3 - 3x^2 + 3x - 1) + C_2\\ &= \frac{1}{3}x^3 - x^2 + x - \frac{1}{3} + C_2\end{aligned}$$

(b) The two answers differ by a constant:

$C_1 = -\frac{1}{3} + C_2$

(c) Answers will vary.

45. (a) $$\begin{aligned}\int x(x^2-1)^2\ dx &= \int (x^5 - 2x^3 + x)\ dx\\ &= \frac{1}{6}x^6 - \frac{1}{2}x^4 + \frac{1}{2}x^2 + C_1\end{aligned}$$

$$\begin{aligned}\int x(x^2-1)^2\ dx &= \frac{1}{2}\int (x^2-1)^2(2x)\ dx\\ &= \frac{1}{2}\frac{(x^2-1)^3}{3} + C_2\\ &= \frac{1}{6}(x^6 - 3x^4 + 3x^2 - 1) + C_2\\ &= \frac{1}{6}x^6 - \frac{1}{2}x^4 + \frac{1}{2}x^2 - \frac{1}{6} + C_2\end{aligned}$$

(b) The two answers differ by a constant:

$C_1 = -\frac{1}{6} + C_2$

(c) Answers will vary.

47. $$\begin{aligned}f(x) &= \int x\sqrt{1-x^2}\ dx\\ &= -\frac{1}{2}\int (1-x^2)^{1/2}(-2x)\ dx\\ &= -\frac{1}{2}\left[\frac{(1-x^2)^{3/2}}{3/2}\right] + C\\ &= -\frac{1}{3}(1-x^2)^{3/2} + C\end{aligned}$$

Because $f(0) = \frac{4}{3}$, it follows that $C = \frac{5}{3}$ and you have

$$\begin{aligned}f(x) &= -\frac{1}{3}(1-x^2)^{3/2} + \frac{5}{3}\\ &= \frac{1}{3}\left[5 - (1-x^2)^{3/2}\right].\end{aligned}$$

49. (a) $W(t) = \int \frac{1280}{(15.08 - 0.128t)^2}\,dt$

$= \int 1280(15.08 - 0.128t)^{-2}\,dt$

$= \frac{10,000}{15.08 - 0.128t} + C$

Because $W(0) = 66.3$, it follows that $C \approx -596.83$, and you have

$W(t) = \frac{10,000}{15.08 - 0.128t} - 596.83.$

(b) $W(5) \approx 95.7$

According to the model, the number of women in the labor force in 2005 was about 95.7 million.

51. (a) $h = \int \frac{17.6t}{\sqrt{17.6t^2 + 1}}\,dt$

$= \frac{1}{2}\int (17.6t^2 + 1)^{-1/2}(35.2t)\,dt$

$= (17.6t^2 + 1)^{1/2} + C$

Because $h(0) = 6$, it follows that $C = 5$, and you have $h(t) = \sqrt{17.6t^2 + 1} + 5.$

(b) $h(5) = 26$ inches

53. (a) $P(t) = \int 0.06t(0.005t^2 + 1)^2\,dt$

$= 6\int 0.01t(0.005t^2 + 1)^2\,dt$

$= \frac{6}{3}(0.005t^2 + 1)^3 + C$

$= 2(0.005t^2 + 1)^3 + C$

Because $P(5) = 3$, it follows that $C = \frac{39}{256}$, and you have $P(t) = 2(0.005t^2 + 1)^3 + \frac{39}{256}.$

(b)

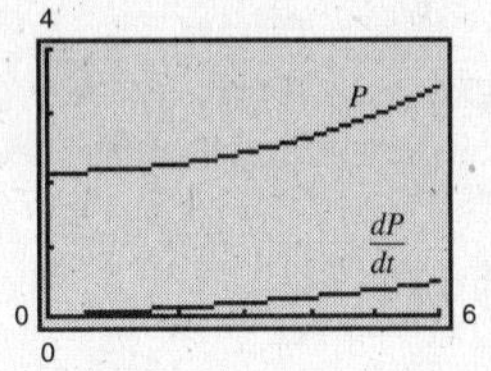

(c) When $t = 3$, $P(3) \approx 2$ and $\frac{dP}{dt} \approx 0.2.$

(d) When $t = 3$:

$P(3) = 2\left[0.005(3)^2 + 1\right]^3 + \frac{39}{256} \approx 2.43$ million

$\frac{dP}{dt} = 0.06(3)\left[0.005(3)^2 + 1\right]^2$

≈ 0.20 million per year

55. Let $u = 3x + 2$, then $du = 3\,dx$ which implies that $\frac{u - 2}{3} = x$ and $dx = \frac{1}{3}\,du.$

$$\int \frac{x}{\sqrt{3x + 2}}\,dx = \int \frac{\frac{u-2}{3}}{\sqrt{u}}\,du$$

$$= \int \frac{u - 2}{3u^{1/2}}\,du$$

$$= \frac{1}{3}\int \left(u^{1/2} - 2u^{-1/2}\right)du$$

$$= \frac{1}{3}\left[\frac{u^{3/2}}{3/2} - \frac{2u^{1/2}}{1/2}\right] + C$$

$$= \frac{2}{27}(3x + 2)^{3/2} - \frac{4}{9}\sqrt{3x + 2} + C$$

$$\frac{d}{dx}\left[\frac{2}{27}(3x + 2)^{3/2} - \frac{4}{9}\sqrt{3x + 2} + C\right] = \frac{2}{27}\left(\frac{3}{2}\right)(3x + 2)^{1/2}(3) - \frac{4}{9}\left(\frac{1}{2}\right)(3x + 2)^{-1/2}(3)$$

$$= \frac{1}{3}\sqrt{3x + 2} - \frac{2}{3\sqrt{3x + 2}}$$

$$= \frac{\sqrt{3x + 2}}{3} \cdot \frac{\sqrt{3x + 2}}{\sqrt{3x + 2}} - \frac{2}{3\sqrt{3x + 2}}$$

$$= \frac{3x + 2 - 2}{3\sqrt{3x + 2}}$$

$$= \frac{x}{\sqrt{3x + 2}}$$

Section 6.3 Exponential and Logarithmic Integrals

Skills Review

1. $y = \ln(2x - 5)$

$2x - 5 > 0$

$x > \frac{5}{2}$

Domain: $\left(\frac{5}{2}, \infty\right)$

2. $y = \ln\left(x^2 - 5x + 6\right)$

$x^2 - 5x + 6 > 0$

$(x - 3)(x - 2) > 0$

Domain: $(-\infty, 2) \cup (3, \infty)$

3. $\dfrac{x^2 + 4x + 2}{x + 2} =$

$$\begin{array}{r|l} & x + 2 - \dfrac{2}{x+2} \\ \hline x + 2 & x^2 + 4x + 2 \\ & -\left(x^2 + 2x\right) \\ & 2x + 2 \\ & -(2x + 4) \\ & -2 \end{array}$$

$$\frac{x^2 + 4x + 2}{x + 2} = x + 2 - \frac{2}{x + 2}$$

4. $\dfrac{x^2 - 6x + 9}{x - 4} =$

$$\begin{array}{r|l} & x - 2 + \dfrac{1}{x-4} \\ \hline x - 4 & x^2 - 6x + 9 \\ & -\left(x^2 - 4x\right) \\ & -2x + 9 \\ & -(-2x + 8) \\ & 1 \end{array}$$

$$\frac{x^2 - 6x + 9}{x - 4} = x - 2 + \frac{1}{x - 4}$$

5. $\dfrac{x^3 + 4x^2 - 30x - 4}{x^2 - 4x} =$

$$\begin{array}{r|l} & x + 8 + \dfrac{2x - 4}{x^2 - 4x} \\ \hline x^2 - 4x & x^3 + 4x^2 - 30x - 4 \\ & -\left(x^3 - 4x^2\right) \\ & 8x^2 - 30x \\ & -\left(8x^2 - 32x\right) \\ & 2x - 4 \end{array}$$

$$\frac{x^3 + 4x^2 - 30x - 4}{x^2 - 4x} = x + 8 + \frac{2x - 4}{x^2 - 4x}$$

6. $\dfrac{x^4 - x^3 + x^2 + 15x + 2}{x^2 + 5} =$

$$\begin{array}{r|l} & x^2 - x - 4 + \dfrac{20x + 22}{x^2 + 5} \\ \hline x^2 + 5 & x^4 - x^3 + x^2 + 15x + 2 \\ & -\left(x^4 + 5x^2\right) \\ & -x^3 - 4x^2 + 15x \\ & -\left(-x^3 - 5x\right) \\ & -4x^2 + 20x + 2 \\ & -\left(-4x^2 - 20\right) \\ & 20x + 22 \end{array}$$

$$\frac{x^4 - x^3 + x^2 + 15x + 2}{x^2 + 5} = x^2 - x - 4 + \frac{20x + 22}{x^2 + 5}$$

Skills Review —*continued*—

7. $\int\left(x^3 + \frac{1}{x^2}\right) dx = \int x^3\, dx + \int \frac{1}{x^2}\, dx$

$= \frac{1}{4}x^4 - \frac{1}{x} + C$

8. $\int \frac{x^2 + 2x}{x}\, dx = \int \frac{x^2}{x}\, dx + \int \frac{2x}{x}\, dx$

$= \int x\, dx + \int 2\, dx$

$= \frac{1}{2}x^2 + 2x + C$

9. $\int \frac{x^3 + 4}{x^2}\, dx = \int \frac{x^3}{x^2}\, dx + \int \frac{4}{x^2}\, dx$

$= \int x\, dx + \int \frac{4}{x^2}\, dx$

$= \frac{1}{2}x^2 - \frac{4}{x} + C$

10. $\int \frac{x + 3}{x^3}\, dx = \int \frac{x}{x^3}\, dx + \int \frac{3}{x^3}\, dx$

$= \int \frac{1}{x^2}\, dx + \int \frac{3}{x^3}\, dx$

$= -\frac{1}{x} - \frac{3}{2x^2} + C$

1. $\int 2e^{2x}\, dx = \int e^{2x}(2)\, dx = e^{2x} + C$

3. $\int e^{4x}\, dx = \frac{1}{4}\int e^{4x}(4)\, dx = \frac{1}{4}e^{4x} + C$

5. $\int 9xe^{-x^2}\, dx = -\frac{9}{2}\int e^{-x^2}(-2x)\, dx = -\frac{9}{2}e^{-x^2} + C$

7. $\int 5x^2e^{x^3}\, dx = \frac{5}{3}\int e^{x^3}(3x^2)\, dx = \frac{5}{3}e^{x^3} + C$

9. $\int (x^2 + 2x)e^{x^3 + 3x^2 - 1}\, dx = \frac{1}{3}\int e^{x^3 + 3x^2 - 1}(3x^2 + 6x)\, dx$

$= \frac{1}{3}e^{x^3 + 3x^2 - 1} + C$

11. $\int 5e^{2-x}\, dx = -5\int e^{2-x}(-1)\, dx = -5e^{2-x} + C$

13. $\int \frac{1}{x + 1}\, dx = \ln|x + 1| + C$

15. $\int \frac{1}{3 - 2x}\, dx = -\frac{1}{2}\int \frac{1(-2)}{3 - 2x}\, dx = -\frac{1}{2}\ln|3 - 2x| + C$

17. $\int \frac{2}{3x + 5}\, dx = \frac{2}{3}\int \frac{3}{3x + 5}\, dx = \frac{2}{3}\ln|3x + 5| + C$

19. $\int \frac{x}{x^2 + 1}\, dx = \frac{1}{2}\int \frac{x(2)}{x^2 + 1}\, dx$

$= \frac{1}{2}\ln(x^2 + 1) + C$

$= \ln(x^2 + 1)^{1/2} + C$

$= \ln\sqrt{x^2 + 1} + C$

21. $\int \frac{x^2}{x^3 + 1}\, dx = \frac{1}{3}\int \frac{3x^2}{x^3 + 1}\, dx = \frac{1}{3}\ln|x^3 + 1| + C$

23. $\int \frac{x + 3}{x^2 + 6x + 7}\, dx = \frac{1}{2}\int \frac{2(x + 3)}{x^2 + 6x + 7}\, dx$

$= \frac{1}{2}\ln|x^2 + 6x + 7| + C$

25. $\int \frac{1}{x\ln x}\, dx = \int \frac{1}{\ln x}\left(\frac{1}{x}\right) dx = \ln|\ln x| + C$

27. $\int \frac{e^{-x}}{1 - e^{-x}}\, dx = \int \frac{1}{1 - e^{-x}}(e^{-x}\, dx) = \ln|1 - e^{-x}| + C$

29. $\int \frac{1}{x^2}e^{2/x}\, dx = -\frac{1}{2}\int e^{2/x}(-2x^{-2})\, dx = -\frac{1}{2}e^{2/x} + C$

31. $\int \frac{1}{\sqrt{x}}e^{\sqrt{x}}dx = 2\int e^{\sqrt{x}}\frac{1}{2\sqrt{x}}\, dx = 2e^{\sqrt{x}} + C$

33. $\int (e^x - 2)^2\, dx = \int (e^{2x} - 4e^x + 4)\, dx$

$= \frac{1}{2}e^{2x} - 4e^x + 4x + C$

35. $\int \frac{e^{-x}}{1 + e^{-x}}\, dx = -\int \frac{1}{1 + e^{-x}}(-e^{-x})\, dx$

$= -\ln(1 + e^{-x}) + C$

37. $\int \frac{4e^{2x}}{5 - e^{2x}}\, dx = -2\int \frac{1}{5 - e^{2x}} - 2e^{2x}\, dx$

$= -2\ln|5 - e^{2x}| + C$

39. $\int \frac{e^{2x} + 2e^x + 1}{e^x}\, dx = \int (e^x + 2 + e^{-x})\, dx$

$= e^x + 2x - e^{-x} + C$

General Power Rule and Exponential Rule

41. $\int e^x\sqrt{1-e^x}\,dx = -\int(1-e^x)^{1/2}(-e^x)\,dx$

$= -\frac{2}{3}(1-e^x)^{3/2} + C$

Exponential Rule

43. $\int\frac{1}{(x-1)^2}\,dx = \int(x-1)^{-2}\,dx$

$= \frac{(x-1)^{-1}}{-1}$

$= -\frac{1}{x-1} + C$

General Power Rule

45. $\int 4e^{2x-1}\,dx = 2\int e^{2x-1}(2)\,dx = 2e^{2x-1} + C$

Exponential Rule

47. $\int\frac{x^3-8x}{2x^2}\,dx = \int\left(\frac{x}{2}-\frac{4}{x}\right)dx = \frac{1}{4}x^2 - 4\ln|x| + C$

General Power Rule, Logarithmic Rule

49. $\int\frac{2}{1+e^{-x}}\,dx = 2\int\frac{1}{e^x+1}e^x\,dx = 2\ln(e^x+1) + C$

Logarithmic Rule

51. $\int\frac{x^2+2x+5}{x-1}\,dx = \int\left(x+3+\frac{8}{x-1}\right)dx$

$= \frac{1}{2}x^2 + 3x + 8\ln|x-1| + C$

General Power Rule, Logarithmic Rule

53. $\int\frac{1+e^{-x}}{1+xe^{-x}}\,dx = \int\frac{e^x+1}{e^x+x}\,dx$

$= \ln|e^x+x| + C$

Logarithmic Rule

55. $f(x) = \int\frac{x^2+4x+3}{x-1}\,dx$

$= \int\left(x+5+\frac{8}{x-1}\right)dx$

$= \frac{x^2}{2} + 5x + 8\ln|x-1| + C$

At $(2, 4)$, $4 = \frac{2^2}{2} + 5(2) + 8\ln|2-1| + C$

$-8 = C$

So, $f(x) = \frac{1}{2}x^2 + 5x + 8\ln|x-1| - 8.$

57. (a) $P = \int\frac{3000}{1+0.25t}\,dt$

$= \frac{3000}{0.25}\int\frac{0.25}{1+0.25t}\,dt$

$= 12{,}000\ln|1+0.25t| + C$

Because $P(0) = 12{,}000\ln 1 + C = 1000$, it follows that $C = 1000$. So,

$P(t) = 12{,}000\ln|1+0.25t| + 1000$

$= 1000[12\ln|1+0.25t| + 1]$

$= 1000\left[1 + \ln(1+0.25t)^{12}\right].$

(b) $P(3) = 1000\left[1+\ln(1+0.75)^{12}\right] \approx 7715$ bacteria

(c) $12{,}000 = 1000\left[1+\ln(1+0.25t)^{12}\right]$

$12 = 1 + \ln(1+0.25t)^{12}$

$11 = 12\ln(1+0.25t)$

$\frac{11}{12} = \ln(1+0.25t)$

$e^{11/12} = 1 + 0.25t$

$\frac{e^{11/12}-1}{0.25} = t$

$t \approx 6$ days

59. (a) $L(t) = \int\left(-\frac{4.28}{t} + 0.14\right)dt$

$= -4.28\int\frac{1}{t}\,dt + 0.14\int dt$

$= -4.28\ln t + 0.14t + C$

Because $L(13) = 6.3$, it follows that $C \approx 15.46$, and you have

$L(t) = -4.28\ln t + 0.14t + 15.46.$

(b) $L(20) \approx 5.4$ million sheep and lambs

61. (a) $S = \int(1724.1e^{-t/4.2})\,dt$

$= -4.2(1724.1)e^{-t/4.2} + C$

$= -7241.22e^{-t/4.2} + C$

When $t = 5$,

$40{,}520 = -7241.22e^{-5/4.2} + C$

$C \approx 42{,}721.88$

$S = -7241.22e^{-t/4.2} + 42{,}721.88$

(b) For 2002, $t = 2$ and

$S = -7241.22e^{-2/4.2} + 42{,}721.88 \approx \$38{,}224.03$

63. False. $(\ln 5)^{1/2} \approx 1.27 \neq \frac{1}{2}\ln 5 \approx 0.80$

$\ln x^{1/2} = \frac{1}{2}\ln x$

Mid-Chapter Quiz Solutions

1. $\int 3\,dx = 3\int dx = 3x + C$

$\frac{d}{dx}[3x + C] = 3$

2. $\int 10x\,dx = 10\int x\,dx = 10\left(\frac{1}{2}x^2 + C\right) = 5x^2 + C$

$\frac{d}{dx}[5x^2 + C] = 10x$

3. $\int \frac{1}{x^5}\,dx = -\frac{1}{4}x^{-4} + C = -\frac{1}{4x^4} + C$

$\frac{d}{dx}\left[-\frac{1}{4x^4} + C\right] = \frac{1}{x^5}$

4. $\int (x^2 - 2x + 15)\,dx = \frac{1}{3}x^3 - x^2 + 15x + C$

$\frac{d}{dx}\left[\frac{1}{3}x^3 - x^2 + 15x + C\right] = x^2 - 2x + 15$

5. $\int x(x + 4)\,dx = \int (x^2 + 4x)\,dx = \frac{1}{3}x^3 + 2x^2 + C$

$\frac{d}{dx}\left[\frac{1}{3}x^3 + 2x^2 + C\right] = x^2 + 4x = x(x + 4)$

6. $\int (6x + 1)^3(6)\,dx = \frac{1}{4}(6x + 1)^4 + C$

$\frac{d}{dx}\left[\frac{1}{4}(6x + 1)^4 + C\right] = (6x + 1)^3(6)$

7. $\int (x^2 - 5x)(2x - 5)\,dx = \int (2x^3 - 15x^2 + 25x)\,dx$

$= \frac{1}{2}x^4 - 5x^3 + \frac{25}{2}x^2 + C$

$\frac{d}{dx}\left[\frac{1}{2}x^4 - 5x^3 + \frac{25}{2}x^2 + C\right] = 2x^3 - 15x^2 + 25x$

$= (x^2 - 5x)(2x - 5)$

8. $\int \frac{3x^2}{(x^3 + 3)^3} = -\frac{1}{2}(x^3 + 3)^{-2} + C$

$\frac{d}{dx}\left[-\frac{1}{2}(x^3 + 3)^{-2} + C\right] = (x^3 + 3)^{-3}(3x^2)$

$= \frac{3x^2}{(x^3 + 3)^3}$

9. $\int \sqrt{5x + 2}\,dx = \frac{1}{5}\int 5\sqrt{5x + 2}\,dx$

$= \frac{2}{15}(5x + 2)^{3/2} + C$

$\frac{d}{dx}\left[\frac{2}{15}(5x + 2)^{3/2} + C\right] = \frac{1}{5}(5x + 2)^{1/2}(5)$

$= \sqrt{5x + 2}$

10. $f'(x) = 16x;\ f(0) = 1$

$f(x) = \int 16x\,dx$

$= 8x^2 + C$

$1 = 8(0)^2 + C$

$C = 1$

$f(x) = 8x^2 + 1$

11. $f'(x) = 9x^2 + 4;\ f(1) = 5$

$f(x) = \int (9x^2 + 4)\,dx$

$= 3x^3 + 4x + C$

$5 = 3(1)^3 + 4(1) + C$

$5 = 7 + C$

$C = -2$

$f(x) = 3x^3 + 4x - 2$

12. $f'(x) = 2x^2 + 1$

$f(x) = \int (2x^2 + 1)\,dx$

$= \frac{2}{3}x^3 + x + C$

$1 = \frac{2}{3}(0)^3 + 0 + C$

$C = 1$

$f(x) = \frac{2}{3}x^3 + x + 1$

13. $\int 5e^{5x+4}\,dx = e^{5x+4} + C$

$\frac{d}{dx}[e^{5x+4} + C] = 5e^{5x+4}$

14. $\int (x + 2e^{2x})\,dx = \frac{1}{2}x^2 + e^{2x} + C$

$\frac{d}{dx}\left[\frac{1}{2}x^2 + e^{2x} + C\right] = x + 2e^{2x}$

15. $\int 3x^2e^{x^5}\,dx = e^{x^3} + C$

$\frac{d}{dx}\left[e^{x^3} + C\right] = 3x^2e^{x^3}$

16. $\int \frac{2}{2x-1}\,dx = \ln|2x-1| + C$

$\frac{d}{dx}\left[\ln|2x-1| + C\right] = \frac{1}{2x-1}(2) = \frac{2}{2x-1}$

17. $\int \frac{-2x}{x^2+3}\,dx = -\ln|x^2+3| + C$

$\frac{d}{dx}\left[-\ln|x^2+3| + C\right] = -\frac{1}{x^2+3}(2x) = -\frac{2x}{x^2+3}$

18. $\int \frac{3(3x^2+4x)}{x^3+2x^2}\,dx = 3\ln|x^3+2x^2| + C$

$\frac{d}{dx}\left[3\ln|x^3+2x^2| + C\right] = 3\left(\frac{1}{x^3+2x^2}\right)(3x^2+4x)$

$= \frac{3(3x^2+4x)}{x^3+2x^2}$

19. (a) $P(t) = \int 3\sqrt{t+2}\,dt$

$= 3\int (t+2)^{1/2}\,dt$

$= \frac{3}{3/2}(t+2)^{3/2} + C = 2(t+2)^{3/2} + C$

Because $P(2) = 25$, it follows that $C = 9$, and you have $P(t) = 2(t+2)^{3/2} + 9$.

(b)

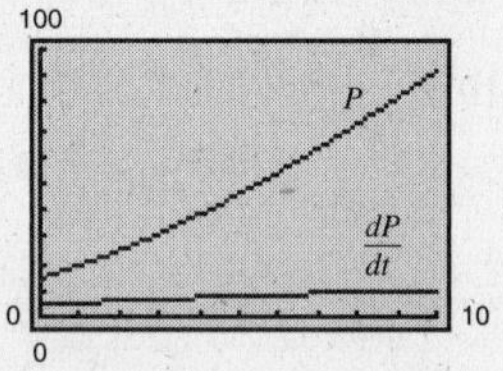

(c) When $t = 7$, $P(7) = 63$ and $\frac{dP}{dt} = 9$.

(d) When $t = 7$:

$P(7) = 2(7+2)^{3/2} + 9 = 54 + 9 = 63 \Rightarrow$ 6300 cells

$\frac{dP}{dt} = 3\sqrt{7+2} = 3 \cdot 3 = 9 \Rightarrow$ 900 cells per day

Section 6.4 Area and the Fundamental Theorem of Calculus

Skills Review

1.

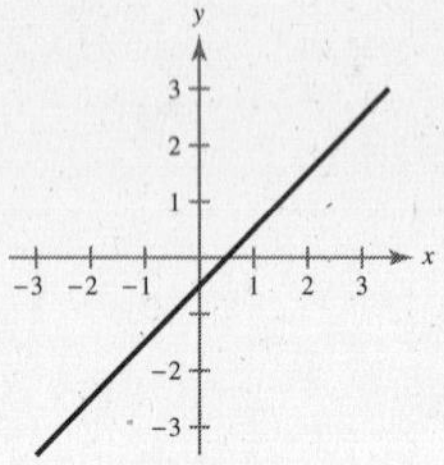

2.

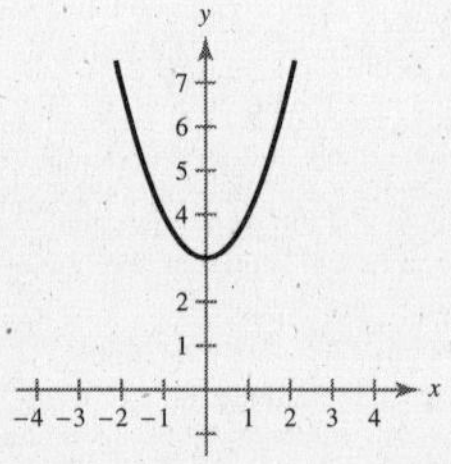

3.

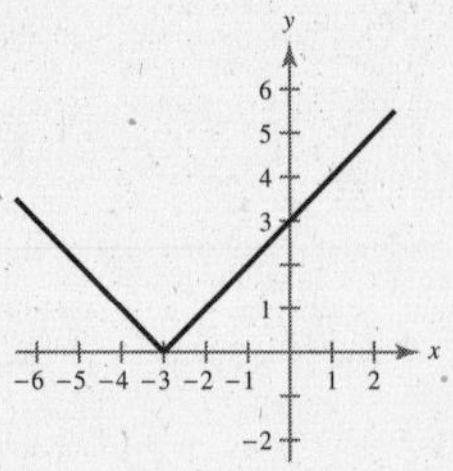

4.

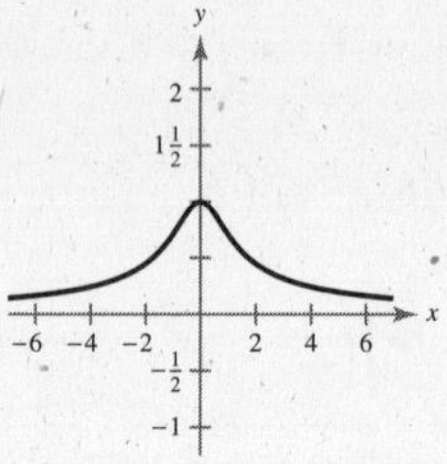

5. $\int (3x+7)\,dx = \frac{3}{2}x^2 + 7x + C$

6. $\int \left(x^{3/2} + 2\sqrt{x}\right)dx = \frac{2}{5}x^{5/2} + \frac{4}{3}x^{3/2} + C$

7. $\int \frac{1}{5x}\,dx = \frac{1}{5}\ln|x| + C$

8. $\int e^{-6x}\,dx = -\frac{1}{6}e^{-6x} + C$

9. $\left(\frac{a}{5} - a\right) - \left(\frac{b}{5} - b\right)$ when $a = 5$ and $b = 3$:

$\left(\frac{5}{5} - 5\right) - \left(\frac{3}{5} - 3\right) = -4 - \left(-\frac{12}{5}\right) = -\frac{8}{5}$

10. $\left(6a - \frac{a^3}{3}\right) - \left(6b - \frac{b^3}{3}\right)$ when $a = 5$ and $b = 3$:

$\left(6(5) - \frac{5^3}{3}\right) - \left(6(3) - \frac{3^3}{3}\right) = -\frac{35}{3} - 9 = -\frac{62}{3}$

1. $\int_0^3 \frac{5x}{x^2+1}\,dx$

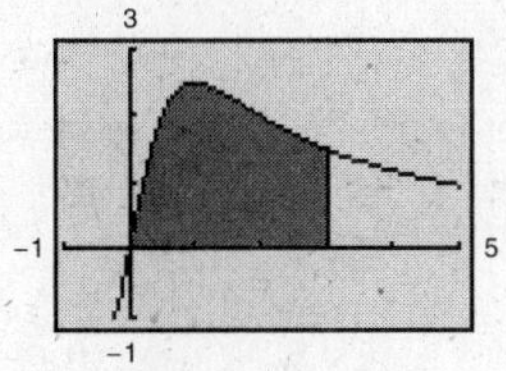

The definite integral is positive.

3. $\int_0^2 3\,dx$

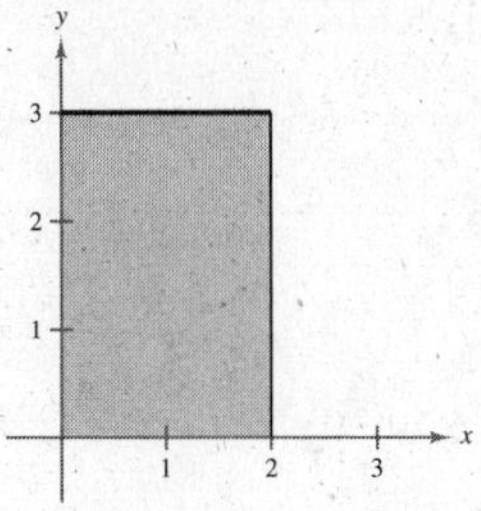

Area $= (\text{base})(\text{height}) = (2)(3) = 6$

5. $\int_0^4 x\,dx$

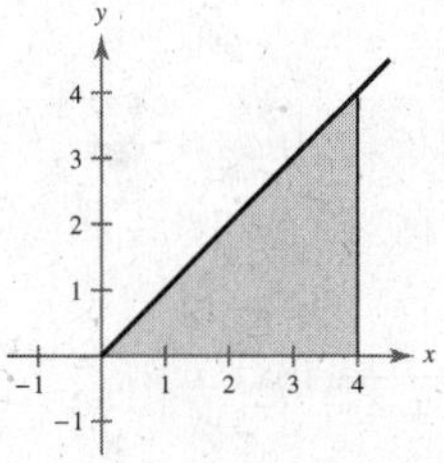

Area $= \frac{1}{2}(\text{base})(\text{height}) = \frac{1}{2}(4)(4) = 8$

7. $\int_0^5 (x+1)\,dx$

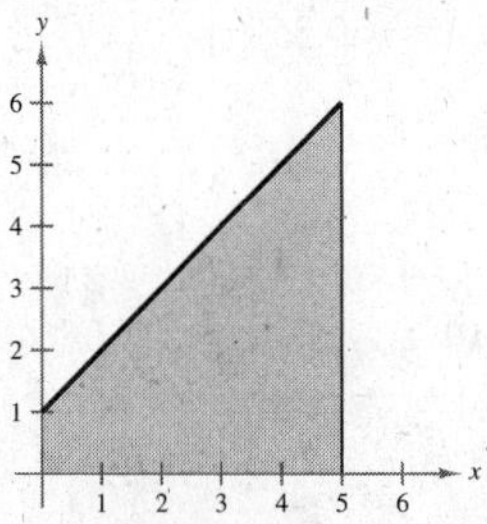

Area $= \frac{1}{2}(h)(b_1 + b_2) = \frac{1}{2}(5)(1+6) = \frac{35}{2}$

9. $\int_{-2}^3 |x-1|\,dx$

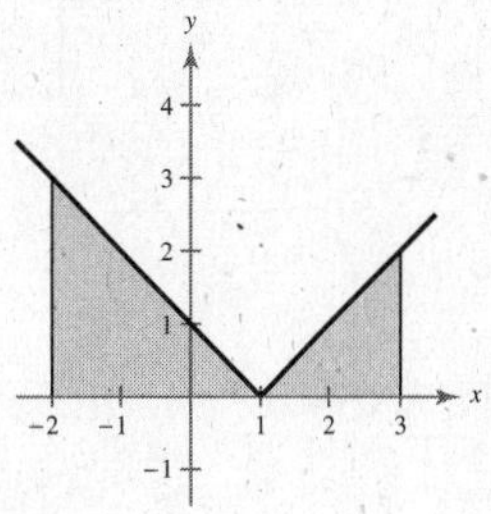

Area $= \frac{1}{2}bh + \frac{1}{2}bh = \frac{1}{2}(3)(3) + \frac{1}{2}(2)(2) = \frac{13}{2}$

11. $\int_{-3}^3 \sqrt{9-x^2}\,dx$

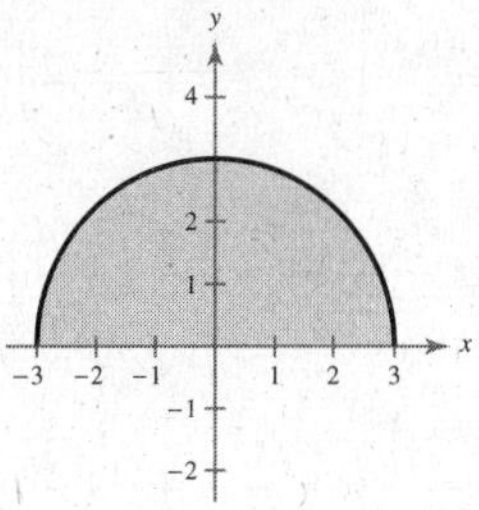

Area $= \frac{1}{2}\pi r^2 = \frac{1}{2}\pi(3)^2 = \frac{9\pi}{2}$

13. (a) $\int_0^5 [f(x) + g(x)]\,dx = \int_0^5 f(x)\,dx + \int_0^5 g(x)\,dx$

$= 6 + 2$

$= 8$

(b) $\int_0^5 [f(x) - g(x)]\,dx = \int_0^5 f(x)\,dx - \int_0^5 g(x)\,dx$

$= 6 - 2$

$= 4$

(c) $\int_0^5 -4f(x)\,dx = -4\int_0^5 f(x)\,dx$

$= -4(6)$

$= -24$

(d) $\int_0^5 [f(x) - 3g(x)] = \int_0^5 f(x)\,dx - 3\int_0^5 g(x)\,dx$

$= 6 - 3(2)$

$= 0$

15. $A = \int_0^1 (1 - x^2)\,dx = \left[\frac{x^2}{2} - \frac{x^3}{3}\right]_0^1 = \frac{1}{2} - \frac{1}{3} = \frac{1}{6}$

17. $A = \int_1^2 \frac{1}{x^2}\,dx = \int_1^2 x^{-2}\,dx = \left.\frac{x^{-1}}{-1}\right]_1^2 = \left.-\frac{1}{x}\right]_1^2 = -\frac{1}{2} + 1 = \frac{1}{2}$

19. $A = \int_0^4 3e^{-x/2}\,dx = 3(-2)\int_0^4 e^{-x/2}\left(-\frac{1}{2}\right)dx = \left.-6e^{-x/2}\right]_0^4 = -6(e^{-2} - 1) = 6(1 - e^{-2}) = 6\left(1 - \frac{1}{e^2}\right)$

21. $A = \int_1^4 \frac{x^2+4}{x}\,dx$

$= \int_1^4 \left[x + 4\left(\frac{1}{x}\right)\right] dx$

$= \left[\frac{x^2}{2} + 4\ln|x|\right]_1^4$

$= (8 + 4\ln 4) - \left(\frac{1}{2} + 4\ln 1\right)$

$= \frac{15}{2} + 8\ln 2$

23. $\int_0^1 2x\,dx = x^2\Big]_0^1 = 1 - 0 = 1$

25. $\int_{-1}^0 (x-2)\,dx = \left[\frac{x^2}{2} - 2x\right]_{-1}^0 = 0 - \left(\frac{1}{2} + 2\right) = -\frac{5}{2}$

27. $\int_{-1}^1 (2t-1)^2\,dt = \frac{1}{6}(2t-1)^3\Big]_{-1}^1 = \frac{1}{6} - \left(\frac{-27}{6}\right) = \frac{14}{3}$

29. $\int_0^3 (x-2)^3\,dx = \frac{(x-2)^4}{4}\Big]_0^3 = \frac{1}{4} - 4 = -\frac{15}{4}$

31. $\int_{-1}^1 \left(\sqrt[3]{t} - 2\right) dt = \left[\frac{3}{4}t^{4/3} - 2t\right]_{-1}^1$

$= \left(\frac{3}{4} - 2\right) - \left(\frac{3}{4} + 2\right)$

$= -\frac{5}{4} - \frac{11}{4}$

$= -4$

33. $\int_1^4 \frac{u-2}{\sqrt{u}}\,du = \int_1^4 \left(u^{1/2} - 2u^{-1/2}\right) du$

$= \left[\frac{2}{3}u^{3/2} - 4u^{1/2}\right]_1^4$

$= \left[\frac{2}{3}(8) - 4(2)\right] - \left(\frac{2}{3} - 4\right)$

$= -\frac{8}{3} + \frac{10}{3} = \frac{2}{3}$

35. $\int_{-1}^0 \left(t^{1/3} - t^{2/3}\right) dt = \left[\frac{3}{4}t^{4/3} - \frac{3}{5}t^{5/3}\right]_{-1}^0$

$= 0 - \left(\frac{3}{4} + \frac{3}{5}\right) = -\frac{27}{20}$

37. $\int_0^4 \frac{1}{\sqrt{2x+1}}\,dx = \frac{1}{2}\int_0^4 (2x+1)^{-1/2}(2)\,dx$

$= \frac{1}{2}(2)(2x+1)^{1/2}\Big]_0^4$

$= \sqrt{2x+1}\Big]_0^4 = 3 - 1 = 2$

39. $\int_0^1 e^{-2x}\,dx = -\frac{1}{2}e^{-2x}\Big]_0^1 = -\frac{e^{-2}}{2} + \frac{1}{2} = \frac{1}{2}\left(1 - e^{-2}\right) \approx 0.43$

41. $\int_1^3 \frac{e^{3/x}}{x^2}\,dx = -\frac{1}{3}\int_1^3 e^{3/x}\left(-\frac{3}{x^2}\right) dx$

$= -\frac{1}{3}e^{3/x}\Big]_1^3$

$= -\frac{1}{3}\left(e - e^3\right)$

$= \frac{e^3 - e}{3} \approx 5.79$

43. $\int_0^1 e^{2x}\sqrt{e^{2x}+1}\,dx = \frac{1}{2}\int_0^1 \left(e^{2x}+1\right)^{1/2} 2e^{2x}\,dx$

$= \frac{1}{3}\left(e^{2x}+1\right)^{3/2}\Big]_0^1$

$= \frac{1}{3}\left[\left(e^2+1\right)^{3/2} - 2\sqrt{2}\right]$

≈ 7.16

45. $\int_0^2 \frac{x}{1+4x^2}\,dx = \frac{1}{8}\int_0^2 \frac{1}{1+4x^2}(8x)\,dx$

$= \frac{1}{8}\ln\left(1+4x^2\right)\Big]_0^2$

$= \frac{1}{8}(\ln 17 - 0)$

$= \frac{1}{8}\ln 17 \approx 0.35$

47. $\int_{-1}^1 |4x|\,dx = 2\int_0^1 |4x|\,dx$

$= 2\int_0^1 4x\,dx$

$= 4x^2\Big]_0^1 = 4$

49. $\int_0^4 \left(2 - |x-2|\right) dx = \int_0^2 \{2 - [-(x-2)]\}\,dx + \int_2^4 [2 - (x-2)]\,dx$

$= \int_0^2 x\,dx + \int_2^4 (4-x)\,dx = \frac{x^2}{2}\Big]_0^2 + \left[4x - \frac{x^2}{2}\right]_2^4 = (2-0) + (8-6) = 4$

51. $\int_{-1}^2 \frac{x}{x^2-9}\,dx = \frac{1}{2}\ln\left|x^2-9\right|\Big]_{-1}^2 = \frac{1}{2}\ln 5 - \frac{1}{2}\ln 8 \approx -0.24$

53. $\displaystyle\int_0^3 \frac{2e^x}{2 + e^x}\,dx = 2\ln\left(2 + e^x\right)\Big]_0^3$

$= 2\ln\left(2 + e^3\right) - 2\ln 3 \approx 3.99$

55. $\displaystyle\int_1^3 (4x - 3)\,dx = \left[2x^2 - 3x\right]_1^3$

$= (18 - 9) - (2 - 3)$

$= 10$

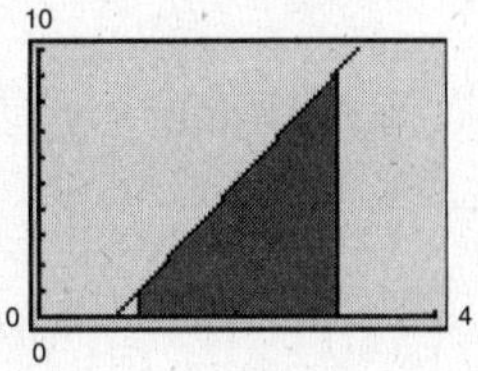

57. $\displaystyle\int_0^1 \left(x - x^3\right)dx = \left[\frac{x^2}{2} - \frac{x^4}{4}\right]_0^1 = \frac{1}{2} - \frac{1}{4} = \frac{1}{4}$

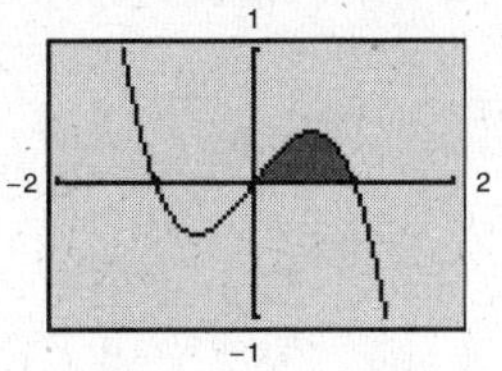

59. $\displaystyle\int_2^4 \frac{3x^2}{x^3 - 1}\,dx = \ln\left(x^3 - 1\right)\Big]_2^4$

$= \ln 63 - \ln 7$

$= \ln 9$

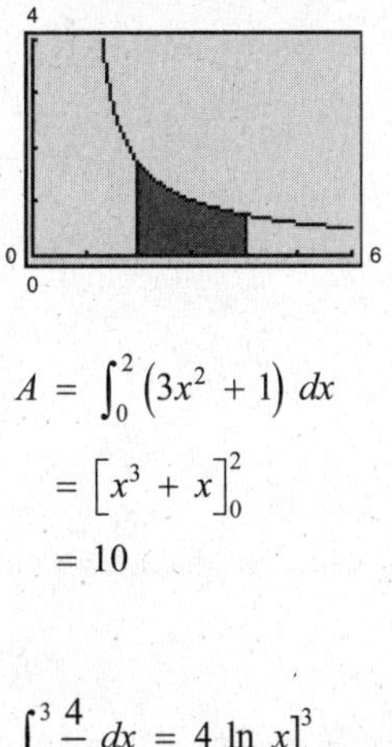

61. $\displaystyle A = \int_0^2 \left(3x^2 + 1\right)dx$

$= \left[x^3 + x\right]_0^2$

$= 10$

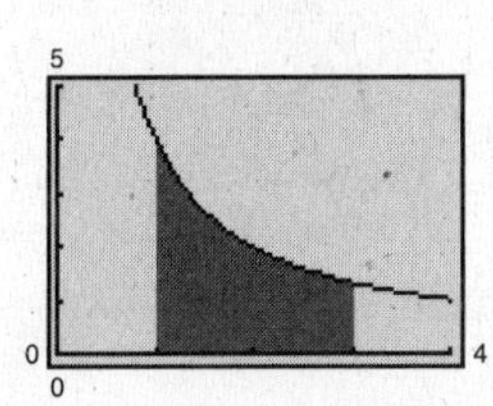

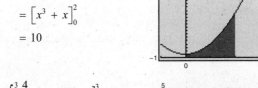

63. $\displaystyle\int_1^3 \frac{4}{x}\,dx = 4\ln x\Big]_1^3$

$= 4\ln 3 - 4\ln 1$

$= 4\ln 3$

≈ 4.39

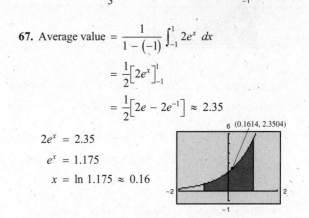

65. Average value $\displaystyle = \frac{1}{2 - (-2)}\int_{-2}^2 \left(4 - x^2\right)dx$

$\displaystyle = \frac{1}{4}\left[4x - \frac{x^3}{3}\right]_{-2}^2$

$\displaystyle = \frac{1}{4}\left[8 - \frac{8}{3} - \left(-8 + \frac{8}{3}\right)\right] = \frac{8}{3}$

$\displaystyle 4 - x^2 = \frac{8}{3}$

$\displaystyle x^2 = \frac{4}{3}$

$\displaystyle x = \pm\frac{2\sqrt{3}}{3} \approx \pm 1.15$

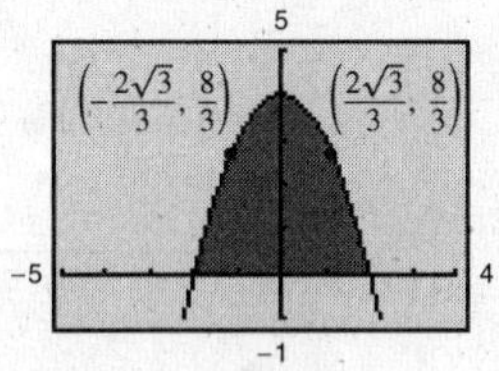

67. Average value $\displaystyle = \frac{1}{1 - (-1)}\int_{-1}^1 2e^x\,dx$

$\displaystyle = \frac{1}{2}\left[2e^x\right]_{-1}^1$

$\displaystyle = \frac{1}{2}\left[2e - 2e^{-1}\right] \approx 2.35$

$2e^x = 2.35$

$e^x = 1.175$

$x = \ln 1.175 \approx 0.16$

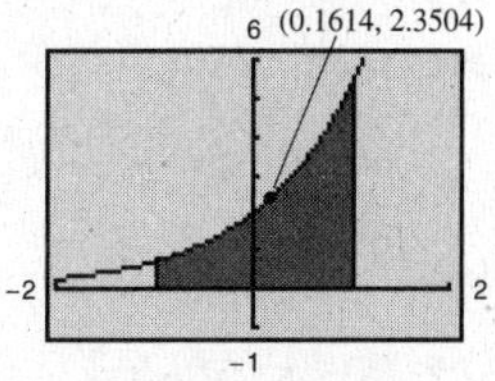

69. Average value $\displaystyle = \frac{1}{2 - 0}\int_0^2 x\sqrt{4 - x^2}\,dx$

$\displaystyle = \frac{1}{2}\left(-\frac{1}{2}\right)\int_0^2 \left(4 - x^2\right)^{1/2}(-2x)\,dx$

$\displaystyle = \frac{1}{2}\left(-\frac{1}{2}\right)\left(\frac{2}{3}\right)\left(4 - x^2\right)^{3/2}\Big]_0^2$

$\displaystyle = -\frac{1}{6}\left(4 - x^2\right)^{3/2}\Big]_0^2 = 0 + \frac{4}{3}$

$\displaystyle = \frac{4}{3}$

$\displaystyle x\sqrt{4 - x^2} = \frac{4}{3}$

$\displaystyle x^2\left(4 - x^2\right) = \frac{16}{9}$

$36x^2 - 9x^4 = 16$

$9x^4 - 36x^2 + 16 = 0$

$\displaystyle x^2 = \frac{36 \pm \sqrt{720}}{18} = 2 \pm \frac{2\sqrt{5}}{3}$

$\displaystyle x = \sqrt{2 \pm \frac{2\sqrt{5}}{3}}$

$x \approx 1.868$ and $x \approx 0.714$

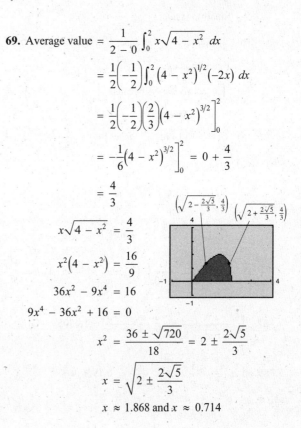

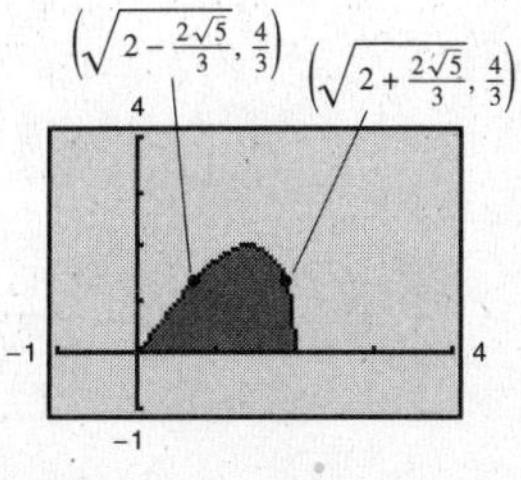

71. Average value $= \frac{1}{7-0}\int_0^7 \frac{6x}{x^2+1}\,dx$

$= \left(\frac{1}{7}\right)3\ln\left(x^2+1\right)\Big]_0^7$

$= \frac{3}{7}\ln 50 - 0 \approx 1.677$

$\frac{6x}{x^2+1} = \frac{3}{7}\ln 50$

Using a graphing utility, you obtain $x \approx 0.306$ and $x \approx 3.273$.

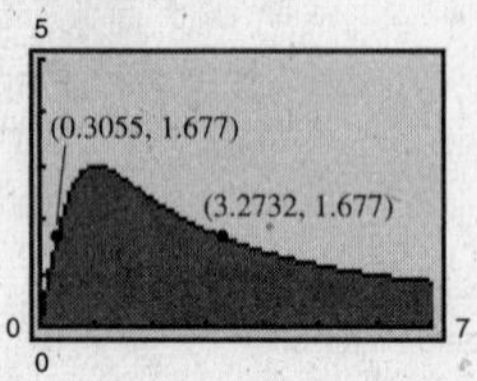

73. Because $f(-x) = 3(-x)^4 = 3x^4 = f(x)$, the function is even.

75. Because $g(-t) = 2(-t)^5 - 3(-t)^2 = -2t^5 - 3t^2$, which is neither $g(t)$ nor $-g(t)$, the function is neither even nor odd.

77. (a) $\int_{-1}^0 x^2\,dx = \frac{1}{3}$ because $\int_0^1 x^2\,dx = \frac{1}{3}$ and the function is even.

(b) $\int_{-1}^1 x^2\,dx = \frac{2}{3}$ because $\int_0^1 x^2\,dx = \frac{1}{3}$ and the function is even.

(c) $\int_0^1 -x^2\,dx = -\frac{1}{3}$ because $\int_0^1 -x^2\,dx = -\int_0^1 x^2\,dx$.

79. $F = \int_0^5 (1.08t - 15.4)\,dt$

$= 0.54t^2 - 15.4t\Big]_0^5$

$= -63.5$

The number of farms decreased by 63.5 thousand from 2000 to 2005.

81. $C(x) = 5000\left(25 + 3\int_0^x t^{1/4}\,dt\right)$

$= 5000\left(25 + \frac{12}{5}t^{5/4}\Big]_0^x\right)$

$= 5000\left(25 + \frac{12}{5}x^{5/4}\right)$

(a) $C(1) = 5000\left[25 + \left(\frac{12}{5}\right)(1)^{5/4}\right] = \$137{,}000.00$

(b) $C(5) = 5000\left[25 + \left(\frac{12}{5}\right)(5)^{5/4}\right] \approx \$214{,}720.93$

(c) $C(10) = 5000\left[25 + \left(\frac{12}{5}\right)(10)^{5/4}\right] \approx \$338{,}393.53$

83. Average $= \frac{1}{R-0}\int_0^R k\left(R^2 - r^2\right)dr$

$= \frac{k}{R}\left(R^2 r - \frac{r^3}{3}\right)\Big]_0^R = \frac{k}{R}\left(R^3 - \frac{R^3}{3}\right)$

$= \frac{k}{R}\left(\frac{2R^3}{3}\right) = \frac{2kR^2}{3}$

85. (a) $C(t) = \int 147e^{-0.42t}\,dt = 147\left(-\frac{1}{0.42}\right)e^{-0.42t} + K = -350e^{-0.42t} + K$

When $t = 2$:

$500 = -350e^{-0.42(2)} + K$

$K \approx 651.1$

$C(t) = -350e^{-0.42t} + 651.1$

(b) Average $= \frac{1}{5-0}\int_0^5\left(-350e^{-0.42t} + 651.1\right)dt = \frac{1}{5}\left[-350\left(\frac{1}{0.42}\right)e^{-0.42t} + 651.1t\right]_0^5 = \frac{1}{5}\left[\frac{2500}{3}e^{-0.42t} + 651.1t\right]_0^5 \approx 505$ coyotes

87. $\int_3^6 \frac{x}{e\sqrt{x^2-8}}\,dx = \frac{2\sqrt{7}-1}{3} \approx 1.4305$

89. $\int_2^5\left(\frac{1}{x^2} - \frac{1}{x^3}\right)dx = \frac{39}{200} = 0.195$

Section 6.5 The Area of a Region Bounded by Two Graphs

Skills Review

1. $(-x^2 + 4x + 3) - (x + 1) = -x^2 + 3x + 2$

2. $(-2x^2 + 3x + 9) - (-x + 5) = -2x^2 + 4x + 4$

3. $(-x^3 + 3x^2 - 1) - (x^2 - 4x + 4) = -x^3 + 2x^2 + 4x - 5$

4. $(3x + 1) - (-x^3 + 9x + 2) = x^3 - 6x - 1$

5. $x^2 - 4x + 4 = 4$
$x(x - 4) = 0$
$x = 0$ or $x - 4 = 0$
$x = 4$
$g(0) = 4$ and $g(4) = 4$
The graphs intersect at $(0, 4)$ and $(4, 4)$.

6. $-3x^2 = 6 - 9x$
$0 = 3x^2 - 9x + 6$
$0 = x^2 - 3x + 2$
$0 = (x - 1)(x - 2)$
$x - 1 = 0$ or $x - 2 = 0$
$x = 1 \quad x = 2$
$f(1) = -3(1)^2 = -3$ and $f(2) = -3(2)^2 = -12$
The graphs intersect at $(1, -3)$ and $(2, -12)$.

7. $x^2 = -x + 6$
$x^2 + x - 6 = 0$
$(x + 3)(x - 2) = 0$
$x + 3 = 0$ or $x - 2 = 0$
$x = -3 \quad x = 2$
$f(-3) = (-3)^2 = 9$ and $f(2) = (2)^2 = 4$
The graphs intersect at $(-3, 9)$ and $(2, 4)$.

8. $\frac{1}{2}x^3 = 2x$
$\frac{1}{2}x^3 - 2x = 0$
$x\left(\frac{1}{2}x^2 - 2\right) = 0$
$x = 0$ or $\frac{1}{2}x^2 - 2 = 0$
$x^2 = 4$
$x = \pm 2$
$g(0) = 2(0) = 0$, $g(-2) = 2(-2) = -4$, and
$g(2) = 2(2) = 4$
The graphs intersect at $(0, 0)$, $(-2, -4)$, and $(2, 4)$.

9. $x^2 - 3x = 3x - 5$
$x^2 - 6x + 5 = 0$
$(x - 1)(x - 5) = 0$
$x - 1 = 0$ or $x - 5 = 0$
$x = 1 \quad x = 5$
$g(1) = 3(1) - 5 = -2$ and $g(5) = 3(5) - 5 = 10$
The graphs intersect at $(1, -2)$ and $(5, 10)$.

10. $e^x = e$
$x = \ln e$
$x = 1$
$g(1) = e$
The graphs intersect at $(1, e)$.

1. $A = \int_0^6 \left[0 - (x^2 - 6x)\right] dx = \left(-\frac{x^3}{3} + 3x^2\right)\Big]_0^6 = 36$

3. $A = \int_0^3 \left[(-x^2 + 2x + 3) - (x^2 - 4x + 3)\right] dx$
$= \int_0^3 (-2x^2 + 6x)\, dx = \left[-\frac{2x^3}{3} + 3x^2\right]_0^3 = 9$

5. $A = 2\int_0^1 \left[0 - 3(x^3 - x)\right] dx = -6\left(\frac{x^4}{4} - \frac{x^2}{2}\right)\Big]_0^1 = \frac{3}{2}$

7. $A = \int_0^1 \left[(e^x - 1) - 0\right] dx$
$= \int_0^1 (e^x - 1)\, dx$
$= \left[e^x - x\right]_0^1$
$= (e - 1) - (1 - 0)$
$= e - 2 \approx 0.718$

9. The region is bounded by the graphs of $y = x + 1$, $y = x/2$, $x = 0$, and $x = 4$, as shown in the figure.

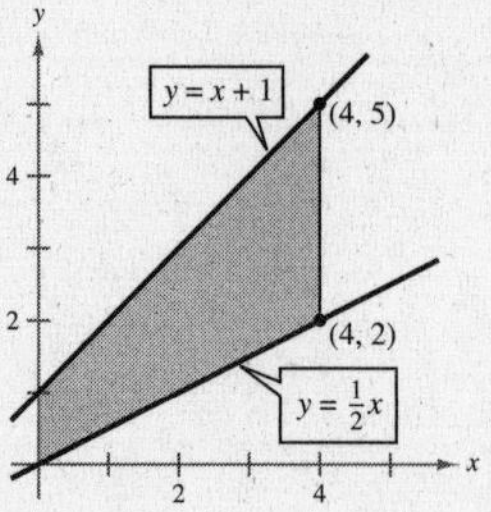

11. The region is bounded by the graphs of $y = 2x^2$ and $y = x^4 - 2x^2$ from $x = -2$ to $x = 2$, as shown in the figure.

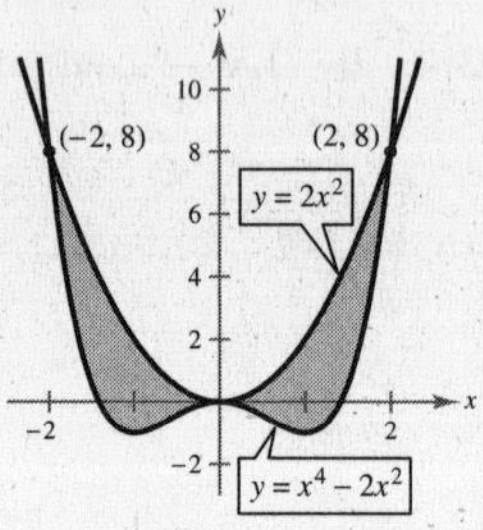

13. The region is bounded by the graphs of $x = 1$ and $x = y^2 + 2$ from $y = -1$ to $y = 2$, as shown in the figure.

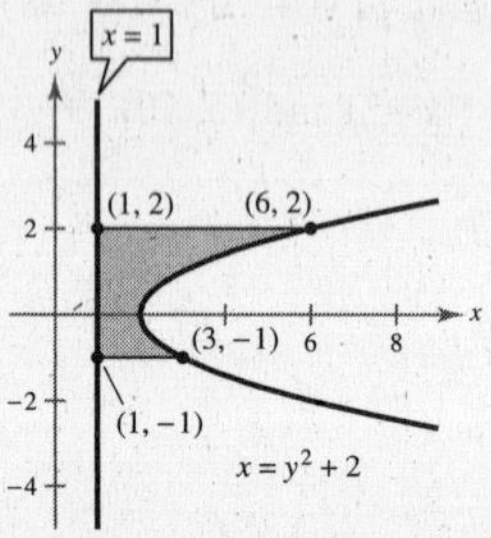

15. $f(x) = x + 1$

$g(x) = (x - 1)^2$

$A \approx 4$

Matches d

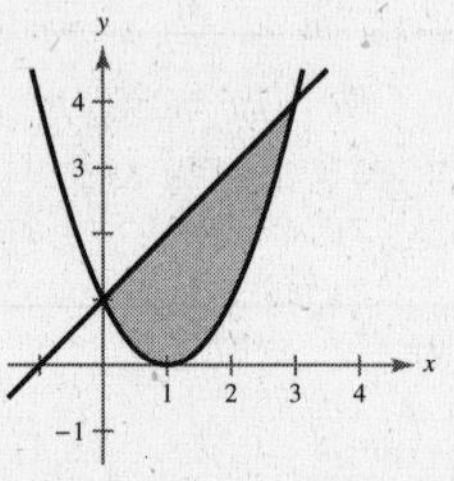

17. $A = \int_1^5 \frac{1}{x^2}\,dx = -\frac{1}{x}\Big]_1^5 = \frac{4}{5}$

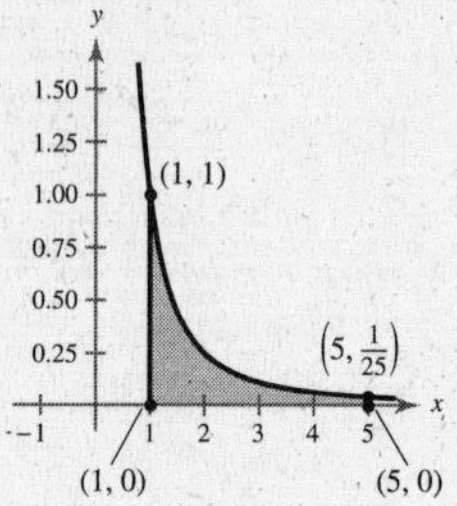

19. $\sqrt[3]{x} = x$

$x = x^3$

$0 = x(x^2 - 1)$

$x = -1, 0, 1$

$A = 2\int_0^1 \left(\sqrt[3]{x} - x\right) dx$

$= 2\left[\frac{3}{4}x^{4/3} - \frac{x^2}{2}\right]_0^1$

$= \frac{1}{2}$

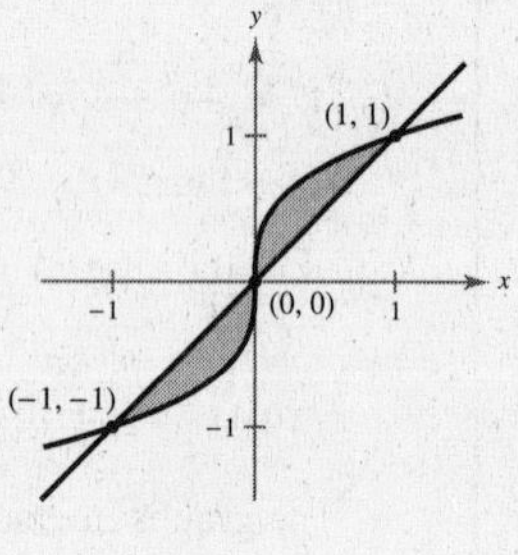

21. $x^2 - 4x + 3 = 3 + 4x - x^2$

$2x^2 - 8x = 0$

$2x(x - 4) = 0$

$x = 0, 4$

$A = \int_0^4 \left[\left(3 + 4x - x^2\right) - \left(x^2 - 4x + 3\right)\right] dx$

$= \int_0^4 \left(-2x^2 + 8x\right) dx$

$= \left[-\frac{2x^3}{3} + 4x^2\right]_0^4$

$= \frac{64}{3}$

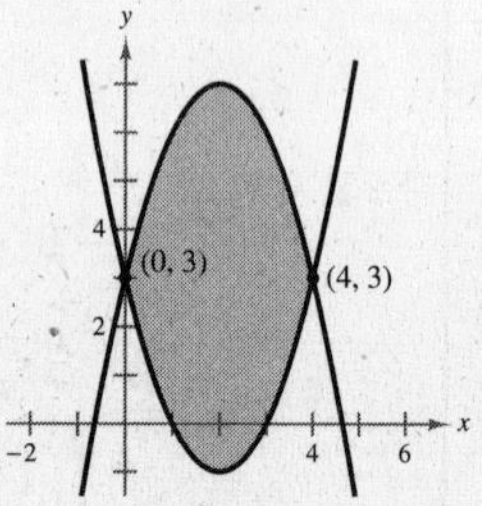

23. $A = \int_0^1 xe^{-x^2}\,dx = -\frac{1}{2}e^{-x^2}\Big]_0^1$

$= -\frac{1}{2}e^{-1} + \frac{1}{2} \approx 0.316$

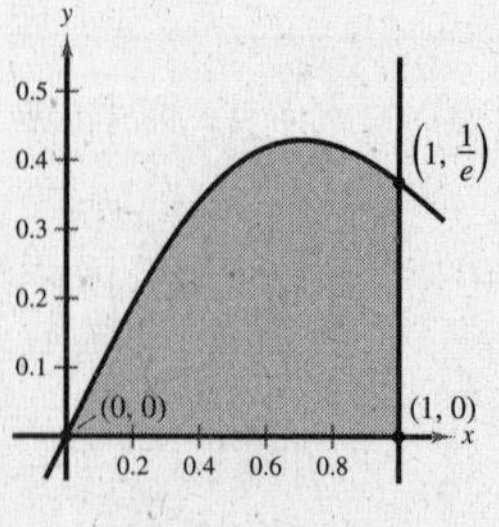

25. $A = \int_1^2 x^2\,dx + \int_2^4 \frac{8}{x}\,dx = \frac{x^3}{3}\Big]_1^2 + 8\ln x\Big]_2^4$

$= \left(\frac{8}{3} - \frac{1}{3}\right) + 8(\ln 4 - \ln 2) = \frac{7}{3} + 8\ln 2 \approx 7.879$

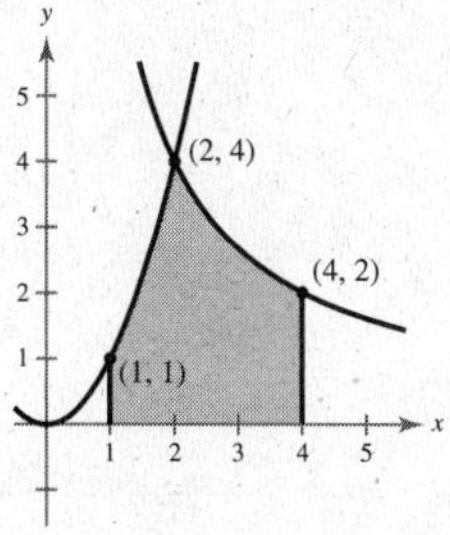

27. $A = \int_1^2 \left[e^{0.5x} - \left(-\frac{1}{x}\right)\right] dx$

$= 2e^{0.5x} + \ln x\Big]_1^2$

$= (2e + \ln 2) - 2e^{0.5}$

≈ 2.832

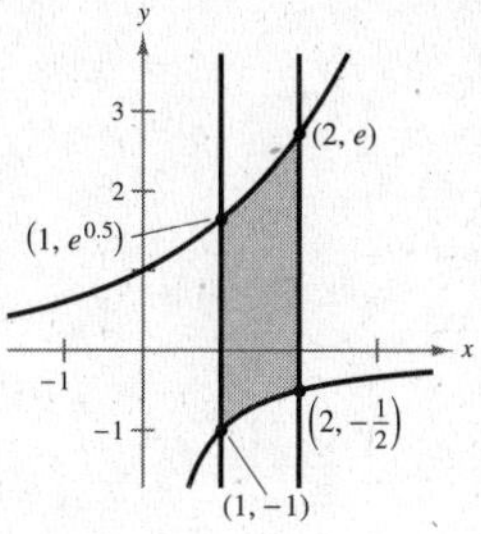

29. $y^2 = y + 2$

$y^2 - y - 2 = 0$

$(y + 1)(y - 2) = 0$

$y = -1, 2$

$A = \int_{-1}^2 \left[(y + 2) - y^2\right] dy$

$= \left[\frac{y^2}{2} + 2y - \frac{y^3}{3}\right]_{-1}^2 = \frac{9}{2}$

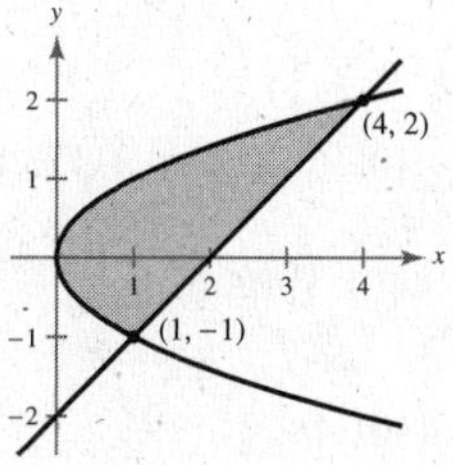

31. $A = \int_0^9 \sqrt{y}\,dy = \frac{2}{3}y^{3/2}\Big]_0^9 = 18$

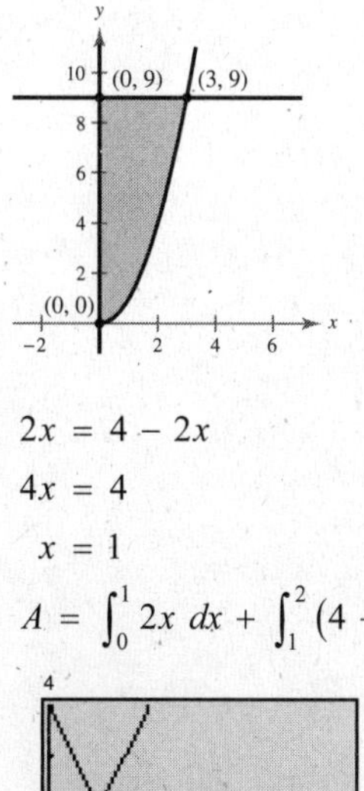

33. $2x = 4 - 2x$

$4x = 4$

$x = 1$

$A = \int_0^1 2x\,dx + \int_1^2 (4 - 2x)\,dx$

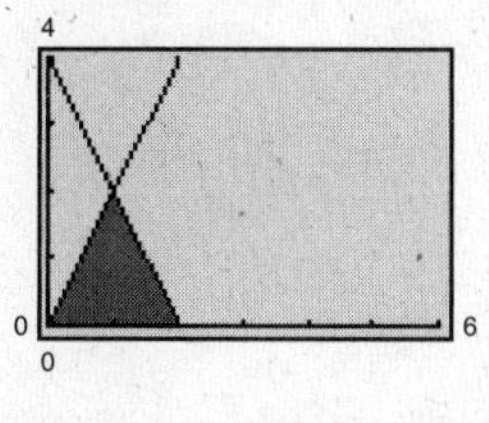

35. $\frac{4}{x} = x$

$4 = x^2$

$x = -2, 2$

$A = \int_1^2 \left(\frac{4}{x} - x\right) dx + \int_2^4 \left(x - \frac{4}{x}\right) dx$

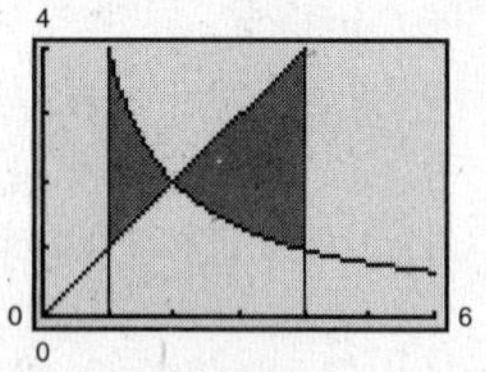

37. $x^2 - 4x = 0$

$x(x - 4) = 0$

$x = 0, 4$

$A = \int_0^4 \left[0 - (x^2 - 4x)\right] dx$

$= -\left(\frac{x^3}{3} - 2x^2\right)\Big]_0^4 = \frac{32}{3}$

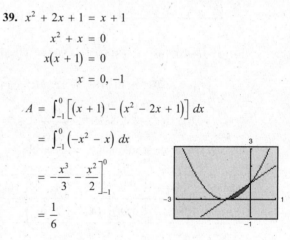

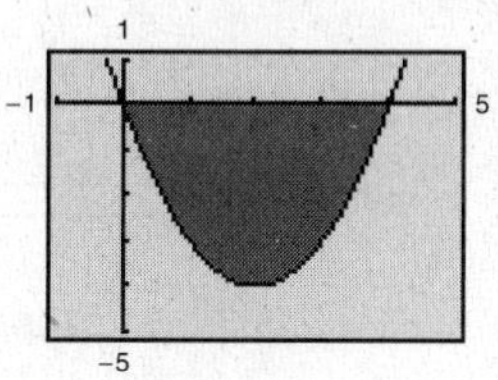

39. $x^2 + 2x + 1 = x + 1$

$x^2 + x = 0$

$x(x + 1) = 0$

$x = 0, -1$

$A = \int_{-1}^0 \left[(x + 1) - (x^2 - 2x + 1)\right] dx$

$= \int_{-1}^0 (-x^2 - x)\,dx$

$= -\frac{x^3}{3} - \frac{x^2}{2}\Big]_{-1}^0$

$= \frac{1}{6}$

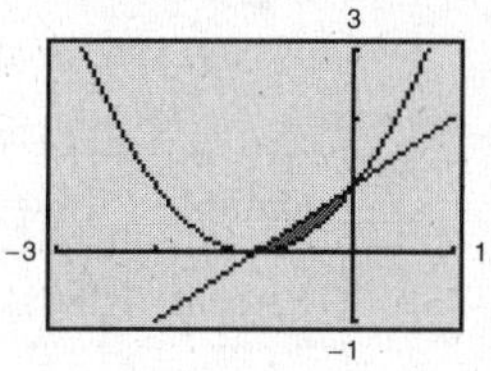

41. The equation of the line passing through $(0, 0)$ and $(4, 4)$ is $y = x$. So, the area is given by

$$A = \int_0^4 x\,dx = \frac{x^2}{2}\Big]_0^4 = 8.$$

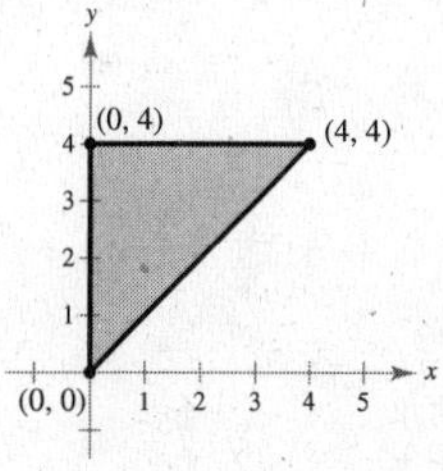

43. Offer 2 is better because the cumulative salary (area under the curve) is greater.

45. $\int_7^{13} (C_1 - C_2)\, dt = \int_7^{13} \left[(568.5 + 7.15t) - (525.6 + 6.43t)\right] dt$

$= \int_7^{13} (42.9 + 0.72t)\, dt$

$= \left[42.9t + 0.36t^2\right]_7^{13}$

$= \$300.6$ million

Explanations will vary.

47. $\int_0^4 (N_1 - N_2)\, dt = \int_0^4 \left[\left(3.07t^2 + 118.2t + 1357\right) - \left(2t^2 + 112t + 1311\right)\right] dt$

$= \int_0^4 \left(1.07t^2 + 6.2t + 46\right) dt$

$= \left[\frac{1.07}{3}t^3 + 3.1t^2 + 46t\right]_0^4$

$\approx \$256.4$ billion

49. (a)

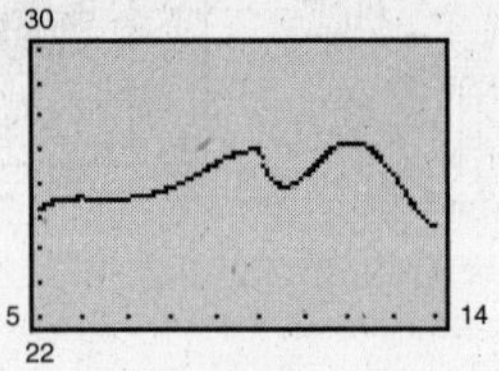

(b) $\int_{10}^{11} \left[\left(-0.02708t^4 + 0.8153t^3 - 8.96t^2 + 42.85t - 50\right) - \left(0.2t^4 - 9.8t^3 + 178.85t^2 - 1440.95t + 4351.4\right)\right]$

$= \int_{10}^{11} \left(-0.22708t^4 + 10.6153t^3 - 187.81t^2 + 1483.8t - 4401.4\right) dt$

$= \left[-0.045416t^5 + 2.653825t^4 - \frac{187.81}{3}t^3 + 741.9t^2 - 4401.4t\right]_{10}^{11}$

≈ 0.506, or about 506 million more pounds

51.

Quintile	Lowest	2nd	3rd	4th	Highest
Percent	2.81	6.98	14.57	27.01	45.73

$y(20) - y(0) \approx 2.81$

$y(40) - y(20) \approx 6.98$

$y(60) - y(40) \approx 14.57$

$y(80) - y(60) \approx 27.01$

$y(100) - y(80) \approx 45.73$

53. Answers will vary.

Section 6.6 Volumes of Solids of Revolution

Skills Review

1. $x^2 = 2x$

$x^2 - 2x = 0$

$x(x - 2) = 0$

$x = 0$

$x - 2 = 0 \Rightarrow x = 2$

2. $-x^2 + 4x = x^2$

$0 = 2x^2 - 4x$

$0 = 2x(x - 2)$

$2x = 0 \Rightarrow x = 0$

$x - 2 = 0 \Rightarrow x = 2$

3. $x = -x^3 + 5x$

$x^3 - 4x = 0$

$x(x^2 - 4) = 0$

$x(x + 2)(x - 2) = 0$

$x = 0$

$x + 2 = 0 \Rightarrow x = -2$

$x - 2 = 0 \Rightarrow x = 2$

4. $x^2 + 1 = x + 3$

$x^2 - x - 2 = 0$

$(x - 2)(x + 1) = 0$

$x - 2 = 0 \Rightarrow x = 2$

$x + 1 = 0 \Rightarrow x = -1$

5. $-x + 4 = \sqrt{4x - x^2}$

$(-x + 4)^2 = 4x - x^2$

$x^2 - 8x + 16 = 4x - x^2$

$2x^2 - 12x + 16 = 0$

$2(x^2 - 6x + 8) = 0$

$2(x - 4)(x - 2) = 0$

$x - 4 = 0 \Rightarrow x = 4$

$x - 2 = 0 \Rightarrow x = 2$

6. $\sqrt{x - 1} = \frac{1}{2}(x - 1)$

$2\sqrt{x - 1} = x - 1$

$4(x - 1) = (x - 1)^2$

$4x - 4 = x^2 - 2x + 1$

$0 = x^2 - 6x + 5$

$0 = (x - 5)(x - 1)$

$x - 5 = 0 \Rightarrow x = 5$

$x - 1 = 0 \Rightarrow x = 1$

7. $\int_0^2 2e^{2x}\,dx = e^{2x}\Big]_0^2 = e^4 - 1$

8. $\int_{-1}^3 \frac{2x + 1}{x^2 + x + 2}\,dx = \Big[\ln\left|x^2 + x + 2\right|\Big]_{-1}^3$

$= \ln 14 - \ln 2 = \ln \frac{14}{2} = \ln 7$

9. $\int_0^2 x\sqrt{x^2 + 1}\,dx = \frac{1}{2}\int_0^2 (x^2 + 1)^{1/2}(2x)\,dx$

$= \left[\frac{1}{2} \cdot \frac{2}{3}(x^2 + 1)^{3/2}\right]_0^2$

$= \left[\frac{1}{3}(x^2 + 1)^{3/2}\right]_0^2$

$= \frac{\sqrt{125}}{3} - \frac{1}{3} = \frac{5\sqrt{5}}{3} - \frac{1}{3}$

10. $\int_1^5 \frac{(\ln x)^2}{x}\,dx = \left[\frac{1}{3}(\ln x)^3\right]_1^5 = \frac{(\ln 5)^3}{3}$

1. $V = \pi \int_0^2 \left(\sqrt{4 - x^2}\right)^2 dx$

$= \pi \int_0^2 (4 - x^2)\,dx$

$= \pi\left(4x - \frac{x^3}{3}\right)\Big]_0^2 = \frac{16\pi}{3}$

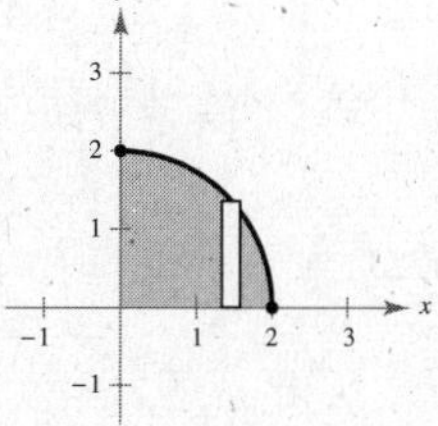

3. $V = \pi \int_1^4 \left(\sqrt{x}\right)^2 dx$

$= \pi \int_1^4 x\,dx$

$= \pi\left(\frac{x^2}{2}\right)\Big]_1^4$

$= 8\pi - \frac{\pi}{2} = \frac{15\pi}{2}$

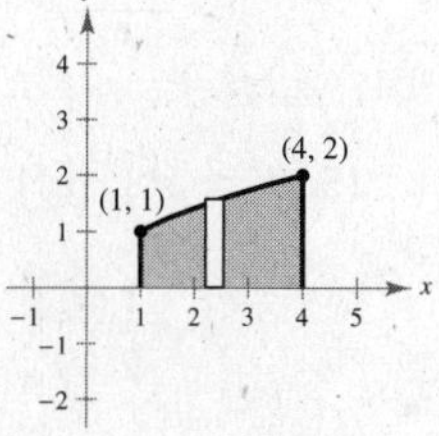

5. $V = \pi\int_{-2}^{2}\left(4 - x^2\right)^2\,dx$

$= \pi\int_{-2}^{2}\left(16 - 8x^2 + x^4\right)dx$

$= \pi\left[16x - \frac{8}{3}x^3 + \frac{x^5}{5}\right]_{-2}^{2}$

$= \frac{512\pi}{15}$

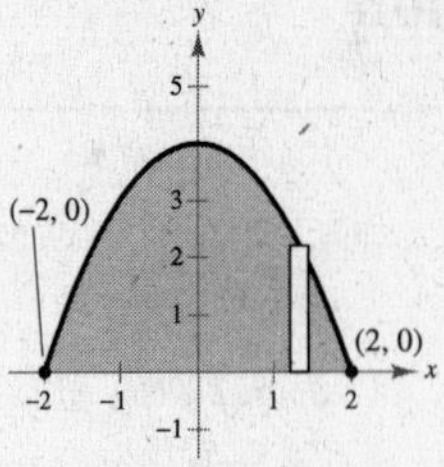

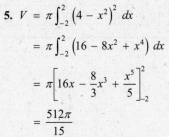

7. $V = \pi\int_{-2}^{2}\left(1 - \frac{x^2}{4}\right)^2 dx$

$= 2\pi\int_{0}^{2}\left(1 - \frac{x^2}{2} + \frac{x^4}{16}\right)dx$

$= 2\pi\left[x - \frac{1}{6}x^3 + \frac{1}{80}x^5\right]_{0}^{2}$

$= \frac{32\pi}{15}$

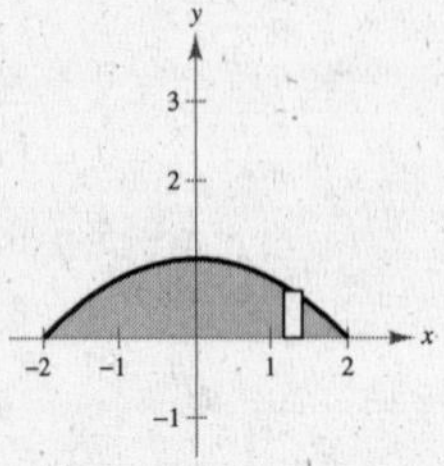

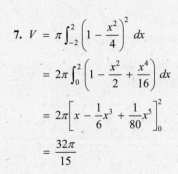

9. $V = \pi\int_{0}^{1}\left(-x + 1\right)^2 dx$

$= \pi\int_{0}^{1}\left(x^2 - 2x + 1\right)dx$

$= \pi\left(\frac{x^3}{3} - x^2 + x\right)\Bigg]_{0}^{1}$

$= \frac{\pi}{3}$

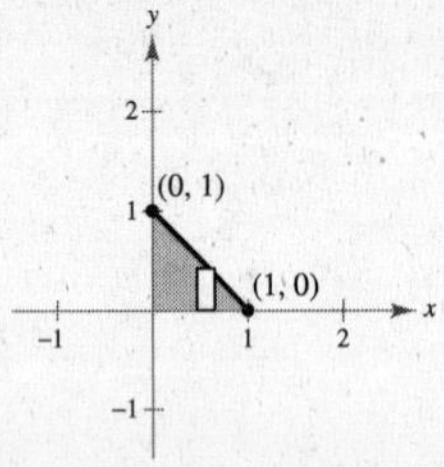

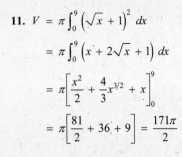

11. $V = \pi\int_{0}^{9}\left(\sqrt{x} + 1\right)^2 dx$

$= \pi\int_{0}^{9}\left(x + 2\sqrt{x} + 1\right)dx$

$= \pi\left[\frac{x^2}{2} + \frac{4}{3}x^{3/2} + x\right]_{0}^{9}$

$= \pi\left[\frac{81}{2} + 36 + 9\right] = \frac{171\pi}{2}$

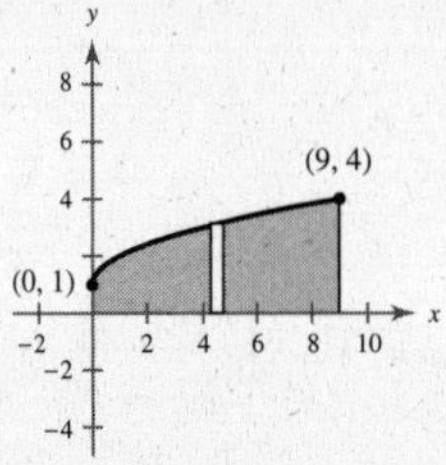

13. $V = \pi\int_{0}^{2}\left(2x^2\right)^2 dx$

$= \pi\int_{0}^{2}4x^4\,dx$

$= \pi\left(\frac{4x^5}{5}\right)\Bigg]_{0}^{2}$

$= \frac{128\pi}{5}$

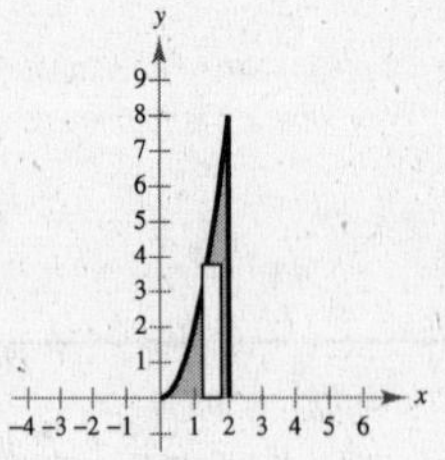

15. $V = \pi\int_{0}^{1}\left(e^x\right)^2 dx$

$= \pi\frac{1}{2}e^{2x}\Big]_{0}^{1}$

$= \frac{\pi}{2}\left(e^2 - 1\right)$

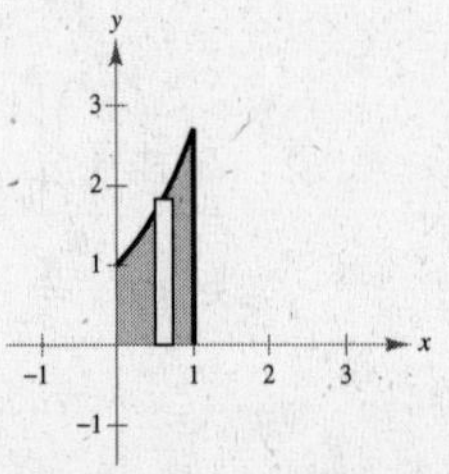

17. $V = \pi\int_{0}^{1}\left(2y\right)^2 dy$

$= \pi\int_{0}^{1}4y^2\,dy$

$= \pi\left[\frac{4y^3}{3}\right]_{0}^{1}$

$= \frac{4\pi}{3}$

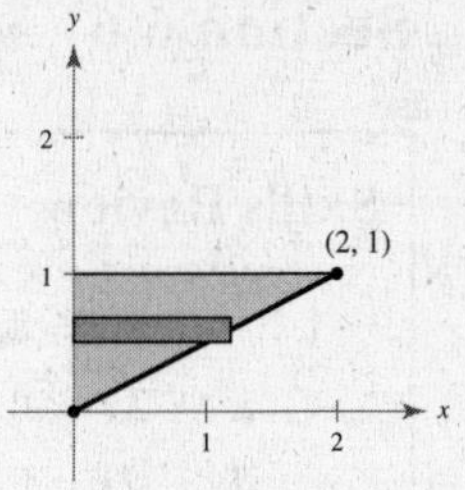

19. $V = \pi\int_{0}^{2}\left(y^3\right)^2 dy$

$= \pi\int_{0}^{2}y^6\,dy$

$= \pi\left[\frac{1}{7}y^7\right]_{0}^{2}$

$= \frac{128\pi}{7}$

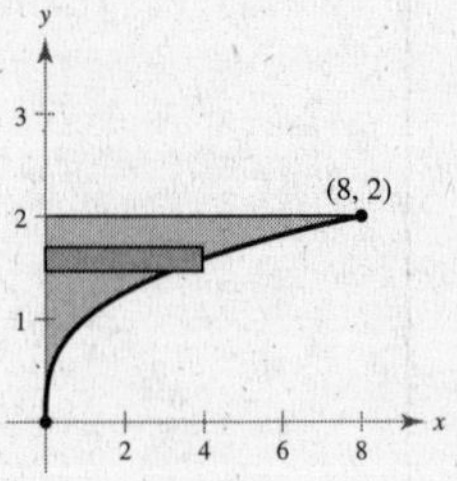

21. $V = \pi\int_{0}^{4}\left(\sqrt{y}\right)^2 dy$

$= \pi\int_{0}^{4}y\,dy$

$= \pi\left(\frac{y^2}{2}\right)\Bigg]_{0}^{4}$

$= 8\pi$

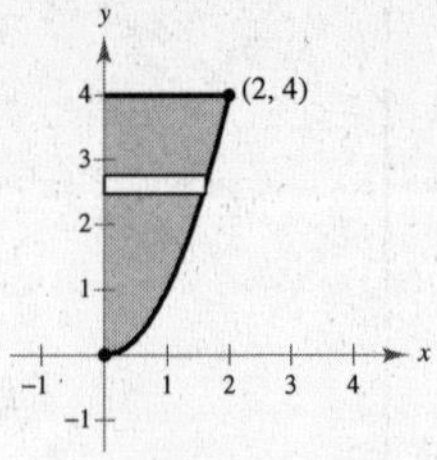

23. $V = \pi\int_{0}^{6}\left(\frac{1}{2}x\right)^2 dx$

$= \frac{\pi}{4}\int_{0}^{6}x^2\,dx$

$= \frac{\pi}{4}\left(\frac{x^3}{3}\right)\Bigg]_{0}^{6}$

$= 18\pi$

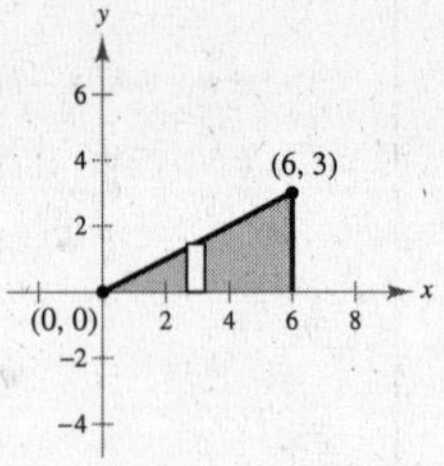

25. A right circular cone can be formed be revolving the region bounded by $y = (r/h)x$, $x = h$, and $y = 0$ about the x-axis.

$V = \pi\int_{0}^{h}\left(\frac{r}{h}x\right)^2 dx$

$= \pi\left(\frac{r^2}{h^2}\cdot\frac{x^3}{3}\right)\Bigg]_{0}^{h}$

$= \pi\frac{r^2}{h^2}\cdot\frac{h^3}{3}$

$= \frac{1}{3}\pi r^2 h$

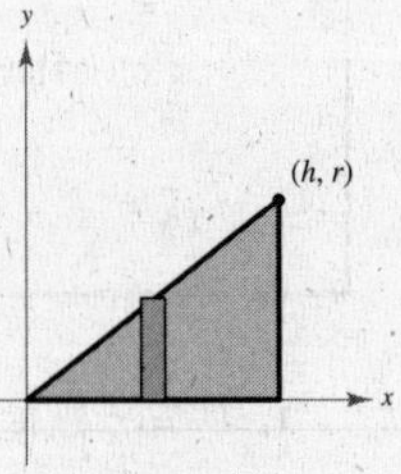

27. The right half of the ellipse is given by $x = 2\sqrt{1 - y^2}$.

$$V = \pi \int_{-1}^{1} \left(2\sqrt{1 - y^2}\right)^2 dy$$
$$= \pi \int_{-1}^{1} 4\left(1 - y^2\right) dy$$
$$= 4\pi \int_{-1}^{1} \left(1 - y^2\right) dy$$
$$= 4\pi\left[y - \frac{y^3}{3}\right]_{-1}^{1}$$
$$= 4\pi\left[\frac{2}{3} + \frac{2}{3}\right] = \frac{16\pi}{3}$$

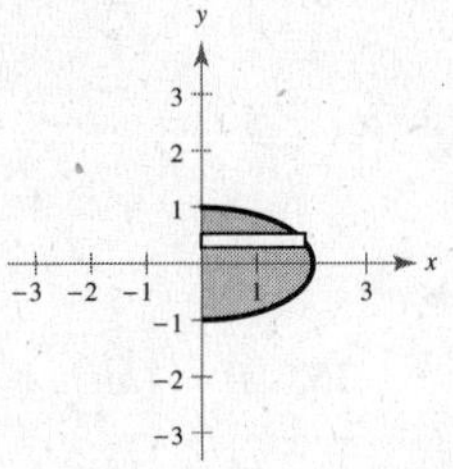

29. The graph has x-intercepts at $(0, 0)$ and $(2, 0)$.

$$V = \pi \int_{0}^{2} \left[\frac{1}{8}x^2\sqrt{2 - x}\right]_0^2 dx$$
$$= \frac{\pi}{64} \int_{0}^{2} \left(2x^4 - x^5\right) dx$$
$$= \frac{\pi}{64}\left[\frac{2x^5}{5} - \frac{x^6}{6}\right]_0^2 = \frac{\pi}{30}$$

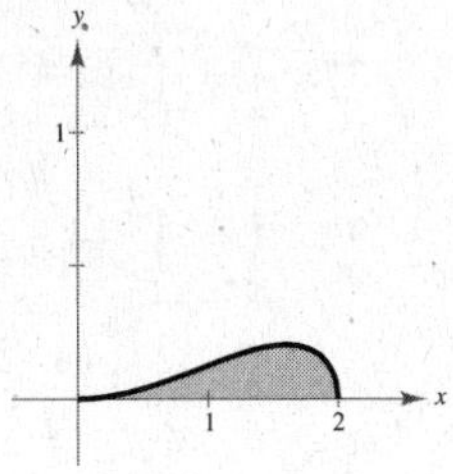

31. (a) $y = 20\left[\left(0.005x\right)^2 - 1\right]$

Solving for x, we obtain

$$x = 200\sqrt{\frac{y}{20} + 1}.$$

$$V = \pi \int_{-20}^{0} \left(200\sqrt{\frac{y}{20} + 1}\right)^2 dy$$
$$= 40{,}000\pi \int_{-20}^{0} \left(\frac{y}{20} + 1\right) dy$$
$$= 40{,}000\pi\left[\frac{y^2}{40} + y\right]_{-20}^{0}$$
$$= 40{,}000\pi(10) \approx 1{,}256{,}637 \text{ cubic feet}$$

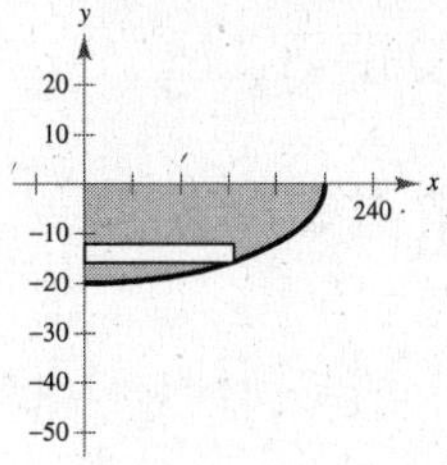

(b) The maximum number of fish that can be supported is

$$\frac{1{,}256{,}637}{500} \approx 2513 \text{ fish.}$$

Review Exercises for Chapter 6

1. $\int 16\, dx = 16x + C$

3. $\int \left(2x^2 + 5x\right) dx = \frac{2}{3}x^3 + \frac{5}{2}x^2 + C$

5. $\int \frac{2}{3\sqrt[3]{x}}\, dx = \int \frac{2}{3}x^{-1/3}\, dx = x^{2/3} + C$

7. $\int \left(3\sqrt{x^4} + 3x\right) dx = \int \left(x^{4/3} + 3x\right) dx$

$$= \frac{3}{7}x^{7/3} + \frac{3}{2}x^2 + C$$

9. $\int \frac{2x^4 - 1}{\sqrt{x}}\, dx = \int \left(2x^{7/2} - x^{-1/2}\right) dx$

$$= \frac{4}{9}x^{9/2} - 2\sqrt{x} + C$$

11. $f(x) = \int (3x + 1)\, dx = \frac{3}{2}x^2 + x + C$

When $f(2) = 6$, $\frac{3}{2}(2)^2 + 2 + C = 6$

$$8 + C = 6$$
$$C = -2$$

So, $f(x) = \frac{3}{2}x^2 + x - 2$.

13. $f'(x) = \int 2x^2\, dx = \frac{2}{3}x^3 + C_1$

When $f'(3) = 10$, $\frac{2}{3}(3)^3 + C_1 = 10$

$$C_1 = -8$$

So, $f'(x) = \frac{2}{3}x^3 - 8$.

$$f(x) = \int \left(\frac{2}{3}x^3 - 8\right) dx = \frac{1}{6}x^4 - 8x + C_2$$

When $f(3) = 6$, $\frac{1}{6}(3)^4 - 8(3) + C_2 = 6$

$$-\frac{21}{2} + C_2 = 6$$
$$C_2 = \frac{33}{2}$$

So, $f(x) = \frac{1}{6}x^4 - 8x + \frac{33}{2}$.

15. (a) $s(t) = -16t^2 + 80t$

$$s'(t) = v(t) = -32t + 80 = 0$$
$$32t = 80$$
$$t = 2.5 \text{ seconds}$$

(b) $s(2.5) = 100$ feet

(c) Net change $= \int_0^1 s'(t)\, dt$

$$= \int_0^1 (-32t + 80)\, dt$$
$$= -16t^2 + 80t\Big]_0^1 = 64 \text{ feet}$$

17. $\int (1+5x)^2\,dx = \int (1+10x+25x^2)\,dx$

$= x + 5x^2 + \frac{25}{3}x^3 + C$

or

$\int (1+5x)^2\,dx = \frac{1}{5}\frac{(1+5x)^3}{3} + C_1$

$= \frac{1}{15}(1+5x)^3 + C_1$

19. $\int \frac{1}{\sqrt{5x-1}}\,dx = \frac{1}{5}\int (5x-1)^{-1/2}(5)\,dx$

$= \frac{1}{5}(2)(5x-1)^{1/2} + C$

$= \frac{2}{5}\sqrt{5x-1} + C$

21. $\int x(1-4x^2)\,dx = \int (x-4x^3)\,dx = \frac{1}{2}x^2 - x^4 + C$

23. $\int (x^4-2x)(2x^3-1)\,dx = \frac{1}{2}\int (x^4-2x)(4x^3-2)\,dx$

$= \frac{1}{2}\left[\frac{(x^4-2x)^2}{2}\right] + C$

$= \frac{1}{4}(x^4-2x)^2 + C$

25. $A(t) = \int 2t(0.001t^2+0.5)^{1/4}\,dt$

$= \frac{1}{0.001}\int (0.001t^2+0.5)^{1/4}(0.002t)\,dt$

$= \frac{1}{0.001}\cdot\frac{4}{5}(0.001t^2+0.5)^{5/4} + C$

$= 800(0.001t^2+0.5)^{5/4} + C$

Because $A(0) = 0$, it follows that $C \approx -336.36$, and you have $A(t) = 800(0.001t^2+0.5)^{5/4} - 336.36$.

(a) $A(3) \approx 7.58$ centimeters

(b) $A(5) \approx 21.15$ centimeters

27. $\int 3e^{-3x}\,dx = -e^{-3x} + C$

29. $\int (x-1)e^{x^2-2x}\,dx = \frac{1}{2}e^{x^2-2x} + C$

31. $\int \frac{x^2}{1-x^3}\,dx = -\frac{1}{3}\int \frac{1}{1-x^3}(-3x^2)\,dx$

$= -\frac{1}{3}\ln\left|1-x^3\right| + C$

33. $\int \frac{(\sqrt{x}+1)^2}{\sqrt{x}}\,dx = \int \frac{x+2\sqrt{x}+1}{\sqrt{x}}\,dx$

$= \int (x^{1/2} + 2 + x^{-1/2})\,dx$

$= \frac{2}{3}x^{3/2} + 2x + 2x^{1/2} + C$

35. $A = \frac{1}{2}bh$

$= \frac{1}{2}(5)(5)$

$= \frac{25}{2}$

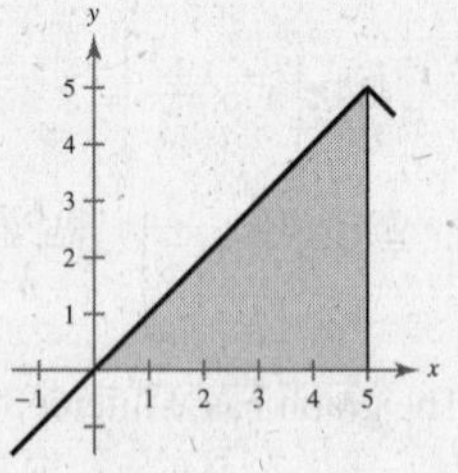

37. $A = \int_0^2 (4-2x)\,dx = \left[4x - x^2\right]_0^2 = 4$

39. $A = \int_{-2}^{2} (4-x^2)\,dx$

$= 2\int_0^2 (4-x^2)\,dx = 2\left[4x - \frac{1}{3}x^3\right]_0^2 = \frac{32}{3}$

41. $A = \int_0^2 (y-2)^2\,dy = \left[\frac{1}{3}y^3 - 2y^2 + 4y - \frac{8}{3}\right]_0^2 = \frac{8}{3}$

43. $A = \int_0^1 \frac{2}{x+1}\,dx = 2\ln(x+1)\Big]_0^1 = 2\ln 2 \approx 1.39$

45. (a) $\int_2^6 [f(x)+g(x)]\,dx = \int_2^6 f(x)\,dx + \int_2^6 g(x)\,dx$

$= 10 + 3 = 13$

(b) $\int_2^6 [f(x)-g(x)]\,dx = \int_2^6 f(x)\,dx - \int_2^6 g(x)\,dx$

$= 10 - 3 = 7$

(c) $\int_2^6 [2f(x)-3g(x)]\,dx = 2\int_2^6 f(x)\,dx - 3\int_2^6 g(x)\,dx$

$= 2(10) - 3(3) = 11$

(d) $\int_2^6 5f(x)\,dx = 5\int_2^6 f(x)\,dx = 5(10) = 50$

47. $\int_0^4 (2+x)\,dx = \left[2x + \frac{1}{2}x^2\right]_0^4 = 16$

49. $\int_4^9 x\sqrt{x}\,dx = \int_4^9 x^{3/2}\,dx = \frac{2}{5}x^{5/2}\Big]_4^9 = \frac{422}{5} = 84.4$

51. $\int_{-1}^{1} (4t^3 - 2t)\,dt = \left[t^4 - t^2\right]_{-1}^{1} = 0$

53. $\int_0^3 \frac{1}{\sqrt{1+x}}\,dx = \int_0^3 (1+x)^{-1/2}\,dx = 2\sqrt{1+x}\Big]_0^3 = 2$

55. $\int_1^2 \left(\frac{1}{x^2} - \frac{1}{x^3}\right)dx = \int_1^2 (x^{-2} - x^{-3})\,dx$

$= \left[-x^{-1} + \frac{1}{2}x^{-2}\right]_1^2 = \frac{1}{8}$

57. $\int_1^3 \frac{3 + \ln x}{x}\,dx = \frac{1}{2}(3 + \ln x)^2\Big]_1^3$

$= \frac{1}{2}\left[6 \ln 3 + (\ln 3)^2\right] \approx 3.899$

59. $\int_{-1}^1 3xe^{x^2-1}\,dx = \frac{3}{2}\int_{-1}^1 e^{x^2-1}(2x)\,dx$

$= \frac{3}{2}e^{x^2-1}\Big]_{-1}^1 = \frac{3}{2}(1 - 1) = 0$

61. $\int_1^3 (2x - 1)\,dx = \left[x^2 - x\right]_1^3 = 6$

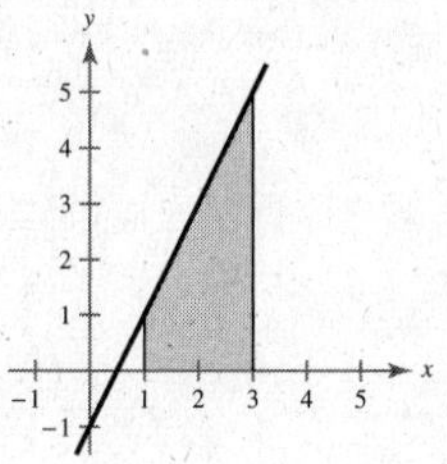

63. $\int_3^4 (x^2 - 9)\,dx = \left[\frac{1}{3}x^3 - 9x\right]_3^4 = \frac{10}{3}$

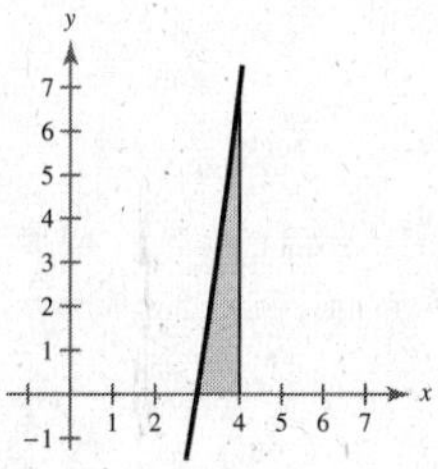

65. Average Value: $\frac{1}{9 - 4}\int_4^9 \frac{1}{\sqrt{x}}\,dx = \frac{1}{5}\left[2\sqrt{x}\right]_4^9 = \frac{2}{5}$

To find the value of x for which $f(x) = \frac{2}{5}$, you solve

$\frac{1}{\sqrt{x}} = \frac{2}{5}$ for x.

$\frac{1}{\sqrt{x}} = \frac{2}{5}$

$2\sqrt{x} = 5$

$x = \frac{25}{4}$

67. Average value $= \frac{1}{5 - 2}\int_2^5 e^{5-x}\,dx$

$= \frac{1}{3}\left[-e^{5-x}\right]_2^5 = \frac{1}{3}\left(-1 + e^3\right) \approx 6.362$

To find the value of x for which $f(x) = \frac{1}{3}\left(-1 + e^3\right)$,

you solve $e^{5-x} = \frac{1}{3}\left(-1 + e^3\right)$ for x.

$e^{5-x} = \frac{1}{3}\left(-1 + e^3\right)$

$5 - x = \ln\left[\frac{1}{3}\left(-1 + e^3\right)\right]$

$x = 5 - \ln\left[\frac{1}{3}\left(-1 + e^3\right)\right]$

$x \approx 3.15$

69. $p = \frac{15{,}000}{33}\int_1^6 \left(0.0782t^2 - 0.352t + 1.75\right)dt$

$= \frac{15{,}000}{33}\left[0.02607t^3 - 0.176t^2 + 1.75t\right]_1^6 \approx \$3724.74$

71. (a) $B = \int(-0.0391t + 0.6108)\,dt = -0.01955t^2 + 0.6108t + C$

$2.95 = -0.01955(16)^2 + 0.6108(16) + C$

$C = -1.818$

$B = -0.01955t^2 + 0.6108t - 1.818$

(b) $3.25 = -0.01955t^2 + 0.6108t - 1.818$

According to the model, the price of beef per pound will never surpass \$3.25. The highest price is approximately \$2.95 per pound in 2005, and after that the prices decrease.

73. $\int_{-2}^2 6x^5\,dx = 0$ (odd function)

75. $\int_{-2}^{-1} \frac{4}{x^2}\,dx = \int_1^2 \frac{4}{x^2}\,dx = 2$ (symmetric about y-axis)

77.

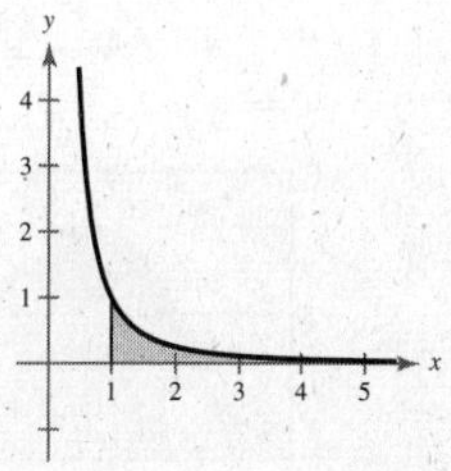

$\int_1^5 \frac{1}{x^2}\,dx = -x^{-1}\Big]_1^5 = \frac{4}{5}$

79.

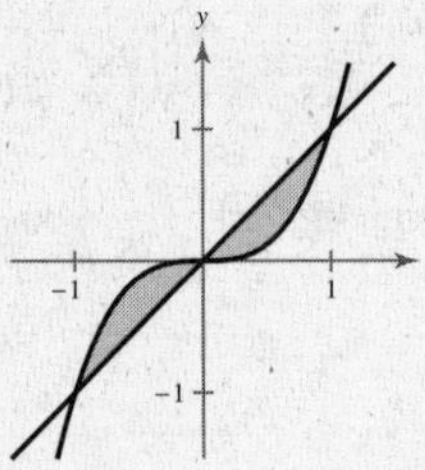

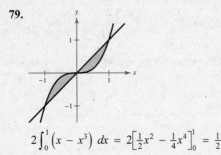

$$2\int_0^1 \left(x - x^3\right) dx = 2\left[\frac{1}{2}x^2 - \frac{1}{4}x^4\right]_0^1 = \frac{1}{2}$$

81.

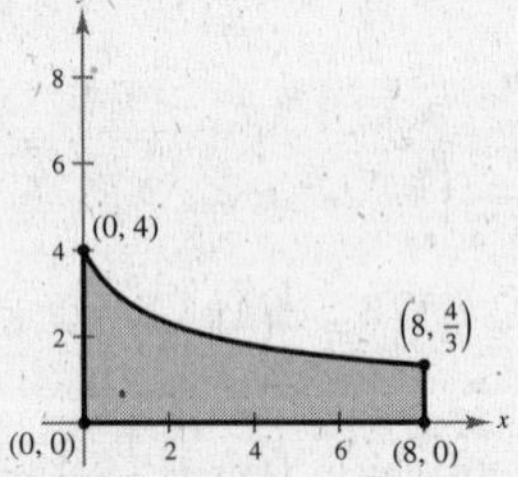

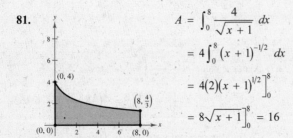

$$\begin{aligned} A &= \int_0^8 \frac{4}{\sqrt{x+1}}\,dx \\ &= 4\int_0^8 (x+1)^{-1/2}\,dx \\ &= 4(2)(x+1)^{1/2}\Big]_0^8 \\ &= 8\sqrt{x+1}\Big]_0^8 = 16 \end{aligned}$$

83.

$$\begin{aligned} (x-3)^2 &= 8 - (x-3)^2 \\ 2(x-3)^2 &= 8 \\ (x-3)^2 &= 4 \\ x - 3 &= \pm 2 \\ x &= 1, 5 \end{aligned}$$

$$\begin{aligned} A &= \int_1^5 \left\{\left[8 - (x-3)^2\right] - (x-3)^2\right\} dx \\ &= \int_1^5 \left[8 - 2(x-3)^2\right] dx \\ &= \left[8x - \frac{2}{3}(x-3)^3\right]_1^5 = \frac{64}{3} \end{aligned}$$

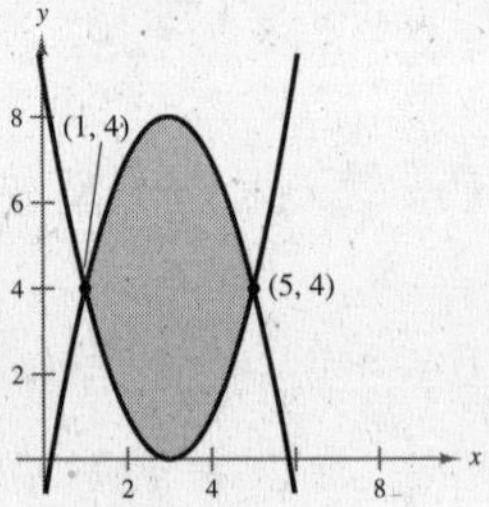

85.

$$\begin{aligned} x &= 2 - x^2 \\ x^2 + x - 2 &= 0 \\ (x+2)(x-1) &= 0 \\ x &= -2, 1 \end{aligned}$$

$$\begin{aligned} A &= \int_{-2}^1 \left[\left(2 - x^2\right) - x\right] dx \\ &= \left[2x - \frac{1}{3}x^3 - \frac{1}{2}x^2\right]_{-2}^1 \\ &= \frac{9}{2} \end{aligned}$$

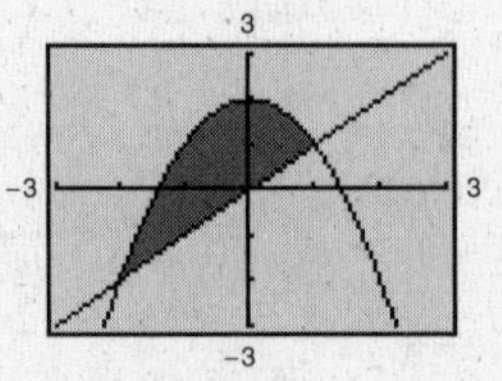

87.

$$\begin{aligned} \int_0^5 (P_1 - P_2)\,dt &= \int_0^5 \left(282.52e^{0.0098t} - 275.69e^{0.0117t}\right) dt \\ &= \left[\frac{282.52}{0.0098}e^{0.0098t} - \frac{275.69}{0.0117}e^{0.0117t}\right]_0^5 \\ &\approx 5293.537 - 5265.324 \\ &= 28.213 \end{aligned}$$

The projection underestimated the population by 28.213 million people.

89.

$$\begin{aligned} V &= \pi\int_1^4 \left(\frac{1}{\sqrt{x}}\right)^2 dx \\ &= \pi\int_1^4 \frac{1}{x}\,dx \\ &= \pi \ln x\Big]_1^4 \\ &= \pi \ln 4 \\ &\approx 4.355 \end{aligned}$$

91.

$$\begin{aligned} V &= \pi\int_0^2 \left(e^{1-x}\right)^2 dx \\ &= \pi\int_0^2 e^{2-2x}\,dx \\ &= -\frac{\pi}{2}\left[e^{2-2x}\right]_0^2 \\ &= -\frac{\pi}{2}\left[e^{-2} - e^2\right] \\ &= \frac{\pi}{2}\left(e^2 - e^{-2}\right) \\ &\approx 11.394 \end{aligned}$$

93.

$$\begin{aligned} V &= \pi\int_0^2 \left[(2x+1)^2 - 1^2\right] dx \\ &= \pi\int_0^2 \left(4x^2 + 4x\right) dx \\ &= \pi\left[\frac{4}{3}x^3 + 2x^2\right]_0^2 \\ &= \pi\left(\frac{32}{3} + 8\right) = \frac{56}{3}\pi \end{aligned}$$

95.

$$\begin{aligned} V &= \pi\int_0^1 \left[\left(x^2\right)^2 - \left(x^3\right)^2\right] dx \\ &= \pi\left[\frac{x^5}{5} - \frac{x^7}{7}\right]_0^1 = \frac{2}{35}\pi \end{aligned}$$

97.

$$\begin{aligned} V &= \pi\int_{-3}^3 \left(\frac{2}{3}\sqrt{9 - x^2}\right)^2 dx \\ &= \frac{4\pi}{9}\int_{-3}^3 \left(9 - x^2\right) dx \\ &= \frac{4\pi}{9}\left[9x - \frac{x^3}{3}\right]_{-3}^3 \\ &= \frac{4\pi}{9}(18 + 18) \\ &= 16\pi \text{ cubic centimeters} \end{aligned}$$

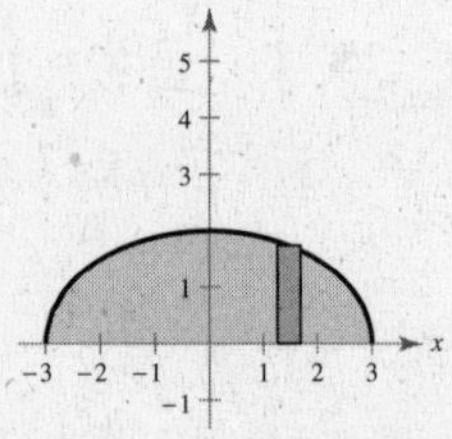

Chapter Test Solutions

1. $\int (9x^2 - 4x + 13)\,dx = 3x^3 - 2x^2 + 13x + C$

2. $\int (x+1)^2\,dx = \frac{1}{3}(x+1)^3 + C$

3. $\int 4x^3\sqrt{x^4 - 7}\,dx = \frac{2}{3}(x^4 - 7)^{3/2} + C$

4. $\int \frac{5x - 6}{\sqrt{x}}\,dx = \int \frac{5x}{\sqrt{x}}\,dx - \int \frac{6}{\sqrt{x}}\,dx$

$= \frac{10}{3}x^{3/2} - 12x^{1/2} + C$

5. $\int 15e^{3x}\,dx = 5e^{3x} + C$

6. $\int \frac{3x^2 - 11}{x^3 - 11x}\,dx = \ln|x^3 - 11x| + C$

7. $\int (e^x + 1)\,dx = e^x + x + C$

$1 = e^0 + 0 + C$

$C = 0$

$f(x) = e^2 + x$

8. $\int \frac{1}{x}\,dx = \ln|x| + C$

$2 = \ln|-1| + C$

$C = 2$

$f(x) = \ln|x| + 2$

9. $\int_0^1 16x\,dx = 8x^2\Big]_0^1 = 8$

10. $\int_{-3}^3 (3 - 2x)\,dx = 3x - x^2\Big]_{-3}^3 = 18$

11. $\int_{-1}^1 (x^3 + x^2)\,dx = \frac{1}{4}x^4 + \frac{1}{3}x^3\Big]_{-1}^1 = \frac{2}{3}$

12. $\int_{-1}^2 \frac{2x}{\sqrt{x^2 + 1}}\,dx = 2(x^2 + 1)^{1/2}\Big]_{-1}^2 = 2\sqrt{5} - 2\sqrt{2} \approx 1.64$

13. $\int_0^3 e^{4e}\,dx = \frac{1}{4}e^{4x}\Big]_0^3 = \frac{e^{12}}{4} - \frac{1}{4} \approx 40{,}688.45$

14. $\int_{-2}^3 \frac{1}{x+3}\,dx = \ln(x+3)\Big]_{-2}^3 = \ln(6) \approx 1.79$

15. (a) Net change $= \int_0^5 (1.714t - 5.06)\,dt$

$= 0.857t^2 - 5.06t\Big]_0^5 = -3.875$

The number of travelers decreased by 3.875 million from 2000 to 2005.

(b) $N(t) = \int (1.714t - 5.06)\,dt$

$= 0.857t^2 - 5.06t + C$

Because $N(0) = 26$, it follows that $C = 26$, and you have $N(t) = 0.857t^2 - 5.06t + 26$.

(c) $N(5) = 22.125$ million travelers

(d) Average $= \frac{1}{5-0}\int_0^5 (0.857t^2 - 5.06t + 26)\,dt$

$= \frac{1}{5}\left[\frac{0.857}{3}t^3 - 2.53t^2 + 26t\right]_0^5$

≈ 20.5 million travelers

16.

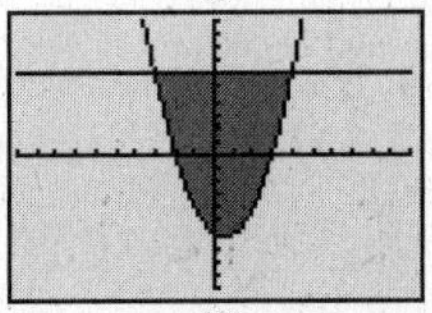

$\int_{-3}^4 6 - (x^2 - x - 6)\,dx = \int_{-3}^4 (6 - x^2 + x + 6)\,dx$

$= -\frac{1}{3}x^3 + \frac{1}{2}x^2 + 12x\Big]_{-3}^4$

$= \frac{104}{3} - (-22.5)$

$= \frac{343}{6} \approx 57.167$

17.

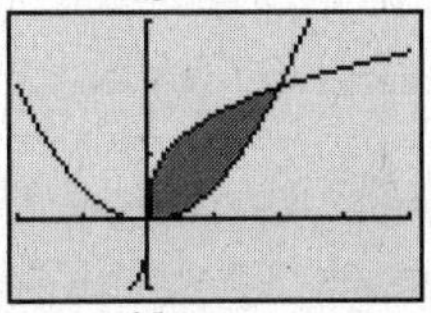

$\int_0^1 (\sqrt[3]{x} - x^2)\,dx = \frac{3}{4}x^{4/3} - \frac{1}{3}x^3\Big]_0^1 = \frac{5}{12}$

18. Volume $= \pi\int_0^3 (3 - x)^2\,dx$

$= \pi\int_0^3 (9 - 6x + x^2)\,dx$

$= \pi\left[9x - 3x^2 + \frac{1}{3}x^3\right]_0^3$

$= 9\pi$

≈ 28.27 cubic inches

Practice Test for Chapter 6

1. Find $\int (3x^2 - 8x + 5)\, dx$.

2. Find $\int (x + 7)(x^2 - 4)\, dx$.

3. Find $\int \frac{x^3 - 9x^2 + 1}{x^2}\, dx$.

4. Find $\int x^3 \sqrt[4]{1 - x^4}\, dx$.

5. Find $\int \frac{3}{\sqrt[3]{7x}}\, dx$.

6. Find $\int \sqrt{6 - 11x}\, dx$.

7. Find $\int \left(\sqrt[4]{x} + \sqrt[6]{x}\right) dx$.

8. Find $\int \left(\frac{1}{x^4} - \frac{1}{x^5}\right) dx$.

9. Find $\int (1 - x^2)^3\, dx$.

10. Find $\int \frac{5x}{(1 + 3x^2)^3}\, dx$.

11. Find $\int e^{7x}\, dx$.

12. Find $\int x e^{4x^2}\, dx$.

13. Find $\int e^x (1 + 4e^x)^3\, dx$.

14. Find $\int (e^x + 2)^2\, dx$.

15. Find $\int \frac{e^{3x} - 4e^x + 1}{e^x}\, dx$.

16. Find $\int \frac{1}{x + 6}\, dx$.

17. Find $\int \frac{x^2}{8 - x^3}\, dx$.

18. Find $\int \frac{e^x}{1 + 3e^x}\, dx$.

19. Find $\int \frac{(\ln x)^6}{x}\, dx$.

20. Find $\int \frac{x^2 + 5}{x - 1}\, dx$.

21. Evaluate $\int_0^3 (x^2 - 4x + 2)\, dx$.

22. Evaluate $\int_1^8 x\sqrt[3]{x}\, dx$.

23. Evaluate $\int_{\sqrt{5}}^{\sqrt{13}} \frac{x}{\sqrt{x^2 - 4}}\, dx$.

24. Sketch the region bounded by the graphs of $f(x) = x^2 - 6x$ and $g(x) = 0$ and find the area of the region.

25. Sketch the region bounded by the graphs of $f(x) = x^3 + 1$ and $g(x) = x + 1$ and find the area of the region.

26. Sketch the region bounded by the graphs of $f(y) = 1/y^2$, $x = 0$, $y = 1$, and $y = 3$ and find the area of the region.

27. Find the volume of the solid generated by revolving the region bounded by the graphs of $f(x) = 1/\sqrt[3]{x}$, $x = 1$, $x = 8$, and $y = 0$ about the x-axis.

28. Find the volume of the solid generated by revolving the region bounded by the graphs of $y = \sqrt{25 - x}$, $y = 0$, and $x = 0$ about the y-axis.

Graphing Calculator Required

29. Use a graphing calculator to sketch the region bounded by $f(x) = 3 - \sqrt{x}$ and $g(x) = 3 - \frac{1}{3}x$.
Based on the graph alone (do no calculations), determine which value best approximates the bounded area.
(a) 13 (b) 3 (c) 5 (d) 6

CHAPTER 7
Techniques of Integration

CHAPTER 7
Techniques of Integration

Section 7.1 Integration by Parts

Skills Review

1. $f(x) = \ln(x + 1)$

   $f'(x) = \dfrac{1}{x + 1}$

2. $f(x) = \ln(x^2 - 1)$

   $f'(x) = \dfrac{2x}{x^2 - 1}$

3. $f(x) = e^{x^3}$

   $f'(x) = 3x^2 e^{x^3}$

4. $f(x) = e^{-x^2}$

   $f'(x) = -2xe^{-x^2}$

5. $f(x) = x^2 e^x$

   $f'(x) = x^2 e^x + 2xe^x$

6. $f(x) = xe^{-2x}$

   $f'(x) = -2xe^{-2x} + e^{-2x}$

7. $A = \int_a^b [f(x) - g(x)]\,dx$

   $= \int_{-2}^{2} [(-x^2 + 4) - (x^2 - 4)]\,dx$

   $= \int_{-2}^{2} (-2x^2 + 8)\,dx$

   $= \left[-\frac{2}{3}x^3 + 8x\right]_{-2}^{2}$

   $= \left(-\frac{16}{3} + 16\right) - \left(\frac{16}{3} - 16\right) = \frac{64}{3}$

8. $A = \int_a^b [f(x) - g(x)]\,dx$

   $= \int_{-1}^{1} [(-x^2 + 2) - 1]\,dx$

   $= \int_{-1}^{1} (-x^2 + 1)\,dx$

   $= \left[-\frac{1}{3}x^3 + x\right]_{-1}^{1}$

   $= \left(-\frac{1}{3} + 1\right) - \left(\frac{1}{3} - 1\right)$

   $= \frac{4}{3}$

9. $A = \int_a^b [f(x) - g(x)]\,dx$

   $= \int_{-1}^{5} [4x - (x^2 - 5)]\,dx$

   $= \int_{-1}^{5} (-x^2 + 4x + 5)\,dx$

   $= \left[-\frac{1}{3}x^3 + 2x^2 + 5x\right]_{-1}^{5}$

   $= \left(-\frac{125}{3} + 50 + 25\right) - \left(\frac{1}{3} + 2 - 5\right)$

   $= 36$

10. $A = \int_a^b [f(x) - g(x)]\,dx$

    $= \int_{-1}^{1} [f(x) - g(x)]\,dx + \int_1^3 [g(x) - f(x)]\,dx$

    $= \int_{-1}^{1} [(x^3 - 3x^2 + 2) - (x - 1)]\,dx + \int_1^3 [(x - 1) - (x^3 - 3x^2 + 2)]\,dx$

    $= \int_{-1}^{1} (x^3 - 3x^2 - x + 3)\,dx + \int_1^3 (-x^3 + 3x^2 + x - 3)\,dx$

    $= \left[\frac{1}{4}x^4 - x^3 - \frac{1}{2}x^2 + 3x\right]_{-1}^{1} + \left[-\frac{1}{4}x^4 + x^3 + \frac{1}{2}x^2 - 3x\right]_1^3$

    $= \left(\frac{1}{4} - 1 - \frac{1}{2} + 3\right) - \left(\frac{1}{4} + 1 - \frac{1}{2} - 3\right) + \left(-\frac{81}{4} + 27 + \frac{9}{2} - 9\right) - \left(-\frac{1}{4} + 1 + \frac{1}{2} - 3\right) = 8$

1. $\int xe^{3x}\,dx$

$u = x,\ dv = e^{3x}\,dx$

3. $\int x \ln 2x\,dx$

$u = \ln 2x,\ dv = x\,dx$

5. Let $u = x$ and $dv = e^{3x}\,dx$. Then $du = dx$ and $v = \frac{1}{3}e^{3x}$.

$$\begin{aligned}\int xe^{3x}\,dx &= \tfrac{1}{3}xe^{3x} - \int \tfrac{1}{3}e^{3x}\,dx \\ &= \tfrac{1}{3}xe^{3x} - \tfrac{1}{9}e^{3x} + C \\ &= \tfrac{1}{9}e^{3x}(3x - 1) + C\end{aligned}$$

7. Let $u = x^2$ and $dv = e^{-x}\,dx$. Then $du = 2x\,dx$ and $v = -e^{-x}$.

$$\int x^2e^{-x}\,dx = -x^2e^{-x} + 2\int xe^{-x}\,dx$$

Let $u = x$ and $dv = e^{-x}\,dx$. Then $du = dx$ and $v = -e^{-x}$.

$$\begin{aligned}\int x^2e^{-x}\,dx &= -x^2e^{-x} + 2\left[-xe^{-x} + \int e^{-x}\,dx\right] \\ &= -x^2e^{-x} - 2xe^{-x} - 2e^{-x} + C \\ &= -e^{-x}\left(x^2 + 2x + 2\right) + C\end{aligned}$$

9. Let $u = \ln 2x$ and $dv = dx$. Then $du = \frac{1}{x}\,dx$ and $v = x$.

$$\begin{aligned}\int \ln 2x\,dx &= x\ln 2x - \int x\left(\frac{1}{x}\right)dx \\ &= x\ln 2x - \int dx \\ &= x\ln 2x - x + C \\ &= x(\ln 2x - 1) + C\end{aligned}$$

11. $\int e^{4x}\,dx = \frac{1}{4}\int e^{4x}(4)\,dx = \frac{1}{4}e^{4x} + C$

13. Let $u = x$ and $dv = e^{4x}\,dx$. Then $du = dx$ and $v = \frac{1}{4}e^{4x}$.

$$\begin{aligned}\int xe^{4x}\,dx &= \frac{1}{4}xe^{4x} - \frac{1}{4}\int e^{4x}\,dx \\ &= \frac{1}{4}xe^{4x} - \frac{1}{16}e^{4x} + C \\ &= \frac{e^{4x}}{16}(4x - 1) + C\end{aligned}$$

15. $\int xe^{x^2}\,dx = \frac{1}{2}\int e^{x^2}(2x)\,dx = \frac{1}{2}e^{x^2} + C$

17. $\int \frac{x}{e^x}\,dx = \int xe^{-x}\,dx$

Let $u = x$ and $dv = e^{-x}\,dx$. Then $du = dx$ and $v = -e^{-x}$.

$$\begin{aligned}\int \frac{x}{e^x}\,dx &= -xe^{-x} - \int -e^{-x}\,dx \\ &= -xe^{-x} - e^{-x} + C \\ &= -e^{-x}(x + 1) + C \\ &= -\frac{x + 1}{e^x} + C\end{aligned}$$

19. Let $u = 2x^2$ and $dv = e^x\,dx$. Then $du = 4x\,dx$ and $v = e^x$.

$$\int 2x^2e^x\,dx = 2x^2e^x - \int 4xe^x\,dx$$

Let $u = 4x$ and $dv = e^x\,dx$. Then $du = 4\,dx$ and $v = e^x$.

$$\begin{aligned}\int 2x^2e^x\,dx &= 2x^2e^x - \left[4xe^x - \int e^x(4\,dx)\right] \\ &= 2x^2e^x - 4xe^x + 4e^x + C \\ &= 2e^x\left(x^2 - 2x + 2\right) + C\end{aligned}$$

21. Let $u = \ln(t + 1)$ and $dv = t\,dt$. Then $du = \frac{1}{(t + 1)}\,dt$ and $v = \frac{t^2}{2}$.

$$\begin{aligned}\int t\ln(t + 1)\,dt &= \frac{t^2}{2}\ln(t + 1) - \frac{1}{2}\int \frac{t^2}{t + 1}\,dt \\ &= \frac{t^2}{2}\ln(t + 1) - \frac{1}{2}\int\left(t - 1 + \frac{1}{t + 1}\right)dt \\ &= \frac{t^2}{2}\ln(t + 1) - \frac{1}{2}\left[\frac{t^2}{2} - t + \ln|t + 1|\right] + C \\ &= \frac{1}{4}\left[2t^2\ln(t + 1) + t(2 - t) - 2\ln|t + 1|\right] + C\end{aligned}$$

23. Let $u = x - 1$ and $dv = e^x$. Then $du = dx$ and $v = e^x$.

$$\int (x - 1)e^x\,dx = (x - 1)e^x - \int e^x\,dx$$
$$= (x - 1)e^x - e^x + C$$
$$= e^x(x - 2) + C$$

25. Let $u = \frac{1}{t}$. Then $du = \left(-\frac{1}{t^2}\right)dt$.

$$\int \frac{e^{1/t}}{t^2}\,dt = -\int e^u\,du$$
$$= -e^u + C$$
$$= -e^{1/t} + C$$

27. Let $u = (\ln x)^2$ and $dv = x\,dx$. Then $du = \frac{(2\ln x)}{x}\,dx$ and $v = \frac{x^2}{2}$.

$$\int x(\ln x)^2\,dx = \frac{x^2}{2}(\ln x)^2 - \int x\ln x\,dx$$

Let $u = \ln x$ and $v = x\,dx$. Then $du = \frac{1}{x}\,dx$ and $v = \frac{x^2}{2}$.

$$\int x(\ln x)^2\,dx = \frac{x^2}{2}(\ln x)^2 - \left[\frac{x^2}{2}\ln x - \int \frac{x}{2}\,dx\right]$$
$$= \frac{x^2}{2}(\ln x)^2 - \frac{x^2}{2}\ln x + \frac{x^2}{4} + C$$
$$= \frac{x^2}{4}\left[2(\ln x)^2 - 2\ln x + 1\right] + C$$

29. $$\int \frac{(\ln x)^2}{x}\,dx = \int (\ln x)^2\left(\frac{1}{x}\right)dx = \frac{(\ln x)^3}{3} + C$$

31. Let $u = \ln x$ and $dv = \frac{1}{x^2}\,dx$. Then $du = \frac{1}{x}\,dx$ and $v = -\frac{1}{x}$.

$$\int \frac{\ln x}{x^2}\,dx = -\frac{\ln x}{x} - \int -\frac{1}{x}\cdot\frac{1}{x}\,dx$$
$$= -\frac{\ln x}{x} + \int \frac{1}{x^2}\,dx$$
$$= -\frac{\ln x}{x} - \frac{1}{x} + C$$
$$= -\frac{\ln x + 1}{x} + C$$

33. Let $u = x$ and $dv = \sqrt{x - 1}\,dx$. Then $du = dx$ and $v = \frac{2}{3}(x - 1)^{3/2}$.

$$\int x\sqrt{x - 1}\,dx = \frac{2}{3}x(x - 1)^{3/2} - \int \frac{2}{3}(x - 1)^{3/2}\,dx$$
$$= \frac{2}{3}x(x - 1)^{3/2} - \frac{4}{15}(x - 1)^{5/2} + C$$
$$= \frac{2}{15}(x - 1)^{3/2}(3x + 2) + C$$

35. $$\int x(x + 1)^2\,dx = \int x(x^2 + 2x + 1)\,dx$$
$$= \int (x^3 + 2x^2 + x)\,dx$$
$$= \frac{x^4}{4} + \frac{2x^3}{3} + \frac{x^2}{2} + C$$

37. Let $u = x^2$ and $dv = e^x\,dx$. Then $du = 2x\,dx$ and $v = e^x$.

$$\int_1^2 x^2e^x\,dx = x^2e^x\Big]_1^2 - 2\int_1^2 xe^x\,dx$$

Let $u = x$ and $dv = e^x\,dx$. Then $du = dx$ and $v = e^x$.

$$\int_1^2 x^2e^x\,dx = 4e^2 - e - 2\left(\left[xe^x\right]_1^2 - \int_1^2 e^x\,dx\right)$$
$$= 4e^2 - e - 2\left(2e^2 - e - \left[e^x\right]_1^2\right)$$
$$= 4e^2 - e - 4e^2 + 2e + 2(e^2 - e)$$
$$= 2e^2 - e$$
$$\approx 12.060$$

39. Let $u = x$ and $dv = e^{-x/2}\,dx$. Then $du = dx$ and $v = -2e^{-x/2}$.

$$\int_0^4 \frac{x}{e^{x/2}}\,dx = -2xe^{-x/2}\Big]_0^4 + 2\int_0^4 e^{-x/2}\,dx$$
$$= -8e^{-2} - \left[4e^{-x/2}\right]_0^4$$
$$= -12e^{-2} + 4$$
$$\approx 2.376$$

41. Let $u = \ln x$ and $dv = x^5\,dx$. Then $du = \left(\frac{1}{x}\right)dx$ and $v = \frac{x^6}{6}$.

$$\int_1^e x^5\ln x\,dx = \frac{x^6}{6}\ln x\Big]_1^e - \int_1^e \frac{x^5}{6}\,dx$$
$$= \frac{e^6}{6} - \left[\frac{x^6}{36}\right]_1^e$$
$$= \frac{e^6}{6} - \frac{e^6}{36} + \frac{1}{36}$$
$$= \frac{5}{36}e^6 + \frac{1}{36} \approx 56.060$$

43. Let $u = \ln(x + 2)$ and $dv = dx$. Then

$du = \dfrac{1}{x+2}\,dx$ and $v = x$.

$$\int_{-1}^{0} \ln(x+2)\,dx = x\ln(x+2)\Big]_{-1}^{0} - \int_{-1}^{0} \frac{x}{x+2}\,dx$$

$$= -\int_{-1}^{0}\left(1 - \frac{2}{x+2}\right)dx$$

$$= \int_{-1}^{0}\left(\frac{2}{x+2} - 1\right)dx$$

$$= 2\ln(x+2) - x\Big]_{-1}^{0}$$

$$= 2\ln 2 - 1 \approx 0.386$$

45. Area $= \int_{0}^{2} x^3 e^x\,dx$

Let $u = x^3$ and $dv = e^x\,dx$. Then $du = 3x^2\,dx$ and $v = e^x$.

$$\int x^3e^x\,dx = x^3e^x - 3\int x^2e^x\,dx$$

Let $u = x^2$ and $dv = e^x\,dx$. Then $du = 2x\,dx$ and $v = e^x$.

$$\int x^3e^x\,dx = x^3e^x - 3\left[x^2e^x - \int 2xe^x\,dx\right]$$

$$= x^3e^x - 3x^2e^x + 6\int xe^x\,dx$$

Let $u = x$ and $dv = e^x\,dx$. Then $du = dx$ and $v = e^x$.

$$\int x^3e^x\,dx = x^3e^x - 3x^2e^x + 6\left[xe^x - \int e^x\,dx\right]$$

$$= \left(x^3 - 3x^2 + 6x - 6\right)e^x + C$$

$$\text{Area} = \int_{0}^{2} x^3e^x\,dx = \left[\left(x^3 - 3x^2 + 6x - 6\right)e^x\right]_{0}^{2}$$

$$= 2e^2 + 6 \approx 20.778$$

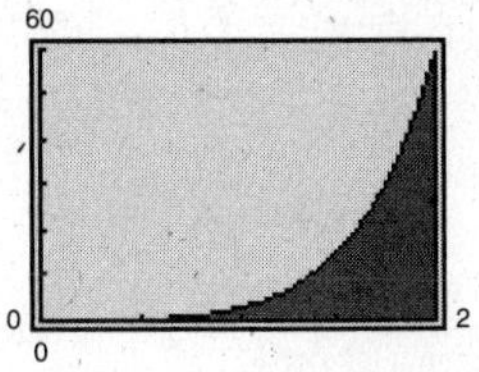

47. Area $= \int_{1}^{e} x^2 \ln x\,dx$

Let $u = \ln x$ and $dv = x^2$. Then $du = \dfrac{1}{x}\,dx$ and $v = \dfrac{x^3}{3}$.

$$\int x^2\ln x\,dx = \frac{x^3}{3}\ln x - \int \frac{x^2}{3}\,dx$$

$$= \frac{x^3}{3}\ln x - \frac{x^3}{9} + C$$

$$\text{Area} = \int_{1}^{e} x^2\ln x\,dx\left[\frac{x^3}{3}\ln x = \frac{x^3}{9}\right]_{1}^{e}$$

$$= \left(\frac{e^3}{3} - \frac{e^3}{9}\right) - \left(-\frac{1}{9}\right) = \frac{2e^3}{9} + \frac{1}{9} = 4.575$$

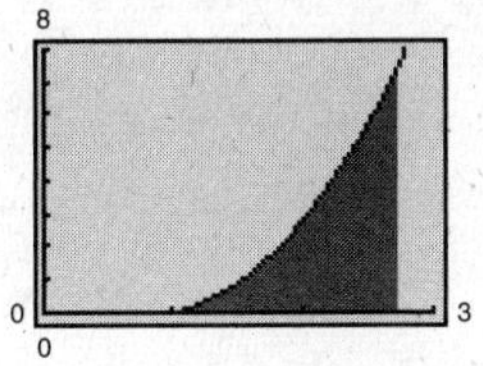

49. Let $u = \ln x$ and $dv = x^n\,dx$. Then $du = \left(\dfrac{1}{x}\right)dx$ and $v = \dfrac{x^{n+1}}{(n+1)}$.

$$\int x^n\ln x\,dx = \frac{x^{n+1}}{n+1}\ln x - \int \frac{1}{x}\cdot\frac{x^{n+1}}{n+1}\,dx$$

$$= \frac{x^{n+1}}{n+1}\ln x - \frac{1}{n+1}\int x^n\,dx$$

$$= \frac{x^{n+1}}{n+1}\ln x - \frac{1}{n+1}\cdot\frac{x^{n+1}}{n+1} + C$$

$$= \frac{x^{n+1}}{(n+1)^2}\left[-1 + (n+1)\ln x\right] + C$$

51. Using $n = 2$ and $a = 5$,

$$\int x^2e^{5x}\,dx = \frac{x^2e^{5x}}{5} - \frac{2}{5}\int xe^{5x}\,dx.$$

Now, using $n = 1$ and $a = 5$,

$$\int x^2e^{5x}\,dx = \frac{x^2e^{5x}}{5} - \frac{2}{5}\left[\frac{xe^{5x}}{5} - \frac{1}{5}\int e^{5x}\,dx\right]$$

$$= \frac{x^2e^{5x}}{5} - \frac{2xe^{5x}}{25} + \frac{2e^{5x}}{125} + C$$

$$= \frac{e^{5x}}{125}\left(25x^2 - 10x + 2\right) + C.$$

53. Using $n = -2$,

$$\int x^{-2} \ln x \, dx = \frac{x^{-2+1}}{(-2+1)^2}\left[-1 + (-2+1)\ln x\right] + C$$
$$= \frac{1}{x}(-1 - \ln x) + C$$
$$= -\frac{1}{x} - \frac{\ln x}{x} + C.$$

55. Let $u = x$ and $dv = e^{-x}\,dx$. Then $du = dx$ and $v = -e^{-x}$.

$$A = \int_0^4 xe^{-x}\,dx$$
$$= -xe^{-x}\Big]_0^4 + \int_0^4 e^{-x}\,dx$$
$$= -4e^{-4} - \left[e^{-x}\right]_0^4$$
$$= -4e^{-4} - e^{-4} + 1$$
$$= 1 - 5e^{-4}$$
$$\approx 0.908.$$

57. Let $u = \ln x$ and $dv = x\,dx$. Then $du = \frac{1}{x}\,dx$ and $v = \frac{x^2}{2}$.

$$A = \int_1^e x \ln x \, dx$$
$$= \frac{x^2}{2}\ln x\Big]_1^e - \int_1^e \frac{x}{2}\,dx$$
$$= \frac{e^2}{2} - \left[\frac{x^2}{4}\right]_1^e$$
$$= \frac{e^2}{4} + \frac{1}{4} \approx 2.097$$

59. $\int_0^2 t^3 e^{-4t}\,dt = \frac{3}{128} - \frac{379}{128}e^{-8} \approx 0.022$

61. $\int_0^5 x^4(25 - x^2)^{3/2}\,dx = \frac{1{,}171{,}875}{256}\pi \approx 14{,}381.070$

63. $P(a \le x \le b) = \int_a^b \frac{e}{e-2}xe^{-x}\,dx,\ 0 \le a \le b \le 1$

Let $u = x$ and $dv = e^{-x}\,dx$. Then $du = dx$ and $v = -e^{-x}$.

$$\frac{e}{e-2}\int_a^b xe^{-x}\,dx = \frac{e}{e-2}\left(\left[-xe^{-x}\right]_a^b - \int_a^b -e^{-x}\,dx\right) = \frac{e}{e-2}\left[-xe^{-x} - e^{-x}\right]_a^b$$

(a) $P(0.40 \le x \le 0.80) = \frac{e}{e-2}\left[-xe^{-x} - e^{-x}\right]_{0.40}^{0.80} \approx 0.491$

(b) $P(0 \le x \le 0.50) = \frac{e}{e-2}\left[-xe^{-x} - e^{-x}\right]_0^{0.50} \approx 0.341$

65. $\int(1.6t \ln t + 1)\,dt = \int dt + \int 1.6\,t \ln t\,dt = t + 1.6\int t \ln t\,dt$

Let $u = \ln t$ and $dv = t\,dt$. Then $du = \frac{1}{t}\,dt$ and $v = \frac{t^2}{2}$.

$$= t + 1.6\left[\frac{t^2}{2}\ln t - \int \frac{t}{2}\,dt\right]$$
$$= t + 1.6\left[\frac{t^2}{2}\ln t - \frac{t^2}{4}\right]$$
$$= t + 0.8t^2 \ln t - 0.4t^2$$

(a) Average value $= \frac{1}{2-1}\int_1^2 (1.6t \ln t + 1)\,dt = \left[t + 0.8t^2 \ln t - 0.4t^2\right]_1^2 = 3.2 \ln 2 - 0.2 \approx 2.018$

(b) Average value $= \frac{1}{4-3}\int_3^4 (1.6t \ln t + 1)\,dt = \left[t + 0.8t^2 \ln t - 0.4t^2\right]_3^4 = 12.8 \ln 4 - 7.2 \ln 3 - 1.8 \approx 8.035$

67. Average $= \frac{1}{5-1}\int_1^5 \left(17.3 + 0.70\sqrt{t}\ln t\right)dt$

Let $u = \ln t$ and $dv = 0.70\sqrt{t}$. Then $du = \frac{1}{t}\,dt$ and $v = \frac{7}{15}t^{3/2}$.

$$\text{Average} = \tfrac{1}{4}\left(\int_1^5 17.3\,dt + \left[\tfrac{7}{15}t^{3/2}\ln t\right]_1^5 - \int_1^5 \tfrac{7}{15}t^{1/2}\,dt\right) = \tfrac{1}{4}\left[17.3t + \tfrac{7}{15}t^{3/2}\ln t - \tfrac{14}{45}t^{3/2}\right]_1^5 \approx 18.61\%$$

Section 7.2 Partial Fractions and Logistic Growth

Skills Review

1. $x^2 - 16 = (x + 4)(x - 4)$

2. $x^2 - 25 = (x + 5)(x - 5)$

3. $x^2 - x - 12 = (x - 4)(x + 3)$

4. $x^2 + x - 6 = (x + 3)(x - 2)$

5. $x^3 - x^2 - 2x = x(x^2 - x - 2)$
$= x(x - 2)(x + 1)$

6. $x^3 - 4x^2 + 4x = x(x^2 - 4x + 4)$
$= x(x - 2)^2$

7. $x^3 - 4x^2 + 5x - 2 = (x - 2)(x^2 - 2x + 1)$
$= (x - 2)(x - 1)^2$

8. $x^3 - 5x^2 + 7x - 3 = (x - 3)(x^2 - 2x + 1)$
$= (x - 3)(x - 1)^2$

9.
$$\begin{array}{r} x \\ x - 2\overline{)\,x^2 - 2x + 1} \\ -(x^2 - 2x) \\ \hline 1 \end{array}$$

$$\frac{x^2 - 2x + 1}{x - 2} = x + \frac{1}{x - 2}$$

10.
$$\begin{array}{r} 2x - 2 \\ x - 1\overline{)\,2x^2 - 4x + 1} \\ -(2x^2 - 2x) \\ \hline -2x + 1 \\ -(-2x + 2) \\ \hline -1 \end{array}$$

$$\frac{2x^2 - 4x + 1}{x - 1} = 2x - 2 - \frac{1}{x - 1}$$

11.
$$\begin{array}{r} x^2 - x - 2 \\ x - 2\overline{)\,x^3 - 3x^2 + 0x + 2} \\ -(x^3 - 2x^2) \\ \hline -x^2 + 0x \\ -(-x^2 + 2x) \\ \hline -2x + 2 \\ -(-2x + 4) \\ \hline -2 \end{array}$$

$$\frac{x^3 - 3x^2 + 2}{x - 2} = x^2 - x - 2 - \frac{2}{x - 2}$$

12.
$$\begin{array}{r} x^2 - x + 3 \\ x + 1\overline{)\,x^3 + 0x^2 + 2x - 1} \\ -(x^3 + x^2) \\ \hline -x^2 + 2x \\ -(-x^2 - x) \\ \hline 3x - 1 \\ -(3x + 3) \\ \hline -4 \end{array}$$

$$\frac{x^3 + 2x - 1}{x + 1} = x^2 - x + 3 - \frac{4}{x + 1}$$

13.
$$\begin{array}{r} x + 4 \\ x^2 - 1\overline{)\,x^3 + 4x^2 + 5x + 2} \\ -(x^3 \qquad - x) \\ \hline 4x^2 + 6x + 2 \\ -(4x^2 \qquad - 4) \\ \hline 6x + 6 \end{array}$$

$$\frac{x^3 + 4x^2 + 5x + 2}{x^2 - 1} = x + 4 + \frac{6x + 6}{x^2 - 1}$$
$$= x + 4 + \frac{6}{x - 1}$$

14.
$$\begin{array}{r} x + 3 \\ x^2 - 1\overline{)\,x^3 + 3x^2 + 0x - 4} \\ -(x^3 \qquad - x) \\ \hline 3x^2 + x - 4 \\ -(3x^2 \qquad - 3) \\ \hline x - 1 \end{array}$$

$$\frac{x^3 + 3x^2 - 4}{x^2 - 1} = x + 3 + \frac{x - 1}{x^2 - 1}$$
$$= x + 3 + \frac{1}{x + 1}$$

1. $\dfrac{2x+40}{(x-5)(x+5)} = \dfrac{A}{x-5} + \dfrac{B}{x+5}$

Basic equation: $2x + 40 = A(x+5) + B(x-5)$

When $x = 5$: $50 = 10A$, $A = 5$

When $x = -5$: $30 = -10B$, $B = -3$

$$\frac{2(x+20)}{x^2-25} = \frac{5}{x-5} - \frac{3}{x+5}$$

3. $\dfrac{8x+3}{x(x-3)} = \dfrac{A}{x} + \dfrac{B}{x-3}$

Basic equation: $8x + 3 = A(x-3) + Bx$

When $x = 0$: $3 = -3A$, $A = -1$

When $x = 3$: $27 = 3B$, $B = 9$

$$\frac{8x+3}{x^2-3x} = \frac{9}{x-3} - \frac{1}{x}$$

5. $\dfrac{4x-13}{(x-5)(x+2)} = \dfrac{A}{x-5} + \dfrac{B}{x+2}$

Basic equation: $4x - 13 = A(x+2) + B(x-5)$

When $x = 5$: $7 = 7A$, $A = 1$

When $x = -2$: $-21 = -7B$, $B = 3$

$$\frac{4x-13}{x^2-3x-10} = \frac{1}{x-5} + \frac{3}{x+2}$$

7. $\dfrac{3x^2-2x-5}{x^2(x+1)} = \dfrac{A}{x} + \dfrac{B}{x^2} + \dfrac{C}{x+1}$

Basic equation:
$3x^2 - 2x - 5 = Ax(x+1) + B(x+1) + Cx^2$

When $x = 0$: $-5 = B$

When $x = -1$: $0 = C$

When $x = 1$: $-4 = 2A + 2B + C$, $A = 3$

$$\frac{3x^2-2x-5}{x^2(x+1)} = \frac{3}{x} - \frac{5}{x^2}$$

9. $\dfrac{1}{3}\left[\dfrac{x+1}{(x-2)^2}\right] = \dfrac{1}{3}\left[\dfrac{A}{x-2} + \dfrac{B}{(x-2)^2}\right]$

Basic equation: $x + 1 = A(x-2) + B$

When $x = 2$: $3 = B$

When $x = 3$: $4 = A + B$, $A = 1$

$$\frac{x+1}{3(x-2)^2} = \frac{1}{3}\left[\frac{1}{x-2} + \frac{3}{(x-2)^2}\right] = \frac{1}{3(x-2)} + \frac{1}{(x-2)^2}$$

11. $\dfrac{8x^2+15x+9}{(x+1)^3} = \dfrac{A}{x+1} + \dfrac{B}{(x+1)^2} + \dfrac{C}{(x+1)^3}$

Basic equation:

$$8x^2 + 15x + 9 = A(x+1)^2 + B(x+1) + C = Ax^2 + (2A+B)x + (A+B+C)$$

So, $A = 8$, $2A + B = 15$, and $A + B + C = 9$. Solving these equations yields $A = 8$, $B = -1$, and $C = 2$.

$$\frac{8x^2+15x+9}{(x+1)^3} = \frac{8}{x+1} - \frac{1}{(x+1)^2} + \frac{2}{(x+1)^3}$$

13. $\dfrac{1}{(x+1)(x-1)} = \dfrac{A}{x+1} + \dfrac{B}{x-1}$

Basic equation: $1 = A(x-1) + B(x+1)$

When $x = -1$: $1 = -2A$, $A = -\dfrac{1}{2}$

When $x = 1$: $1 = 2B$, $B = \dfrac{1}{2}$

$$\int \frac{1}{x^2-1}\,dx = -\frac{1}{2}\int\frac{1}{x+1}\,dx + \frac{1}{2}\int\frac{1}{x-1}\,dx = -\frac{1}{2}\ln|x+1| + \frac{1}{2}\ln|x-1| + C = \frac{1}{2}\ln\left|\frac{x-1}{x+1}\right| + C$$

15. $\dfrac{-2}{(x+4)(x-4)} = \dfrac{A}{x+4} + \dfrac{B}{x-4}$

Basic equation: $-2 = A(x-4) + B(x+4)$

When $x = -4$: $-2 = -8A$, $A = \dfrac{1}{4}$

When $x = 4$: $-2 = 8B$, $B = -\dfrac{1}{4}$

$$\int \frac{-2}{x^2-16}\,dx = \frac{1}{4}\int\frac{1}{x+4}\,dx - \frac{1}{4}\int\frac{1}{x-4}\,dx = \frac{1}{4}\ln|x+4| - \frac{1}{4}\ln|x-4| + C = \frac{1}{4}\ln\left|\frac{x+4}{x-4}\right| + C$$

17. $\dfrac{1}{x(2x-1)} = \dfrac{A}{x} + \dfrac{B}{2x-1}$

Basic equation: $1 = A(2x-1) + Bx$

When $x = 0$: $1 = -A$, $A = -1$

When $x = \dfrac{1}{2}$: $1 = \dfrac{1}{2}B$, $B = 2$

$$\int \frac{1}{2x^2 - x}\,dx = -\int \frac{1}{x}\,dx + \int \frac{2}{2x-1}\,dx$$
$$= -\ln|x| + \ln|2x-1| + C$$
$$= \ln\left|\frac{2x-1}{x}\right| + C$$

19. $\dfrac{10}{x(x-10)} = \dfrac{A}{x} + \dfrac{B}{x-10}$

Basic equation: $10 = A(x-10) + Bx$

When $x = 0$: $10 = -10A$, $A = -1$

When $x = 10$: $10 = 10B$, $B = 1$

$$\int \frac{10}{x^2 - 10x}\,dx = -\int \frac{1}{x}\,dx + \int \frac{1}{x-10}\,dx$$
$$= -\ln|x| + \ln|x-10| + C$$
$$= \ln\left|\frac{x-10}{x}\right| + C$$

21. $\dfrac{3}{(x-1)(x+2)} = \dfrac{A}{x-1} + \dfrac{B}{x+2}$

Basic equation: $3 = A(x+2) + B(x-1)$

When $x = 1$: $3 = 3A$, $A = 1$

When $x = -2$: $3 = -3B$, $B = -1$

$$\int \frac{3}{x^2 + x - 2}\,dx = \int \frac{1}{x-1}\,dx - \int \frac{1}{x+2}\,dx$$
$$= \ln|x-1| - \ln|x+2| + C$$
$$= \ln\left|\frac{x-1}{x+2}\right| + C$$

23. $\dfrac{5-x}{(2x-1)(x+1)} = \dfrac{A}{2x-1} + \dfrac{B}{x+1}$

Basic equation: $5 - x = A(x+1) + B(2x-1)$

When $x = \dfrac{1}{2}$: $4.5 = 1.5A$, $A = 3$

When $x = -1$: $6 = -3B$, $B = -2$

$$\int \frac{5-x}{2x^2 + x - 1}\,dx = 3\int \frac{1}{2x-1}\,dx - 2\int \frac{1}{x+1}\,dx$$
$$= \frac{3}{2}\ln|2x-1| - 2\ln|x+1| + C$$

25. $\dfrac{x^2 - 4x - 4}{x(x+2)(x-2)} = \dfrac{A}{x} + \dfrac{B}{x+2} + \dfrac{C}{x-2}$

Basic equation: $x^2 - 4x - 4 = A(x+2)(x-2) + Bx(x-2) + Cx(x+2)$

When $x = 0$: $-4 = -4A$, $A = 1$

When $x = -2$: $8 = 8B$, $B = 1$

When $x = 2$: $-8 = 8C$, $C = -1$

$$\int \frac{x^2 - 4x - 4}{x^3 - 4x}\,dx = \int \frac{1}{x}\,dx + \int \frac{1}{x+2}\,dx - \int \frac{1}{x-2}\,dx$$
$$= \ln|x| + \ln|x+2| - \ln|x-2| + C = \ln\left|\frac{x(x+2)}{x-2}\right| + C$$

27. $\dfrac{x+2}{x(x-4)} = \dfrac{A}{x-4} + \dfrac{B}{x}$

Basic equation: $x + 2 = Ax + B(x-4)$

When $x = 4$: $6 = 4A$, $A = \dfrac{3}{2}$

When $x = 0$: $2 = -4B$, $B = -\dfrac{1}{2}$

$$\int \frac{x+2}{x^2 - 4x}\,dx = \frac{1}{2}\left[3\int \frac{1}{x-4}\,dx - \int \frac{1}{x}\,dx\right]$$
$$= \frac{1}{2}\left[3\ln|x-4| - \ln|x|\right] + C$$

29. $\dfrac{2x-3}{(x-1)^2} = \dfrac{A}{x-1} + \dfrac{B}{(x-1)^2}$

Basic equation: $2x - 3 = A(x-1) + B$

When $x = 1$: $-1 = B$, $B = -1$

When $x = 0$: $-3 = -A + B$, $A = 2$

$$\int \frac{2x-3}{(x-1)^2}\,dx = 2\int \frac{1}{x-1}\,dx - \int \frac{1}{(x-1)^2}\,dx$$
$$= 2\ln|x-1| + \frac{1}{x-1} + C$$

31. $\dfrac{3x^2 + 3x + 1}{x(x + 1)^2} = \dfrac{A}{x} + \dfrac{B}{x + 1} + \dfrac{C}{(x + 1)^2}$

Basic equation: $3x^2 + 3x + 1 = A(x + 1)^2 + Bx(x + 1) + Cx$

When $x = 0$: $1 = A$, $A = 1$

When $x = -1$: $1 = -C$, $C = 1$

When $x = 1$: $7 = 4A + 2B + C$, $B = 2$

$$\int \frac{3x^2 + 3x + 1}{x(x^2 + 2x + 1)}\,dx = \int \frac{1}{x}\,dx + 2\int \frac{1}{x + 1}\,dx - 1\int \frac{1}{(x + 1)^2}\,dx = \ln|x| + 2\ln|x + 1| + \frac{1}{x + 1} + C$$

33. $\dfrac{1}{9 - x^2} = \dfrac{-1}{(x - 3)(x + 3)} = \dfrac{A}{x - 3} + \dfrac{B}{x + 3}$

Basic equation: $-1 = A(x + 3) + B(x - 3)$

When $x = 3$: $-1 = 6A$, $A = -\dfrac{1}{6}$

When $x = -3$: $-1 = -6B$, $B = \dfrac{1}{6}$

$$\int_4^5 \frac{1}{9 - x^2}\,dx = -\frac{1}{6}\int_4^5 \frac{1}{x - 3}\,dx + \frac{1}{6}\int_4^5 \frac{1}{x + 3}\,dx$$

$$= \frac{1}{6}\Big[-\ln(x - 3) + \ln(x + 3)\Big]_4^5 = \frac{1}{6}\big[-\ln 2 + \ln 8 - \ln 7\big] = \frac{1}{6}\ln\frac{4}{7} \approx -0.093$$

35. $\dfrac{x - 1}{x^2(x + 1)} = \dfrac{A}{x} + \dfrac{B}{x^2} + \dfrac{C}{x + 1}$

Basic equation: $x - 1 = Ax(x + 1) + B(x + 1) + Cx^2$

When $x = 0$: $-1 = B$, $B = -1$

When $x = -1$: $-2 = C$, $C = -2$

When $x = 1$: $0 = 2A + 2B + C$, $A = 2$

$$\int_1^5 \frac{x - 1}{x^2(x + 1)}\,dx = 2\int_1^5 \frac{1}{x}\,dx - \int_1^5 \frac{1}{x^2}\,dx - 2\int_1^5 \frac{1}{x + 1}\,dx$$

$$= \left[2\ln|x| + \frac{1}{x} - 2\ln|x + 1|\right]_1^5$$

$$= \left[2\ln\left|\frac{x}{x + 1}\right| + \frac{1}{x}\right]_1^5$$

$$= \left(2\ln\frac{5}{6} + \frac{1}{5}\right) - \left(2\ln\frac{1}{2} + 1\right)$$

$$= 2\ln\left(\frac{5}{3}\right) - \frac{4}{5} \approx 0.222$$

37. $\dfrac{x^3}{x^2 - 2} = x + \dfrac{2x}{x^2 - 2}$

$$\int_0^1 \left(x + \frac{2x}{x^2 - 2}\right)dx = \left[\frac{x^2}{2} + \ln|x^2 - 2|\right]_0^1 = \frac{1}{2} - \ln 2 \approx -0.193$$

39. $\dfrac{x^3 - 4x^2 - 3x + 3}{x^2 - 3x} = x - 1 - \dfrac{6x - 3}{x^2 - 3x}$

$\dfrac{6x - 3}{x(x - 3)} = \dfrac{A}{x} + \dfrac{B}{x - 3}$

Basic equation: $6x - 3 = A(x - 3) + Bx$

When $x = 3$: $15 = 3B$, $B = 5$

When $x = 0$: $-3 = -3A$, $A = 1$

$$\int_1^2 \frac{x^3 - 4x^2 - 3x + 3}{x^2 - 3x}\,dx = \int_1^2 \left(x - 1 - \frac{1}{x} - \frac{5}{x - 3}\right) dx$$

$$= \left[\frac{x^2}{2} - x - \ln|x| - 5\ln|x - 3|\right]_1^2 = (-\ln 2) - \left(-\frac{1}{2} - 5\ln 2\right) = \frac{1}{2} + 4\ln 2 \approx 3.273$$

41. To find the limits of integration, solve:

$$2 = \frac{14}{16 - x^2}$$

$$32 - 2x^2 = 14$$

$$18 = 2x^2$$

$$x = \pm 3$$

$$A = \int_{-3}^{3}\left(2 - \frac{14}{16 - x^2}\right) dx = \int_{-3}^{3}\left(2 - \frac{7}{4}\cdot\frac{1}{x + 4} + \frac{7}{4}\cdot\frac{1}{x - 4}\right) dx = \left[2x - \frac{7}{4}\ln|x + 4| + \frac{7}{4}\ln|x - 4|\right]_{-3}^{3}$$

$$= \left(6 - \frac{7}{4}\ln 7\right) - \left(-6 + \frac{7}{4}\ln 7\right) = 12 - \frac{7}{2}\ln 7 \approx 5.189$$

43. $A = \displaystyle\int_2^5 \frac{x + 1}{x^2 - x}\,dx$

$= \displaystyle\int_2^5 \left(-\frac{1}{x} + \frac{2}{x - 1}\right) dx$

$= \left[-\ln x + 2\ln(x - 1)\right]_2^5$

$= (-\ln 5 + 2\ln 4) - (-\ln 2)$

$= 5\ln 2 - \ln 5 \approx 1.856$

45. $A = \displaystyle\int_0^1 \frac{12}{x^2 + 5x + 6}\,dx$

$= \displaystyle\int_0^1 \left(\frac{12}{x + 2} - \frac{12}{x + 3}\right) dx$

$= \left[12\ln(x + 2) - 12\ln(x + 3)\right]_0^1$

$= (12\ln 3 - 12\ln 4) - (12\ln 2 - 12\ln 3)$

$= 24\ln 3 - 36\ln 2 \approx 1.413$

47. $\dfrac{1}{(a + x)(a - x)} = \dfrac{A}{a + x} + \dfrac{B}{a - x}$

Basic equation: $1 = A(a - x) + B(a + x)$

When $x = -a$: $1 = 2aA$, $A = \dfrac{1}{2a}$

When $x = a$: $1 = 2aB$, $B = \dfrac{1}{2a}$

$\dfrac{1}{a^2 - x^2} = \dfrac{1}{2a}\left(\dfrac{1}{a + x} + \dfrac{1}{a - x}\right)$

49. $\dfrac{1}{x(a - x)} = \dfrac{A}{x} + \dfrac{B}{a - x}$

Basic equation: $1 = A(a - x) + Bx$

When $x = a$: $1 = Ba$, $B = \dfrac{1}{a}$

When $x = 0$: $1 = Aa$, $A = \dfrac{1}{a}$

$\dfrac{1}{x(a - x)} = \dfrac{1}{a}\left(\dfrac{1}{x} + \dfrac{1}{a - x}\right)$

51. $B(t) = \displaystyle\int \frac{24{,}361e^{-0.193t}}{\left(1 + 784e^{-0.193t}\right)^2}\,dt$

$= -\dfrac{24{,}361}{151{,}312}\displaystyle\int \frac{-151{,}312e^{-0.193t}}{\left(1 + 784e^{-0.193t}\right)^2}\,dt$

$= \dfrac{24{,}361}{151{,}312}\left(\dfrac{1}{1 + 784e^{-0.193t}}\right) + C$

Because $B(8) = 0.04$, it follows that $C \approx -0.916$, and

you have $B(t) = \dfrac{24{,}361}{151{,}312}\left(\dfrac{1}{1 + 784e^{-0.193t}}\right) - 0.916.$

$B(25) \approx 21.16$ ounces

53. $N(t) = \int \frac{125e^{-0.125t}}{\left(1 + 9e^{-0.125t}\right)^2}\,dt$

$= -\frac{1000}{9}\int \frac{-1.125e^{-0.125t}}{\left(1 + 9e^{-0.125t}\right)^2}\,dt$

$= \frac{1000}{9}\left(\frac{1}{1 + 9e^{-0.125t}}\right) + C$

(a) Because $N(0) = 100$, it follows that $C = \frac{800}{9}$. So,

you have $N(t) = \frac{1000}{9}\left(\frac{1}{1 + 9e^{-0.125t}}\right) + \frac{800}{9}$.

$N(24) \approx 166$ animals

(b) As t increases without bound, $N(t)$ approaches 200. So, the limiting size of the population is 200 animals.

55. $t = 5010\int \frac{1}{(x + 1)(500 - x)}\,dx$

$= \int \left(\frac{10}{x + 1} - \frac{10}{x - 500}\right)dx$

$t = 10 \ln|x + 1| - 10 \ln|500 - x| + C$

Because $x = 1$ when $t = 0$:

$0 = 10 \ln 2 - 10 \ln 499 + C,\ C \approx 55.1946$.

So, $t = 10 \ln(x + 1) - 10 \ln(500 - x) + 55.1946$.

(a) If $x = (0.75)(500) = 375$, then $t \approx 66.2$ hours.

(b) When $t = 100$, you can use a graphing utility to solve for x:

$100 = 10 \ln(x + 1) - 10 \ln(500 - x) + 55.1946$,

$x \approx 494$ individuals

57. Both species have positive growth rates. The rate of growth is increasing on $[0, 3]$ and decreasing on $[3, \infty)$ for P. aurelia. The rate of growth is increasing on $[0, 2]$ and decreasing on $[2, \infty)$ for P. caudatum. Hence, the former species has a greater limiting population.

59. (a) $\frac{dy}{dt} = ky\left(1 - \frac{y}{839.1}\right)$

(b) $y = \frac{839.1}{1 + be^{-kt}}$

When $t = 0$, $y = 76$:

$$76 = \frac{839.1}{1 + be^{-k(0)}}$$
$$76 = \frac{839.1}{1 + b}$$
$$76(1 + b) = 839.1$$
$$1 + b = \frac{839.1}{76}$$
$$b \approx 10.04$$

When $t = 106$, $y = 300$:

$$300 = \frac{839.1}{1 + 10.04e^{-106k}}$$
$$300\left(1 + 10.04e^{-106k}\right) = 839.1$$
$$1 + 10.04e^{-106k} = 2.797$$
$$10.04e^{-106k} = 1.797$$
$$e^{-106k} \approx 0.17898$$
$$-106k = \ln(0.17898)$$
$$k \approx -0.01623$$

$$y = \frac{839.1}{1 + 10.04e^{-0.01623t}}$$

(c)

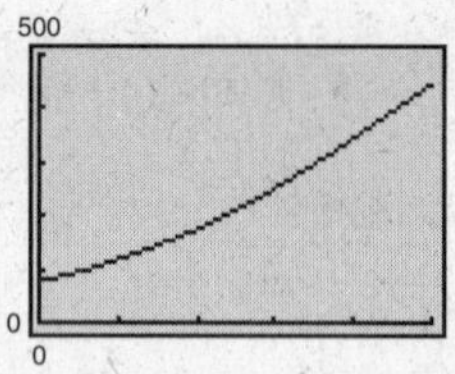

The population will reach 400 million in about 2036 $(t \approx 136)$.

61. (a) Log Rule; The derivative of the denominator is $2x + 1$, which is the expression in the numerator.

(b) Partial Fractions; The fraction can be rewritten as the sum of two more simple rational functions.

Section 7.3 Integrals of Trigonometric Functions

Skills Review

1. $\cos\frac{5\pi}{4} = -\frac{\sqrt{2}}{2}$

2. $\sin\frac{7\pi}{6} = -\frac{1}{2}$

3. $\sin\left(-\frac{\pi}{3}\right) = -\frac{\sqrt{3}}{2}$

4. $\cos\left(-\frac{\pi}{6}\right) = \frac{\sqrt{3}}{2}$

5. $\tan\frac{5\pi}{6} = -\frac{\sqrt{3}}{3}$

6. $\cot\frac{5\pi}{3} = -\frac{\sqrt{3}}{3}$

Skills Review —*continued*—

7. $\sec \pi = -1$

8. $\cos \frac{\pi}{2} = 0$

9. $\sin x \sec x = \sin x \cdot \frac{1}{\cos x} = \tan x$

10. $\csc x \cos x = \frac{1}{\sin x} \cdot \cos x = \cot x$

11. $\cos^2 x(\sec^2 x - 1) = \cos^2 x \tan^2 x$
$= \cos^2 x \cdot \frac{\sin^2 x}{\cos^2 x} = \sin^2 x$

12. $\sin^2 x(\csc^2 x - 1) = \sin^2 x(\cot^2 x)$
$= \sin^2 x \cdot \frac{\cos^2 x}{\sin^2 x} = \cos^2 x$

13. $\sec x \sin\left(\frac{\pi}{2} - x\right) = \sec x \cos x = \frac{1}{\cos x} \cdot \cos x = 1$

14. $\cot x \cos\left(\frac{\pi}{2} - x\right) = \cot x \sin x$
$= \frac{\cos x}{\sin x} \cdot \sin x = \cos x$

15. $\cot x \sec x = \frac{\cos x}{\sin x} \cdot \frac{1}{\cos x} = \frac{1}{\sin x} = \csc x$

16. $\cot x(\sin^2 x) = \frac{\cos x}{\sin x} \cdot \sin^2 x = \cos x \sin x$

17. $\int_0^4 (x^2 + 3x - 4)\,dx = \left[\frac{1}{3}x^3 + \frac{3}{2}x^2 - 4x\right]_0^4$
$= \frac{64}{3} + 24 - 16 = \frac{88}{3}$

18. $\int_{-1}^1 (1 - x^2)\,dx = \left[x - \frac{1}{3}x^3\right]_{-1}^1$
$= \left(1 - \frac{1}{3}\right) - \left(-1 + \frac{1}{3}\right) = \frac{4}{3}$

19. $\int_0^2 x(4 - x^2)\,dx = \int_0^2 (4x - x^3)\,dx$
$= \left[2x^2 - \frac{1}{4}x^4\right]_0^2$
$= 8 - 4 = 4$

20. $\int_0^1 x(9 - x^2)\,dx = \int_0^1 (9x - x^3)\,dx$
$= \left[\frac{9}{2}x^2 - \frac{1}{4}x^4\right]_0^1$
$= \frac{9}{2} - \frac{1}{4} = \frac{17}{4}$

1. $\int (2 \sin x + 3 \cos x)\,dx = -2 \cos x + 3 \sin x + C$

3. $\int (1 - \csc t \cot t)\,dt = t + \csc t + C$

5. $\int (\csc^2 \theta - \cos \theta)\,d\theta = -\cot \theta - \sin \theta + C$

7. $\int \sin 2x\,dx = \frac{1}{2}\int 2 \sin 2x\,dx = -\frac{1}{2} \cos 2x + C$

9. $\int 2x \cos x^2\,dx = \sin x^2 + C$

11. $\int \sec^2 \frac{x}{2}\,dx = 2\int \frac{1}{2} \sec^2 \frac{x}{2}\,dx = 2 \tan \frac{x}{2} + C$

13. $\int \tan 3x\,dx = \frac{1}{3}\int 3 \tan 3x\,dx = -\frac{1}{3} \ln|\cos 3x| + C$

15. $\int \tan^3 x \sec^2 x\,dx = \frac{1}{4} \tan^4 x + C$

17. $\int \cot \pi x\,dx = \frac{1}{\pi}\int \pi \cot \pi x\,dx = \frac{1}{\pi} \ln|\sin \pi x| + C$

19. $\int \csc 2x\,dx = \frac{1}{2}\int 2 \csc 2x\,dx$
$= \frac{1}{2} \ln|\csc 2x - \cot 2x| + C$

21. $\int \frac{\sec^2 x}{\tan x}\,dx = \ln|\tan x| + C$

23. $\int \frac{\sec x \tan x}{\sec x - 1}\,dx = \ln|\sec x - 1| + C$

25. $\int \frac{\sin x}{1 + \cos x}\,dx = -\ln|1 + \cos x| + C$

27. $\int \frac{\csc^2 x}{\cot^3 x}\,dx = -\int \cot^{-3} x(-\csc^2 x)\,dx$
$= -\frac{\cot^{-2} x}{-2} + C$
$= \frac{1}{2} \tan^2 x + C$

29. $\int e^x \sin e^x\,dx = -\cos e^x + C$

31. $\int e^{\sin x} \cos x\,dx = e^{\sin x} + C$

33. $\int (\sin x + \cos x)^2\, dx = \int (\sin^2 x + 2 \sin x \cos x + \cos^2 x)\, dx$

$= \int (1 + \sin 2x)\, dx$

$= \int dx + \frac{1}{2}\int 2 \sin 2x\, dx$

$= x - \frac{1}{2}\cos 2x + C$

$= \sin^2 x + x + C$ or $-\cos^2 x + x + C$

35. Using integration by parts, you let $u = x$ and $dv = \cos x\, dx$. Then $du = dx$ and $v = \sin x$.

$$\int x \cos x\, dx = x \sin x - \int \sin x\, dx$$
$$= x \sin x + \cos x + C$$

37. Using integration by parts, you let $u = x$ and $dv = \sec^2 x\, dx$. Then $du = dx$ and $v = \tan x$.

$$\int x \sec^2 x\, dx = x \tan x - \int \tan x\, dx$$
$$= x \tan x + \ln|\cos x| + C$$

39. $\int_0^{\pi/4} \cos \frac{4x}{3}\, dx = \frac{3}{4} \sin \frac{4x}{3}\Big]_0^{\pi/4}$

$= \frac{3}{4}\left(\sin \frac{\pi}{3}\right) = \frac{3\sqrt{3}}{8} \approx 0.6495$

41. $\int_{\pi/2}^{2\pi/3} \sec^2 \frac{x}{2}\, dx = 2 \tan \frac{x}{2}\Big]_{\pi/2}^{2\pi/3} = 2\left(\sqrt{3} - 1\right) \approx 1.4641$

43. $\int_{\pi/12}^{\pi/4} \csc 2x \cot 2x\, dx = -\frac{1}{2} \csc 2x\Big]_{\pi/12}^{\pi/4}$

$= -\frac{1}{2}\left[\csc \frac{\pi}{2} - \csc \frac{\pi}{6}\right]$

$= -\frac{1}{2}[1 - 2] = \frac{1}{2}$

45. $\int_0^1 \tan(1 - x)\, dx = \ln\left|\cos(1 - x)\right|\Big]_0^1$

$= \ln(\cos 0) - \ln(\cos 1) \approx 0.6156$

47. Area $= \int_0^{2\pi} \cos \frac{x}{4}\, dx = 4 \sin \frac{x}{4}\Big]_0^{2\pi} = 4$ square units

49. Area $= \int_0^{\pi} (x + \sin x)\, dx$

$= \left[\frac{x^2}{2} - \cos x\right]_0^{\pi}$

$= \frac{\pi^2}{2} + 2 \approx 6.9348$ square units

51. Area $= \int_0^{\pi} (\sin x + \cos 2x)\, dx$

$= \left[-\cos x + \frac{1}{2} \sin 2x\right]_0^{\pi}$

$= 1 + 1 = 2$ square units

53. $Q = 588 + 390 \cos(0.46t - 0.25),\ 0 \le t \le 12$

($t = 1$ corresponds to January.)

Average $= \frac{1}{12 - 0}\int_0^{12} \left[588 + 390 \cos(0.46t - 0.25)\right] dt$

$= \frac{1}{12}\left[588t + \frac{390}{0.46} \sin(0.46t - 0.25)\right]_0^{12}$

≈ 545.5 trillion Btu

55. $W = 7594 + 455.2 \sin(0.41t - 1.713)$

($t = 1$ corresponds to January.)

(a) Average $= \frac{1}{3 - 0}\int_0^3 \left[7594 + 455.2 \sin(0.41t - 1.713)\right] dt$

$= \frac{1}{3}\left[7594t - \frac{455.2}{0.41} \cos(0.41t - 1.713)\right]_0^3 \approx 7213.8$ thousand workers

(b) Average $= \frac{1}{6 - 3}\int_3^6 \left[7594 + 455.2 \sin(0.41t - 1.713)\right] dt$

$= \frac{1}{3}\left[7594t - \frac{455.2}{0.41} \cos(0.41t - 1.713)\right]_3^6 \approx 7650.2$ thousand workers

(c) Average $= \frac{1}{12 - 0}\int_0^{12} \left[7594 + 455.2 \sin(0.41t - 1.713)\right] dt$

$= \frac{1}{12}\left[7594t - \frac{455.2}{0.41} \cos(0.41t - 1.713)\right]_0^{12} \approx 7673.2$ thousand workers

57. Total $= \int_0^{12} \left[2.47 \sin(0.40t + 1.80) + 2.08\right] dt$

$= \left[-6.175 \cos(0.40t + 1.80) + 2.08t\right]_0^{12}$

≈ 17.69 inches

59. (a) $C = 0.3\int_8^{20} \left[72 + 12 \sin \frac{\pi(t-8)}{12} - 72\right] dt$

$\approx \$27.50$

(b) $C = 0.3\int_{10}^{18} \left[72 + 12 \sin \frac{\pi(t-8)}{12} - 78\right] dt$

$\approx \$9.42$

Savings $= \$27.50 - \$9.42 = \$18.08$

61. The volume is $V = \int_0^3 0.9 \sin \frac{\pi t}{3}\, dt$

$= -\frac{3}{\pi}(0.9) \cos \frac{\pi t}{3}\Big]_0^3$

$= -\frac{2.7}{\pi}(-1 - 1) \approx 1.7189$ liters.

63. $V = \pi \int_0^{\pi} \left(\sqrt{\sin x}\right)^2 dx$

$= \pi \int_0^{\pi} \sin x\, dx = \pi\left[-\cos x\right]_0^{\pi} = 2\pi$

65. False. $4\int \sin x \cos x\, dx = \int 2 \sin 2x\, dx = -\cos 2x + C$

Mid-Chapter Quiz Solutions

1. Let $u = x$ and $dv = e^{5x}\, dx$. Then $du = dx$ and $v = \frac{1}{5}e^{5x}.$

$\int xe^{5x}\, dx = \frac{1}{5}xe^{5x} - \frac{1}{5}\int e^{5x}\, dx$

$= \frac{1}{5}xe^{5x} - \frac{1}{25}e^{5x} + C$

$= \frac{1}{5}e^{5x}\left(x - \frac{1}{5}\right) + C$

2. $\int \ln x^3\, dx = 3\int \ln x\, dx$

Let $u = \ln x$ and $dv = dx$. Then $du = \frac{1}{x}\, dx$ and $v = x$.

$\int \ln x^3\, dx = 3\left[x \ln x - \int dx\right]$

$= 3x \ln x - 3x + C = 3x(\ln x - 1) + C$

3. Let $u = \ln x$ and $dv = (x + 1)\, dx$. Then $du = \frac{1}{x}\, dx$ and $v = \frac{1}{2}x^2 + x$.

$\int (x+1) \ln x\, dx = \ln x\left(\frac{1}{2}x^2 + x\right) - \int\left(\frac{1}{2}x^2 + x\right)\left(\frac{1}{x}\right) dx = \ln x\left(\frac{1}{2}x^2 + x\right) - \int\left(\frac{x}{2} + 1\right) dx$

$= \ln x\left(\frac{1}{2}x^2 + x\right) - \frac{1}{4}x^2 - x + C = \frac{1}{2}x^2 \ln x + x \ln x - \frac{1}{4}x^2 - x + C$

4. $\frac{10}{(x+5)(x-5)} = \frac{A}{x+5} + \frac{B}{x-5}$

Basic equation: $10 = A(x - 5) + B(x + 5)$

When $x = -5$: $10 = -10A,\ A = -1$

When $x = 5$: $10 = 10B,\ B = 1$

$\int \frac{10}{x^2 - 25}\, dx = -\int \frac{1}{x+5}\, dx + \int \frac{1}{x-5}\, dx$

$= -\ln|x + 5| + \ln|x - 5| + C$

$= \ln\left|\frac{x-5}{x+5}\right| + C$

5. $\frac{x - 14}{(x+4)(x-2)} = \frac{A}{x+4} + \frac{B}{x-2}$

Basic equation: $x - 14 = A(x - 2) + B(x + 4)$

When $x = -4$: $-18 = -6A,\ A = 3$

When $x = 2$: $-12 = 6B,\ B = -2$

$\int \frac{x - 14}{x^2 + 2x - 8}\, dx = 3\int \frac{1}{x+4}\, dx - 2\int \frac{1}{x-2}\, dx$

$= 3 \ln|x + 4| - 2 \ln|x - 2| + C$

6. $\frac{5x - 1}{(x+1)^2} = \frac{A}{x+1} + \frac{B}{(x+1)^2}$

Basic equation: $5x - 1 = A(x + 1) + B$

When $x = -1$: $-6 = B$

When $x = 0$: $-1 = A + B,\ A = 5$

$\int \frac{5x - 1}{(x+1)^2}\, dx = 5\int \frac{1}{x+1}\, dx - 6\int \frac{1}{(x+1)^2}\, dx$

$= 5 \ln|x + 1| + \frac{6}{x+1} + C$

7. $\int \sin 5x\, dx = \frac{1}{5}\int 5 \sin 5x\, dx = -\frac{1}{5} \cos 5x + C$

8. $\int x \csc x^2\, dx = \frac{1}{2}\int 2x \csc x^2\, dx$

$= \frac{1}{2} \ln\left|\csc x^2 - \cot x^2\right| + C$

9. $\int \frac{\cos \sqrt{x}}{2\sqrt{x}}\, dx = \sin \sqrt{x} + C$

10. Let $u = x$ and $dv = e^{x/2}\,dx$. Then $du = dx$ and $v = 2e^{x/2}$.

$$\int_{-2}^{0} xe^{x/2}\,dx = 2xe^{x/2}\Big]_{-2}^{0} - \int_{-2}^{0} 2e^{x/2}\,dx$$
$$= 4e^{-1} - \Big[4e^{x/2}\Big]_{-2}^{0}$$
$$= 4e^{-1} - \left(4 - 4e^{-1}\right)$$
$$= \frac{8}{e} - 4 \approx -1.057$$

11. Let $u = (\ln x)^2$ and $dv = dx$. Then $du = \dfrac{2}{x}(\ln x)\,dx$ and $v = x$.

$$\int_1^e (\ln x)^2\,dx = x(\ln x)^2\Big]_1^e - 2\int_1^e \ln x\,dx$$

Let $u = \ln x$ and $dv = dx$. Then $du = \dfrac{1}{x}\,dx$ and $v = x$.

$$\int_1^e (\ln x)^2\,dx = x(\ln x)^2\Big]_1^e - 2x\ln x\Big]_1^e - 2\int_1^e dx = \Big[x(\ln x)^2 - 2x\ln x + 2x\Big]_1^e = e - 2 \approx 0.718$$

12. Let $u = x$ and $dv = (1 + 2x)^{-1/2}\,dx$. Then $du = dx$ and $v = (1 + 2x)^{1/2}$.

$$\int_0^1 \frac{x}{\sqrt{1 + 2x}}\,dx = x(1 + 2x)^{1/2}\Big]_0^1 - \int_0^1 (1 + 2x)^{1/2}\,dx = \left[x(1 + 2x)^{1/2} - \frac{1}{3}(1 + 2x)^{3/2}\right]_0^1 = \frac{1}{3}$$

13. $\dfrac{3x + 1}{x(x + 1)} = \dfrac{A}{x} + \dfrac{B}{x + 1}$

Basic equation: $3x + 1 = A(x + 1) + Bx$

When $x = 0$: $1 = A$

When $x = -1$: $-2 = -B$, $B = 2$

$$\int_1^4 \frac{3x + 1}{x(x + 1)}\,dx = \int_1^4 \frac{1}{x}\,dx + 2\int_1^4 \frac{1}{x + 1}\,dx = \Big[\ln x + 2\ln|x + 1|\Big]_1^4 = (\ln 4 + 2\ln 5) - 2\ln 2 \approx 3.219$$

14. $\dfrac{120}{(x - 3)(x + 5)} = \dfrac{A}{x - 3} + \dfrac{B}{x + 5}$

Basic equation: $120 = A(x + 5) + B(x - 3)$

When $x = 3$: $120 = 8A$, $A = 15$

When $x = -5$: $120 = -8B$, $B = -15$

$$\int_4^5 \frac{120}{(x - 3)(x + 5)}\,dx = 15\int_4^5 \frac{1}{x - 3}\,dx - 15\int_4^5 \frac{1}{x + 5}\,dx$$
$$= 15\Big[\ln|x - 3| - \ln|x + 5|\Big]_4^5$$
$$= 15\big[(\ln 2 - \ln 10) + \ln 9\big]$$
$$\approx 8.817$$

15. $\dfrac{1}{x(0.1+0.2x)} = \dfrac{A}{x} + \dfrac{B}{0.1+0.2x}$

Basic equation: $1 = A(0.1+0.2x) + Bx$

When $x = 0$: $A = 10$

When $x = -0.5$: $B = -2$

$$\int_1^2 \frac{1}{x(0.1+0.2x)}\,dx = 10\int_1^2 \frac{1}{x}\,dx - 2\int_1^2 \frac{1}{0.1+0.2x}\,dx$$
$$= \Big[10 \ln x - 10 \ln|0.1+0.2x|\Big]_1^2$$
$$= 10 \ln 2 - 10 \ln 0.5 + 10 \ln 0.3$$
$$\approx 1.823$$

16. $\displaystyle\int_0^\pi \sec^2 \frac{x}{3} \tan \frac{x}{3}\,dx = 3\int_0^\pi \frac{1}{3}\sec^2 \frac{x}{3} \tan \frac{x}{3}\,dx = \frac{3}{2}\sec^2 \frac{x}{3}\Big]_0^\pi = \frac{9}{2}$

17. $\displaystyle\int_{1/4}^{1/2} \cos \pi x\,dx = \frac{1}{11}\int_{1/4}^{1/2} \pi \cos \pi x\,dx = \frac{1}{\pi}\sin \pi x\Big]_{1/4}^{1/2} = \frac{1}{\pi} - \frac{\sqrt{2}}{2\pi} \approx 0.093$

18. $\displaystyle\int_{\pi/4}^{\pi/2} \frac{e^{\cot x}}{\sin^2 x}\,dx = \int_{\pi/4}^{\pi/2} \csc^2 x e^{\cot x}\,dx = -\int_{\pi/4}^{\pi/2} -\csc^2 x e^{\cot x}\,dx = -e^{\cot x}\Big]_{\pi/4}^{\pi/2} = -1 + e \approx 1.718$

19. $P(a \le x \le b) = \displaystyle\int_a^b 2x^3 e^{x^2}\,dx$

Let $u = x^2$. Then $du = 2x\,dx$. Rewrite the integral as $\int_a^b x^2 e^{x^2}\,2x\,dx = \int_a^b ue^u\,du$. Using integration by parts, let $w = u$ and $dv = e^u\,du$. Then $dw = du$ and $v = e^u$.

$$\int_a^b x^2 e^{x^2}\,2x\,dx = \int_a^b ue^u\,du = ue^u\Big]_a^b - \int_a^b e^u\,du = \Big[ue^u - e^u\Big]_a^b = \Big[x^2 e^{x^2} - e^{x^2}\Big]_a^b$$

(a) $P(0 \le x \le 0.40) = \Big[x^2 e^{x^2} - e^{x^2}\Big]_0^{0.40} \approx 0.014$

(b) $P(0.80 \le x \le 1) = \Big[x^2 e^{x^2} - e^{x^2}\Big]_{0.80}^{1} \approx 0.683$

20. $\dfrac{dy}{dt} = ky\left(1 - \dfrac{y}{100{,}000}\right)$

$y = \dfrac{100{,}000}{1 + be^{-kt}}$

When $t = 0$,

$y = 25{,}000$: $25{,}000 = \dfrac{100{,}000}{1 + be^{-k(0)}}$, $b = 3$

When $t = 13$,

$y = 28{,}000$: $28{,}000 = \dfrac{100{,}000}{1 + 3e^{-k(13)}}$, $k \approx 0.01186$

$y = \dfrac{100{,}000}{1 + 3e^{-0.01186t}}$

21. $W = 139.8 + 37.33\sin(0.612t - 2.66)$

($t = 1$ corresponds to January.)

(a) $\text{Average} = \dfrac{1}{3-0}\int_0^3 [139.8 + 37.33\sin(0.612 - 2.66)]\,dt$

$= \dfrac{1}{3}\left[139.8t - \dfrac{37.33}{0.612}\cos(0.612t - 2.66)\right]_0^3 \approx 108.0$ thousand workers

(b) $\text{Average} = \dfrac{1}{6-3}\int_3^6 [139.8 + 37.33\sin(0.612t - 2.66)]\,dt$

$= \dfrac{1}{3}\left[139.8t - \dfrac{37.33}{0.612}\cos(0.612t - 2.66)\right]_3^6 \approx 142.8$ thousand workers

(c) $\text{Average} = \dfrac{1}{12-0}\int_0^{12} [139.8 + 37.33\sin(0.612t - 2.66)]\,dt$

$= \dfrac{1}{12}\left[139.8t - \dfrac{37.33}{0.612}\cos(0.612t - 2.66)\right]_0^{12} \approx 135.4$ thousand workers

Section 7.4 The Definite Integral as the Limit of a Sum

Skills Review

1. $\left[0, \dfrac{1}{3}\right]$ midpoint: $\dfrac{0 + \frac{1}{3}}{2} = \dfrac{1}{6}$

2. $\left[\dfrac{1}{10}, \dfrac{2}{10}\right]$ midpoint: $\dfrac{\frac{1}{10} + \frac{2}{10}}{2} = \dfrac{3}{20}$

3. $\left[\dfrac{3}{20}, \dfrac{4}{20}\right]$ midpoint: $\dfrac{\frac{3}{20} + \frac{4}{20}}{2} = \dfrac{7}{40}$

4. $\left[1, \dfrac{7}{6}\right]$ midpoint: $\dfrac{1 + \frac{7}{6}}{2} = \dfrac{13}{12}$

5. $\left[2, \dfrac{31}{15}\right]$ midpoint: $\dfrac{2 + \frac{31}{15}}{2} = \dfrac{61}{30}$

6. $\left[\dfrac{26}{9}, 3\right]$ midpoint: $\dfrac{\frac{26}{9} + 3}{2} = \dfrac{53}{18}$

7. $\displaystyle\lim_{x\to\infty} \frac{2x^2 + 4x - 1}{3x^2 - 2x} = \lim_{x\to\infty} \frac{2 + \frac{4}{x} - \frac{1}{x^2}}{3 - \frac{2}{x}} = \frac{2}{3}$

8. $\displaystyle\lim_{x\to\infty} \frac{4x + 5}{7x - 5} = \lim_{x\to\infty} \frac{4 + \frac{5}{x}}{7 - \frac{5}{x}} = \frac{4}{7}$

9. $\displaystyle\lim_{x\to\infty} \frac{x - 7}{x^2 + 1} = \lim_{x\to\infty} \frac{\frac{1}{x} - \frac{7}{x^2}}{1 + \frac{1}{x^2}} = 0$

10. $\displaystyle\lim_{x\to\infty} \frac{5x^3 + 1}{x^3 + x^2 + 4} = \lim_{x\to\infty} \frac{5 + \frac{1}{x^3}}{1 + \frac{1}{x} + \frac{4}{x^3}} = 5$

1. The midpoints of the four intervals are $\frac{1}{8}$, $\frac{3}{8}$, $\frac{5}{8}$, and $\frac{7}{8}$. The approximate area is

$$A \approx \frac{1-0}{4}\left[f\left(\frac{1}{8}\right) + f\left(\frac{3}{8}\right) + f\left(\frac{5}{8}\right) + f\left(\frac{7}{8}\right)\right]$$

$$= \frac{1}{4}\left[\frac{11}{4} + \frac{9}{4} + \frac{7}{4} + \frac{5}{4}\right] = 2.$$

The exact area is

$$A = \int_0^1 (-2x + 3)\,dx = \left[-x^2 + 3x\right]_0^1 = 2.$$

3. The midpoints of the four intervals are $\frac{1}{8}$, $\frac{3}{8}$, $\frac{5}{8}$, and $\frac{7}{8}$.

The approximate area is

$$A \approx \frac{1-0}{4}\left[f\left(\frac{1}{8}\right)+f\left(\frac{3}{8}\right)+f\left(\frac{5}{8}\right)+f\left(\frac{7}{8}\right)\right]$$
$$=\frac{1}{4}\left[\sqrt{\frac{1}{8}}+\sqrt{\frac{3}{8}}+\sqrt{\frac{5}{8}}+\sqrt{\frac{7}{8}}\right]$$
$$=\frac{1}{4}\left[\frac{\sqrt{2}}{4}+\frac{\sqrt{6}}{4}+\frac{\sqrt{10}}{4}+\frac{\sqrt{14}}{4}\right]\approx 0.6730.$$

The exact area is

$$A=\int_0^1 \sqrt{x}\,dx=\frac{2}{3}x^{3/2}\Big]_0^1=\frac{2}{3}\approx 0.6667.$$

5. The midpoints are $\frac{1}{4}$, $\frac{3}{4}$, $\frac{5}{4}$, and $\frac{7}{4}$.

The approximate area is

$$A \approx \frac{2-0}{4}\left[f\left(\frac{1}{4}\right)+f\left(\frac{3}{4}\right)+f\left(\frac{5}{4}\right)+f\left(\frac{7}{4}\right)\right]$$
$$=\frac{2}{4}\left[\frac{63}{16}+\frac{55}{16}+\frac{39}{16}+\frac{15}{16}\right]$$
$$=\frac{43}{8}=5.375.$$

The exact area is

$$A=\int_0^2 \left(4-x^2\right)dx$$
$$=\left[4x-\frac{x^3}{3}\right]_0^2$$
$$=\frac{16}{3}$$
$$\approx 5.333.$$

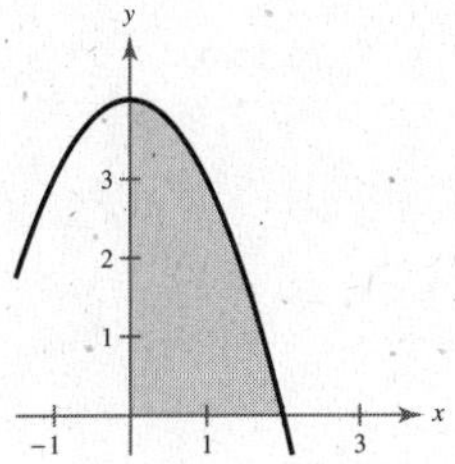

7. The midpoints are $-\frac{3}{4}$, $-\frac{1}{4}$, $\frac{1}{4}$, and $\frac{3}{4}$.

The approximate area is

$$A \approx \frac{1-(-1)}{4}\left[f\left(-\frac{3}{4}\right)+f\left(-\frac{1}{4}\right)+f\left(\frac{1}{4}\right)+f\left(\frac{3}{4}\right)\right]$$
$$=\frac{1}{2}\left(\frac{57}{16}+\frac{49}{16}+\frac{49}{16}+\frac{57}{16}\right)$$
$$=\frac{53}{8}$$
$$=6.625.$$

The exact area is

$$A=\int_{-1}^1 \left(x^2+3\right)dx$$
$$=\left[\frac{1}{3}x^3+3x\right]_{-1}^1$$
$$=\frac{20}{3}\approx 6.67.$$

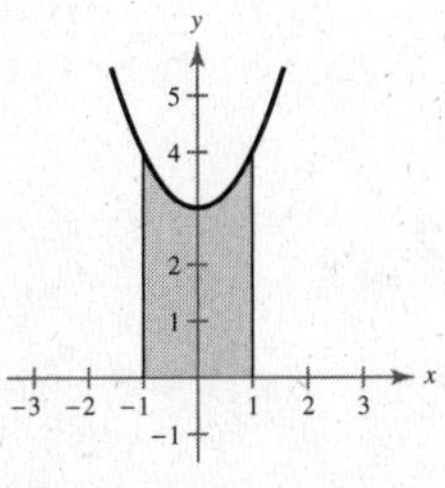

9. The midpoints are $\frac{5}{4}$, $\frac{7}{4}$, $\frac{9}{4}$, and $\frac{11}{4}$.

The approximate area is

$$A \approx \frac{3-1}{4}\left[f\left(\frac{5}{4}\right)+f\left(\frac{7}{4}\right)+f\left(\frac{9}{4}\right)+f\left(\frac{11}{4}\right)\right]$$
$$=\frac{1}{2}\left[\frac{25}{8}+\frac{49}{8}+\frac{81}{8}+\frac{121}{8}\right]=\frac{69}{4}=17.25.$$

The exact area is

$$A=\int_1^3 2x^2\,dx$$
$$=\frac{2x^3}{3}\Big]_1^3$$
$$=\frac{52}{3}$$
$$\approx 17.3333.$$

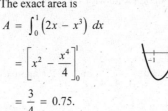

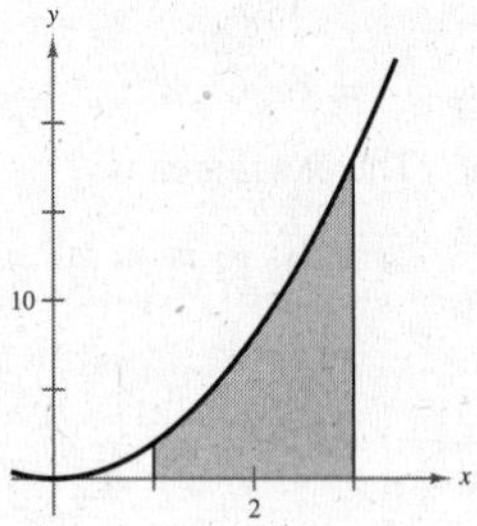

11. The midpoints are $\frac{1}{8}$, $\frac{3}{8}$, $\frac{5}{8}$, and $\frac{7}{8}$.

The approximate area is

$$A \approx \frac{1-0}{4}\left[f\left(\frac{1}{8}\right)+f\left(\frac{3}{8}\right)+f\left(\frac{5}{8}\right)+f\left(\frac{7}{8}\right)\right]$$
$$=\frac{1}{4}\left[\frac{127}{512}+\frac{357}{512}+\frac{515}{512}+\frac{553}{512}\right]$$
$$=\frac{97}{128}\approx 0.758.$$

The exact area is

$$A=\int_0^1 \left(2x-x^3\right)dx$$
$$=\left[x^2-\frac{x^4}{4}\right]_0^1$$
$$=\frac{3}{4}=0.75.$$

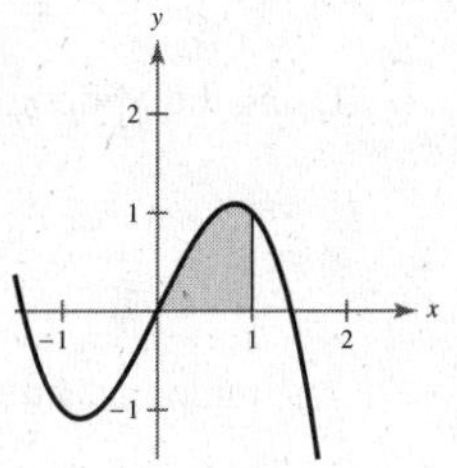

13. The midpoints are $-\frac{7}{8}$, $-\frac{5}{8}$, $-\frac{3}{8}$, and $-\frac{1}{8}$.

The approximate area is

$$A \approx \frac{0-(-1)}{4}\left[f\left(-\frac{7}{8}\right)+f\left(-\frac{5}{8}\right)+f\left(-\frac{3}{8}\right)+f\left(-\frac{1}{8}\right)\right]$$
$$=\left[\frac{735}{512}+\frac{325}{512}+\frac{99}{512}+\frac{9}{512}\right]$$
$$=\frac{73}{178}\approx 0.5703.$$

The exact area is

$$A=\int_{-1}^0 \left(x^2-x^3\right)dx$$
$$=\left[\frac{x^3}{3}-\frac{x^4}{4}\right]_{-1}^0$$
$$=\frac{7}{12}\approx 0.5833.$$

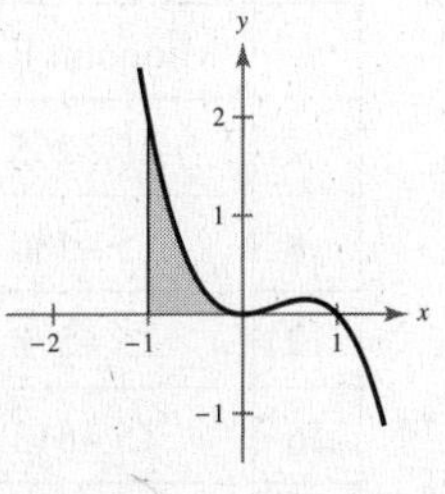

15. The midpoints are $\frac{3}{8}, \frac{9}{8}, \frac{15}{8}$, and $\frac{21}{8}$.

The approximate area is

$$A \approx \frac{3-0}{4}\left[f\left(\frac{3}{8}\right) + f\left(\frac{9}{8}\right) + f\left(\frac{15}{8}\right) + f\left(\frac{21}{8}\right)\right]$$

$$= \frac{3}{4}\left[\frac{189}{512} + \frac{1215}{512} + \frac{2025}{512} + \frac{1323}{512}\right]$$

$$= \frac{891}{128}$$

$$\approx 6.961.$$

The exact area is

$$A = \int_0^3 \left(3x^2 - x^3\right) dx$$

$$= \left[x^3 - \frac{x^4}{4}\right]_0^3$$

$$= \frac{27}{4} = 6.75.$$

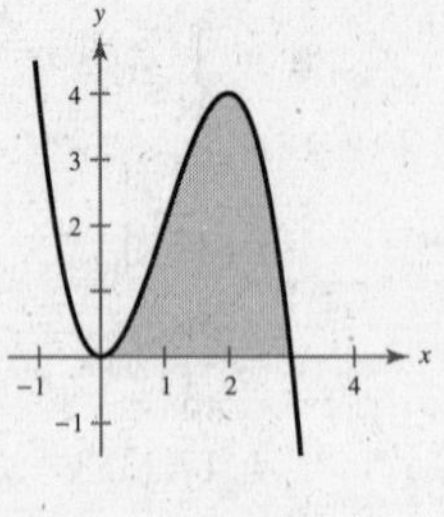

17. $\int_0^4 \left(2x^2 + 3\right) dx = \left[\frac{2x^3}{3} + 3x\right]_0^4 = \frac{164}{3} \approx 54.6667$

Using the Midpoint Rule with $n = 31$, you get 54.66.

19. $\int_1^2 \left(2x^2 - x + 1\right) dx = \left[\frac{2x^3}{3} - \frac{x^2}{2} + x\right]_1^2 = \frac{25}{6} \approx 4.166$

Using the Midpoint Rule with $n = 5$, you get 4.16.

21. $\int_1^4 \frac{1}{x+1}\, dx = \ln(x+1)\Big]_1^4 = \ln 5 - \ln 2 \approx 0.916$

Using the Midpoint Rule with $n = 5$, you get 0.913.

23. The midpoints are $\frac{9}{4}, \frac{11}{4}, \frac{13}{4}$, and $\frac{15}{4}$.

The approximate area is

$$A \approx \frac{4-2}{4}\left[f\left(\frac{9}{4}\right) + f\left(\frac{11}{4}\right) + f\left(\frac{13}{4}\right) + f\left(\frac{15}{4}\right)\right]$$

$$= \frac{1}{2}\left[\frac{9}{16} + \frac{11}{16} + \frac{13}{16} + \frac{15}{16}\right] = 1.5.$$

The exact area is $A = \int_2^4 \frac{1}{4} y\, dy = \frac{y^2}{8}\Big]_2^4 = 1.5.$

25. The midpoints are $\frac{1}{2}, \frac{3}{2}, \frac{5}{2}$, and $\frac{7}{2}$.

The approximate area is

$$A \approx \frac{4-0}{4}\left[f\left(\frac{1}{2}\right) + f\left(\frac{3}{2}\right) + f\left(\frac{5}{2}\right) + f\left(\frac{7}{2}\right)\right]$$

$$= \frac{5}{4} + \frac{13}{4} + \frac{29}{4} + \frac{53}{4}$$

$$= 25.$$

The exact area is $A = \int_0^4 \left(y^2 + 1\right) dy$

$$= \left[\frac{y^3}{3} + y\right]_0^4$$

$$= \frac{76}{3} = 25.33.$$

27. The midpoints are $\frac{1}{4}, \frac{3}{4}, \frac{5}{4}$, and $\frac{7}{4}$. The approximation is

$$\int_0^2 \frac{1}{x+1}\, dx \approx \frac{2-0}{4}\left[f\left(\frac{1}{4}\right) + f\left(\frac{3}{4}\right) + f\left(\frac{5}{4}\right) + f\left(\frac{7}{4}\right)\right] = \frac{1}{2}\left[\frac{4}{5} + \frac{4}{7} + \frac{4}{9} + \frac{4}{11}\right] = \frac{3776}{3465} \approx 1.0898$$

29. The midpoints are $-\frac{3}{4}, -\frac{1}{4}, \frac{1}{4}$, and $\frac{3}{4}$.

The approximation is

$$\int_{-1}^1 \frac{1}{x^2+1}\, dx \approx \frac{1-(-1)}{4}\left[f\left(-\frac{3}{4}\right) + f\left(-\frac{1}{4}\right) + f\left(\frac{1}{4}\right) + f\left(\frac{3}{4}\right)\right] = \frac{8}{25} + \frac{8}{17} + \frac{8}{17} + \frac{8}{25} = \frac{672}{425} \approx 1.5812.$$

31. $\int_0^4 \sqrt{2 + 3x^2}\, dx$

n	Midpoint Rule
4	15.3965
8	15.4480
12	15.4578
16	15.4613
20	15.4628

33. The midpoints are $\frac{3}{20}, \frac{9}{20}, \frac{3}{4}, \frac{21}{20}, \frac{27}{20}, \frac{33}{20}, \frac{39}{20}, \frac{9}{4}, \frac{51}{20}$, and $\frac{57}{20}$.

The approximate area is

$$A \approx \frac{3-0}{10}\left[f\left(\frac{3}{20}\right) + f\left(\frac{9}{20}\right) + f\left(\frac{3}{4}\right) + f\left(\frac{21}{20}\right) + f\left(\frac{27}{20}\right) + f\left(\frac{33}{20}\right) + f\left(\frac{39}{20}\right) + f\left(\frac{9}{4}\right) + f\left(\frac{51}{20}\right) + f\left(\frac{57}{20}\right)\right]$$

$$\approx \frac{3}{10}\left[0.0296 + 0.1602 + 0.3603 + 0.6264 + 0.9636 + 1.3826 + 1.9018 + 2.5513 + 3.3816 + 4.4866\right]$$

$$= 4.7532$$

35. The midpoints for velocity are 14.65, 40.3, 58.65, and 69.65.

Distance $\approx \frac{20}{4}(14.65 + 40.3 + 58.65 + 69.65) = 916.25$ feet

37. (a) Area $\approx \frac{160}{8}(25 + 52 + 68 + 82 + 77.5 + 74 + 77.5 + 40) = 9920 \text{ ft}^2$

Section 7.5 Numerical Integration

Skills Review

1. $f(x) = \frac{1}{x}$

$f'(x) = -\frac{1}{x^2}$

$f''(x) = \frac{2}{x^3}$

2. $f(x) = \ln(2x + 1)$

$f'(x) = \frac{2}{2x + 1}$

$f''(x) = -\frac{4}{(2x + 1)^2}$

$f'''(x) = \frac{16}{(2x + 1)^3}$

$f^{(4)}(x) = -\frac{96}{(2x + 1)^4}$

3. $f(x) = 2 \ln x$

$f'(x) = \frac{2}{x}$

$f''(x) = -\frac{2}{x^2}$

$f'''(x) = \frac{4}{x^3}$

$f^{(4)}(x) = -\frac{12}{x^4}$

4. $f(x) = x^3 - 2x^2 + 7x - 12$

$f'(x) = 3x^2 - 4x + 7$

$f''(x) = 6x - 4$

5. $f(x) = e^{2x}$

$f'(x) = 2e^{2x}$

$f''(x) = 4e^{2x}$

$f'''(x) = 8e^{2x}$

$f^{(4)}(x) = 16e^{2x}$

6. $f(x) = e^{x^2}$

$f'(x) = 2xe^{x^2}$

$f''(x) = 4x^2e^{x^2} + 2e^{x^2}$

7. $f(x) = -x^2 + 6x + 9 \quad [0, 4]$

$f'(x) = -2x + 6 = 0$ when $x = 3$

Interval	$\infty < x < 3$	$3 < x < \infty$
Test value	$x = 0$	$x = 4$
Sign of $f'(x)$	$f'(x) > 0$	$f'(x) < 0$
Conclusion	Increasing	Decreasing

$f(0) = 9$, $f(3) = 18$, and $f(4) = 17$, so $(3, 18)$ is the absolute maximum.

Skills Review —continued—

8. $f(x) = \frac{8}{x^3}$ $[1, 2]$

$f'(x) = -\frac{24}{x^4}$

Because $-\frac{24}{x^4} \neq 0$ for any value of x, there are no critical numbers. $f(1) = 8$ and $f(2) = 1$, so $(1, 8)$ is the absolute maximum.

9.
$$\frac{1}{4n^2} < 0.001$$
$$\frac{1}{0.001} < 4n^2$$
$$1000 < 4n^2$$
$$250 < n^2$$
$$n^2 - 250 = 0$$
$$n^2 = 250$$
$$n = \pm 5\sqrt{10}$$

So, $n < -5\sqrt{10}$ or $n > 5\sqrt{10}$.

Test interval	$(-\infty, -5\sqrt{10})$	$(-5\sqrt{10}, 5\sqrt{10})$	$(5\sqrt{10}, \infty)$
n-value	-16	1	16
Function value	0.00098	0.25	0.00098
Conclusion	Does satisfy	Does not satisfy	Does satisfy

10.
$$\frac{1}{16n^4} < 0.0001$$
$$\frac{1}{0.0001} < 16n^4$$
$$10{,}000 < 16n^4$$
$$625 < n^4$$
$$n^4 = 625$$
$$n = \pm 5$$

So, $n < -5$ or $n > 5$.

Test interval	$(-\infty, -5)$	$(-5, 5)$	$(5, \infty)$
n-value	-6	1	6
Function value	0.000045	0.0625	0.000045
Conclusion	Does satisfy	Does not satisfy	Does satisfy

1. Exact: $\int_0^2 x^2\,dx = \frac{1}{3}x^3\Big]_0^2 = \frac{8}{3} \approx 2.6667$

Trapezoidal Rule: $\int_0^2 x^2\,dx \approx \frac{1}{4}\left[0 + 2\left(\frac{1}{2}\right)^2 + 2(1)^2 + 2\left(\frac{3}{2}\right)^2 + (2)^2\right] = \frac{11}{4} = 2.75$

Simpson's Rule: $\int_0^2 x^2\,dx \approx \frac{1}{6}\left[0 + 4\left(\frac{1}{2}\right)^2 + 2(1)^2 + 4\left(\frac{3}{2}\right)^2 + (2)^2\right] = \frac{8}{3} \approx 2.6667$

3. Exact: $\int_0^2 (x^4 + 1)\,dx = \frac{x^5}{5} + x\Big]_0^2 = \frac{42}{5} = 8.4$

Trapezoidal Rule: $\int_0^2 (x^4 + 1)\,dx \approx \frac{1}{4}\left[1 + 2\left(\frac{1}{16} + 1\right) + 2(1 + 1) + 2\left(\frac{81}{16} + 1\right) + 17\right] = \frac{36.25}{4} = 9.0625$

Simpson's Rule: $\int_0^2 (x^4 + 1)\,dx \approx \frac{1}{6}\left[1 + 4\left(\frac{1}{16} + 1\right) + 2(1 + 1) + 4\left(\frac{81}{16} + 1\right) + 17\right] = \frac{50.5}{6} \approx 8.4167$

5. Exact: $\int_0^2 x^3\,dx = \frac{1}{4}x^4\Big]_0^2 = 4$

Trapezoidal Rule: $\int_0^2 x^3\,dx \approx \frac{1}{8}\left[0 + 2\left(\frac{1}{4}\right)^3 + 2\left(\frac{1}{2}\right)^3 + 2\left(\frac{3}{4}\right)^3 + 2(1)^3 + 2\left(\frac{5}{4}\right)^3 + 2\left(\frac{3}{2}\right)^3 + 2\left(\frac{7}{4}\right)^3 + 8\right] = \frac{65}{16} = 4.0625$

Simpson's Rule: $\int_0^2 x^3\,dx \approx \frac{1}{12}\left[0 + 4\left(\frac{1}{4}\right)^3 + 2\left(\frac{1}{2}\right)^3 + 4\left(\frac{3}{4}\right)^3 + 2(1)^3 + 4\left(\frac{5}{4}\right)^3 + 2\left(\frac{3}{2}\right)^3 + 4\left(\frac{7}{4}\right)^3 + 8\right] = 4$

7. Exact: $\int_1^2 \frac{1}{x}\,dx = \ln|x|\Big]_1^2 = \ln 2 \approx 0.6931$

Trapezoidal Rule: $\int_1^2 \frac{1}{x}\,dx \approx \frac{1}{16}\left[1 + 2\left(\frac{8}{9}\right) + 2\left(\frac{4}{5}\right) + 2\left(\frac{8}{11}\right) + 2\left(\frac{2}{3}\right) + 2\left(\frac{8}{13}\right) + 2\left(\frac{4}{7}\right) + 2\left(\frac{8}{15}\right) + \frac{1}{2}\right] \approx 0.6941$

Simpson's Rule: $\int_1^2 \frac{1}{x}\,dx \approx \frac{1}{24}\left[1 + 4\left(\frac{8}{9}\right) + 2\left(\frac{4}{5}\right) + 4\left(\frac{8}{11}\right) + 2\left(\frac{2}{3}\right) + 4\left(\frac{8}{13}\right) + 2\left(\frac{4}{7}\right) + 4\left(\frac{8}{15}\right) + \frac{1}{2}\right] \approx 0.6932$

9. Exact: $\int_0^4 \sqrt{x}\,dx = \frac{2}{3}x^{3/2}\Big]_0^4 = \frac{16}{3} \approx 5.3333$

Trapezoidal Rule: $\int_0^4 \sqrt{x}\,dx \approx \frac{1}{4}\left(0 + 2\sqrt{\frac{1}{2}} + 2 + 2\sqrt{\frac{3}{2}} + 2\sqrt{2} + 2\sqrt{\frac{5}{2}} + 2\sqrt{3} + 2\sqrt{\frac{7}{2}} + 2\right) \approx 5.2650$

Simpson's Rule: $\int_0^4 \sqrt{x}\,dx \approx \frac{1}{6}\left(0 + 4\sqrt{\frac{1}{2}} + 2 + 4\sqrt{\frac{3}{2}} + 2\sqrt{2} + 4\sqrt{\frac{5}{2}} + 2\sqrt{3} + 4\sqrt{\frac{7}{2}} + 2\right) \approx 5.3046$

11. Exact: $\int_4^9 \sqrt{x}\,dx = \frac{2}{3}x^{3/2}\Big]_4^9 = \frac{38}{3} \approx 12.6667$

Trapezoidal Rule: $\int_4^9 \sqrt{x}\,dx \approx \frac{5}{16}\left[2 + 2\sqrt{\frac{37}{8}} + 2\sqrt{\frac{21}{4}} + 2\sqrt{\frac{47}{8}} + 2\sqrt{\frac{13}{2}} + 2\sqrt{\frac{57}{8}} + 2\sqrt{\frac{31}{4}} + 2\sqrt{\frac{67}{8}} + 3\right] \approx 12.6640$

Simpson's Rule: $\int_4^9 \sqrt{x}\,dx \approx \frac{5}{24}\left[2 + 4\sqrt{\frac{37}{8}} + 2\sqrt{\frac{21}{4}} + 4\sqrt{\frac{47}{8}} + 2\sqrt{\frac{13}{2}} + 4\sqrt{\frac{57}{8}} + 2\sqrt{\frac{31}{4}} + 4\sqrt{\frac{67}{8}} + 3\right] \approx 12.6667$

13. Exact: $\int_0^1 \frac{1}{1+x}\,dx = \ln|1+x|\Big]_0^1 = \ln 2 \approx 0.6931$

Trapezoidal Rule: $\int_0^1 \frac{1}{1+x}\,dx \approx \frac{1}{8}\left[1 + 2\left(\frac{4}{5}\right) + 2\left(\frac{2}{3}\right) + 2\left(\frac{4}{7}\right) + \frac{1}{2}\right] \approx 0.6970$

Simpson's Rule: $\int_0^1 \frac{1}{1+x}\,dx \approx \frac{1}{12}\left[1 + 4\left(\frac{4}{5}\right) + 2\left(\frac{2}{3}\right) + 4\left(\frac{4}{7}\right) + \frac{1}{2}\right] \approx 0.6933$

15. (a) Trapezoidal Rule: $\int_0^1 \frac{1}{1+x^2}\,dx \approx \frac{1}{8}\left[1 + 2\left(\frac{1}{1+\frac{1}{16}}\right) + 2\left(\frac{1}{1+\frac{1}{4}}\right) + 2\left(\frac{1}{1+\frac{9}{16}}\right) + \frac{1}{2}\right] \approx 0.783$

(b) Simpson's Rule: $\int_0^1 \frac{1}{1+x^2}\,dx \approx \frac{1}{12}\left[1 + 4\left(\frac{1}{1+\frac{1}{16}}\right) + 2\left(\frac{1}{1+\frac{1}{4}}\right) + 4\left(\frac{1}{1+\frac{9}{16}}\right) + \frac{1}{2}\right] \approx 0.785$

17. (a) Trapezoidal Rule: $\int_0^2 \sqrt{1+x^3}\,dx \approx \frac{1}{4}\left[1 + 2\sqrt{\frac{9}{8}} + 2\sqrt{2} + 2\sqrt{\frac{35}{8}} + 3\right] \approx 3.283$

(b) Simpson's Rule: $\int_0^2 \sqrt{1+x^3}\,dx \approx \frac{1}{6}\left[1 + 4\sqrt{\frac{9}{8}} + 2\sqrt{2} + 4\sqrt{\frac{35}{8}} + 3\right] \approx 3.240$

19. (a) Trapezoidal Rule: $\int_0^1 \sqrt{1-x^2}\,dx \approx \frac{1}{8}\left[1 + 2\sqrt{\frac{15}{16}} + 2\sqrt{\frac{3}{4}} + 2\sqrt{\frac{7}{16}} + 0\right] \approx 0.749$

(b) Simpson's Rule: $\int_0^1 \sqrt{1-x^2}\,dx \approx \frac{1}{12}\left[1 + 4\sqrt{\frac{15}{16}} + 2\sqrt{\frac{3}{4}} + 4\sqrt{\frac{7}{16}} + 0\right] \approx 0.771$

21. (a) Trapezoidal Rule: $\int_0^2 e^{-x^2}\,dx \approx \frac{1}{2}\left[e^0 + 2e^{-1} + e^{-4}\right] \approx 0.877$

(b) Simpson's Rule: $\int_0^2 e^{-x^2}\,dx \approx \frac{1}{3}\left[e^0 + 4e^{-1} + e^{-4}\right] \approx 0.830$

23. (a) Trapezoidal Rule: $\int_0^3 \frac{1}{2 - 2x + x^2}\,dx \approx \frac{1}{4}\left[\frac{1}{2} + 2\left(\frac{4}{5}\right) + 2(1) + 2\left(\frac{4}{5}\right) + 2\left(\frac{1}{2}\right) + 2\left(\frac{4}{13}\right) + \frac{1}{5}\right] \approx 1.879$

(b) Simpson's Rule: $\int_0^3 \frac{1}{2 - 2x + x^2}\,dx \approx \frac{1}{6}\left[\frac{1}{2} + 4\left(\frac{4}{5}\right) + 2(1) + 4\left(\frac{4}{5}\right) + 2\left(\frac{1}{2}\right) + 4\left(\frac{4}{13}\right) + \frac{1}{5}\right] \approx 1.888$

25. (a) Trapezoidal Rule:

$$\int_0^{\pi/4} x\tan x\,dx \approx \frac{\pi}{32}\left[0 + 2(0.0391) + 2(0.1627) + 2(0.3936) + \frac{\pi}{4}\right] \approx 0.194$$

(b) Simpson's Rule:

$$\int_0^{\pi/4} x\tan x\,dx \approx \frac{\pi}{48}\left[0 + 4(0.0391) + 2(0.1627) + 4(0.3936) + \frac{\pi}{4}\right] \approx 0.186$$

27. (a) Trapezoidal Rule: $A \approx \frac{1}{2}\left[3 + 2(7) + 2(9) + 2(7) + 0\right] = 24.5$

(b) Simpson's Rule: $A \approx \frac{1}{3}\left[3 + 4(7) + 2(9) + 4(7) + 0\right] = \frac{77}{3} \approx 25.667$

29. $P(0 \le x \le 1) = \int_0^1 \frac{1}{\sqrt{2\pi}} e^{-x^2/2}\,dx$

Using a program similar to the Simpson's Rule program, when $n = 6$, $P \approx 0.3413 = 34.13\%$.

31. $P(0 \le x \le 4) = \int_0^4 \frac{1}{\sqrt{2\pi}} e^{-x^2/2}\,dx$

Using a program similar to the Simpson's Rule program, when $n = 6$, $P \approx 0.499958 = 49.996\%$.

33. $A \approx \frac{1000}{3(10)}\left[125 + 4(125) + 2(120) + 4(112) + 2(90) + 4(90) + 2(95) + 4(88) + 2(75) + 4(35) + 0\right] = 89{,}500$ square feet

35.

$$f(x) = x^3$$
$$f'(x) = 3x^2$$
$$f''(x) = 6x$$
$$f'''(x) = 6$$
$$f^{(4)}(x) = 0$$

(a) Trapezoidal Rule: Because $|f''(x)|$ is maximum in $[0, 2]$ when $x = 2$ and $|f''(2)| = 12$, you have

$$|\text{Error}| \le \frac{(2 - 0)^3}{12(4)^2}(12) = 0.5.$$

(b) Simpson's Rule: Because $f^{(4)}(x) = 0$ for all x, you have

$$|\text{Error}| \le \frac{(2 - 0)^5}{180(4)^4}(0) = 0.$$

37.

$$f(x) = e^{x^3}$$
$$f'(x) = 3x^2e^{x^3}$$
$$f''(x) = 3(3x^4 + 2x)e^{x^3}$$
$$f'''(x) = 3(9x^6 + 18x^3 + 2)e^{x^3}$$
$$f^{(4)}(x) = 9(9x^8 + 36x^5 + 20x^2)e^{x^3}$$

(a) Trapezoidal Rule: Because $|f''(x)|$ is maximum in $[0, 1]$ when $x = 1$ and $|f''(1)| = 15e$, you have

$$|\text{Error}| \le \frac{(1 - 0)^3}{12(4)^2}(15e) = \frac{5e}{64} \approx 0.212.$$

(b) Simpson's Rule: Because $|f^{(4)}(x)|$ is maximum in $[0, 1]$ when $x = 1$ and $|f^{(4)}(1)| = 585e$, you have

$$|\text{Error}| \le \frac{(1 - 0)^5}{180(4)^4}(585e) = \frac{13e}{1024} \approx 0.035.$$

39. $f(x) = x^3$

$f'(x) = 3x^2$

$f''(x) = 6x$

$f'''(x) = 6$

$f^{(4)}(x) = 0$

(a) Trapezoidal Rule: Because $|f''(x)|$ is maximum in $[0, 1]$ when $x = 1$ and $|f''(1)| = 6$, you have

$$|\text{Error}| \le \frac{(1-0)^3}{12n^2}(6) \le 0.0001$$

$$\frac{1}{2n^2} \le 0.0001$$

$$n^2 > 5000$$

$$n > 70.7.$$

Let $n = 71$.

(b) Simpson's Rule: Because $f^{(4)}(x) = 0$ in $[0, 1]$ for all x, n can be any value and will result in zero error. Let $n = 1$.

41. $f(x) = e^{2x}$

$f'(x) = 2e^{2x}$

$f''(x) = 4e^{2x}$

$f'''(x) = 8e^{2x}$

$f^{(4)}(x) = 16e^{2x}$

(a) Trapezoidal Rule: Because $|f''(x)|$ is maximum in $[1, 3]$ when $x = 3$ and $|f''(3)| = 4e^6$, you have

$$|\text{Error}| \le \frac{(3-1)^3}{12n^2}(4e^6) < 0.0001$$

$$\frac{8e^6}{3n^2} < 0.0001$$

$$n^2 > 10{,}758{,}101.16$$

$$n > 3279.95.$$

Let $n = 3280$.

(b) Simpson's Rule: Because $|f^{(4)}(x)|$ is maximum in $[1, 3]$ when $x = 3$ and $|f^{(4)}(3)| = 16e^6$, you have

$$|\text{Error}| \le \frac{(3-1)^5}{180n^4}(16e^6) < 0.0001$$

$$\frac{128e^6}{45n^4} < 0.0001$$

$$n^4 > 11{,}475{,}307.90$$

$$n > 58.2.$$

Let $n = 60$ (n must be even).

43. Using a program similar to the Simpson's Rule program, when $n = 100$,

$$\int_1^4 x\sqrt{x+4}\,dx \approx 19.5215.$$

45. Using a program similar to the Simpson's Rule program, when $n = 100$,

$$\int_2^5 10xe^{-x}\,dx \approx 3.6558.$$

47. Using a program similar to the Simpson's Rule program, when $n = 4$,

$$A = \int_1^5 x\sqrt[3]{x+4}\,dx \approx 23.3750.$$

49. $y = \dfrac{x^2}{800}$

$y' = \dfrac{x}{400}$

Arc length $= \displaystyle\int_{-200}^{200} \sqrt{1 + \frac{x^2}{160{,}000}}\, dx$

For $n = 12$, the length of the cable is about 416.1 feet.

51. (a) $\displaystyle\int_7^{15} f(x)\, dx \approx \frac{1}{3}\left[15.1 + 4(14.7) + 2(15.1) + 4(16.4) + 2(17.0) + 4(17.8) + 2(18.3) + 4(20.0) + 20.6\right] = \frac{4121}{30}$

≈ 137.37 billion board feet

Average $= \dfrac{4121}{30}\left(\dfrac{1}{8}\right) \approx 17.17$ billion board feet per year

(b) $\displaystyle\int_7^{15} \left(6.613 + 0.93t + 2095.7e^{-t}\right) dt$

$\approx \dfrac{1}{3}\left[15.03 + 4(14.76) + 2(15.24) + 4(16.01) + 2(16.88) + 4(17.79) + 2(18.71) + 4(19.63) + 20.56\right] = \dfrac{41{,}001}{300}$

≈ 136.67 billion board feet

Average $= \dfrac{136.67}{8} \approx 17.08$ billion board feet per year

(c) The results of parts (a) and (b) are approximately the same.

53. $C = \displaystyle\int_0^{12} \left[8 - \ln\left(t^2 - 2t + 4\right)\right] dt$

$\approx \frac{1}{2}\left[6.61 + 4(6.82) + 2(6.05) + 4(5.28) + 2(4.67) + 4(4.19) + 2(3.80) + 4(3.46) + 3.18\right] \approx 58.915$ milligrams

55. $S = \displaystyle\int_0^6 1000t^2e^{-t}\, dt \approx \frac{1}{6}[0 + 4(151.63) + 2(367.88) + 4(502.04) + 2(541.34) + 4(513.03) + 2(448.08)$

$+ 4(369.92) + 2(293.05) + 4(224.96) + 2(168.45) + 4(123.62) + 89.24]$

≈ 1878 subscribers

Section 7.6 Improper Integrals

Skills Review

1. $\displaystyle\lim_{x\to 2} (2x + 5) = 2(2) + 5 = 9$

2. $\displaystyle\lim_{x\to 1} \left(\frac{1}{x} + 2x^2\right) = \frac{1}{1} + 2\left(1^2\right) = 3$

3. $\displaystyle\lim_{x\to -4} \frac{x + 4}{x^2 - 16} = \lim_{x\to -4} \frac{x + 4}{(x + 4)(x - 4)}$

$\displaystyle= \lim_{x\to -4} \frac{1}{x - 4}$

$= \dfrac{1}{-4 - 4}$

$= -\dfrac{1}{8}$

4. $\displaystyle\lim_{x\to 0} \frac{x^2 - 2x}{x^3 + 3x^2} = \lim_{x\to 0} \frac{x(x - 2)}{x^2(x + 3)} = \lim_{x\to 0} \frac{x - 2}{x(x + 3)}$

$\displaystyle\lim_{x\to 0^-} \frac{x - 2}{x(x + 3)} = \infty$

$\displaystyle\lim_{x\to 0^+} \frac{x - 2}{x(x + 3)} = -\infty$

Limit does not exist.

5. $\displaystyle\lim_{x\to 1} \frac{1}{\sqrt{x - 1}} = \infty$

Limit does not exist.

6. $\displaystyle\lim_{x\to -3} \frac{x^2 + 2x - 3}{x + 3} = \lim_{x\to -3} \frac{(x + 3)(x - 1)}{x + 3}$

$\displaystyle= \lim_{x\to -3} x - 1 = -3 - 1 = -4$

Skills Review —*continued*—

7. $\frac{4}{3}(2x-1)^3$

(a) $\frac{4}{3}(2b-1)^3$

(b) $\frac{4}{3}(2\cdot 0-1)^3 = -\frac{4}{3}$

8. $\dfrac{1}{x-5} + \dfrac{3}{(x-2)^2}$

(a) $\dfrac{1}{b-5} + \dfrac{3}{(b-2)^2}$

(b) $\dfrac{1}{0-5} + \dfrac{3}{(0-2)^2} = -\dfrac{1}{5} + \dfrac{3}{4} = \dfrac{11}{20}$

9. $\ln(5-3x^2) - \ln(x+1)$

(a) $\ln(5-3b^2) - \ln(b+1)$

(b) $\ln(5-3\cdot 0^2) - \ln(0+1) = \ln 5 - \ln 1$
$= \ln 5$
≈ 1.609

10. $e^{3x^2} + e^{-3x^2}$

(a) $e^{3b^2} + e^{-3b^2} = e^{-3b^2}\left(e^{6b^2} + 1\right)$

(b) $e^{3(0)^2} + e^{-3(0)^2} = e^0 + e^0$
$= 1 + 1$
$= 2$

1. The integral is improper because the function has an infinite discontinuity in $[0, 1]$.

3. The integral is not improper.

5. The integral is improper because it has an infinite discontinuity in $[0, 4]$.

This integral converges because

$$\int_0^4 \frac{1}{\sqrt{x}}\,dx = \lim_{b\to 0^+}\left[2\sqrt{x}\right]_b^4 = 4.$$

7. The integral is improper because it has an infinite discontinuity in $[0, 2]$.

This integral converges because

$$\int_0^2 \frac{1}{(x-1)^{2/3}}\,dx = \int_0^1 \frac{1}{(x-1)^{2/3}}\,dx + \int_1^2 \frac{1}{(x-1)^{2/3}}\,dx$$
$$= \lim_{b\to 1^-}\left[3(x-1)^{1/3}\right]_0^b + \lim_{a\to 1^+}\left[3(x-1)^{1/3}\right]_a^2 = 3 + 3 = 6.$$

9. The integral is improper because one of the limits of integration is infinite.

This integral converges because

$$\int_0^\infty e^{-x}\,dx = \lim_{b\to\infty}\left[-e^{-x}\right]_0^b = 0 + 1 = 1.$$

$-1 - (-e^{-(-\infty)}) = -1 - (-e^{\infty}) = -1 - (-\infty)$
$= -1 + \infty$

11. This integral converges because

$$\int_1^\infty \frac{1}{x^2}\,dx = \lim_{b\to\infty}\left[-\frac{1}{x}\right]_1^b = 0 + 1 = 1.$$

13. This integral diverges because

$$\int_0^\infty e^{x/3}\,dx = \lim_{b\to\infty}\left[3e^{x/3}\right]_0^b = \infty.$$

15. This integral diverges because

$$\int_5^\infty \frac{x}{\sqrt{x^2-16}}\,dx = \lim_{b\to\infty}\left[\sqrt{x^2-16}\right]_5^b = \infty - 3 = \infty.$$

17. This integral diverges because

$$\int_{-\infty}^0 e^{-x}\,dx = \lim_{a\to-\infty}\left[-e^{-x}\right]_a^0 = -1 + \infty = \infty.$$

19. This integral diverges because

$$\int_1^\infty \frac{e^{\sqrt{x}}}{\sqrt{x}}\,dx = \lim_{b\to\infty}\left[2e^{\sqrt{x}}\right]_1^b = \lim_{b\to\infty}\left[2e^{\sqrt{b}} - 2e\right] = \infty.$$

21. This integral converges because

$$\int_{-\infty}^{\infty} 2xe^{-3x^2}\,dx = \int_{-\infty}^{0} 2xe^{-3x^2}\,dx + \int_{0}^{\infty} 2xe^{-3x^2}\,dx$$
$$= \lim_{a\to-\infty}\left[-\tfrac{1}{3}e^{-3x^2}\right]_a^0 + \lim_{b\to\infty}\left[-\tfrac{1}{3}e^{-3x^2}\right]_0^b$$
$$= \left(-\tfrac{1}{3} + 0\right) + \left(0 + \tfrac{1}{3}\right) = 0.$$

23. This integral diverges because

$$\int_0^1 \frac{1}{1-x}\,dx = \lim_{b\to1^-}\left[-\ln|1-x|\right]_0^b = \infty - 0 = \infty.$$

25. This integral converges because

$$\int_0^9 \frac{1}{\sqrt{9-x}}\,dx = \lim_{b\to9^-}\left[-2\sqrt{9-x}\right]_0^b$$
$$= 0 - \left(-2\sqrt{9}\right)$$
$$= 6.$$

27. This integral diverges because

$$\int_0^1 \frac{1}{x^2}\,dx = \lim_{b\to0^+}\left[-\frac{1}{x}\right]_b^1$$
$$= -1 + \infty$$
$$= \infty.$$

29. This integral converges because

$$\int_0^2 \frac{1}{\sqrt[3]{x-1}}\,dx = \int_0^1 \frac{1}{\sqrt[3]{x-1}}\,dx + \int_1^2 \frac{1}{\sqrt[3]{x-1}}\,dx$$
$$= \lim_{b\to1^-}\left[\frac{3}{2}(x-1)^{2/3}\right]_0^b + \lim_{a\to1^+}\left[\frac{3}{2}(x-1)^{2/3}\right]_a^2$$
$$= -\frac{3}{2} + \frac{3}{2} = 0.$$

31. This integral converges because

$$\int_3^4 \frac{1}{\sqrt{x^2-9}}\,dx = \lim_{a\to3^+}\left[\ln\left(x + \sqrt{x^2-9}\right)\right]_a^4$$
$$= \ln\left(4 + \sqrt{7}\right) - \ln 3 \approx 0.7954.$$

33. $A = \int_1^\infty \frac{1}{x^2}\,dx = \lim_{b\to\infty}\left[-\frac{1}{x}\right]_1^b = 0 + 1 = 1$

35. $a = 1, n = 1: \lim_{x\to\infty} xe^{-x}$

x	1	10	25	50
xe^{-x}	0.3679	0.0005	0.0000	0.0000

37. $a = \frac{1}{2}, n = 2: \lim_{x\to\infty} x^2e^{-(1/2)x}$

x	1	10	25	50
$x^2e^{-(1/2)x}$	0.6065	0.6738	0.0023	0.0000

39. $\int_0^\infty x^2e^{-x}\,dx = \lim_{b\to\infty}\left[-x^2e^{-x} - 2xe^{-x} - 2e^{-x}\right]_0^b$
$$= (0 - 0 - 0) - (0 + 0 - 2) = 2$$

41. $\int_0^\infty xe^{-2x}\,dx = \lim_{b\to\infty}\left[-\tfrac{1}{2}xe^{-2x} - \tfrac{1}{4}e^{-2x}\right]_0^b$
$$= (0 - 0) - \left(0 - \tfrac{1}{4}\right) = \tfrac{1}{4}$$

43. $\mu = 64.5,\ \sigma = 2.7$

$$f(x) = \frac{1}{\sigma\sqrt{2\pi}}e^{-(x-\mu)^2/2\sigma^2} = 0.147756e^{-(x-64.5)^2/14.58}$$

Using a graphing utility:

(a) $\int_{60}^{72} f(x)\,dx \approx 0.9495$

(b) $\int_{68}^{\infty} f(x)\,dx \approx 0.0974$

(c) $\int_{72}^{\infty} f(x)\,dx \approx 0.0027$

45. Answers will vary.

Review Exercises for Chapter 7

1. Let $u = \ln x$ and $dv = \left(1/\sqrt{x}\right)dx$. Then $du = (1/x)\,dx$ and $v = 2\sqrt{x}$.

$$\int \frac{\ln x}{\sqrt{x}}\,dx = 2\sqrt{x}\ln x - \int 2\sqrt{x}\left(\frac{1}{x}\right)dx$$
$$= 2\sqrt{x}\ln x - \int 2x^{-1/2}\,dx$$
$$= 2\sqrt{x}\ln x - 4\sqrt{x} + C$$

3. Let $u = x + 1$ and $dv = e^x\,dx$. Then $du = dx$ and $v = e^x$.

$$\int (x+1)e^x\,dx = (x+1)e^x - \int e^x\,dx$$
$$= (x+1)e^x - e^x + C$$
$$= xe^x + C$$

5. Let $u = x$ and $dv = (x+1)^{-1/2}\,dx$. Then $du = dx$ and $v = 2\sqrt{x+1}$.

$$\int \frac{x}{\sqrt{x+1}}\,dx = 2x\sqrt{x+1} - \int 2\sqrt{x+1}\,dx$$

$$= 2x\sqrt{x+1} - \frac{4}{3}(x+1)^{3/2} + C = 2(x+1)^{1/2}\left[x - \frac{2}{3}(x+1)\right] + C = \frac{2}{3}\sqrt{x+1}(x-2) + C$$

7. Let $u = 2x^2$ and $dv = e^{2x}$. Then $du = 4x\,dx$ and $v = \frac{1}{2}e^{2x}$.

$$\int 2x^2e^{2x}\,dx = (2x^2)\left(\tfrac{1}{2}e^{2x}\right) - \int \tfrac{1}{2}e^{2x}(4x\,dx) = x^2e^{2x} - \int 2xe^{2x}\,dx$$

Let $u = 2x$ and $dv = e^{2x}$. Then $du = 2\,dx$ and $v = \frac{1}{2}e^{2x}$.

$$\int 2x^2e^{2x}\,dx = x^2e^{2x} - \left[(2x)\left(\tfrac{1}{2}e^{2x}\right) - \int \tfrac{1}{2}e^{2x}(2)\,dx\right] = x^2e^{2x} - xe^{2x} + \int e^{2x}\,dx = x^2e^{2x} - xe^{2x} + \tfrac{1}{2}e^{2x} + C$$

9. Let $u = x$ and $dv = (9+16x)^{-1/2}\,dx$. Then $du = dx$ and $v = \frac{1}{8}\sqrt{9+16x}$.

$$P(a \le x \le b) = \frac{96}{11}\left[\frac{1}{8}x\sqrt{9+16x} - \frac{1}{192}(9+16x)^{3/2}\right]_a^b = \frac{96}{11}\left[\frac{2(8x-9)}{384}\sqrt{9+16x}\right]_a^b = \frac{1}{22}\left[(8x-9)\sqrt{9+16x}\right]_a^b$$

(a) $P(0 \le x \le 0.8) = \frac{1}{22}\left[(8x-9)\sqrt{9+16x}\right]_0^{0.8} = \frac{1}{22}\left(-2.6\sqrt{21.8} + 27\right) \approx 0.675$

(b) $P(0 \le x \le 0.5) = \frac{1}{22}\left[(8x-9)\sqrt{9+16x}\right]_0^{0.5} = \frac{1}{22}\left(-5\sqrt{17} + 27\right) \approx 0.290$

11. $\dfrac{1}{x(x+5)} = \dfrac{A}{x} + \dfrac{B}{x+5}$

Basic equation: $1 = A(x+5) + Bx$

When $x = 0$: $1 = 5A$, $A = \dfrac{1}{5}$

When $x = -5$: $1 = -5B$, $B = -\dfrac{1}{5}$

$$\int \frac{1}{x(x+5)}\,dx = \frac{1}{5}\int\left[\frac{1}{x} - \frac{1}{x+5}\right]dx$$

$$= \frac{1}{5}\left[\ln|x| - \ln|x+5|\right] + C$$

$$= \frac{1}{5}\ln\left|\frac{x}{x+5}\right| + C$$

13. $\dfrac{x-28}{(x-3)(x+2)} = \dfrac{A}{x-3} + \dfrac{B}{x+2}$

Basic equation: $x - 28 = A(x+2) + B(x-3)$

When $x = 3$: $-25 = 5A$, $A = -5$

When $x = -2$: $-30 = -5B$, $B = 6$

$$\int \frac{x-28}{x^2-x-6}\,dx = \int \frac{-5}{x-3}\,dx + \int \frac{6}{x+2}\,dx$$

$$= -5\ln|x-3| + 6\ln|x+2| + C$$

$$= 6\ln|x+2| - 5\ln|x-3| + C$$

15. $\dfrac{x^2}{x^2+2x-15} = 1 - \dfrac{2x-15}{(x+5)(x-3)}$

$\dfrac{2x-15}{(x+5)(x-3)} = \dfrac{A}{x+5} + \dfrac{B}{x-3}$

Basic equation: $2x - 15 = A(x-3) + B(x+5)$

When $x = -5$: $-25 = -8A$, $A = \dfrac{25}{8}$

When $x = 3$: $-9 = 8B$, $B = -\dfrac{9}{8}$

$$\int \frac{x^2}{x^2+2x-15}\,dx = \int\left[1 - \frac{25}{8}\left(\frac{1}{x+5}\right) + \frac{9}{8}\left(\frac{1}{x-3}\right)\right]dx = x - \frac{25}{8}\ln|x+5| + \frac{9}{8}\ln|x-3| + C$$

17. $\dfrac{x}{(2+3x)^2} = \dfrac{A}{2+3x} + \dfrac{B}{(2+3x)^2}$

Basic equation: $x = A(2+3x) + B$

When $x = -\frac{2}{3}$: $-\frac{2}{3} = B$

When $x = 0$: $0 = 2A - \frac{2}{3}$, $A = \frac{1}{3}$

$$\int \frac{x}{(2+3x)^2}\,dx = \frac{1}{3}\int \frac{1}{2+3x}\,dx - \frac{2}{3}\int \frac{1}{(2+3x)^2}\,dx$$
$$= \frac{1}{9}\ln|2+3x| + \frac{2}{9(2+3x)} + C$$
$$= \frac{1}{9}\left(\ln|2+3x| + \frac{2}{2+3x}\right) + C$$

19. (a) $y = \dfrac{L}{1+be^{-kt}} = \dfrac{10{,}000}{1+be^{-kt}}$

Because $y = 1250$ when $t = 0$,

$\dfrac{10{,}000}{1+b} = 1250$, $b = 7$.

When $t = 24$, $y = 6500 = \dfrac{10{,}000}{1+7e^{-k(24)}}$.

Solving for k,

$$1 + 7e^{-24k} = \frac{20}{13}$$
$$7e^{-24k} = \frac{7}{13}$$
$$e^{-24k} = \frac{1}{13}$$
$$k = -\frac{1}{24}\ln\left(\frac{1}{13}\right) \approx 0.106873.$$

So, $y = \dfrac{10{,}000}{1+7e^{-0.106873t}}$.

(b)

t	0	3	6	12	24
y	1250	1645	2134	3400	6500

(c) The sales will be 7500 when $t \approx 28$ weeks.

21. $\int(3\sin x - 2\cos x)\,dx = -3\cos x - 2\sin x + C$

23. $\int \sin^3 x \cos x\,dx = \frac{1}{4}\sin^4 x + C$

25. $\int_0^{\pi}(1+\sin x)\,dx = \left[x - \cos x\right]_0^{\pi}$

$= \left[\pi - (-1)\right] - (-1) = \pi + 2$

27. $\int_{-\pi/6}^{\pi/6}\sec^2 x\,dx = \tan x\Big]_{-\pi/6}^{\pi/6}$

$= \dfrac{1}{\sqrt{3}} - \left(-\dfrac{1}{\sqrt{3}}\right) = \dfrac{2}{\sqrt{3}} = \dfrac{2\sqrt{3}}{3}$

29. $\int_{-\pi/3}^{\pi/3} 4\sec x\tan x\,dx = 4\sec x\Big]_{-\pi/3}^{\pi/3}$

$= 4(2) - 4(2) = 0$

31. $\int_{-\pi/2}^{\pi/2}(2x + \cos x)\,dx = \left[x^2 + \sin x\right]_{-\pi/2}^{\pi/2}$

$= \dfrac{\pi^2}{4} + 1 - \left(\dfrac{\pi^2}{4} - 1\right) = 2$

33. $A = \int_0^{\pi/3}\sin 3x\,dx$

$= \frac{1}{3}\int_0^{\pi/3} 3\sin 3x\,dx$

$= \left[-\frac{1}{3}\cos 3x\right]_0^{\pi/3}$

$= \frac{1}{3} + \frac{1}{3} = \frac{2}{3}$ square unit

35. $A = \int_0^{\pi/2}(2\sin x + \cos 3x)\,dx$

$= \left[-2\cos x + \frac{1}{3}\sin 3x\right]_0^{\pi/2}$

$= \left(-\frac{1}{3}\right) - (-2) = \frac{5}{3}$ square units

37. $P = 2.91\sin(0.4t + 1.81) + 2.38$, $0 \le t \le 12$

($t = 1$ corresponds to January.)

Total $= \int_0^{12}\left[2.91\sin(0.4t + 1.81) + 2.38\right]dt$

$= \left[-\dfrac{2.91}{0.4}\cos(0.4t + 1.81) + 2.38t\right]_6^{12}$

≈ 19.95 inches

The total annual precipitation for San Francisco is about 19.95 inches.

39. $\int_0^2 (x^2+1)^2\,dx$

$n = 4$: 13.3203

$n = 20$: 13.7167 (exact $13.7\overline{3}$)

41. $\int_0^1 \dfrac{1}{x^2+1}\,dx$

$n = 4$: 0.7867

$n = 20$: 0.7855

43. $\int_1^3 \dfrac{1}{x^2}\,dx \approx \dfrac{1}{4}\left[1 + 2\left(\dfrac{4}{9}\right) + 2\left(\dfrac{1}{4}\right) + 2\left(\dfrac{4}{25}\right) + \dfrac{1}{9}\right] = 0.705$

45. $\int_1^2 \frac{1}{1+\ln x}\,dx \approx \frac{1}{8}\left(1 + 2\left[\frac{1}{1+\ln(5/4)}\right] + 2\left[\frac{1}{1+\ln(3/2)}\right] + 2\left[\frac{1}{1+\ln(7/4)}\right] + \frac{1}{1+\ln 2}\right) \approx 0.741$

47. $\int_1^2 \frac{1}{x^3}\,dx \approx \frac{1}{12}\left[1 + 4\left(\frac{4}{5}\right)^3 + 2\left(\frac{2}{3}\right)^3 + 4\left(\frac{4}{7}\right)^3 + \frac{1}{8}\right] \approx 0.376$

49. $\int_0^1 \frac{x^{3/2}}{2-x^2}\,dx \approx \frac{1}{12}\left[0 + 4\left(\frac{2}{31}\right) + 2\left(\frac{\sqrt{2}}{7}\right) + 4\left(\frac{6\sqrt{3}}{23}\right) + 1\right] \approx 0.289$

51. $f(x) = e^{2x}$

$f'(x) = 2e^{2x}$

$f''(x) = 4e^{2x}$

$\left|f''(x)\right| \le 4e^{2(2)} = 4e^4$ on $[0, 2]$.

$|E| \le \frac{(2-0)^3}{12(4)^2}\left(4e^4\right) \le 9.0997$

53. $f(x) = \frac{1}{x-1} = (x-1)^{-1}$

$f'(x) = -(x-1)^{-2}$

$f''(x) = 2(x-1)^{-3}$

$f'''(x) = -6(x-1)^{-4}$

$f^{(4)}(x) = 24(x-1)^{-5} = \frac{24}{(x-1)^5}$

$\left|f^{(4)}(x)\right| \le 24$ on $[2, 4]$.

$|E| \le \frac{(4-2)^5}{180(4)^4}(24) \approx 0.017$

55. This integral converges because

$$\int_0^\infty 4xe^{-2x^2}\,dx = \lim_{b\to\infty} -\int_0^b e^{-2x^2}(-4x\,dx) = \lim_{b\to\infty}\left[-e^{-2x^2}\right]_0^b = 1.$$

57.
$$\int_{-\infty}^0 \frac{1}{3x^2}\,dx = \lim_{b\to-\infty}\int_b^{-1}\frac{1}{3}x^{-2}\,dx + \lim_{a\to 0^-}\int_{-1}^a \frac{1}{3}x^{-2}\,dx$$
$$= \lim_{b\to-\infty}\left[\frac{-1}{3x}\right]_b^{-1} + \lim_{a\to 0^-}\left[\frac{-1}{3x}\right]_{-1}^0$$
$$= \left(\frac{1}{3} - 0\right) + \left(\infty - \frac{1}{3}\right) = \infty.$$

59. This integral converges because

$$\int_0^4 \frac{1}{\sqrt{4x}}\,dx = \lim_{a\to 0^+}\int_a^4 \frac{1}{2}x^{-1/2}\,dx = \lim_{a\to 0^+}\left[x^{1/2}\right]_a^4 = 2.$$

61. This integral converges because

$$\int_2^3 \frac{1}{\sqrt{x-2}}\,dx = \lim_{b\to 2^+}\left[2(x-2)^{1/2}\right]_b^3 = 2.$$

63. (a) $P(500 \le x \le 650) = \int_{500}^{600} 0.0035e^{-(x-518)^2/26{,}450}\,dx \approx 0.441$

(b) $P(650 \le x \le 800) = \int_{650}^{800} 0.0035e^{-(x-518)^2/26{,}450}\,dx \approx 0.119$

(c) $P(750 \le x \le 800) = \int_{750}^{800} 0.0035e^{-(x-518)^2/26{,}450}\,dx \approx 0.015$

Chapter Test Solutions

1. Let $u = x$ and $dv = e^{x+1}\,dx$. Then $du = dx$ and $v = e^{x+1}$.

$$\int xe^{x+1}\,dx = xe^{x+1} - \int e^{x+1}\,dx$$
$$= xe^{x+1} - e^{x+1} + C = (x-1)e^{x+1} + C$$

2. Let $u = \ln x$ and $dv = 9x^2\,dx$. Then $du = \frac{1}{x}\,dx$ and $v = 3x^3$.

$$\int 9x^2 \ln x\,dx = 3x^3\ln x - 3\int x^2\,dx$$
$$= 3x^3\ln x - x^3 + C = x^3(3\ln x - 1) + C$$

3. Let $u = x^2$ and $dv = e^{-x/3}\,dx$. Then $du = 2x\,dx$ and $v = -3e^{-x/3}$.

$$\int x^2e^{-x/3}\,dx = -3x^2e^{-x/3} + \int 6xe^{-x/3}\,dx$$

Let $u = 6x$ and $dv = e^{-x/3}\,dx$. Then $du = 6\,dx$ and $v = -3e^{-x/3}$.

$$\int x^2e^{-x/3}\,dx = -3x^2e^{-x/3} - 18xe^{-x/3} + \int 18e^{-x/3}\,dx$$
$$= -3x^2e^{-x/3} - 18xe^{-x/3} - 54e^{-x/3} + C$$
$$= -3e^{-x/3}\left(x^2 + 6x + 18\right) + C$$

4. $\frac{18}{(x+9)(x-9)} = \frac{A}{x+9} + \frac{B}{x-9}$

Basic equation: $18 = A(x-9) + B(x+9)$

When $x = -9$: $18 = -18A,\ A = -1$

When $x = 9$: $18 = 18B,\ B = 1$

$$\int \frac{18}{x^2-81}\,dx = -\int \frac{1}{x+9}\,dx + \int \frac{1}{x-9}\,dx$$
$$= -\ln|x+9| + \ln|x-9| + C$$
$$= \ln\left|\frac{x-9}{x+9}\right| + C$$

5. $\frac{3x}{(3x+1)^2} = \frac{A}{3x+1} + \frac{B}{(3x+1)^2}$

Basic equation: $3x = A(3x+1) + B$

When $x = -\frac{1}{3}$: $-1 = B$

When $x = 0$: $0 = A + B,\ A = 1$

$$\int \frac{3x}{(3x+1)^2}\,dx = \int \frac{1}{3x+1}\,dx - \int \frac{1}{(3x+1)^2}\,dx$$
$$= \frac{1}{3}\ln|3x+1| + \frac{1}{3(3x+1)} + C$$

6. $\frac{x+4}{x(x+2)} = \frac{A}{x} + \frac{B}{x+2}$

Basic equation: $x + 4 = A(x+2) + Bx$

When $x = 0$: $4 = 2A,\ A = 2$

When $x = -2$: $2 = -2B,\ B = -1$

$$\int \frac{x+4}{x^2+2x}\,dx = 2\int \frac{1}{x}\,dx - \int \frac{1}{x+2}\,dx$$
$$= 2\ln|x| - \ln|x+2| + C$$

7. $\int \frac{\cos\theta}{\sin\theta}\,d\theta = \int \cot\theta\,d\theta = \ln|\sin\theta| + C$

8. $\int \tan 5\theta\,d\theta = \frac{1}{5}\int 5\tan 5\theta\,d\theta = -\frac{1}{5}\ln|\cos 5\theta| + C$

9. $\int \csc 2\theta\,d\theta = \frac{1}{2}\int 2\csc 2\theta\,d\theta$

$$= \frac{1}{2}\ln|\csc 2\theta - \cot 2\theta| + C$$

10. The midpoints are $\frac{1}{8}, \frac{3}{8}, \frac{5}{8}$, and $\frac{7}{8}$. The approximate area is

$$A \approx \frac{1-0}{4}\left[f\left(\frac{1}{8}\right) + f\left(\frac{3}{8}\right) + f\left(\frac{5}{8}\right) + f\left(\frac{7}{8}\right)\right]$$
$$= \frac{1}{4}\left[\frac{3}{64} + \frac{27}{64} + \frac{75}{64} + \frac{147}{64}\right]$$
$$= \frac{63}{64} \approx 0.984.$$

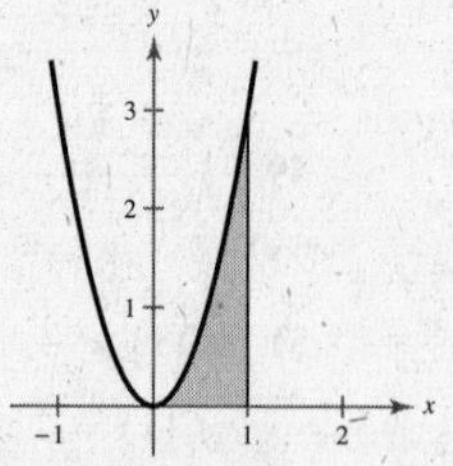

The exact area is

$$A = \int_0^1 3x^2\,dx = x^3\Big]_0^1 = 1.$$

11. The midpoints are $-\frac{3}{4}, -\frac{1}{4}, \frac{1}{4}$, and $\frac{3}{4}$. The approximate area is

$$A \approx \frac{1-(-1)}{4}\left[f\left(-\frac{3}{4}\right) + f\left(-\frac{1}{4}\right) + f\left(\frac{1}{4}\right) + f\left(\frac{3}{4}\right)\right]$$
$$= \frac{1}{2}\left[\frac{25}{16} + \frac{17}{16} + \frac{17}{16} + \frac{25}{16}\right] = \frac{21}{8} = 2.625.$$

The exact area is

$$A = \int_{-1}^1 (x^2+1)\,dx$$
$$= \frac{1}{3}x^3 + x\Big]_{-1}^1$$
$$= \frac{4}{3} - \left(-\frac{4}{3}\right)$$
$$= \frac{8}{3}.$$

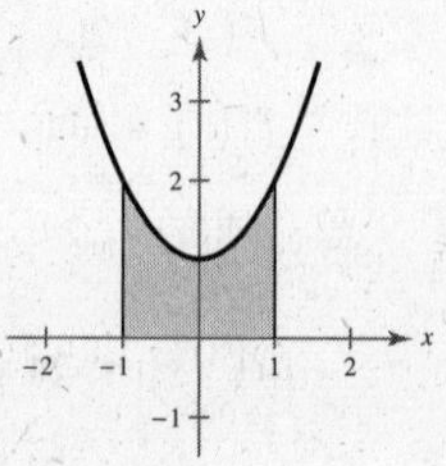

12. Let $u = \ln(3-2x)$ and $dv = dx$. Then

$du = -\frac{2}{3-2x}\,dx$ and $v = x$.

$$\int_0^1 \ln(3-2x)\,dx = x\ln(3-2x)\Big]_0^1 + \int_0^1 \frac{2x}{3-2x}\,dx$$
$$= \int_0^1 \left(\frac{3}{3-2x} - 1\right)dx$$
$$= \left[-\frac{3}{2}\ln|3-2x| - x\right]_0^1$$
$$= -1 + \frac{3}{2}\ln 3 \approx 0.648$$

13. $\frac{28}{(x-4)(x+3)} = \frac{A}{x-4} + \frac{B}{x+3}$

Basic Equation: $28 = A(x+3) + B(x-4)$

When $x = 4$: $28 = 7A,\ A = 4$

When $x = 3$: $28 = -7B,\ B = -4$

$$\int_5^{10} \frac{28}{x^2-x-12}\,dx = 4\int_5^{10} \frac{1}{x-4}\,dx - 4\int_5^{10} \frac{1}{x+3}$$
$$= \Big[4\ln|x-4| - 4\ln|x+3|\Big]_5^{10}$$
$$= 4(\ln 6 - \ln 13 + \ln 8) \approx 5.2250$$

14. $\int_0^{\pi/2} \cos 2x\,dx = \frac{1}{2}\int_0^{\pi/2} 2\cos 2x\,dx = \frac{1}{2}\sin 2x\Big]_0^{\pi/2} = 0$

15. Trapezoidal Rule: $\int_1^2 \frac{1}{x^2\sqrt{x^2+4}}\,dx \approx \frac{1}{8}\left[\frac{1}{\sqrt{5}} + 2\left(\frac{16}{25\sqrt{89/16}}\right) + 2\left(\frac{4}{9\sqrt{25/4}}\right) + 2\left(\frac{16}{49\sqrt{113/16}}\right) + \frac{1}{4\sqrt{8}}\right] \approx 0.2100$

16. Exact: Let $u = 9x$ and $dv = e^{3x}\,dx$. Then $du = 9\,dx$ and $v = \frac{1}{3}e^{3x}$.

$$\begin{aligned}\int_0^1 9xe^{3x}\,dx &= 3xe^{3x}\Big]_0^1 - \int 3e^{3x}\,dx\\ &= 3e^3 - \Big[e^{3x}\Big]_0^1\\ &= 3e^3 - \left(e^3 - 1\right)\\ &= 2e^3 + 1 \approx 41.1711\end{aligned}$$

Simpson's Rule: $\int_0^1 9xe^{3x}\,dx = \frac{1}{12}\left[0 + 4\left(\frac{9}{4}e^{3/4}\right) + 2\left(\frac{9}{2}e^{3/2}\right) + 4\left(\frac{27}{4}e^{9/4}\right) + 9e^3\right] \approx 41.3606$

17. This integral converges because $\int_0^{\infty} e^{-3x}\,dx = \lim_{b\to\infty}\left[-\frac{1}{3}e^{-3x}\right]_0^b = 0 + \frac{1}{3} = \frac{1}{3}$.

18. This integral converges because $\int_0^9 \frac{2}{\sqrt{x}}\,dx = \lim_{a\to 0^+}\left[4\sqrt{x}\right]_b^9 = 12$.

19. This integral diverges because $\int_{-\infty}^0 \frac{1}{(4x-1)^{2/3}}\,dx = \frac{3}{4}\lim_{a\to-\infty}\left[(4x-1)^{1/3}\right]_a^0 = -\frac{3}{4} + \infty = \infty$.

20. Area $\approx \frac{180}{9}(27.5 + 62.5 + 65 + 60 + 66 + 68 + 60.5 + 55.5 + 27)$

$\approx 9840\ \text{ft}^2$

Practice Test for Chapter 7

1. Find $\int xe^{2x}\,dx$.

2. Find $\int x^3 \ln x\,dx$.

3. Find $\int x^2\sqrt{x-6}\,dx$.

4. Find $\int x^2 e^{4x}\,dx$.

5. Find $\int \frac{-5}{x^2+x-6}\,dx$.

6. Find $\int \frac{x+12}{x^2+4x}\,dx$.

7. Find $\int \frac{5x+3}{(x+2)^2}\,dx$.

8. Find $\int \frac{3x^3+9x^2-x+3}{x(x+3)}\,dx$.

9. Find $\int \csc^2 \frac{x}{8}\,dx$.

10. Find $\int \sin^5 x \cos x\,dx$.

11. Evaluate $\int_0^{\pi/4} (2x - \cos x)\,dx$.

12. Approximate the definite integral by the Midpoint Rule using $n = 4$.

$$\int_0^1 \sqrt{x^3+2}\,dx$$

13. Approximate the definite integral by the Midpoint Rule using $n = 4$.

$$\int_3^4 \frac{1}{x^2-5}\,dx$$

14. Approximate the integral using (a) the Trapezoidal Rule and (b) Simpson's Rule.

$$\int_0^4 \sqrt{3+x^3}\,dx,\ n = 8$$

15. Approximate the integral using (a) the Trapezoidal Rule and (b) Simpson's Rule.

$$\int_0^2 e^{-x^2/2}\,dx,\ n = 4$$

16. Determine the divergence or convergence of the integral. Evaluate the integral if it converges.

$$\int_0^9 \frac{1}{\sqrt{x}}\,dx$$

17. Determine the divergence or convergence of the integral. Evaluate the integral if it converges.

$$\int_1^\infty \frac{1}{x-3}\,dx$$

18. Determine the divergence or convergence of the integral. Evaluate the integral if it converges.

$$\int_{-\infty}^0 e^{-3x}\,dx$$

19. Complete the following chart using Simpson's Rule to determine the convergence or divergence of

$$\int_0^1 \frac{1}{\sqrt{1-x^2}}\,dx.$$

n	100	1000	10,000
$\int_1^{0.9999999} \frac{1}{\sqrt{1-x^2}}\,dx$			

Note: The upper limit cannot equal 1 to avoid division by zero. Let $b = 0.9999999$.

Graphing Calculator Required

20. Use a program similar to that on page 457 of the textbook to approximate the following integral for $n = 50$ and $n = 100$.

$$\int_0^4 \sqrt{1+x^4}\,dx$$

21. Use a program similar to that on page 466 of the textbook to approximate

$$\int_0^3 \frac{1}{\sqrt{x^3+1}}\,dx$$

when $n = 50$ and $n = 100$.

CHAPTER 8
Matrices

CHAPTER 8
Matrices

Section 8.1 Systems of Equations in Two Variables

Skills Review

1. $3x + (x - 5) = 15 + 4$

$3x + x - 5 = 19$

$4x - 5 = 19$

$4x = 24$

$x = 6$

2. $y^2 + (y - 2)^2 = 2$

$y^2 + y^2 - 4y + 4 = 2$

$2y^2 - 4y + 4 = 2$

$2y^2 - 4y + 2 = 0$

$2(y^2 - 2y + 1) = 0$

$2(y - 1)(y - 1) = 0$

$y - 1 = 0$

$y = 1$

3. Line 1: $2x - 3y = -10$

$-3y = -2x - 10$

$y = \frac{2}{3}x + \frac{10}{3}$

Slope $= \frac{2}{3}$

Line 2: $3x + 2y = 11$

$2y = -3x + 11$

$y = -\frac{3}{2}x + \frac{11}{2}$

Slope $= -\frac{3}{2}$

Because the slopes are negative reciprocals of each other, the lines are perpendicular.

4. Line 1: $4x - 12y = 5$

$-12y = -4x + 5$

$y = \frac{1}{3}x + \frac{1}{12}$

Slope $= \frac{1}{3}$

Line 2: $-2x + 6y = 3$

$6y = 2x + 3$

$y = \frac{1}{3}x + \frac{1}{2}$

Slope $= \frac{1}{3}$

Because the slopes are equal, the lines are parallel.

1. $\begin{cases} 4x - y = 1 \\ 6x + y = -6 \end{cases}$

(a) $4(0) - (-3) \neq 1$

$(0, -3)$ *is not* a solution.

(b) $4(-1) - (-4) \neq 1$

$(-1, -4)$ *is not* a solution.

(c) $4\left(-\frac{3}{2}\right) - (-2) \neq 1$

$\left(-\frac{3}{2}, -2\right)$ *is not* a solution.

(d) $4\left(-\frac{1}{2}\right) - (-3) = 1$

$6\left(-\frac{1}{2}\right) + (-3) = -6$

$\left(-\frac{1}{2}, -3\right)$ *is* a solution.

3. $\begin{cases} y = -2e^x \\ 3y - y = 2 \end{cases}$

(a) $0 \neq -2e^{-2}$

$(-2, 0)$ *is not* a solution.

(b) $-2 = -2e^0$

$3(0) - (-2) = 2$

$(0, -2)$ *is* a solution.

(c) $-3 \neq -2e^0$

$(0, -3)$ *is not* a solution.

(d) $2 \neq -2e^{-1}$

$(-1, 2)$ *is not* a solution.

5. $\begin{cases} x - y = 0 & \text{Equation 1} \\ 5x - 3y = 10 & \text{Equation 2} \end{cases}$

Solve for y in Equation 1: $y = x$

Solve for y in Equation 2: $5x - 3x = 10$

Solve for x: $2x = 10 \Rightarrow x = 5$

Back-substitute in Equation 1: $y = x = 5$

Solution: $(5, 5)$

7. $\begin{cases} 2x - y + 2 = 0 & \text{Equation 1} \\ 4x + y - 5 = 0 & \text{Equation 2} \end{cases}$

Solve for y in Equation 1: $y = 2x + 2$

Substitute for y in Equation 2: $4x + (2x + 2) - 5 = 0$

Solve for x: $6x - 3 = 0 \Rightarrow x = \frac{1}{2}$

Back-substitute $x = \frac{1}{2}$: $y = 2x + 2 = 2\left(\frac{1}{2}\right) + 2 = 3$

Solution: $\left(\frac{1}{2}, 3\right)$

9. $\begin{cases} 1.5x + 0.8y = 2.3 & \text{Equation 1} \\ 0.3x + 0.2y = 0.1 & \text{Equation 2} \end{cases}$

Multiply the equations by 10.

$15x + 8y = 23$ Revised Equation 1

$3x - 2y = 1$ Revised Equation 2

Solve for y in revised Equation 2: $y = \frac{3}{2}x - \frac{1}{2}$

Substitute for y in revised Equation 1:

$15x + 8\left(\frac{3}{2}x - \frac{1}{2}\right) = 23$

Solve for x:

$15x + 12x - 4 = 23 \Rightarrow 27x = 27 \Rightarrow x = 1$

Back-substitute $x = 1$: $y = \frac{3}{2}(1) - \frac{1}{2} = 1$

Solution: $(1, 1)$

11. $\begin{cases} \frac{1}{5}x + \frac{1}{2}y = 8 & \text{Equation 1} \\ x + y = 20 & \text{Equation 2} \end{cases}$

Solve for x in Equation 2: $x = 20 - y$

Substitute for x in Equation 1: $\frac{1}{5}(20 - y) + \frac{1}{2}y = 8$

Solve for y: $4 + \frac{3}{10}y = 8 \Rightarrow y = \frac{40}{3}$

Back-substitute $y = \frac{40}{3}$: $x = 20 - y = 20 - \frac{40}{3} = \frac{20}{3}$

Solution: $\left(\frac{20}{3}, \frac{40}{3}\right)$

13. $\begin{cases} 6x + 5y = -3 & \text{Equation 1} \\ -x - \frac{5}{6}y = -7 & \text{Equation 2} \end{cases}$

Solve for x in Equation 2: $x = 7 - \frac{5}{6}y$

Substitute for x in Equation 1: $6\left(7 - \frac{5}{6}y\right) + 5y = -3$

Solve for y: $42 - 5y + 5y = -3 \Rightarrow 42 = -3$ (False)

No solution

15. $\begin{cases} x^2 - y = 0 & \text{Equation 1} \\ 2x + y = 0 & \text{Equation 2} \end{cases}$

Solve for y in Equation 2: $y = -2x$

Substitute for y in Equation 1: $x^2 - (-2x) = 0$

Solve for x: $x^2 + 2x = 0 \Rightarrow x(x + 2) = 0$

$\Rightarrow x = 0, -2$

Back-substitute $x = 0$: $y = -2(0) = 0$

Back-substitute $x = -2$: $y = -2(-2) = 4$

Solutions: $(0, 0), (-2, 4)$

17. $\begin{cases} x - y = -1 & \text{Equation 1} \\ x^2 - y = -4 & \text{Equation 2} \end{cases}$

Solve for y in Equation 1: $y = x + 1$

Substitute for y in Equation 2: $x^2 - (x + 1) = -4$

Solve for x: $x^2 - x - 1 = -4 \Rightarrow x^2 - x + 3 = 0$

The Quadratic Formula yields no real solutions.

19. $\begin{cases} x + 2y = 4 & \text{Equation 1} \\ x - 2y = 1 & \text{Equation 2} \end{cases}$

Add to eliminate y:

$2x = 5$

$x = \frac{5}{2}$

Substitute $x = \frac{5}{2}$ in Equation 1:

$\frac{5}{2} + 2y = 4 \Rightarrow y = \frac{3}{4}$

Solution: $\left(\frac{5}{2}, \frac{3}{4}\right)$

21. $\begin{cases} -4x + 5y = -12 & \text{Equation 1} \\ 2x - y = 3 & \text{Equation 2} \end{cases}$

Multiply Equation 2 by 2: $4x - 2y = 6$

Add this to Equation 1 to eliminate x:

$$\begin{array}{r} -4x + 5y = -12 \\ 4x - 2y = 6 \\ \hline 3y = -6 \end{array} \Rightarrow y = -2$$

Substitute $y = -2$ in Equation 1:

$-4x + 5(-2) = -12 \Rightarrow x = \frac{1}{2}$

Solution: $\left(\frac{1}{2}, -2\right)$

23. $\begin{cases} 3x + 2y = 10 & \text{Equation 1} \\ 2x + 5y = 3 & \text{Equation 2} \end{cases}$

Multiply Equation 1 by 2 and Equation 2 by −3:

$\begin{cases} 6x + 4y = 20 \\ -6x - 15y = -9 \end{cases}$

Add to eliminate x: $-11y = 11 \Rightarrow y = -1$

Substitute $y = -1$ in Equation 1:

$3x - 2 = 10 \Rightarrow x = 4$

Solution: $(4, -1)$

25. $\begin{cases} 5u + 6v = 24 & \text{Equation 1} \\ 3u + 5v = 18 & \text{Equation 2} \end{cases}$

Multiply Equation 1 by 5 and Equation 2 by −6:

$\begin{cases} 25u + 30v = 120 \\ -18u - 30v = -108 \end{cases}$

Add to eliminate v: $7u = 12 \Rightarrow u = \frac{12}{7}$

Substitute $u = \frac{12}{7}$ in Equation 1:

$5\left(\frac{12}{7}\right) + 6v = 24 \Rightarrow 6v = \frac{108}{7} \Rightarrow v = \frac{18}{7}$

Solution: $\left(\frac{12}{7}, \frac{18}{7}\right)$

27. $\begin{cases} \frac{9}{5}x + \frac{6}{5}y = 4 & \text{Equation 1} \\ 9x + 6y = 3 & \text{Equation 2} \end{cases}$

Multiply Equation 1 by 10 and Equation 2 by −2:

$\begin{cases} 18x + 12y = 40 \\ -18x - 12y = -6 \end{cases}$

Add to eliminate x and y: $0 = 34$

Inconsistent

No solution

29. $\begin{cases} \dfrac{x}{4} + \dfrac{y}{6} = 1 & \text{Equation 1} \\ x - y = 3 & \text{Equation 2} \end{cases}$

Multiply Equation 1 by 6: $\dfrac{3}{2}x + y = 6$

Add this to Equation 2 to eliminate y:

$\dfrac{5}{2}x = 9 \Rightarrow x = \dfrac{18}{5}$

Substitute $x = \dfrac{18}{5}$ in Equation 2:

$\dfrac{18}{5} - y = 3 \Rightarrow y = \dfrac{3}{5}$

Solution: $\left(\dfrac{18}{5}, \dfrac{3}{5}\right)$

31. $\begin{cases} -5x + 6y = -3 & \text{Equation 1} \\ 20x - 24y = 12 & \text{Equation 2} \end{cases}$

Multiply Equation 1 by 4:

$\begin{cases} -20x + 24y = -12 \\ 20x - 24y = 12 \end{cases}$

Add to eliminate x and y: $0 = 0$

The equations are dependent. There are infinitely many solutions.

Let $x = a$, then

$-5a + 6y = -3 \Rightarrow y = \dfrac{5a - 3}{6} = \dfrac{5}{6}a - \dfrac{1}{2}.$

Solution: $\left(a, \dfrac{5}{6}a - \dfrac{1}{2}\right)$ where a is any real number

33. $\begin{cases} 0.05x - 0.03y = 0.21 & \text{Equation 1} \\ 0.07x + 0.02y = 0.16 & \text{Equation 2} \end{cases}$

Multiply Equation 1 by 200 and Equation 2 by 300:

$\begin{cases} 10x - 6y = 42 \\ 21x + 6y = 48 \end{cases}$

Add to eliminate y: $31x = 90 \Rightarrow x = \frac{90}{31}$

Substitute $x = \frac{90}{31}$ in Equation 2:

$0.07\left(\frac{90}{31}\right) + 0.02y = 0.16 \Rightarrow y = -\frac{67}{31}$

Solution: $\left(\frac{90}{31}, -\frac{67}{31}\right)$

35. $\begin{cases} 4b + 3m = 3 & \text{Equation 1} \\ 3b + 11m = 13 & \text{Equation 2} \end{cases}$

Multiply Equation 1 by 3 and Equation 2 by −4:

$\begin{cases} 12b + 9m = 9 \\ -12b - 44m = -52 \end{cases}$

Add to eliminate b: $-35m = -43 \Rightarrow m = \frac{43}{35}$

Substitute $m = \frac{43}{35}$ in Equation 1:

$4b + 3\left(\frac{43}{35}\right) = 3 \Rightarrow b = -\frac{6}{35}$

Solution: $\left(-\frac{6}{35}, \frac{43}{35}\right)$

37. $\begin{cases} \dfrac{x+3}{4} + \dfrac{y-1}{3} = 1 & \text{Equation 1} \\ 2x - y = 12 & \text{Equation 2} \end{cases}$

Multiply Equation 1 by 12 and Equation 2 by 4:

$\begin{cases} 3x + 4y = 7 \\ 8x - 4y = 48 \end{cases}$

Add to eliminate y: $11x = 55 \Rightarrow x = 5$

Substitute $x = 5$ into Equation 2:

$2(5) - y = 12 \Rightarrow y = -2$

Solution: $(5, -2)$

39. $\begin{cases} 3x - 5y = 7 & \text{Equation 1} \\ 2x + y = 9 & \text{Equation 2} \end{cases}$

Multiply Equation 2 by 5: $10x + 5y = 45$

Add this to Equation 1: $13x = 52 \Rightarrow x = 4$

Back-substitute $x = 4$ into Equation 2:

$2(4) + y = 9 \Rightarrow y = 1$

Solution: $(4, 1)$

41. $\begin{cases} y = 2x - 5 & \text{Equation 1} \\ y = 5x - 11 & \text{Equation 2} \end{cases}$

Because both equations are solved for y, set them equal to one another and solve for x.

$$\begin{aligned} 2x - 5 &= 5x - 11 \\ 6 &= 3x \\ 2 &= x \end{aligned}$$

Back-substitute $x = 2$ into Equation 1:

$y = 2(2) - 5 = -1$

Solution: $(2, -1)$

43. $\begin{cases} x - 5y = 21 & \text{Equation 1} \\ 6x + 5y = 21 & \text{Equation 2} \end{cases}$

Add the equations: $7x = 42 \Rightarrow x = 6$

Back-substitute $x = 6$ into Equation 1:

$6 - 5y = 21 \Rightarrow -5y = 15 \Rightarrow y = -3$

Solution: $(6, -3)$

45. $\begin{cases} -2x + 8y = 19 & \text{Equation 1} \\ y = x - 3 & \text{Equation 2} \end{cases}$

Substitute the expression for y from Equation 2 into Equation 1.

$$\begin{aligned} -2x + 8(x - 3) = 19 \Rightarrow -2x + 8x - 24 &= 19 \\ 6x &= 43 \\ x &= \tfrac{43}{6} \end{aligned}$$

Back-substitute $x = \frac{43}{6}$ into Equation 2:

$y = \frac{43}{6} - 3 \Rightarrow y = \frac{25}{6}$

Solution: $\left(\frac{43}{6}, \frac{25}{6}\right)$

47. Let r_1 = the air speed of the plane and

r_2 = the wind air speed.

$\begin{cases} 3.6(r_1 - r_2) = 1800 & \text{Equation 1} \\ 3(r_1 + r_2) = 1800 & \text{Equation 2} \end{cases}$

Multiply Equation 1 by $\frac{5}{18}$ and Equation 2 by $\frac{1}{3}$; then add the equations:

$$\begin{aligned} r_1 - r_2 &= 500 \\ r_1 + r_2 &= 600 \\ \hline 2r_1 &= 1100 \Rightarrow r_1 = 550 \end{aligned}$$

Back-substitute $r_1 = 550$ into revised Equation 2:

$550 + r_2 = 600 \Rightarrow r_2 = 50$

The air speed of the plane is 550 miles per hour and the speed of the wind is 50 miles per hour.

49. Let x = number of calories in cheeseburger

y = number of calories in a small order of french fries

$\begin{cases} 2x + y = 850 & \text{Equation 1} \\ 3x + 2y = 1390 & \text{Equation 2} \end{cases}$

Multiply Equation 1 by -2; then add the equations

$$\begin{aligned} -4x - 2y &= -1700 \\ 3x + 2y &= 1390 \\ \hline -x &= -310 \Rightarrow x = 310 \end{aligned}$$

Back-substitute $x = 310$ into Equation 1:

$2(310) + y = 850 \Rightarrow y = 230$

The cheeseburger contains 310 calories and the fries contain 230 calories.

51. $\begin{cases} B = 0.73a + 11 & \text{Males} \\ B = 0.61a + 12.8 & \text{Females} \end{cases}$

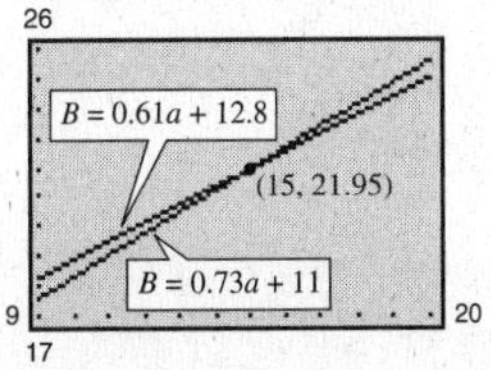

The curves intersect when $a = 15$, which corresponds to age 15. According to these models, the BMI for males exceeds the BMI for females after age 15.

53.

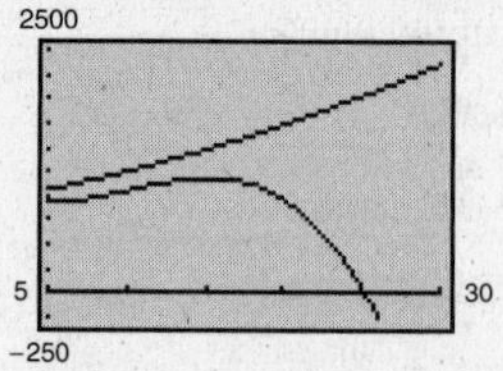

SAT: $y = 0.68t^2 + 28.1t + 903$

ACT: $y = -0.485t^3 + 14.88t^2 - 115.1t + 1201$

According to the model, ACT testing will not overtake SAT testing. The model for SAT participants may continue to be accurate, because it continues to increase. The ACT model eventually decreases and becomes negative. It is unlikely that this model will continue to be accurate.

55. Answers will vary.

Section 8.2 Systems of Linear Equations in More than Two Variables

Skills Review

1. $\begin{cases} x + y = 25 & \text{Equation 1} \\ y = 10 & \text{Equation 2} \end{cases}$

Substitute $y = 10$ into Equation 1:

$x + 10 = 25 \Rightarrow x = 15$

Solution: $(15, 10)$

2. $\begin{cases} 2x - 3y = 4 & \text{Equation 1} \\ 6x = -12 & \text{Equation 2} \end{cases}$

Multiply Equation 2 by $\frac{1}{6}$: $x = -2$

Substitute $x = -2$ into Equation 1:

$2(-2) - 3y = 4$

$-4 - 3y = 4$

$-3y = 8$

$y = -\frac{8}{3}$

Solution: $\left(-2, -\frac{8}{3}\right)$

3. $\begin{cases} x + y = 32 & \text{Equation 1} \\ x - y = 24 & \text{Equation 2} \end{cases}$

Add to eliminate y

$2x = 56$

$x = 28$

Substitute $x = 28$ into Equation 1:

$28 + y = 32 \Rightarrow y = 4$

Solution: $(28, 4)$

4. $\begin{cases} 2r - s = 5 & \text{Equation 1} \\ r + 2s = 10 & \text{Equation 2} \end{cases}$

Multiply Equation 1 by 2: $4r - 2s = 10$

Add this to Equation 2 to eliminate s:

$$\begin{array}{rcl} 4r - 2s &=& 10 \\ r + 2s &=& 10 \\ \hline 5r &=& 20 \Rightarrow r = 4 \end{array}$$

Substitute $r = 4$ into Equation 1:

$2(4) - s = 5$

$8 - s = 5$

$3 = s$

Solution: $(4, 3)$

5. $5x - 3y + 4z = 2$; $(-1, -2, 1)$

$5(-1) - 3(-2) + 4(1) \stackrel{?}{=} 2$

$-5 + 6 + 4 \stackrel{?}{=} 2$

$5 \neq 2$

$(-1, -2, 1)$ *is not* a solution.

6. $x - 2y + 12z = 9$; $(6, 3, 2)$

$6 - 2(3) + 12(2) \stackrel{?}{=} 9$

$6 - 6 + 24 \stackrel{?}{=} 9$

$24 \neq 9$

$(6, 3, 2)$ *is not* a solution.

Skills Review —*continued*—

7. $2x - 5y + 3z = -9;\ (a - 2, a + 1, a)$

$$2(a - 2) - 5(a + 1) + 3a \stackrel{?}{=} -9$$

$$2a - 4 - 5a - 5 + 3a \stackrel{?}{=} -9$$

$$-9 = -9$$

$(a - 2, a + 1, a)$ *is* a solution.

8. $-5x + y + z = 21;\ (a - 4, 4a + 1, a)$

$$-5(a - 4) + 4a + 1 + a \stackrel{?}{=} 21$$

$$-5a + 20 + 4a + 1 + a \stackrel{?}{=} 21$$

$$21 = 21$$

$(a - 4, 4a + 1, a)$ *is* a solution.

9. $x + 2y - 3z = 4$

Substitute $y = 1 - a$ and $z = a$ into the original equation.

$$x + 2(1 - a) - 3a = 4$$

$$x + 2 - 2a - 3a = 4$$

$$x + 2 - 5a = 4$$

$$x = 5a + 2$$

10. $x - 3y + 5z = 4$

Substitute $y = 2a + 3$ and $z = a$ into the original equation.

$$x - 3(2a + 3) + 5a = 4$$

$$x - 6a - 9 + 5a = 4$$

$$x - a - 9 = 4$$

$$x = a + 13$$

1. $\begin{cases} 3x - y + z = 1 \\ 2x - 3z = -14 \\ 5y + 2z = 8 \end{cases}$

(a) $3(2) - (0) + (-3) \neq 1$

$(2, 0, -3)$ *is not* a solution.

(b) $3(-2) - (0) + 8 \neq 1$

$(-2, 0, 8)$ *is not* a solution

(c) $3(0) - (-1) + 3 \neq 1$

$(0, -1, 3)$ *is not* a solution.

(d) $3(-1) - (0) + 4 = 1$

$2(-1) - 3(4) = -14$

$5(0) + 2(4) = 8$

$(-1, 0, 4)$ *is* a solution.

3. $\begin{cases} 4x + y - z = 0 \\ -8x - 6y + z = -\frac{7}{4} \\ 3x - y = -\frac{9}{4} \end{cases}$

(a) $4\left(\frac{1}{2}\right) + \left(-\frac{3}{4}\right) - \left(-\frac{7}{4}\right) \neq 0$

$\left(\frac{1}{2}, -\frac{3}{4}, -\frac{7}{4}\right)$ *is not* a solution

(b) $4\left(-\frac{3}{2}\right) + \left(\frac{5}{4}\right) - \left(-\frac{5}{4}\right) \neq 0$

$\left(-\frac{3}{2}, \frac{5}{4}, -\frac{5}{4}\right)$ *is not* a solution.

(c) $4\left(-\frac{1}{2}\right) + \left(\frac{3}{4}\right) - \left(-\frac{5}{4}\right) = 0$

$-8\left(-\frac{1}{2}\right) - 6\left(\frac{3}{4}\right) + \left(-\frac{5}{4}\right) = -\frac{7}{4}$

$3\left(-\frac{1}{2}\right) - \left(\frac{3}{4}\right) = -\frac{9}{4}$

$\left(-\frac{1}{2}, \frac{3}{4}, -\frac{5}{4}\right)$ *is* a solution

(d) $4\left(-\frac{1}{2}\right) + \left(\frac{1}{6}\right) - \left(-\frac{3}{4}\right) \neq 0$

$\left(-\frac{1}{2}, \frac{1}{6}, -\frac{3}{4}\right)$ *is not* a solution.

5. $\begin{cases} 2x - y + 5z = 24 & \text{Equation 1} \\ y + 2z = 6 & \text{Equation 2} \\ z = 4 & \text{Equation 3} \end{cases}$

Back-substitute $z = 4$ into Equation 2:

$$y + 2(4) = 6$$

$$y = -2$$

Back-substitute $y = -2$ and $z = 4$ into Equation 1:

$$2x - (-2) + 5(4) = 24$$

$$2x + 22 = 24$$

$$x = 1$$

Solution: $(1, -2, 4)$

7. $\begin{cases} 2x + y - 3z = 10 & \text{Equation 1} \\ y + z = 12 & \text{Equation 2} \\ z = 2 & \text{Equation 3} \end{cases}$

Substitute $z = 2$ into Equation 2:
$y + (2) = 12 \Rightarrow y = 10$

Substitute $y = 10$ and $z = 2$ into Equation 1:

$$\begin{aligned} 2x + (10) - 3(2) &= 10 \\ 2x + 4 &= 10 \\ 2x &= 6 \\ x &= 3 \end{aligned}$$

Solution: $(3, 10, 2)$

9. $\begin{cases} 4x - 2y + z = 8 & \text{Equation 1} \\ -y + z = 4 & \text{Equation 2} \\ z = 2 & \text{Equation 3} \end{cases}$

Substitute $z = 2$ into Equation 2:

$-y + (2) = 4 \Rightarrow y = -2$

Substitute $y = -2$ and $z = 2$ into Equation 1:

$$\begin{aligned} 4x - 2(-2) + (2) &= 8 \\ 4x + 6 &= 8 \\ 4x &= 2 \\ x &= \tfrac{1}{2} \end{aligned}$$

Solution: $\left(\frac{1}{2}, -2, 2\right)$

11. $\begin{cases} x - 2y + 3z = 5 & \text{Equation 1} \\ -x + 3y - 5z = 4 & \text{Equation 2} \\ 2x \quad - 3z = 0 & \text{Equation 3} \end{cases}$

Add Equation 1 to Equation 2:

$\begin{cases} x - 2y + 3z = 5 \\ y - 2z = 9 \\ 2x \quad -3z = 0 \end{cases}$

This operation eliminated the x-variable from Equation 2.

13. $\begin{cases} x + y + z = 3 \\ x - 2y + 4z = 5 \\ 3y + 4z = 5 \end{cases}$

$\begin{cases} x + y + z = 3 \\ -3y + 3z = 2 & (-1)\text{Eq. 1} + \text{Eq. 2} \\ 3y + 4z = 5 \end{cases}$

$\begin{cases} x + y + z = 3 \\ -3y + 3z = 2 \\ 7z = 7 & \text{Eq. 2} + \text{Eq. 3} \end{cases}$

$\begin{cases} x + y + z = 3 \\ y - z = -\frac{2}{3} & \left(-\frac{1}{3}\right)\text{Eq. 2} \\ z = 1 & \left(\frac{1}{7}\right)\text{Eq. 3} \end{cases}$

$y - 1 = -\frac{2}{3} \Rightarrow y = \frac{1}{3}$

$x + \frac{1}{3} + 1 = 3 \Rightarrow x = \frac{5}{3}$

Solution: $\left(\frac{5}{3}, \frac{1}{3}, 1\right)$

15. $\begin{cases} x + y - z = -1 & \text{Interchange equations.} \\ 2x + 4y + z = 1 \\ x - 2y - 3z = 2 \end{cases}$

$\begin{cases} x + y - z = -1 \\ 2y + 3z = 3 & (-2)\text{Eq.1} + \text{Eq.2} \\ -3y - 2z = 3 & (-1)\text{Eq.1} + \text{Eq.3} \end{cases}$

$\begin{cases} x + y - z = -1 \\ 2y + 3z = 3 \\ -6y - 4z = 6 & 2\text{Eq.3} \end{cases}$

$\begin{cases} x + y - z = -1 \\ 2y + 3z = 3 \\ 5z = 15 & 3\text{Eq.2} + \text{Eq.3} \end{cases}$

$\begin{cases} x + y - z = -1 \\ y + \frac{3}{2}z = \frac{3}{2} & \left(\frac{1}{2}\right)\text{Eq.2} \\ z = 3 & \left(\frac{1}{5}\right)\text{Eq.3} \end{cases}$

$y + \frac{3}{2}(3) = \frac{3}{2} \Rightarrow y = -3$

$x - 3 - 3 = -1 \Rightarrow x = 5$

Solution: $(5, -3, 3)$

17. $\begin{cases} x + 4y + z = 0 \\ 2x + 4y - z = 7 \\ 2x - 4y + 2z = -6 \end{cases}$ Interchange equations.

$\begin{cases} x + 4y + z = 0 \\ -4y - 3z = 7 & (-2)\text{Eq.1} + \text{Eq.2} \\ -12 = -6 & (-2)\text{Eq.1} + \text{Eq.3} \end{cases}$

$\begin{cases} x + 4y + z = 0 \\ -4y - 3z = 7 \\ 9z = -27 & (-3)\text{Eq.2} + \text{Eq.3} \end{cases}$

$\begin{cases} x + 4y + z = 0 \\ y + \frac{3}{4}z = -\frac{7}{4} & \left(-\frac{1}{4}\right)\text{Eq.2} \\ z = -3 & \left(\frac{1}{9}\right)\text{Eq.3} \end{cases}$

$y + \frac{3}{4}(-3) = -\frac{7}{4} \Rightarrow y = \frac{1}{2}$

$x + 4\left(\frac{1}{2}\right) + (-3) = 0 \Rightarrow x = 1$

Solution: $\left(1, \frac{1}{2}, -3\right)$

19. $\begin{cases} 2x + y - 3z = 4 \\ 4x + 2z = 10 \\ -2x + 3y - 13z = -8 \end{cases}$

$\begin{cases} 2x + y - 3z = 4 & (-2)\text{Eq.1} + \text{Eq.2} \\ -2y + 8z = 2 & \text{Eq.1} + \text{Eq.3} \\ 4y - 16z = -4 \end{cases}$

$\begin{cases} 2x + y - 3z = 4 \\ -2y + 8z = 2 \\ 0 = 0 & 2\text{Eq.2} + \text{Eq.3} \end{cases}$

$\begin{cases} 2x + z = 5 & \left(\frac{1}{2}\right)\text{Eq.2} + \text{Eq.1} \\ -2y + 8z = 2 \end{cases}$

$\begin{cases} x + \frac{1}{2}z = \frac{5}{2} & \left(\frac{1}{2}\right)\text{Eq.1} \\ y - 4z = -1 & \left(-\frac{1}{2}\right)\text{Eq.2} \end{cases}$

Let $z = a$, then:

$y - 4a = -1 \Rightarrow y = 4a - 1$

$x + \frac{1}{2}a = \frac{5}{2} \Rightarrow x = -\frac{1}{2}a + \frac{5}{2}$

Solution: $\left(-\frac{1}{2}a + \frac{5}{2}, 4a - 1, a\right)$

21. $\begin{cases} x + 2z = 5 \\ 3x - y - z = 1 \\ 6x - y + 5z = 16 \end{cases}$

$\begin{cases} x + 2z = 5 \\ -y - 7z = -14 & (-3)\text{Eq.1} + \text{Eq.2} \\ -y - 7z = -14 & (-6)\text{Eq.1} + \text{Eq.3} \end{cases}$

$\begin{cases} x + 2z = 5 \\ -y - 7z = -14 \\ 0 = 0 & (-1)\text{Eq.2} + \text{Eq.3} \end{cases}$

$\begin{cases} x + 2z = 5 \\ y + 7z = 14 & (-1)\text{Eq.2} \end{cases}$

Let $z = a$, then:

$y + 7a = 14 \Rightarrow y = -7a + 14$

$x + 2a = 5 \Rightarrow x = -2a + 5$

Solution: $(-2a + 5, -7a + 14, a)$

23. $\begin{cases} 2x - 3y + z = -2 \\ -4x + 9y = 7 \end{cases}$

$\begin{cases} 2x - 3y + z = -2 \\ 3y + 2z = 3 & 2\text{Eq.1} + \text{Eq.2} \end{cases}$

$\begin{cases} 2x + 3z = 1 & \text{Eq.2} + \text{Eq.1} \\ 3y + 2z = 3 \end{cases}$

Let $z = a$, then:

$y = -\frac{2}{3}a + 1$

$x = -\frac{3}{2}a + \frac{1}{2}$

Solution: $\left(-\frac{3}{2}a + \frac{1}{2}, -\frac{2}{3}a + 1, a\right)$

25. $\begin{cases} x + 3w = 4 \\ 2y - z - w = 0 \\ 3y - 2w = 1 \\ 2x - y + 4z = 5 \end{cases}$

$\begin{cases} x + 3w = 4 \\ 2y - z - w = 0 \\ 3y - 2w = 1 \\ -y + 4z - 6w = -3 & -2\text{Eq.1} + \text{Eq.4} \end{cases}$

$\begin{cases} x + 3w = 4 \\ y - 4z + 6w = 3 & -\text{Eq.4 and interchange} \\ 2y - z - w = 0 & \text{the equations.} \\ 3y - 2w = 1 \end{cases}$

$\begin{cases} x + 3w = 4 \\ y - 4z + 6w = 3 \\ 7z - 13w = -6 & -\text{Eq.2} + \text{Eq.3} \\ 12z - 20w = -8 & -3\text{Eq.2} + \text{Eq.4} \end{cases}$

$\begin{cases} x + 3w = 4 \\ y - 4z + 6w = 3 \\ z - 3w = -2 & -\frac{1}{2}\text{Eq.4} + \text{Eq.3} \\ 12z - 20w = -8 \end{cases}$

$\begin{cases} x + 3w = 4 \\ y - 4z + 6w = 3 \\ z - 3w = -2 \\ 16w = 16 & -12\text{Eq.3} + \text{Eq.4} \end{cases}$

$16w = 16 \Rightarrow w = 1$

$z - 3(1) = -2 \Rightarrow z = 1$

$y - 4(1) + 6(1) = 3 \Rightarrow y = 1$

$x + 3(1) = 4 \Rightarrow x = 1$

Solution: $(1, 1, 1, 1)$

27. $\begin{cases} x + 4z = 1 \\ x + y + 10z = 10 \\ 2x - y + 2z = -5 \end{cases}$

$\begin{cases} x + 4z = 1 \\ y + 6z = 9 & -\text{Eq.1} + \text{Eq.2} \\ -y - 6z = -7 & -2\text{Eq.1} + \text{Eq.3} \end{cases}$

$\begin{cases} x + 4z = 1 \\ y + 6z = 9 \\ 0 = 2 & \text{Eq.2} + \text{Eq.3} \end{cases}$

No solution, inconsistent

29. $\begin{cases} 2x + 3y = 0 \\ 4x + 3y - z = 0 \\ 8x + 3y + 3z = 0 \end{cases}$

$\begin{cases} 2x + 3y = 0 \\ -3y - z = 0 & -2\text{Eq.1} + \text{Eq.2} \\ -9y + 3z = 0 & -4\text{Eq.1} + \text{Eq.3} \end{cases}$

$\begin{cases} 2x + 3y = 0 \\ -3y - z = 0 \\ 6z = 0 & -3\text{Eq.2} + \text{Eq.3} \end{cases}$

$6z = 0 \Rightarrow z = 0$

$-3y - 0 = 0 \Rightarrow y = 0$

$2x + 3(0) = 0 \Rightarrow x = 0$

Solution: $(0, 0, 0)$

31. $\begin{cases} 12x + 5y + z = 0 \\ 23x + 4y - z = 0 \end{cases}$

$\begin{cases} 24x + 10y + 2z = 0 & 2\text{Eq.1} \\ 23x + 4y - z = 0 \end{cases}$

$\begin{cases} x + 6y + 3z = 0 & -\text{Eq.2} + \text{Eq.1} \\ 23x + 4y - z = 0 \end{cases}$

$\begin{cases} x + 6y + 3z = 0 \\ -134y - 70z = 0 & -23\text{Eq.1} + \text{Eq.2} \end{cases}$

$\begin{cases} x + 6y + 3z = 0 \\ -67y - 35z = 0 & \frac{1}{2}\text{Eq.2} \end{cases}$

To avoid fractions, let $z = 67a$, then:

$-67y - 35(67a) = 0 \Rightarrow y = -35a$

$x + 6(-35a) + 3(67a) = 0 \Rightarrow x = 9a$

Solution: $(9a, -35a, 67a)$

33. $s = \frac{1}{2}at^2 + v_0t + s_0$

$(1, 452), (2, 372), (3, 260)$

$$\begin{cases} 452 = \frac{1}{2}a + v_0 + S_0 \\ 372 = 2a + 2v_0 + S_0 \\ 260 = \frac{9}{2}a + 3v_0 + S_0 \end{cases}$$

$$\begin{cases} a + 2v_0 + 2s_0 = 904 & 2\text{Eq.1} \\ 2a + 2v_0 + s_0 = 372 & \\ 9a + 6v_0 + 2s_0 = 520 & 2\text{Eq.3} \end{cases}$$

$$\begin{cases} a + 2v_0 + 2s_0 = 904 & \\ -2v_0 - 3s_0 = -1436 & -2\text{Eq.1} + \text{Eq.2} \\ -12v_0 - 16s_0 = -7616 & -9\text{Eq.1} + \text{Eq.3} \end{cases}$$

$$\begin{cases} a + 2v_0 + 2s_0 = 904 & \\ -2v_0 - 3s_0 = -1436 & \\ 2s_0 = 1000 & (-6)\text{Eq.2} + \text{Eq.3} \end{cases}$$

$$\begin{cases} a + 2v_0 + 2s_0 = 904 & \\ v_0 + 1.5s_0 = 718 & (-0.5)\text{Eq.2} \\ s_0 = 500 & (0.5)\text{Eq.3} \end{cases}$$

$v_0 + 1.5(500) = 718 \Rightarrow v_0 = -32$

$a + 2(-32) + 2(500) = 904 \Rightarrow a = -32$

So, $s = \frac{1}{2}(-32)t^2 + (-32)t + 500$

$= -16t^2 - 32t + 500.$

35. $y = ax^2 + bx + c$ passing through $(0, 0)$, $(2, -2)$, $(4, 0)$

$(0, 0)$: $0 = c$

$(2, -2)$: $-2 = 4a + 2b + c \Rightarrow -1 = 2a + b$

$(4, 0)$: $0 = 16a + 4b + c \Rightarrow 0 = 4a + b$

$$\begin{cases} 2a + b = -1 \\ 4a + b = 0 \end{cases}$$

$$\begin{cases} 2a + b = -1 & \\ -b = 2 & -2\text{Eq.1} + \text{Eq.2} \end{cases}$$

Solution: $a = \frac{1}{2}$, $b = -2$, $c = 0$

The equation of the parabola is $y = \frac{1}{2}x^2 - 2x$.

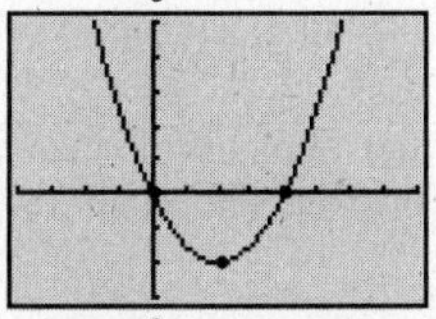

37. $y = ax^2 + bx + c$ passing through $(2, 0)$, $(3, -1)$, $(4, 0)$

$(2, 0)$: $0 = 4a + 2b + c$

$(3, -1)$: $-1 = 9a + 3b + c$

$(4, 0)$: $0 = 16a + 4b + c$

$$\begin{cases} 0 = 4a + 2b + c & \\ -1 = 5a + b & -\text{Eq.1} + \text{Eq.2} \\ 0 = 12a + 2b & -\text{Eq.1} + \text{Eq.3} \end{cases}$$

$$\begin{cases} 0 = 4a + 2b + c & \\ -1 = 5a + b & \\ 2 = 2a & -2\text{Eq.2} + \text{Eq.3} \end{cases}$$

Solution: $a = 1$, $b = -6$, $c = 8$

The equation of the parabola is $y = x^2 - 6x + 8$.

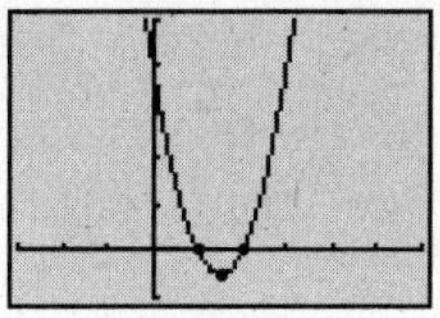

39. $x^2 + y^2 + Dx + Ey + F = 0$ passing through $(0, 0)$, $(2, 2)$, $(4, 0)$

$(0, 0)$: $F = 0$

$(2, 2)$: $8 + 2D + 2E + F = 0 \Rightarrow D + E = -4$

$(4, 0)$: $16 + 4D + F = 0 \Rightarrow D = -4$ and $E = 0$

The equation of the circle is $x^2 + y^2 - 4x = 0$.

To graph, let $y_1 = \sqrt{4x - x^2}$ and $y_2 = -\sqrt{4x - x^2}$.

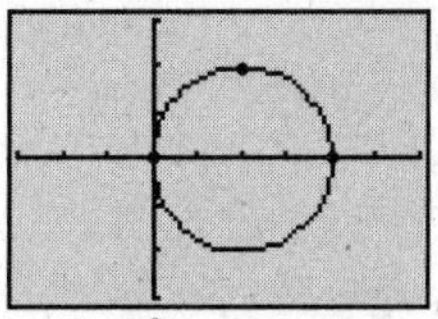

41. $x^2 + y^2 + Dx + Ey + F = 0$ passing through $(-3, -1)$, $(2, 4)$, $(-6, 8)$

$(-3, -1)$: $10 - 3D - E + F = 0$

$3D + E - F = 10$

$(2, 4)$: $20 + 2D + 4E + F = 0$

$-2D - 4E - F = 20$

$(-6, 8)$: $100 - 6D + 8E + F = 0$

$6D - 8E - F = 100$

$$\begin{cases} 3D + E - F = 10 \\ -2D - 4E - F = 20 \\ 6D - 8E - F = 100 \end{cases}$$

$$\begin{cases} 6D - 8E - F = 100 \\ -2D - 4E - F = 20 \\ 3D + E - F = 10 \end{cases} \quad \text{Interchange Eq.1 and Eq.3.}$$

$$\begin{cases} 6D - 8E - F = 100 \\ 20E + 4F = -160 & (-3)\text{Eq.2} - \text{Eq.1} \\ 10E - F = -80 & 2\text{Eq.3} - \text{Eq.1} \end{cases}$$

$$\begin{cases} 6D - 8E - F = 100 \\ 20E + 4F = -160 \\ 6F = 0 & (-2)\text{Eq.3} + \text{Eq.2} \end{cases}$$

Solution: $D = 6$, $E = -8$, $F = 0$

The equation of the circle is $x^2 + y^2 + 6x - 8y = 0$. To graph, complete the squares first, then solve for y.

$$\left(x^2 + 6x + 9\right) + \left(y^2 - 8y + 16\right) = 0 + 9 + 16$$

$$(x + 3)^2 + (y - 4)^2 = 25$$

$$(y - 4)^2 = 25 - (x + 3)^2$$

$$y - 4 = \pm\sqrt{25 - (x + 3)^2}$$

$$y = 4 \pm \sqrt{25 - (x + 3)^2}$$

Let $y_1 = 4 + \sqrt{25 - (x + 3)^2}$ and

$y_2 = 4 - \sqrt{25 - (x + 3)^2}$.

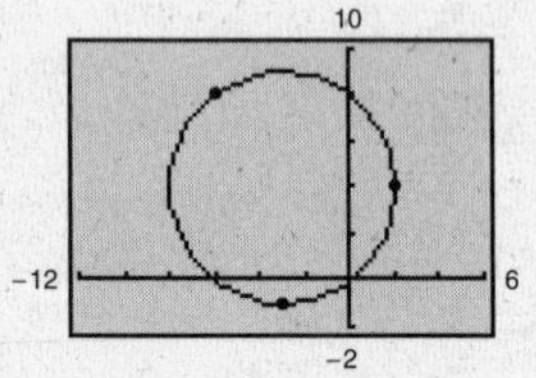

43. Let x = pounds of brand X.

Let y = pounds of brand Y.

Let z = pounds of brand Z.

Fertilizer A: $\frac{1}{3}y + \frac{2}{9}z = 5$

Fertilizer B: $\frac{1}{2}x + \frac{2}{3}y + \frac{5}{9}z = 13$

Fertilizer C: $\frac{1}{2}x + \frac{2}{9}z = 4$

$$\begin{cases} \frac{1}{2}x + \frac{2}{3}y + \frac{5}{9}z = 13 \\ \frac{1}{3}y + \frac{2}{9}z = 5 \\ \frac{1}{2}x + \frac{2}{9}z = 4 \end{cases} \quad \text{Interchange Eq.1 and Eq.2.}$$

$$\begin{cases} \frac{1}{2}x + \frac{2}{3}y + \frac{5}{9}z = 13 \\ \frac{1}{3}y + \frac{2}{9}z = 5 \\ -\frac{2}{3}y - \frac{1}{3}z = -9 & -\text{Eq.1} + \text{Eq.3} \end{cases}$$

$$\begin{cases} \frac{1}{2}x + \frac{2}{3}y + \frac{5}{9}z = 13 \\ \frac{1}{3}y + \frac{2}{9}z = 5 \\ \frac{1}{9}z = 1 & 2\text{Eq.2} + \text{Eq.3} \end{cases}$$

$z = 9$

$\frac{1}{3}y + \frac{2}{9}(9) = 5 \Rightarrow y = 9$

$\frac{1}{2}x + \frac{2}{3}(9) + \frac{5}{9}(9) = 13 \Rightarrow x = 4$

4 pounds of brand X, 9 pounds of brand Y, and 9 pounds of brand Z are needed to obtain the desired mixture.

45. Let x = pounds of Vanilla coffee.

Let y = pounds of Hazelnut coffee.

Let z = pounds of Mocha coffee.

$$\begin{cases} x + y + z = 10 \\ 2x + 2.50y + 3z = 26 \\ y - z = 0 \end{cases}$$

$$\begin{cases} x + y + z = 10 \\ 0.5y + z = 6 & -2\text{Eq.1} + \text{Eq.2} \\ y - z = 0 \end{cases}$$

$$\begin{cases} x + y + z = 10 \\ 0.5y + z = 6 \\ -3z = -12 & -2\text{Eq.2} + \text{Eq.3} \end{cases}$$

$z = 4$

$0.5y + 4 = 6 \Rightarrow y = 4$

$x + 4 + 4 = 10 \Rightarrow x = 2$

2 pounds of Vanilla coffee, 4 pounds of Hazelnut coffee, and 4 pounds of Mocha coffee are needed to obtain the desired mixture.

47. Let x = number of television ads.

Let y = number of radio ads.

Let z = number of local newspaper ads.

$$\begin{cases} x + y + z = 60 \\ 1000x + 200y + 500z = 42{,}000 \\ x - y - z = 0 \end{cases}$$

$$\begin{cases} x + y + z = 60 & \\ -800y - 500z = -18{,}000 & -1000\text{Eq.1} + \text{Eq.2} \\ -2y - 2z = -60 & -\text{Eq.1} + \text{Eq.3} \end{cases}$$

$$\begin{cases} x + y + z = 60 & \\ -2y - 2z = -60 & \text{Interchange} \\ -800y - 500z = -18{,}000 & \text{Eq.2 and Eq.3} \end{cases}$$

$$\begin{cases} x + y + z = 60 & \\ -2y - 2z = -60 & \\ 300z = 6000 & -400\text{Eq.2} + \text{Eq.3} \end{cases}$$

$z = 20$

$-2y - 2(20) = -60 \Rightarrow y = 10$

$x + 10 + 20 = 60 \Rightarrow x = 30$

30 television ads, 10 radio ads, and 20 newspaper ads can be run each month.

49. (a) To use 2 liters of the 50% solution:

Let x = amount of 10% solution.

Let y = amount of 20% solution.

$x + y = 8 \Rightarrow y = 8 - x$

$$x(0.10) + y(0.20) + 2(0.50) = 10(0.25)$$
$$0.10x + 0.20(8 - x) + 1 = 2.5$$
$$0.10x + 1.6 - 0.20x + 1 = 2.5$$
$$-0.10x = -0.1$$
$$x = 1$$
$$y = 7$$

2 liters of 50% solution, 1 liter of 10% solution, and 7 liters of 20% solution.

(b) To use as little of the 50% solution as possible, the chemist should use no 10% solution.

Let x = amount of 20% solution.

Let y = amount of 50% solution.

$x + y = 10 \Rightarrow y = 10 - x$

$$x(0.20) + y(0.50) = 10(0.25)$$
$$x(0.20) + (10 - x)(0.50) = 10(0.25)$$
$$x(0.20) + 5 - 0.50x = 2.5$$
$$-0.30x = -2.5$$
$$x = 8\tfrac{1}{3}$$
$$y = 1\tfrac{2}{3}$$

$8\frac{1}{3}$ liters of 20% solution, $1\frac{2}{3}$ liters of 50% solution, and no liters of 10% solution.

(c) To use as much of the 50% solution as possible, the chemist should use no 20% solution.

Let x = amount of 10% solution.

Let y = amount of 50% solution.

$x + y = 10 \Rightarrow y = 10 - x$

$$x(0.10) + y(0.50) = 10(0.25)$$
$$0.10x + 0.50(10 - x) = 2.5$$
$$0.10x + 5 - 0.50x = 2.5$$
$$-0.40x = -2.5$$
$$x = 6\tfrac{1}{4}$$
$$y = 3\tfrac{3}{4}$$

$6\frac{1}{4}$ liters of 10% solution, $3\frac{3}{4}$ liters of 50% solution, and no liters of 20% solution.

51. There are an infinite number of linear systems that have $(4, -1, 2)$ as their solution. Two such systems are as follows:

$$\begin{cases} 3x + y - z = 9 \\ x + 2y - z = 0 \\ -x + y + 3z = 1 \end{cases} \qquad \begin{cases} x + y + z = 5 \\ x \quad\quad - 2z = 0 \\ 2y + z = 0 \end{cases}$$

53. There are an infinite number of linear systems that have $\left(3, -\frac{1}{2}, \frac{7}{4}\right)$ as their solution. Two such systems are as follows:

$$\begin{cases} x + 2y - 4z = -5 \\ -x - 4y + 8z = 13 \\ x + 6y + 4z = 7 \end{cases} \qquad \begin{cases} x + 2y + 4z = 9 \\ y + 2z = 3 \\ x \quad\quad - 4z = -4 \end{cases}$$

55. False. Equation 2 does not have a leading coefficient of 1.

57. Answers will vary.

Section 8.3 Matrices and Systems of Equations

Skills Review

1. $\begin{cases} 4x - 2y + 3z = -5 & \text{Equation 1} \\ x + 3y - z = 11 & \text{Equation 2} \\ -x + 2y \quad\quad = 5 & \text{Equation 3} \end{cases}$

$4(1) - 2(3) + 3(-1) = -5$

$1 + 3(3) - (-1) = 11$

$-1 + 2(3) = 5$

$(1, 3, -1)$ *is* a solution.

2. $\begin{cases} -x + 2y + z = 4 & \text{Equation 1} \\ 2x \quad\quad - 3z = 5 & \text{Equation 2} \\ 3x + 5y - 2z = 21 & \text{Equation 3} \end{cases}$

$-(1) + 2(3) + (-1) = 4$

$2(1) - 3(-1) = 5$

$3(1) + 5(3) - 2(-1) = 20$

$(1, 3, -1)$ *is not* a solution

3. $\begin{cases} 2x - 3y = 4 & \text{Equation 1} \\ y = 2 & \text{Equation 2} \end{cases}$

Back-substitute $y = 2$ into Equation 1:

$2x - 3(2) = 4$

$2x = 10$

$x = 5$

Solution: $(5, 2)$

4. $\begin{cases} 5x + 4y = 0 & \text{Equation 1} \\ y = -3 & \text{Equation 2} \end{cases}$

Back-substitute $y = -3$ into Equation 1:

$5x + 4(-3) = 0$

$5x = 12$

$x = \frac{12}{5}$

Solution: $\left(\frac{12}{5}, -3\right)$

5. $\begin{cases} x - 3y + z = 0 & \text{Equation 1} \\ y - 3z = 8 & \text{Equation 2} \\ z = 2 & \text{Equation 3} \end{cases}$

Back-substitute $z = 2$ into Equation 2:

$y - 3(2) = 8$

$y = 14$

Back-substitute $z = 2$ and $y = 14$ into Equation 1:

$x - 3(14) + 2 = 0$

$x = 40$

Solution: $(40, 14, 2)$

6. $\begin{cases} 2x - 5y + 3z = -2 & \text{Equation 1} \\ y - 4z = 0 & \text{Equation 2} \\ z = 1 & \text{Equation 3} \end{cases}$

Back-substitute $z = 1$ into Equation 2:

$y - 4(1) = 0$

$y = 4$

Back-substitute $z = 1$ and $y = 4$ into equation 1:

$2x - 5(4) + 3(1) = -2$

$2x = 15$

$x = \frac{15}{2}$

Solution: $\left(\frac{15}{2}, 4, 1\right)$

1. Because the matrix has one row and two columns, its order is 1×2.

3. Because the matrix has three rows and one column, its order is 3×1.

5. Because the matrix has two rows and two columns, its order is 2×2.

7. $\begin{cases} 4x - 3y = -5 \\ -x + 3y = 12 \end{cases}$

$$\left[\begin{array}{cc|c} 4 & -3 & -5 \\ -1 & 3 & 12 \end{array}\right]$$

9. $\begin{cases} x + 10y - 2z = 2 \\ 5x - 3y + 4z = 0 \\ 2x + y \quad = 6 \end{cases}$

$$\left[\begin{array}{ccc|c} 1 & 10 & -2 & 2 \\ 5 & -3 & 4 & 0 \\ 2 & 1 & 0 & 6 \end{array}\right]$$

11. $\begin{cases} 7x - 5y + z = 13 \\ 19x \quad - 8z = 10 \end{cases}$

$$\left[\begin{array}{ccc|c} 7 & -5 & 1 & 13 \\ 19 & 0 & -8 & 10 \end{array}\right]$$

13. $\left[\begin{array}{cc|c} 1 & 2 & 7 \\ 2 & -3 & 4 \end{array}\right]$

$\begin{cases} x + 2y = 7 \\ 2x - 3y = 4 \end{cases}$

15. $\left[\begin{array}{ccc|c} 2 & 0 & 5 & -12 \\ 0 & 1 & -2 & 7 \\ 6 & 3 & 0 & 2 \end{array}\right]$

$\begin{cases} 2x \quad + 5z = -12 \\ \quad y - 2z = 7 \\ 6x + 3y \quad = 2 \end{cases}$

17. $\left[\begin{array}{cccc|c} 9 & 12 & 3 & 0 & 0 \\ -2 & 18 & 5 & 2 & 10 \\ 1 & 7 & -8 & 0 & -4 \\ 3 & 0 & 2 & 0 & -10 \end{array}\right]$

$\begin{cases} 9x + 12y + 3z \quad = 0 \\ -2x + 18y + 5z + 2w = 10 \\ x + 7y - 8z \quad = -4 \\ 3x \quad + 2z \quad = -10 \end{cases}$

19. $\begin{bmatrix} 1 & 4 & 3 \\ 2 & 10 & 5 \end{bmatrix}$

$$-2R_1 + R_2 \rightarrow \begin{bmatrix} 1 & 4 & 3 \\ 0 & \boxed{2} & -1 \end{bmatrix}$$

21. $\begin{bmatrix} 1 & 1 & 4 & -1 \\ 3 & 8 & 10 & 3 \\ -2 & 1 & 12 & 6 \end{bmatrix}$

$$\begin{array}{r} \\ -3R_1 + R_2 \rightarrow \\ 2R_1 + R_3 \rightarrow \end{array} \begin{bmatrix} 1 & 1 & 4 & -1 \\ 0 & 5 & \boxed{-2} & \boxed{6} \\ 0 & 3 & \boxed{20} & \boxed{4} \end{bmatrix}$$

$$\tfrac{1}{5}R_2 \rightarrow \begin{bmatrix} 1 & 1 & 4 & -1 \\ 0 & 1 & -\frac{2}{5} & \frac{6}{5} \\ 0 & 3 & \boxed{20} & \boxed{4} \end{bmatrix}$$

23. $\begin{bmatrix} -2 & 5 & 1 \\ 3 & -1 & -8 \end{bmatrix} \rightarrow \begin{bmatrix} 13 & 0 & -39 \\ 3 & -1 & -8 \end{bmatrix}$

Add 5 times Row 2 to Row 1.

25. $\begin{bmatrix} 0 & -1 & -5 & 5 \\ -1 & 3 & -7 & 6 \\ 4 & -5 & 1 & 3 \end{bmatrix} \rightarrow \begin{bmatrix} -1 & 3 & -7 & 6 \\ 0 & -1 & -5 & 5 \\ 0 & 7 & -27 & 27 \end{bmatrix}$

Interchange Row 1 and Row 2. Then add 4 times the new Row 1 to Row 3.

27. $\begin{bmatrix} 1 & 2 & 3 \\ 2 & -1 & -4 \\ 3 & 1 & -1 \end{bmatrix}$

(a) $\begin{bmatrix} 1 & 2 & 3 \\ 0 & -5 & -10 \\ 3 & 1 & -1 \end{bmatrix}$ (b) $\begin{bmatrix} 1 & 2 & 3 \\ 0 & -5 & -10 \\ 0 & -5 & -10 \end{bmatrix}$

(c) $\begin{bmatrix} 1 & 2 & 3 \\ 0 & -5 & -10 \\ 0 & 0 & 0 \end{bmatrix}$ (d) $\begin{bmatrix} 1 & 2 & 3 \\ 0 & 1 & 2 \\ 0 & 0 & 0 \end{bmatrix}$

(e) $\begin{bmatrix} 1 & 0 & -1 \\ 0 & 1 & 2 \\ 0 & 0 & 0 \end{bmatrix}$

This matrix is in reduced row-echelon form.

29. $\begin{bmatrix} 1 & 0 & 0 & 0 \\ 0 & 1 & 1 & 5 \\ 0 & 0 & 0 & 0 \end{bmatrix}$

This matrix is in reduced row-echelon form.

31. $\begin{bmatrix} 2 & 0 & 4 & 0 \\ 0 & -1 & 3 & 6 \\ 0 & 0 & 1 & 5 \end{bmatrix}$

The first nonzero entries in Rows 1 and 2 are not 1. The matrix is not in row-echelon form.

33. $\begin{bmatrix} 1 & 1 & 0 & 5 \\ -2 & -1 & 2 & -10 \\ 3 & 6 & 7 & 14 \end{bmatrix}$

$\begin{matrix} \\ 2R_1 + R_2 \rightarrow \\ -3R_1 + R_3 \rightarrow \end{matrix} \begin{bmatrix} 1 & 1 & 0 & 5 \\ 0 & 1 & 2 & 0 \\ 0 & 3 & 7 & -1 \end{bmatrix}$

$\begin{matrix} \\ \\ -3R_2 + R_3 \rightarrow \end{matrix} \begin{bmatrix} 1 & 1 & 0 & 5 \\ 0 & 1 & 2 & 0 \\ 0 & 0 & 1 & -1 \end{bmatrix}$

35. $\begin{bmatrix} 1 & -1 & -1 & 1 \\ 5 & -4 & 1 & 8 \\ -6 & 8 & 18 & 0 \end{bmatrix}$

$\begin{matrix} \\ -5R_1 + R_2 \rightarrow \\ 6R_1 + R_3 \rightarrow \end{matrix} \begin{bmatrix} 1 & -1 & -1 & 1 \\ 0 & 1 & 6 & 3 \\ 0 & 2 & 12 & 6 \end{bmatrix}$

$\begin{matrix} \\ \\ -2R_2 + R_3 \rightarrow \end{matrix} \begin{bmatrix} 1 & -1 & -1 & 1 \\ 0 & 1 & 6 & 3 \\ 0 & 0 & 0 & 0 \end{bmatrix}$

37. Use the reduced row-echelon form feature of a graphing utility.

$\begin{bmatrix} 3 & 3 & 3 \\ -1 & 0 & -4 \\ 2 & 4 & -2 \end{bmatrix} \Rightarrow \begin{bmatrix} 1 & 0 & 0 \\ 0 & 1 & 0 \\ 0 & 0 & 1 \end{bmatrix}$

39. Use the reduced row-echelon form feature of a graphing utility.

$\begin{bmatrix} 1 & 2 & 3 & -5 \\ 1 & 2 & 4 & -9 \\ -2 & -4 & -4 & 3 \\ 4 & 8 & 11 & -14 \end{bmatrix} \Rightarrow \begin{bmatrix} 1 & 2 & 0 & 0 \\ 0 & 0 & 1 & 0 \\ 0 & 0 & 0 & 1 \\ 0 & 0 & 0 & 0 \end{bmatrix}$

41. Use the reduced row-echelon form feature of a graphing utility.

$\begin{bmatrix} -3 & 5 & 1 & 12 \\ 1 & -1 & 1 & 4 \end{bmatrix} \Rightarrow \begin{bmatrix} 1 & 0 & 3 & 16 \\ 0 & 1 & 2 & 12 \end{bmatrix}$

43. $\begin{cases} x - 2y = 4 \\ y = -3 \end{cases}$

$x - 2(-3) = 4$

$x = -2$

Solution: $(-2, -3)$

45. $\begin{cases} x - y + 2z = 4 \\ y - z = 2 \\ z = -2 \end{cases}$

$y - (-2) = 2$

$y = 0$

$x - 0 + 2(-2) = 4$

$x = 8$

Solution: $(8, 0, -2)$

47. $\begin{cases} 2x + 6y = 16 \\ 2x + 3y = 7 \end{cases}$

$\left[\begin{array}{cc:c} 2 & 6 & 16 \\ 2 & 3 & 7 \end{array}\right]$

$\begin{matrix} \\ -R_1 + R_2 \rightarrow \end{matrix} \left[\begin{array}{cc:c} 2 & 6 & 16 \\ 0 & -3 & -9 \end{array}\right]$

$\begin{matrix} \frac{1}{2}R_1 \rightarrow \\ -\frac{1}{3}R_2 \rightarrow \end{matrix} \left[\begin{array}{cc:c} 1 & 3 & 8 \\ 0 & 1 & 3 \end{array}\right]$

$\begin{cases} x + 3y = 8 \\ y = 3 \end{cases}$

$y = 3$

$x + 3(3) = 8 \Rightarrow x = -1$

Solution: $(-1, 3)$

49. $\begin{cases} -x + y = 4 \\ 2x - 4y = -34 \end{cases}$

$\left[\begin{array}{cc:c} -1 & 1 & 4 \\ 2 & -4 & -34 \end{array}\right]$

$\begin{matrix} (-1)R_1 \rightarrow \\ \left(\frac{1}{2}\right)R_2 \rightarrow \end{matrix} \left[\begin{array}{cc:c} 1 & -1 & -4 \\ 1 & -2 & -17 \end{array}\right]$

$\begin{matrix} \\ -R_1 + R_2 \rightarrow \end{matrix} \left[\begin{array}{cc:c} 1 & -1 & -4 \\ 0 & -1 & -13 \end{array}\right]$

$\begin{matrix} \\ (-1)R_2 \rightarrow \end{matrix} \left[\begin{array}{cc:c} 1 & -1 & -4 \\ 0 & 1 & 13 \end{array}\right]$

$\begin{cases} x - y = -4 \\ y = 13 \end{cases}$

$y = 13$

$x - 13 = -4 \Rightarrow x = 9$

Solution: $(9, 13)$

51. $\begin{cases} 5x - 5y = -5 \\ -2x - 3y = 7 \end{cases}$

$$\begin{bmatrix} 5 & -5 & \vdots & -5 \\ -2 & -3 & \vdots & 7 \end{bmatrix}$$

$$\tfrac{1}{5}R_1 \rightarrow \begin{bmatrix} 1 & -1 & \vdots & -1 \\ -2 & -3 & \vdots & 7 \end{bmatrix}$$

$$2R_1 + R_2 \rightarrow \begin{bmatrix} 1 & -1 & \vdots & -1 \\ 0 & -5 & \vdots & 5 \end{bmatrix}$$

$$-\tfrac{1}{5}R_2 \rightarrow \begin{bmatrix} 1 & -1 & \vdots & -1 \\ 0 & 1 & \vdots & -1 \end{bmatrix}$$

$\begin{cases} x - y = -1 \\ y = -1 \end{cases}$

$y = -1$

$x - (-1) = -1 \Rightarrow x = -2$

Solution: $(-2, -1)$

53. $\begin{cases} 2x - y + 3z = 24 \\ 2y - z = 14 \\ 7x - 5y = 6 \end{cases}$

$$\begin{bmatrix} 2 & -1 & 3 & \vdots & 24 \\ 0 & 2 & -1 & \vdots & 14 \\ 7 & -5 & 0 & \vdots & 6 \end{bmatrix}$$

$$R_3 + (-3)R_1 \rightarrow \begin{bmatrix} 1 & -2 & -9 & \vdots & -66 \\ 0 & 2 & -1 & \vdots & 14 \\ 7 & -5 & 0 & \vdots & 6 \end{bmatrix}$$

$$\begin{matrix} \\ \\ -7R_1 + R_3 \rightarrow \end{matrix} \begin{bmatrix} 1 & -2 & -9 & \vdots & -66 \\ 0 & 2 & -1 & \vdots & 14 \\ 0 & 9 & 63 & \vdots & 468 \end{bmatrix}$$

$$4R_2 \rightarrow \begin{bmatrix} 1 & -2 & -9 & \vdots & -66 \\ 0 & 8 & -4 & \vdots & 56 \\ 0 & 9 & 63 & \vdots & 468 \end{bmatrix}$$

$$-R_3 + R_2 \rightarrow \begin{bmatrix} 1 & -2 & -9 & \vdots & -66 \\ 0 & -1 & -67 & \vdots & -412 \\ 0 & 9 & 63 & \vdots & 468 \end{bmatrix}$$

$$\begin{matrix} \\ \\ 9R_2 + R_3 \rightarrow \end{matrix} \begin{bmatrix} 1 & -2 & -9 & \vdots & -66 \\ 0 & -1 & -67 & \vdots & -412 \\ 0 & 0 & -540 & \vdots & -3240 \end{bmatrix}$$

$$\begin{matrix} \\ -R_2 \rightarrow \\ -\tfrac{1}{540}R_3 \rightarrow \end{matrix} \begin{bmatrix} 1 & -2 & -9 & \vdots & -66 \\ 0 & 1 & 67 & \vdots & 412 \\ 0 & 0 & 1 & \vdots & 6 \end{bmatrix}$$

$\begin{cases} x - 2y - 9z = -66 \\ y + 67z = 412 \\ z = 6 \end{cases}$

$z = 6$

$y + 67(6) = 412 \Rightarrow y = 10$

$x - 2(10) - 9(6) = -66 \Rightarrow x = 8$

Solution: $(8, 10, 6)$

55. $\begin{cases} 2x + 2y - z = 2 \\ x - 3y + z = -28 \\ -x + y = 14 \end{cases}$

$$\begin{bmatrix} 2 & 2 & -1 & \vdots & 2 \\ 1 & -3 & 1 & \vdots & -28 \\ -1 & 1 & 0 & \vdots & 14 \end{bmatrix}$$

$$\begin{matrix} R_2 \\ R_1 \\ \end{matrix}\begin{bmatrix} 1 & -3 & 1 & \vdots & -28 \\ 2 & 2 & -1 & \vdots & 2 \\ -1 & 1 & 0 & \vdots & 14 \end{bmatrix}$$

$$\begin{matrix} \\ R_3 \\ R_2 \end{matrix}\begin{bmatrix} 1 & -3 & 1 & \vdots & -28 \\ -1 & 1 & 0 & \vdots & 14 \\ 2 & 2 & -1 & \vdots & 2 \end{bmatrix}$$

$$\begin{matrix} \\ R_1 + R_2 \rightarrow \\ -2R_1 + R_3 \rightarrow \end{matrix}\begin{bmatrix} 1 & -3 & 1 & \vdots & -28 \\ 0 & -2 & 1 & \vdots & -14 \\ 0 & 8 & -3 & \vdots & 58 \end{bmatrix}$$

$$\begin{matrix} \\ \\ 4R_2 + R_3 \rightarrow \end{matrix}\begin{bmatrix} 1 & -3 & 1 & \vdots & -28 \\ 0 & -2 & 1 & \vdots & -14 \\ 0 & 0 & 1 & \vdots & 2 \end{bmatrix}$$

$$\begin{matrix} \\ -\frac{1}{2}R_2 \rightarrow \\ \\ \end{matrix}\begin{bmatrix} 1 & -3 & 1 & \vdots & -28 \\ 0 & 1 & -\frac{1}{2} & \vdots & 7 \\ 0 & 0 & 1 & \vdots & 2 \end{bmatrix}$$

$\begin{cases} x - 3y + z = -28 \\ y - \frac{1}{2}z = 7 \\ z = 2 \end{cases}$

$z = 2$

$y - \frac{1}{2}(2) = 7 \Rightarrow y = 8$

$x - 3(8) + 2 = -28 \Rightarrow x = -6$

Solution: $(-6, 8, 2)$

57. $\begin{cases} 3x - 2y + z = 15 \\ -x + y + 2z = -10 \\ x - y - 4z = 14 \end{cases}$

$$\begin{bmatrix} 3 & -2 & 1 & \vdots & 15 \\ -1 & 1 & 2 & \vdots & -10 \\ 1 & -1 & -4 & \vdots & 14 \end{bmatrix}$$

$$\begin{matrix} R_3 \\ \\ R_1 \end{matrix}\begin{bmatrix} 1 & -1 & -4 & \vdots & 14 \\ -1 & 1 & 2 & \vdots & -10 \\ 3 & -2 & 1 & \vdots & 15 \end{bmatrix}$$

$$\begin{matrix} \\ R_1 + R_2 \rightarrow \\ -3R_1 + R_3 \rightarrow \end{matrix}\begin{bmatrix} 1 & -1 & -4 & \vdots & 14 \\ 0 & 0 & -2 & \vdots & 4 \\ 0 & 1 & 13 & \vdots & -27 \end{bmatrix}$$

$$\begin{matrix} \\ R_3 \\ R_2 \end{matrix}\begin{bmatrix} 1 & -1 & -4 & \vdots & 14 \\ 0 & 1 & 13 & \vdots & -27 \\ 0 & 0 & -2 & \vdots & 4 \end{bmatrix}$$

$$\begin{matrix} \\ \\ -\frac{1}{2}R_3 \rightarrow \end{matrix}\begin{bmatrix} 1 & -1 & -4 & \vdots & 14 \\ 0 & 1 & 13 & \vdots & -27 \\ 0 & 0 & 1 & \vdots & -2 \end{bmatrix}$$

$\begin{cases} x - y - 4z = 14 \\ y + 13z = -27 \\ z = -2 \end{cases}$

$z = -2$

$y + 13(-2) = -27 \Rightarrow y = -1$

$x - (-1) - 4(-2) = 14 \Rightarrow x = 5$

Solution: $(5, -1, -2)$

59. $\begin{cases} x + 2y + z + 2w = 8 \\ 3x + 7y + 6z + 9w = 26 \end{cases}$

$$\begin{bmatrix} 1 & 2 & 1 & 2 & \vdots & 8 \\ 3 & 7 & 6 & 9 & \vdots & 26 \end{bmatrix}$$

$$\begin{matrix} \\ -3R_1 + R_2 \rightarrow \end{matrix}\begin{bmatrix} 1 & 2 & 1 & 2 & \vdots & 8 \\ 0 & 1 & 3 & 3 & \vdots & 2 \end{bmatrix}$$

$$\begin{matrix} -2R_2 + R_1 \rightarrow \\ \\ \end{matrix}\begin{bmatrix} 1 & 0 & -5 & -4 & \vdots & 4 \\ 0 & 1 & 3 & 3 & \vdots & 2 \end{bmatrix}$$

$\begin{cases} x - 5z - 4w = 4 \\ y + 3z + 3w = 2 \end{cases}$

Let $w = a$ and $z = b$, then:

$y + 3b + 3a = 2 \Rightarrow y = 2 - 3b - 3a$

$x - 5b - 4a = 4 \Rightarrow x = 4 + 5b + 4a$

Solution: $(4 + 5b + 4a, 2 - 3b - 3a, b, a)$ where a and b are real numbers

61. $\begin{cases} -x + y = -22 \\ 3x + 4y = 4 \\ 4x - 8y = 32 \end{cases}$

$$\begin{bmatrix} -1 & 1 & \vdots & -22 \\ 3 & 4 & \vdots & 4 \\ 4 & -8 & \vdots & 32 \end{bmatrix}$$

$$-R_1 \rightarrow \begin{bmatrix} 1 & -1 & \vdots & 22 \\ 3 & 4 & \vdots & 4 \\ 4 & -8 & \vdots & 32 \end{bmatrix}$$

$$\begin{matrix} \\ -3R_1 + R_2 \rightarrow \\ -4R_1 + R_3 \rightarrow \end{matrix} \begin{bmatrix} 1 & -1 & \vdots & 22 \\ 0 & 7 & \vdots & -62 \\ 0 & -4 & \vdots & -56 \end{bmatrix}$$

$$\begin{matrix} \\ \frac{1}{7}R_2 \rightarrow \\ -\frac{1}{4}R_3 \rightarrow \end{matrix} \begin{bmatrix} 1 & -1 & \vdots & 22 \\ 0 & 1 & \vdots & -\frac{62}{7} \\ 0 & 1 & \vdots & 14 \end{bmatrix}$$

$$\begin{matrix} \\ \\ -R_2 + R_3 \rightarrow \end{matrix} \begin{bmatrix} 1 & -1 & \vdots & 22 \\ 0 & 1 & \vdots & -\frac{62}{7} \\ 0 & 0 & \vdots & \frac{160}{7} \end{bmatrix}$$

The system is inconsistent and there is no solution.

63. (a) $\begin{cases} x - 2y + z = -6 \\ y - 5z = 16 \\ z = -3 \end{cases}$

$$y - 5(-3) = 16$$
$$y = 1$$
$$x - 2(1) + (-3) = -6$$
$$x = -1$$

Solution: $(-1, 1, -3)$

(b) $\begin{cases} x + y - 2z = 6 \\ y + 3z = -8 \\ z = -3 \end{cases}$

$$y + 3(-3) = -8$$
$$y = 1$$
$$x + (1) - 2(-3) = 6$$
$$x = -1$$

Solution: $(-1, 1, -3)$

Both systems yield the same solution, namely $(-1, 1, -3)$.

65. (a) $\begin{cases} x - 4y + 5z = 27 \\ y - 7z = -54 \\ z = 8 \end{cases}$

$$y - 7(8) = -54$$
$$y = 2$$
$$x - 4(2) + 5(8) = 27$$
$$x = -5$$

Solution: $(-5, 2, 8)$

(b) $\begin{cases} x - 6y + z = 15 \\ y + 5z = 42 \\ z = 8 \end{cases}$

$$y + 5(8) = 42$$
$$y = 2$$
$$x - 6(2) + (8) = 15$$
$$x = 19$$

Solution: $(19, 2, 8)$

The systems do *not* yield the same solution.

67. Let x = amount at 8%.

Let y = amount at 9%.

Let z = amount at 12%.

$$\begin{aligned} x + y + z &= 2{,}200{,}000 \\ 0.08x + 0.09y + 0.12z &= 204{,}000 \\ -x + 2z &= 0 \end{aligned}$$

$$\left[\begin{array}{ccc:c} 1 & 1 & 1 & 2{,}200{,}000 \\ 0.08 & 0.09 & 0.12 & 204{,}000 \\ -1 & 0 & 2 & 0 \end{array}\right]$$

$$\begin{array}{r} \\ 100R_2 \\ R_1 + R_3 \end{array}\left[\begin{array}{ccc:c} 1 & 1 & 1 & 2{,}200{,}000 \\ 8 & 9 & 12 & 20{,}400{,}000 \\ 0 & 1 & 3 & 2{,}200{,}000 \end{array}\right]$$

$$\begin{array}{r} \\ -8R_1 + R_2 \\ \\ \end{array}\left[\begin{array}{ccc:c} 1 & 1 & 1 & 2{,}200{,}000 \\ 0 & 1 & 4 & 2{,}800{,}000 \\ 0 & 1 & 3 & 2{,}200{,}000 \end{array}\right]$$

$$\begin{array}{r} \\ R_3 \\ R_2 \end{array}\left[\begin{array}{ccc:c} 1 & 1 & 1 & 2{,}200{,}000 \\ 0 & 1 & 3 & 2{,}200{,}000 \\ 0 & 1 & 4 & 2{,}800{,}000 \end{array}\right]$$

$$\begin{array}{r} \\ \\ -R_2 + R_3 \end{array}\left[\begin{array}{ccc:c} 1 & 1 & 1 & 2{,}200{,}000 \\ 0 & 1 & 3 & 2{,}200{,}000 \\ 0 & 0 & 1 & 600{,}000 \end{array}\right]$$

$$\begin{array}{r} \\ -3R_3 + R_2 \\ \\ \end{array}\left[\begin{array}{ccc:c} 1 & 1 & 1 & 2{,}200{,}000 \\ 0 & 1 & 0 & 400{,}000 \\ 0 & 0 & 1 & 600{,}000 \end{array}\right]$$

$z = 600{,}000$

$y = 400{,}000$

$x + 400{,}000 + 600{,}000 = 2{,}200{,}000$

$\Rightarrow x = 1{,}200{,}000$

Solution: (1,200,000, 400,000, 600,000)

So, $1,200,000 was borrowed at 8%, $400,000 was borrowed at 9%, and $600,000 was borrowed at 12%.

69. $12b + 66a = 831$

$66b + 506a = 5643$

$$\left[\begin{array}{cc:c} 12 & 66 & 831 \\ 66 & 506 & 5643 \end{array}\right]$$

$$\begin{array}{r} \frac{1}{12}R_1 \\ \\ \end{array}\left[\begin{array}{cc:c} 1 & 5.5 & 69.25 \\ 66 & 506 & 5643 \end{array}\right]$$

$$\begin{array}{r} \\ -66R_1 + R_2 \end{array}\left[\begin{array}{cc:c} 1 & 5.5 & 69.25 \\ 0 & 143 & 1072.5 \end{array}\right]$$

$143a = 1072.5 \Rightarrow a \approx 7.5$

$b + 5.5(7.5) = 69.25 \Rightarrow b \approx 28$

$y = 7.5t + 28$

$t = 14$ corresponds to 2011.

$y = 7.5(14) + 28 = 133$

According to the model, the total number of new cases in 2011 would be 133. Because the data values increased in a linear pattern, this estimate seems reasonable.

71. False. The rows are in the wrong order. To change this matrix to reduced row-echelon form, interchange Row 1 and Row 4, and interchange Row 2 and Row 3.

Mid-Chapter Quiz Solutions

1. $\begin{cases} 2.5x - y = 6 & \text{Equation 1} \\ 3x + 4y = 2 & \text{Equation 2} \end{cases}$

Multiply Equation 1 by 4; then add the equations:

$$\begin{aligned} 10x - 4y &= 6 \\ 3x + 4y &= 2 \\ \hline 13x &= 8 \\ x &= 2 \end{aligned}$$

Back-substitute $x = 2$ into Equation 2:

$3(2) + 4y = 2$

$4y = -4$

$y = -1$

Solution: $(2, -1)$

2. $\begin{cases} \frac{1}{2}x + \frac{1}{3}y = 1 & \text{Equation 1} \\ x - 2y = -2 & \text{Equation 2} \end{cases}$

Multiply Equation 1 by 6; then add the equations:

$$\begin{aligned} 3x + 2y &= 6 \\ x - 2y &= -2 \\ \hline 4x &= 4 \\ x &= 1 \end{aligned}$$

Back-substitute $x = 1$ into Equation 2:

$1 - 2y = -2$

$-2y = -3$

$y = \frac{3}{2}$

Solution: $\left(1, \frac{3}{2}\right)$

3. $\begin{cases} x \qquad\quad + 3z = 10 \\ 2x + 3y - z = -7 \\ \qquad 2y + z = -1 \end{cases}$ Interchange the equations.

$\begin{cases} x \qquad\quad + 3z = 10 \\ \qquad 3y - 7z = -27 \\ \qquad 2y + z = -1 \end{cases}$ $\begin{matrix} \\ -2\text{Eq.1} + \text{Eq.2} \\ \\ \end{matrix}$

$\begin{cases} x \qquad\quad + 3z = 10 \\ \qquad y - 8z = -26 \\ \qquad\quad 17z = 51 \end{cases}$ $\begin{matrix} \\ -1\text{Eq.3} + \text{Eq.2} \\ -2\text{Eq.2} + \text{Eq.3} \end{matrix}$

$z = 3$

$y - 8(3) = -26 \Rightarrow y = -2$

$x + 3(3) = 10 \Rightarrow x = 1$

Solution: $(1, -2, 3)$

4. $\begin{cases} x + y - 2z = 12 \\ 2x - y - z = 6 \\ \quad y - z = 6 \end{cases}$

$\begin{cases} x + \quad y - 2z = 12 \\ \quad -3y + 3z = 18 \\ \qquad y - z = 6 \end{cases}$ $\begin{matrix} \\ -2\text{Eq.1} + \text{Eq.2} \\ \\ \end{matrix}$

$\begin{cases} x + \quad y - 2z = 12 \\ \quad -3y + 3z = -18 \\ \qquad\qquad 0 = 0 \end{cases}$ $\begin{matrix} \\ \\ \text{Eq.2} + 3\text{Eq.3} \end{matrix}$

Let $z = a$, then:

$y = a + 6$

$x = a + 6$

Solution: $(a + 6, a + 6, a)$; a is any real number.

5. $\begin{cases} 3x + 2y + z = -17 \\ -x + \quad y + z = 4 \\ \quad x - \quad y - z = 3 \end{cases}$ $\begin{matrix} \text{Equation 1} \\ \text{Equation 2} \\ \text{Equation 3} \end{matrix}$

$\begin{cases} 3x + 2y + z = -17 \\ -x + \quad y + z = 4 \\ \qquad\qquad 0 = 7 \end{cases}$ $\begin{matrix} \\ \\ \text{Eq.2} + \text{Eq.3} \end{matrix}$

Inconsistent, no solution

6. $\begin{cases} 2x + \quad y + \quad 3z = 1 \\ 2x + 6y + \quad 8z = 3 \\ 6x + 8y + 18z = 5 \end{cases}$ $\begin{matrix} \text{Equation 1} \\ \text{Equation 2} \\ \text{Equation 3} \end{matrix}$

$\begin{cases} 2x + \quad y + 3z = 1 \\ \qquad 5y + 5z = 2 \\ \qquad 5y + 9z = 2 \end{cases}$ $\begin{matrix} \\ (-1)\text{Eq.1} + \text{Eq.2} \\ (-3)\text{Eq.1} + \text{Eq.3} \end{matrix}$

$\begin{cases} 2x + \quad y + 3z = 1 \\ \qquad 5y + 5z = 2 \\ \qquad\qquad 4z = 0 \end{cases}$ $\begin{matrix} \\ \\ (-1)\text{Eq.2} + \text{Eq.3} \end{matrix}$

$\begin{cases} x + \frac{1}{2}y + \frac{3}{2}z = \frac{1}{2} \\ \qquad y + \quad z = \frac{2}{5} \\ \qquad\qquad z = 0 \end{cases}$ $\begin{matrix} \left(\frac{1}{2}\right)\text{Eq.1} \\ \left(\frac{1}{5}\right)\text{Eq.2} \\ \left(\frac{1}{4}\right)\text{Eq.3} \end{matrix}$

$z = 0$

$y + 0 = \frac{2}{5} \Rightarrow y = \frac{2}{5}$

$x + \frac{1}{2}\left(\frac{2}{5}\right) + \frac{3}{2}(0) = \frac{1}{2} \Rightarrow x = \frac{3}{10}$

Solution: $\left(\frac{3}{10}, \frac{2}{5}, 0\right)$

7. Any matrix with 4 rows and 3 columns will satisfy this requirement. Some examples are

$$\begin{bmatrix} 1 & 0 & 0 \\ 0 & 1 & 0 \\ 0 & 1 & 1 \\ 1 & 0 & 1 \end{bmatrix}, \begin{bmatrix} 1 & 2 & 3 \\ 4 & 5 & 6 \\ 7 & 8 & 9 \\ 10 & 11 & 12 \end{bmatrix}, \begin{bmatrix} -2 & 1 & 1 \\ 7 & -1 & 0 \\ 13 & 5 & 6 \\ 0 & 0 & 0 \end{bmatrix}$$

8. Any matrix with 3 rows and 1 column will satisfy this requirement. Some examples are

$$\begin{bmatrix} 1 \\ 2 \\ 3 \end{bmatrix}, \begin{bmatrix} 0 \\ -1 \\ 5 \end{bmatrix}, \begin{bmatrix} a \\ b \\ c \end{bmatrix}$$

9. $\begin{bmatrix} 3 & 2 & \vdots & -2 \\ 5 & -1 & \vdots & 19 \end{bmatrix}$

10. $\begin{bmatrix} 1 & 0 & 3 & \vdots & -5 \\ 1 & 2 & -1 & \vdots & 3 \\ 3 & 0 & 4 & \vdots & 0 \end{bmatrix}$

11. $\begin{bmatrix} 3 & 2 & \vdots & -2 \\ 5 & -1 & \vdots & 19 \end{bmatrix}$

$$\begin{array}{r} \frac{1}{3}R_1 \rightarrow \\ -5R_1 + R_2 \rightarrow \end{array} \begin{bmatrix} 1 & \frac{2}{3} & \vdots & -\frac{2}{3} \\ 0 & -\frac{13}{3} & \vdots & \frac{67}{3} \end{bmatrix}$$

$$-\tfrac{3}{13}R_2 \rightarrow \begin{bmatrix} 1 & \frac{2}{3} & \vdots & -\frac{2}{3} \\ 0 & 1 & \vdots & -\frac{67}{13} \end{bmatrix}$$

$$y = -\frac{67}{13}$$

$$x + \frac{2}{3}\left(-\frac{67}{13}\right) = -\frac{2}{3} \Rightarrow x = \frac{36}{13}$$

Solution: $\left(\frac{36}{13}, -\frac{67}{13}\right)$

12. $\begin{bmatrix} 1 & 0 & 3 & \vdots & -5 \\ 1 & 2 & -1 & \vdots & 3 \\ 3 & 0 & 4 & \vdots & 0 \end{bmatrix}$

$$\begin{array}{r} \\ -R_1 + R_2 \rightarrow \\ -3R_1 + R_3 \rightarrow \end{array} \begin{bmatrix} 1 & 0 & 3 & \vdots & -5 \\ 0 & 2 & -4 & \vdots & 8 \\ 0 & 0 & -5 & \vdots & 15 \end{bmatrix}$$

$$\begin{array}{r} \\ \frac{1}{2}R_2 \rightarrow \\ -\frac{1}{5}R_3 \rightarrow \end{array} \begin{bmatrix} 1 & 0 & 3 & \vdots & -5 \\ 0 & 1 & -2 & \vdots & 4 \\ 0 & 0 & 1 & \vdots & -3 \end{bmatrix}$$

$$\begin{array}{r} -3R_3 + R_1 \rightarrow \\ 2R_3 + R_2 \rightarrow \\ \\ \end{array} \begin{bmatrix} 1 & 0 & 0 & \vdots & 4 \\ 0 & 1 & 0 & \vdots & -2 \\ 0 & 0 & 1 & \vdots & -3 \end{bmatrix}$$

So, $x = 4$, $y = -2$, and $z = -3$.

Solution: $(4, -2, -3)$

13. $-5R_1 + R_2 \begin{bmatrix} 1 & 1 & 1 \\ 0 & -7 & -1 \end{bmatrix}$

14. $-\frac{1}{3}R_1 \begin{bmatrix} 1 & -1 & -4 \\ 18 & -8 & 4 \end{bmatrix}$

15. Let x = liters of spray X.

Let y = liters of spray Y.

Let z = liters of spray Z.

Chemical A: $\frac{1}{2}x + \frac{1}{2}y = 15$

Chemical B: $\frac{1}{2}x + \frac{1}{3}y = 11$

Chemical C: $\frac{1}{6}y + z = 20$

$$\begin{cases} \frac{1}{2}x + \frac{1}{2}y = 15 \\ -\frac{1}{6}y = -4 & -\text{Eq.1} + \text{Eq.2} \\ \frac{1}{6}y + z = 20 \end{cases}$$

$$\begin{cases} x + y = 30 & 2\text{Eq.1} \\ -\frac{1}{6}y = -4 \\ z = 16 & \text{Eq.2} + \text{Eq.3} \end{cases}$$

$$z = 16$$

$$-\frac{1}{6}y = -4 \Rightarrow y = 24$$

$$x + (24) = 30 \Rightarrow x = 6$$

To get the desired mixture, you need 6 liters of spray X, 24 liters of spray Y, and 16 liters of spray Z.

Section 8.4 Operations with Matrices

Skills Review

1. $\begin{bmatrix} 0 & 1 & 0 & -5 \\ 1 & 0 & 3 & 2 \\ 0 & 0 & 1 & 0 \end{bmatrix}$

This matrix is not in reduced row-echelon form.

2. $\begin{bmatrix} 1 & 0 & 0 & 2 & 3 \\ 0 & 0 & 0 & 0 & 0 \\ 0 & 1 & 1 & 3 & 10 \end{bmatrix}$

This matrix is not in reduced row-echelon form.

3. $\begin{cases} -5x + 10y = 12 \\ 7x - 3y = 0 \end{cases}$

$$\begin{bmatrix} -5 & 10 & \vdots & 12 \\ 7 & -3 & \vdots & 0 \end{bmatrix}$$

4. $\begin{cases} 10x + 15y - 9z = 42 \\ 6x - 5y = 0 \end{cases}$

$$\begin{bmatrix} 10 & 15 & -9 & \vdots & 42 \\ 6 & -5 & 0 & \vdots & 0 \end{bmatrix}$$

5. $\begin{bmatrix} 1 & 0 & \vdots & 0 \\ 0 & 1 & \vdots & 2 \end{bmatrix}$

$$\begin{cases} x = 0 \\ y = 2 \end{cases}$$

Solution: $(0, 2)$

Skills Review —*continued*—

6. $\begin{bmatrix} 1 & 0 & -1 & \vdots & 2 \\ 0 & 1 & 1 & \vdots & 3 \end{bmatrix}$

$\begin{cases} x \quad\quad -z = 2 \\ \quad y + \; z = 3 \end{cases}$

Let $z = a$, then:

$y + a = 3 \Rightarrow y = -a + 3$

$x - a = 2 \Rightarrow x = \; a + 2$

Solution: $(a + 2, -a + 3, a)$, where a is a real number

7. $\begin{bmatrix} 1 & -1 & 0 & \vdots & 3 \\ 0 & 1 & -2 & \vdots & 1 \\ 0 & 0 & 1 & \vdots & -1 \end{bmatrix}$

$\begin{cases} x - y \quad\quad = 3 \\ \quad\; y - 2z = 1 \\ \quad\quad\quad\; z = -1 \end{cases}$

$z = -1$

$y - 2(-1) = 1 \Rightarrow y = -1$

$x - (-1) = 3 \Rightarrow x = \; 2$

Solution: $(2, -1, -1)$

1. $x = -2,\ y = 4$

3. $2x + 1 = 5 \Rightarrow x = 2$

$3x = 6 \Rightarrow x = 2$

$3y - 5 = 4 \Rightarrow y = 3$

5. (a) $A + B = \begin{bmatrix} 1 & -1 \\ 2 & -1 \end{bmatrix} + \begin{bmatrix} 2 & -1 \\ -1 & 8 \end{bmatrix} = \begin{bmatrix} 1+2 & -1-1 \\ 2-1 & -1+8 \end{bmatrix} = \begin{bmatrix} 3 & -2 \\ 1 & 7 \end{bmatrix}$

(b) $A - B = \begin{bmatrix} 1 & -1 \\ 2 & -1 \end{bmatrix} - \begin{bmatrix} 2 & -1 \\ -1 & 8 \end{bmatrix} = \begin{bmatrix} 1-2 & -1+1 \\ 2+1 & -1-8 \end{bmatrix} = \begin{bmatrix} -1 & 0 \\ 3 & -9 \end{bmatrix}$

(c) $3A = 3\begin{bmatrix} 1 & -1 \\ 2 & -1 \end{bmatrix} = \begin{bmatrix} 3(1) & 3(-1) \\ 3(2) & 3(-1) \end{bmatrix} = \begin{bmatrix} 3 & -3 \\ 6 & -3 \end{bmatrix}$

(d) $3A - 2B = \begin{bmatrix} 3 & -3 \\ 6 & -3 \end{bmatrix} - 2\begin{bmatrix} 2 & -1 \\ -1 & 8 \end{bmatrix} = \begin{bmatrix} 3 & -3 \\ 6 & -3 \end{bmatrix} + \begin{bmatrix} -4 & 2 \\ 2 & -16 \end{bmatrix} = \begin{bmatrix} -1 & -1 \\ 8 & -19 \end{bmatrix}$

7. (a) $A + B = \begin{bmatrix} 1 & -1 \\ -3 & 1 \\ 4 & 1 \end{bmatrix} + \begin{bmatrix} 4 & -2 \\ 0 & 1 \\ 2 & 3 \end{bmatrix} = \begin{bmatrix} 1+4 & -1+(-2) \\ -3+0 & 1+1 \\ 4+2 & 1+3 \end{bmatrix} = \begin{bmatrix} 5 & -3 \\ -3 & 2 \\ 6 & 4 \end{bmatrix}$

(b) $A - B = \begin{bmatrix} 1 & -1 \\ -3 & 1 \\ 4 & 1 \end{bmatrix} - \begin{bmatrix} 4 & -2 \\ 0 & 1 \\ 2 & 3 \end{bmatrix} = \begin{bmatrix} 1-4 & -1-(-2) \\ -3-0 & 1-1 \\ 4-2 & 1-3 \end{bmatrix} = \begin{bmatrix} -3 & 1 \\ -3 & 0 \\ 2 & -2 \end{bmatrix}$

(c) $3A = 3\begin{bmatrix} 1 & -1 \\ -3 & 1 \\ 4 & 1 \end{bmatrix} = \begin{bmatrix} 3(1) & 3(-1) \\ 3(-3) & 3(1) \\ 3(4) & 3(1) \end{bmatrix} = \begin{bmatrix} 3 & -3 \\ -9 & 3 \\ 12 & 3 \end{bmatrix}$

(d) $3A - 2B = 3\begin{bmatrix} 1 & -1 \\ -3 & 1 \\ 4 & 1 \end{bmatrix} - 2\begin{bmatrix} 4 & -2 \\ 0 & 1 \\ 2 & 3 \end{bmatrix} = \begin{bmatrix} 3 & -3 \\ -9 & 3 \\ 12 & 3 \end{bmatrix} + \begin{bmatrix} -8 & 4 \\ 0 & -2 \\ -4 & -6 \end{bmatrix} = \begin{bmatrix} -5 & 1 \\ -9 & 1 \\ 8 & -3 \end{bmatrix}$

9. $A = \begin{bmatrix} 6 & 0 & 3 \\ -1 & -4 & 0 \end{bmatrix},\ B = \begin{bmatrix} 8 & -1 \\ 4 & -3 \end{bmatrix}$

(a) $A + B$ is not possible. A and B do not have the same order.

(b) $A - B$ is not possible. A and B do not have the same order.

(c) $3A = \begin{bmatrix} 18 & 0 & 9 \\ -3 & -12 & 0 \end{bmatrix}$

(d) $3A - 2B$ is not possible. A and B do not have the same order.

11. $\begin{bmatrix} -5 & 0 \\ 3 & -6 \end{bmatrix} + \begin{bmatrix} 7 & 1 \\ -2 & -1 \end{bmatrix} + \begin{bmatrix} -10 & -8 \\ 14 & 6 \end{bmatrix} = \begin{bmatrix} -5 + 7 + (-10) & 0 + 1 + (-8) \\ 3 + (-2) + 14 & -6 + (-1) + 6 \end{bmatrix} = \begin{bmatrix} -8 & -7 \\ 15 & -1 \end{bmatrix}$

13. $\frac{1}{2}([5 \ \ -2 \ \ 4 \ \ 0] + [14 \ \ 6 \ \ -18 \ \ 9]) = \frac{1}{2}[5 + 14 \ \ -2 + 6 \ \ 4 + (-18) \ \ 0 + 9]$

$= \frac{1}{2}[19 \ \ 4 \ \ -14 \ \ 9]$

$= \left[\frac{19}{2} \ \ 2 \ \ -7 \ \ \frac{9}{2}\right]$

15. $\frac{3}{7}\begin{bmatrix} 2 & 5 \\ -1 & -4 \end{bmatrix} + 6\begin{bmatrix} -3 & 0 \\ 2 & 2 \end{bmatrix} \approx \begin{bmatrix} -17.143 & 2.143 \\ 11.571 & 10.286 \end{bmatrix}$

17. $-\begin{bmatrix} 3.211 & 6.829 \\ -1.004 & 4.914 \\ 0.055 & -3.889 \end{bmatrix} - \begin{bmatrix} -1.630 & -3.090 \\ 5.256 & 8.335 \\ -9.768 & 4.251 \end{bmatrix} = \begin{bmatrix} -1.581 & -3.739 \\ -4.252 & -13.249 \\ 9.713 & -0.362 \end{bmatrix}$

19. $X = 3\begin{bmatrix} -2 & -1 \\ 1 & 0 \\ 3 & -4 \end{bmatrix} - 2\begin{bmatrix} 0 & 3 \\ 2 & 0 \\ -4 & -1 \end{bmatrix} = \begin{bmatrix} -6 & -3 \\ 3 & 0 \\ 9 & -12 \end{bmatrix} - \begin{bmatrix} 0 & 6 \\ 4 & 0 \\ -8 & -2 \end{bmatrix} = \begin{bmatrix} -6 & -9 \\ -1 & 0 \\ 17 & -10 \end{bmatrix}$

21. $X = -\frac{3}{2}A + \frac{1}{2}B = -\frac{3}{2}\begin{bmatrix} -2 & -1 \\ 1 & 0 \\ 3 & -4 \end{bmatrix} + \frac{1}{2}\begin{bmatrix} 0 & 3 \\ 2 & 0 \\ -4 & -1 \end{bmatrix} = \begin{bmatrix} 3 & \frac{3}{2} \\ -\frac{3}{2} & 0 \\ -\frac{9}{2} & 6 \end{bmatrix} + \begin{bmatrix} 0 & \frac{3}{2} \\ 1 & 0 \\ -2 & -\frac{1}{2} \end{bmatrix} = \begin{bmatrix} 3 & 3 \\ -\frac{1}{2} & 0 \\ -\frac{13}{2} & \frac{11}{2} \end{bmatrix}$

23. A is 3×2 and B is 3×3. AB is not possible.

25. A is 3×3, B is $3 \times 2 \Rightarrow AB$ is 3×2.

$$\begin{bmatrix} 0 & -1 & 0 \\ 4 & 0 & 2 \\ 8 & -1 & 7 \end{bmatrix}\begin{bmatrix} 2 & 1 \\ -3 & 4 \\ 1 & 6 \end{bmatrix} = \begin{bmatrix} (0)(2) + (-1)(-3) + (0)(1) & (0)(1) + (-1)(4) + (0)(6) \\ (4)(2) + (0)(-3) + (2)(1) & (4)(1) + (0)(4) + (2)(6) \\ (8)(2) + (-1)(-3) + (7)(1) & (8)(1) + (-1)(4) + (7)(6) \end{bmatrix} = \begin{bmatrix} 3 & -4 \\ 10 & 16 \\ 26 & 46 \end{bmatrix}$$

27. A is 3×3, B is $3 \times 3 \Rightarrow AB$ is 3×3.

$$\begin{bmatrix} 0 & 0 & 5 \\ 0 & 0 & -3 \\ 0 & 0 & 4 \end{bmatrix}\begin{bmatrix} 6 & -11 & 4 \\ 8 & 16 & 4 \\ 0 & 0 & 0 \end{bmatrix} = \begin{bmatrix} (0)(6) + (0)(8) + (5)(0) & (0)(-11) + (0)(16) + (5)(0) & (0)(4) + (0)(4) + (5)(0) \\ (0)(6) + (0)(8) + (-3)(0) & (0)(-11) + (0)(16) + (-3)(0) & (0)(4) + (0)(4) + (-3)(0) \\ (0)(6) + (0)(8) + (4)(0) & (0)(-11) + (0)(16) + (4)(0) & (0)(4) + (0)(4) + (4)(0) \end{bmatrix}$$

$$= \begin{bmatrix} 0 & 0 & 0 \\ 0 & 0 & 0 \\ 0 & 0 & 0 \end{bmatrix}$$

29. $\begin{bmatrix} 5 & 6 & -3 \\ -2 & 5 & 1 \\ 10 & -5 & 5 \end{bmatrix}\begin{bmatrix} 1 & -1 & 2 \\ 8 & 1 & 4 \\ 4 & -2 & 9 \end{bmatrix} = \begin{bmatrix} 41 & 7 & 7 \\ 42 & 5 & 25 \\ -10 & -25 & 45 \end{bmatrix}$

31. $\begin{bmatrix} -3 & 8 & -6 & 8 \\ -12 & 15 & 9 & 6 \\ 5 & -1 & 1 & 5 \end{bmatrix}\begin{bmatrix} 3 & 1 & 6 \\ 24 & 15 & 14 \\ 16 & 10 & 21 \\ 8 & -4 & 10 \end{bmatrix} = \begin{bmatrix} 151 & 25 & 48 \\ 516 & 279 & 387 \\ 47 & -20 & 87 \end{bmatrix}$

33. (a) $AB = \begin{bmatrix} 1 & 2 \\ 4 & 2 \end{bmatrix}\begin{bmatrix} 2 & -1 \\ -1 & 8 \end{bmatrix} = \begin{bmatrix} (1)(2)+(2)(-1) & (1)(-1)+(2)(8) \\ (4)(2)+(2)(-1) & (4)(-1)(2)(8) \end{bmatrix} = \begin{bmatrix} 0 & 15 \\ 6 & 12 \end{bmatrix}$

(b) $BA = \begin{bmatrix} 2 & -1 \\ -1 & 8 \end{bmatrix}\begin{bmatrix} 1 & 2 \\ 4 & 2 \end{bmatrix} = \begin{bmatrix} (2)(1)+(-1)(4) & (2)(2)+(-1)(2) \\ (-1)(1)+(8)(4) & (-1)(2)+(8)(2) \end{bmatrix} = \begin{bmatrix} -2 & 2 \\ 31 & 14 \end{bmatrix}$

(c) $A^2 = \begin{bmatrix} 1 & 2 \\ 4 & 2 \end{bmatrix}\begin{bmatrix} 1 & 2 \\ 4 & 2 \end{bmatrix} = \begin{bmatrix} (1)(1)+(2)(4) & (1)(2)+(2)(2) \\ (4)(1)+(2)(4) & (4)(2)+(2)(2) \end{bmatrix} = \begin{bmatrix} 9 & 6 \\ 12 & 12 \end{bmatrix}$

35. (a) $AB = \begin{bmatrix} 6 \\ 9 \\ -2 \end{bmatrix}\begin{bmatrix} 2 & 2 & 1 \end{bmatrix} = \begin{bmatrix} 6(2) & 6(2) & 6(1) \\ 9(2) & 9(2) & 9(1) \\ -2(2) & -2(2) & -2(1) \end{bmatrix} = \begin{bmatrix} 12 & 12 & 6 \\ 18 & 18 & 9 \\ -4 & -4 & -2 \end{bmatrix}$

(b) $BA = \begin{bmatrix} 2 & 2 & 1 \end{bmatrix}\begin{bmatrix} 6 \\ 9 \\ -2 \end{bmatrix} = \begin{bmatrix} 2(6)+2(9)+1(-2) \end{bmatrix} = \begin{bmatrix} 28 \end{bmatrix}$

(c) The number of columns of A does not equal the number of rows of A; the multiplication is not possible.

37. $\begin{bmatrix} 3 & 1 \\ 0 & -2 \end{bmatrix}\begin{bmatrix} 1 & 0 \\ -2 & 2 \end{bmatrix}\begin{bmatrix} 1 & 0 \\ 2 & 4 \end{bmatrix} = \begin{bmatrix} 1 & 2 \\ 4 & -4 \end{bmatrix}\begin{bmatrix} 1 & 0 \\ 2 & 4 \end{bmatrix} = \begin{bmatrix} 5 & 8 \\ -4 & -16 \end{bmatrix}$

39. $\begin{bmatrix} 0 & 2 & -2 \\ 4 & 1 & 2 \end{bmatrix}\left(\begin{bmatrix} 4 & 0 \\ 0 & -1 \\ -1 & 2 \end{bmatrix} + \begin{bmatrix} -2 & 3 \\ -3 & 5 \\ 0 & -3 \end{bmatrix}\right) = \begin{bmatrix} 0 & 2 & -2 \\ 4 & 1 & 2 \end{bmatrix}\begin{bmatrix} 2 & 3 \\ -3 & 4 \\ -1 & -1 \end{bmatrix} = \begin{bmatrix} -4 & 10 \\ 3 & 14 \end{bmatrix}$

41. (a) $\begin{bmatrix} -1 & 1 \\ -2 & 1 \end{bmatrix}\begin{bmatrix} x_1 \\ x_2 \end{bmatrix} = \begin{bmatrix} 4 \\ 0 \end{bmatrix}$

(b) $\begin{bmatrix} -1 & 1 & \vdots & 4 \\ -2 & 1 & \vdots & 0 \end{bmatrix}$

$-R_2 + R_1 \to \begin{bmatrix} 1 & 0 & \vdots & 4 \\ -2 & 1 & \vdots & 0 \end{bmatrix}$

$2R_1 + R_2 \to \begin{bmatrix} 1 & 0 & \vdots & 4 \\ 0 & 1 & \vdots & 8 \end{bmatrix}$

$X = \begin{bmatrix} 4 \\ 8 \end{bmatrix}$

43. (a) $\begin{bmatrix} -2 & -3 \\ 6 & 1 \end{bmatrix}\begin{bmatrix} x_1 \\ x_2 \end{bmatrix} = \begin{bmatrix} -4 \\ -36 \end{bmatrix}$

(b) $\begin{bmatrix} -2 & -3 & \vdots & -4 \\ 6 & 1 & \vdots & -36 \end{bmatrix}$

$3R_1 + R_2 \to \begin{bmatrix} -2 & -3 & \vdots & -4 \\ 0 & -8 & \vdots & -48 \end{bmatrix}$

$\begin{matrix} -\frac{1}{2}R_1 \to \\ -\frac{1}{8}R_2 \to \end{matrix} \begin{bmatrix} 1 & \frac{3}{2} & \vdots & 2 \\ 0 & 1 & \vdots & 6 \end{bmatrix}$

$-\frac{3}{2}R_2 + R_1 \to \begin{bmatrix} 1 & 0 & \vdots & -7 \\ 0 & 1 & \vdots & 6 \end{bmatrix}$

$X = \begin{bmatrix} -7 \\ 6 \end{bmatrix}$

45. (a) $A = \begin{bmatrix} 1 & -2 & 3 \\ -1 & 3 & -1 \\ 2 & -5 & 5 \end{bmatrix}\begin{bmatrix} x_1 \\ x_2 \\ x_3 \end{bmatrix} = \begin{bmatrix} 9 \\ -6 \\ 17 \end{bmatrix}$

(b) $\begin{bmatrix} 1 & -2 & 3 & \vdots & 9 \\ -1 & 3 & -1 & \vdots & -6 \\ 2 & -5 & 5 & \vdots & 17 \end{bmatrix}$

$\begin{matrix} \\ R_1 + R_2 \to \\ -2R_2 + R_3 \to \end{matrix} \begin{bmatrix} 1 & -2 & 3 & \vdots & 9 \\ 0 & 1 & 2 & \vdots & 3 \\ 0 & -1 & -1 & \vdots & -1 \end{bmatrix}$

$\begin{matrix} 2R_2 + R_1 \to \\ \\ R_2 + R_3 \to \end{matrix} \begin{bmatrix} 1 & 0 & 7 & \vdots & 15 \\ 0 & 1 & 2 & \vdots & 3 \\ 0 & 0 & 1 & \vdots & 2 \end{bmatrix}$

$\begin{matrix} -7R_3 + R_1 \to \\ -2R_3 + R_2 \to \\ \end{matrix} \begin{bmatrix} 1 & 0 & 0 & \vdots & 1 \\ 0 & 1 & 0 & \vdots & -1 \\ 0 & 0 & 1 & \vdots & 2 \end{bmatrix}$

$X = \begin{bmatrix} 1 \\ -1 \\ 2 \end{bmatrix}$

47. $P^2 = \begin{bmatrix} 0.6 & 0.1 & 0.1 \\ 0.2 & 0.7 & 0.1 \\ 0.2 & 0.2 & 0.8 \end{bmatrix}\begin{bmatrix} 0.6 & 0.1 & 0.1 \\ 0.2 & 0.7 & 0.1 \\ 0.2 & 0.2 & 0.8 \end{bmatrix} = \begin{bmatrix} 0.40 & 0.15 & 0.15 \\ 0.28 & 0.53 & 0.17 \\ 0.32 & 0.32 & 0.68 \end{bmatrix}$

The P^2 matrix gives the proportion of the voting population that changed parties or remained loyal to their party from the first election to the third.

49. (a) Bicycled, Jogged, Walked: $B = \begin{bmatrix} 2 & 0.5 & 3 \end{bmatrix}$ 20-minute time periods

(b) $BA = \begin{bmatrix} 2 & 0.5 & 3 \end{bmatrix}\begin{bmatrix} 109 & 136 \\ 127 & 159 \\ 64 & 79 \end{bmatrix} = \begin{bmatrix} 473.5 & 588.5 \end{bmatrix}$ Calories burned (120-pound person, 150-pound person)

The first entry represents the total calories burned by the 120-pound person and the second entry represents the total calories burned by the 150-pound person.

51. $X_2 = LX_1 = \begin{bmatrix} 0 & 2 \\ \frac{1}{2} & 0 \end{bmatrix}\begin{bmatrix} 10 \\ 10 \end{bmatrix} = \begin{bmatrix} 0(10) + 2(10) \\ \frac{1}{2}(10) + 0(10) \end{bmatrix} = \begin{bmatrix} 20 \\ 5 \end{bmatrix}$

$X_3 = LX_2 = \begin{bmatrix} 0 & 2 \\ \frac{1}{2} & 0 \end{bmatrix}\begin{bmatrix} 20 \\ 5 \end{bmatrix} = \begin{bmatrix} 0(20) + 2(5) \\ \frac{1}{2}(20) + 0(5) \end{bmatrix} = \begin{bmatrix} 10 \\ 10 \end{bmatrix}$

53. $X_2 = LX_1 = \begin{bmatrix} 0 & 3 & 4 \\ 1 & 0 & 0 \\ 0 & \frac{1}{2} & 0 \end{bmatrix}\begin{bmatrix} 12 \\ 10 \\ 12 \end{bmatrix} = \begin{bmatrix} 0(12) + 3(10) + 4(12) \\ 1(12) + 0(10) + 0(12) \\ 0(12) + \frac{1}{2}(10) + 0(12) \end{bmatrix} = \begin{bmatrix} 78 \\ 12 \\ 5 \end{bmatrix}$

$X_3 = LX_2 = \begin{bmatrix} 0 & 3 & 4 \\ 1 & 0 & 0 \\ 0 & \frac{1}{2} & 0 \end{bmatrix}\begin{bmatrix} 78 \\ 12 \\ 5 \end{bmatrix} = \begin{bmatrix} 0(78) + 3(12) + 4(5) \\ 1(78) + 0(12) + 0(5) \\ 0(78) + \frac{1}{2}(12) + 0(5) \end{bmatrix} = \begin{bmatrix} 56 \\ 78 \\ 5 \end{bmatrix}$

Section 8.5 The Inverse of a Square Matrix

Skills Review

1. $4\begin{bmatrix} 1 & 6 \\ 0 & -4 \\ 12 & 2 \end{bmatrix} = \begin{bmatrix} 4(1) & 4(6) \\ 4(0) & 4(-4) \\ 4(12) & 4(2) \end{bmatrix} = \begin{bmatrix} 4 & 24 \\ 0 & -16 \\ 48 & 8 \end{bmatrix}$

2. $\frac{1}{2}\begin{bmatrix} 11 & 10 & 48 \\ 1 & 0 & 16 \\ 0 & 2 & 8 \end{bmatrix} = \begin{bmatrix} \frac{1}{2}(11) & \frac{1}{2}(10) & \frac{1}{2}(48) \\ \frac{1}{2}(1) & \frac{1}{2}(0) & \frac{1}{2}(16) \\ \frac{1}{2}(0) & \frac{1}{2}(2) & \frac{1}{2}(8) \end{bmatrix} = \begin{bmatrix} \frac{11}{2} & 5 & 24 \\ \frac{1}{2} & 0 & 8 \\ 0 & 1 & 4 \end{bmatrix}$

3. $\begin{bmatrix} 5 & 20 \\ -7 & 15 \end{bmatrix} - 3\begin{bmatrix} 6 & 3 \\ 4 & -2 \end{bmatrix} = \begin{bmatrix} 5 & 20 \\ -7 & 15 \end{bmatrix} + \begin{bmatrix} -18 & -9 \\ -12 & 6 \end{bmatrix} = \begin{bmatrix} -13 & 11 \\ -19 & 21 \end{bmatrix}$

4. $\begin{bmatrix} 1 & 0 \\ 0 & 1 \end{bmatrix}\begin{bmatrix} 6 & 5 \\ 3 & -2 \end{bmatrix} = \begin{bmatrix} 1(6) + 0(3) & 1(5) + 0(-2) \\ 0(6) + 1(3) & 0(5) + 1(-2) \end{bmatrix} = \begin{bmatrix} 6 & 5 \\ 3 & -2 \end{bmatrix}$

Skills Review —*continued*—

5. $\begin{bmatrix} 2 & 0 & 0 \\ 0 & -1 & 0 \\ 0 & 0 & 3 \end{bmatrix}\begin{bmatrix} \frac{1}{2} & 0 & 0 \\ 0 & -1 & 0 \\ 0 & 0 & \frac{1}{3} \end{bmatrix} = \begin{bmatrix} 2\left(\frac{1}{2}\right) + 0(0) + 0(0) & 2(0) + 0(-1) + 0(0) & 2(0) + 0(0) + 0\left(\frac{1}{3}\right) \\ 0\left(\frac{1}{2}\right) + (-1)(0) + 0(0) & 0(0) + (-1)(-1) + 0(0) & 0(0) + (-1)(0) + 0\left(\frac{1}{3}\right) \\ 0\left(\frac{1}{2}\right) + 0(0) + 3(0) & 0(0) + 0(-1) + 3(0) & 0(0) + 0(0) + 3\left(\frac{1}{3}\right) \end{bmatrix} = \begin{bmatrix} 1 & 0 & 0 \\ 0 & 1 & 0 \\ 0 & 0 & 1 \end{bmatrix}$

6. $\begin{bmatrix} 1 & -1 & 0 \\ 1 & 0 & -1 \\ 6 & -2 & -3 \end{bmatrix}\begin{bmatrix} -2 & -3 & 1 \\ -3 & -3 & 1 \\ -2 & -4 & 1 \end{bmatrix} = \begin{bmatrix} 1(-2) + (-1)(-3) + 0(-2) & 1(-3) + (-1)(-3) + 0(-4) & 1(1) + (-1)(1) + 0(1) \\ 1(-2) + 0(-3) + (-1)(-2) & 1(-3) + 0(-3) + (-1)(-4) & 1(1) + 0(1) + (-1)(1) \\ 6(-2) + (-2)(-3) + (-3)(-2) & 6(-3) + (-2)(-3) + (-3)(-4) & 6(1) + (-2)(1) + (-3)(1) \end{bmatrix}$

$= \begin{bmatrix} 1 & 0 & 0 \\ 0 & 1 & 0 \\ 0 & 0 & 1 \end{bmatrix}$

7. $\begin{bmatrix} 3 & -2 & 1 & 0 \\ 4 & -3 & 0 & 1 \end{bmatrix}$

$\frac{1}{3}R_1 \rightarrow \begin{bmatrix} 1 & -\frac{2}{3} & \frac{1}{3} & 0 \\ 4 & -3 & 0 & 1 \end{bmatrix}$

$-4R_1 + R_2 \rightarrow \begin{bmatrix} 1 & -\frac{2}{3} & \frac{1}{3} & 0 \\ 0 & -\frac{1}{3} & -\frac{4}{3} & 1 \end{bmatrix}$

$-3R_2 \rightarrow \begin{bmatrix} 1 & -\frac{2}{3} & \frac{1}{3} & 0 \\ 0 & 1 & 4 & -3 \end{bmatrix}$

$\frac{2}{3}R_2 + R_1 \rightarrow \begin{bmatrix} 1 & 0 & 3 & -2 \\ 0 & 1 & 4 & -3 \end{bmatrix}$

8. $\begin{bmatrix} 1 & 1 & 2 & 1 & 0 & 0 \\ -1 & 0 & 3 & 0 & 1 & 0 \\ 1 & 2 & 8 & 0 & 0 & 1 \end{bmatrix}$

$\begin{matrix} \\ R_1 + R_2 \rightarrow \\ -R_1 + R_3 \rightarrow \end{matrix} \begin{bmatrix} 1 & 1 & 2 & 1 & 0 & 0 \\ 0 & 1 & 5 & 1 & 1 & 0 \\ 0 & 1 & 6 & -1 & 0 & 1 \end{bmatrix}$

$\begin{matrix} \\ \\ -R_2 + R_3 \rightarrow \end{matrix} \begin{bmatrix} 1 & 1 & 2 & 1 & 0 & 0 \\ 0 & 1 & 5 & 1 & 1 & 0 \\ 0 & 0 & 1 & -2 & -1 & 1 \end{bmatrix}$

$\begin{matrix} -R_2 + R_1 \rightarrow \\ \\ \\ \end{matrix} \begin{bmatrix} 1 & 0 & -3 & 0 & -1 & 0 \\ 0 & 1 & 5 & 1 & 1 & 0 \\ 0 & 0 & 1 & -2 & -1 & 1 \end{bmatrix}$

$\begin{matrix} 3R_3 + R_1 \rightarrow \\ -5R_3 + R_2 \rightarrow \\ \\ \end{matrix} \begin{bmatrix} 1 & 0 & 0 & -6 & -4 & 3 \\ 0 & 1 & 0 & 11 & 6 & -5 \\ 0 & 0 & 1 & -2 & -1 & 1 \end{bmatrix}$

1. $AB = \begin{bmatrix} 5 & 3 \\ 3 & 2 \end{bmatrix}\begin{bmatrix} 2 & -3 \\ -3 & 5 \end{bmatrix}\begin{bmatrix} 10 - 9 & -15 + 15 \\ 6 - 6 & -9 + 10 \end{bmatrix} = \begin{bmatrix} 1 & 0 \\ 0 & 1 \end{bmatrix}$

$BA = \begin{bmatrix} 2 & -3 \\ -3 & 5 \end{bmatrix}\begin{bmatrix} 5 & 3 \\ 3 & 2 \end{bmatrix} = \begin{bmatrix} 10 - 9 & 6 - 6 \\ -15 + 15 & -9 + 10 \end{bmatrix} = \begin{bmatrix} 1 & 0 \\ 0 & 1 \end{bmatrix}$

3. $AB = \begin{bmatrix} 1 & 2 \\ 3 & 4 \end{bmatrix}\begin{bmatrix} -2 & 1 \\ \frac{3}{2} & -\frac{1}{2} \end{bmatrix} = \begin{bmatrix} -2 + 3 & 1 - 1 \\ -6 + 6 & 3 - 2 \end{bmatrix} = \begin{bmatrix} 1 & 0 \\ 0 & 1 \end{bmatrix}$

$BA = \begin{bmatrix} -2 & 1 \\ \frac{3}{2} & -\frac{1}{2} \end{bmatrix}\begin{bmatrix} 1 & 2 \\ 3 & 4 \end{bmatrix} = \begin{bmatrix} -2 + 3 & -4 + 4 \\ \frac{3}{2} - \frac{3}{2} & 3 - 2 \end{bmatrix} = \begin{bmatrix} 1 & 0 \\ 0 & 1 \end{bmatrix}$

5. $AB = \begin{bmatrix} 2 & -17 & 11 \\ -1 & 11 & -7 \\ 0 & 3 & -2 \end{bmatrix}\begin{bmatrix} 1 & 1 & 2 \\ 2 & 4 & -3 \\ 3 & 6 & -5 \end{bmatrix} = \begin{bmatrix} 2 - 34 + 33 & 2 - 68 + 66 & 4 + 51 - 55 \\ -1 + 22 - 21 & -1 + 44 - 42 & -2 - 33 + 35 \\ 6 - 6 & 12 - 12 & -9 + 10 \end{bmatrix} = \begin{bmatrix} 1 & 0 & 0 \\ 0 & 1 & 0 \\ 0 & 0 & 1 \end{bmatrix}$

$BA = \begin{bmatrix} 1 & 1 & 2 \\ 2 & 4 & -3 \\ 3 & 6 & -5 \end{bmatrix}\begin{bmatrix} 2 & -17 & 11 \\ -1 & 11 & -7 \\ 0 & 3 & -2 \end{bmatrix} = \begin{bmatrix} 2 - 1 & -17 + 11 + 6 & 11 - 7 - 4 \\ 4 - 4 & -34 + 44 - 9 & 22 - 28 + 6 \\ 6 - 6 & -51 + 66 - 15 & 33 - 42 + 10 \end{bmatrix} = \begin{bmatrix} 1 & 0 & 0 \\ 0 & 1 & 0 \\ 0 & 0 & 1 \end{bmatrix}$

7. $AB = \begin{bmatrix} 2 & 0 & 1 & 1 \\ 3 & 0 & 0 & 1 \\ -1 & 1 & -2 & 1 \\ 4 & -1 & 1 & 0 \end{bmatrix}\begin{bmatrix} -1 & 2 & -1 & -1 \\ -4 & 9 & -5 & -6 \\ 0 & 1 & -1 & -1 \\ 3 & -5 & 3 & 3 \end{bmatrix} = \begin{bmatrix} -2+3 & 4+1-5 & -2-1+3 & -2-1+3 \\ 0 & 6-5 & 0 & 0 \\ 1-4+3 & -2+9-2-5 & 1-5+2+3 & 1-6+2+3 \\ 0 & 8-9+1 & -4+5-1 & -4+6-1 \end{bmatrix} = \begin{bmatrix} 1 & 0 & 0 & 0 \\ 0 & 1 & 0 & 0 \\ 0 & 0 & 1 & 0 \\ 0 & 0 & 0 & 1 \end{bmatrix}$

$BA = \begin{bmatrix} -1 & 2 & -1 & -1 \\ -4 & 9 & -5 & -6 \\ 0 & 1 & -1 & -1 \\ 3 & -5 & 3 & 3 \end{bmatrix}\begin{bmatrix} 2 & 0 & 1 & 1 \\ 3 & 0 & 0 & 1 \\ -1 & 1 & -2 & 1 \\ 4 & -1 & 1 & 0 \end{bmatrix} = \begin{bmatrix} -2+6+1-4 & 0 & -1+2-1 & -1+2-1 \\ -8+27+5-24 & -5+6 & -4+10-6 & -4+9-5 \\ 3+1-4 & 0 & 2-1 & 0 \\ 6-15-3+12 & 0 & 3-6+3 & 3-5+3 \end{bmatrix} = \begin{bmatrix} 1 & 0 & 0 & 0 \\ 0 & 1 & 0 & 0 \\ 0 & 0 & 1 & 0 \\ 0 & 0 & 0 & 1 \end{bmatrix}$

9. $[A \;\vdots\; I] = \left[\begin{array}{cc:cc} 2 & 0 & 1 & 0 \\ 0 & 3 & 0 & 1 \end{array}\right]$

$\begin{array}{l} \frac{1}{2}R_1 \rightarrow \\ \frac{1}{3}R_2 \rightarrow \end{array} \left[\begin{array}{cc:cc} 1 & 0 & \frac{1}{2} & 0 \\ 0 & 1 & 0 & \frac{1}{3} \end{array}\right] = [I \;\vdots\; A^{-1}]$

$A^{-1} = \begin{bmatrix} \frac{1}{2} & 0 \\ 0 & \frac{1}{3} \end{bmatrix}$

11. $[A \;\vdots\; I] = \left[\begin{array}{cc:cc} 1 & -2 & 1 & 0 \\ 2 & -3 & 0 & 1 \end{array}\right]$

$-2R_1 + R_2 \rightarrow \left[\begin{array}{cc:cc} 1 & -2 & 1 & 0 \\ 0 & 1 & -2 & 1 \end{array}\right]$

$2R_2 + R_1 \rightarrow \left[\begin{array}{cc:cc} 1 & 0 & -3 & 2 \\ 0 & 1 & -2 & 1 \end{array}\right] = [I \;\vdots\; A^{-1}]$

$A^{-1} = \begin{bmatrix} -3 & 2 \\ -2 & 1 \end{bmatrix}$

13. $[A \;\vdots\; I] = \left[\begin{array}{cc:cc} -1 & 1 & 1 & 0 \\ -2 & 1 & 0 & 1 \end{array}\right]$

$-R_2 + R_1 \rightarrow \left[\begin{array}{cc:cc} 1 & 0 & 1 & -1 \\ -2 & 1 & 0 & 1 \end{array}\right]$

$2R_1 + R_2 \rightarrow \left[\begin{array}{cc:cc} 1 & 0 & 1 & -1 \\ 0 & 1 & 2 & -1 \end{array}\right] = [I \;\vdots\; A^{-1}]$

$A^{-1} = \begin{bmatrix} 1 & -1 \\ 2 & -1 \end{bmatrix}$

15. $[A \;\vdots\; I] = \left[\begin{array}{cc:cc} 2 & 4 & 1 & 0 \\ 4 & 8 & 0 & 1 \end{array}\right]$

$-2R_1 + R_2 \rightarrow \left[\begin{array}{cc:cc} 2 & 4 & 1 & 0 \\ 0 & 0 & -2 & 1 \end{array}\right]$

The two zeros in the second row imply that the inverse does not exist.

17. $A = \begin{bmatrix} 2 & 7 & 1 \\ -3 & -9 & 2 \end{bmatrix}$ A has no inverse because it is not square.

19. $[A \;\vdots\; I] = \left[\begin{array}{ccc:ccc} 1 & 1 & 1 & 1 & 0 & 0 \\ 3 & 5 & 4 & 0 & 1 & 0 \\ 3 & 6 & 5 & 0 & 0 & 1 \end{array}\right]$

$\begin{array}{l} \\ -3R_1 + R_2 \rightarrow \\ -3R_1 + R_3 \rightarrow \end{array} \left[\begin{array}{ccc:ccc} 1 & 1 & 1 & 1 & 0 & 0 \\ 0 & 2 & 1 & -3 & 1 & 0 \\ 0 & 3 & 2 & -3 & 0 & 1 \end{array}\right]$

$\frac{1}{2}R_2 \rightarrow \left[\begin{array}{ccc:ccc} 1 & 1 & 1 & 1 & 0 & 0 \\ 0 & 1 & \frac{1}{2} & -\frac{3}{2} & \frac{1}{2} & 0 \\ 0 & 3 & 2 & -3 & 0 & 1 \end{array}\right]$

$\begin{array}{l} -R_2 + R_1 \rightarrow \\ \\ -3R_2 + R_3 \rightarrow \end{array} \left[\begin{array}{ccc:ccc} 1 & 0 & \frac{1}{2} & \frac{5}{2} & -\frac{1}{2} & 0 \\ 0 & 1 & \frac{1}{2} & -\frac{3}{2} & \frac{1}{2} & 0 \\ 0 & 0 & \frac{1}{2} & \frac{3}{2} & -\frac{3}{2} & 1 \end{array}\right]$

$\begin{array}{l} -R_3 + R_1 \rightarrow \\ -R_3 + R_2 \rightarrow \\ \\ \end{array} \left[\begin{array}{ccc:ccc} 1 & 0 & 0 & 1 & 1 & -1 \\ 0 & 1 & 0 & -3 & 2 & -1 \\ 0 & 0 & \frac{1}{2} & \frac{3}{2} & -\frac{3}{2} & 1 \end{array}\right]$

$\begin{array}{l} \\ \\ 2R_3 \rightarrow \end{array} \left[\begin{array}{ccc:ccc} 1 & 0 & 0 & 1 & 1 & -1 \\ 0 & 1 & 0 & -3 & 2 & -1 \\ 0 & 0 & 1 & 3 & -3 & 2 \end{array}\right] = [I \;\vdots\; A^{-1}]$

$A^{-1} = \begin{bmatrix} 1 & 1 & -1 \\ -3 & 2 & -1 \\ 3 & -3 & 2 \end{bmatrix}$

21. $[A \ \vdots \ I] = \left[\begin{array}{ccc:ccc} 1 & 0 & 0 & 1 & 0 & 0 \\ 3 & 4 & 0 & 0 & 1 & 0 \\ 2 & 5 & 5 & 0 & 0 & 1 \end{array}\right]$

$$\begin{array}{r} \\ -3R_1 + R_2 \rightarrow \\ -2R_1 + R_3 \rightarrow \end{array} \left[\begin{array}{ccc:ccc} 1 & 0 & 0 & 1 & 0 & 0 \\ 0 & 4 & 0 & -3 & 1 & 0 \\ 0 & 5 & 5 & -2 & 0 & 1 \end{array}\right]$$

$$\begin{array}{r} \\ \\ -\frac{5}{4}R_2 + R_3 \rightarrow \end{array} \left[\begin{array}{ccc:ccc} 1 & 0 & 0 & 1 & 0 & 0 \\ 0 & 4 & 0 & -3 & 1 & 0 \\ 0 & 0 & 5 & \frac{7}{4} & -\frac{5}{4} & 1 \end{array}\right]$$

$$\begin{array}{r} \\ \frac{1}{4}R_2 \rightarrow \\ \frac{1}{5}R_3 \rightarrow \end{array} \left[\begin{array}{ccc:ccc} 1 & 0 & 0 & 1 & 0 & 0 \\ 0 & 1 & 0 & -\frac{3}{4} & \frac{1}{4} & 0 \\ 0 & 0 & 1 & \frac{7}{20} & -\frac{1}{4} & \frac{1}{5} \end{array}\right] = [I \ \vdots \ A^{-1}]$$

$$A^{-1} = \begin{bmatrix} 1 & 0 & 0 \\ -\frac{3}{4} & \frac{1}{4} & 0 \\ \frac{7}{20} & -\frac{1}{4} & \frac{1}{5} \end{bmatrix}$$

23. $[A \ \vdots \ I] = \left[\begin{array}{cccc:cccc} -8 & 0 & 0 & 0 & 1 & 0 & 0 & 0 \\ 0 & 1 & 0 & 0 & 0 & 1 & 0 & 0 \\ 0 & 0 & 4 & 0 & 0 & 0 & 1 & 0 \\ 0 & 0 & 0 & -5 & 0 & 0 & 0 & 1 \end{array}\right]$

$$\begin{array}{r} -\frac{1}{8}R_1 \rightarrow \\ \\ \frac{1}{4}R_3 \rightarrow \\ -\frac{1}{5}R_4 \rightarrow \end{array} \left[\begin{array}{cccc:cccc} 1 & 0 & 0 & 0 & -\frac{1}{8} & 0 & 0 & 0 \\ 0 & 1 & 0 & 0 & 0 & 1 & 0 & 0 \\ 0 & 0 & 1 & 0 & 0 & 0 & \frac{1}{4} & 0 \\ 0 & 0 & 0 & 1 & 0 & 0 & 0 & -\frac{1}{5} \end{array}\right] = [I \ \vdots \ A^{-1}]$$

$$A^{-1} = \begin{bmatrix} -\frac{1}{8} & 0 & 0 & 0 \\ 0 & 1 & 0 & 0 \\ 0 & 0 & \frac{1}{4} & 0 \\ 0 & 0 & 0 & -\frac{1}{5} \end{bmatrix}$$

25. $A = \begin{bmatrix} 1 & 2 & -1 \\ 3 & 7 & -10 \\ -5 & -7 & -15 \end{bmatrix}$

$$A^{-1} = \begin{bmatrix} -175 & 37 & -13 \\ 95 & -20 & 7 \\ 14 & -3 & 1 \end{bmatrix}$$

27. $A = \begin{bmatrix} 1 & 1 & 2 \\ 3 & 1 & 0 \\ -2 & 0 & 3 \end{bmatrix}$

$$A^{-1} = \frac{1}{2}\begin{bmatrix} -3 & 3 & 2 \\ 9 & -7 & -6 \\ -2 & 2 & 2 \end{bmatrix} = \begin{bmatrix} -1.5 & 1.5 & 1 \\ 4.5 & -3.5 & -3 \\ -1 & 1 & 1 \end{bmatrix}$$

29. $A = \begin{bmatrix} -\frac{1}{2} & \frac{3}{4} & \frac{1}{4} \\ 1 & 0 & -\frac{3}{2} \\ 0 & -1 & \frac{1}{2} \end{bmatrix}$

$$A^{-1} = \begin{bmatrix} -12 & -5 & -9 \\ -4 & -2 & -4 \\ -8 & -4 & -6 \end{bmatrix}$$

31. $A = \begin{bmatrix} 0.1 & 0.2 & 0.3 \\ -0.3 & 0.2 & 0.2 \\ 0.5 & 0.4 & 0.4 \end{bmatrix}$

$$A^{-1} = \frac{5}{11}\begin{bmatrix} 0 & -4 & 2 \\ -22 & 11 & 11 \\ 22 & -6 & -8 \end{bmatrix} = \begin{bmatrix} 0 & -1.\overline{81} & 0.\overline{90} \\ -10 & 5 & 5 \\ 10 & -2.\overline{72} & -3.\overline{63} \end{bmatrix}$$

33. $A = \begin{bmatrix} -1 & 0 & 1 & 0 \\ 0 & 2 & 0 & -1 \\ 2 & 0 & -1 & 0 \\ 0 & -1 & 0 & 1 \end{bmatrix}$

$A^{-1} = \begin{bmatrix} 1 & 0 & 1 & 0 \\ 0 & 1 & 0 & 1 \\ 2 & 0 & 1 & 0 \\ 0 & 1 & 0 & 2 \end{bmatrix}$

35. $A = \begin{bmatrix} a & b \\ c & d \end{bmatrix}, A^{-1} = \dfrac{1}{ad - bc}\begin{bmatrix} d & -b \\ -c & a \end{bmatrix}$

$A = \begin{bmatrix} 1 & -3 \\ 5 & 4 \end{bmatrix}$

$ad - bc = (1)(4) - (-3)(5) = 19$

$A^{-1} = \frac{1}{19}\begin{bmatrix} 4 & 3 \\ -5 & 1 \end{bmatrix} = \begin{bmatrix} \frac{4}{19} & \frac{3}{19} \\ -\frac{5}{19} & \frac{1}{19} \end{bmatrix}$

37. $A = \begin{bmatrix} -4 & -6 \\ 2 & 3 \end{bmatrix}$

$ad - bc = (-4)(3) - (-2)(-6) = 0$

Because $ad - bc = 0$, A^{-1} does not exist.

39. $A = \begin{bmatrix} \frac{7}{2} & -\frac{3}{4} \\ \frac{1}{5} & \frac{4}{5} \end{bmatrix}$

$ad - bc = \left(\frac{7}{2}\right)\left(\frac{4}{5}\right) - \left(-\frac{3}{4}\right)\left(\frac{1}{5}\right) = \frac{28}{10} + \frac{3}{20} = \frac{59}{20}$

$A^{-1} = \dfrac{1}{59/20}\begin{bmatrix} \frac{4}{5} & \frac{3}{4} \\ -\frac{1}{5} & \frac{7}{2} \end{bmatrix} = \frac{20}{59}\begin{bmatrix} \frac{4}{5} & \frac{3}{4} \\ -\frac{1}{5} & \frac{7}{2} \end{bmatrix} = \begin{bmatrix} \frac{16}{59} & \frac{15}{59} \\ -\frac{4}{59} & \frac{70}{59} \end{bmatrix}$

41. $\begin{bmatrix} x \\ y \end{bmatrix} = \begin{bmatrix} -3 & 2 \\ -2 & 1 \end{bmatrix}\begin{bmatrix} 5 \\ 10 \end{bmatrix} = \begin{bmatrix} 5 \\ 0 \end{bmatrix}$

Solution: $(5, 0)$

43. $\begin{bmatrix} x \\ y \end{bmatrix} = \begin{bmatrix} -3 & 2 \\ -2 & 1 \end{bmatrix}\begin{bmatrix} 4 \\ 2 \end{bmatrix} = \begin{bmatrix} -8 \\ -6 \end{bmatrix}$

Solution: $(-8, -6)$

45. $\begin{bmatrix} x \\ y \\ z \end{bmatrix} = \begin{bmatrix} 1 & 1 & -1 \\ -3 & 2 & -1 \\ 3 & -3 & 2 \end{bmatrix}\begin{bmatrix} 0 \\ 5 \\ 2 \end{bmatrix} = \begin{bmatrix} 3 \\ 8 \\ -11 \end{bmatrix}$

Solution: $(3, 8, -11)$

47. $\begin{bmatrix} x_1 \\ x_2 \\ x_3 \\ x_4 \end{bmatrix} = \begin{bmatrix} -24 & 7 & 1 & -2 \\ -10 & 3 & 0 & -1 \\ -29 & 7 & 3 & -2 \\ 12 & -3 & -1 & 1 \end{bmatrix}\begin{bmatrix} 0 \\ 1 \\ -1 \\ 2 \end{bmatrix} = \begin{bmatrix} 2 \\ 1 \\ 0 \\ 0 \end{bmatrix}$

Solution: $(2, 1, 0, 0)$

49. $A = \begin{bmatrix} 3 & 4 \\ 5 & 3 \end{bmatrix}$

$A^{-1} = \dfrac{1}{9 - 20}\begin{bmatrix} 3 & -4 \\ -5 & 3 \end{bmatrix}$

$\begin{bmatrix} x \\ y \end{bmatrix} = -\frac{1}{11}\begin{bmatrix} 3 & -4 \\ -5 & 3 \end{bmatrix}\begin{bmatrix} -2 \\ 4 \end{bmatrix} = -\frac{1}{11}\begin{bmatrix} -22 \\ 22 \end{bmatrix} = \begin{bmatrix} 2 \\ -2 \end{bmatrix}$

Solution: $(2, -2)$

51. $A = \begin{bmatrix} -0.4 & 0.8 \\ 2 & -4 \end{bmatrix}$

$A^{-1} = \dfrac{1}{1.6 - 1.6}\begin{bmatrix} -4 & -0.8 \\ -2 & -0.4 \end{bmatrix}$

A^{-1} does not exist.

This implies that there is no unique solution; that is, either the system is inconsistent *or* there are infinitely many solutions.

Find the reduced row-echelon form of the matrix corresponding to the system.

$$\begin{bmatrix} -0.4 & 0.8 & \vdots & 1.6 \\ 2 & -4 & \vdots & 5 \end{bmatrix}$$

$$-2.5R_1 \rightarrow \begin{bmatrix} 1 & -2 & \vdots & -4 \\ 2 & -4 & \vdots & 5 \end{bmatrix}$$

$$-2R_1 + R_2 \rightarrow \begin{bmatrix} 1 & -2 & \vdots & -4 \\ 0 & 0 & \vdots & 13 \end{bmatrix}$$

The given system is inconsistent and there is no solution.

53. $A = \begin{bmatrix} -\frac{1}{4} & \frac{3}{8} \\ \frac{3}{2} & \frac{3}{4} \end{bmatrix}$

$A^{-1} = \dfrac{1}{-\frac{3}{16} - \frac{9}{16}}\begin{bmatrix} \frac{3}{4} & -\frac{3}{8} \\ -\frac{3}{2} & -\frac{1}{4} \end{bmatrix} = -\frac{4}{3}\begin{bmatrix} \frac{3}{4} & -\frac{3}{8} \\ -\frac{3}{2} & -\frac{1}{4} \end{bmatrix} = \begin{bmatrix} -1 & \frac{1}{2} \\ 2 & \frac{1}{3} \end{bmatrix}$

$\begin{bmatrix} x \\ y \end{bmatrix} = \begin{bmatrix} -1 & \frac{1}{2} \\ 2 & \frac{1}{3} \end{bmatrix}\begin{bmatrix} -2 \\ -12 \end{bmatrix} = \begin{bmatrix} -4 \\ -8 \end{bmatrix}$

Solution: $(-4, -8)$

55. $A = \begin{bmatrix} 4 & -1 & 1 \\ 2 & 2 & 3 \\ 5 & -2 & 6 \end{bmatrix}$

Find A^{-1}.

$$[A \ \vdots \ I] = \left[\begin{array}{ccc:ccc} 4 & -1 & 1 & 1 & 0 & 0 \\ 2 & 2 & 3 & 0 & 1 & 0 \\ 5 & -2 & 6 & 0 & 0 & 1 \end{array}\right]$$

$$\begin{array}{l} R_1 \\ \\ R_3 \end{array} \left[\begin{array}{ccc:ccc} 5 & -2 & 6 & 0 & 0 & 1 \\ 2 & 2 & 3 & 0 & 1 & 0 \\ 4 & -1 & 1 & 1 & 0 & 0 \end{array}\right]$$

$$\begin{array}{r} -R_3 + R_1 \rightarrow \\ \\ \\ \end{array} \left[\begin{array}{ccc:ccc} 1 & -1 & 5 & -1 & 0 & 1 \\ 2 & 2 & 3 & 0 & 1 & 0 \\ 4 & -1 & 1 & 1 & 0 & 0 \end{array}\right]$$

$$\begin{array}{r} \\ -2R_1 + R_2 \rightarrow \\ -4R_1 + R_3 \rightarrow \end{array} \left[\begin{array}{ccc:ccc} 1 & -1 & 5 & -1 & 0 & 1 \\ 0 & 4 & -7 & 2 & 1 & -2 \\ 0 & 3 & -19 & 5 & 0 & -4 \end{array}\right]$$

$$\begin{array}{r} \\ -R_3 + R_2 \rightarrow \\ \\ \end{array} \left[\begin{array}{ccc:ccc} 1 & -1 & 5 & -1 & 0 & 1 \\ 0 & 1 & 12 & -3 & 1 & 2 \\ 0 & 3 & -19 & 5 & 0 & -4 \end{array}\right]$$

$$\begin{array}{r} R_2 + R_1 \rightarrow \\ \\ -3R_2 + R_3 \rightarrow \end{array} \left[\begin{array}{ccc:ccc} 1 & 0 & 17 & -4 & 1 & 3 \\ 0 & 1 & 12 & -3 & 1 & 2 \\ 0 & 0 & -55 & 14 & -3 & -10 \end{array}\right]$$

$$\begin{array}{r} \\ \\ -\frac{1}{55}R_3 \rightarrow \end{array} \left[\begin{array}{ccc:ccc} 1 & 0 & 17 & -4 & 1 & 3 \\ 0 & 1 & 12 & -3 & 1 & 2 \\ 0 & 0 & 1 & -\frac{14}{55} & \frac{3}{55} & \frac{2}{11} \end{array}\right]$$

$$\begin{array}{r} -17R_3 + R_1 \rightarrow \\ -12R_3 + R_2 \rightarrow \\ \\ \end{array} \left[\begin{array}{ccc:ccc} 1 & 0 & 0 & \frac{18}{55} & \frac{4}{55} & -\frac{1}{11} \\ 0 & 1 & 0 & \frac{3}{55} & \frac{19}{55} & -\frac{2}{11} \\ 0 & 0 & 1 & -\frac{14}{55} & \frac{3}{55} & \frac{2}{11} \end{array}\right]$$

$$= [I \ \vdots \ A^{-1}]$$

$$A^{-1} = \frac{1}{55}\begin{bmatrix} 18 & 4 & -5 \\ 3 & 19 & -10 \\ -14 & 3 & 10 \end{bmatrix}$$

$$\begin{bmatrix} x \\ y \\ z \end{bmatrix} = \frac{1}{55}\begin{bmatrix} 18 & 4 & -5 \\ 3 & 19 & -10 \\ -14 & 3 & 10 \end{bmatrix}\begin{bmatrix} -5 \\ 10 \\ 1 \end{bmatrix} = \frac{1}{55}\begin{bmatrix} -55 \\ 165 \\ 110 \end{bmatrix} = \begin{bmatrix} -1 \\ 3 \\ 2 \end{bmatrix}$$

Solution: $(-1, 3, 2)$

57. $A = \begin{bmatrix} 5 & -3 & 2 \\ 2 & 2 & -3 \\ 1 & -7 & 8 \end{bmatrix}$

A^{-1} does not exist. This implies that there is no unique solution; that is, either the system is inconsistent *or* the system has infinitely many solutions. Use a graphing utility to find the reduced row-echelon form of the matrix corresponding to the system.

$$\left[\begin{array}{ccc:c} 5 & -3 & 2 & 2 \\ 2 & 2 & -3 & 3 \\ 1 & -7 & 8 & -4 \end{array}\right]$$

$$\left[\begin{array}{ccc:c} 1 & 0 & -\frac{5}{16} & \frac{13}{16} \\ 0 & 1 & -\frac{19}{16} & \frac{11}{16} \\ 0 & 0 & 0 & 0 \end{array}\right]$$

$$\begin{cases} x - \frac{5}{16}z = \frac{13}{16} \\ y - \frac{19}{16}z = \frac{11}{16} \end{cases}$$

Let $z = a$. Then $x = \frac{5}{16}a + \frac{13}{16}$ and $y = \frac{19}{16}a + \frac{11}{16}$.

Solution: $\left(\frac{5}{16}a + \frac{13}{16}, \frac{19}{16}a + \frac{11}{16}, a\right)$ where a is a real number.

59. $A = \begin{bmatrix} 3 & -2 & 1 \\ -4 & 1 & -3 \\ 1 & -5 & 1 \end{bmatrix}$

$$A^{-1} = \begin{bmatrix} 0.56 & 0.12 & -0.2 \\ -0.04 & -0.08 & -0.2 \\ -0.76 & -0.52 & 0.2 \end{bmatrix}$$

$$\begin{bmatrix} x \\ y \\ z \end{bmatrix} = \begin{bmatrix} 0.56 & 0.12 & -0.2 \\ -0.04 & -0.08 & -0.2 \\ -0.76 & -0.52 & 0.2 \end{bmatrix}\begin{bmatrix} -29 \\ 37 \\ -24 \end{bmatrix} = \begin{bmatrix} -7 \\ 3 \\ -2 \end{bmatrix}$$

Solution: $(-7, 3, -2)$

61. $A = \begin{bmatrix} 7 & -3 & 0 & 2 \\ -2 & 1 & 0 & -1 \\ 4 & 0 & 1 & -2 \\ -1 & 1 & 0 & -1 \end{bmatrix}$

$$A^{-1} = \begin{bmatrix} 0 & -1 & 0 & 1 \\ -1 & -5 & 0 & 3 \\ -2 & -4 & 1 & -2 \\ -1 & -4 & 0 & 1 \end{bmatrix}$$

$$\begin{bmatrix} x \\ y \\ z \\ w \end{bmatrix} = \begin{bmatrix} 0 & -1 & 0 & 1 \\ -1 & -5 & 0 & 3 \\ -2 & -4 & 1 & -2 \\ -1 & -4 & 0 & 1 \end{bmatrix}\begin{bmatrix} 41 \\ -13 \\ 12 \\ -8 \end{bmatrix} = \begin{bmatrix} 5 \\ 0 \\ -2 \\ 3 \end{bmatrix}$$

Solution: $(5, 0, -2, 3)$

63. $\begin{cases} 2x + y + 2z = 500 \\ x + 2y + 2z = 500 \\ x + y + 2z = 400 \end{cases}$

$$A = \begin{bmatrix} 2 & 1 & 2 \\ 1 & 2 & 2 \\ 1 & 1 & 2 \end{bmatrix};\ A^{-1} = \begin{bmatrix} 1 & 0 & -1 \\ 0 & 1 & -1 \\ -0.5 & -0.5 & 1.5 \end{bmatrix};\ B = \begin{bmatrix} 500 \\ 500 \\ 400 \end{bmatrix}$$

$$\begin{bmatrix} x \\ y \\ z \end{bmatrix} = A^{-1}B = \begin{bmatrix} 1 & 0 & -1 \\ 0 & 1 & -1 \\ -0.5 & -0.5 & 1.5 \end{bmatrix}\begin{bmatrix} 500 \\ 500 \\ 400 \end{bmatrix} = \begin{bmatrix} 100 \\ 100 \\ 100 \end{bmatrix}$$

100 bags of potting soil for seedings, 100 bags of potting soil for general potting, and 100 bags of potting soil for hardwood plants can be produced.

65. $\begin{cases} 2x + y + 2z = 350 \\ x + 2y + 2z = 445 \\ x + y + 2z = 345 \end{cases}$

$$A = \begin{bmatrix} 2 & 1 & 2 \\ 1 & 2 & 2 \\ 1 & 1 & 2 \end{bmatrix};\ A^{-1} = \begin{bmatrix} 1 & 0 & -1 \\ 0 & 1 & -1 \\ -0.5 & -0.5 & 1.5 \end{bmatrix};\ B = \begin{bmatrix} 350 \\ 445 \\ 345 \end{bmatrix}$$

$$\begin{bmatrix} x \\ y \\ z \end{bmatrix} = A^{-1}B = \begin{bmatrix} 1 & 0 & -1 \\ 0 & 1 & -1 \\ -0.5 & -0.5 & 1.5 \end{bmatrix}\begin{bmatrix} 350 \\ 445 \\ 345 \end{bmatrix}\begin{bmatrix} 5 \\ 100 \\ 120 \end{bmatrix}$$

5 bags of potting soil for seedings, 100 bags of potting soil for general potting, and 120 bags of potting soil for hardwood plants can be produced.

67. (a) $\begin{bmatrix} 6 & 3 \\ 3 & 19 \end{bmatrix}^{-1} = \begin{bmatrix} \frac{19}{105} & -\frac{1}{35} \\ -\frac{1}{35} & \frac{2}{35} \end{bmatrix}$

(b) $$\begin{bmatrix} \frac{19}{105} & -\frac{1}{35} \\ -\frac{1}{35} & \frac{2}{35} \end{bmatrix}\begin{bmatrix} 123.1 \\ 79.2 \end{bmatrix} = \begin{bmatrix} \left(\frac{19}{105}\right)(123.1) + \left(-\frac{1}{35}\right)(79.2) \\ \left(-\frac{1}{35}\right)(123.1) + \left(\frac{2}{35}\right)(79.2) \end{bmatrix} = \begin{bmatrix} 20.0 \\ 1.0 \end{bmatrix}$$

$b = 20.0,\ a = 1.0$

The least squares regression line is $y = t + 20.0$.

(c) For 2007, $t = 5$:

$y = 5 + 20.0 = 25.0$

There will be $25 billion of child support collections in 2007.

69. (a) $A = \begin{bmatrix} 2 & 0 & 4 \\ 0 & 1 & 4 \\ 1 & 1 & -1 \end{bmatrix}$

$$\begin{bmatrix} A & \vdots & I \end{bmatrix} = \left[\begin{array}{ccc|ccc} 2 & 0 & 4 & 1 & 0 & 0 \\ 0 & 1 & 4 & 0 & 1 & 0 \\ 1 & 1 & -1 & 0 & 0 & 1 \end{array}\right]$$

$$\begin{array}{c} R_1 \\ \\ R_3 \end{array}\left[\begin{array}{ccc|ccc} 1 & 1 & -1 & 0 & 0 & 1 \\ 0 & 1 & 4 & 0 & 1 & 0 \\ 2 & 0 & 4 & 1 & 0 & 0 \end{array}\right]$$

$$\begin{array}{r} \\ \\ -2R_1 + R_3 \rightarrow \end{array}\left[\begin{array}{ccc|ccc} 1 & 1 & -1 & 0 & 0 & 1 \\ 0 & 1 & 4 & 0 & 1 & 0 \\ 0 & -2 & 6 & 1 & 0 & -2 \end{array}\right]$$

$$\begin{array}{r} -R_2 + R_1 \rightarrow \\ \\ 2R_2 + R_3 \rightarrow \end{array}\left[\begin{array}{ccc|ccc} 1 & 0 & -5 & 0 & -1 & 1 \\ 0 & 1 & 4 & 0 & 1 & 0 \\ 0 & 0 & 14 & 1 & 2 & -2 \end{array}\right]$$

$$\begin{array}{r} \\ \\ \frac{1}{14}R_3 \rightarrow \end{array}\left[\begin{array}{ccc|ccc} 1 & 0 & -5 & 0 & -1 & 1 \\ 0 & 1 & 4 & 0 & 1 & 0 \\ 0 & 0 & 1 & \frac{1}{14} & \frac{1}{7} & -\frac{1}{7} \end{array}\right]$$

$$\begin{array}{r} 5R_3 + R_1 \rightarrow \\ -4R_3 + R_2 \rightarrow \\ \\ \end{array}\left[\begin{array}{ccc|ccc} 1 & 0 & 0 & \frac{5}{14} & -\frac{2}{7} & \frac{2}{7} \\ 0 & 1 & 0 & -\frac{2}{7} & \frac{3}{7} & \frac{4}{7} \\ 0 & 0 & 1 & \frac{1}{14} & \frac{1}{7} & -\frac{1}{7} \end{array}\right]$$

$$= \begin{bmatrix} I & \vdots & A^{-1} \end{bmatrix}$$

$$A^{-1} = \frac{1}{14}\begin{bmatrix} 5 & -4 & 4 \\ -4 & 6 & 8 \\ 1 & 2 & -2 \end{bmatrix}$$

$$\begin{bmatrix} I_1 \\ I_2 \\ I_3 \end{bmatrix} = \frac{1}{14}\begin{bmatrix} 5 & -4 & 4 \\ -4 & 6 & 8 \\ 1 & 2 & -2 \end{bmatrix}\begin{bmatrix} 14 \\ 28 \\ 0 \end{bmatrix} = \begin{bmatrix} -3 \\ 8 \\ 5 \end{bmatrix}$$

Solution: $I_1 = -3$ amperes, $I_2 = 8$ amperes, $I_3 = 5$ amperes

(b) $$\begin{bmatrix} I_1 \\ I_2 \\ I_3 \end{bmatrix} = \frac{1}{14}\begin{bmatrix} 5 & -4 & 4 \\ -4 & 6 & 8 \\ 1 & 2 & -2 \end{bmatrix}\begin{bmatrix} 24 \\ 23 \\ 0 \end{bmatrix} = \begin{bmatrix} 2 \\ 3 \\ 5 \end{bmatrix}$$

Solution: $I_1 = 2$ amperes, $I_2 = 3$ amperes, $I_3 = 5$ amperes

71. True. If A and B are both square matrices and $AB = I_n$, it can be shown that $BA = I_n$.

73. (a) Given $A = \begin{bmatrix} a_{11} & 0 \\ 0 & a_{22} \end{bmatrix}$, $A^{-1} = \begin{bmatrix} \frac{1}{a_{11}} & 0 \\ 0 & \frac{1}{a_{22}} \end{bmatrix}$.

Given $A = \begin{bmatrix} a_{11} & 0 & 0 \\ 0 & a_{22} & 0 \\ 0 & 0 & a_{33} \end{bmatrix}$,

$$A^{-1} = \begin{bmatrix} \frac{1}{a_{11}} & 0 & 0 \\ 0 & \frac{1}{a_{22}} & 0 \\ 0 & 0 & \frac{1}{a_{33}} \end{bmatrix}.$$

(b) In general, the inverse of a matrix in the form of A is

$$\begin{bmatrix} \frac{1}{a_{11}} & 0 & 0 & \cdots & 0 \\ 0 & \frac{1}{a_{22}} & 0 & \cdots & 0 \\ 0 & 0 & \frac{1}{a_{33}} & \cdots & 0 \\ \vdots & \vdots & \vdots & \cdots & \vdots \\ 0 & 0 & 0 & \cdots & \frac{1}{a_{nn}} \end{bmatrix}.$$

Review Exercises for Chapter 8

1. $\begin{cases} x + 3y = 10 & \text{Equation 1} \\ 4x - 5y = -28 & \text{Equation 2} \end{cases}$

Solve for x in Equation 1: $x = 10 - 3y$

Substitute for x in Equation 2; then solve for y:

$4(10 - 3y) - 5y = -28 \Rightarrow -17y = -68 \Rightarrow y = 4$

Back-substitute $y = 4$: $x = 10 - 3(4) = -2$

Solution: $(-2, 4)$

3. $\begin{cases} \frac{1}{2}x + \frac{3}{5}y = -2 & \text{Equation 1} \\ 2x + y = 6 & \text{Equation 2} \end{cases}$

Solve for y in Equation 2: $y = 6 - 2x$

Substitute for y in Equation 1; then solve for x:

$5x + 6(6 - 2x) = -20 \Rightarrow -7x = 56 \Rightarrow x = 8$

Back-substitute $x = 8$: $y = 6 - 2(8) = -10$

Solution: $(8, -10)$

5. $\begin{cases} x^2 + y^2 = 100 & \text{Equation 1} \\ x + 2y = 20 & \text{Equation 2} \end{cases}$

Solve for x in Equation 2: $x = 20 - 2y$

Substitute for x in Equation 1; then solve for y:

$(20 - 2y)^2 + y^2 = 100$

$5y^2 - 80y + 300 = 0$

$5(y - 6)(y - 10) = 0 \Rightarrow y = 6, 10$

Back-substitute $y = 6$: $x = 20 - 2(6) = 8$

Back-substitute $y = 10$: $x = 20 - 2(10) = 0$

Solutions: $(8, 6)$, $(0, 10)$

7. $\begin{cases} 2x - 3y = 21 & \text{Equation 1} \\ 3x + y = 4 & \text{Equation 2} \end{cases}$

Multiply Equation 2 by 3; then add the equations:

$$\begin{aligned} 2x - 3y &= 21 \\ 9x + 3y &= 12 \\ \hline 11x &= 33 \\ x &= 3 \end{aligned}$$

Back-substitute $x = 3$ into Equation 2:

$3(3) + y = 4 \Rightarrow y = -5$

Solution: $(3, -5)$

9. $\begin{cases} 4x - 3y = 10 & \text{Equation 1} \\ 8x - 6y = 20 & \text{Equation 2} \end{cases}$

Multiply Equation 1 by -2; then add the equations:

$$\begin{aligned} -8x + 6y &= -20 \\ 8x - 6y &= 20 \\ \hline 0 &= 0 \end{aligned}$$

The equations are dependent. There are infinitely many solutions.

Let $y = a$, then

$4x - 3a = 10 \Rightarrow x = \frac{3}{4}a + \frac{5}{2}$

Solution: $\left(\frac{3}{4}a + \frac{5}{2}, a\right)$ where a is any real number

11. $\begin{cases} 1.25x - 2y = 3.5 & \text{Equation 1} \\ 5x - 8y = 14 & \text{Equation 2} \end{cases}$

Multiply Equation 1 by -4; then add the equations:

$$\begin{aligned} -5x + 8y &= -14 \\ 5x - 8y &= 14 \\ \hline 0 &= 0 \end{aligned}$$

There are infinitely many solutions.

Let $y = a$, then

$5x - 8a = 14 \Rightarrow x = \frac{8}{5}a + \frac{14}{5}$

Solution: $\left(\frac{8}{5}a + \frac{14}{5}, a\right)$ where a is any real number

13. $\begin{cases} \dfrac{x-2}{3} + \dfrac{y+3}{4} = 5 & \text{Equation 1} \\ 2x - y = 7 & \text{Equation 2} \end{cases}$

Multiply Equation 1 by 12 and Equation 2 by 3; then add the equations:

$$\begin{aligned} 4x + 3y &= 59 \\ 6x - 3y &= 21 \\ \hline 10x &= 80 \\ x &= 8 \end{aligned}$$

Back-substitute $x = 8$ into Equation 2:

$2(8) - y = 7 \Rightarrow y = 9$

Solution: $(8, 9)$

15. $\begin{cases} 2x + y = -1 \\ 3x - 2y = -5 \end{cases}$

$2x + y = -1 \Rightarrow y = -2x - 1$

$$\begin{aligned} 3x - 2(-2x - 1) &= -5 \\ 7x &= -7 \\ x &= -1 \end{aligned}$$

$y = -2(-1) - 1 = 1$

The graph intersects at the point $(-1, 1)$.

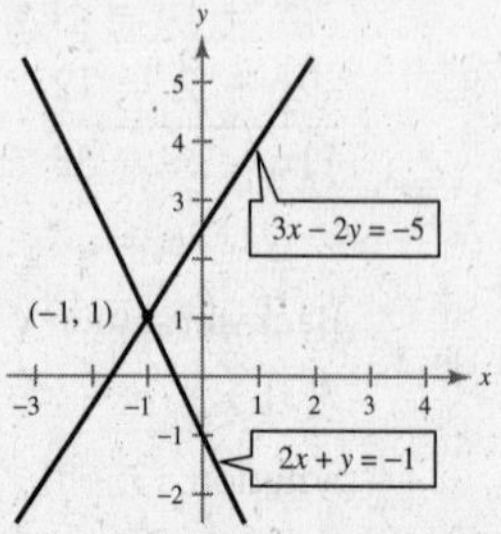

17. $\begin{cases} y = 950x + 10{,}000 & \text{Station A} \\ y = -875x + 18{,}000 & \text{Station B} \end{cases}$

$$\begin{aligned} 950x + 10{,}000 &= -875x + 18{,}000 \\ 1825x &= 8000 \\ x &\approx 4.384 \end{aligned}$$

The goal will be achieved during the fourth month of the new format.

19. Let x = amount of 10% solution.

Let y = amount of 50% solution.

(a) $\begin{cases} x + y = 12 \Rightarrow y = 12 - x \\ 0.10x + 0.50y = 0.25(12) \end{cases}$

$$\begin{aligned} 0.10x + 0.50(12 - x) &= 3 \\ -0.40x &= -3 \\ x &= 7.5 \\ y &= 12 - 7.5 = 4.5 \end{aligned}$$

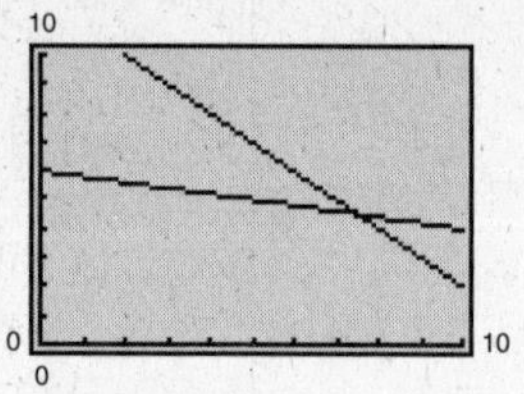

(b) 7.5 gallons of 10% solution, 4.5 gallons of 50% solution.

21. $\begin{cases} 4x - 3y + 2z = 1 \\ 2y - 4z = 2 \\ z = 2 \end{cases}$

$2y - 4(2) = 2 \Rightarrow y = 5$

$4x - 3(5) + 2(2) = 1 \Rightarrow x = 3$

Solution: $(3, 5, 2)$

23. $\begin{cases} 2x + y + z = 6 \\ x - 4y - z = 3 \\ x + y + z = 4 \end{cases}$

$\begin{cases} 2x + y + z = 6 & \\ 9y + 3z = 0 & \text{Eq.1} + (-2)\text{Eq.2} \\ -y - z = -2 & \text{Eq.1} + (-2)\text{Eq.3} \end{cases}$

$\begin{cases} 2x + y + z = 6 & \\ 9y + 3z = 0 & \\ -6z = -18 & \text{Eq.2} + 9\text{Eq.3} \end{cases}$

$$\begin{aligned} -6z &= -18 \Rightarrow z = 3 \\ 9y + 3(3) &= 0 \Rightarrow y = -1 \\ 2x + (-1) + 3 &= 6 \Rightarrow x = 2 \end{aligned}$$

Solution: $(2, -1, 3)$

25. $\begin{cases} x - 3y + z = 2 \\ 2x + 6y - z = 1 \\ 3x + 3y = 12 \end{cases}$ Interchange Eq.1 and Eq.2. 2Eq.3

$$\begin{cases} x - 3y + z = 2 \\ 12y - 3z = -3 & -2\text{Eq.1} + \text{Eq.2} \\ 12y - 3z = 6 & -3\text{Eq.1} + \text{Eq.3} \end{cases}$$

$$\begin{cases} x - 3y + z = 2 \\ 12y - 3z = -3 \\ 0 = 9 & -\text{Eq.2} + \text{Eq.3} \end{cases}$$

Inconsistent system

There is no solution.

27. $\begin{cases} x + y + z = 10 \Rightarrow 3x + 3y + 3z = 30 \\ -2x + 3y + 4z = 22 \Rightarrow 2x - 3y - 4z = -22 \end{cases}$

$$5x - z = 8 \Rightarrow x = \frac{8 + z}{5}$$

Let $z = a$. Then $x = \dfrac{8 + a}{5}$ and $\dfrac{8 + a}{5} = y + a = 10 \Rightarrow y = \dfrac{42 - 6a}{5}$.

Solution: $\left(\dfrac{8 + a}{5}, \dfrac{42 - 6a}{5}, a\right)$ where a is any real number

29. From the following chart you obtain the system of equations.

	A	B	C
Mixture X	$\frac{1}{5}$	$\frac{2}{5}$	$\frac{2}{5}$
Mixture Y	0	0	1
Mixture Z	$\frac{1}{3}$	$\frac{1}{3}$	$\frac{1}{3}$
Desired Mixture	$\frac{6}{27}$	$\frac{8}{27}$	$\frac{13}{27}$

$$\left.\begin{aligned} \tfrac{1}{5}x + \tfrac{1}{3}z &= \tfrac{6}{27} \\ \tfrac{2}{5}x + \tfrac{1}{3}z &= \tfrac{8}{27} \end{aligned}\right\} \Rightarrow x = \tfrac{10}{27},\ z = \tfrac{12}{27}$$

$$\tfrac{2}{5}x + y + \tfrac{1}{3}z = \tfrac{13}{27} \Rightarrow y = \tfrac{5}{27}$$

To obtain the desired mixture, use 10 gallons of spray X, 5 gallons of spray Y, and 12 gallons of spray Z.

31. $y = ax^2 + bx + c$

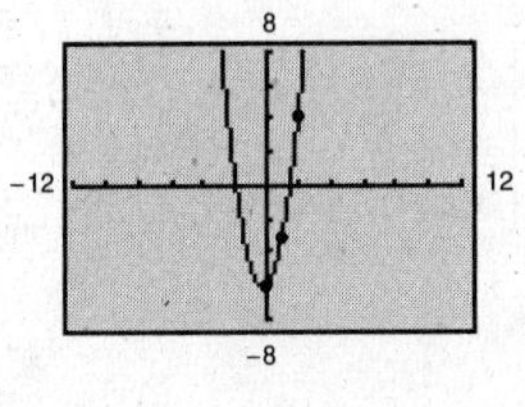

At $(0, -6)$: $-6 = c$

At $(1, -3)$: $-3 = a + b + c \Rightarrow a + b = 3 \Rightarrow -2a - 2b = -6$

At $(2, 4)$: $4 = 4a + 2b + c \Rightarrow 4a + 2b = 10 \Rightarrow 4a + 2b = 10$

$$2a = 4$$

$$a = 2 \Rightarrow b = 1$$

Equation of parabola: $y = 2x^2 + x - 6$

33. $x^2 + y^2 + Dx + Ey + F = 0$

At $(2, 2)$: $8 + 2D + 2E + F = 0 \Rightarrow 2D + 2E + F = -8$ Equation 1

At $(5, -1)$: $26 + 5D - E + F = 0 \Rightarrow 5D - E + F = -26$ Equation 2

At $(-1, -1)$: $2 - D - E + F = 0 \Rightarrow D + E - F = 2$ Equation 3

Equation 1 + Equation 3: $3D + 3E = -6 \Rightarrow D + E = -2$

Equation 2 + Equation 3: $6D = -24 \Rightarrow D = -4 \Rightarrow E = 2$

Equation 3: $(-4) + 2 - F = 2 \Rightarrow F = -4$

Equation of circle: $x^2 + y^2 - 4x + 2y - 4 = 0$

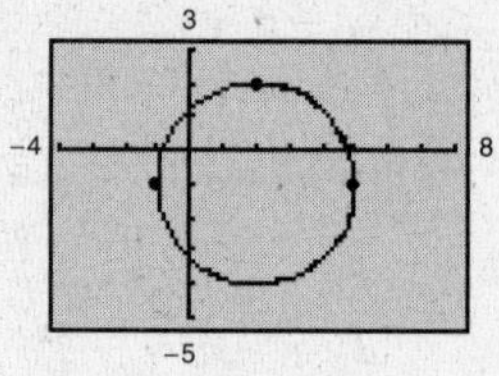

35. 2×4

37. $\begin{bmatrix} 1 & 3 & 0 & 2 \\ 3 & 10 & 1 & 8 \\ 2 & 3 & 3 & 10 \end{bmatrix}$

$\begin{array}{r} \\ -3R_1 + R_2 \\ -2R_1 + R_3 \end{array}\begin{bmatrix} 1 & 3 & 0 & 2 \\ 0 & 1 & 1 & 2 \\ 0 & -3 & 3 & 6 \end{bmatrix}$

$\begin{array}{r} \\ \\ 3R_2 + R_3 \end{array}\begin{bmatrix} 1 & 3 & 0 & 2 \\ 0 & 1 & 1 & 2 \\ 0 & 0 & 6 & 12 \end{bmatrix}$

$\begin{array}{r} \\ \\ \frac{1}{6}R_3 \end{array}\begin{bmatrix} 1 & 3 & 0 & 2 \\ 0 & 1 & 1 & 2 \\ 0 & 0 & 1 & 2 \end{bmatrix}$

39. $\begin{bmatrix} 1 & 2 & 3 \\ -2 & 0 & 2 \\ 2 & 1 & 2 \end{bmatrix}$

$\begin{array}{r} \\ 2R_1 + R_2 \\ -2R_1 + R_3 \end{array}\begin{bmatrix} 1 & 2 & 3 \\ 0 & 4 & 8 \\ 0 & -3 & -4 \end{bmatrix}$

$\begin{array}{r} -2R_1 + R_2 \\ \frac{1}{4}R_2 \\ \frac{3}{4}R_2 + R_3 \end{array}\begin{bmatrix} 1 & 0 & 2 \\ 0 & 1 & 2 \\ 0 & 0 & 2 \end{bmatrix}$

$\begin{array}{r} R_1 - R_3 \\ R_2 - R_3 \\ \frac{1}{2}R_3 \end{array}\begin{bmatrix} 1 & 0 & 0 \\ 0 & 1 & 0 \\ 0 & 0 & 1 \end{bmatrix}$

41. $\begin{cases} 4x - 3y = 18 \\ x + y = 1 \end{cases}$

$\begin{array}{r} R_2 \leftrightarrow R_1 \\ R_1 \leftrightarrow R_2 \end{array}\left[\begin{array}{rr:r} 1 & 1 & 1 \\ 4 & -3 & 18 \end{array}\right] \Rightarrow \begin{array}{r} \\ -4R_1 + R_2 \end{array}\left[\begin{array}{rr:r} 1 & 1 & 1 \\ 0 & -7 & 14 \end{array}\right]$

$-7y = 14 \Rightarrow y = -2$

$x + (-2) = 1 \Rightarrow x = 3$

Solution: $(3, -2)$

43. $\begin{cases} 2x + 3y - z = 13 \\ 3x + z = 8 \\ x - 2y + 3z = -4 \end{cases}$

$\begin{array}{r} R_3 \leftrightarrow R_1 \\ R_1 \leftrightarrow R_2 \\ R_2 \leftrightarrow R_3 \end{array}\left[\begin{array}{rrr:r} 1 & -2 & 3 & -4 \\ 2 & 3 & -1 & 13 \\ 3 & 0 & 1 & 8 \end{array}\right] \Rightarrow \begin{array}{r} \\ -2R_1 + R_2 \\ -3R_1 + R_3 \end{array}\left[\begin{array}{rrr:r} 1 & -2 & 3 & -4 \\ 0 & 7 & -7 & 21 \\ 0 & 6 & -8 & 20 \end{array}\right] \Rightarrow \begin{array}{r} \\ \frac{1}{7}R_2 \\ -\frac{6}{7}R_2 + R_3 \end{array}\left[\begin{array}{rrr:r} 1 & 0 & 1 & 2 \\ 0 & 1 & -1 & 3 \\ 0 & 0 & -2 & 2 \end{array}\right]$

$\Rightarrow \begin{array}{r} \frac{1}{2}R_3 + R_1 \\ -\frac{1}{2}R_3 + R_2 \\ -\frac{1}{2}R_3 \end{array}\left[\begin{array}{rrr:r} 1 & 0 & 0 & 3 \\ 0 & 1 & 0 & 2 \\ 0 & 0 & 1 & -1 \end{array}\right]$

$x = 3$

$y = 2$

$z = -1$

Solution: $(3, 2, -1)$

45. $\begin{cases} x + 2y + 2z = 10 \\ 2x + 3y + 5z = 20 \end{cases}$

$$\begin{bmatrix} 1 & 2 & 2 & \vdots & 10 \\ 2 & 3 & 5 & \vdots & 20 \end{bmatrix} \Rightarrow \begin{matrix} \\ -2R_1 + R_2 \end{matrix}\begin{bmatrix} 1 & 2 & 2 & \vdots & 10 \\ 0 & -1 & 1 & \vdots & 0 \end{bmatrix}$$

$$\Rightarrow \begin{matrix} 2R_2 + R_1 \\ -1R_2 \end{matrix}\begin{bmatrix} 1 & 0 & 4 & \vdots & 10 \\ 0 & 1 & -1 & \vdots & 0 \end{bmatrix}$$

$x + 4z = 10$

$y - z = 0$

Let $z = a$. Then $y = a$ and $x = 10 - 4a$, for all a.

Solution: $(10 - 4a, a, a)$, where a is any real number

47. $\begin{cases} 2x + y - 3z = 4 \\ x + 2y + 2z = 10 \\ x - 2z = 12 \\ x + y + z = 6 \end{cases}$

$$\begin{matrix} R_4 \leftrightarrow R_1 \\ R_1 \leftrightarrow R_2 \\ R_2 \leftrightarrow R_3 \\ R_3 \leftrightarrow R_4 \end{matrix}\begin{bmatrix} 1 & 1 & 1 & \vdots & 6 \\ 2 & 1 & -3 & \vdots & 4 \\ 1 & 2 & 2 & \vdots & 10 \\ 1 & 0 & -2 & \vdots & 12 \end{bmatrix} \Rightarrow \begin{matrix} \\ -2R_1 + R_2 \\ R_3 - R_4 \\ \\ \end{matrix}\begin{bmatrix} 1 & 1 & 1 & \vdots & 6 \\ 0 & -1 & -5 & \vdots & -8 \\ 0 & 2 & 4 & \vdots & -2 \\ 1 & 0 & -2 & \vdots & 12 \end{bmatrix}$$

$$\Rightarrow \begin{matrix} \\ -R_2 \\ 2R_2 + R_3 \\ R_4 - R_1 \end{matrix}\begin{bmatrix} 1 & 1 & 1 & \vdots & 6 \\ 0 & 1 & 5 & \vdots & 8 \\ 0 & 0 & -6 & \vdots & -18 \\ 0 & -1 & -3 & \vdots & 6 \end{bmatrix}$$

$$\Rightarrow \begin{matrix} \\ \\ -\frac{1}{6}R_3 \\ R_4 + R_2 \end{matrix}\begin{bmatrix} 1 & 1 & 1 & \vdots & 6 \\ 0 & 1 & 5 & \vdots & 8 \\ 0 & 0 & 1 & \vdots & 3 \\ 0 & 0 & 2 & \vdots & 14 \end{bmatrix}$$

$z = 3$ and $2z = 14$

$z = 7$

Inconsistent

No solution

49. (a) $A + B = \begin{bmatrix} -1 & 5 \\ 2 & 1 \end{bmatrix} + \begin{bmatrix} 4 & 2 \\ -6 & 3 \end{bmatrix} = \begin{bmatrix} 3 & 7 \\ -4 & 4 \end{bmatrix}$

(b) $A - B = \begin{bmatrix} -1 & 5 \\ 2 & 1 \end{bmatrix} - \begin{bmatrix} 4 & 2 \\ -6 & 3 \end{bmatrix} = \begin{bmatrix} -5 & 3 \\ 8 & -2 \end{bmatrix}$

(c) $4A = 4\begin{bmatrix} -1 & 5 \\ 2 & 1 \end{bmatrix} = \begin{bmatrix} -4 & 20 \\ 8 & 4 \end{bmatrix}$

(d) $4A - 3B = \begin{bmatrix} -4 & 20 \\ 8 & 4 \end{bmatrix} - \begin{bmatrix} 12 & 6 \\ -18 & 9 \end{bmatrix} = \begin{bmatrix} -16 & 14 \\ 26 & -5 \end{bmatrix}$

51. (a) $A + B = \begin{bmatrix} 1 & 3 & -2 & 6 \\ 0 & 1 & 3 & 2 \end{bmatrix} + \begin{bmatrix} 2 & 1 & 4 & -5 \\ 3 & -6 & 3 & -2 \end{bmatrix} = \begin{bmatrix} 3 & 4 & 2 & 1 \\ 3 & -5 & 6 & 0 \end{bmatrix}$

(b) $A - B = \begin{bmatrix} 1 & 3 & -2 & 6 \\ 0 & 1 & 3 & 2 \end{bmatrix} - \begin{bmatrix} 2 & 1 & 4 & -5 \\ 3 & -6 & 3 & -2 \end{bmatrix} = \begin{bmatrix} -1 & 2 & -6 & 11 \\ -3 & 7 & 0 & 4 \end{bmatrix}$

(c) $4A = 4\begin{bmatrix} 1 & 3 & -2 & 6 \\ 0 & 1 & 3 & 2 \end{bmatrix} = \begin{bmatrix} 4 & 12 & -8 & 24 \\ 0 & 4 & 12 & 8 \end{bmatrix}$

(d) $4A - 3B = \begin{bmatrix} 4 & 12 & -8 & 24 \\ 0 & 4 & 12 & 8 \end{bmatrix} - \begin{bmatrix} 6 & 3 & 12 & -15 \\ 9 & -18 & 9 & -6 \end{bmatrix} = \begin{bmatrix} -2 & 9 & -20 & 39 \\ -9 & 22 & 3 & 14 \end{bmatrix}$

53. $AB = \begin{bmatrix} 1 & 4 \\ -2 & -1 \\ 3 & 2 \end{bmatrix}\begin{bmatrix} -4 \\ 3 \end{bmatrix} = \begin{bmatrix} 8 \\ 5 \\ -6 \end{bmatrix}$

55. $AB = \begin{bmatrix} 4 & 0 & 0 \\ 0 & 3 & 0 \\ 0 & 0 & -2 \end{bmatrix}\begin{bmatrix} \frac{1}{4} & 0 & 0 \\ 0 & \frac{1}{3} & 0 \\ 0 & 0 & -\frac{1}{2} \end{bmatrix} = \begin{bmatrix} 1 & 0 & 0 \\ 0 & 1 & 0 \\ 0 & 0 & 1 \end{bmatrix}$

57. A is 2×5 and B is 2×2. So, AB is *not* defined. (Note: BA *is* defined.)

59. (a) $AB = \begin{bmatrix} 1 & -3 & 4 \end{bmatrix}\begin{bmatrix} 2 \\ -2 \\ -1 \end{bmatrix} = \begin{bmatrix} 4 \end{bmatrix}$

(b) $BA = \begin{bmatrix} 2 \\ -2 \\ -1 \end{bmatrix}\begin{bmatrix} 1 & -3 & 4 \end{bmatrix} = \begin{bmatrix} 2 & -6 & 8 \\ -2 & 6 & -8 \\ -1 & 3 & -4 \end{bmatrix}$

(c) A^2 is not possible.

61. $X_2 = LX_1 = \begin{bmatrix} 0 & 2 & 2 & 0 \\ \frac{1}{4} & 0 & 0 & 0 \\ 0 & 1 & 0 & 0 \\ 0 & 0 & \frac{1}{2} & 0 \end{bmatrix}\begin{bmatrix} 100 \\ 100 \\ 100 \\ 100 \end{bmatrix} = \begin{bmatrix} 0(100) + 2(100) + 2(100) + 0(100) \\ \frac{1}{4}(100) + 0(100) + 0(100) + 0(100) \\ 0(100) + 1(100) + 0(100) + 0(100) \\ 0(100) + 0(100) + \frac{1}{2}(100) + 0(100) \end{bmatrix} = \begin{bmatrix} 400 \\ 25 \\ 100 \\ 50 \end{bmatrix}$

$X_3 = LX_2 = \begin{bmatrix} 0 & 2 & 2 & 0 \\ \frac{1}{4} & 0 & 0 & 0 \\ 0 & 1 & 0 & 0 \\ 0 & 0 & \frac{1}{2} & 0 \end{bmatrix}\begin{bmatrix} 400 \\ 25 \\ 100 \\ 50 \end{bmatrix} = \begin{bmatrix} 0(400) + 2(25) + 2(100) + 0(50) \\ \frac{1}{4}(400) + 0(25) + 0(100) + 0(50) \\ 0(400) + 1(25) + 0(100) + 0(50) \\ 0(400) + 0(25) + \frac{1}{2}(100) + 0(50) \end{bmatrix} = \begin{bmatrix} 250 \\ 100 \\ 25 \\ 50 \end{bmatrix}$

63. $X = 4A - 3B = 4\begin{bmatrix} 1 & -2 \\ 0 & 1 \\ 2 & 3 \end{bmatrix} - 3\begin{bmatrix} 0 & 1 \\ 1 & 1 \\ 3 & 5 \end{bmatrix} = \begin{bmatrix} 4 & -11 \\ -3 & 1 \\ -1 & -3 \end{bmatrix}$

65. $2X - 3A = B \Rightarrow X = \frac{1}{2}(B + 3A)$

$= \frac{1}{2}\left(\begin{bmatrix} 0 & 1 \\ 1 & 1 \\ 3 & 5 \end{bmatrix} + 3\begin{bmatrix} 1 & -2 \\ 0 & 1 \\ 2 & 3 \end{bmatrix}\right) = \frac{1}{2}\begin{bmatrix} 3 & -5 \\ 1 & 4 \\ 9 & 14 \end{bmatrix}$

67. (a) $\begin{bmatrix} 5 & 4 \\ -1 & 1 \end{bmatrix}\begin{bmatrix} x_1 \\ x_2 \end{bmatrix} = \begin{bmatrix} 2 \\ -22 \end{bmatrix}$

(b) $\begin{matrix} R_2 \\ R_1 \end{matrix}\begin{bmatrix} -1 & 1 & \vdots & -22 \\ 5 & 4 & \vdots & 2 \end{bmatrix}$

$-R_1 \to \begin{bmatrix} 1 & -1 & \vdots & 22 \\ 5 & 4 & \vdots & 2 \end{bmatrix}$

$-5R_1 + R_2 \to \begin{bmatrix} 1 & -1 & \vdots & 22 \\ 0 & 9 & \vdots & -108 \end{bmatrix}$

$\frac{1}{9}R_2 \to \begin{bmatrix} 1 & -1 & \vdots & 22 \\ 0 & 1 & \vdots & -12 \end{bmatrix}$

$R_2 + R_1 \to \begin{bmatrix} 1 & 0 & \vdots & 10 \\ 0 & 1 & \vdots & -12 \end{bmatrix}$

$X = \begin{bmatrix} 10 \\ -12 \end{bmatrix}$

69. (a) $\begin{bmatrix} 2 & 3 & 1 \\ 2 & -3 & -3 \\ 4 & -2 & 3 \end{bmatrix}\begin{bmatrix} x_1 \\ x_2 \\ x_3 \end{bmatrix} = \begin{bmatrix} 10 \\ 22 \\ -2 \end{bmatrix}$

(b)
$$\left[\begin{array}{ccc:c} 2 & 3 & 1 & 10 \\ 2 & -3 & -3 & 22 \\ 4 & -2 & 3 & -2 \end{array}\right]$$

$$\tfrac{1}{2}R \rightarrow \left[\begin{array}{ccc:c} 1 & \frac{3}{2} & \frac{1}{2} & 5 \\ 2 & -3 & -3 & 22 \\ 4 & -2 & 3 & -2 \end{array}\right]$$

$$\begin{array}{r} \\ -2R_1 + R_2 \rightarrow \\ -4R_1 + R_3 \rightarrow \end{array} \left[\begin{array}{ccc:c} 1 & \frac{3}{2} & \frac{1}{2} & 5 \\ 0 & -6 & -4 & 12 \\ 0 & -8 & 1 & -22 \end{array}\right]$$

$$\begin{array}{r} \\ -\frac{1}{6}R_2 \rightarrow \\ -\frac{1}{8}R_3 \rightarrow \end{array} \left[\begin{array}{ccc:c} 1 & \frac{3}{2} & \frac{1}{2} & 5 \\ 0 & 1 & \frac{2}{3} & -2 \\ 0 & 1 & -\frac{1}{8} & \frac{11}{4} \end{array}\right]$$

$$\begin{array}{r} -\frac{3}{2}R_2 + R_1 \rightarrow \\ \\ -R_2 + R_3 \rightarrow \end{array} \left[\begin{array}{ccc:c} 1 & 0 & -\frac{1}{2} & 8 \\ 0 & 1 & \frac{2}{3} & -2 \\ 0 & 0 & -\frac{19}{24} & \frac{19}{4} \end{array}\right]$$

$$\begin{array}{r} \\ \\ -\frac{24}{19}R_3 \rightarrow \end{array} \left[\begin{array}{ccc:c} 1 & 0 & -\frac{1}{2} & 8 \\ 0 & 1 & \frac{2}{3} & -2 \\ 0 & 0 & 1 & -6 \end{array}\right]$$

$$\begin{array}{r} \frac{1}{2}R_3 + R_1 \rightarrow \\ -\frac{2}{3}R_3 + R_2 \rightarrow \\ \\ \end{array} \left[\begin{array}{ccc:c} 1 & 0 & 0 & 5 \\ 0 & 1 & 0 & 2 \\ 0 & 0 & 1 & -6 \end{array}\right]$$

$$X = \begin{bmatrix} 5 \\ 2 \\ -6 \end{bmatrix}$$

71. $AB = \begin{bmatrix} -4 & -1 \\ 7 & 2 \end{bmatrix}\begin{bmatrix} -2 & -1 \\ 7 & 4 \end{bmatrix}$

$= \begin{bmatrix} -4(-2) + (-1)(7) & -4(-1) + (-1)(4) \\ 7(-2) + 2(7) & 7(-1) + 2(4) \end{bmatrix}$

$= \begin{bmatrix} 1 & 0 \\ 0 & 1 \end{bmatrix} = I$

$BA = \begin{bmatrix} -2 & -1 \\ 7 & 4 \end{bmatrix}\begin{bmatrix} -4 & -1 \\ 7 & 2 \end{bmatrix}$

$= \begin{bmatrix} -2(-4) + (-1)(7) & -2(-1) + (-1)(2) \\ 7(-4) + 4(7) & 7(-1) + 4(2) \end{bmatrix}$

$= \begin{bmatrix} 1 & 0 \\ 0 & 1 \end{bmatrix} = I$

73. $\begin{bmatrix} 1 & 2 & 1 \\ 3 & 6 & 4 \\ 0 & 1 & 3 \end{bmatrix}\begin{bmatrix} -14 & 5 & -2 \\ 9 & -3 & 1 \\ -3 & 1 & 0 \end{bmatrix} = \begin{bmatrix} -14 & 5 & -2 \\ 9 & -3 & 1 \\ -3 & 1 & 0 \end{bmatrix}\begin{bmatrix} 1 & 2 & 1 \\ 3 & 6 & 4 \\ 0 & 1 & 3 \end{bmatrix} = \begin{bmatrix} 1 & 0 & 0 \\ 0 & 1 & 0 \\ 0 & 0 & 1 \end{bmatrix}$

$AB = BA = I$

75. $[A : I] = \left[\begin{array}{ccc:ccc} -1 & 0 & 0 & 1 & 0 & 0 \\ 0 & 2 & 0 & 0 & 1 & 0 \\ 0 & 0 & 4 & 0 & 0 & 1 \end{array}\right] \Rightarrow \begin{array}{r} -1R_1 \\ \frac{1}{2}R_2 \\ \frac{1}{4}R_3 \end{array}\left[\begin{array}{ccc:ccc} 1 & 0 & 0 & -1 & 0 & 0 \\ 0 & 1 & 0 & 0 & \frac{1}{2} & 0 \\ 0 & 0 & 1 & 0 & 0 & \frac{1}{4} \end{array}\right]; A^{-1} = \begin{bmatrix} -1 & 0 & 0 \\ 0 & \frac{1}{2} & 0 \\ 0 & 0 & \frac{1}{4} \end{bmatrix}$

77. $A^{-1} = \dfrac{1}{ad - bc}\begin{bmatrix} d & -b \\ -c & a \end{bmatrix} = \dfrac{1}{5 - 6}\begin{bmatrix} 5 & -3 \\ -2 & 1 \end{bmatrix}$

$= \begin{bmatrix} -5 & 3 \\ 2 & -1 \end{bmatrix}$

$A^{-1} = \begin{bmatrix} -5 & 3 \\ 2 & -1 \end{bmatrix}$

79. $\begin{bmatrix} 1 & 3 \\ 2 & 5 \end{bmatrix}\begin{bmatrix} x \\ y \end{bmatrix} = \begin{bmatrix} 15 \\ 26 \end{bmatrix}$

$\begin{bmatrix} x \\ y \end{bmatrix} = \begin{bmatrix} -5 & 3 \\ 2 & -1 \end{bmatrix}\begin{bmatrix} 15 \\ 26 \end{bmatrix} = \begin{bmatrix} 3 \\ 4 \end{bmatrix}$

Solution: $(3, 4)$

81. $\begin{bmatrix} 3 & 2 & 2 \\ 0 & 2 & 1 \\ 1 & 0 & 1 \end{bmatrix}\begin{bmatrix} x \\ y \\ z \end{bmatrix} = \begin{bmatrix} 13 \\ 4 \\ 5 \end{bmatrix}$

$$\begin{bmatrix} x \\ y \\ z \end{bmatrix} = \frac{1}{4}\begin{bmatrix} 2 & -2 & -2 \\ 1 & 1 & -3 \\ -2 & 2 & 6 \end{bmatrix}\begin{bmatrix} 13 \\ 4 \\ 5 \end{bmatrix} = \begin{bmatrix} 2 \\ \frac{1}{2} \\ 3 \end{bmatrix}$$

Solution: $\left(2, \frac{1}{2}, 3\right)$

83. $\begin{cases} -3x + 10y = 8 \\ 5x - 17y = -13 \end{cases}$

$$A = \begin{bmatrix} -3 & 10 \\ 5 & -17 \end{bmatrix}, X = \begin{bmatrix} x \\ y \end{bmatrix}, B = \begin{bmatrix} 8 \\ -13 \end{bmatrix},$$

$$A^{-1} = \begin{bmatrix} -17 & -10 \\ -5 & -3 \end{bmatrix}$$

$$X = A^{-1}B = \begin{bmatrix} -6 \\ -1 \end{bmatrix}$$

Solution: $(-6, -1)$

85. $\begin{cases} 3x + 2y - z = 6 \\ x - y + 2z = -1 \\ 5x + y + z = 7 \end{cases}$

$$A = \begin{bmatrix} 3 & 2 & -1 \\ 1 & -1 & 2 \\ 5 & 1 & 1 \end{bmatrix}, X = \begin{bmatrix} x \\ y \\ z \end{bmatrix}, B = \begin{bmatrix} 6 \\ -1 \\ 7 \end{bmatrix},$$

$$A^{-1} = \frac{1}{3}\begin{bmatrix} -3 & -3 & 3 \\ 9 & 8 & -7 \\ 6 & 7 & -5 \end{bmatrix}, X = A^{-1}B = \begin{bmatrix} 2 \\ -1 \\ -2 \end{bmatrix}$$

Solution: $(2, -1, -2)$

87. $\begin{cases} 9x + 10y + 14z = 240 \\ x + 3y + 2z = 44 \\ x + y + 2z = 28 \end{cases}$

$$A = \begin{bmatrix} 9 & 10 & 14 \\ 1 & 3 & 2 \\ 1 & 1 & 2 \end{bmatrix}, X = \begin{bmatrix} x \\ y \\ z \end{bmatrix}, B = \begin{bmatrix} 240 \\ 44 \\ 28 \end{bmatrix},$$

$$A^{-1} = \begin{bmatrix} \frac{1}{2} & -\frac{3}{4} & -\frac{11}{4} \\ 0 & \frac{1}{2} & -\frac{1}{2} \\ -\frac{1}{4} & \frac{1}{8} & \frac{17}{8} \end{bmatrix}$$

$$X = A^{-1}B = \begin{bmatrix} \frac{1}{2} & -\frac{3}{4} & -\frac{11}{4} \\ 0 & \frac{1}{2} & -\frac{1}{2} \\ -\frac{1}{4} & \frac{1}{8} & \frac{17}{8} \end{bmatrix}\begin{bmatrix} 240 \\ 44 \\ 28 \end{bmatrix} = \begin{bmatrix} 10 \\ 8 \\ 5 \end{bmatrix}$$

You can produce 10 units of fluid A, 8 units of fluid B, and 5 units of fluid C.

89. (a) $\begin{bmatrix} 6 & 3 \\ 3 & 19 \end{bmatrix}^{-1} = \begin{bmatrix} \frac{19}{105} & -\frac{1}{35} \\ -\frac{1}{35} & \frac{2}{35} \end{bmatrix}$

(b) $\begin{bmatrix} \frac{19}{105} & -\frac{1}{35} \\ -\frac{1}{35} & \frac{2}{35} \end{bmatrix}\begin{bmatrix} 2691 \\ 1188 \end{bmatrix} = \begin{bmatrix} \left(\frac{19}{105}\right)(2691) + \left(-\frac{1}{35}\right)(1188) \\ \left(-\frac{1}{35}\right)(2691) + \left(\frac{2}{35}\right)(1188) \end{bmatrix}$

$$= \begin{bmatrix} 453 \\ -9 \end{bmatrix}$$

$b = 453, a = -9$

The least squares regression line is $y = -9t + 453$.

(c) For 2007, $t = 5$:

$y = -9(5) + 453 = 408$

The amount of lead produced in 2007 is 408 thousand metric tons.

Chapter Test Solutions

1. $5x - 7y = -18 \Rightarrow y = \dfrac{5x + 18}{7}$

$$4x + 3y = 20 \Rightarrow 4x + 3\left(\frac{5x + 18}{7}\right) = 20$$

$$4x + \frac{15}{7}x + \frac{54}{7} = 20$$

$$\frac{43}{7}x = \frac{86}{7}$$

$$x = 2$$

$$y = \frac{5(2) + 18}{7} = 4$$

Solution: (2, 4)

2. $x + y = 3 \Rightarrow y = 3 - x$

$$x^2 + y = 9 \Rightarrow x^2 + (3 - x) = 9$$

$$x^2 - x - 6 = 0$$

$$(x - 3)(x + 2) = 0$$

$x = 3$ or $x = -2$

$y = 0$ $\quad y = 5$

Solutions: (3, 0), (–2, 5)

3. $3x - 2y = 6 \Rightarrow y = \frac{3}{2}x - 3$

$$2x^2 + 2y = 8 \Rightarrow 2x^2 + 2\left(\frac{3}{2}x - 3\right) = 8$$
$$2x^2 + 3x - 14 = 0$$
$$(2x + 7)(x - 2) = 0$$
$$x = -\frac{7}{2}, 2$$

$x = -\frac{7}{2} \Rightarrow y = -\frac{33}{4}$

$x = 2 \Rightarrow y = 0$

Solutions: $\left(-\frac{7}{2}, -\frac{33}{4}\right)$, $(2, 0)$

4. $\begin{cases} 1.5x - 2y = 8 \\ 2.5x + 2y = 5.75 \end{cases}$

$$4x = 13.75$$
$$x = 3.4375$$

$1.5(3.4375) - 2y = 8 \Rightarrow y = -1.421875$

Solution: $(3.4375, -1.421875)$

5. $\begin{cases} x + 2y + 3z = 9 \\ 2x - 4y + z = 11 \\ 3y + 5z = 12 \end{cases}$ Interchange Eq.1 and Eq.2.

$\begin{cases} x + 2y + 3z = 9 \\ -8y - 5z = -7 \\ 3y + 5z = 12 \end{cases}$ $-$Eq.1 + Eq.2

$\begin{cases} x + 2y + 3z = 9 \\ -8y - 5z = -7 \\ -5y = 5 \end{cases}$ Eq.2 + Eq.3

$y = -1$

$-8(-1) - 5z = -7 \Rightarrow z = 3$

$x + 2(-1) + 3(3) = 9 \Rightarrow x = 2$

Solution: $(2, -1, 3)$

6. $\begin{cases} 3x - 2y + z = 16 \\ 2x - y - z = 3 \\ 5x - z = 6 \end{cases}$ Interchange Eq.2 and Eq.3.

$\begin{cases} -x + 3z = 10 \\ 2x - y - z = 3 \\ 5x - z = 6 \end{cases}$ Eq.1 $-$ 2Eq.2

$\begin{cases} -x + 3z = 10 \\ 2x - y - z = 3 \\ \frac{14}{3}x = \frac{28}{3} \end{cases}$ $\frac{1}{3}$Eq.1 + Eq.3

$\begin{cases} -x + 3z = 10 \\ 2x - y - z = 3 \\ 14x = 28 \end{cases}$ 3Eq.3

$x = 2$

$-(2) + 3z = 10 \Rightarrow z = 4$

$2(2) - y - (4) = 3 \Rightarrow y = -3$

Solution: $(2, -3, 4)$

7. $\begin{cases} 2x + y + 4z = 2 \\ x + 4y - z = 0 \\ -x + 3y + 3z = -1 \end{cases} \Rightarrow \left[\begin{array}{ccc|c} 2 & 1 & 4 & 2 \\ 1 & 4 & -1 & 0 \\ -1 & 3 & 3 & -1 \end{array}\right]$

8. $\begin{cases} 3x + 4y + 2z = 4 \\ 2x + 3y = -2 \\ 2y - 3z = -13 \end{cases} \Rightarrow \left[\begin{array}{ccc|c} 3 & 4 & 2 & 4 \\ 2 & 3 & 0 & -2 \\ 0 & 2 & -3 & -13 \end{array}\right]$

9. $\left[\begin{array}{ccc|c} 1 & 2 & 3 & 16 \\ 5 & 4 & -1 & 22 \end{array}\right]$

$-5R_1 + R_2 \left[\begin{array}{ccc|c} 1 & 2 & 3 & 16 \\ 0 & -6 & -16 & -58 \end{array}\right]$

$-\frac{1}{6}R_2 \left[\begin{array}{ccc|c} 1 & 2 & 3 & 16 \\ 0 & 1 & \frac{8}{3} & \frac{29}{3} \end{array}\right]$

$-2R_2 + R_1 \left[\begin{array}{ccc|c} 1 & 0 & -\frac{7}{3} & -\frac{10}{3} \\ 0 & 1 & \frac{8}{3} & \frac{29}{3} \end{array}\right]$

Dependent System

Solution: $\left(\frac{7}{3}a - \frac{10}{3}, -\frac{8}{3}a + \frac{29}{3}, a\right)$ a is any real number.

10. $\begin{bmatrix} 1 & -2 & 1 & \vdots & 14 \\ 0 & 1 & -3 & \vdots & 2 \\ 0 & 0 & 1 & \vdots & -6 \end{bmatrix}$

$\begin{matrix} 2R_2 + R_1 \\ \\ \\ \end{matrix} \begin{bmatrix} 1 & 0 & -5 & \vdots & 18 \\ 0 & 1 & -3 & \vdots & 2 \\ 0 & 0 & 1 & \vdots & -6 \end{bmatrix}$

$\begin{matrix} 5R_3 + R_1 \\ 3R_3 + R_2 \\ \\ \end{matrix} \begin{bmatrix} 1 & 0 & 0 & \vdots & -12 \\ 0 & 1 & 0 & \vdots & -16 \\ 0 & 0 & 1 & \vdots & -6 \end{bmatrix}$

$x = -12,\ y = -16,\ z = -6$

Solution: $(-12, -16, -6)$

11. $\begin{bmatrix} 2 & -3 & 1 & \vdots & 14 \\ 1 & 2 & 0 & \vdots & -4 \\ 0 & 1 & -1 & \vdots & -4 \end{bmatrix}$

$\begin{matrix} -R_2 + R_1 \\ \\ \\ \end{matrix} \begin{bmatrix} 1 & -5 & 1 & \vdots & 18 \\ 1 & 2 & 0 & \vdots & -4 \\ 0 & 1 & -1 & \vdots & -4 \end{bmatrix}$

$\begin{matrix} \\ -R_1 + R_2 \\ \\ \end{matrix} \begin{bmatrix} 1 & -5 & 1 & \vdots & 18 \\ 0 & 7 & -1 & \vdots & -22 \\ 0 & 1 & -1 & \vdots & -4 \end{bmatrix}$

$\begin{matrix} \\ R_3 \leftrightarrow R_2 \\ R_2 \leftrightarrow R_3 \end{matrix} \begin{bmatrix} 1 & -5 & 1 & \vdots & 18 \\ 0 & 1 & -1 & \vdots & -4 \\ 0 & 7 & -1 & \vdots & -22 \end{bmatrix}$

$\begin{matrix} 5R_2 + R_1 \\ \\ -7R_2 + R_3 \end{matrix} \begin{bmatrix} 1 & 0 & -4 & \vdots & -2 \\ 0 & 1 & -1 & \vdots & -4 \\ 0 & 0 & 6 & \vdots & 6 \end{bmatrix}$

$\begin{matrix} \frac{2}{3}R_3 + R_1 \\ \frac{1}{6}R_3 + R_2 \\ \frac{1}{6}R_3 \end{matrix} \begin{bmatrix} 1 & 0 & 0 & \vdots & 2 \\ 0 & 1 & 0 & \vdots & -3 \\ 0 & 0 & 1 & \vdots & 1 \end{bmatrix}$

$x = 2,\ y = -3,\ z = 1$

Solution: $(2, -3, 1)$

12. $2A + C = 2\begin{bmatrix} 1 & 3 \\ 2 & 4 \end{bmatrix} + \begin{bmatrix} 0 & -2 \\ 3 & 5 \end{bmatrix} = \begin{bmatrix} 2 & 6 \\ 4 & 8 \end{bmatrix} + \begin{bmatrix} 0 & -2 \\ 3 & 5 \end{bmatrix}$

$= \begin{bmatrix} 2 & 4 \\ 7 & 13 \end{bmatrix}$

13. $CA = \begin{bmatrix} 0 & -2 \\ 3 & 5 \end{bmatrix}\begin{bmatrix} 1 & 3 \\ 2 & 4 \end{bmatrix} = \begin{bmatrix} -4 & -8 \\ 13 & 29 \end{bmatrix}$

14. $BD = \begin{bmatrix} 2 & -1 & 3 \\ 4 & 0 & 1 \end{bmatrix}\begin{bmatrix} 3 \\ 2 \\ -1 \end{bmatrix} = \begin{bmatrix} 1 \\ 11 \end{bmatrix}$

15. $A^2 = \begin{bmatrix} 1 & 3 \\ 2 & 4 \end{bmatrix}\begin{bmatrix} 1 & 3 \\ 2 & 4 \end{bmatrix} = \begin{bmatrix} 7 & 15 \\ 10 & 22 \end{bmatrix}$

16. $A^{-1} = \dfrac{1}{2(4) - (-1)(-3)}\begin{bmatrix} 4 & -(-1) \\ -(-3) & 2 \end{bmatrix} = \dfrac{1}{5}\begin{bmatrix} 4 & 1 \\ 3 & 2 \end{bmatrix}$

17. Because $A = I$, $A^{-1} = A = \begin{bmatrix} 1 & 0 \\ 0 & 1 \end{bmatrix}$.

18. $[A \,\vdots\, I] = \begin{bmatrix} 3 & 4 & 2 & \vdots & 1 & 0 & 0 \\ 2 & 3 & 0 & \vdots & 0 & 1 & 0 \\ 0 & 2 & -3 & \vdots & 0 & 0 & 1 \end{bmatrix} \Rightarrow \begin{matrix} R_1 - R_2 \\ \\ \\ \end{matrix} \begin{bmatrix} 1 & 1 & 2 & \vdots & 1 & -1 & 0 \\ 2 & 3 & 0 & \vdots & 0 & 1 & 0 \\ 0 & 2 & -3 & \vdots & 0 & 0 & 1 \end{bmatrix} \Rightarrow \begin{matrix} \\ -2R_1 + R_2 \\ \\ \end{matrix} \begin{bmatrix} 1 & 1 & 2 & \vdots & 1 & -1 & 0 \\ 0 & 1 & -4 & \vdots & -2 & 3 & 0 \\ 0 & 2 & -3 & \vdots & 0 & 0 & 1 \end{bmatrix}$

$\Rightarrow \begin{matrix} \\ \\ -2R_2 + R_3 \end{matrix} \begin{bmatrix} 1 & 1 & 2 & \vdots & 1 & -1 & 0 \\ 0 & 1 & -4 & \vdots & -2 & 3 & 0 \\ 0 & 0 & 5 & \vdots & 4 & -6 & 1 \end{bmatrix} \Rightarrow \begin{matrix} R_1 - R_2 \\ \frac{4}{5}R_3 + R_2 \\ \frac{1}{5}R_3 \end{matrix} \begin{bmatrix} 1 & 0 & 6 & \vdots & 3 & -4 & 0 \\ 0 & 1 & 0 & \vdots & \frac{6}{5} & -\frac{9}{5} & \frac{4}{5} \\ 0 & 0 & 1 & \vdots & \frac{4}{5} & -\frac{6}{5} & \frac{1}{5} \end{bmatrix}$

$\Rightarrow \begin{matrix} -6R_3 + R_1 \\ \\ \\ \end{matrix} \begin{bmatrix} 1 & 0 & 0 & \vdots & -\frac{9}{5} & \frac{16}{5} & -\frac{6}{5} \\ 0 & 1 & 0 & \vdots & \frac{6}{5} & -\frac{9}{5} & \frac{4}{5} \\ 0 & 0 & 1 & \vdots & \frac{4}{5} & -\frac{6}{5} & \frac{1}{5} \end{bmatrix}$

$$A^{-1} \begin{bmatrix} -\frac{9}{5} & \frac{16}{5} & -\frac{6}{5} \\ \frac{6}{5} & -\frac{9}{5} & \frac{4}{5} \\ \frac{4}{5} & -\frac{6}{5} & \frac{1}{5} \end{bmatrix} = \frac{1}{5}\begin{bmatrix} -9 & 16 & -6 \\ 6 & -9 & 4 \\ 4 & -6 & 1 \end{bmatrix}$$

19. Let x = the number of liters at 60%.

Let y = the number of liters at 20%.

$$\begin{cases} x + y = 100 & \text{Equation 1} \\ 0.6x + 0.2y = 0.5(100) & \text{Equation 2} \end{cases}$$

Multiply Equation 1 by −6 and multiply Equation 2 by 10:

$$\begin{cases} -6x - 6y = -600 \\ \underline{6x + 2y = 500} \end{cases}$$

$-4y = -100$ Add equations

$y = 25$

$x + 25 = 100 \Rightarrow x = 75$

In order to obtain the desired mixture, 75 liters of the 60% solution and 25 liters of the 20% solution are required.

20. B = total votes cast for Bush

K = total votes cast for Kerry

N = Total votes cast for Nader

$$\begin{cases} B + K + N = 120.884 \text{ million} \\ B - K = 2.978 \text{ million} \\ N = 0.00096(120.884) \text{ million} \end{cases}$$

So, $N \approx 0.11604864$ million or 116,049

$$\begin{cases} B + K + 116{,}049 = 120{,}884{,}000 \\ B - K = 2{,}978{,}000 \end{cases}$$

$$\begin{cases} B + K = 120{,}767{,}951 \\ \underline{B - K = 2{,}978{,}000} \end{cases}$$

$2B = 123{,}745{,}951$ Add equations

$B = 61{,}872{,}975.5$

$61{,}872{,}976 + K = 120{,}767{,}951 \Rightarrow K = 58{,}894{,}975$

Bush: 61,872,976 votes

Kerry: 58,894,975 votes

Nader: 116,049 votes

Practice Test for Chapter 8

1. Use the substitution method to solve $\begin{cases} x + y = 1 \\ 3x - y = 15 \end{cases}$.

2. Use the substitution method to solve $\begin{cases} x - 3y = -3 \\ x^2 + 6y = 5 \end{cases}$.

3. Use the substitution method to solve $\begin{cases} x + y + z = 6 \\ 2x - y + 3z = 0 \\ 5x + 2y - z = -3 \end{cases}$.

4. Find the two numbers whose sum is 110 and product is 2800.

5. Find the dimensions of a rectangle if its perimeter is 170 feet and its area is 1500 square feet.

6. Use the elimination method to solve $\begin{cases} 2x + 15y = 4 \\ x - 3y = 23 \end{cases}$.

7. Use the elimination method to solve $\begin{cases} x + y = 2 \\ 38x - 19y = 7 \end{cases}$.

8. Use the elimination method to solve $\begin{cases} 0.4x + 0.5y = 0.112 \\ 0.3x - 0.7y = -0.131 \end{cases}$.

9. Herbert invests \$17,000 in two funds that pay 11% and 13% simple interest, respectively. If he receives \$2080 in yearly interest, how much is invested in each fund?

10. Find the least squares regression line for the points $(4, 3)$, $(1, 1)$, $(-1, -2)$, and $(-2, -1)$.

11. Solve $\begin{cases} x + y = -2 \\ 2x - y + z = 11 \\ 4y - 3z = -20 \end{cases}$.

12. Solve $\begin{cases} 3x + 2y - z = 5 \\ 6x - y + 5z = 2 \end{cases}$.

13. Find the equation of the parabola $y = ax^2 + bx + c$ passing through the points $(0, -1)$, $(1, 4)$ and $(2, 13)$.

14. Put $\begin{bmatrix} 1 & -2 & 4 \\ 3 & -5 & 9 \end{bmatrix}$ in reduced row-echelon form.

15. Use a matrix to solve $\begin{cases} 3x + 5y = 3 \\ 2x - y = -11 \end{cases}$.

16. Use a matrix to solve $\begin{cases} 2x + 3y = -3 \\ 3x + 2y = 8 \\ x + y = 1 \end{cases}$.

17. Use a matrix to solve $\begin{cases} x + 3z = -5 \\ 2x + y = 0. \\ 3x + y - z = 3 \end{cases}$

18. Multiply $\begin{bmatrix} 1 & 4 & 5 \\ 2 & 0 & -3 \end{bmatrix}\begin{bmatrix} 1 & 6 \\ 0 & -7 \\ -1 & 2 \end{bmatrix}$.

19. Given $A = \begin{bmatrix} 9 & 1 \\ -4 & 8 \end{bmatrix}$ and $B = \begin{bmatrix} 6 & -2 \\ 3 & 5 \end{bmatrix}$, find $3A - 5B$.

20. Find $f(A)$ when $f(x) = x^2 - 7x + 8$ and $A = \begin{bmatrix} 3 & 0 \\ 7 & 1 \end{bmatrix}$.

21. True or false: $(A + B)(A + 3B) = A^2 + 4AB + 3B^2$ where A and B are matrices. (Assume that A^2, AB, and B^2 exist.)

22. Find the inverse, if it exists, of $\begin{bmatrix} 1 & 2 \\ 3 & 5 \end{bmatrix}$.

23. Find the inverse, if it exists, of $\begin{bmatrix} 1 & 1 & 1 \\ 3 & 6 & 5 \\ 6 & 10 & 8 \end{bmatrix}$.

24. Use an inverse matrix to solve each system.

(a) $x + 2y = 4$
$3x + 5y = 1$

(b) $x + 2y = 3$
$3x + 5y = -2$

CHAPTER 9
Functions of Several Variables

CHAPTER 9
Functions of Several Variables

Section 9.1 The Three-Dimensional Coordinate System

Skills Review

1. $(5, 1), (3, 5)$

$$d = \sqrt{(3-5)^2 + (5-1)^2} = \sqrt{4+16} = \sqrt{20} = 2\sqrt{5}$$

2. $(2, 3), (-1, -1)$

$$d = \sqrt{(-1-2)^2 + (-1-3)^2} = \sqrt{9+16} = \sqrt{25} = 5$$

3. $(-5, 4), (-5, -4)$

$$d = \sqrt{(-5-(-5))^2 + (-4-4)^2} = \sqrt{64} = 8$$

4. $(-3, 6), (-3, -2)$

$$d = \sqrt{(-3-(-3))^2 + (-2-6)^2} = \sqrt{64} = 8$$

5. $(2, 5), (6, 9)$

$$\text{Midpoint} = \left(\frac{2+6}{2}, \frac{5+9}{2}\right) = (4, 7)$$

6. $(-1, -2), (3, 2)$

$$\text{Midpoint} = \left(\frac{-1+3}{2}, \frac{-2+2}{2}\right) = (1, 0)$$

7. $(-6, 0), (6, 6)$

$$\text{Midpoint} = \left(\frac{-6+6}{2}, \frac{0+6}{2}\right) = (0, 3)$$

8. $(-4, 3), (2, -1)$

$$\text{Midpoint} = \left(\frac{-4+2}{2}, \frac{3+(-1)}{2}\right) = (-1, 1)$$

9. $c: (2, 3),\ r = 2$

$$(x-2)^2 + (y-3)^2 = 2^2$$
$$(x-2)^2 + (y-3)^2 = 4$$

10. $C = \left(\frac{4+(-2)}{2}, \frac{0+8}{2}\right) = (1, 4)$

$$r = \frac{1}{2}\sqrt{(-2-4)^2 + (8-0)^2} = \frac{1}{2}\sqrt{36+64} = \frac{1}{2}\sqrt{100} = 5$$

$$(x-1)^2 + (y-4)^2 = 5^2$$
$$(x-1)^2 + (y-4)^2 = 25$$

1. (a) and (b)

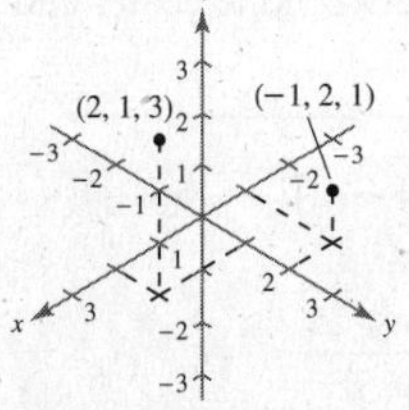

3. (a) and (b)

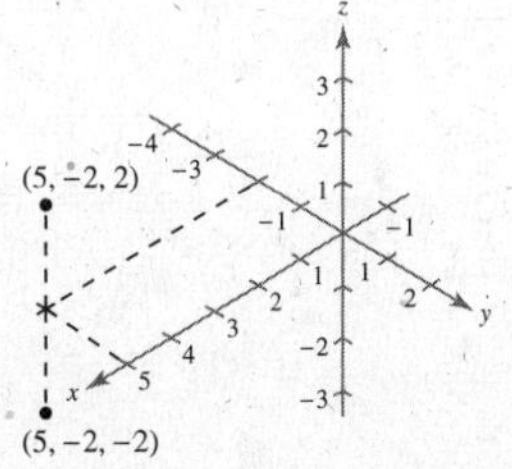

5. $A(2, 3, 4)$

$B(-1, -2, 2)$

7. $x = -3,\ y = 4,\ z = 5: (-3, 4, 5)$

9. $y = z = 0,\ x = 10: (10, 0, 0)$

11. The z-coordinate is 0.

13. $(4, 1, 5), (8, 2, 6)$

$$\begin{aligned} d &= \sqrt{(8-4)^2 + (2-1)^2 + (6-5)^2} \\ &= \sqrt{18} \\ &= 3\sqrt{2} \end{aligned}$$

15. $(-1, -5, 7), (-3, 4, -4)$

$$\begin{aligned} d &= \sqrt{(-3+1)^2 + (4+5)^2 + (-4-7)^2} \\ &= \sqrt{206} \end{aligned}$$

17. $(6, -9, 1), (-2, -1, 5)$

$$\begin{aligned} \text{Midpoint} &= \left(\frac{6+(-2)}{2}, \frac{-9+(-1)}{2}, \frac{1+5}{2}\right) \\ &= (2, -5, 3) \end{aligned}$$

19. $(-5, -2, 5), (6, 3, -7)$

$$\begin{aligned} \text{Midpoint} &= \left(\frac{-5+6}{2}, \frac{-2+3}{2}, \frac{5+(-7)}{2}\right) \\ &= \left(\frac{1}{2}, \frac{1}{2}, -1\right) \end{aligned}$$

21. $(2, -1, 3) = \left(\frac{x+(-2)}{2}, \frac{y+1}{2}, \frac{z+1}{2}\right)$

$$\begin{array}{lll} 2 = \dfrac{x-2}{2} & -1 = \dfrac{y+1}{2} & 3 = \dfrac{z+1}{2} \\ 4 = x - 2 & -2 = y + 1 & 6 = z + 1 \\ x = 6 & y = -3 & z = 5 \end{array}$$

$(x, y, z) = (6, -3, 5)$

23. $\left(\frac{3}{2}, 1, 2\right) = \left(\frac{x+2}{2}, \frac{y+0}{2}, \frac{z+3}{2}\right)$

$$\begin{array}{lll} \dfrac{3}{2} = \dfrac{x+2}{2} & 1 = \dfrac{y}{2} & 2 = \dfrac{z+3}{2} \\ 3 = x + 2 & & 4 = z + 3 \\ x = 1 & y = 2 & z = 1 \end{array}$$

$(x, y, z) = (1, 2, 1)$

25. Let $A = (0, 0, 0)$, $B = (2, 2, 1)$, and $C = (2, -4, 4)$.

Then you have

$$\begin{aligned} d(AB) &= \sqrt{(2-0)^2 + (2-0)^2 + (1-0)^2} = 3 \\ d(AC) &= \sqrt{(2-0)^2 + (-4-0)^2 + (4-0)^2} = 6 \\ d(BC) &= \sqrt{(2-2)^2 + (-4-2)^2 + (4-1)^2} = 3\sqrt{5}. \end{aligned}$$

The triangle is a right triangle because

$$\begin{aligned} d^2(AB) + d^2(AC) &= (3)^2 + (6)^2 \\ &= 45 \\ &= \left(3\sqrt{5}\right)^2 = d^2(BC). \end{aligned}$$

27. Let $A = (-2, 2, 4)$, $B = (-2, 2, 6)$, and $C = (-2, 4, 8)$. Then you have

$$\begin{aligned} d(AB) &= \sqrt{[-2-(-2)]^2 + (2-2)^2 + (6-4)^2} \\ &= 2 \\ d(AC) &= \sqrt{[-2-(-2)]^2 + (4-2)^2 + (8-4)^2} \\ &= 2\sqrt{5} \\ d(BC) &= \sqrt{[-2-(-2)]^2 + (4-2)^2 + (8-6)^2} \\ &= 2\sqrt{2}. \end{aligned}$$

The triangle is not a right triangle because

$$\begin{aligned} d^2(AB) + d^2(BC) &= (2)^2 + \left(2\sqrt{2}\right)^2 \\ &= 12 \\ &\neq \left(2\sqrt{5}\right)^2 = 8 = d^2(AC). \end{aligned}$$

The triangle is neither right nor isosceles.

29. Each z-coordinate is increased by 5 units:

$(0, 0, 5), (2, 2, 6), (2, -4, 9)$

31. $x^2 + (y-2)^2 + (z-2)^2 = 4$

33. The midpoint of the diameter is the center.

$$\text{Center} = \left(\frac{2+1}{2}, \frac{1+3}{2}, \frac{3+(-1)}{2}\right) = \left(\frac{3}{2}, 2, 1\right)$$

The radius is the distance between the center and either endpoint.

$$\begin{aligned} \text{Radius} &= \sqrt{\left(2 - \frac{3}{2}\right)^2 + (1-2)^2 + (3-1)^2} \\ &= \sqrt{\frac{1}{4} + 1 + 4} = \frac{\sqrt{21}}{2} \end{aligned}$$

$$\left(x - \frac{3}{2}\right)^2 + (y-2)^2 + (z-1)^2 = \frac{21}{4}$$

35. $(x-1)^2 + (y-1)^2 + (z-5)^2 = 9$

37. The midpoint of the diameter is the center.

$$\text{Center} = \left(\frac{2+0}{2}, \frac{0+6}{2}, \frac{0+0}{2}\right) = (1, 3, 0)$$

The radius is the distance from the center to either endpoint.

$$\text{Radius} = \sqrt{(1-2)^2 + (3-0)^2 + (0-0)^2} = \sqrt{10}$$

$$(x-1)^2 + (y-3)^2 + z^2 = 10$$

39. The distance from $(-2, 1, 1)$ to the xy-coordinate plane is the radius $r = 1$.

$$(x+2)^2 + (y-1)^2 + (z-1)^2 = 1$$

41.

$$\left(x^2 - 5x + \tfrac{25}{4}\right) + y^2 + z^2 = \tfrac{25}{4}$$

$$x^2 + y^2 + z^2 - 5x = 0$$

$$\left(x - \tfrac{5}{2}\right)^2 + (y-0)^2 + (z-0)^2 = \tfrac{25}{4}$$

Center: $\left(\frac{5}{2}, 0, 0\right)$

Radius: $\frac{5}{2}$

43.

$$x^2 + y^2 + z^2 - 2x + 6y + 8z + 1 = 0$$

$$\left(x^2 - 2x + 1\right) + \left(y^2 + 6y + 9\right) + \left(z^2 + 8z + 16\right) = -1 + 1 + 9 + 16$$

$$(x-1)^2 + (y+3)^2 + (z+4)^2 = 25$$

Center: $(1, -3, -4)$

Radius: 5

45.

$$2x^2 + 2y^2 + 2z^2 - 4x - 12y - 8z + 3 = 0$$

$$\left(x^2 - 2x + 1\right) + \left(y^2 - 6y + 9\right) + \left(z^2 - 4z + 4\right) = -\frac{3}{2} + 1 + 9 + 4$$

$$(x-1)^2 + (y-3)^2 + (z-2)^2 = \frac{25}{2}$$

Center: $(1, 3, 2)$

Radius: $\dfrac{5}{\sqrt{2}} = \dfrac{5\sqrt{2}}{2}$

47. $(x-1)^2 + (y-3)^2 + (z-2)^2 = 25$

To find the xy-trace, let $z = 0$.

$$(x-1)^2 + (y-3)^2 + (0-2)^2 = 25$$

$$(x-1)^2 + (y-3)^2 = 21$$

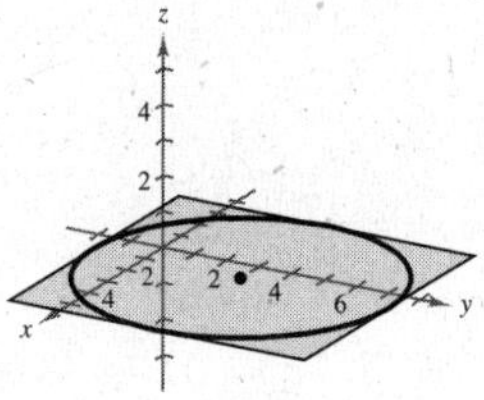

49. $x^2 + y^2 + z^2 - 6x - 10y + 6z + 30 = 0$

To find the xy-trace, let $z = 0$.

$$x^2 + y^2 + (0)^2 - 6x - 10y + 6(0) + 30 = 0$$

$$\left(x^2 - 6x + 9\right) + \left(y^2 - 10y + 25\right) = -30 + 9 + 25$$

$$(x-3)^2 + (y-5)^2 = 4$$

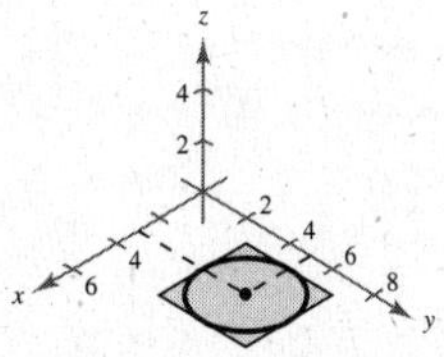

51. $x^2 + (y + 3)^2 + z^2 = 25$

To find the *yz*-trace, let $x = 0$.

$0^2 + (y + 3)^2 + z^2 = 25$

$(y + 3)^2 + z^2 = 25$

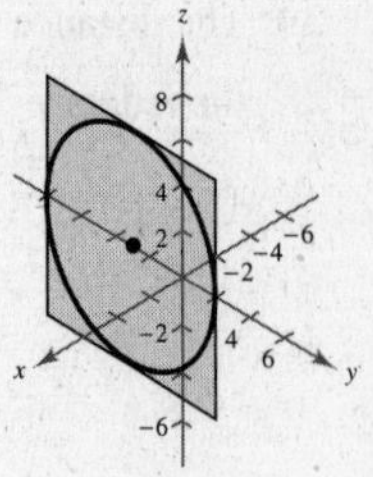

53. $x^2 + y^2 + z^2 - 4x - 4y - 6z - 12 = 0$

To find the *yz*-trace, let $x = 0$.

$(0)^2 + y^2 + z^2 - 4(0) - 4y - 6z - 12 = 0$

$(y^2 - 4y + 4) + (z^2 - 6z + 9) = 12 + 4 + 9$

$(y - 2)^2 + (z - 3)^2 = 25$

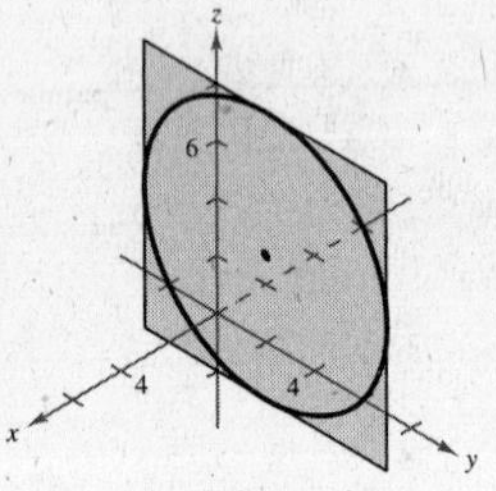

55. $x^2 + y^2 + z^2 = 25$

(a) To find the trace, let $z = 3$.

$x^2 + y^2 + 3^2 = 25$

$x^2 + y^2 = 16$

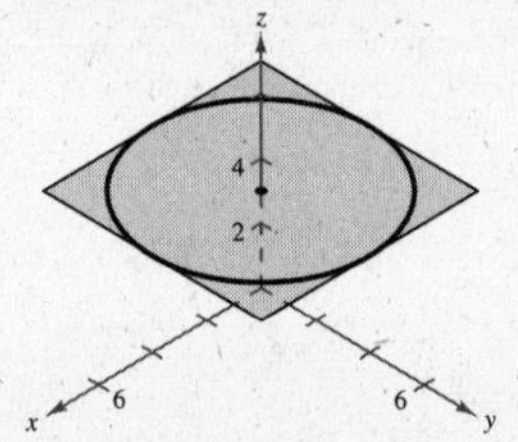

(b) To find the trace, let $x = 4$.

$4^2 + y^2 + z^2 = 25$

$y^2 + z^2 = 9$

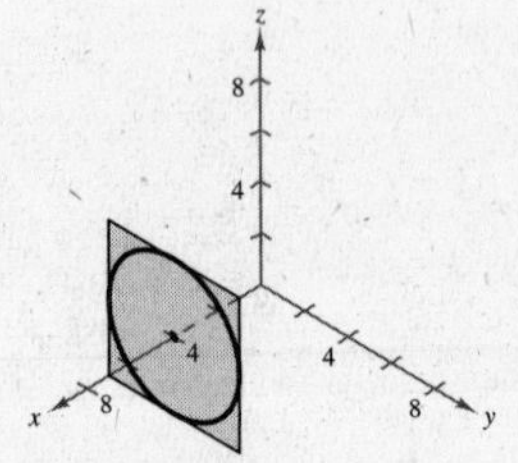

57. $x^2 + y^2 + z^2 - 4x - 6y + 9 = 0$

(a) To find the trace, let $x = 2$.

$2^2 + y^2 + z^2 - 4(2) - 6y + 9 = 0$

$(y^2 - 6y + 9) + z^2 = -9 - 4 + 8 + 9$

$(y - 3)^2 + z^2 = 2^2$

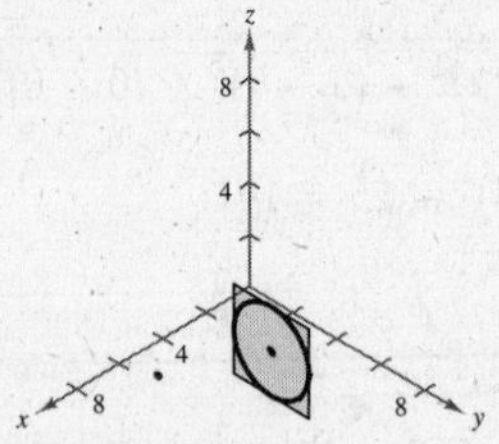

(b) To find the trace, let $y = 3$.

$x^2 + 3^2 + z^2 - 4x - 6(3) + 9 = 0$

$(x^2 - 4x + 4) + z^2 = -9 + 18 - 9 + 4$

$(x - 2)^2 + z^2 = 2^2$

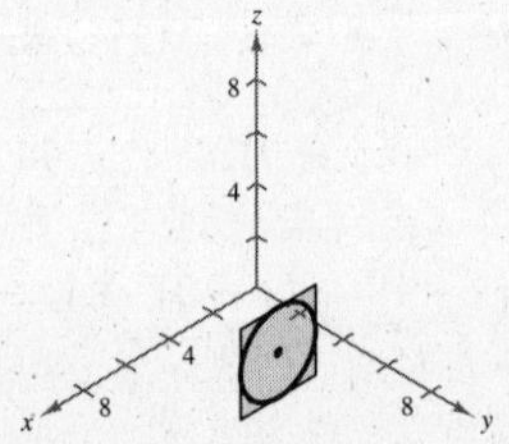

59. Because the crystal is a cube, $A = (3, 3, 0)$. So,

$(x, y, z) = (3, 3, 3)$.

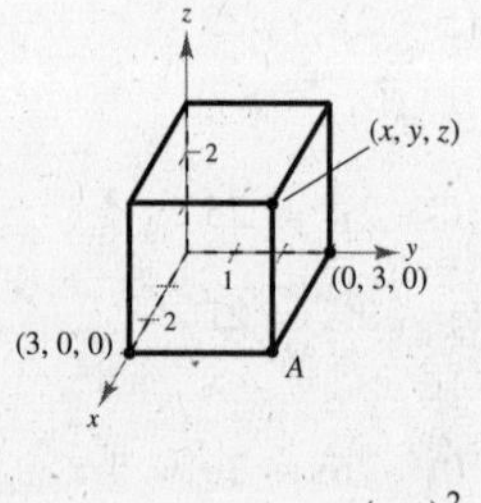

61. $x^2 + y^2 + z^2 = \left(\frac{165}{2}\right)^2$

$x^2 + y^2 + z^2 = 6806.25$

Section 9.2 Surfaces in Space

Skills Review

1. $3x + 4y = 12$

Let $x = 0$ to find the y-intercept.

$$3(0) + 4y = 12$$
$$y = 3$$

y-intercept: $(0, 3)$

Let $y = 0$ to find the x-intercept.

$$3x + 4(0) = 12$$
$$x = 4$$

x-intercept: $(4, 0)$

2. $6x + y = -8$

Let $x = 0$ to find the y-intercept.

$$6(0) + y = -8$$
$$y = -8$$

y-intercept: $(0, -8)$

Let $y = 0$ to find the x-intercept.

$$6x + 0 = -8$$
$$x = -\frac{4}{3}$$

x-intercept: $\left(-\frac{4}{3}, 0\right)$

3. $-2x + y = -2$

Let $x = 0$ to find the y-intercept.

$$-2(0) + y = -2$$
$$y = -2$$

y-intercept: $(0, -2)$

Let $y = 0$ to find the x-intercept.

$$-2x + 0 = -2$$
$$x = 1$$

x-intercept: $(1, 0)$

4. $-x - y = 5$

Let $x = 0$ to find the y-intercept.

$$-0 - y = 5$$
$$y = -5$$

y-intercept: $(0, -5)$

Let $y = 0$ to find the x-intercept.

$$-x - 0 = 5$$
$$x = -5$$

x-intercept: $(-5, 0)$

5.
$$x^2 + y^2 + z^2 - 2x - 4y - 6z + 15 = 0$$
$$(x^2 - 2x + 1) + (y^2 - 4y + 4) + (z^2 - 6z + 9) = -15 + 1 + 4 + 9$$
$$(x - 1)^2 + (y - 2)^2 + (z - 3)^2 = -1$$

6.
$$x^2 + y^2 - z^2 - 8x + 4y - 6z + 11 = 0$$
$$(x^2 - 8x + 16) + (y^2 + 4y + 4) - (z^2 + 6z + 9) = -11 + 16 + 4 - 9$$
$$(x - 4)^2 + (y + 2)^2 - (z + 3)^2 = 0$$

7.
$$z - 2 = x^2 + y^2 + 2x - 2y$$
$$z - 2 + 1 + 1 = (x^2 + 2x + 1) + (y^2 - 2y + 1)$$
$$z = (x + 1)^2 + (y - 1)^2$$

8.
$$x^2 + y^2 + z^2 - 6x + 10y + 26z = -202$$
$$(x^2 - 6x + 9) + (y^2 + 10y + 25) + (z^2 + 26z + 169) = -202 + 9 + 25 + 169$$
$$(x - 3)^2 + (y + 5)^2 + (z + 13)^2 = 1$$

Skills Review —*continued*—

9. $16x^2 + 16y^2 + 16z^2 = 4$

$x^2 + y^2 + z^2 = \frac{1}{4}$

10. $9x^2 + 9y^2 + 9z^2 = 36$

$x^2 + y^2 + z^2 = 4$

1. $4x + 2y + 6z = 12$

To find the x-intercept, let $y = 0$ and $z = 0$.

$4x = 12 \Rightarrow x = 3$

To find the y-intercept, let $x = 0$ and $z = 0$.

$2y = 12 \Rightarrow y = 6$

To find the z-intercept, let $x = 0$ and $y = 0$.

$6z = 12 \Rightarrow z = 2$

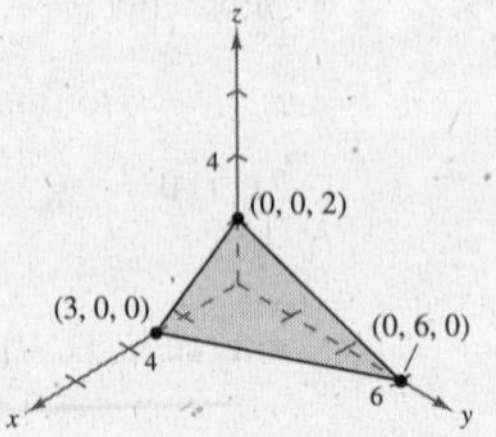

3. $3x + 3y + 5z = 15$

To find the x-intercept, let $y = 0$ and $z = 0$.

$3x = 15 \Rightarrow x = 5$

To find the y-intercept, let $x = 0$ and $z = 0$.

$3y = 15 \Rightarrow y = 5$

To find the z-intercept, let $x = 0$ and $y = 0$.

$5z = 15 \Rightarrow z = 3$

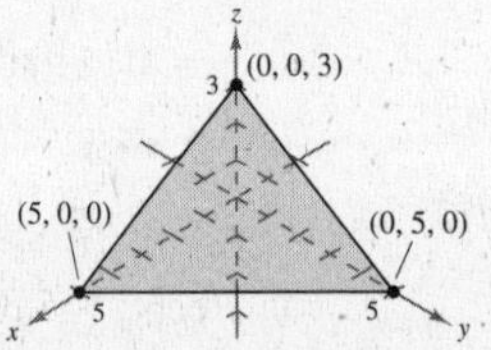

5. $2x - y + 3z = 4$

To find the x-intercept, let $y = 0$ and $z = 0$.

$2x = 4 \Rightarrow x = 2$

To find the y-intercept, let $x = 0$ and $z = 0$.

$-y = 4 \Rightarrow y = -4$

To find the z-intercept, let $x = 0$ and $y = 0$.

$3z = 4 \Rightarrow z = \frac{4}{3}$

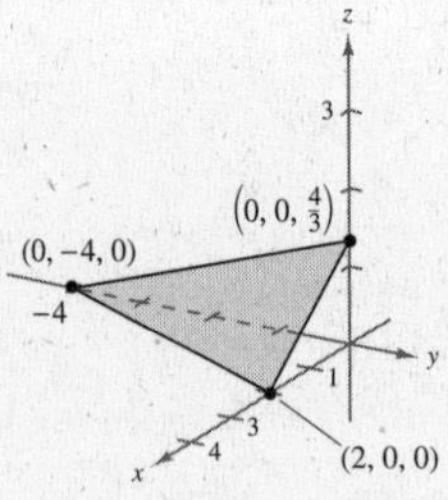

7. $z = 8$

Because the coefficients of x and y are zero, the only intercept is the z-intercept of 8. The plane is parallel to the xy-plane.

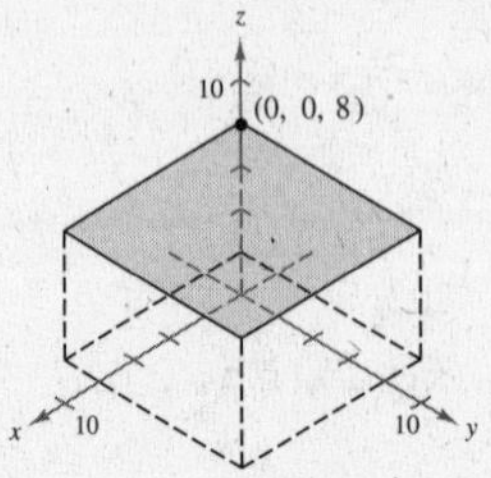

9. $y + z = 5$

Because the coefficient of x is zero, there is no x-intercept.

To find the y-intercept, let $z = 0$.

$y = 5$

To find the z-intercept, let $y = 0$.

$z = 5$

The plane is parallel to the x-axis.

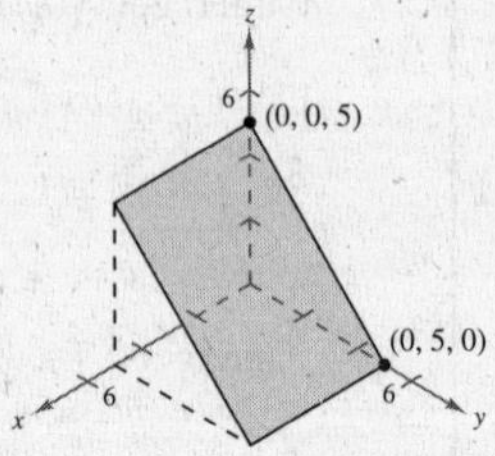

11. $x + y - z = 0$

To find the x-intercept, let $y = 0$ and $z = 0$.

$x = 0$

To find the y-intercept, let $x = 0$ and $z = 0$.

$y = 0$

To find the z-intercept, let $x = 0$ and $y = 0$.

$z = 0$

The only intercept is the origin. The xy-trace is the line $x + y = 0$.

The xz-trace is the line $x - z = 0$.

The yz-trace is the line $y - z = 0$.

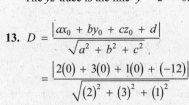

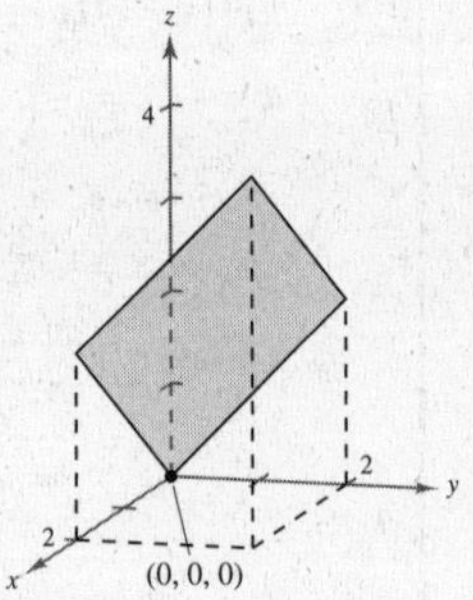

13. $$D = \frac{|ax_0 + by_0 + cz_0 + d|}{\sqrt{a^2 + b^2 + c^2}}$$

$$= \frac{|2(0) + 3(0) + 1(0) + (-12)|}{\sqrt{(2)^2 + (3)^2 + (1)^2}}$$

$$= \frac{12}{\sqrt{14}}$$

$$= \frac{6\sqrt{14}}{7} \approx 3.207$$

15. $$D = \frac{|ax_0 + by_0 + cz_0 + d|}{\sqrt{a^2 + b^2 + c^2}}$$

$$= \frac{|(3)(1) + (-1)(5) + 2(-4) - 6|}{\sqrt{(3)^2 + (-1)^2 + (2)^2}}$$

$$= \frac{16}{\sqrt{14}} = \frac{8\sqrt{14}}{7} \approx 4.276$$

17. $D = \dfrac{|ax_0 + by_0 + cz_0 + d|}{\sqrt{a^2 + b^2 + c^2}}$

$= \dfrac{|2(1) + (-4)(0) + 3(-1) + (-12)|}{\sqrt{2^2 + (-4)^2 + 3^2}}$

$= \dfrac{13}{\sqrt{29}} = \dfrac{13\sqrt{29}}{29} \approx 2.414$

19. $D = \dfrac{|ax_0 + by_0 + cz_0 + d|}{\sqrt{a^2 + b^2 + c^2}}$

$= \dfrac{|2(3) + (-3)(2) + 4(-1) + (-24)|}{\sqrt{2^2 + (-3)^2 + 4^2}}$

$= \dfrac{28}{\sqrt{29}} = \dfrac{28\sqrt{29}}{29} \approx 5.199$

21. For the first plane, $5x - 3y + z = 4$, $a_1 = 5$, $b_1 = -3$, and $c_1 = 1$. For the second plane, $x + 4y + 7z = 1$, $a_2 = 1$, $b_2 = 4$, and $c_2 = 7$. So you have

$a_1a_2 + b_1b_2 + c_1c_2 = (5)(1) + (-3)(4) + (1)(7)$

$= 5 - 12 + 7 = 0.$

The planes are perpendicular.

23. For the first plane, $x - 5y - z = 1$, $a_1 = 1$, $b_1 = -5$, and $c_1 = -1$. For the second plane, $5x - 25y - 5z = -3$, $a_2 = 5$, $b_2 = -25$, and $c_2 = -5$. So you have $a_2 = 5a_1$, $b_2 = 5b_1$, and $c_2 = 5c_1$. The planes are parallel.

25. For the first plane, $x + 2y = 3$, $a_1 = 1$, $b_1 = 2$, and $c_1 = 0$. For the second plane, $4x + 8y = 5$. $a_2 = 4$, $b_2 = 8$, and $c_2 = 0$. So you have $a_2 = 4a_1$, $b_2 = 4b_1$, and $c_2 = 4c_1$. The planes are parallel.

27. For the first plane, $2x + y = 3$, $a_1 = 2$, $b_1 = 1$, and $c_1 = 0$. For the second plane, $3x - 5z = 0$, $a_2 = 3$, $b_2 = 0$, and $c_2 = -5$. The planes are not parallel because $3a_1 = 2a_2$ and $3b_1 \neq 2b_2$. The planes are not perpendicular because

$a_1a_2 + b_1b_2 + c_1c_2 = (2)(3) + (1)(0) + (0)(-5) = 6 \neq 0.$

29. For the first plane, $x = 6$, $a_1 = 1$, $b_1 = 0$, and $c_1 = 0$. For the second plane, $y = -1$, $a_2 = 0$, $b_2 = 1$, and $c_2 = 0$. So you have

$a_1a_2 + b_1b_2 + c_1c_2 = (1)(0) + (0)(1) + (0)(0) = 0.$

The planes are perpendicular.

31. $x^2 - y - z^2 = 0$

Trace in xy-plane $(z = 0)$:	$y = x^2$	Parabola
Trace in plane $y = 1$:	$x^2 - z^2 = 1$	Hyperbola
Trace in yz-plane $(x = 0)$:	$y = -z^2$	Parabola

33. $\dfrac{x^2}{4} + y^2 + z^2 = 1$

Trace in xy-plane $(z = 0)$:	$\dfrac{x^2}{4} + y^2 = 1$	Ellipse
Trace in xz-plane $(y = 0)$:	$\dfrac{x^2}{4} + z^2 = 1$	Ellipse
Trace in yz-plane $(x = 0)$:	$y^2 + z^2 = 1$	Circle

35. $\dfrac{x^2}{9} + \dfrac{y^2}{16} + \dfrac{z^2}{9} = 1$ is an ellipsoid.
Matches graph (c)

37. $4x^2 - y^2 + 4z^2 = 4$ is a hyperboloid of one sheet.
Matches graph (f)

39. $4x^2 - 4y + z^2 = 0$ is an elliptic paraboloid.
Matches graph (d)

41. The graph of $x^2 + \dfrac{y^2}{4} + z^2 = 1$ is an ellipsoid.

43. $25x^2 + 25y^2 - z^2 = 5$

Standard form: $\dfrac{x^2}{1/5} + \dfrac{y^2}{1/5} - \dfrac{z^2}{5} = 1$

The graph is a hyperboloid of one sheet.

45. $x^2 - y + z^2 = 0$

Standard form: $y = x^2 + z^2$

The graph is an elliptic paraboloid.

47. $x^2 - y^2 + z = 0$

Standard form: $z = y^2 - x^2$

The graph is a hyperbolic paraboloid.

49. $2x^2 - y^2 + 2z^2 = -4$

Standard form: $-\dfrac{x^2}{2} + \dfrac{y^2}{4} - \dfrac{z^2}{2} = 1$

The graph is a hyperboloid of two sheets.

51. $z^2 = 9x^2 + y^2$

Standard form: $x^2 + \dfrac{y^2}{9} - \dfrac{z^2}{9} = 0$

The graph is an elliptic cone.

53. You are viewing the paraboloid from the x-axis: (20, 0, 0)

55. You are viewing the paraboloid from the z-axis: (0, 0, 20)

57. (a) $z = 1.25x - 0.125y + 0.95$

Year	1999	2000	2001	2002	2003	2004
x	6.2	6.1	5.9	5.8	5.6	5.5
y	7.3	7.1	7.0	7.0	6.9	6.9
z(actual)	7.8	7.7	7.4	7.3	7.2	6.9
z(model)	7.7875	7.6875	7.45	7.325	7.0875	6.9625

(b) Rewrite the model as an equation giving x in terms of y and z.

$$x = \frac{0.125y + z - 0.95}{1.25}$$

From the equation, you can see that according to the model, an increase in y and in z corresponds to an increase in x, the per capita consumption of reduced fat (1%) and skim milk.

Section 9.3 Functions of Several Variables

Skills Review

1. $f(x) = 5 - 2x,\ x = -3$

$f(-3) = 5 - 2(-3) = 11$

2. $f(x) = -x^2 + 4x + 5,\ x = -3$

$f(-3) = -(-3)^2 + 4(-3) + 5 = -16$

3. $y = \sqrt{4x^2 - 3x + 4},\ x = -3$

$y = \sqrt{4(-3)^2 - 3(-3) + 4} = \sqrt{49} = 7$

4. $y = \sqrt[3]{34 - 4x + 2x^2},\ x = -3$

$y = \sqrt[3]{34 - 4(-3) + 2(-3)^2} = \sqrt[3]{64} = 4$

5. $f(x) = 5x^2 + 3x - 2$

Domain: $(-\infty, \infty)$

6. $g(x) = \dfrac{1}{2x} - \dfrac{2}{x + 3}$

Domain: $(-\infty, -3) \cup (-3, 0) \cup (0, \infty)$

7. $h(y) = \sqrt{y - 5}$

Domain: $[5, \infty)$

8. $f(y) = \sqrt{y^2 - 5}$

Domain: $\left(-\infty, -\sqrt{5}\right] \cup \left[\sqrt{5}, \infty\right)$

9. $(476)^{0.65} \approx 55.0104$

10. $(251)^{0.35} \approx 6.9165$

1. $f(x, y) = \dfrac{x}{y}$

(a) $f(3, 2) = \dfrac{3}{2}$ (b) $f(-1, 4) = -\dfrac{1}{4}$

(c) $f(30, 5) = \dfrac{30}{5} = 6$ (d) $f(5, y) = \dfrac{5}{y}$

(e) $f(x, 2) = \dfrac{x}{2}$ (f) $f(5, t) = \dfrac{5}{t}$

3. $f(x, y) = xe^y$

(a) $f(5, 0) = 5e^0 = 5$ (b) $f(3, 2) = 3e^2$

(c) $f(2, -1) = 2e^{-1} = \dfrac{2}{e}$ (d) $f(5, y) = 5e^y$

(e) $f(x, 2) = xe^2$ (f) $f(t, t) = te^t$

5. $h(x, y, z) = \dfrac{xy}{z}$

(a) $h(2, 3, 9) = \dfrac{(2)(3)}{9} = \dfrac{2}{3}$

(b) $h(1, 0, 1) = \dfrac{(1)(0)}{1} = 0$

7. $V(r, h) = \pi r^2 h$

(a) $V(3, 10) = \pi(3)^2(10) = 90\pi$

(b) $V(5, 2) = \pi(5)^2(2) = 50\pi$

9. $A(P, r, t) = P\left[\left(1 + \frac{r}{12}\right)^{12t} - 1\right]\left(1 + \frac{12}{r}\right)$

(a) $A(100, 0.10, 10) = 100\left[\left(1 + \frac{0.10}{12}\right)^{120} - 1\right]\left(1 + \frac{12}{0.10}\right) \approx \$20{,}655.20$

(b) $A(275, 0.0925, 40) = 275\left[\left(1 + \frac{0.0925}{12}\right)^{480} - 1\right]\left(1 + \frac{12}{0.0925}\right) \approx \$1{,}397{,}672.67$

11. $f(x, y) = \int_x^y (2t - 3)\,dt$

(a) $f(1, 2) = \int_1^2 (2t - 3)\,dt = \left[(t^2 - 3t)\right]_1^2 = (-2) - (-2) = 0$

(b) $f(1, 4) = \int_1^4 (2t - 3)\,dt = \left[(t^2 - 3t)\right]_1^4 = 4 - (-2) = 6$

13. $f(x, y) = x^2 - 2y$

(a) $f(x + \Delta x, y) = (x + \Delta x)^2 - 2y = x^2 + 2x\Delta x + (\Delta x)^2 - 2y$

(b) $\frac{f(x, y + \Delta y) - f(x, y)}{\Delta y} = \frac{\left[x^2 - 2(y + \Delta y)\right] - (x^2 - 2y)}{\Delta y} = \frac{x^2 - 2y - 2\Delta y - x^2 + 2y}{\Delta y} = -\frac{2\Delta y}{\Delta y} = -2,\ \Delta y \neq 0$

15. $f(x, y) = \sqrt{16 - x^2 - y^2}$

The domain is the set of all points inside and on the circle $x^2 + y^2 = 16$ because $16 - x^2 - y^2 \geq 0$. The range is $[0, 4]$.

17. $f(x, y) = e^{x/y}$

The domain is the set of all points above or below the x-axis because $y \neq 0$. The range is $(0, \infty)$.

19. $z = \sqrt{4 - x^2 - y^2}$

The domain is the set of all points on or inside the circle $x^2 + y^2 = 4$ because $4 - x^2 - y^2 > 0$.

21. $f(x, y) = x^2 + y^2$

The domain is the set of all points in the xy-plane.

23. $f(x, y) = \frac{1}{xy}$

The domain is the set of all points in the xy-plane except those on the x- and y-axes.

25. $h(x, y) = x\sqrt{y}$

The domain is the set of all points in the xy-plane such that $y \geq 0$.

27. $g(x, y) = \ln(4 - x - y)$

The domain is the half plane below the line $y = -x + 4$ because $4 - x - y > 0$.

29. $f(x, y) = x^2 + \frac{y^2}{4}$

The contour map consists of ellipses

$x^2 + \frac{y^2}{4} = C.$

Matches (b)

31. $f(x, y) = e^{1-x^2-y^2}$

The contour map consists of curves $e^{1-x^2-y^2} = C$, or

$1 - x^2 - y^2 = \ln C \Rightarrow x^2 + y^2 = 1 - \ln C$, circles.

Matches (a)

33.

$c = -1,$	$-1 = x + y,$	$y = -x - 1$
$c = 0,$	$0 = x + y,$	$y = -x$
$c = 2,$	$2 = x + y,$	$y = -x + 2$
$c = 4,$	$4 = x + y,$	$y = -x + 4$

The level curves are parallel lines.

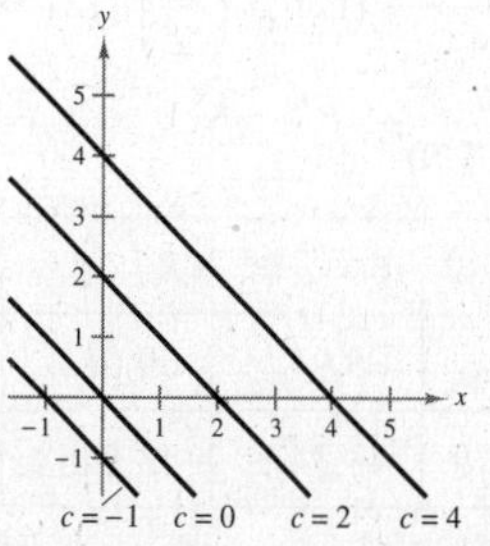

35. $c = 0, \quad 0 = \sqrt{25 - x^2 - y^2}, \quad x^2 + y^2 = 25$

$c = 1, \quad 1 = \sqrt{24 - x^2 - y^2}, \quad x^2 + y^2 = 24$

$c = 2, \quad 2 = \sqrt{21 - x^2 - y^2}, \quad x^2 + y^2 = 21$

$c = 3, \quad 3 = \sqrt{16 - x^2 - y^2}, \quad x^2 + y^2 = 16$

$c = 4, \quad 4 = \sqrt{9 - x^2 - y^2}, \quad x^2 + y^2 = 9$

$c = 5, \quad 5 = \sqrt{0 - x^2 - y^2}, \quad x^2 + y^2 = 0$

The level curves are circles.

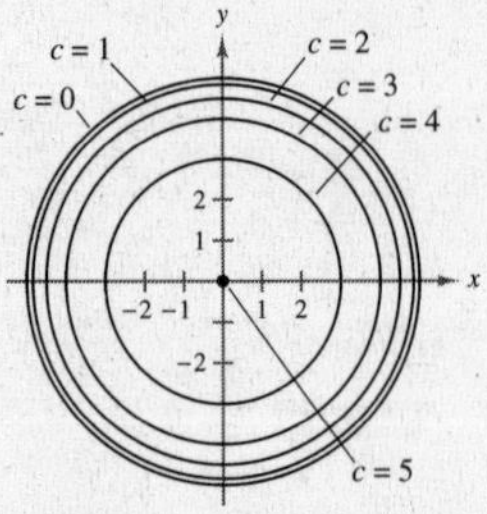

37. $c = \pm 1, \quad xy = \pm 1$

$c = \pm 2, \quad xy = \pm 2$

$c = \pm 3, \quad xy = \pm 3$

$c = \pm 4, \quad xy = \pm 4$

$c = \pm 5, \quad xy = \pm 5$

$c = \pm 6, \quad xy = \pm 6$

The level curves are hyperbolas.

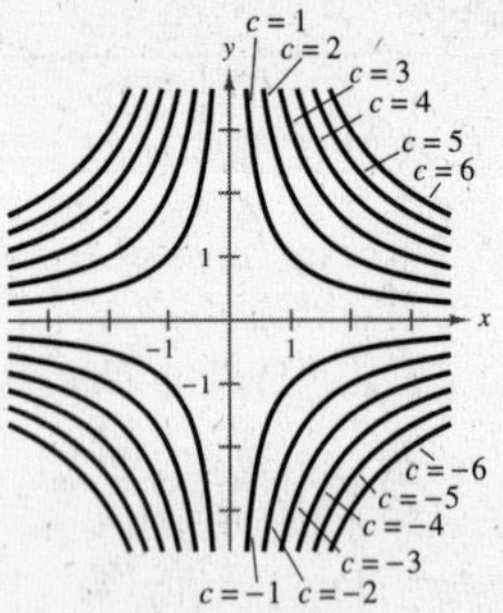

39. $c = \frac{1}{2}, \quad \frac{1}{2} = \frac{x}{x^2 + y^2}, \quad (x - 1)^2 + y^2 = 1$

$c = -\frac{1}{2}, \quad -\frac{1}{2} = \frac{x}{x^2 + y^2}, \quad (x + 1)^2 + y^2 = 1$

$c = 1, \quad 1 = \frac{x}{x^2 + y^2}, \quad \left(x - \frac{1}{2}\right)^2 + y^2 = \frac{1}{4}$

$c = -1, \quad -1 = \frac{x}{x^2 + y^2}, \quad \left(x + \frac{1}{2}\right)^2 + y^2 = \frac{1}{4}$

$c = \frac{3}{2}, \quad \frac{3}{2} = \frac{x}{x^2 + y^2}, \quad \left(x - \frac{1}{3}\right)^2 + y^2 = \frac{1}{9}$

$c = -\frac{3}{2}, \quad -\frac{3}{2} = \frac{x}{x^2 + y^2}, \quad \left(x + \frac{1}{3}\right)^2 + y^2 = \frac{1}{9}$

$c = 2, \quad 2 = \frac{x}{x^2 + y^2}, \quad \left(x - \frac{1}{4}\right)^2 + y^2 = \frac{1}{16}$

$c = -2, \quad -2 = \frac{x}{x^2 + y^2}, \quad \left(x + \frac{1}{4}\right)^2 + y^2 = \frac{1}{16}$

The level curves are circles.

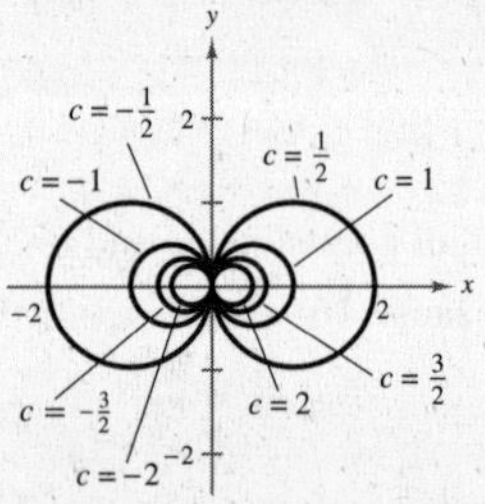

41. Volume = Volume of hemispheres + Volume of cylinder

$V = \frac{4}{3}\pi r^3 + \pi r^2 l$

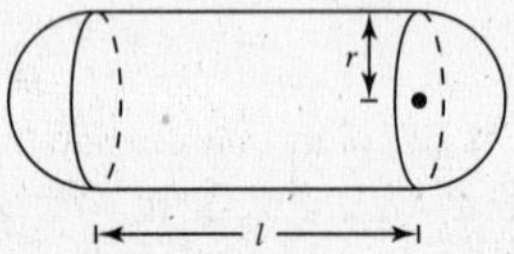

43. (a) $N(d, L) = \left(\frac{d - 4}{4}\right)^2 L$

$N(22, 12) = \left(\frac{22 - 4}{4}\right)^2 (12) = \left(\frac{9}{2}\right)^2 (12) = 243$ board-feet

(b) $N(30, 12) = \left(\frac{30 - 4}{4}\right)^2 (12) = \left(\frac{13}{2}\right)^2 (12) = 507$ board-feet

45. $c = 0.58l - 0.36a + 6.70$

a	2	4	6	8	10	12
l	57.40	62.57	66.91	69.01	72.00	75.83
c	39.272	41.5506	43.3478	43.8458	44.86	46.3614

47. (a) C, highest pressure

(b) A, lowest pressure

(c) B, highest wind velocity

49. $w = -0.59h + 7.25a + 47.37$

(a) For $h = 117$ and $a = 6$:

$$w = -0.59(117) + 7.25(6) + 47.37 = 21.84 \text{ kilograms}$$

(b) For $w = 47.9$ and $a = 13$:

$$47.9 = -0.59h + 7.25(13) + 47.37$$
$$47.9 = -0.59h + 141.62$$
$$-93.72 = -0.59h$$
$$158.8 \approx h$$

The mean height is about 158.8 centimeters.

(c) The variable representing age has the greater influence on weight. When h is held constant and a changes, w changes more than when a is held constant and h changes.

Section 9.4 Partial Derivatives

Skills Review

1. $f(x) = \sqrt{x^2 + 3}$

$$f'(x) = \frac{1}{2}(x^2 + 3)^{-1/2}(2x) = \frac{x}{\sqrt{x^2 + 3}}$$

2. $g(x) = (3 - x^2)^3$

$$g'(x) = 3(3 - x^2)^2(-2x) = -6x(3 - x^2)^2$$

3. $g(t) = te^{2t+1}$

$$g'(t) = te^{2t+1}(2) + e^{2t+1}(1) = e^{2t+1}(2t + 1)$$

4. $f(x) = e^{2x}\sqrt{1 - e^{2x}}$

$$f'(x) = e^{2x}\left(\frac{1}{2}\right)(1 - e^{2x})^{-1/2}(-e^{2x})(2) + \sqrt{1 - e^{2x}}\,e^{2x}(2)$$

$$= \frac{e^{2x}(-e^{2x})}{\sqrt{1 - e^{2x}}} + 2e^{2x}\sqrt{1 - e^{2x}} = \frac{e^{2x}}{\sqrt{1 - e^{2x}}}\left(-e^{2x} + 2(1 - e^{2x})\right) = \frac{e^{2x}}{\sqrt{1 - e^{2x}}}(2 - 3e^{2x})$$

5. $f(x) = \ln(3 - 2x)$

$$f'(x) = -\frac{2}{3 - 2x}$$

6. $u(t) = \ln\sqrt{t^3 - 6t}$

$$u'(t) = \frac{1}{\sqrt{t^3 - 6t}}\left(\frac{1}{2}\right)(t^3 - 6t)^{-1/2}(3t^2 - 6)$$

$$= \frac{3(t^2 - 2)}{2t(t^2 - 6)}$$

7. $g(x) = \dfrac{5x^2}{(4x - 1)^2}$

$$g'(x) = \frac{(4x - 1)^2(10x) - 5x^2(2)(4x - 1)(4)}{(4x - 1)^4} = \frac{(4x - 1)10x - 40x^2}{(4x - 1)^3} = -\frac{10x}{(4x - 1)^3}$$

Skills Review —*continued*—

8. $f(x) = \dfrac{(x+2)^3}{(x^2-9)^2}$

$$f'(x) = \frac{(x^2-9)^2(3)(x+2)^2 - (x+2)^3(2)(x^2-9)(2x)}{(x^2-9)^4}$$

$$= \frac{3(x+2)^2(x^2-9) - 4x(x+2)^3}{(x^2-9)^3} = \frac{(x+2)^2\left[3(x^2-9) - 4x(x+2)\right]}{(x^2-9)^3} = -\frac{(x+2)^2(x^2+8x+27)}{(x^2-9)^3}$$

9. $f(x) = x^2e^{x-2}$

$f'(x) = x^2e^{x-2} + e^{x-2}(2x)$

$f'(2) = (2)^2e^{2-2} + e^{2-2}(2(2)) = 4 + 4 = 8$

10. $g(x) = x\sqrt{x^2 - x + 2}$

$$g'(x) = x\left(\frac{1}{2}\right)(x^2 - x + 2)^{-1/2}(2x-1) + \sqrt{x^2 - x + 2} = \frac{x^2 - x/2}{\sqrt{x^2 - x + 2}} + \sqrt{x^2 - x + 2}$$

$$g'(2) = \frac{2^2 - 2/2}{\sqrt{2^2 - 2 + 2}} + \sqrt{2^2 - 2 + 2} = \frac{3}{2} + 2 = \frac{7}{2}$$

1. $\dfrac{\partial z}{\partial x} = 3$

$\dfrac{\partial z}{\partial y} = 5$

3. $f_x(x, y) = 3$

$f_y(x, y) = -12y$

5. $f_x(x, y) = \dfrac{1}{y}$

$f_y(x, y) = -xy^{-2} = -\dfrac{x}{y^2}$

7. $f_x(x, y) = \dfrac{1}{2}(x^2 + y^2)^{-1/2}(2x) = \dfrac{x}{\sqrt{x^2 + y^2}}$

$f_y(x, y) = \dfrac{1}{2}(x^2 + y^2)^{-1/2}(2y) = \dfrac{y}{\sqrt{x^2 + y^2}}$

9. $\dfrac{\partial z}{\partial x} = 2xe^{2y}$

$\dfrac{\partial z}{\partial y} = 2x^2e^{2y}$

11. $h_x(x, y) = -2xe^{-(x^2+y^2)}$

$h_y(x, y) = -2ye^{-(x^2+y^2)}$

13. $z = \ln\dfrac{x+y}{x-y} = \ln(x+y) - \ln(x-y)$

$\dfrac{\partial z}{\partial x} = \dfrac{1}{x+y} - \dfrac{1}{x-y} = -\dfrac{2y}{x^2 - y^2}$

$\dfrac{\partial z}{\partial y} = \dfrac{1}{x+y} + \dfrac{1}{x-y} = \dfrac{2x}{x^2 - y^2}$

15. $f_x(x, y) = 6xye^{x-y} + 3x^2ye^{x-y} = 3xye^{x-y}(2 + x)$

17. $g_x(x, y) = 3y^2e^{y-x} - 3xy^2e^{y-x} = 3y^2e^{y-x}(1 - x)$

19. Using the solution from Exercise 15, $f_x(1, 1) = 9$.

21. $f_x(x, y) = 6x + y \qquad f_x(2, 1) = 13$

$f_y(x, y) = x - 2y \qquad f_y(2, 1) = 0$

23. $f_x(x, y) = 3ye^{3xy} \qquad f_x(0, 4) = 3(4)e^0 = 12$

$f_y(x, y) = 3xe^{3xy} \qquad f_y(0, 4) = 0$

25. $f_x(x, y) = \dfrac{(x - y)y - xy}{(x - y)^2} = -\dfrac{y^2}{(x - y)^2}$ $\quad f_x(2, -2) = -\dfrac{4}{16} = -\dfrac{1}{4}$

$f_y(x, y) = \dfrac{(x - y)x - xy(-1)}{(x - y)^2} = \dfrac{x^2}{(x - y)^2}$ $\quad f_y(2, -2) = \dfrac{4}{16} = \dfrac{1}{4}$

27. $f_x(x, y) = \dfrac{1}{x^2 + y^2}(2x) = \dfrac{2x}{x^2 + y^2}$ $\quad f_x(1, 0) = \dfrac{2}{1 + 0} = 2$

$f_y(x, y) = \dfrac{1}{x^2 + y^2}(2y) = \dfrac{2y}{x^2 + y^2}$ $\quad f_y(1, 0) = 0$

29. $w_x = yz,\ w_y = xz,\ w_z = xy$

31. $w_x = \dfrac{(x + y)(0) - 2z(1)}{(x + y)^2} = -\dfrac{2z}{(x + y)^2}$

$w_y = \dfrac{(x + y)(0) - 2z(1)}{(x + y)^2} = -\dfrac{2z}{(x + y)^2}$

$w_z = \dfrac{2}{x + y}$

33. $w_x = \dfrac{x}{\sqrt{x^2 + y^2 + z^2}}$ $\quad w_x(2, -1, 2) = \dfrac{2}{3}$

$w_y = \dfrac{y}{\sqrt{x^2 + y^2 + z^2}}$ $\quad w_y(2, -1, 2) = -\dfrac{1}{3}$

$w_z = \dfrac{z}{\sqrt{x^2 + y^2 + z^2}}$ $\quad w_z(2, -1, 2) = \dfrac{2}{3}$

35. $w = \ln\sqrt{x^2 + y^2 + z^2} = \dfrac{1}{2}\ln(x^2 + y^2 + z^2)$

$w_x = \dfrac{x}{x^2 + y^2 + z^2}$ $\quad w_x(3, 0, 4) = \dfrac{3}{25}$

$w_y = \dfrac{y}{x^2 + y^2 + z^2}$ $\quad w_y(3, 0, 4) = 0$

$w_z = \dfrac{z}{x^2 + y^2 + z^2}$ $\quad w_z(3, 0, 4) = \dfrac{4}{25}$

37. $w_x = 2z^2 + 3yz$ $\quad w_x(1, -1, 2) = 2$

$w_y = 3xz - 12yz$ $\quad w_y(1, -1, 2) = 30$

$w_z = 4xz + 3xy - 6y^2$ $\quad w_z(1, -1, 2) = -1$

39.
$$\begin{aligned} f_x(x, y) &= 2x + 4y - 4 = 0 \Rightarrow & -4x - 8y &= -8\\ f_y(x, y) &= 4x + 2y + 16 = 0 \Rightarrow & 4x + 2y &= -16\\ && -6y &= -24\\ && y &= 4\\ && x &= -6 \end{aligned}$$

Solution: $(-6, 4)$

41.
$$\left.\begin{aligned} f_x(x, y) &= -\frac{1}{x^2} + y = 0 \Rightarrow x^2y = 1\\ f_y(x, y) &= -\frac{1}{y^2} + x = 0 \Rightarrow y^2x = 1 \end{aligned}\right\} x = y = 1$$

Solution: $(1, 1)$

43. (a) $\dfrac{\partial z}{\partial x} = y$ $\quad \dfrac{\partial z}{\partial x}(1, 2, 2) = 2$

(b) $\dfrac{\partial z}{\partial y} = x$ $\quad \dfrac{\partial z}{\partial y}(1, 2, 2) = 1$

45. (a) $\dfrac{\partial z}{\partial x} = -2x$ $\quad \dfrac{\partial z}{\partial x}(1, 1, 2) = -2$

(b) $\dfrac{\partial z}{\partial y} = -2y$ $\quad \dfrac{\partial z}{\partial y}(1, 1, 2) = -2$

47. (a) $\dfrac{\partial z}{\partial x} = -e^{-x}\cos y$

$\dfrac{\partial z}{\partial x}(0, 0, 1) = -1$

(b) $\dfrac{\partial z}{\partial y} = -e^{-x}\sin y$

$\dfrac{\partial z}{\partial y}(0, 0, 1) = 0$

49. $\dfrac{dz}{dx} = 2x - 2y$

$\dfrac{dz}{dy} = -2x + 6y$

$\dfrac{d^2z}{dx^2} = 2$

$\dfrac{d^2z}{dxdy} = -2$

$\dfrac{d^2z}{dydx} = -2$

$\dfrac{d^2z}{dy^2} = 6$

51. $\dfrac{dz}{dx} = \dfrac{4xe^{2xy}(2y) - e^{2xy}(4)}{16x^2} = \dfrac{e^{2xy}(2xy - 1)}{4x^2}$

$\dfrac{dz}{dy} = \dfrac{1}{2}e^{2xy}$

$$\frac{d^2z}{dx^2} = \frac{4x^2\left[2ye^{2xy}(2xy - 1) + 2ye^{2xy}\right] - 8xe^{2xy}(2xy - 1)}{16x^4}$$

$$= \frac{e^{2xy}\left[xy(2xy - 1) + xy - (2xy - 1)\right]}{2x^3} = \frac{e^{2xy}(2x^2y^2 - 2xy + 1)}{2x^3}$$

$$\frac{d^2z}{dxdy} = \frac{1}{4x^2}\left[e^{2xy}(2x) + e^{2xy}(2xy - 1)(2x)\right] = ye^{2xy}$$

$\dfrac{d^2z}{dydx} = ye^{2xy}$

$\dfrac{d^2z}{dy^2} = xe^{2xy}$

53. $\dfrac{dz}{dx} = 3x^2$

$\dfrac{dz}{dy} = -8y$

$\dfrac{d^2z}{dx^2} = 6x$

$\dfrac{d^2z}{dxdy} = 0$

$\dfrac{d^2z}{dydx} = 0$

$\dfrac{d^2z}{dy^2} = -8$

55. $\dfrac{dz}{dx} = -\dfrac{1}{(x - y)^2}$

$\dfrac{dz}{dy} = \dfrac{1}{(x - y)^2}$

$\dfrac{d^2z}{dx^2} = \dfrac{2}{(x - y)^3}$

$\dfrac{d^2z}{dxdy} = -\dfrac{2}{(x - y)^3}$

$\dfrac{d^2z}{dydx} = -\dfrac{2}{(x - y)^3}$

$\dfrac{d^2z}{dy^2} = \dfrac{2}{(x - y)^3}$

57. $\dfrac{dz}{dx} = e^{-y^2}$

$\dfrac{dz}{dy} = x\left(e^{-y^2}\right)(-2y) = -2xye^{-y^2}$

$\dfrac{d^2z}{dx^2} = 0$

$\dfrac{d^2z}{dxdy} = -2ye^{-y^2}$

$\dfrac{d^2z}{dydx} = e^{-y^2}(-2y) = -2ye^{-y^2}$

$\dfrac{d^2z}{dy^2} = -2xy\left(e^{-y^2}\right)(-2y) + e^{-y^2}(-2x) = 4xy^2e^{-y^2} - 2xe^{-y^2}$

59. $f_x(x, y) = 4x^3 - 6xy^2$ $\quad f_y(x, y) = -6x^2y + 2y$

$f_{xx}(x, y) = 12x^2 - 6y^2$ $\quad f_{xx}(1, 0) = 12$

$f_{xy}(x, y) = -12xy$ $\quad f_{xy}(1, 0) = 0$

$f_{yx}(x, y) = -12xy$ $\quad f_{yx}(1, 0) = 0$

$f_{yy}(x, y) = -6x^2 + 2$ $\quad f_{yy}(1, 0) = -4$

61. $f_x(x, y) = \dfrac{1}{x - y}$ $\quad f_y(x, y) = -\dfrac{1}{x - y}$

$f_{xx}(x, y) = -\dfrac{1}{(x - y)^2}$ $\quad f_{xx}(2, 1) = -1$

$f_{xy}(x, y) = \dfrac{1}{(x - y)^2}$ $\quad f_{xy}(2, 1) = 1$

$f_{yx}(x, y) = \dfrac{1}{(x - y)^2}$ $\quad f_{yx}(2, 1) = 1$

$f_{yy}(x, y) = -\dfrac{1}{(x - y)^2}$ $\quad f_{yy}(2, 1) = -1$

63. (a) $\dfrac{\partial z}{\partial x} = 1.25$ $\quad \dfrac{\partial z}{\partial y} = -0.125$

(b) For every gallon increase in the per capita consumption of reduced fat (1%) and skim milk, consumption of whole milk increases by 1.25 gallons. For every gallon increase in the per capita consumption of reduced fat (2%) milk, consumption of whole milk decreases by 0.125 gallon.

65. $T = 500 - 0.6x^2 - 1.5y^2$

$T_x(x, y) = -1.2x$

$T_x(2, 3) = -1.2(2) = -2.4$ degrees per meter

$T_y(x, y) = -3y$

$T_y(2, 3) = -3(3) = -9$ degrees per meter

67. Since both first partials are negative, an increase in the charge for food and housing or tuition will cause a decrease in the number of applicants.

69. Answers will vary.

Mid-Chapter Quiz Solutions

1. (a)

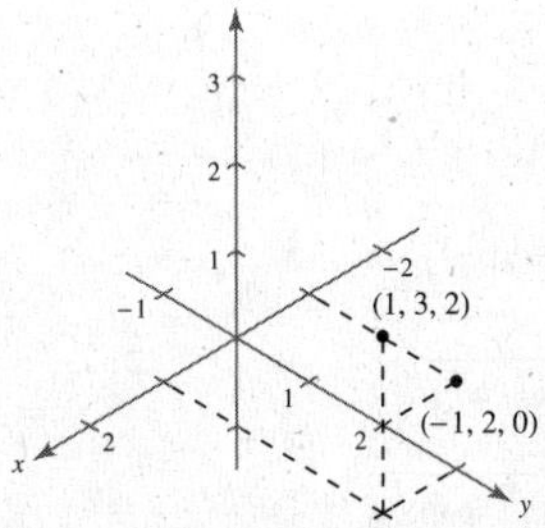

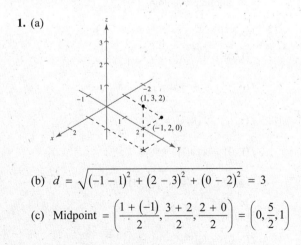

(b) $d = \sqrt{(-1 - 1)^2 + (2 - 3)^2 + (0 - 2)^2} = 3$

(c) Midpoint $= \left(\dfrac{1 + (-1)}{2}, \dfrac{3 + 2}{2}, \dfrac{2 + 0}{2}\right) = \left(0, \dfrac{5}{2}, 1\right)$

2. (a)

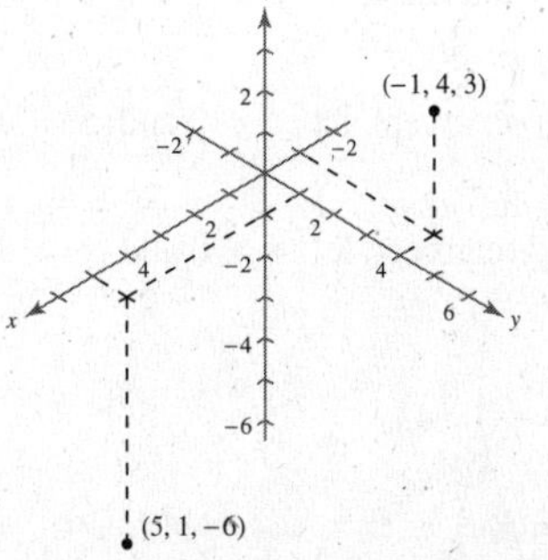

(b) $d = \sqrt{(5 + 1)^2 + (1 - 4)^2 + (-6 - 3)^2}$

$= \sqrt{126} = 3\sqrt{14}$

(c) Midpoint $= \left(\dfrac{-1 + 5}{2}, \dfrac{4 + 1}{2}, \dfrac{3 - 6}{2}\right) = \left(2, \dfrac{5}{2}, -\dfrac{3}{2}\right)$

3. (a)

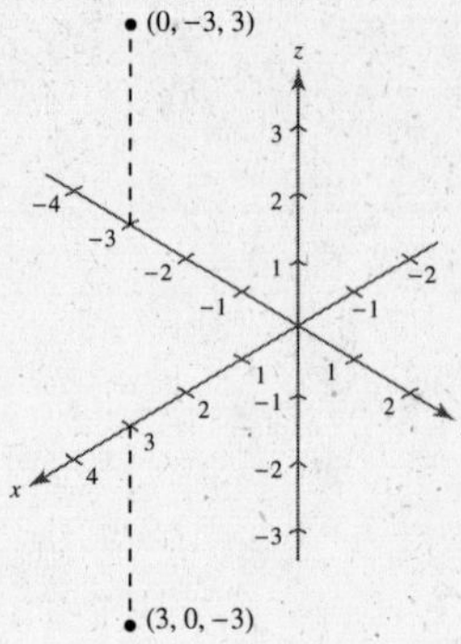

(b) $d = \sqrt{(3-0)^2 + (0+3)^2 + (-3-3)^2}$

$= \sqrt{54} = 3\sqrt{6}$

(c) $\text{Midpoint} = \left(\dfrac{0+3}{2}, \dfrac{-3+0}{2}, \dfrac{3+(-3)}{2}\right)$

$= \left(\dfrac{3}{2}, -\dfrac{3}{2}, 0\right)$

4. $(x-2)^2 + (y+1)^2 + (z-3)^2 = 16$

5. Center: $\left(\dfrac{0+2}{2}, \dfrac{3+5}{2}, \dfrac{1+(-5)}{2}\right) = (1, 4, -2)$

Radius $= \sqrt{(1-0)^2 + (4-3)^2 + (-2-1)^2} = \sqrt{11}$

Standard Form: $(x-1)^2 + (y-4)^2 + (z+2)^2 = 11$

6. $x^2 + y^2 + z^2 - 8x - 2y - 6z - 23 = 0$

$(x^2 - 8x + 16) + (y^2 - 2y + 1) + (z^2 - 6z + 9)$

$= 23 + 16 + 1 + 9$

$(x-4)^2 + (y-1)^2 + (z-3)^2 = 49$

Center: $(4, 1, 3)$ Radius: $\sqrt{49} = 7$

7. $2x + 3y + z = 6$

To find the x-intercept, let $y = 0$ and $z = 0$.

$2x = 6 \Rightarrow x = 3$

To find the y-intercept, let $x = 0$ and $z = 0$.

$3y = 6 \Rightarrow y = 2$

To find the z-intercept, let $x = 0$ and $y = 0$.

$z = 6$

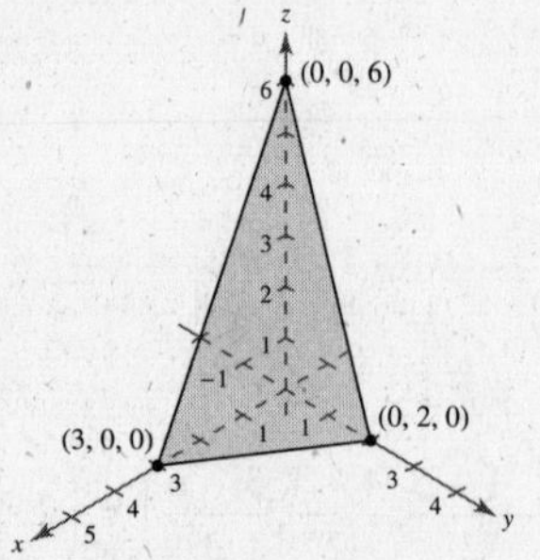

8. $x - 2z = 4$

To find the x-intercept, let $z = 0$.

$x = 4$

Because the y-coefficient is 0, there is no y-intercept. The plane is parallel to the y-axis.

To find the z-intercept, let $x = 0$.

$-2z = 4 \Rightarrow z = -2$

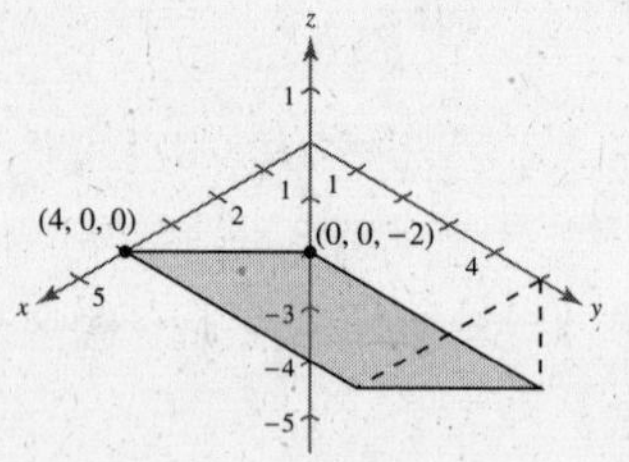

9. The only intercept is $z = -5$. The plane is parallel to the xy-plane.

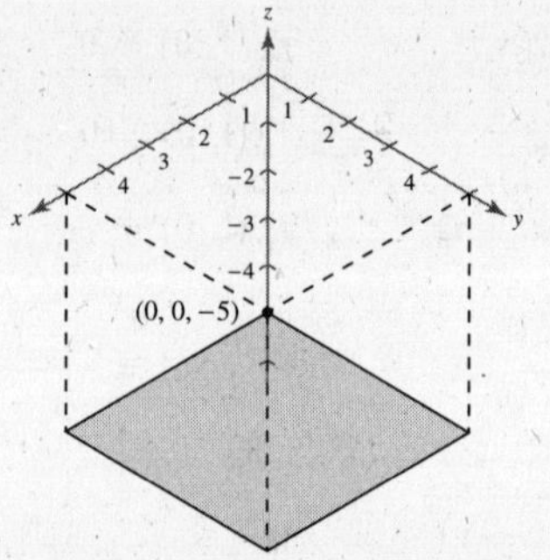

10. The graph of $\dfrac{x^2}{4} + \dfrac{y^2}{9} + \dfrac{z^2}{16} = 1$ is an ellipsoid.

11. The graph of $z^2 - x^2 - y^2 = 25$ or

$\dfrac{z^2}{25} - \dfrac{x^2}{25} - \dfrac{y^2}{25} = 1$ is a hyperboloid of two sheets.

12. The graph of $81z - 9x^2 - y^2 = 0$ or $z = \dfrac{x^2}{9} + \dfrac{y^2}{81}$ is an elliptic paraboloid.

13. $f(x, y) = x - 9y^2$

$f(1, 0) = 1 - 9(0)^2 = 1$

$f(4, -1) = 4 - 9(-1)^2 = -5$

14. $f(x, y) = \sqrt{4x^2 + y}$

$f(1, 0) = \sqrt{4(1)^2 + 0} = 2$

$f(4, -1) = \sqrt{4(4)^2 + (-1)} = \sqrt{63} = 3\sqrt{7}$

15. $f(x, y) = \ln(x + 3y)$

$f(1, 0) = \ln(1 + 3(0)) = 0$

$f(4, -1) = \ln(4 + 3(-1)) = 0$

16. (a) The temperatures in the Great Lakes region range from about 30° to about 50°.

(b) The temperatures in the United States range from 40° to 80°.

(c) The temperatures in Mexico range from about 70° to almost 90°.

17. $f(x, y) = x^2 + 2y^2 - 3x - y + 1$

$f_x(x, y) = 2x - 3 \qquad f_x(-2, 3) = 2(-2) - 3 = -7$

$f_y(x, y) = 4y - 1 \qquad f_y(-2, 3) = 4(3) - 1 = 11$

18. $f(x, y) = \dfrac{3x - y^2}{x + y}$

$$f_x(x, y) = \frac{(x + y)(3) - (3x - y^2)}{(x + y)^2} = \frac{y^2 + 3y}{(x + y)^2} = \frac{y(y + 3)}{(x + y)^2} \qquad f_x(-2, 3) = \frac{3^2 + 3(3)}{(-2 + 3)^2} = 18$$

$$f_y(x, y) = \frac{(x + y)(-2y) - (3x - y^2)}{(x + y)^2} = \frac{-y^2 - 2xy - 3x}{(x + y)^2} \qquad f_y(-2, 3) = \frac{-3^2 - 2(-2)(3) - 3(-2)}{(-2 + 3)^2} = 9$$

19. Equation of sphere: $x^2 + y^2 + z^2 = 3963^2$

Lines of longitude could be represented by traces in planes passing through the z-axis. The traces are circles.
Lines of latitude could be represented by traces in the xy-plane and planes parallel to it. The traces are circles.

Section 9.5 Extrema of Functions of Two Variables

Skills Review

1. $\begin{cases} 5x = 15 \\ 3x - 2y = 5 \end{cases}$

$5x = 15$

$x = 3$

Substitute in the other equation.

$3(3) - 2y = 5$

$-2y = -4$

$y = 2$

The solution is (3, 2).

2. $\begin{cases} \frac{1}{2}y = 3 \\ -x + 5y = 19 \end{cases}$

$\frac{1}{2}y = 3$

$y = 6$

Substitute in the other equation.

$-x + 5(6) = 19$

$-x = -11$

$x = 11$

The solution is $(11, 6)$.

3. $\begin{cases} x + y = 5 \\ x - y = -3 \end{cases}$

Adding the two equations gives $2x = 2$, so $x = 1$.

Substitute.

$1 + y = 5$

$y = 4$

The solution is $(1, 4)$.

4. $\begin{cases} x + y = 8 \\ 2x - y = 4 \end{cases}$

Adding the two equations gives $3x = 12$, so $x = 4$.

Substitute.

$4 + y = 8$

$y = 4$

The solution is (4, 4).

5. $\begin{cases} 2x - y = 8 & \text{Equation 1} \\ 3x - 4y = 7 & \text{Equation 2} \end{cases}$

Multiply Equation 1 by −4: $-8x + 4y = -32$

Add the new equation to Equation 2: $-5x = -25$

Simplify: $x = 5$

Substitute 5 for x in Equation 1: $2(5) - y = 8$

Solve for y: $y = 2$

The solution is $(5, 2)$.

Skills Review —continued—

6. $\begin{cases} 2x - 4y = 14 & \text{Equation 1} \\ 3x + y = 7 & \text{Equation 2} \end{cases}$

Multiply Equation 2 by 4: $12x + 4y = 28$

Add new Equation to Equation 1: $14x = 42$

Simplify: $x = 3$

Substitute 3 for x in Equation 2: $3(3) + y = 7$

Simplify: $y = -2$

The solution is $(3, -2)$.

7. $\begin{cases} x^2 + x = 0 & \text{Equation 1} \\ 2yx + y = 0 & \text{Equation 2} \end{cases}$

Factor Equation 1: $x(x + 1) = 0$

Solve Equation 1 for x: $x = -1$ or $x = 0$

Substitute -1 for x in Equation 2: $2y(-1) + y = 0$

Solve for y: $y = 0$

Substitute 0 for x in Equation 2: $2y(0) + y = 0$

Solve for y: $y = 0$

The solutions are $(-1, 0)$ and $(0, 0)$.

8. $\begin{cases} 3y^2 + 6y = 0 & \text{Equation 1} \\ xy + x + 2 = 0 & \text{Equation 2} \end{cases}$

Factor Equation 1: $y(3y + 6) = 0$

Solve for y: $y = 0$

$3y + 6 = 0$

$y = -2$

Substitute 0 for y in Equation 2: $x(0) + x + 2 = 0$

Solve for x: $x = -2$

Substitute -2 for y in Equation 2: $x(-2) + x + 2 = 0$

Solve for x: $x = 2$

The solutions are $(-2, 0)$ and $(2, -2)$.

9. $z = 4x^3 - 3y^2$

$\dfrac{\partial z}{\partial x} = 12x^2 \qquad \dfrac{\partial z}{\partial y} = -6y$

$\dfrac{\partial^2 z}{\partial x^2} = 24x \qquad \dfrac{\partial^2 z}{\partial y^2} = -6$

$\dfrac{\partial^2 z}{\partial x \partial y} = 0 \qquad \dfrac{\partial^2 z}{\partial y \partial x} = 0$

10. $z = 2x^5 - y^3$

$\dfrac{\partial z}{\partial x} = 10x^4 \qquad \dfrac{\partial z}{\partial y} = -3y^2$

$\dfrac{\partial^2 z}{\partial x^2} = 40x^3 \qquad \dfrac{\partial^2 z}{\partial y^2} = -6y$

$\dfrac{\partial^2 z}{\partial x \partial y} = 0 \qquad \dfrac{\partial^2 z}{\partial y \partial x} = 0$

11. $z = x^4 - \sqrt{xy} + 2y$

$\dfrac{\partial z}{\partial x} = 4x^3 - \dfrac{y}{2\sqrt{xy}} = 4x^3 - \dfrac{\sqrt{xy}}{2x}$

$\dfrac{\partial z}{\partial y} = -\dfrac{x}{2\sqrt{xy}} + 2 = -\dfrac{\sqrt{xy}}{2y} + 2$

$\dfrac{\partial^2 z}{\partial x^2} = 12x^2 - \dfrac{y}{4x\sqrt{xy}} + \dfrac{2\sqrt{xy}}{4x^2} = 12x^2 + \dfrac{\sqrt{xy}}{4x^2}$

$\dfrac{\partial^2 z}{\partial y^2} = -\dfrac{x}{4y\sqrt{xy}} + \dfrac{\sqrt{xy}}{2y^2} = \dfrac{\sqrt{xy}}{4y^2}$

$\dfrac{\partial^2 z}{\partial x \partial y} = -\dfrac{x}{4x\sqrt{xy}} = -\dfrac{\sqrt{xy}}{4xy}$

$\dfrac{\partial^2 z}{\partial y \partial x} = -\dfrac{x}{4x\sqrt{xy}} - \dfrac{\sqrt{xy}}{4xy}$

12. $z = 2x^2 - 3xy + y^2$

$\dfrac{\partial z}{\partial x} = 4x - 3y \qquad \dfrac{\partial z}{\partial y} = -3x + 2y$

$\dfrac{\partial^2 z}{\partial x^2} = 4 \qquad \dfrac{\partial^2 z}{\partial y^2} = 2$

$\dfrac{\partial^2 z}{\partial x \partial y} = -3 \qquad \dfrac{\partial^2 z}{\partial y \partial x} = -3$

13. $z = ye^{xy^2}$

$\dfrac{\partial z}{\partial x} = y(y^2)e^{xy^2} = y^3e^{xy^2}$

$\dfrac{\partial z}{\partial y} = y(2xy)e^{xy^2} + e^{xy^2}(1) = 2xy^2e^{xy^2} + e^{xy^2}$

$\dfrac{\partial^2 z}{\partial x^2} = y^3(y^2)e^{xy^2} = y^5e^{xy^2}$

$\dfrac{\partial^2 z}{\partial y^2} = 2xy^2(2xy)e^{xy^2} + e^{xy^2}(4xy) + 2xye^{xy^2}$

$= 4x^2y^3e^{xy^2} + 6xye^{xy^2}$

$\dfrac{\partial^2 z}{\partial x \partial y} = 2xy^4e^{xy^2} + 3y^2e^{xy^2}$

$\dfrac{\partial^2 z}{\partial y \partial x} = y^3(2xy)e^{xy^2} + e^{xy^2}3y^2 = 2xy^4e^{xy^2} + 3y^2e^{xy^2}$

Skills Review —*continued*—

14. $z = xe^{xy}$

$$\frac{\partial z}{\partial x} = xye^{xy} + e^{xy} = e^{xy}(xy + 1)$$

$$\frac{\partial z}{\partial y} = x^2e^{xy}$$

$$\frac{\partial^2 z}{\partial x^2} = e^{xy}(y) + (xy + 1)ye^{xy} = ye^{xy}(xy + 2)$$

$$\frac{\partial^2 z}{\partial y^2} = x^3e^{xy}$$

$$\frac{\partial^2 z}{\partial x \partial y} = e^{xy}(x) + (xy + 1)xe^{xy} = xe^{xy}(xy + 2)$$

$$\frac{\partial^2 z}{\partial y \partial x} = x^2ye^{xy} + e^{xy}(2x) = xe^{xy}(xy + 2)$$

1. $f(x, y) = x^2 - y^2 + 4x - 8y - 11$

The first partial derivatives of f, $f_x(x, y) = 2x + 4$ and $f_y(x, y) = -2y - 8$, are zero at the point $(-2, -4)$. Moreover, because

$f_{xx}(x, y) = 2$, $f_{yy}(x, y) = -2$, and $f_{xy}(x, y) = 0$,

it follows that $f_{xx}(-2, -4) > 0$ and

$f_{xx}(-2, -4)f_{yy}(-2, -4) - [f_{xy}(-2, -4)]^2 = -4 < 0.$

So, $(-2, -4, 1)$ is a saddle point. There are no relative extrema.

3. $f(x, y) = \sqrt{x^2 + y^2 + 1}$

The first partial derivatives of f,

$f_x(x, y) = \dfrac{x}{\sqrt{x^2 + y^2 + 1}}$ and

$f_y(x, y) = \dfrac{y}{\sqrt{x^2 + y^2 + 1}}$, are zero at the point $(0, 0)$.

Moreover, because

$f_{xx}(x, y) = \dfrac{y^2 + 1}{(x^2 + y^2 + 1)^{3/2}}$,

$f_{yy}(x, y) = \dfrac{x^2 + 1}{(x^2 + y^2 + 1)^{3/2}}$, and

$f_{xy}(x, y) = \dfrac{-xy}{(x^2 + y^2 + 1)^{3/2}}$,

it follows that $f_{xx}(0, 0) = 1 > 0$ and

$f_{xx}(0, 0)f_{yy}(0, 0) - [f_{xy}(0, 0)]^2 = 1 > 0.$

So, $(0, 0, 1)$ is a relative minimum.

5. $f(x, y) = (x - 1)^2 + (y - 3)^2$

The first partial derivatives of f, $f_x(x, y) = 2(x - 1)$ and $f_y(x, y) = 2(y - 3)$, are zero at the point $(1, 3)$. Moreover, because

$f_{xx}(x, y) = 2$, $f_{yy}(x, y) = 2$, and $f_{xy}(x, y) = 0$,

it follows that $f_{xx}(1, 3) > 0$ and

$f_{xx}(1, 3)f_{yy}(1, 3) - [f_{xy}(1, 3)]^2 = 4 > 0.$

So, $(1, 3, 0)$ is a relative minimum.

7. $f(x, y) = 2x^2 + 2xy + y^2 + 2x - 3$

The first partial derivatives of f, $f_x(x, y) = 4x + 2y + 2$ and $f_y(x, y) = 2x + 2y$, are zero at the point $(-1, 1)$.

Moreover, because $f_{xx}(x, y) = 4$, $f_{yy}(x, y) = 2$, and $f_{xy}(x, y) = 2$, it follows that $f_{xx}(-1, 1) > 0$ and

$f_{xx}(-1, 1)f_{yy}(-1, 1) - [f_{xy}(-1, 0)]^2 = 4 > 0.$

So, $(-1, 1, -4)$ is a relative minimum.

9. $f(x, y) = -5x^2 + 4xy - y^2 + 16x + 10$

The first partial derivatives of f, $f_x(x, y) = -10x + 4y + 16$ and $f_y(x, y) = 4x - 2y$, are zero at the point $(8, 16)$.

Moreover, because $f_{xx}(x, y) = -10$, $f_{yy}(x, y) = -2$, and $f_{xy}(x, y) = 4$, it follows that $f_{xx}(8, 16) < 0$ and

$f_{xx}(8, 16)f_{yy}(8, 16) - [f_{xy}(8, 16)]^2 = 4 > 0.$

So, $(8, 16, 74)$ is a relative maximum.

11. $f(x, y) = 3x^2 + 2y^2 - 12x - 4y + 7$

The first partial derivatives of f, $f_x(x, y) = 6x - 12$ and $f_y(x, y) = 4y - 4$, are zero at the point (2, 1).

Moreover, because $f_{xx}(x, y) = 6$, $f_{yy}(x, y) = 4$, and $f_{xy}(x, y) = 0$, it follows that $f_{xx}(2, 1) > 0$ and $f_{xx}(2, 1)f_{yy}(2, 1) - [f_{xy}(2, 1)]^2 = 24 > 0$.

So, $(2, 1, -7)$ is a relative minimum.

13. $f(x, y) = x^2 - y^2 + 4x - 4y - 8$

The first partial derivatives of f, $f_x(x, y) = 2x + 4 = 2(x + 2)$ and $f_y(x, y) = -2y - 4 = -2(y + 2)$, are zero at the point $(-2, -2)$.

Moreover, because $f_{xx}(x, y) = 2$, $f_{yy}(x, y) = -2$, and $f_{xy}(x, y) = 0$, it follows that $f_{xx}(-2, -2) > 0$ and $f_{xx}(1, -2)f_{yy}(1, -2) - [f_{xy}(1, -2)]^2 = -4 < 0$.

So, $(-2, -2, -8)$ is a saddle point.

15. $f(x, y) = \frac{1}{2}xy$

The first partial derivatives of f, $f_x(x, y) = \frac{1}{2}y$ and $f_y(x, y) = \frac{1}{2}x$, are zero at the point $(0, 0)$.

Moreover, because $f_{xx}(x, y) = 0$, $f_{yy}(x, y) = 0$, and $f_{xy}(x, y) = \frac{1}{2}$, it follows that $f_{xx}(0, 0) = 0$ and $f_{xx}(0, 0)f_{yy}(0, 0) - [f_{xy}(0, 0)]^2 = -\frac{1}{4} < 0$.

So, (0, 0, 0) is a saddle point.

17. $f(x, y) = (x + y)e^{1-x^2-y^2}$

The first partial derivatives of f, $f_x(x, y) = (-2x^2 - 2xy + 1)e^{1-x^2-y^2}$ and $f_y(x, y) = (-2y^2 - 2xy + 1)e^{1-x^2-y^2}$, are zero at $\left(\frac{1}{2}, \frac{1}{2}\right)$ and $\left(-\frac{1}{2}, -\frac{1}{2}\right)$. Moreover, because

$f_{xx}(x, y) = (4x^3 + 4x^2y - 6x - 2y)e^{1-x^2-y^2}$,

$f_{yy}(x, y) = (4y^3 + 4xy^2 - 6y - 2x)e^{1-x^2-y^2}$, and

$f_{xy}(x, y) = (4x^2y + 4xy^2 - 2y - 2x)e^{1-x^2-y^2}$,

it follows that $f_{xx}\left(\frac{1}{2}, \frac{1}{2}\right) = -3e^{1/2} < 0$,

$f_{xx}\left(\frac{1}{2}, \frac{1}{2}\right)f_{yy}\left(\frac{1}{2}, \frac{1}{2}\right) - \left[f_{xy}\left(\frac{1}{2}, \frac{1}{2}\right)\right]^2 = 0$,

$f_{xx}\left(-\frac{1}{2}, -\frac{1}{2}\right) = 3e^{1/2} > 0$, and

$f_{xx}\left(-\frac{1}{2}, -\frac{1}{2}\right)f_{yy}\left(-\frac{1}{2}, -\frac{1}{2}\right) - \left[f_{xy}\left(-\frac{1}{2}, -\frac{1}{2}\right)\right]^2 = 0$.

So, $\left(\frac{1}{2}, \frac{1}{2}, e^{1/2}\right)$ is a relative maximum and $\left(-\frac{1}{2}, -\frac{1}{2}, -e^{1/2}\right)$ is a relative minimum.

19. $f(x, y) = 4e^{xy}$

The first partial derivatives of f, $f_x(x, y) = 4ye^{xy}$ and $f_y(x, y) = 4xe^{xy}$, are zero at the point (0, 0).

Moreover, because

$f_{xx}(x, y) = 4y^2e^{xy}$, $f_{yy}(x, y) = 4x^2e^{xy}$, and $f_{xy}(x, y) = 4e^{xy}(xy + 1)$, it follows that $f_{xx}(0, 0) = 0$ and $f_{xx}(0, 0)f_{yy}(0, 0) - [f_{xy}(0, 0)]^2 = -16 < 0$.

So, (0, 0, 4) is a saddle point.

21. $f_{xx} > 0$ and $f_{xx}f_{yy} - (f_{xy})^2 = (9)(4) - 6^2 = 0$

Insufficient information

23. $f_{xx} < 0$ and $f_{xx}f_{yy} - (f_{xy})^2 = (-9)(6) - 10^2 < 0$

f has a saddle point at (x_0, y_0).

25. $f(x, y) = (xy)^2$

The first partial derivatives of f, $f_x(x, y) = 2xy^2$ and $f_y(x, y) = 2x^2y$, are zero at the points $(a, 0)$ and $(0, b)$ where a and b are any real numbers. Because

$f_{xx}(x, y) = 2y^2$, $f_{yy}(x, y) = 2x^2$, and $f_{xy}(x, y) = 4xy$, it follows that

$f_{xx}(a, 0)f_{yy}(a, 0) - [f_{xy}(a, 0)]^2 = 0$ and

$f_{xx}(0, b)f_{yy}(0, b) - [f_{xy}(0, b)]^2 = 0$ and the Second-Derivative Test fails. You can note that $f(x, y) = (xy)^2$ is nonnegative for all $(a, 0, 0)$ and $(0, b, 0)$ where a and b are real numbers.

So, $(a, 0, 0)$ and $(0, b, 0)$ are relative minima.

27. $f(x, y) = x^3 + y^3$

The first partial derivatives of f, $f_x(x, y) = 3x^2$ and $f_y(x, y) = 3y^2$, are zero at $(0, 0)$. Because $f_{xx}(x, y) = 6x$, $f_{yy}(x, y) = 6y$, $f_{xy}(x, y) = 0$, and $f_{xx}(0, 0)f_{yy}(0, 0) - [f_{xy}(0, 0)]^2 = 0$, the Second-Partials Test fails. By testing "nearby" points, you can conclude that (0, 0, 0) is a saddle point.

29. $f(x, y) = x^{2/3} + y^{2/3}$

The first partial derivatives of f,

$f_x(x, y) = \frac{2}{3\sqrt[3]{x}}$ and $f_y(x, y) = \frac{2}{3\sqrt[3]{y}}$, are undefined

at the point (0, 0). Because $f_{xx}(x, y) = -\frac{2}{9x^{4/3}}$,

$f_{yy}(x, y) = -\frac{2}{9y^{4/3}}$, $f_{xy}(x, y) = 0$ and $f_{xx}(0, 0)$ is undefined, the Second-Derivative Test fails. Because $f(x, y) \geq 0$ for all points in the xy-coordinate plane, (0, 0, 0) is a relative minimum.

31. $f(x, y, z) = (x - 1)^2 + (y + 3)^2 + z^2$

Critical point: $(x, y, z) = (1, -3, 0)$

Relative minimum

33. The sum is $x + y + z = 30$ and $z = 30 - x - y$, and the product is $P = xyz = 30xy - x^2y - xy^2$. The first partial derivatives of P are

$P_x = 30y - 2xy - y^2 = y(30 - 2x - y)$

$P_y = 30x - x^2 - 2xy = x(30 - x - 2y)$.

Setting these equal to zero produces the system

$2x + 2y = 30$

$x + 2y = 30$.

Solving the system, you have $x = 10$, $y = 10$, and $z = 10$.

35. The sum is

$x + y + z = 30$

$z = 30 - x - y$

and the sum of the squares is

$S = x^2 + y^2 + z^2 = x^2 + y^2 + (30 - x - y)^2$. The first partial derivatives of S are

$S_x = 2x - 2(30 - x - y) = 4x + 2y - 60$ and

$S_y = 2y - 2(30 - x - y) = 2x + 4y - 60$.

Setting these equal to zero produces the system

$2x + y = 30$

$x + 2y = 30$.

Solving this system, you have $x = 10$, $y = 10$.

So, $z = 10$.

37. The duration function is

$D(x, y) = x^2 + 2y^2 - 18x - 24y + 2xy + 120$ and the first partial derivatives are $D_x = 2x - 18 + 2y$ and $D_y = 4y - 24 + 2x$.

Setting these equal to zero produces the system

$2x + 2y = 18$

$2x + 4y = 24$.

Solving this system, you have $x = 6$ and $y = 3$.

So, to minimize the duration of the infection you should take 600 mg of the first drug and 300 mg of the second drug.

39. Let x = length, y = width, and z = height.

The sum of length and girth is

$x + (2y + 2z) = 96$

$x = 96 - 2y - 2z$.

The volume is $V = xyz = 96yz - 2zy^2 - 2yz^2$ and the first partial derivatives are

$V_y = 96z - 4zy - 2z^2 = 2z(48 - 2y - z)$ and

$V_z = 96y - 2y^2 - 4yz = 2y(48 - y - 2z)$.

Setting these equal to zero produces the system

$2y + z = 48$

$y + 2z = 48$.

Solving this system, you have $y = 16$ and $z = 16$.

So, $x = 32$.

The volume is a maximum when the length is 32 inches and the width and height are each 16 inches.

41. Let x = length, y = width, and z = height.

The volume is $4608 = xyz$.

The surface area of the room is $S = xy + 2yz + 2xz$.

Because the walls will have three coats and the ceiling will have two coats, the surface area is $S = 2xy + 6yz + 6xz$.

Solve the volume function for x and substitute the result into S.

$$x = \frac{4608}{yz}$$

$$S = 2\left(\frac{4608}{yz}\right)(y) + 6yz + 6\left(\frac{4608}{yz}\right)(z)$$

$$= \frac{9216}{z} + 6yz + \frac{27{,}648}{y}$$

The first partial derivatives are

$$S_y = 6z - \frac{27{,}648}{y^2} \text{ and } S_z = -\frac{9216}{z^2} + 6y.$$

Setting these equal to zero produces the following.

$$z = \frac{4608}{y^2} \text{ and } y = \frac{1536}{z^2}$$

Substitute the expression for z into y and solve for y.

$$y = \frac{1536}{\left(\frac{4608}{y^2}\right)^2} = \frac{1}{13{,}824}y^4$$

$$0 = \frac{1}{13{,}824}y^4 - y$$

$y = 0$ or $y = 24$

Because $y = 0$ doesn't make sense, you can conclude $y = 24$ and $z = 8$. Substitute these values into the equation for x.

$$x = \frac{4608}{(24)(8)} = 24$$

To minimize the amount of paint, the dimensions of the room are 24 feet by 24 feet by 8 feet.

The surface area is

$S = 2(24)(24) + 6(24)(8) + 6(24)(8) = 3456$ square feet

Gallons needed: $\frac{3456}{500} \approx 6.91$

So, the contractor should buy 7 gallons of paint.

43. (a) Total area: $A = x(12 - x) + y(12 - x - y)$

$$= 12x - x^2 + 12y - xy - y^2$$

First partial derivatives:

$A_x = 12 - 2x - y \qquad A_y = 12 - x - 2y$

Setting these equal to zero produces the system

$2x + y = 12$

$x + 2y = 12.$

Solving this system, you have $x = 4$ and $y = 4$.

So, the area is a maximum when $x = 4$ meters and $y = 4$ meters.

(b) Maximum area:

$A = 12(4) - 4^2 + 12(4) - 4(4) - 4^2$

$= 48$ square meters

(c) Total perimeter:

$P = 2(4) + 2(12 - 4) + 2(4) + 2(12 - 4 - 4)$

$= 40$ meters

So, 40 meters of fencing are needed.

45. The population function is

$P(p, q, r) = 2pq + 2pr + 2qr.$

Because $p + q + r = 1$, $r = 1 - p - q$.

So, $P = 2pq + 2p(1 - p - q) + 2q(1 - p - q)$

$= -2p^2 + 2p - 2q^2 + 2q - 2pq$

and the first partial derivatives are

$P_p = -4p + 2 - 2q$

$P_q = -4q + 2 - 2p.$

Setting these equal to zero produces the system

$4p + 2q = 2$

$2p + 4q = 2.$

Solving this system, you have $p = \frac{1}{3}$ and $q = \frac{1}{3}$. So, $r = \frac{1}{3}$.

The proportion is a maximum when $p = \frac{1}{3}$, $q = \frac{1}{3}$, and $r = \frac{1}{3}$.

The maximum proportion is

$P = 2\left(\frac{1}{3}\right)\left(\frac{1}{3}\right) + 2\left(\frac{1}{3}\right)\left(\frac{1}{3}\right) + 2\left(\frac{1}{3}\right)\left(\frac{1}{3}\right) = \frac{6}{9} = \frac{2}{3}.$

47. $S = 12xyz$

$0.13 = 12xyz$

$\dfrac{0.13}{12xy} = z$

Total cost:

$$C = x + 2y + 3z = x + 2y + 3\left(\frac{0.13}{12xy}\right) = x + 2y + \frac{0.0325}{xy}$$

First partial derivatives:

$$C_x = 1 - \frac{0.0325}{x^2y}$$

$$C_y = 2 - \frac{0.0325}{xy^2}$$

Setting these equal to zero produces the system

$$1 - \frac{0.0325}{x^2y} = 0$$

$$2 - \frac{0.0325}{xy^2} = 0.$$

Solving this system, you have $x \approx 0.402$ and $y \approx 0.201$.

So, $z = \dfrac{0.13}{12(0.402)(0.201)} \approx 0.134.$

To minimize the cost, use 0.402 liter of nutrient solution x, 0.201 liter of nutrient solution y, and 0.134 liter of nutrient solution z.

49. Let x = length, y = width, and z = height.

$V = xyz = 1000$

$z = \dfrac{1000}{xy}$

Heat loss: $H = 5xy + xy + 3(2yz) + 3(2xz)$

$$= 6xy + 6yz + 6xz$$

$$= 6xy + 6y\left(\frac{1000}{xy}\right) + 6x\left(\frac{1000}{xy}\right)$$

$$= 6xy + \frac{6000}{x} + \frac{6000}{y}$$

First partial derivatives:

$$H_x = 6y - \frac{6000}{x^2} \qquad H_y = 6x - \frac{6000}{y^2}$$

Setting these equal to zero produces the system

$$6y - \frac{6000}{x^2} = 0$$

$$6x - \frac{6000}{y^2} = 0.$$

Solving this system, you have $x = 10$ and $y = 10$. So, $z = \dfrac{1000}{(10)(10)} = 10$. To minimize heat loss, the length of the room is 10 feet, the width is 10 feet, and the height is 10 feet.

51. True

Section 9.6 Least Squares Regression Analysis

Skills Review

1. $(2.5 - 1)^2 + (3.25 - 2)^2 + (4.1 - 3)^2 = (1.5)^2 + (1.25)^2 + (1.1)^2 = 2.25 + 1.5625 + 1.21 = 5.0225$

2. $(1.1 - 1)^2 + (2.08 - 2)^2 + (2.95 - 3)^2 = (0.1)^2 + (0.08)^2 + (-0.05)^2 = 0.01 + 0.0064 + 0.0025 = 0.0189$

3. $S = a^2 + 6b^2 - 4a - 8b - 4ab + 6$

$$\frac{\partial S}{\partial a} = 2a - 4 - 4b$$

$$\frac{\partial S}{\partial b} = 12b - 8 - 4a$$

4. $S = 4a^2 + 9b^2 - 6a - 4b - 2ab + 8$

$$\frac{\partial S}{\partial a} = 8a - 6 - 2b$$

$$\frac{\partial S}{\partial b} = 18b - 4 - 2a$$

5. $\displaystyle\sum_{i=1}^{5} i = 1 + 2 + 3 + 4 + 5 = 15$

6. $\displaystyle\sum_{i=1}^{6} 2i = 2(1) + 2(2) + 2(3) + 2(4) + 2(5) + 2(6) = 42$

7. $\displaystyle\sum_{i=1}^{4} \frac{1}{i} = \frac{1}{1} + \frac{1}{2} + \frac{1}{3} + \frac{1}{4} = \frac{25}{12}$

8. $\displaystyle\sum_{i=1}^{3} i^2 = 1^2 + 2^2 + 3^2 = 14$

Skills Review —*continued*—

9. $\sum_{i=1}^{6}(2-i)^2 = (2-1)^2 + (2-2)^2 + (2-3)^2 + (2-4)^2 + (2-5)^2 + (2-6)^2$

$= 1^2 + 0^2 + (-1)^2 + (-2)^2 + (-3)^2 + (-4)^2 = 31$

10. $\sum_{i=1}^{5}(30-i^2) = (30-1) + (30-(2^2)) + (30-(3^2)) + (30-(4^2)) + (30-(5^2))$

$= 29 + 26 + 21 + 14 + 5 = 95$

1. (a) $\sum x_i = 0$

$\sum y_i = 4$

$\sum x_i y_i = 6$

$\sum x_i^2 = 8$

$$a = \frac{3(6) - 0(4)}{3(8) - 0^2} = \frac{3}{4}$$

$$b = \frac{1}{3}\left[4 - \frac{3}{4}(0)\right] = \frac{4}{3}$$

The regression line is $y = \frac{3}{4}x + \frac{4}{3}$.

(b) $S = \left(-\frac{3}{2} + \frac{4}{3} - 0\right)^2 + \left(\frac{4}{3} - 1\right)^2 + \left(\frac{3}{2} + \frac{4}{3} - 3\right)^2$

$= \frac{1}{6}$

3. (a) $\sum x_i = 4$

$\sum y_i = 8$

$\sum x_i y_i = 4$

$\sum x_i^2 = 6$

$$a = \frac{4(4) - 4(8)}{4(6) - 4^2} = -2$$

$$b = \frac{1}{4}\left[8 + 2(4)\right] = 4$$

The regression line is $y = -2x + 4$.

(b) $S = (4-4)^2 + (2-3)^2 + (2-1)^2 + (0-0)^2$

$= 2$

5. (a) $\sum x_i = 11$

$\sum y_i = 15$

$\sum x_i y_i = 47$

$\sum x_i^2 = 39$

$$a = \frac{5(47) - (11)(15)}{5(39) - 11^2} = \frac{35}{37}$$

$$b = \frac{1}{5}\left[15 - \frac{35}{37}(11)\right] = \frac{34}{37}$$

The regression line is $y = \frac{35}{37}x + \frac{34}{37}$.

(b) $S = \left(\frac{34}{37} - 2\right)^2 + \left(\frac{69}{37} - 1\right)^2 + \left(\frac{104}{37} - 2\right)^2 + \left(\frac{139}{37} - 4\right)^2 + \left(\frac{209}{37} - 6\right)^2 = \frac{102}{37}$

7. The sum of the squared errors is as follows.

$$S = (-2a + b + 1)^2 + (0a + b)^2 + (2a + b - 3)^2$$

$$\frac{\partial S}{\partial a} = 2(-2a + b + 1)(-2) + 2(2a + b - 3)(2) = 16a - 16$$

$$\frac{\partial S}{\partial b} = 2(-2a + b + 1) + 2b + 2(2a + b - 3) = 6b - 4$$

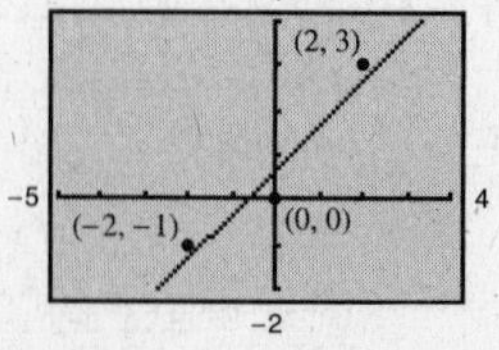

Setting these partial derivatives equal to zero produces $a = 1$ and $b = \frac{2}{3}$. So, $y = x + \frac{2}{3}$.

9. The sum of the squared errors is as follows.

$$S = (-2a + b - 4)^2 + (-a + b - 1)^2 + (b + 1)^2 + (a + b + 3)^2$$

$$\frac{\partial S}{\partial a} = -4(-2a + b - 4) - 2(-a + b - 1) + 2(a + b + 3) = 12a - 4b + 24$$

$$\frac{\partial S}{\partial b} = 2(-2a + b - 4) + 2(-a + b - 1) + 2(b + 1) + 2(a + b + 3) = -4a + 8b - 2$$

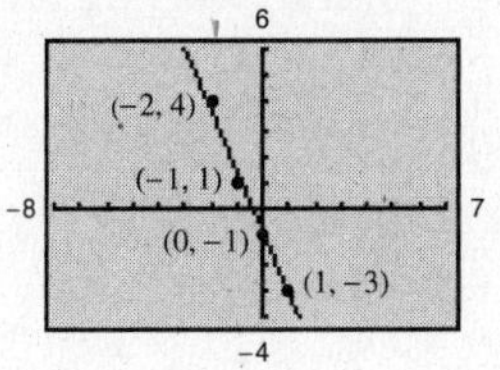

Setting these partial derivatives equal to zero produces:

$$12a - 4b = -24$$
$$-4a + 8b = 2$$

So, $a = -2.3$ and $b = -0.9$, and $y = -2.3x - 0.9$.

11. $y = 0.7x + 1.4$

13. $y = x + 4$

15. $y = -0.65x + 1.75$

17. $y = 0.8605x + 0.1628$

19. $y = -1.1824x + 6.3851$

21. $y = 0.4286x^2 + 1.2x + 0.7429$

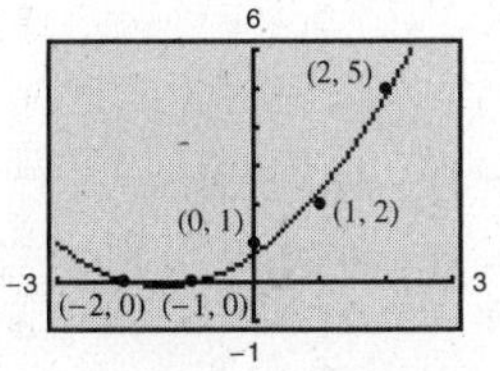

23. $y = x^2 - x$

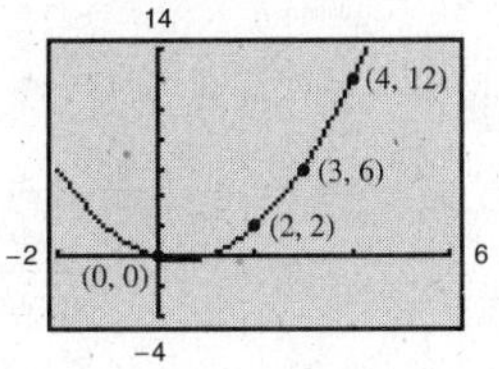

25. Linear: $y = 1.4286x + 6$

Quadratic: $y = 0.1190x^2 + 1.6667x + 5.6429$

The quadratic model is a better fit.

27. Linear: $y = -68.9143x + 753.9524$

Quadratic: $y = 2.8214x^2 - 83.0214x + 763.3571$

The quadratic model is a better fit.

29. $y = 71.2t + 40$

For 2005, $t = 15$.

When $t = 15$, $y = 1108$ thousand 45-54 years old.

31. The least squares regression quadratic is

$y = -54.9464x^2 + 4959.4643x - 50{,}050$.

When $x = 28$, $y \approx \$45{,}737$.

33. (a) The least squares regression line is $y = -0.2383t + 11.9286$.

For 2010, $t = 30$. When $t = 30$,

$y \approx 4.8$ deaths per 1000 live births.

(b) The least squares regression quadratic is $y = 0.0088t^2 - 0.4579t + 12.6607$.

When $t = 30$, $y \approx 6.8$ deaths per 1000 live births.

35. (a)

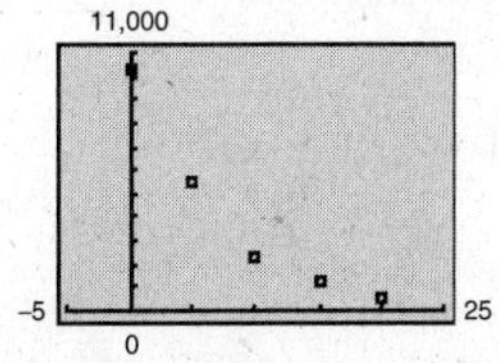

(b) Sample answer: $P = 28.92h^2 - 1057.9h + 10{,}250$

(c) Sample answer: The r-value obtained for the quadratic model, $r \approx 0.998$, is closer to 1 than that of the linear model, $r \approx -0.940$.

37. Linear: $y = 3.7569x + 9.0347$

Quadratic: $y = 0.0063x^2 + 3.6252x + 9.4282$

Either model is a good fit for the data.

39. Quadratic: $y = -0.08715x^2 + 2.8159x + 0.3975$

41. Positive correlation $r \approx 0.9981$

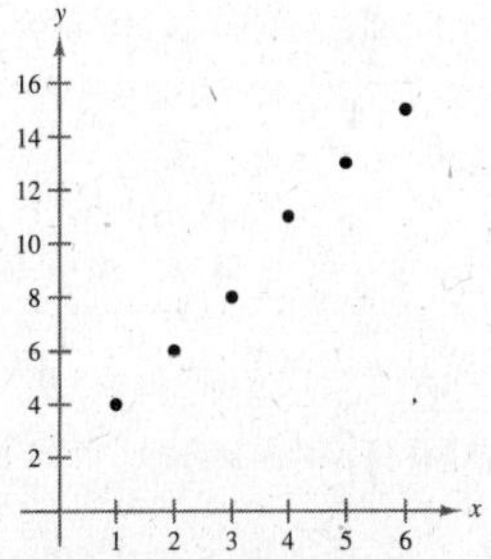

43. No correlation

$r = 0$

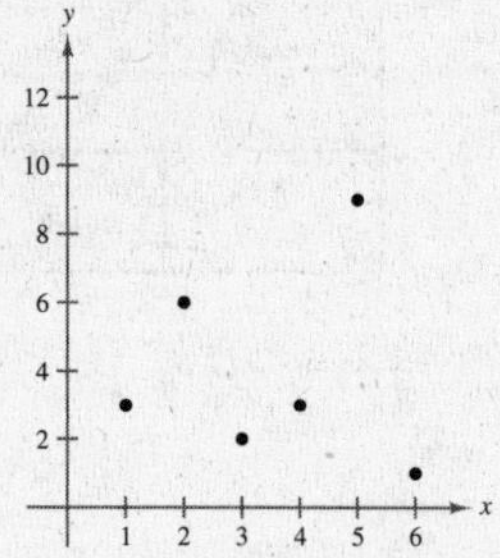

45. No correlation

$r \approx 0.0750$

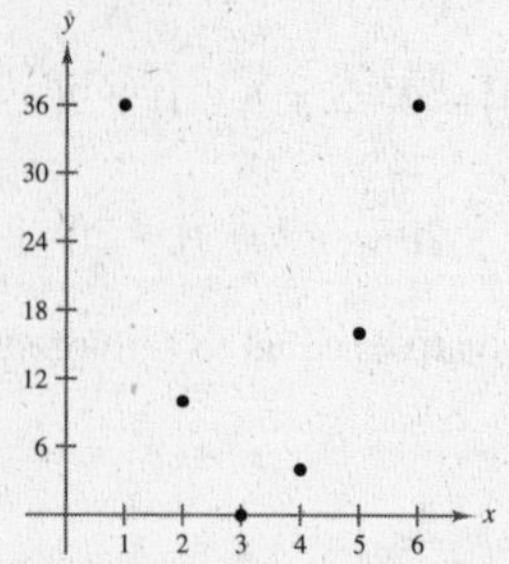

47. False, the slope is positive, which means there is a positive correlation.

49. True

51. True

53. Answers will vary.

Section 9.7 Double Integrals and Area in a Plane

Skills Review

1. $\int_0^1 dx = x\Big]_0^1 = 1$

2. $\int_0^2 3dy = 3y\Big]_0^2 = 6$

3. $\int_1^4 2x^2\,dx = \frac{2}{3}x^3\Big]_1^4 = \left[\frac{2}{3}(4^3) - \frac{2}{3}(1^3)\right] = 42$

4. $\int_0^1 2x^3\,dx = \frac{1}{2}x^4\Big]_0^1 = \frac{1}{2}$

5. $\int_1^2 \left(x^3 - 2x + 4\right) dx = \left[\frac{1}{4}x^4 - x^2 + 4x\right]_1^2$

$= (4 - 4 + 8) - \left(\frac{1}{4} - 1 + 4\right) = \frac{19}{4}$

6. $\int_0^2 \left(4 - y^2\right) dy = \left[4y - \frac{1}{3}y^3\right]_0^2 = \frac{16}{3}$

7. $\int_1^2 \frac{2}{7x^2}\,dx = -\frac{2}{7x}\Big]_1^2 = -\frac{2}{14} + \frac{2}{7} = \frac{1}{7}$

8. $\int_1^4 \frac{2}{\sqrt{x}}\,dx = 4\sqrt{x}\Big]_1^4 = 8 - 4 = 4$

9. $\int_0^2 \frac{2x}{x^2 + 1}\,dx = \ln\left(x^2 + 1\right)\Big]_0^2$

$= \ln 5 - \ln 1 = \ln 5 \approx 1.609$

10. $\int_2^e \frac{1}{y - 1}\,dy = \ln(y - 1)\Big]_2^e$

$= \ln(e - 1) - \ln(2 - 1)$

$= \ln(e - 1) \approx 0.541$

11. $\int_0^2 xe^{x^2+1}\,dx = \frac{1}{2}e^{x^2+1}\Big]_0^2 = \frac{1}{2}e^5 - \frac{1}{2}e \approx 72.847$

12. $\int_0^1 e^{-2y}\,dy = -\frac{1}{2}e^{-2y}\Big]_0^1 = -\frac{1}{2}e^{-2} + \frac{1}{2} \approx 0.432$

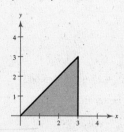

13. $y = x,\ y = 0,\ x = 3$

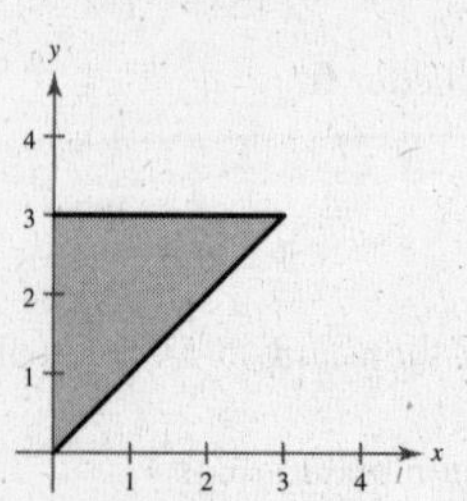

14. $y = x,\ y = 3,\ x = 0$

Skills Review —continued—

15. $y = 4 - x^2,\ y = 0$

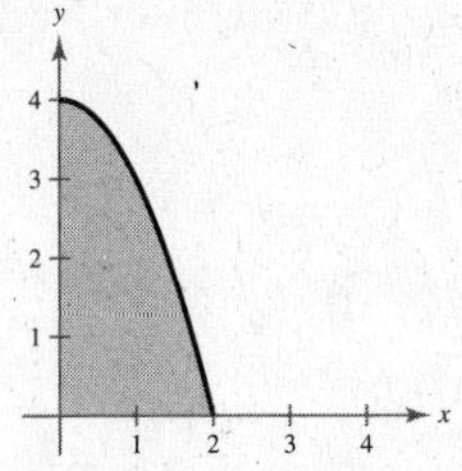

16. $y = x^2,\ y = 4x$

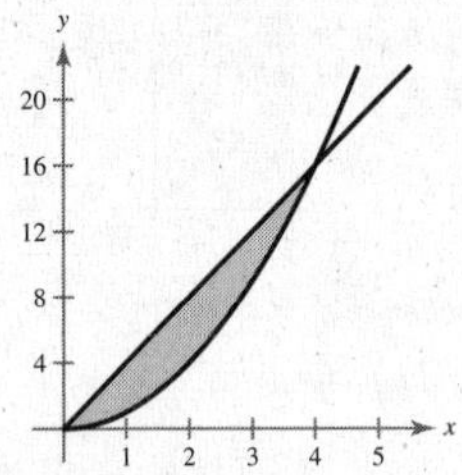

1. $\int_0^x (2x - y)\, dy = \left[2xy - \frac{y^2}{2}\right]_0^x = \frac{3x^2}{2}$

3. $\int_1^{2y} \frac{y}{x}\, dx = y \ln|x|\Big]_1^{2y} = y \ln|2y|$

5. $\int_0^{\sqrt{4-x^2}} x^2 y\, dy = \frac{x^2y^2}{2}\Big]_0^{\sqrt{4-x^2}} = \frac{x^2(4 - x^2)}{2} = \frac{4x^2 - x^4}{2}$

7. $\int_1^{e^y} \frac{y \ln x}{x}\, dx = \frac{y(\ln x)^2}{2}\Big]_1^{e^y}$

$= \frac{y(\ln e^y)^2}{2} - \frac{y(\ln 1)^2}{2} = \frac{y^3}{2}$

9. Using Formula 35,

$\int_0^x ye^{xy}\, dy = \frac{1}{x^2}\int_0^x xye^{xy}x\, dy$

$= \frac{1}{x^2}(xy - 1)e^{xy}\Big]_0^x$

$= \frac{1}{x^2}(x^2 - 1)e^{x^2} - \frac{1}{x^2}(-1)(1)$

$= e^{x^2} - \frac{e^{x^2}}{x^2} + \frac{1}{x^2}$

11. $\int_0^1\int_0^2 (x + y)\, dy\, dx = \int_0^1\left[xy + \frac{y^2}{2}\right]_0^2 dx$

$= \int_0^1 (2x + 2)\, dx$

$= \left[x^2 + 2x\right]_0^1$

$= 3$

13. $\int_0^4\int_0^3 xy\, dy\, dx = \int_0^4 \frac{xy^2}{2}\Big]_0^3 dx$

$= \frac{9}{2}\int_0^4 x\, dx = \frac{9}{2}\left[\frac{x^2}{2}\right]_0^4 = 36$

15. $\int_0^1\int_0^y (x + y)dx\, dy = \int_0^1\left[\frac{x^2}{2} + xy\right]_0^y dy$

$= \int_0^1 \frac{3y^2}{2}\, dy = \left[\frac{y^3}{2}\right]_0^1 = \frac{1}{2}$

17. $\int_1^2\int_0^4 (3x^2 - 2y^2 + 1)\, dx\, dy = \int_1^2 \left[x^3 - 2xy^2 + x\right]_0^4 dy$

$= \int_1^2 (68 - 8y^2)\, dy$

$= \left[68y - \frac{8y^3}{3}\right]_1^2$

$= \frac{148}{3}$

19. $\int_0^2\int_0^{\sqrt{1-y^2}} -5xy\, dx\, dy = -5\int_0^2 \frac{x^2}{2}y\Big]_0^{\sqrt{1-y^2}} dy$

$= \frac{-5}{2}\int_0^2 (y - y^3)\, dy$

$= -\frac{5}{2}\left[\frac{y^2}{2} - \frac{y^4}{4}\right]_0^2$

$= -\frac{5}{2}(2 - 4) = 5$

21. $\int_0^2\int_0^{6x^2} x^3\, dy\, dx = \int_0^2 x^3y\Big]_0^{6x^2} dx$

$= \int_0^2 6x^5\, dx = x^6\Big]_0^2 = 64$

23. Because (for a fixed x)

$\lim_{b\to\infty} -2e^{-(x+y)/2}\Big]_0^b = 2e^{-x/2}$,

you have

$\int_0^\infty\int_0^\infty e^{-(x+y)/2}\, dy\, dx = \int_0^\infty 2e^{-x/2}\, dx$

$= \lim_{b\to\infty} -4e^{-x/2}\Big]_0^b = 4.$

25. $\int_0^1 \int_0^2 dy\,dx = \int_0^1 2\,dx = 2$

$\int_0^2 \int_0^1 dx\,dy = \int_0^2 dy = 2$

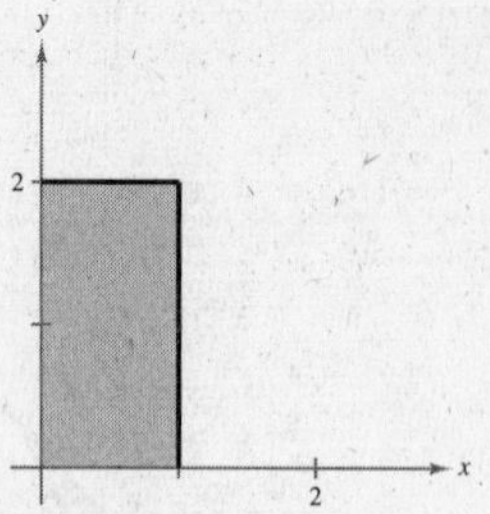

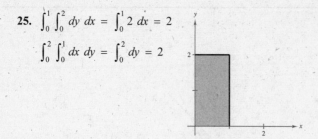

27. $\int_0^1 \int_{2y}^2 dx\,dy = \int_0^1 (2 - 2y)\,dy = \Big[2y - y^2\Big]_0^1 = 1$

$\int_0^2 \int_0^{x/2} dy\,dx = \int_0^2 \frac{x}{2}\,dx = \frac{x^2}{4}\Big]_0^2 = 1$

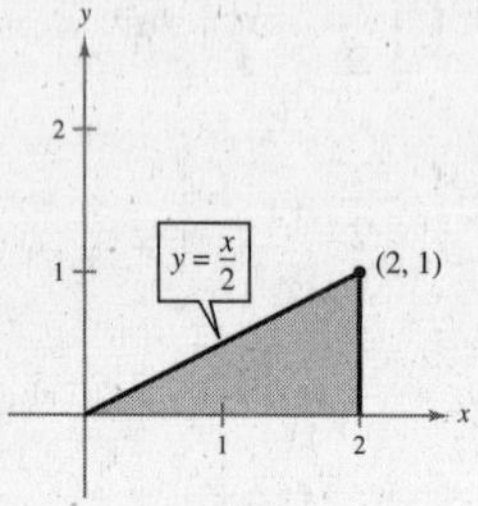

29. $\int_0^2 \int_{x/2}^1 dy\,dx = \int_0^2 \left(1 - \frac{x}{2}\right) dx = \left[x - \frac{x^2}{4}\right]_0^2 = 1$

$\int_0^1 \int_0^{2y} dx\,dy = \int_0^1 2y\,dy = y^2\Big]_0^1 = 1$

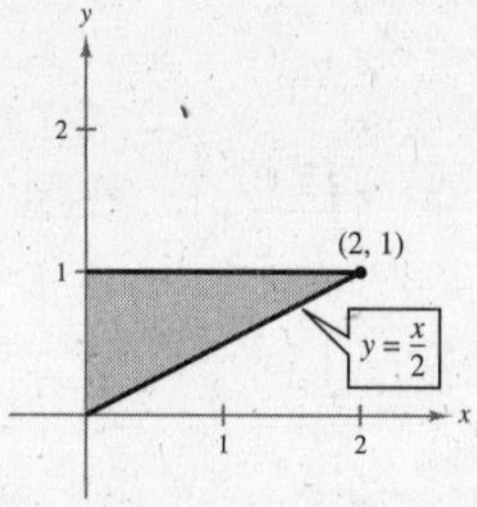

31. $\int_0^1 \int_{y^2}^{\sqrt[3]{y}} dx\,dy = \int_0^1 \left(\sqrt[3]{y} - y^2\right) dy$

$= \left[\frac{3}{4}y^{4/3} - \frac{y^3}{3}\right]_0^1 = \frac{5}{12}$

$\int_0^1 \int_{x^3}^{\sqrt{x}} dy\,dx = \int_0^1 \left(\sqrt{x} - x^3\right) dx$

$= \left[\frac{2}{3}x^{3/2} - \frac{x^4}{4}\right]_0^1 = \frac{5}{12}$

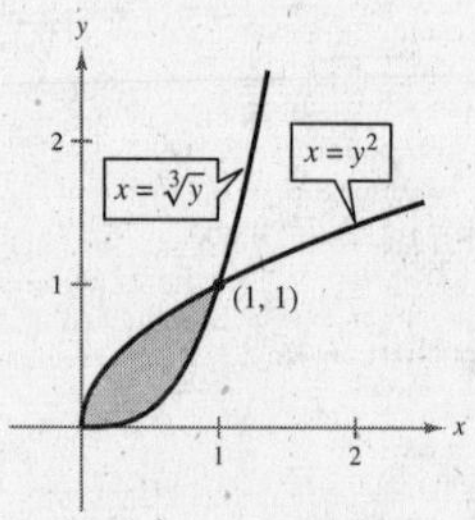

33. $\int_0^3 \int_y^3 e^{x^2}\,dx\,dy = \int_0^3 \int_0^x e^{x^2}\,dy\,dx$

$= \int_0^3 ye^{x^2}\Big]_0^x\,dx$

$= \int_0^3 xe^{x^2}\,dx$

$= \frac{1}{2}e^{x^2}\Big]_0^3$

$= \frac{1}{2}\left(e^9 - 1\right)$

≈ 4051.042

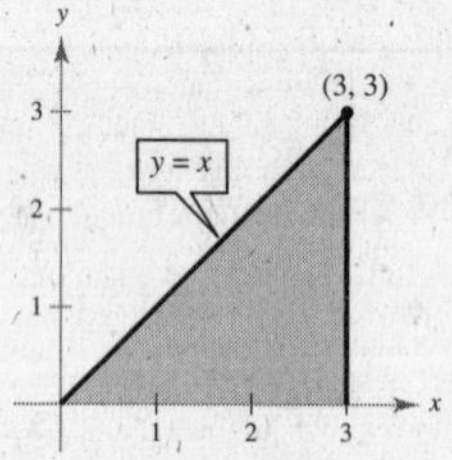

35. $A = \int_0^8 \int_0^3 dy\,dx$

$= \int_0^8 3\,dx$

$= 3x\Big]_0^8$

$= 24$

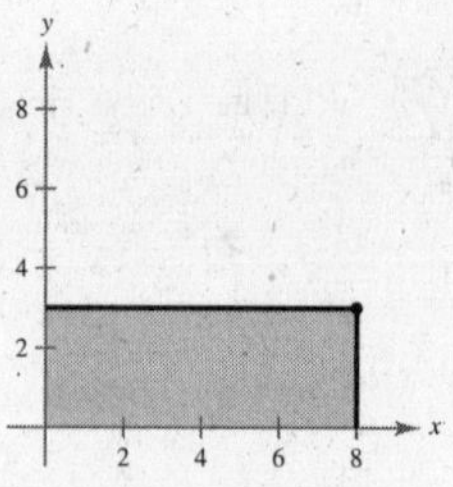

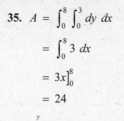

37. $A = \int_0^2 \int_0^{4-x^2} dy\,dx$

$= \int_0^2 \left(4 - x^2\right) dx$

$= \left[4x - \frac{x^3}{3}\right]_0^2$

$= \frac{16}{3}$

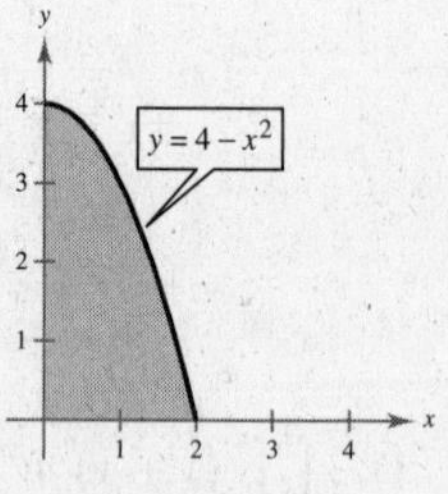

39. $A = \int_0^4 \int_0^{\left(2-\sqrt{x}\right)^2} dy\,dx$

$= \int_0^4 \left(4 - 4\sqrt{x} + x\right) dx$

$= \left[4x - \frac{8}{3}x^{3/2} + \frac{x^2}{2}\right]_0^4$

$= \frac{8}{3}$

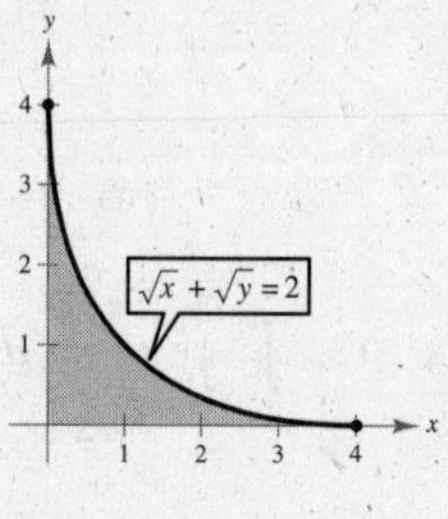

41. $A = \int_{-3}^{3}\int_{0}^{9-x^2} dy\,dx$

$= \int_{-3}^{3} (9 - x^2)\,dx$

$= \left[9x - \frac{x^3}{3}\right]_{-3}^{3}$

$= 36$

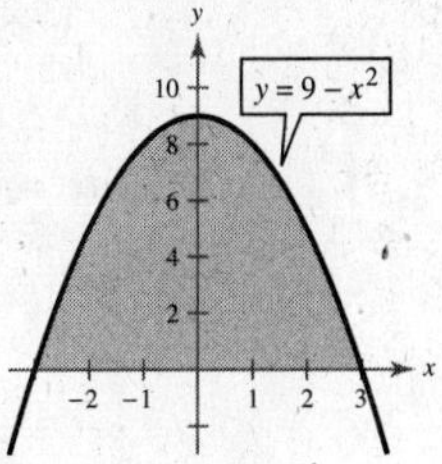

43. $A = \int_{0}^{2}\int_{\frac{3}{2}y}^{-y+5} dx\,dy$

$= \int_{0}^{2}\left(-y + 5 - \frac{3}{2}y\right) dy$

$= \int_{0}^{2}\left(-\frac{5}{2}y + 5\right) dy$

$= \left[-\frac{5y^2}{4} + 5y\right]_{0}^{2} = 5$

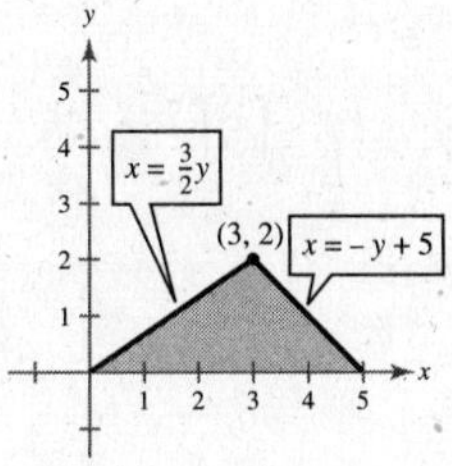

45. The point of intersection of the two graphs is found by equating $y = 2x$ and $y = x$, which yields $x = y = 0$.

$A = \int_{0}^{2}\int_{x}^{2x} dy\,dx$

$= \int_{0}^{2} (2x - x)\,dx$

$= \int_{0}^{2} x\,dx$

$= \left.\frac{x^2}{2}\right]_{0}^{2} = 2$

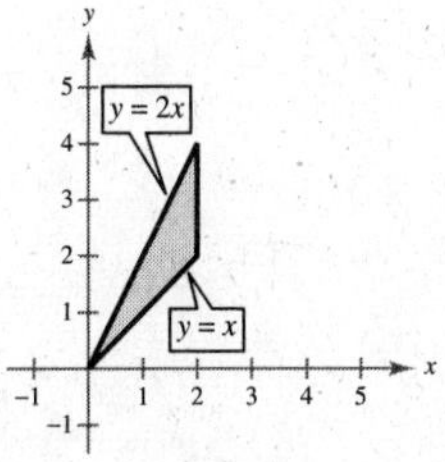

47. $\int_{0}^{1}\int_{0}^{2} e^{-x^2-y^2}\,dx\,dy \approx 0.6588$

49. $\int_{1}^{2}\int_{0}^{x} e^{xy}\,dy\,dx \approx 8.1747$

51. $\int_{0}^{1}\int_{x}^{1} \sqrt{1 - x^2}\,dy\,dx \approx 0.4521$

53. $\int_{0}^{2}\int_{\sqrt{4-x^2}}^{4-x^2/4} \frac{xy}{x^2 + y^2 + 1}\,dy\,dx \approx 1.1190$

55. True

$\int_{-1}^{1}\int_{-2}^{2} y\,dy\,dx = \int_{-1}^{1} \left.\frac{y^2}{2}\right]_{-2}^{2} dx = \int_{-1}^{1} 0\,dx = 0$

$\int_{-1}^{1}\int_{-2}^{2} y\,dx\,dy = \int_{-1}^{1} xy\Big]_{-2}^{2} dy = \int_{-1}^{1} 4y\,dy = 2y^2\Big]_{-1}^{1} = 0$

Section 9.8 Applications of Double Integrals

Skills Review

1.

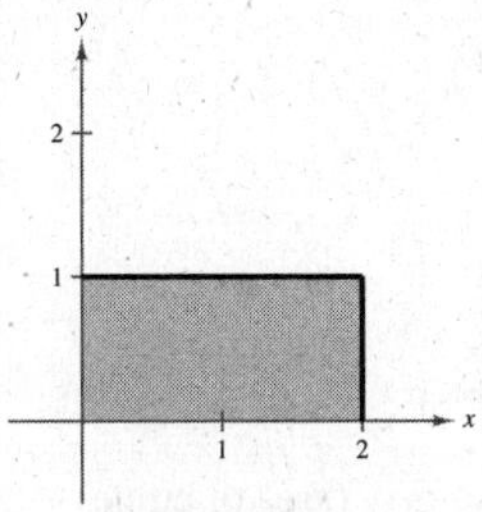

2.

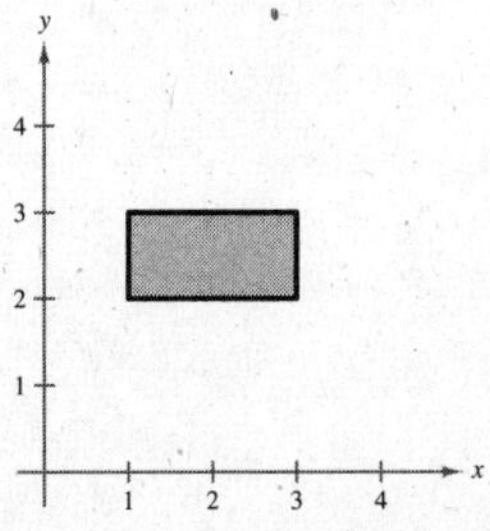

3.

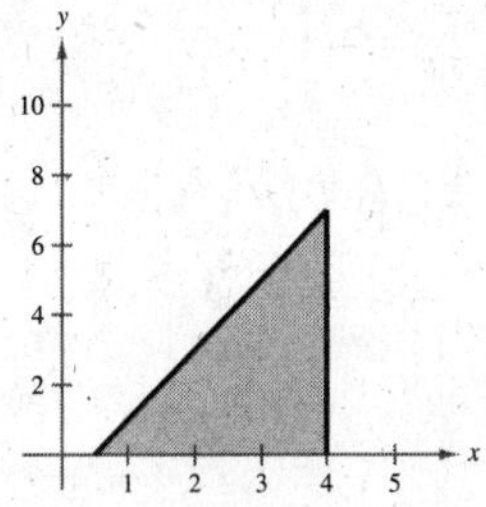

4.

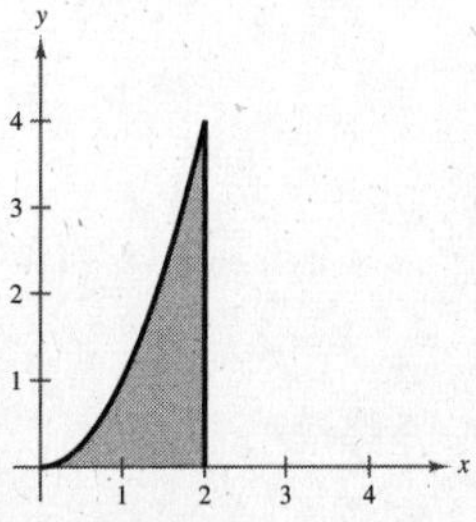

Skills Review —*continued*—

5. $\int_0^1 \int_1^2 dy\,dx = \int_0^1 y\Big]_1^2 dx = \int_0^1 dx = x\Big]_0^1 = 1$

6. $\int_0^3 \int_1^3 dx\,dy = \int_0^3 x\Big]_1^3 dy = \int_0^3 2\,dy = 2y\Big]_0^3 = 6$

7. $\int_0^1 \int_0^x x\,dy\,dx = \int_0^1 xy\Big]_0^x dx$

$= \int_0^1 x^2\,dx = \frac{1}{3}x^3\Big]_0^1 = \frac{1}{3}$

8. $\int_0^4 \int_1^y y\,dx\,dy = \int_0^4 xy\Big]_1^y dy = \int_0^4 \left(y^2 - y\right) dy$

$= \left[\frac{1}{3}y^3 - \frac{1}{2}y^2\right]_0^4 = \frac{40}{3}$

9. $\int_1^3 \int_x^{x^2} 2\,dy\,dx = \int_1^3 2y\Big]_x^{x^2} dx$

$= \int_1^3 \left(2x^2 - 2x\right) dx$

$= \left[\frac{2}{3}x^3 - x^2\right]_1^3$

$= 9 + \frac{1}{3} = \frac{28}{3}$

10. $\int_0^1 \int_x^{-x^2+2} dy\,dx = \int_0^1 y\Big]_x^{-x^2+2} dx$

$= \int_0^1 \left(-x^2 + 2 - x\right) dx$

$= \left[-\frac{1}{3}x^3 + 2x - \frac{1}{2}x^2\right]_0^1 = \frac{7}{6}$

1. $\int_0^2 \int_0^1 (3x + 4y)\,dy\,dx = \int_0^2 \left[3xy + 2y^2\right]_0^1 dx$

$= \int_0^2 (3x + 2)\,dx$

$= \left[\frac{3}{2}x^2 + 2x\right]_0^2$

$= 10$

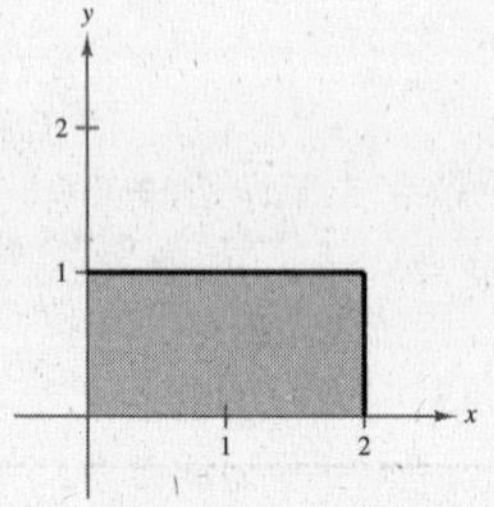

3. $\int_0^1 \int_y^{\sqrt{y}} x^2y^2\,dx\,dy = \int_0^1 \frac{x^3y^2}{3}\Big]_y^{\sqrt{y}} dy$

$= \frac{1}{3}\int_0^1 \left(y^{7/2} - y^5\right) dy$

$= \frac{1}{3}\left[\frac{2}{9}y^{9/2} - \frac{1}{6}y^6\right]_0^1$

$= \frac{1}{54}$

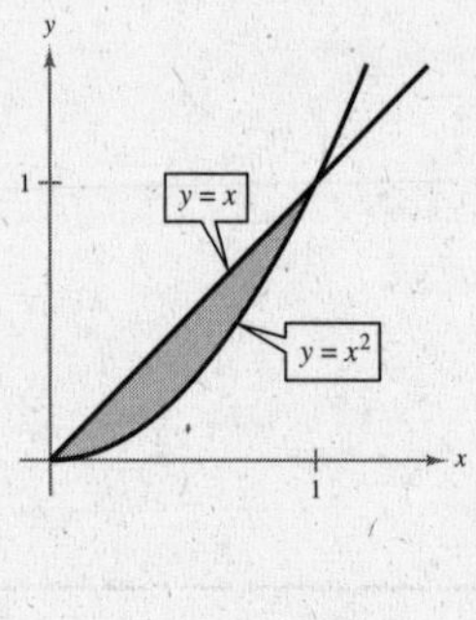

5. $\int_0^1 \int_0^{\sqrt{1-x^2}} y\,dy\,dx = \int_0^1 \frac{y^2}{2}\Big]_0^{\sqrt{1-x^2}} dx$

$= \frac{1}{2}\int_0^1 \left(1 - x^2\right) dx$

$= \frac{1}{2}\left[x - \frac{x^3}{3}\right]_0^1$

$= \frac{1}{3}$

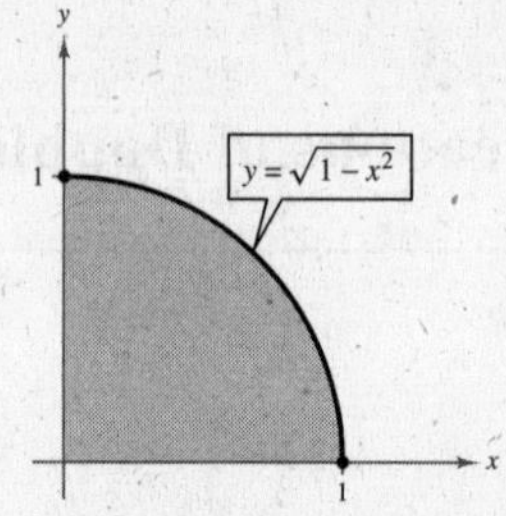

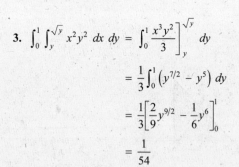

7. $\int_{-a}^a \int_{-\sqrt{a^2-x^2}}^{\sqrt{a^2-x^2}} dy\,dx = \int_{-a}^a y\Big]_{-\sqrt{a^2-x^2}}^{\sqrt{a^2-x^2}} dx$

$= \int_{-a}^a 2\sqrt{a^2 - x^2}\,dx$

$= \pi a^2$ (area of circle)

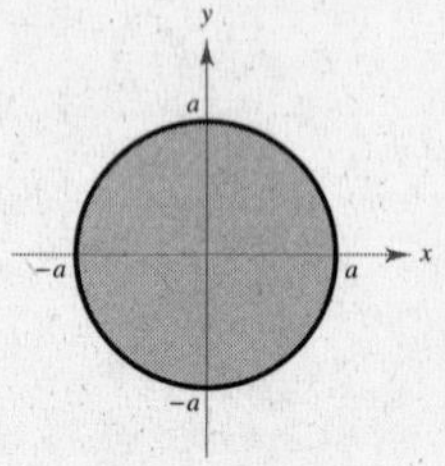

9. $\int_0^3 \int_0^5 xy\, dy\, dx = \int_0^5 \int_0^3 xy\, dx\, dy$

$$\int_0^3 \int_0^5 xy\, dy\, dx = \int_0^3 \frac{xy^2}{2}\bigg]_0^5 dx = \int_0^3 \frac{25}{2}x\, dx = \frac{25}{4}x^2\bigg]_0^3 = \frac{225}{4}$$

11. $\int_0^2 \int_x^{2x} \frac{y}{x^2+y^2}\, dy\, dx = \int_0^2 \int_{y/2}^{y} \frac{y}{x^2+y^2}\, dx\, dy + \int_2^4 \int_{y/2}^{2} \frac{y}{x^2+y^2}\, dx\, dy$

$$\int_0^2 \int_x^{2x} \frac{y}{x^2+y^2}\, dy\, dx = \int_0^2 \frac{1}{2}\ln\left(x^2+y^2\right)\bigg]_x^{2x} dx$$

$$= \frac{1}{2}\int_0^2 \left[\ln\left(5x^2\right) - \ln\left(2x^2\right)\right] dx = \frac{1}{2}\int_0^2 \ln\frac{5}{2}\, dx = \left(\frac{1}{2}\ln\frac{5}{2}\right)x\bigg]_0^2 = \ln\frac{5}{2} \approx 0.916$$

13. $\int_0^1 \int_{y/2}^{1/2} e^{-x^2}\, dx\, dy = \int_0^{1/2} \int_0^{2x} e^{-x^2}\, dy\, dx$

$$= \int_0^{1/2} \left[e^{-x^2} y\right]_0^{2x} dx$$

$$= \int_0^{1/2} 2xe^{-x^2}\, dx$$

$$= \left[-e^{-x^2}\right]_0^{1/2}$$

$$= -e^{-1/4} + 1$$

$$= 1 - e^{-1/4} \approx 0.221$$

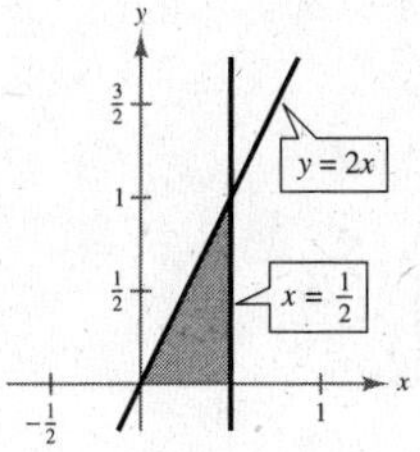

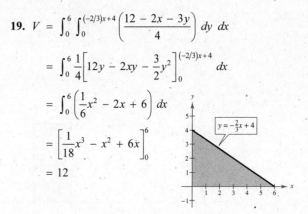

15. $V = \int_0^2 \int_0^4 \frac{y}{2}\, dx\, dy$

$$= \int_0^2 \frac{xy}{2}\bigg]_0^4 dy$$

$$= \int_0^2 2y\, dy$$

$$= y^2\bigg]_0^2$$

$$= 4$$

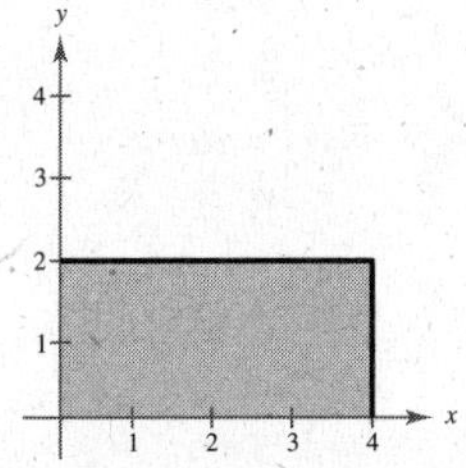

17. $V = \int_0^2 \int_0^y (4 - x - y)\, dx\, dy$

$$= \int_0^2 \left[4x - \frac{x^2}{2} - xy\right]_0^y dy$$

$$= \int_0^2 \left(4y - \frac{3y^2}{2}\right) dy$$

$$= \left[2y^2 - \frac{y^3}{2}\right]_0^2 = 4$$

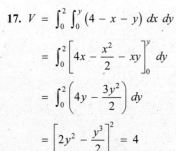

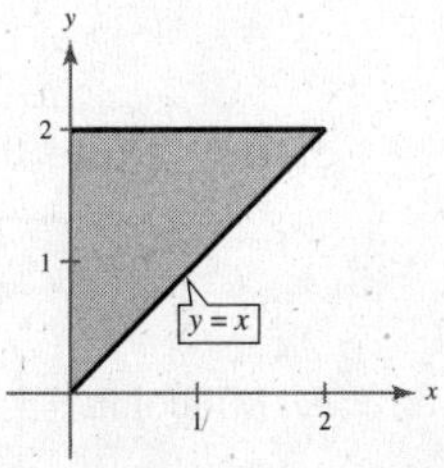

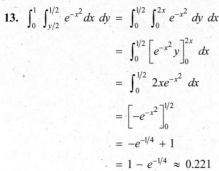

19. $V = \int_0^6 \int_0^{(-2/3)x+4} \left(\frac{12 - 2x - 3y}{4}\right) dy\, dx$

$$= \int_0^6 \frac{1}{4}\left[12y - 2xy - \frac{3}{2}y^2\right]_0^{(-2/3)x+4} dx$$

$$= \int_0^6 \left(\frac{1}{6}x^2 - 2x + 6\right) dx$$

$$= \left[\frac{1}{18}x^3 - x^2 + 6x\right]_0^6$$

$$= 12$$

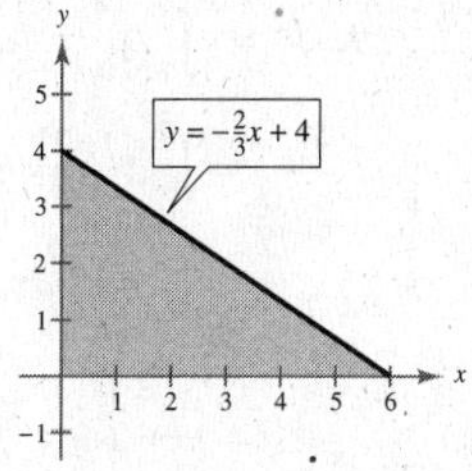

21. $V = \int_0^1 \int_0^y (1 - xy)\, dx\, dy$

$$= \int_0^1 \left[x - \frac{x^2 y}{2}\right]_0^y dy$$

$$= \int_0^1 \left(y - \frac{y^3}{2}\right) dy$$

$$= \left[\frac{y^2}{2} - \frac{y^4}{8}\right]_0^1 = \frac{3}{8}$$

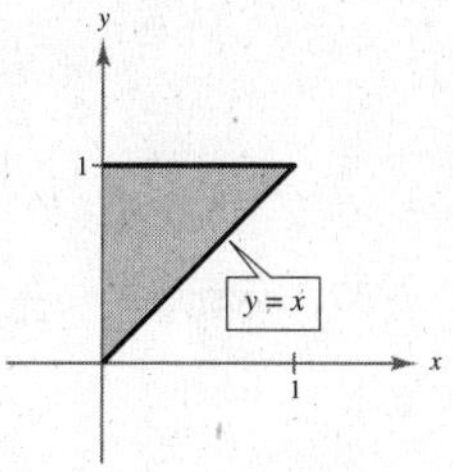

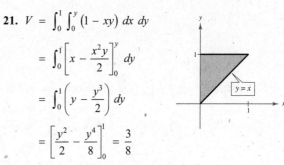

23. $V = 4\int_0^1 \int_0^1 \left(4 - x^2 - y^2\right) dy\, dx$

$$= 4\int_0^1 \left[4y - x^2 y - \frac{y^3}{3}\right]_0^1 dx$$

$$= 4\int_0^1 \left[4 - x^2 - \frac{1}{3}\right] dx$$

$$= 4\int_0^1 \left(\frac{11}{3} - x^2\right) dx$$

$$= 4\left[\frac{11x}{3} - \frac{x^3}{3}\right]_0^1 = \frac{40}{3}$$

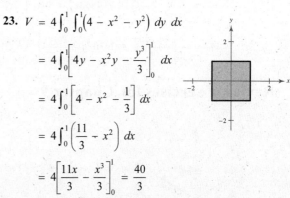

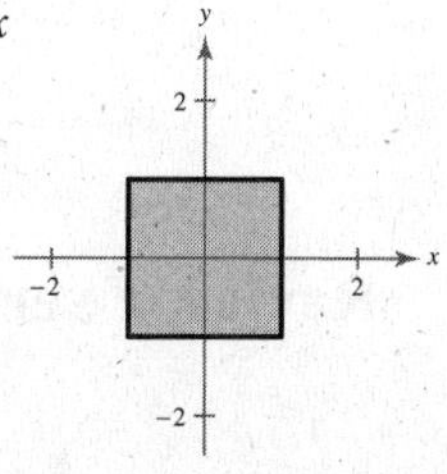

25. $V = \int_0^1 \int_0^x \sqrt{1 - x^2}\, dy\, dx$

$= \int_0^1 x\sqrt{1 - x^2}\, dx$

$= -\frac{1}{3}\left(1 - x^2\right)^{3/2}\Big]_0^1 = \frac{1}{3}$

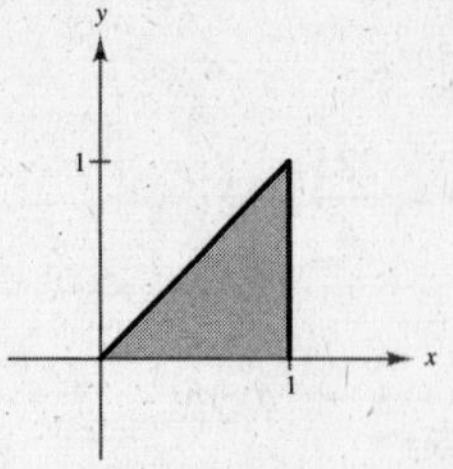

27. $V = \int_0^4 \int_0^1 xy\, dx\, dy$

$= \int_0^4 \frac{x^2 y}{2}\Big]_0^1 dy$

$= \frac{1}{2}\int_0^4 y\, dy$

$= \frac{y^2}{4}\Big]_0^4 = 4$

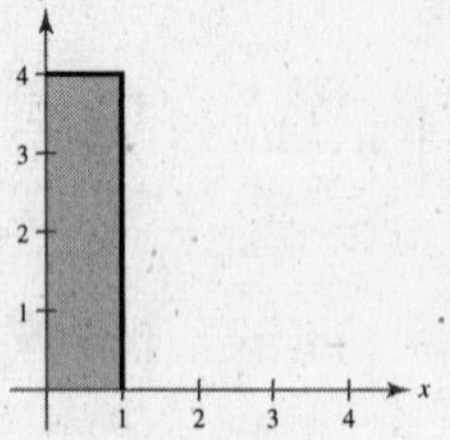

29. $V = \int_0^2 \int_0^4 x^2\, dy\, dx$

$= \int_0^2 x^2 y\Big]_0^4 dx$

$= \int_0^2 4x^2\, dx$

$= \frac{4x^3}{3}\Big]_0^2 = \frac{32}{3}$

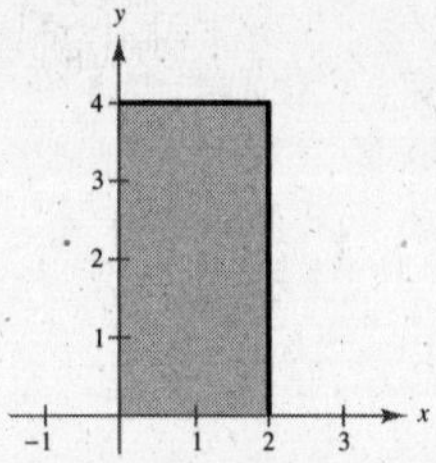

31. $P = \int_0^2 \int_0^2 \frac{120{,}000}{(2 + x + y)^3}\, dy\, dx$

$= \int_0^2 -60{,}000(2 + x + y)^{-2}\Big]_0^2 dx$

$= -60{,}000 \int_0^2 \left(\frac{1}{(4 + x)^2} - \frac{1}{(2 + x)^2}\right) dx$

$= 60{,}000\left[\frac{1}{4 + x} - \frac{1}{2 + x}\right]_0^2$

$= 10{,}000$ people

33. $\text{Average} = \frac{1}{8}\int_0^4 \int_0^2 x\, dy\, dx$

$= \frac{1}{8}\int_0^4 xy\Big]_0^2 dx$

$= \frac{1}{8}\int_0^4 2x\, dx$

$= \frac{x^2}{8}\Big]_0^4 = 2$

35. $\text{Average} = \frac{1}{4}\int_0^2 \int_0^2 \left(x^2 + y^2\right) dx\, dy$

$= \frac{1}{4}\int_0^2 \left[\frac{x^3}{3} + xy^2\right]_0^2 dy$

$= \frac{1}{4}\int_0^2 \left(\frac{8}{3} + 2y^2\right) dy$

$= \frac{1}{4}\left[\frac{8}{3}y + \frac{2}{3}y^3\right]_0^2 = \frac{8}{3}$

37. $\text{Average} = \frac{1}{50}\int_{45}^{50} \int_{40}^{50} \left(192x_1 + 576x_2 - x_1^2 - 5x_2^2 - 2x_1x_2 - 5000\right) dx_1\, dx_2$

$= \frac{1}{50}\int_{45}^{50} \left[96x_1^2 + 576x_1x_2 - \frac{x_1^3}{3} - 5x_1x_2^2 - x_1^2x_2 - 5000x_1\right]_{40}^{50} dx_2$

$= \frac{1}{50}\int_{45}^{50} \left(\frac{48{,}200}{3} + 4860x_2 - 50x_2^2\right) dx_2$

$= \frac{1}{50}\left[\frac{48{,}200x}{3} + 2430x_2^2 - \frac{50x_2^3}{3}\right]_{45}^{50} = \$13{,}400$

Review Exercises for Chapter 9

1.

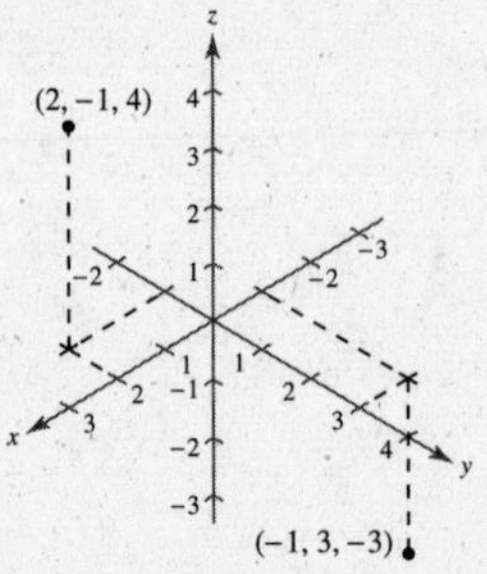

3. $d = \sqrt{(2 - 0)^2 + (5 - 0)^2 + (9 - 0)^2}$

$= \sqrt{4 + 25 + 81} = \sqrt{110}$

5. $\text{Midpoint} = \left(\frac{2 + (-4)}{2}, \frac{6 + 2}{2}, \frac{4 + 8}{2}\right) = (-1, 4, 6)$

7. $(x - 0)^2 + (y - 1)^2 + (z - 0)^2 = 5^2$

$x^2 + (y - 1)^2 + z^2 = 25$

9. Center $= \left(\frac{0+4}{2}, \frac{0+6}{2}, \frac{4+0}{2}\right) = (2, 3, 2)$

Radius $= \sqrt{(2-0)^2 + (3-0)^2 + (2-4)^2} = \sqrt{4+9+4} = \sqrt{17}$

Sphere: $(x-2)^2 + (y-3)^2 + (z-2)^2 = 17$

11.
$$x^2 + y^2 + z^2 + 4x - 2y - 8z + 5 = 0$$
$$(x^2 + 4x + 4) + (y^2 - 2y + 1) + (z^2 - 8z + 16) = -5 + 4 + 1 + 16$$
$$(x+2)^2 + (y-1)^2 + (z-4)^2 = 16$$

Center: $(-2, 1, 4)$

Radius: 4

13. Let $z = 0$.

$$(x+2)^2 + (y-1)^2 + (0-3)^2 = 25$$
$$(x+2)^2 + (y-1)^2 = 16$$

Circle of radius 4

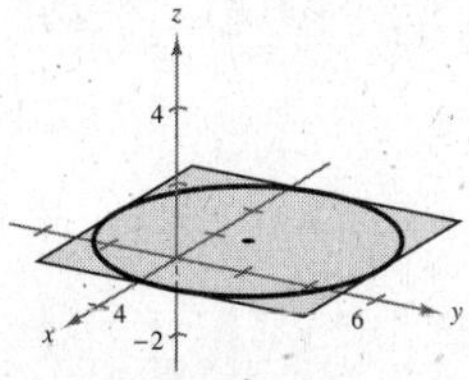

15. Because the crystal is a rectangular prisim, $(x, y, z) = (2, 2, 4)$.

17. $x + 2y + 3z = 6$

To find the x-intercept, let $y = 0$ and $z = 0$.

$x = 6$

To find the y-intercept, let $x = 0$ and $z = 0$.

$2y = 6 \Rightarrow y = 3$

To find the z-intercept, let $x = 0$ and $y = 0$.

$3z = 6 \Rightarrow z = 2$

x-intercept: $(6, 0, 0)$

y-intercept: $(0, 3, 0)$

z-intercept: $(0, 0, 2)$

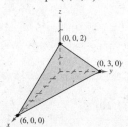

19. $3x - 6z = 12$

To find the x-intercept, let $z = 0$.

$3x = 12 \Rightarrow x = 4$

Because the coefficient of y is zero, there is no y-intercept.

To find the z-intercept, let $x = 0$.

$-6z = 12 \Rightarrow z = -2$

x-intercept: $(4, 0, 0)$

z-intercept: $(0, 0, -2)$

The plane is parallel to the y-axis.

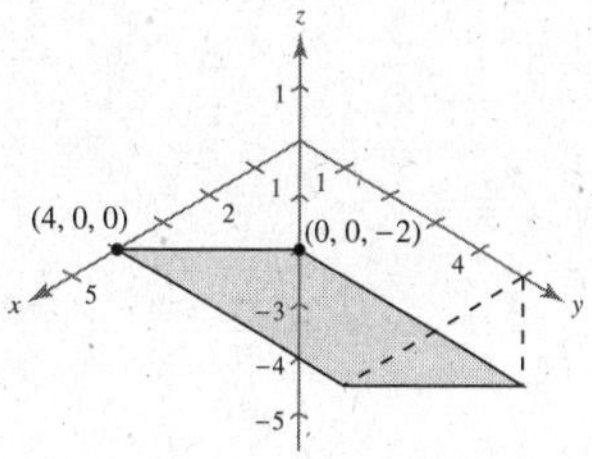

21. The graph of $x^2 + y^2 + z^2 - 2x + 4y - 6z + 5 = 0$ is a sphere whose standard equation is

$$(x-1)^2 + (y+2)^2 + (z-3)^2 = 9.$$

23. The graph of $x^2 + \frac{y^2}{16} + \frac{z^2}{9} = 1$ is an ellipsoid.

25. The graph of $z = \frac{x^2}{9} + y^2$ is an elliptic paraboloid.

27. The graph of $z = \sqrt{x^2 + y^2}$ is the top half of a circular cone whose standard equation is $x^2 + y^2 - z^2 = 0$.

29. $f(x, y) = xy^2$

(a) $f(2, 3) = 2(3)^2 = 18$

(b) $f(0, 1) = 0(1)^2 = 0$

(c) $f(-5, 7) = -5(7)^2 = -245$

(d) $f(-2, -4) = -2(-4)^2 = -32$

31. The domain of $f(x, y) = \sqrt{1 - x^2 - y^2}$ is the set of all points inside or on the circle $x^2 + y^2 = 1$. The range is $[0, 1]$.

33. $z = 10 - 2x - 5y$

$c = 0: 10 - 2x - 5y = 0,\ 2x + 5y = 10$

$c = 2: 10 - 2x - 5y = 2,\ 2x + 5y = 8$

$c = 4: 10 - 2x - 5y = 4,\ 2x + 5y = 6$

$c = 5: 10 - 2x - 5y = 5,\ 2x + 5y = 5$

$c = 10: 10 - 2x - 5y = 10,\ 2x + 5y = 0$

The level curves are lines of slope $-\frac{2}{5}$.

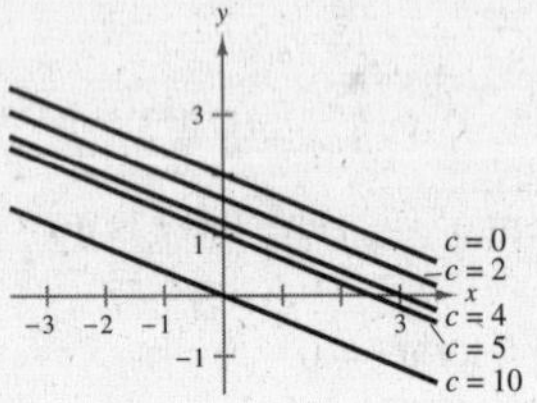

35. $z = (xy)^2$

$c = 1: (xy)^2 = 1,\ y = \pm\frac{1}{x}$

$c = 4: (xy)^2 = 4,\ y = \pm\frac{2}{x}$

$c = 9: (xy)^2 = 9,\ y = \pm\frac{3}{x}$

$c = 12: (xy)^2 = 12,\ y = \pm\frac{2\sqrt{3}}{x}$

$c = 16: (xy)^2 = 16,\ y = \pm\frac{4}{x}$

The level curves are hyperbolas.

37. (a) The level curves represent lines of equal rainfall and separate the four colors.
As the color darkens from light green to dark green, the average yearly precipitation increases.

(b) The small eastern portion containing Davenport

(c) The northwestern portion containing Sioux City

39. Southwest to northeast

41. $w = 0.06h + 3.09a - 2.32$

(a) For $h = 103$ and $a = 4$:

$w = 0.06(103) + 3.09(4) - 2.32 = 16.22$ kilograms

(b) For $w = 42.4$ and $a = 11.5$:

$42.4 = 0.06h + 3.09(11.5) - 2.32$

$42.4 = 0.06h + 33.215$

$9.185 = 0.06h$

$153.1 \approx h$

The mean height is about 153.1 centimeters.

(c) The variable representing age has the greater influence on weight. When h is held constant and a changes, w changes more than when a is held constant and h changes.

43. $f(x, y) = x^2y + 3xy + 2x - 5y$

$f_x = 2xy + 3y + 2$

$f_y = x^2 + 3x - 5$

45. $z = \frac{x^2}{y^2}$

$\frac{\partial z}{\partial x} = \frac{2x}{y^2}$

$\frac{\partial z}{\partial y} = -\frac{2x^2}{y^3}$

47. $f(x, y) = \ln(2x + 3y)$

$f_x = \frac{2}{2x + 3y}$

$f_y = \frac{3}{2x + 3y}$

49. $f(x, y) = xe^y + ye^x$

$f_x = ye^x + e^y$

$f_y = xe^y + e^x$

51. $w = xyz^2$

$\frac{\partial w}{\partial x} = yz^2$

$\frac{\partial w}{\partial y} = xz^2$

$\frac{\partial w}{\partial z} = 2xyz$

53. $z = 3x - 4y + 9,\ (3, 2, 10)$

(a) $\frac{\partial z}{\partial x} = 3$

(b) $\frac{\partial z}{\partial y} = -4$

55. $z = 8 - x^2 - y^2, (1, 2, 3)$

(a) $\dfrac{\partial z}{\partial x} = -2x$; At $(1, 2, 3)$, $\dfrac{\partial z}{\partial x} = -2.$

(b) $\dfrac{\partial z}{\partial y} = -2y$; At $(1, 2, 3)$, $\dfrac{\partial z}{\partial y} = -4.$

57. $f(x, y) = 3x^2 - xy + 2y^3$

$f_x = 6x - y \quad f_y = -x + 6y^2$

$f_{xx} = 6 \quad f_{yy} = 12y \quad f_{xy} = f_{yx} = -1$

59. $f(x, y) = \sqrt{1 + x + y}$

$f_x = \dfrac{1}{2\sqrt{1 + x + y}} \quad f_y = \dfrac{1}{2\sqrt{1 + x + y}}$

$f_{xx} = f_{yy} = f_{xy} = f_{yx} = \dfrac{-1}{4(1 + x + y)^{3/2}}$

61. $A(w, h) = 101.4w^{0.425}h^{0.725}$

(a) $\dfrac{\partial A}{\partial w} = 43.095w^{-0.575}h^{0.725}$

$\dfrac{\partial A}{\partial h} = 73.515w^{0.425}h^{-0.275}$

(b) $\dfrac{\partial A}{\partial w}(180, 70) = 43.095(180)^{-0.575}(70)^{0.725} \approx 47.35$

The surface area increases approximately 47 cm^2 per pound for a human weighing 180 pounds and is 70 inches tall.

63. $f(x, y) = x^2 + 2y^2$

The first partial derivatives of f, $f_x(x, y) = 2x$ and $f_y(x, y) = 4y$, are zero at the point (0, 0).
Moreover, because $f_{xx}(xy) = 2$, $f_{yy}(xy) = 4$, and $f_{xy}(x, y) = 0$, it follows that $f_{xx}(0, 0) > 0$ and

$f_{xx}(0, 0)f_{yy}(0, 0) - [f_{xy}(x, y)]^2 = 8 > 0.$

So, (0, 0, 0) is a relative minimum.

65. $f(x, y) = 1 - (x + 2)^2 + (y - 3)^2$

The first partial derivatives of f, $f_x(x, y) = -2(x + 2)$ and $f_y(x, y) = 2(y - 3)$, are zero at the point $(-2, 3)$.
Moreover, because $f_{xx}(x, y) = -2$, $f_{yy}(x, y) = 2$, and $f_{xy}(x, y) = 0$, it follows that $f_{xx}(-2, 3) > 0$ and $f_{xx}(-2, 3)f_{yy}(-2, 3) - [f_{xy}(-2, 3)]^2 = -4 < 0.$
So, $(-2, 3, 1)$ is a saddle point.

67. $f(x, y) = x^3 + y^2 - xy$

The first partial derivatives of f, $f_x(x, y) = 3x^2 - y$ and $f_y(x, y) = 2y - x$, are zero at the points $(0, 0)$ and $\left(\frac{1}{6}, \frac{1}{12}\right)$.
Moreover, because $f_{xx}(x, y) = 6x$, $f_{yy}(x, y) = 2$, and $f_{xy}(x, y) = -1$, it follows that $f_{xx}(0, 0) = 0$,

$f_{xx}(0, 0)f_{yy}(0, 0) - [f_{xy}(0, 0)]^2 = -1 < 0,$

$f_{xx}\left(\frac{1}{6}, \frac{1}{12}\right) = 1 > 0$, and

$f_{xx}\left(\frac{1}{6}, \frac{1}{12}\right)f_{yy}\left(\frac{1}{6}, \frac{1}{12}\right) - \left[f_{xy}\left(\frac{1}{6}, \frac{1}{12}\right)\right]^2 = 1 > 0.$

So, $(0, 0, 0)$ is a saddle point and $\left(\frac{1}{6}, \frac{1}{12}, -\frac{1}{432}\right)$ is a relative minimum.

69. $f(x, y) = x^3 + y^3 - 3x - 3y + 2$

The first partial derivatives of f, $f_x(x, y) = 3x^2 - 3$ and $f_y(x, y) = 3y^2 - 3$, are zero at the points $(1, 1)$, $(-1, -1)$, $(1, -1)$, and $(-1, 1)$.
Moreover, because $f_{xx}(x, y) = 6x$, $f_{yy}(x, y) = 6y$, and $f_{xy}(x, y) = 0$, it follows that

$f_{xx}(1, 1) = 6 > 0,$

$f_{xx}(1, 1)f_{yy}(1, 1) - [f_{xy}(1, 1)]^2 = 36 > 0,$

$f_{xx}(-1, -1) = -6 < 0,$

$f_{xx}(-1, -1)f_{yy}(-1, -1) - [f_{xy}(-1, -1)]^2 = 36 > 0,$

$f_{xx}(1, -1) = 6 > 0,$

$f_{xx}(1, -1)f_{yy}(1, -1) - [f_{xy}(1, -1)]^2 = -36 < 0,$

$f_{xx}(-1, 1) = -6 < 0$, and

$f_{xx}(-1, 1)f_{yy}(-1, 1) - [f_{xy}(-1, 1)]^2 = -36 < 0.$

So, $(1, 1, -2)$ is a relative minimum, $(-1, -1, 6)$ is a relative maximum, $(1, -1, 2)$ is a saddle point, and $(-1, 1, 2)$ is a saddle point.

71. (a) $\sum x_i = 1$

$\sum y_i = 0$

$\sum x_i^2 = 15$

$\sum x_i y_i = 15$

$$a = \frac{4(15) - (1)(0)}{4(15) - (1)^2} = \frac{60}{59}$$

$$b = \frac{1}{4}\left(0 - \frac{60}{59}(1)\right) = -\frac{15}{59}$$

$$y = \frac{60}{59}x - \frac{15}{59}$$

(b) $\left(\frac{60}{59}(-2) - \frac{15}{59} + 3\right)^2 + \left(\frac{60}{59}(-1) - \frac{15}{59} + 1\right)^2 + \left(\frac{60}{59}(1) - \frac{15}{59} - 2\right)^2 + \left(\frac{60}{59}(3) - \frac{15}{59} - 2\right)^2 \approx 2.746$

73. (a) $y = 0.456t^2 - 9.21t + 100.4$

(b) For 2008, $t = 18$.

When $t = 18$, $y = 82.364$ billions of board-feet.

75. (a) $y = 0.96t^2 - 5.6t + 87$

(b) For 2008, $t = 8$.

When $t = 8$, $y = 103.64$ thousand workers.

77. The least squares quadratic is

$y = 1.71x^2 - 2.57x + 5.56$.

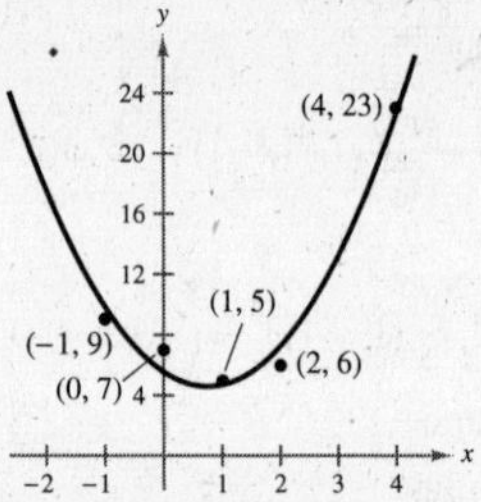

79. $$\int_0^1 \int_0^{1+x} (4x - 2y)\,dy\,dx = \int_0^1 \Big[4xy - y^2\Big]_0^{1+x} dx$$
$$= \int_0^1 \Big[4x(1 + x) - (1 + x)^2\Big]\,dx$$
$$= \int_0^1 \left(3x^2 + 2x - 1\right) dx$$
$$= \Big[x^3 + x^2 - x\Big]_0^1$$
$$= 1$$

81. $$\int_1^2 \int_1^{2y} \frac{x}{y^2}\,dx\,dy = \int_1^2 \frac{x^2}{2y^2}\Big]_1^{2y} dy$$
$$= \int_1^2 \left(\frac{4y^2}{2y^2} - \frac{1}{2y^2}\right) dy$$
$$= \int_1^2 \left(2 - \frac{1}{2}y^{-2}\right) dy$$
$$= \left[2y + \frac{1}{2y}\right]_1^2$$
$$= \left(4 + \frac{1}{4}\right) - \left(2 + \frac{1}{2}\right) = \frac{7}{4}$$

83. $$A = \int_{-2}^2 \int_5^{9-x^2} dy\,dx = \int_5^9 \int_{-\sqrt{9-y}}^{\sqrt{9-y}} dx\,dy$$

$$\int_{-2}^2 \int_5^{9-x^2} dy\,dx = \int_{-2}^2 \Big[\left(9 - x^2\right) - 5\Big]\,dx = \int_{-2}^2 \left(4 - x^2\right) dx = \left[4x - \frac{x^3}{3}\right]_{-2}^2 = \left(8 - \frac{8}{3}\right) - \left(-8 + \frac{8}{3}\right) = \frac{32}{3}$$

85. $$A = \int_{-3}^6 \int_{1/3(x+3)}^{\sqrt{x+3}} dy\,dx = \int_{-3}^6 \left(\sqrt{x + 3} - \frac{1}{3}(x + 3)\right) dx = \left[\frac{2}{3}(x + 3)^{3/2} - \frac{x^2}{6} - x\right]_{-3}^6 = (18 - 6 - 6) - \left(0 - \frac{3}{2} + 3\right) = \frac{9}{2}$$

87. $$V = \int_0^4 \int_0^4 (xy)^2\,dy\,dx = \int_0^4 \int_0^4 x^2y^2\,dy\,dx = \int_0^4 \frac{x^2y^3}{3}\Big]_0^4 dx = \int_0^4 \frac{64x^2}{3}\,dx = \frac{64x^3}{9}\Big]_0^4 = \frac{4096}{9}$$

89. Average $= \dfrac{\int_0^{10} \int_0^{25-2.5x} (0.25 - 0.025x - 0.01y)\, dy\, dx}{\text{area}}$

$$= \frac{\int_0^{10} \left[0.25y - 0.025xy - 0.005y^2\right]_0^{25-2.5x} dx}{125}$$

$$= \frac{\int_0^{10} \left[0.25(25 - 2.5x) - 0.025x(25 - 2.5x) - 0.005(25 - 2.5x)^2\right] dx}{125}$$

$$= \frac{\int_0^{10} \left(0.03125x^2 - 0.625x + 3.125\right) dx}{125}$$

$$= \frac{1}{125}\left[\frac{0.03125}{3}x^3 - 0.3125x^2 + 3.125x\right]_0^{10} = \frac{1}{125}\left[\frac{0.03125}{3}(10)^3 - 0.3125(10)^2 + 3.125(10)\right] = \frac{1}{12}$$

The average elevation is $\frac{1}{12}$ mile.

Chapter Test Solutions

1. (a)

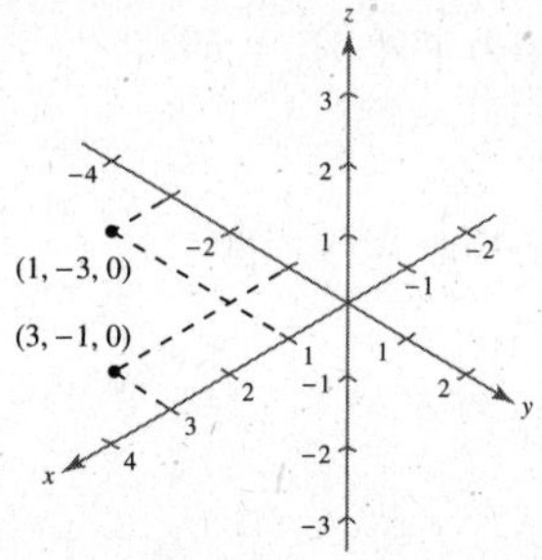

(b) $d = \sqrt{(3-1)^2 + (-1+3)^2 + (0-0)^2}$

$= \sqrt{4+4+0} = 2\sqrt{2}$

(c) Midpoint $= \left(\dfrac{1+3}{2}, \dfrac{-3-1}{2}, \dfrac{0+0}{2}\right) = (2, -2, 0)$

2. (a)

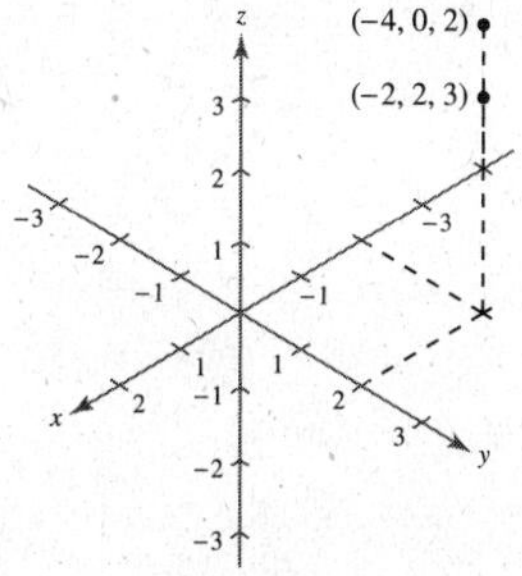

(b) $d = \sqrt{(-4+2)^2 + (0-2)^2 + (2-3)^2}$

$= \sqrt{4+4+1} = 3$

(c) Midpoint $= \left(\dfrac{-2-4}{2}, \dfrac{2+0}{2}, \dfrac{3+2}{2}\right) = \left(-3, 1, \dfrac{5}{2}\right)$

3. (a)

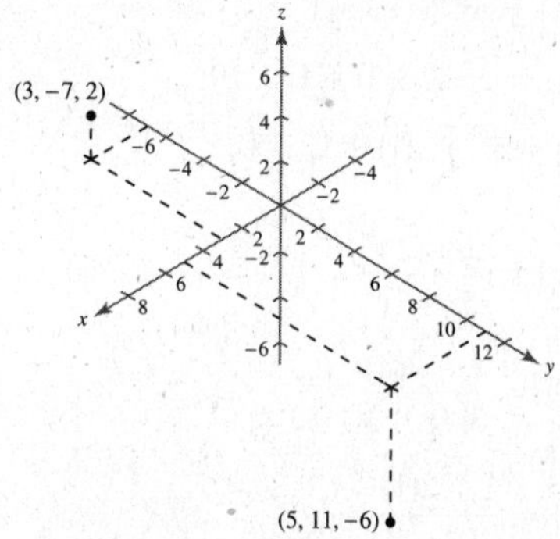

(b) $d = \sqrt{(5-3)^2 + (11+7)^2 + (-6-2)^2}$

$= \sqrt{4 + 324 + 64} = 14\sqrt{2}$

(c) Midpoint $= \left(\dfrac{3+5}{2}, \dfrac{-7+11}{2}, \dfrac{2-6}{2}\right)$

$= (4, 2, -2)$

4. $x^2 + y^2 + z^2 - 20x + 10y - 10z + 125 = 0$

$$\left(x^2 - 20x + 100\right) + \left(y^2 + 10y + 25\right) + \left(z^2 - 10z + 25\right) = -125 + 100 + 25 + 25$$

$$(x - 10)^2 + (y + 5)^2 + (z - 5)^2 = 25$$

Center: $(10, -5, 5)$

Radius: 5

5. The graph of $3x - y - z = 0$ is a plane.

6. The graph of $36x^2 + 9y^2 - 4z^2 = 0$ is an elliptic cone whose standard equation is $x^2 + \dfrac{y^2}{4} - \dfrac{z^2}{9} = 0$

7. The graph of $4x^2 - y^2 - 16z = 0$ is a hyperbolic paraboloid whose standard equation is $z = \dfrac{x^2}{4} - \dfrac{y^2}{16}$

8. $f(x, y) = x^2 + xy + 1$

$f(3, 3) = 3^2 + 3(3) + 1 = 19$

$f(1, 1) = 1^2 + 1(1) + 1 = 3$

9. $f(x, y) = \dfrac{x + 2y}{3x - y}$

$f(3, 3) = \dfrac{3 + 2(3)}{3(3) - 3} = \dfrac{3}{2}$

$f(1, 1) = \dfrac{1 + 2(1)}{3(1) - 1} = \dfrac{3}{2}$

10. $f(x, y) = xy \ln \dfrac{x}{y}$

$f(3, 3) = 3(3)\ln\left(\dfrac{3}{3}\right) = 0$

$f(1, 1) = 1(1)\ln\left(\dfrac{1}{1}\right) = 0$

11. $f(x, y) = 3x^2 + 9xy^2 - 2$

$f_x(x, y) = 6x + 9y^2$

$f_x(10, -1) = 6(10) + 9(-1)^2 = 69$

$f_y(x, y) = 18xy$

$f_y(10, -1) = 18(10)(-1) = -180$

12. $f(x, y) = x\sqrt{x + y}$

$$f_x(x, y) = x\left(\frac{1}{2}\right)(x + y)^{-1/2}(1) + \sqrt{x + y}(1) = \frac{x}{2\sqrt{x + y}} + \sqrt{x + y}$$

$$f_x(10, -1) = \frac{10}{2\sqrt{10 + (-1)}} + \sqrt{10 - 1} = \frac{14}{3}$$

$$f_y(x, y) = x\left(\frac{1}{2}\right)(x + y)^{-1/2}(1) + \sqrt{x + y}(0) = \frac{x}{2\sqrt{x + y}}$$

$$f_y(10, -1) = \frac{10}{2\sqrt{10 + (-1)}} = \frac{5}{3}$$

13. $f(x, y) = 3x^2 + 4y^2 - 6x + 16y - 4$

The first partial derivatives of f, $f_x(x, y) = 6x - 6$ and $f_y(x, y) = 8y + 16$, are zero at the point $(1, -2)$. Moreover, because $f_{xx}(x, y) = 6$, $f_{yy}(x, y) = 8$, and $f_{xy}(x, y) = 0$, it follows that $f_{xx}(1, -2) > 0$ and $f_{xx}(1, -2)f_{yy}(1, -2) - \left[f_{xy}(1, -2)\right]^2 = 48 > 0$.

So, $(1, -2, -23)$ is a relative minimum.

14. $f(x, y) = 4xy - x^4 - y^4$

The first partial derivatives of f, $f_x(x, y) = 4y - 4x^3$ and $f_y(x, y) = 4x - 4y^3$, are zero at the points $(0, 0)$, $(1, 1)$, and $(-1, -1)$.

Moreover, because $f_{xx}(x, y) = -12x^2$, $f_{yy}(x, y) = -12y^2$, and $f_{xy}(x, y) = 4$, it follows that $f_{xx}(0, 0) = 0$, $f_{xx}(0, 0)f_{yy}(0, 0) - [f_{xy}(0, 0)]^2 = -16 < 0$, $f_{xx}(1, 1) = -12 < 0$, $f_{xx}(1, 1)f_{yy}(1, 1) - [f_{xy}(1, 1)]^2 = 128 > 0$, $f_{xx}(-1, -1) = -12 < 0$, and $f_{xx}(-1, -1)f_{yy}(-1, -1) - [f_{xy}(-1, -1)]^2 = 128 > 0$.

So, $(0, 0, 0)$ is a saddle point, $(1, 1, 2)$ is a relative maximum, and $(-1, -1, 2)$ is a relative maximum.

15. $y = -1.839x^2 + 31.70x + 73.6$

16.
$$\begin{aligned}\int_0^1 \int_x^1 (30x^2y - 1)\,dy\,dx &= \int_0^1 \left[15x^2y^2 - y\right]_x^1 dx \\ &= \int_0^1 \left[(15x^2 - 1) - (15x^2(x)^2 - x)\right] dx \\ &= \int_0^1 (-15x^4 + 15x^2 + x - 1)\,dx \\ &= \left[-3x^5 + 5x^3 + \frac{x}{2} - x\right]_0^1 = \frac{3}{2}\end{aligned}$$

17.
$$\begin{aligned}\int_0^{\sqrt{e-1}} \int_0^{2y} \frac{1}{y^2 + 1}\,dx\,dy &= \int_0^{\sqrt{e-1}} \left[\frac{x}{y^2 + 1}\right]_0^{2y} dy \\ &= \int_0^{\sqrt{e-1}} \frac{2y}{y^2 + 1}\,dy \\ &= \left[\ln\left|y^2 + 1\right|\right]_0^{\sqrt{e-1}} \\ &= \ln e = 1\end{aligned}$$

18.
$$\begin{aligned}\int_0^2 \int_{x^2-2x+3}^3 dy\,dx &= \int_0^2 \left[3 - (x^2 - 2x + 3)\right] dx \\ &= \int_0^2 (-x^2 + 2x)\,dx \\ &= \left[-\frac{x^3}{3} + x^2\right]_0^2 \\ &= \frac{4}{3} \text{ square units}\end{aligned}$$

19.
$$\begin{aligned}\text{Average} &= \frac{\int_0^1 \int_0^3 (x^2 + y)\,dy\,dx}{\text{area}} \\ &= \frac{\int_0^1 \left[x^2y + \frac{y^2}{2}\right]_0^3 dx}{3} \\ &= \frac{\int_0^1 \left(3x^2 + \frac{9}{2}\right) dx}{3} \\ &= \frac{\left[x^3 + \frac{9}{2}x\right]_0^1}{3} = \frac{11}{6}\end{aligned}$$

Practice Test for Chapter 9

1. Find the distance between the points $(3, -7, 2)$ and $(5, 11, -6)$ and find the midpoint of the line segment joining the two points.

2. Find the standard form of the equation of the sphere whose center is $(1, -3, 0)$ and whose radius is $\sqrt{5}$.

3. Find the center and radius of the sphere whose equation is $x^2 + y^2 + z^2 - 4x + 2y + 8z = 0$.

4. Sketch the graph of the plane.
   (a) $3x + 8y + 6z = 24$ (b) $y = 2$

5. Identify the surface.
   (a) $\dfrac{x^2}{16} + \dfrac{y^2}{4} - \dfrac{z^2}{9} = 1$ (b) $z = \dfrac{x^2}{25} + y^2$

6. Find the domain of the function.
   (a) $f(x, y) = \ln(3 - x - y)$ (b) $f(x, y) = \dfrac{1}{x^2 + y^2}$

7. Find the first partial derivatives of $f(x, y) = 3x^2 + 9xy^2 + 4y^3 - 3x - 6y + 1$.

8. Find the first partial derivatives of $f(x, y) = \ln(x^2 + y^2 + 5)$.

9. Find the first partial derivatives of $f(x, y, z) = x^2y^3\sqrt{z}$.

10. Find the second partial derivatives of $z = \dfrac{x}{x^2 + y^2}$.

11. Find the relative extrema of $f(x, y) = 3x^2 + 4y^2 - 6x + 16y - 4$.

12. Find the relative extrema of $f(x, y) = 4xy - x^4 - y^4$.

13. Find the least squares regression line for the points $(-3, 7)$, $(1, 5)$, $(8, -2)$, and $(4, 4)$.

14. Find the least squares regression quadratic for the points $(-5, 8)$, $(-1, 2)$, $(1, 3)$, and $(5, 5)$.

15. Evaluate $\int_0^3 \int_0^{\sqrt{x}} xy^3\, dy\, dx$.

16. Evaluate $\int_{-1}^2 \int_0^{3y} (x^2 - 4xy)\, dx\, dy$.

17. Set up a double integral to find the area of the indicated region.

(a)

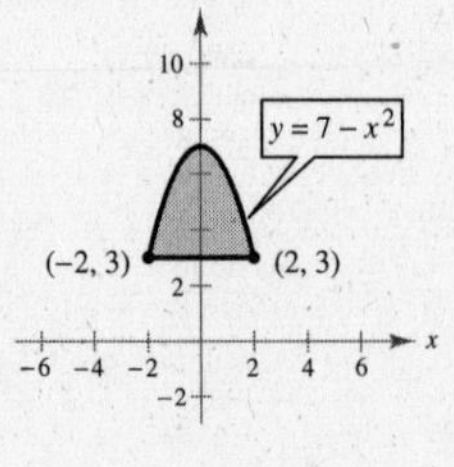

(b)

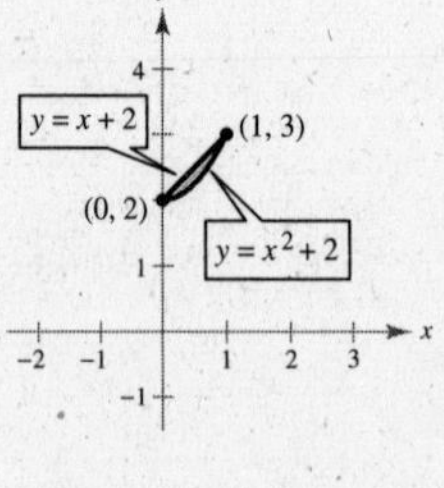

18. Find the volume of the solid bounded by the first octant and the plane $x + y + z = 4$.

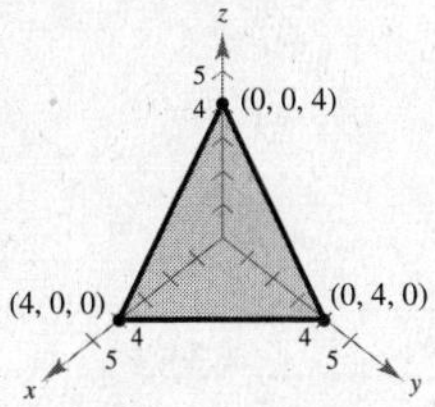

Graphing Calculator Required

19. Use a graphing utility to find the least squares regression line and the correlation coefficient for the given data points.

$(0, 20.4)$, $(1, 21.3)$, $(2, 22.9)$, $(3, 23.4)$, $(4, 24)$, $(5, 24.2)$, $(6, 24.9)$, $(7, 25.3)$, $(8, 26.2)$, $(9, 27.5)$, $(10, 29)$, $(11, 30.3)$, and $(12, 31.1)$

20. Use a graphing calculator or a computer algebra system to approximate $\int_2^3 \int_0^{x+2} e^{x^2 y}\, dx$.

CHAPTER 10
Differential Equations

CHAPTER 10
Differential Equations

Section 10.1 Solutions of Differential Equations

Skills Review

1. $y = 3x^2 + 2x + 1$

$y' = 6x + 2$

$y'' = 6$

2. $y = -2x^3 - 8x + 4$

$y' = -6x^2 - 8$

$y'' = -12x$

3. $y = -3e^{2x}$

$y' = -6e^{2x}$

$y'' = -12e^{2x}$

4. $y = -3e^{x^2}$

$y' = -6xe^{x^2}$

$y'' = -6x\left(2xe^{x^2}\right) + e^{x^2}(-6) = -12x^2e^{x^2} - 6e^{x^2}$

5.
$$\begin{aligned} 0.5 &= 9 - 9e^{-k} \\ -8.5 &= -9e^{-k} \\ \tfrac{17}{18} &= e^{-k} \\ \ln \tfrac{17}{18} &= -k \\ -\ln \tfrac{17}{18} &= k \\ 0.0572 &\approx k \end{aligned}$$

6.
$$\begin{aligned} 14.75 &= 25 - 25e^{-2k} \\ -10.25 &= -25e^{-2k} \\ 0.41 &= e^{-2k} \\ \ln 0.41 &= -2k \\ -\tfrac{1}{2}\ln 0.41 &= k \\ 0.4458 &\approx k \end{aligned}$$

1. $y = x^3 + 5$

$y' = 3x^2$

3. $y = e^{-2x}$

$y' = -2e^{-2x}$ and $y' + 2y = -2e^{-2x} + 2\left(e^{-2x}\right) = 0$

5. $y = 2x^3$

$y' = 6x^2$ and $y' - \dfrac{3}{x}y = 6x^2 - \dfrac{3}{x}\left(2x^3\right) = 0$

7. $y = x^2$

$y' = 2x$

$y'' = 2$ and $x^2y'' - 2y = x^2(2) - 2\left(x^2\right) = 0$

9. $y = 2e^{2x}$

$y' = 4e^{2x}$

$y'' = 8e^{2x}$ and

$y'' - y' - 2y = 8e^{2x} - 4e^{2x} - 2\left(2x^{2x}\right) = 0$

11. By differentiation, you have $\dfrac{dy}{dx} = -\dfrac{1}{x^2}$.

13. By differentiation, you have

$$\frac{dy}{dx} = 4Ce^{4x} = 4y.$$

15. Because $\dfrac{dy}{dt} = -\left(\dfrac{1}{3}\right)Ce^{-t/3}$, you have

$$3\frac{dy}{dt} + y - 7 = 3\left(-\frac{1}{3}Ce^{-t/3}\right) + \left(Ce^{-t/3} + 7\right) - 7 = 0.$$

17. Because $y' = 2Cx - 3$, you have

$$xy' - 3x - 2y = x(2Cx - 3) - 3x - 2\left(Cx^2 - 3x\right) = 0.$$

19. Because $y' = C_2e^x$ and $y'' = C_2e^x$, you have

$$y'' - y' = C_2e^x - C_2e^x = 0.$$

21. Because $y' = \dfrac{3}{5}x^2 - 1 + \dfrac{C}{2\sqrt{x}}$, you have $2xy' - y = 2x\left(\dfrac{3}{5}x^2 - 1 + \dfrac{C}{2\sqrt{x}}\right) - \left(\dfrac{1}{5}x^3 - x + C\sqrt{x}\right) = x^3 - x$.

23. Because $y' = \ln x + 1 + C$, you have $x(y' - 1) - (y - 4) = x(\ln x + 1 + C - 1) - (x \ln x + Cx + 4 - 4) = 0.$

25.
$$\begin{aligned} y &= e^{-2x} \\ y' &= -2e^{-2x} \\ y'' &= 4e^{-2x} \\ y''' &= -8e^{-2x} \\ y^{(4)} &= 16e^{-2x} \\ y^{(4)} - 16y &= 16e^{-2x} - 16\left(e^{-2x}\right) = 0 \end{aligned}$$

So, it is a solution of the differential equation.

27.
$$\begin{aligned} y &= 4x^{-1} \\ y' &= -4x^{-2} \\ y'' &= 8x^{-3} \\ y''' &= -24x^{-4} \\ y^{(4)} &= 96x^{-5} \\ y^{(4)} - 16y &= 96x^{-5} - 16\left(4x^{-1}\right) \neq 0 \end{aligned}$$

So, it is *not* a solution of the differential equation.

29.
$$\begin{aligned} y &= \tfrac{2}{9}xe^{-2x} \\ y' &= -\tfrac{4}{9}xe^{-2x} + \tfrac{2}{9}e^{-2x} \\ y'' &= \tfrac{8}{9}xe^{-2x} - \tfrac{8}{9}e^{-2x} \\ y''' &= -\tfrac{16}{9}xe^{-2x} + \tfrac{24}{9}e^{-2x} \end{aligned}$$

$$y''' - 3y' + 2y = \left(-\tfrac{16}{9}xe^{-2x} + \tfrac{24}{9}e^{-2x}\right) - 3\left(-\tfrac{4}{9}xe^{-2x} + \tfrac{2}{9}e^{-2x}\right) + 2\left(\tfrac{2}{9}xe^{-2x}\right) = 2e^{-2x}$$

So, it is *not* a solution of the differential equation.

31.
$$\begin{aligned} y &= xe^{x} \\ y' &= xe^{x} + e^{x} \\ y'' &= xe^{x} + 2e^{x} \\ y''' &= xe^{x} + 3e^{x} \end{aligned}$$

$$y''' - 3y' + 2y = \left(xe^{x} + 3e^{x}\right) - 3\left(xe^{x} + e^{x}\right) + 2\left(xe^{x}\right) = 0$$

So, it is a solution of the differential equation.

33. Because $y' = -2Ce^{-2x} = -2y$, it follows that $y' + 2y = 0$. To find the particular solution, use the fact that $y = 3$ when $x = 0$. That is, $3 = Ce^{0} = C$. So, $C = 3$ and the particular solution is $y = 3e^{-2x}$.

35. Because $y' = 4C_1e^{4x} - 3C_2e^{-3x}$ and $y'' = 16C_1e^{4x} + 9C_2e^{-3x}$, it follows that $y'' - y' - 12y = 0$. To find the particular solution, use the fact that $y = 5$ and $y' = 6$ when $x = 0$. That is,

$$C_1 + C_2 = 5$$
$$4C_1 - 3C_2 = 6$$

which implies that $C_1 = 3$ and $C_2 = 2$. So, the particular solution is $y = 3e^{4x} + 2e^{-3x}$.

37. When $C = 0$, the graph is a straight line.

When $C = 1$, the graph is a parabola opening upward with a vertex at $(-2, 0)$.

When $C = -1$, the graph is a parabola opening downward with a vertex at $(-2, 0)$.

When $C = 2$, the graph is a parabola opening upward with a vertex at $(-2, 0)$.

When $C = -2$, the graph is a parabola opening downward with a vertex at $(-2, 0)$.

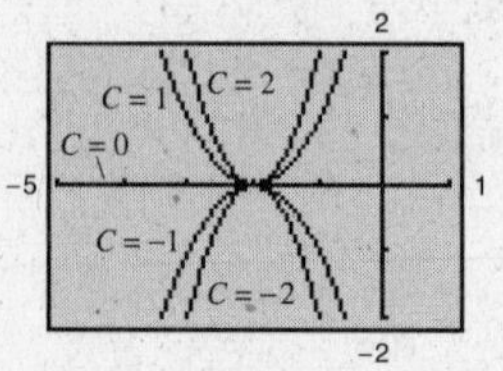

39. $y = \int 3x^2 \, dx = x^3 + C$

41. $y = \int \dfrac{x + 3}{x} \, dx = \int \left(1 + \dfrac{3}{x}\right) dx = x + 3 \ln|x| + C$

43. By using partial fractions:

$$\frac{1}{x^2-1} = \frac{A}{x-1} + \frac{B}{x+1}$$
$$1 = A(x+1) + B(x-1)$$

When $x = 1$: $1 = 2A \Rightarrow A = \frac{1}{2}$.

When $x = -1$: $1 = -2B \Rightarrow B = -\frac{1}{2}$.

You have $y = \int \frac{1}{x^2-1}\,dx$

$$= \frac{1}{2}\int\left[\frac{1}{x-1} - \frac{1}{x+1}\right]dx$$
$$= \frac{1}{2}\left[\ln|x-1| - \ln|x+1|\right] + C$$
$$= \frac{1}{2}\ln\left|\frac{x-1}{x+1}\right| + C.$$

45. $y = \int \cos 4x\,dx = \frac{1}{4}\sin 4x + C$

47. Because $y = 4$ when $x = 4$, you have $4^2 = C4^3$, which implies that $C = \frac{1}{4}$, and the particular solution is $y^2 = \frac{1}{4}x^3$.

49. Because $y = 3$ when $x = 0$, you have $3 = Ce^0$, which implies that $C = 3$, and the particular solution is $y = 3e^x$.

51. (a) Because $N = 100$ when $t = 0$, it follows that $C = 650$. So, the population function is $N = 750 - 650e^{-kt}$. Also, because $N = 160$ when $t = 2$, it follows that

$$160 = 750 - 650e^{-2k}$$
$$e^{-2k} = \frac{59}{65}$$
$$k = -\frac{1}{2}\ln\frac{59}{65}$$
$$\approx -0.0484.$$

So, the population function is $N = 750 - 650e^{-0.0484t}$.

(b) See accompanying graph.

(c) When $t = 4$, $N \approx 214$.

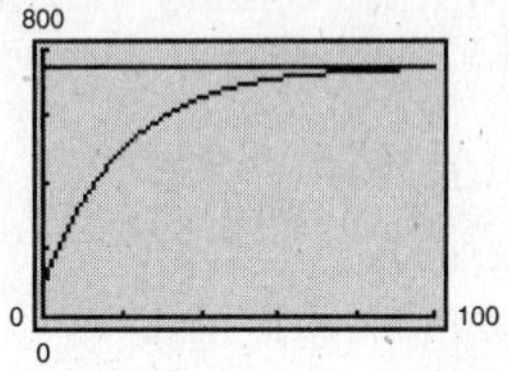

53.
$$y = a + Ce^{k(1-b)t}$$
$$\frac{dy}{dt} = Ck(1-b)e^{k(1-b)t}$$
$$a + b(y-a) + \frac{1}{k}\frac{dy}{dt} = a + b\left[\left(a + Ce^{k(1-b)t}\right) - a\right] + \frac{1}{k}\left[Ck(1-b)e^{k(1-b)t}\right]$$
$$= a + bCe^{k(1-b)t} + C(1-b)e^{k(1-b)t} = a + Ce^{k(1-b)t}\left[b + (1-b)\right] = a + Ce^{k(1-b)t} = y$$

55. True

Section 10.2 Separation of Variables

Skills Review

1. $\int x^{3/2}\,dx = \frac{2}{5}x^{5/2} + C$

$$\frac{d}{dx}\left(\frac{2}{5}x^{5/2} + C\right) = x^{3/2}$$

2. $\int\left(t^3 - t^{1/3}\right)dt = \frac{1}{4}t^4 - \frac{3}{4}t^{4/3} + C$

$$\frac{d}{dt}\left(\frac{1}{4}t^4 - \frac{3}{4}t^{4/3} + C\right) = t^3 - t^{1/3}$$

3. $\int \frac{2}{x-5}\,dx = 2\ln|x-5| + C$

$$\frac{d}{dx}\left(2\ln|x-5| + C\right) = \frac{2}{x-5}$$

4. $\int \frac{y}{2y^2+1}\,dy = \frac{1}{4}\int\frac{4y}{2y^2+1}\,dy$

$$= \frac{1}{4}\ln\left|2y^2+1\right| + C$$
$$= \frac{1}{4}\ln\left(2y^2+1\right) + C$$
$$\frac{d}{dy}\left(\frac{1}{4}\ln\left(2y^2+1\right) + C\right) = \frac{1}{4}\,\frac{4y}{2y^2+1} = \frac{y}{2y^2+1}$$

Skills Review —continued—

5. $\int e^{2y}\,dy = \frac{1}{2}\int 2e^{2y}\,dy = \frac{1}{2}e^{2y} + C$

$\frac{d}{dy}\left(\frac{1}{2}e^{2y} + C\right) = \frac{1}{2}(2)e^{2y} = e^{2y}$

6. $\int xe^{1-x^2}\,dx = -\frac{1}{2}\int -2xe^{1-x^2}\,dx = -\frac{1}{2}e^{1-x^2} + C$

$\frac{d}{dx}\left(-\frac{1}{2}e^{1-x^2} + C\right) = -\frac{1}{2}(-2x)e^{1-x^2} = xe^{1-x^2}$

7. $(3)^2 - 6(3) = 1 + C$

$-9 = 1 + C$

$-10 = C$

8. $(-1)^2 + (-2)^2 = C$

$1 + 4 = C$

$5 = C$

9. $10 = 2e^{2k}$

$5 = e^{2k}$

$\ln 5 = 2k$

$\frac{1}{2}\ln 5 = k$

$0.8047 \approx k$

10. $(6)^2 - 3(6) = e^{-k}$

$18 = e^{-k}$

$\ln 18 = -k$

$-\ln 18 = k$

$-2.8904 \approx k$

1. Yes, $\frac{dy}{dx} = \frac{x}{y+3}$

$(y + 3)\,dy = x\,dx$

3. Yes, $\frac{dy}{dx} = \frac{1}{x} + 1$

$dy = \left(\frac{1}{x} + 1\right)dx$

5. No. the variables cannot be separated.

7. $\frac{dy}{dx} = 2x$

$\int dy = \int 2x\,dx$

$y = x^2 + C$

9. $\frac{dr}{ds} = 0.05r$

$\int \frac{dr}{r} = \int 0.05\,ds$

$\ln|r| = 0.05s + C_1$

$r = e^{0.05s + C_1} = e^{C_1}e^{0.05s} = Ce^{0.05s}$

11. $\frac{dy}{dx} = \frac{3x}{2y}$

$\int y\,dy = \frac{3}{2}\int x\,dx$

$\frac{1}{2}y^2 = \frac{3}{4}x^2 + C_1$

$y^2 = \frac{3}{2}x^2 + C$

13. $3y^2\frac{dy}{dx} = 1$

$\int 3y^2\,dy = \int dx$

$y^3 = x + C$

$y = \sqrt[3]{x + C}$

15. $(y + 1)\frac{dy}{dx} = 2x$

$\int (y + 1)\,dy = \int 2x\,dx$

$\frac{(y+1)^2}{2} = x^2 + C_1$

$C = 2x^2 - (y + 1)^2$

17. $y' - xy = 0$

$\frac{dy}{dx} = xy$

$\int \frac{1}{y}\,dy = \int x\,dx$

$\ln|y| = \frac{1}{2}x^2 + C_1$

$y = e^{(x^2/2) + C_1} = e^{C_1}e^{x^2/2} = Ce^{x^2/2}$

19. $\frac{dy}{dt} = \frac{e^t}{4y}$

$\int 4y\,dy = \int e^t\,dt$

$2y^2 = e^t + C_1$

$y^2 = \frac{1}{2}e^t + C$

21.
$$\frac{dy}{dx} = \sqrt{1-y}$$
$$\int (1-y)^{-1/2}\,dy = \int dx$$
$$-2(1-y)^{1/2} = x + C_1$$
$$\sqrt{1-y} = \frac{-x}{2} + C$$
$$1 - y = \left(C - \frac{x}{2}\right)^2$$
$$y = 1 - \left(C - \frac{x}{2}\right)^2$$

23. $(2+x)y' = 2y$
$$(2+x)\frac{dy}{dx} = 2y$$
$$\int \frac{1}{2y}\,dy = \int \frac{1}{2+x}\,dx$$
$$\frac{1}{2}\ln|y| = \ln|C_1(2+x)|$$
$$\sqrt{y} = C_1(2+x)$$
$$y = C(2+x)^2$$

25. $xy' = y$
$$x\frac{dy}{dx} = y$$
$$\int \frac{1}{y}\,dy = \int \frac{1}{x}\,dx$$
$$\ln|y| = \ln|x| + \ln|C|$$
$$\ln|y| = \ln|Cx|$$
$$y = Cx$$

27.
$$y' = \frac{x}{y} - \frac{x}{1+y}$$
$$\frac{dy}{dx} = \frac{x}{y} - \frac{x}{1+y} = x\left(\frac{1}{y+y^2}\right)$$
$$\int (y + y^2)\,dy = \int x\,dx$$
$$\frac{y^2}{2} + \frac{y^3}{3} = \frac{x^2}{2} + C_1$$
$$3y^2 + 2y^3 = 3x^2 + C$$

29. $e^x(y'+1) = 1$
$$\left(\frac{dy}{dx} + 1\right) = e^{-x}$$
$$\frac{dy}{dx} = e^{-x} - 1$$
$$\int dy = \int (e^{-x} - 1)\,dx$$
$$y = -e^{-x} - x + C$$

31.
$$y\frac{dy}{dx} = \sin x$$
$$\int y\,dy = \int \sin x\,dx$$
$$\tfrac{1}{2}y^2 = -\cos x + C_1$$
$$y^2 = -2\cos x + C$$

33. (a)
$$\frac{dy}{dx} = x$$
$$\int dy = \int x\,dx$$
$$y = \frac{1}{2}x^2 + C$$

(b)

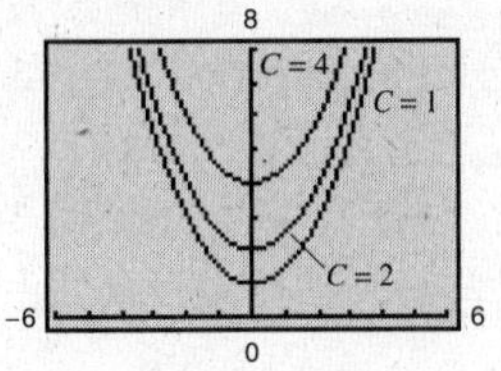

35. (a)
$$\frac{dy}{dx} = 4 - y$$
$$\int \frac{1}{4-y}\,dy = \int dx$$
$$-\ln|4-y| = x + C_1$$
$$\ln|4-y| = -x + C_2$$
$$4 - y = e^{-x+C_2}$$
$$y = 4 - Ce^{-x}$$

(b)

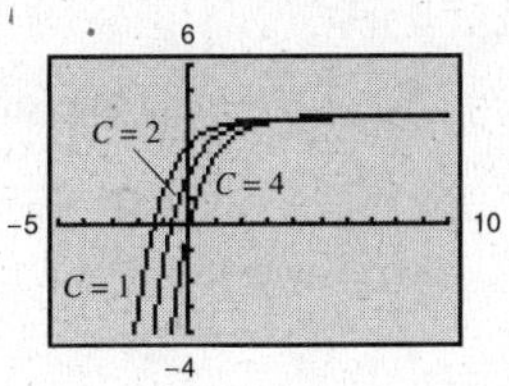

37. $yy' - e^x = 0$
$$y\frac{dy}{dx} = e^x$$
$$\int y\,dy = \int e^x\,dx$$
$$\frac{y^2}{2} = e^x + C$$

When $x = 0$, $y = 4$. So, $C = 7$ and the particular solution is $y^2 = 2e^x + 14$.

39. $x(y + 4) + y' = 0$

$$\frac{dy}{dx} = -x(y + 4)$$

$$\int \frac{1}{y + 4}\, dy = \int -x\, dx$$

$$\ln|y + 4| = -\frac{x^2}{2} + C_1$$

$$|y + 4| = e^{-x^2/2 + C_1}$$

$$y = -4 + Ce^{-x^2/2}$$

When $x = 0$, $y = -5$. So, $C = -1$ and the particular solution is $y = -4 - e^{-x^2/2}$.

41.

$$dP = 6P\, dt$$

$$\int \frac{1}{P}\, dP = \int 6\, dt$$

$$\ln|P| = 6t + C_1$$

$$P = Ce^{6t}$$

When $t = 0$, $P = 5$. So, $C = 5$ and the particular solution is $P = 5e^{6t}$.

43.

$$\frac{dy}{dx} = y \cos x$$

$$\int \frac{1}{y}\, dy = \int \cos x\, dx$$

$$\ln|y| = \sin x + C_1$$

$$y = e^{\sin x + C_1}$$

$$y = Ce^{\sin x}$$

When $x = 0$, $y = 1$. So, $C = 1$ and the particular solution is $y = e^{\sin x}$.

45.

$$\frac{dy}{dx} = \frac{6x}{5y}$$

$$\int 5y\, dy = \int 6x\, dx$$

$$\frac{5}{2}y^2 = 3x^2 + C_1 \Rightarrow 5y^2 = 6x^2 + C$$

When $x = -1$, $y = 1$. So, $C = -1$ and the equation is $5y^2 = 6x^2 - 1$ or $6x^2 - 5y^2 = 1$ (hyperbola).

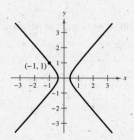

47.

$$\frac{dv}{dt} = 3.456 - 0.1v$$

$$\int \frac{dv}{3.456 - 0.1v} = \int dt$$

$$-10 \ln|3.456 - 0.1v| = t + C_1$$

$$(3.456 - 0.1v)^{-10} = C_2 e^t$$

$$3.456 - 0.1v = Ce^{-0.1t}$$

$$v = -10Ce^{-0.1t} + 34.56$$

When $t = 0$, $v = 0$. So, $C = 3.456$ and the solution is $v = 34.56(1 - e^{-0.1t})$.

49.

$$\frac{dV}{dt} = kV^{2/3}$$

$$\int V^{-2/3}\, dV = \int k\, dt$$

$$3V^{1/3} = kt + C_1$$

$$V^{1/3} = \frac{1}{3}kt + C$$

$$V = \left(\frac{1}{3}kt + C\right)^3$$

51.

$$\frac{dw}{dt} = k(1200 - w)$$

$$\int \frac{1}{1200 - w}\, dw = \int k\, dt$$

$$-\ln|1200 - w| = kt + C_1$$

$$\ln|1200 - w| = -kt + C_2$$

$$1200 - w = e^{-kt + C_2}$$

$$w = 1200 - Ce^{-kt}$$

When $t = 0$, $w = 60$. So, $C = 1140$ and the particular solution is $w = 1200 - 1140e^{-kt}$.

(a)

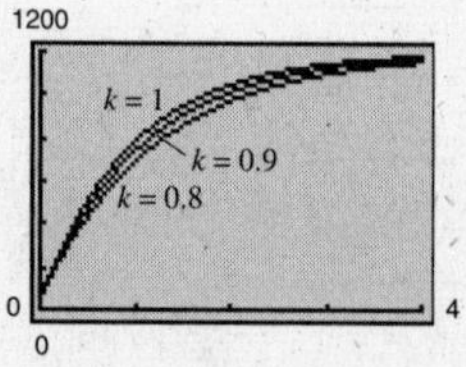

(b) For $k = 0.8$ and $w = 800$:

$$800 = 1200 - 1140e^{-0.8t}$$
$$-400 = 1140e^{-0.8t}$$
$$\tfrac{20}{57} = e^{-0.8t}$$
$$\ln \tfrac{20}{57} = -0.8t$$
$$-1.25 \ln \tfrac{20}{57} = t$$
$$1.31 \text{ years} \approx t$$

For $k = 1$ and $w = 800$:

$$800 = 1200 - 1140e^{-t}$$
$$-400 = -1140e^{-t}$$
$$\tfrac{20}{57} = e^{-t}$$
$$\ln \tfrac{20}{57} = -t$$
$$-\ln \tfrac{20}{57} = t$$
$$1.05 \text{ years} \approx t$$

For $k = 0.9$ and $w = 800$:

$$800 = 1200 - 1140e^{-0.9t}$$
$$-400 = -1140e^{-0.9t}$$
$$\tfrac{20}{57} = e^{-0.9t}$$
$$\ln \tfrac{20}{57} = -0.9t$$
$$-\tfrac{10}{9} \ln \tfrac{20}{57} = t$$
$$1.16 \text{ years} \approx t$$

(c) The weight of the calf approaches 1200 pounds as t increases, for each of the models.

53. Let y represent the mass of the radium and let t represent the time (in years).

$$\frac{dy}{dt} = ky$$
$$\int \frac{1}{y}\,dy = \int k\,dt$$
$$\ln y = kt + C_1$$
$$y = e^{kt + C_1}$$
$$y = Ce^{kt}$$

The half-life is 1599 years.

$$\tfrac{1}{2}C = Ce^{k(1599)}$$
$$\tfrac{1}{2} = e^{1599k}$$
$$\ln \tfrac{1}{2} = 1599k$$
$$\frac{1}{1599} \ln \frac{1}{2} = k$$
$$-0.000433 \approx k$$

The equation is $y = Ce^{-0.000433t}$. After $t = 25$ years, $y \approx 0.989C$. So, about 98.9% of a present amount will remain.

55. Answers will vary.

Mid-Chapter Quiz Solutions

1. Because $y' = -\left(\tfrac{1}{2}\right)Ce^{-x/2}$, you have $2y' + y = 2\left(-\tfrac{1}{2}Ce^{-x/2}\right) + Ce^{-x/2} = 0$.

2. Because $y' = -C_1 \sin x + C_2 \cos x$ and $y'' = -C_1 \cos x - C_2 \sin x$, you have $y'' + y = -C_1 \cos x - C_2 \sin x + C_1 \cos x + C_2 \sin x = 0$.

3. Because $y' = 3C_1 \cos 3x - 3C_2 \sin 3x$ and $y'' = -9C_1 \sin 3x - 9C_2 \cos 3x$, it follows that $y'' + 9y = -9C_1 \sin 3x - 9C_2 \cos 3x + 9(C_1 \sin 3x + C_2 \cos 3x) = 0$. To find the particular solution, use the fact that $y = 2$ and $y' = 1$ when $x = \dfrac{\pi}{6}$. That is,

$$C_1 \sin \frac{3\pi}{6} + C_2 \cos \frac{3\pi}{6} = 2 \quad \Rightarrow C_1 = 2$$
$$3C_1 \cos \frac{3\pi}{6} - 3C_2 \sin \frac{3\pi}{6} = 1 \Rightarrow C_2 = -\frac{1}{3}.$$

So, the particular solution is $y = 2 \sin 3x - \tfrac{1}{3} \cos 3x$.

4. Because $y' = C_1 + 3C_2x^2$ and $y'' = 6C_2x$, it follows that $x^2y'' - 3xy' + 3y = 6C_2x^3 - 3x(C_1 + 3C_2x^2) + 3(C_1x + C_2x^3) = 0$. To find the particular solution, use the fact that $y = 0$ and $y' = 4$ when $x = 2$.

That is, $2C_1 + 8C_2 = 0$
$C_1 + 12C_2 = 4$.

Solving this system, you obtain $C_1 = -2$ and $C_2 = \tfrac{1}{2}$. So, the particular solution is $y = -2x + \tfrac{1}{2}x^3$.

5. $\dfrac{dy}{dx} = -4x + 4$

$\int dy = \int(-4x + 4)\,dx$

$y = -2x^2 + 4x + C$

6. $y' = (x + 2)(y - 1)$

$\dfrac{1}{y - 1}\dfrac{dy}{dx} = x + 2$

$\int \dfrac{dy}{y - 1} = \int (x + 2)\,dx$

$\ln|y - 1| = \dfrac{(x + 2)^2}{2} + C_1$

$y - 1 = e^{(x+2)^2/2 + C_1}$

$y = 1 + Ce^{(x+2)^2/2}$

7. $y\dfrac{dy}{dx} = \dfrac{1}{x^2 - 1}$

$\int y\,dy = \int \dfrac{1}{x^2 - 1}\,dx$

$\int y\,dy = \dfrac{1}{2}\int\left(\dfrac{1}{x - 1} + \dfrac{1}{x + 1}\right)dx$

$\dfrac{1}{2}y^2 = \dfrac{1}{2}\ln\left|\dfrac{x - 1}{x + 1}\right| + C$

$y^2 = \ln\left|\dfrac{x - 1}{x + 1}\right| + C$

8. (a) $\dfrac{dy}{dx} = \dfrac{x^2 + 1}{2y}$

$\int 2y\,dy = \int (x^2 + 1)\,dx$

$y^2 = \dfrac{1}{3}x^3 + x + C$

$y = \pm\sqrt{\dfrac{1}{3}x^3 + x + C}$

(b)

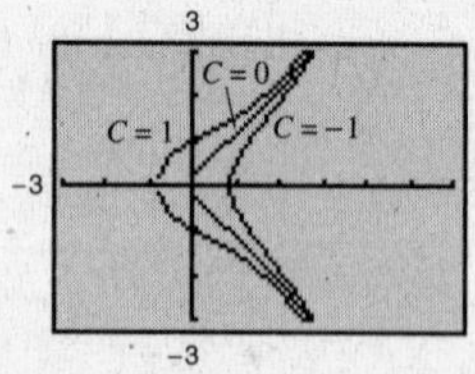

9. (a) $\dfrac{dy}{dx} = \dfrac{y}{x - 3}$

$\int \dfrac{1}{y}\,dy = \int \dfrac{1}{x - 3}\,dx$

$\ln|y| = \ln|x - 3| + \ln|C_1|$

$\ln|y| = \ln|C_1(x - 3)|$

$y = C(x - 3)$

(b)

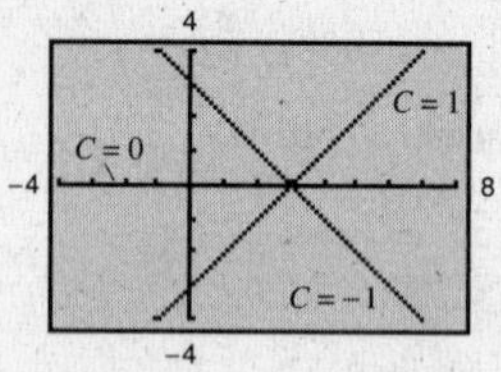

10. $y' + 2y - 1 = 0$

$\dfrac{dy}{dx} = 1 - 2y$

$\int \dfrac{1}{1 - 2y}\,dy = \int dx$

$-\dfrac{1}{2}\ln|1 - 2y| = x + C_1$

$\ln|1 - 2y| = -2x + C_2$

$1 - 2y = e^{-2x + C_2}$

$-2y = e^{-2x + C_2} - 1$

$y = -\tfrac{1}{2}(Ce^{-2x} - 1)$

When $x = 0$, $y = 1$. So, $C = -1$ and the particular solution is $y = \tfrac{1}{2}(e^{-2x} + 1)$.

11. $\dfrac{dy}{dx} = y \sin \pi x$

$\int \dfrac{1}{y}\,dy = \int \sin \pi x\,dx$

$\ln y = -\dfrac{1}{\pi}\cos \pi x + C_1$

$y = e^{-1/\pi \cos \pi x + C_1}$

$y = Ce^{-(\cos \pi x)/\pi}$

When $x = \dfrac{1}{2}$, $y = -3$. So, $C = -3$ and the particular solution is $y = -3e^{-(\cos \pi x)/\pi}$.

12. $\dfrac{dy}{dx} = 3x^2y$

$$\int \frac{1}{y}\,dy = \int 3x^2\,dx$$

$$\ln|y| = x^3 + C_1$$

$$y = e^{x^3 + C_1}$$

$$y = Ce^{x^3}$$

When $x = 0$, $y = 2$. So, $C = 2$ and the equation is $y = 2e^{x^3}$.

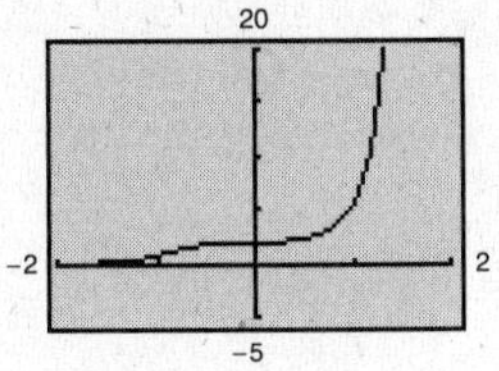

13. $\dfrac{dv}{dt} = k(20 - v)$

$$\int \frac{1}{20 - v}\,dv = \int k\,dt$$

$$-\ln|20 - v| = kt + C_1$$

$$\ln|20 - v| = -kt + C_2$$

$$20 - v = e^{-kt + C_2}$$

$$20 - v = Ce^{-kt}$$

$$v = 20 - Ce^{-kt}$$

Because the boat is starting at rest, you have $t = 0$ and $v = 0$. So, $C = 20$.

When $t = 0.5$ and $v = 10$, you have

$$10 = 20 - 20e^{-0.5k}$$

$$e^{-0.5k} = \tfrac{1}{2}$$

$$-0.5k = \ln \tfrac{1}{2}$$

$$k \approx 1.386.$$

So, $v = 20 - 20e^{-1.386t}$.

Section 10.3 First-Order Linear Differential Equations

Skills Review

1. $e^{-x}\left(e^{2x} + e^{x}\right) = e^{-x}e^{2x} + e^{-x}e^{x} = e^{x} + 1$

2. $\dfrac{1}{e^{-x}}\left(e^{-x} + e^{2x}\right) = \dfrac{e^{-x}}{e^{-x}} + \dfrac{e^{2x}}{e^{-x}} = 1 + e^{3x}$

3. $e^{-\ln x^3} = \dfrac{1}{e^{\ln x^3}} = \dfrac{1}{x^3}$

4. $e^{2\ln x + x} = e^{2\ln x}e^{x} = e^{\ln x^2}e^{x} = x^2e^{x}$

5. $\int 4e^{2x}\,dx = 4\left(\frac{1}{2}\right)\int e^{2x}(2)\,dx = 2e^{2x} + C$

6. $\int xe^{3x^2}\,dx = \frac{1}{6}\int e^{3x^2}(6x)\,dx = \frac{1}{6}e^{3x^2} + C$

7. $\displaystyle\int \frac{1}{2x + 5}\,dx = \frac{1}{2}\int \frac{2}{2x + 5}\,dx = \frac{1}{2}\ln|2x + 5| + C$

8. $\displaystyle\int \frac{x + 1}{x^2 + 2x + 3}\,dx = \frac{1}{2}\int \frac{2x + 2}{x^2 + 2x + 3}\,dx$

$$= \frac{1}{2}\ln\left|x^2 + 2x + 3\right| + C$$

9. $\int (4x - 3)^2\,dx = \frac{1}{4}\int 4(4x - 3)^2\,dx$

$$= \tfrac{1}{4}\left(\tfrac{1}{3}\right)(4x - 3)^3 + C$$

$$= \tfrac{1}{12}(4x - 3)^3 + C$$

10. $\int x\left(1 - x^2\right)^2\,dx = -\frac{1}{2}\int \left(1 - x^2\right)^2(-2x)\,dx$

$$= -\tfrac{1}{6}\left(1 - x^2\right)^3 + C = \tfrac{1}{6}\left(x^2 - 1\right)^3 + C$$

1. $x^3 - 2x^2y' + 3y = 0$

$$-2x^2y' + 3y = -x^3$$

$$y' - \frac{3}{2x^2}y = \frac{x}{2}$$

3. $xy' + y = xe^{x}$

$$y' + \frac{1}{x}y = e^{x}$$

5. $y + 1 = (x - 1)y'$

$$(1 - x)y' + y = -1$$

$$y' + \frac{1}{1 - x}y = \frac{1}{x - 1}$$

7. For this linear differential equation, you have $P(x) = 3$ and $Q(x) = 6$. So, the integrating factor is $u(x) = e^{\int 3\,dx} = e^{3x}$ and the general solution is

$$y = \frac{1}{u(x)}\int Q(x)u(x)\,dx = e^{-3x}\int 6e^{3x}\,dx$$
$$= e^{-3x}\left(2e^{3x} + C\right) = 2 + Ce^{-3x}.$$

9. For this linear differential equation, you have $P(x) = 1$ and $Q(x) = e^{-x}$. So, the integrating factor is $u(x) = e^{\int dx} = e^{x}$ and the general solution is

$$y = \frac{1}{u(x)}\int Q(x)u(x)\,dx = e^{-x}\int e^{-x}e^{x}\,dx = e^{-x}(x + C).$$

11. $$\frac{dy}{dx} = \frac{x^2 + 3}{x}$$
$$\int dy = \int\left(x + \frac{3}{x}\right)dx$$
$$y = \frac{1}{2}x^2 + 3\ln|x| + C$$

13. For this linear differential equation, you have $P(x) = 5x$ and $Q(x) = x$. So, the integrating factor is $u(x) = e^{\int 5x\,dx} = e^{(5/2)x^2}$ and the general solution is

$$y = \frac{1}{u(x)}\int Q(x)u(x)\,dx = \frac{1}{e^{(5/2)x^2}}\int xe^{(5/2)x^2}\,dx$$
$$= \frac{1}{e^{(5/2)x^2}}\left(\frac{1}{5}e^{(5/2)x^2} + C\right) = \frac{1}{5} + Ce^{-(5/2)x^2}.$$

15. For this linear differential equation, $y' + y\left(\frac{1}{x-1}\right) = x + 1$, you have $P(x) = \frac{1}{x-1}$ and $Q(x) = x + 1$. So, the integrating factor is $u(x) = e^{\int 1/(x-1)\,dx} = e^{\ln(x-1)} = x - 1$ and the general solution is

$$y = \frac{1}{u(x)}\int Q(x)u(x)\,dx = \frac{1}{x-1}\int (x+1)(x-1)\,dx$$
$$= \frac{1}{x-1}\left(\frac{x^3}{3} - x + C_1\right) = \frac{x^3 - 3x + C}{3(x-1)}.$$

17. For this linear differential equation, $y' + \frac{2}{x^3}y = \frac{1}{x^3}e^{1/x^2}$, you have $P(x) = \frac{2}{x^3}$ and $Q(x) = \frac{1}{x^3}e^{1/x^2}$. So, the integrating factor is $u(x) = e^{\int(2/x^3)\,dx} = e^{-1/x^2}$ and the general solution is

$$y = \frac{1}{u(x)}\int Q(x)u(x)\,dx = e^{1/x^2}\int \frac{1}{x^3}e^{1/x^2}e^{-1/x^2}\,dx$$
$$= e^{1/x^2}\int\frac{1}{x^3}\,dx = e^{1/x^2}\left(-\frac{1}{2x^2} + C\right).$$

19. Separation of Variables:

$$\frac{dy}{dx} = 4 - y$$
$$\int\frac{dy}{4-y} = \int dx$$
$$-\ln|4 - y| = x + C_1$$
$$4 - y = e^{-(x+C_1)}$$
$$4 - y = C_2e^{-x}$$
$$y = 4 - C_2e^{x} = 4 + Ce^{-x}$$

First-Order Linear:

$$P(x) = 1,\ Q(x) = 4$$
$$u(x) = e^{\int 1\,dx} = e^{x}$$
$$y = \frac{1}{e^x}\int 4e^x\,dx = \frac{1}{e^x}\left[4e^x + C\right] = 4 + Ce^{-x}$$

21. Separation of Variables:

$$\frac{dy}{dx} = 2x(1 + y)$$
$$\int\frac{dy}{1+y} = \int 2x\,dx$$
$$\ln|1 + y| = x^2 + C_1$$
$$1 + y = e^{x^2 + C_1}$$
$$y = Ce^{x^2} - 1$$

First-Order Linear:

$$P(x) = -2x,\ Q(x) = 2x$$
$$u(x) = e^{\int -2x\,dx} = e^{-x^2}$$
$$y = \frac{1}{e^{-x^2}}\int 2xe^{-x^2}\,dx$$
$$= e^{x^2}\left[-e^{-x^2} + C\right] = Ce^{x^2} - 1$$

23. $y' - 2x = 0$ matches (c) because $y = x^2 + C \Rightarrow y' = 2x \Rightarrow y' - 2x = 0$.

25. $y' - 2xy = 0$ matches (a) because $y = Ce^{x^2} \Rightarrow y' = Ce^{x^2}(2x) = 2xy \Rightarrow y' - 2xy = 0$.

27. Because $P(x) = 1$ and $Q(x) = 6e^x$, the integrating factor is $u(x) = e^{\int dx} = e^x$ and the general solution is

$$y = \frac{1}{e^x}\int 6e^x\left(e^x\right)dx = \frac{1}{e^x}\left[3e^{2x} + C\right] = 3e^x + Ce^{-x}.$$

Because $y = 3$ when $x = 0$, it follows that $C = 0$ and the particular solution is $y = 3e^x$.

29. Because $P(x) = \frac{1}{x}$ and $Q(x) = 0$, the integrating factor is $u(x) = e^{\int 1/x\,dx} = e^{\ln x} = x$ and the general solution is

$y = \frac{1}{x}\int 0\,dx = \frac{C}{x}$. Because $y = 2$ when $x = 2$, it follows that $C = 4$ and the particular solution is $y = \frac{4}{x}$ or $xy = 4$.

31. Because $P(x) = 3x^2$ and $Q(x) = 3x^2$, the integrating factor is $u(x) = e^{\int 3x^2\,dx} = e^{x^3}$ and the general solution is

$y = e^{-x^3}\int 3x^2 e^{x^3}\,dx = e^{-x^3}\left(e^{x^3} + C\right) = 1 + Ce^{-x^3}$. Because $y = 6$ when $x = 0$, it follows that $C = 5$ and the particular solution is $y = 1 + 5e^{-x^3}$.

33. Because $P(x) = -\frac{2}{x}$ and $Q(x) = -x$, the integrating factor is $u(x) = e^{\int -2/x\,dx} = e^{-2\ln x} = \frac{1}{x^2}$ and the general solution is

$y = x^2\int(-x)\left(\frac{1}{x^2}\right)dx = x^2\left(-\ln|x| + C\right)$. Because $y = 5$ when $x = 1$, it follows that $C = 5$ and the particular solution is

$y = x^2\left(5 - \ln|x|\right)$.

35. $\frac{dS}{dt} = 0.2(100 - S) + 0.2t$

$S' + 0.2S = 20 + 0.2t$

Because $P(t) = 0.2$ and $Q(t) = 20 + 0.2t$, the integrating factor is $u(t) = e^{\int 0.2\,dt} = e^{t/5}$ and the general solution is

$S = e^{-t/5}\int e^{t/5}\left(20 + \frac{t}{5}\right)dt$. Using integration by parts, the integral is

$S = e^{-t/5}\left(100e^{t/5} + te^{t/5} - 5e^{t/5} + C\right) = 100 + t - 5 + Ce^{-t/5} = 95 + t + Ce^{-t/5}$. Because $S = 0$ when $t = 0$, it follows that $C = -95$ and the particular solution is $S = t + 95\left(1 - e^{-t/5}\right)$. During the first 10 years, the sales are as follows.

t	0	1	2	3	4	5	6	7	8	9	10
S	0	18.22	33.32	45.86	56.31	65.05	72.39	78.57	83.82	88.30	92.14

37. $\frac{dv}{dt} = kv - 9.8$

$v' - kv = -9.8$

Because $P(t) = -k$ and $Q(t) = -9.8$, the integrating factor is $u(t) = e^{\int -k\,dt} = e^{-kt}$ and the general solution is

$v = e^{kt}\int -9.8e^{-kt}\,dt = e^{kt}\left(\frac{9.8}{k}e^{-kt} + C_1\right) = \frac{9.8}{k} + C_1e^{kt}$

or $v = \frac{1}{k}\left(9.8 + Ce^{kt}\right)$.

39. (a) $\frac{dN}{dt} = k(40 - N)$

$N' = 40k - kN$

$N' + kN = 40k$

Because $P(t) = k$ and $Q(t) = 40k$, the integrating factor is $u(t) = e^{\int k\,dt} = e^{kt}$ and the general solution is

$N = e^{-kt}\int 40ke^{kt}\,dt = e^{-kt}\left(40e^{kt} + C\right) = 40 + Ce^{-kt}$.

(b) For $t = 1$, $N = 10$:

$10 = 40 + Ce^{-kt} \Rightarrow -30 = Ce^{-k}$

For $t = 20$, $N = 19$:

$19 = 40 + Ce^{-20k} \Rightarrow -21 = Ce^{-20k}$

$\frac{30}{21} = \frac{e^{-k}}{e^{-20k}}$

$\frac{10}{7} = e^{19k}$

$\ln\left(\frac{10}{7}\right) = 19k \Rightarrow k = \frac{1}{19}\ln\left(\frac{10}{7}\right) \approx 0.0188$

$-30 = Ce^{-k} \Rightarrow C = -30e^{k} \approx -30.57$

$N = 40 - 30.57e^{-0.0188t}$

41. (a) $\frac{dy}{dt} = \frac{1-y}{4}$

$\frac{4}{1-y}\,dy = dt$

$-4\ln(1-y) = t + C_1$

$y(0) = 0 \Rightarrow C_1 = 0$

So, $\ln(1-y) = -\frac{1}{4}t$

$1 - y = e^{(-1/4)t}$

$y = 1 - e^{(-1/4)t}$

(b) $\frac{1}{2} = 1 - e^{(-1/4)t}$

$e^{(-1/4)t} = \frac{1}{2}$

$-\frac{1}{4}t = \ln\frac{1}{2} = -\ln 2$

$t = 4\ln 2 \approx 2.77$ years

(c) $y = 1 - e^{(-1/4)(4)} = 1 - e^{-1} \approx 0.632$ or 63.2%

Section 10.4 Applications of Differential Equations

Skills Review

1. $\frac{dy}{dx} = 3x$

$\int dy = \int 3x\,dx$

$y = \frac{3}{2}x^2 + C$

2. $2y\frac{dy}{dx} = 3$

$\int y\,dy = \int \frac{3}{2}\,dx$

$\frac{1}{2}y^2 = \frac{3}{2}x + C_1$

$y^2 = 3x + C$

3. $\frac{dy}{dx} = 2xy$

$\int \frac{1}{y}\,dy = \int 2x\,dx$

$\ln|y| = x^2 + C_1$

$y = e^{x^2 + C_1}$

$y = Ce^{x^2}$

4. $\frac{dy}{dx} = \frac{x-4}{4y^3}$

$\int 4y^3\,dy = \int (x-4)\,dx$

$y^4 = \frac{1}{2}(x-4)^2 + C$

5. For this linear differential equation, you have $P(x) = 2$ and $Q(x) = 4$. So, the integrating factor is $u(x) = e^{\int 2\,dx} = e^{2x}$ and the general solution is

$y = \frac{1}{u(x)}\int Q(x)u(x)\,dx = e^{-2x}\int 4e^{2x}\,dx$

$= e^{-2x}(2e^{2x} + C) = 2 + Ce^{-2x}.$

6. For this linear differential equation, you have $P(x) = 2$ and $Q(x) = e^{-2x}$. So, the integrating factor is $u(x) = e^{\int 2\,dx} = e^{2x}$ and the general solution is

$y = \frac{1}{u(x)}\int Q(x)u(x)\,dx = e^{-2x}\int e^{-2x}e^{2x}\,dx$

$= e^{-2x}\int dx = e^{-2x}(x + C).$

Skills Review —continued—

7. For this linear differential equation, you have $P(x) = x$ and $Q(x) = x$. So, the integrating factor is $u(x) = e^{\int x\,dx} = e^{x^2/2}$ and the general solution is

$$y = \frac{1}{u(x)}\int Q(x)u(x)\,dx = e^{-x^2/2}\int xe^{x^2/2}\,dx$$
$$= e^{-x^2/2}\left(e^{x^2/2} + C\right) = 1 + Ce^{-x^2/2}.$$

8. For this linear differential equation, you have $P(x) = \frac{2}{x}$ and $Q(x) = x$. So, the integrating factor is $u(x) = e^{\int 2/x\,dx} = e^{2\ln x} = e^{\ln x^2} = x^2$ and the general solution is

$$y = \frac{1}{u(x)}\int Q(x)u(x)\,dx = \frac{1}{x^2}\int x^3\,dx$$
$$= \frac{1}{x^2}\left(\frac{1}{4}x^4 + C\right) = \frac{1}{4}x^2 + \frac{C}{x^2}.$$

9. $\frac{dy}{dx} = kx^2$

10. $\frac{dx}{dt} = k(x - t)$

1. The general solution is $y = Ce^{kx}$. Because $y = 1$ when $x = 0$, it follows that $C = 1$. So, $y = e^{kx}$. Because $y = 2$ when $x = 3$, it follows that $2 = e^{3k}$, which implies that $k = \frac{\ln 2}{3} \approx 0.2310$. So, the particular solution is $y \approx e^{0.2310x}$.

3. The general solution is $y = Ce^{kx}$. Because $y = 4$ when $x = 0$, it follows that $C = 4$. So, $y = 4e^{kx}$. Because $y = 1$ when $x = 4$, it follows that $\frac{1}{4} = e^{4k}$, which implies that $k = \frac{1}{4}\ln\frac{1}{4} \approx -0.3466$. So, the particular solution is $y \approx 4e^{-0.3466x}$.

5. The general solution is $y = Ce^{kx}$. Because $y = 2$ when $x = 2$ and $y = 4$ when $x = 3$, it follows that $2 = Ce^{2k}$ and $4 = Ce^{3k}$. By equating C-values from these two equations, you have

$$2e^{-2k} = 4e^{-3k}$$
$$\frac{1}{2} = e^{-k} \Rightarrow k = \ln 2 \approx 0.6931.$$

This implies that

$$C = 2e^{-2\ln 2} = 2e^{\ln(1/4)} = 2\left(\frac{1}{4}\right) = \frac{1}{2}.$$

So, the particular solution is $y = \frac{1}{2}e^{x\ln 2} \approx \frac{1}{2}e^{0.6931x}$.

7. Let $t = 0$ represent the year 1998. The general solution is $P = Ce^{kt}$ where $C = 400{,}000$ and $k = 0.015$. So, the particular solution is $P = 400{,}000e^{0.015t}$. In the year 2005, you have $t = 7$ and the population of the city is estimated to be $P = 400{,}000e^{0.015(7)} \approx 444{,}284$.

9. The general solution is $y = Ce^{20kx}(20 - y)$.

Because $y = 1$ when $x = 0$, it follows that $C = \frac{1}{19}$.

So, $y = \frac{1}{19}e^{20kx}(20 - y)$.

Because $y = 10$ when $x = 5$, it follows that

$$19 = e^{100k}$$
$$20k = \frac{\ln 19}{5} \approx 0.5889.$$

So, the particular solution is

$$y = \frac{1}{19}e^{0.5889x}(20 - y)$$
$$y\left(19 + e^{0.5889x}\right) = 20e^{0.5889x}$$
$$y = \frac{20e^{0.5889x}}{19 + e^{0.5889x}} = \frac{20}{1 + 19e^{-0.5889x}}.$$

11. The general solution is $y = Ce^{5000kx}(5000 - y)$.

Because $y = 250$ when $x = 0$, it follows that

$C = \frac{1}{19}$. So, $y = \frac{1}{19}e^{5000kx}(5000 - y)$.

Because $y = 2000$ when $x = 25$, it follows that

$$\frac{38}{3} = e^{125{,}000k}$$
$$5000k = \frac{\ln(38/3)}{25} \approx 0.10156.$$

So, the particular solution is

$$y = \frac{1}{19}e^{0.10156x}(5000 - y)$$
$$y\left(19 + e^{0.10156x}\right) = 5000e^{0.10156x}$$
$$y = \frac{5000e^{0.10156x}}{19 + e^{0.10156x}} = \frac{5000}{1 + 19e^{-0.10156x}}.$$

13. (a)

$$\frac{dN}{dt} = kN(500 - N)$$

$$\int \frac{dN}{N(500 - N)} = \int k\, dt$$

$$\frac{1}{500}\int\left[\frac{1}{N} + \frac{1}{500 - N}\right] dN = \int k\, dt$$

$$\ln|N| - \ln|500 - N| = 500(kt + C_1)$$

$$\frac{N}{500 - N} = e^{500kt + C_2} = Ce^{500kt}$$

$$N = \frac{500Ce^{500kt}}{1 + Ce^{500kt}}$$

When $t = 0$, $N = 100$.

So,

$$100 = \frac{500C}{1 + C} \Rightarrow C = 0.25.$$

So,

$$N = \frac{125e^{500kt}}{1 + 0.25e^{500kt}}.$$

When $t = 4$, $N = 200$.

So,

$$200 = \frac{125e^{2000k}}{1 + 0.25e^{2000k}} \Rightarrow k = \frac{\ln(8/3)}{2000} \approx 0.00049.$$

So,

$$N = \frac{125e^{0.2452t}}{1 + 0.25e^{0.2452t}} = \frac{500}{1 + 4e^{-0.2452t}}.$$

(b) When $t = 1$:

$$N = \frac{500}{1 + 4e^{-0.2452}} \approx 121 \text{ deer}$$

(c) When $N = 350$:

$$350 = \frac{500}{1 + 4e^{-0.2452t}}$$

$$1 + 4e^{-0.2452t} = \frac{10}{7}$$

$$4e^{-0.2452t} = \frac{3}{7}$$

$$e^{-0.2452t} = \frac{3}{28}$$

$$-0.2452t = \ln\left(\frac{3}{28}\right)$$

$$t = \frac{\ln\left(\frac{3}{28}\right)}{-0.2452}$$

$$t \approx 9.1 \text{ years}$$

15. The differential equation is given by

$$\frac{dP}{dn} = kP(L - P)$$

$$\int \frac{1}{P(L - P)}dP = \int k\, dn$$

$$\frac{1}{L}\left[\ln|P| - \ln|L - P|\right] = kn + C_1$$

$$\frac{P}{L - P} = Ce^{Lkn}$$

$$P = \frac{CLe^{Lkn}}{1 + Ce^{Lkn}} = \frac{CL}{e^{-Lkn} + C}.$$

17. The general solution is $y = \dfrac{-1}{kt + C}$. Because $y = 45$ when $t = 0$, it follows that $45 = \dfrac{-1}{C}$ and $C = \dfrac{-1}{45}$.

So, $y = -\dfrac{1}{kt - (1/45)} = \dfrac{45}{1 - 45kt}$.

Because $y = 4$ when $t = 2$, you have

$$4 = \frac{45}{1 - 45k(2)} \Rightarrow k = -\frac{45}{360}.$$

So, $y = \dfrac{45}{1 + (41/8)t} = \dfrac{360}{8 + 41t}$.

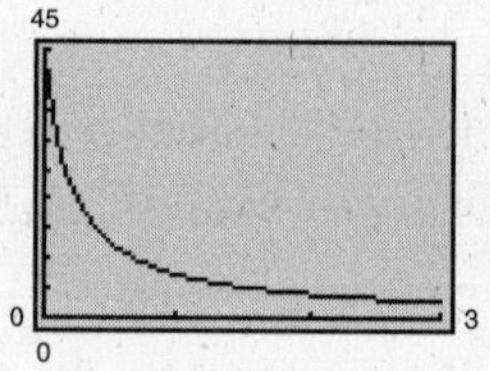

19.

$$\frac{dy}{dt} = ky$$

$$\int \frac{1}{y}\, dy = \int k\, dt$$

$$\ln|y| = kt + C_1$$

$$y = e^{kt + C_1}$$

$$y = Ce^{kt}$$

Because the initial amount is 20 grams, you know that $C = 20$. Because $y = 16$ when $t = 1$, it follows that

$$16 = 20e^k$$

$$\frac{4}{5} = e^k$$

$$\ln\left(\frac{4}{5}\right) = k$$

$$-0.2231 \approx k.$$

So, the particular solution is $y = 20e^{-0.2231t}$.

21. $$y' = k(y - 70)$$
$$\int \frac{1}{y - 70}\,dy = \int k\,dt$$
$$\ln|y - 70| = kt + C_1$$
$$y - 70 = e^{kt + C_1}$$
$$y = 70 + Ce^{kt}$$

Because $y = 350$ when $t = 0$, it follows that $C = 280$. So, $y = 70 + 280e^{kt}$.

Because $y = 150$ when $t = 45$, it follows that
$$150 = 70 + 280e^{45k}$$
$$80 = 280e^{45k}$$
$$\tfrac{2}{7} = e^{45k}$$
$$\ln\left(\tfrac{2}{7}\right) = 45k$$
$$\tfrac{1}{45}\ln\left(\tfrac{2}{7}\right) = k$$
$$-0.0278 \approx k.$$

So, the particular solution is $y = 70 + 280e^{-0.02784t}$.

23. When $y = 80$:
$$y = 70 + 280e^{-0.02784t}$$
$$80 = 70 + 280e^{-0.02784t}$$
$$10 = 280e^{-0.02784t}$$
$$\tfrac{1}{28} = e^{-0.02784t}$$
$$\ln\left(\tfrac{1}{28}\right) = -0.02784t$$
$$119.69 \approx t$$

So, it will require about $119.69 - 45 = 74.69$ more minutes for the object to cool to a temperature of 80°F.

25. Because $y = 100$ when $t = 0$, it follows that $100 = 500e^{-C}$, which implies that $C = \ln 5$.

So you have $y = 500e^{(-\ln 5)e^{-kt}}$.

Because $y = 150$ when $t = 2$, it follows that
$$150 = 500e^{(-\ln 5)e^{-2k}}$$
$$e^{-2k} = \frac{\ln 0.3}{\ln 0.2}$$
$$k = -\frac{1}{2}\ln\frac{\ln 0.3}{\ln 0.2}$$
$$\approx 0.1452.$$

So, $y = 500e^{-1.6904e^{-0.1451t}}$.

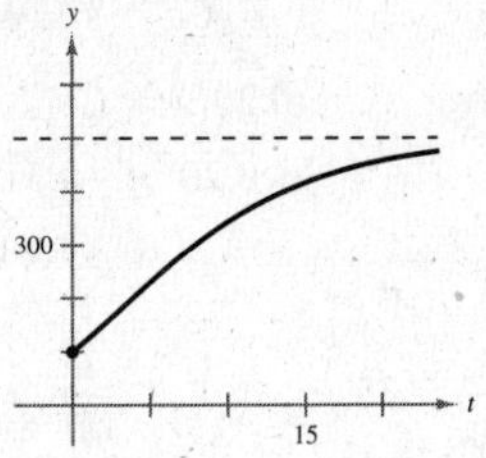

27. From Example 3, the general solution is $y = 60e^{-Ce^{-kt}}$.

Because $y = 8$ when $t = 0$,
$$8 = 60e^{-C} \Rightarrow C = \ln\frac{15}{2} \approx 2.0149.$$

Because $y = 15$ when $t = 3$,
$$15 = 60e^{-2.0149e^{-3k}}$$
$$\frac{1}{4} = e^{-2.0149e^{-3k}}$$
$$\ln\frac{1}{4} = -2.0149e^{-3k}$$
$$k = -\frac{1}{3}\ln\left(\frac{\ln 1/4}{-2.0149}\right) \approx 0.1246.$$

So, $y = 60e^{-2.0149e^{-0.1246t}}$.

When $t = 10$, $y \approx 34$ beavers.

29. Following Example 4, the differential equation is $\frac{dy}{dt} = ky(1 - y)(2 - y)$ and its general solution is $\frac{y(2 - y)}{(1 - y)^2} = Ce^{2kt}$.

$$y = \frac{1}{2} \text{ when } t = 0 \Rightarrow \frac{(1/2)(3/2)}{(1/2)^2} = C \Rightarrow C = 3$$

$$y = 0.75 = \frac{3}{4} \text{ when } t = 4 \Rightarrow \frac{(3/4)(5/4)}{(1/4)^2} = 15 = 3e^{2k(4)} \Rightarrow k = \frac{1}{8}\ln 5 \approx 0.2012.$$

So, the particular solution is $\frac{y(2 - y)}{(1 - y)^2} = 3e^{0.4024t}$.

Using a symbolic algebra utility or graphing utility, you find that when $t = 10$, $y \approx 0.92$, or 92%.

31. (a)
$$\frac{dQ}{dt} = -\frac{Q}{20}$$
$$\int \frac{dQ}{Q} = \int -\frac{1}{20}\,dt$$
$$\ln|Q| = -\frac{1}{20}t + C_1$$
$$Q = e^{-(1/20)t + C_1} = Ce^{-(1/20)t}$$

Because $Q = 25$ when $t = 0$, you have $25 = C$.

So, the particular solution is $Q = 25e^{-(1/20)t}$.

(b) When $Q = 15$, you have $15 = 25e^{-(1/20)t}$.
$$\frac{3}{5} = e^{-(1/20)t}$$
$$\ln\left(\frac{3}{5}\right) = -\frac{1}{20}t$$
$$-20\ln\left(\frac{3}{5}\right) = t$$
$$t \approx 10.217 \text{ minutes}$$

33. The general solution is $y = Ce^{kt}$. Because $y = 0.60C$ when $t = 1$, you have

$0.60C = Ce^{k} \Rightarrow k = \ln 0.60 \approx -0.5108.$

So,

$y = Ce^{-0.5108t}$. When $y = 0.20C$, you have

$$0.20C = Ce^{-0.5108t}$$
$$\ln 0.20 = -0.5108t$$
$$t \approx 3.15 \text{ hours.}$$

35.
$$\frac{dP}{dt} = kP + N$$
$$\int \frac{1}{kP + N}\,dP = \int dt$$
$$\frac{1}{k}\ln|kP + N| = t + C_1$$
$$kP + N = C_2e^{kt}$$
$$P = Ce^{kt} - \frac{N}{k}$$

37.
$$\frac{dA}{dt} = rA + P$$
$$\int \frac{1}{rA + P}\,dA = \int dt$$
$$\frac{1}{r}\ln|rA + P| = t + C_1$$
$$rA + P = Ce^{rt}$$
$$A = \frac{1}{r}\left(Ce^{rt} - P\right)$$

Because $A = 0$ when $t = 0$, it follows that $C = P$.

So, you have $A = \dfrac{P}{r}\left(e^{rt} - 1\right)$.

39. (a)
$$\frac{dC}{dt} = \left(-\frac{R}{V}\right)C$$
$$\int \frac{dC}{C} = \int -\frac{R}{V}\,dt$$
$$\ln|C| = -\frac{R}{V}t + K_1$$
$$C = Ke^{-Rt/V}$$

Because $C = C_0$ when $t = 0$, it follows that $K = C_0$ and the function is $C = C_0e^{-Rt/V}$.

(b) Finally, as $t \to \infty$, you have

$$\lim_{t\to\infty} C = \lim_{t\to\infty} C_0e^{-Rt/V} = 0.$$

41. (a)
$$\frac{dC}{dt} = \frac{Q}{V} - \left(\frac{R}{V}\right)C$$
$$\int \frac{1}{Q - RC}\,dC = \int \frac{1}{V}\,dt$$
$$-\frac{1}{R}\ln|Q - RC| = \frac{t}{V} + K_1$$
$$Q - RC = e^{-R[(t/V)+K_1]}$$
$$C = \frac{1}{R}\left(Q - e^{-R[(t/V)+K_1]}\right)$$
$$= \frac{1}{R}\left(Q - Ke^{-Rt/V}\right)$$

Because $C = 0$ when $t = 0$, it follows that $K = Q$ and you have $C = \dfrac{Q}{R}\left(1 - e^{-Rt/V}\right)$.

(b) As $t \to \infty$, the limit of C is Q/R.

Review Exercises for Chapter 10

1. By differentiation, you have

$$\frac{dy}{dx} = \frac{1}{2}Ce^{x/2} \Rightarrow 2\frac{dy}{dx} = 2\left(\frac{1}{2}Ce^{x/2}\right) = Ce^{x/2} = y$$

3. Because $y' = C_1e^{x} - C_2e^{-x}$ and $y'' = C_1e^{x} + C_2e^{-x}$, you have

$$y'' - y = \left(C_1e^{x} + C_2e^{-x}\right) - \left(C_1e^{x} + C_2e^{-x}\right) = 0.$$

5. $y = \int\left(2x^2 + 5\right)dx = \frac{2}{3}x^3 + 5x + C$

7. $y = \int \cos 2x\,dx = \frac{1}{2}\sin 2x + C$

9. Let $u = x - 7$. Then $x = u + 7$ and $du = dx$.

$$y = \int 2x\sqrt{x - 7}\,dx = 2\int (u + 7)u^{1/2}\,du$$
$$= 2\int\left(u^{3/2} + 7u^{1/2}\right)du$$
$$= \tfrac{4}{5}u^{5/2} + \tfrac{28}{3}u^{3/2} + C$$
$$= \tfrac{4}{5}(x - 7)^{5/2} + \tfrac{28}{3}(x - 7)^{3/2} + C$$
$$= \tfrac{4}{15}(x - 7)^{3/2}\left[3(x - 7) + 35\right] + C$$
$$= \tfrac{4}{15}(x - 7)^{3/2}(3x + 14) + C$$

11. Because $y = e^{4x}$ and $y' = 4e^{4x}$, then you know $y' = 4e^{4x} = 4y$.

13. Because $y = x^2$ and $y' = 2x$, then

$$\frac{1}{2}y' - \frac{1}{x}y = \frac{1}{2}(2x) - \frac{1}{x}\left(x^2\right) = 0.$$

15. $y = x^2$

$y' = 2x$

$xy' - 2y = x(2x) - 2(x^2) = 0 \neq x^3e^x$

So, the function is *not* a solution of the differential equation.

17. $y = x^2e^x$

$y' = x^2e^x + 2xe^x$

$xy' - 2y = x(x^2e^x + 2xe^x) - 2x^2e^x = x^3e^x$

So, the function is a solution of the differential equation.

19. Because $y' = -5Ce^{-5x} = -5y$, it follows that $y' + 5y = 0$. To find the particular solution, use the fact that $y = 1$ when $x = 0$. That is, $1 = Ce^0 \Rightarrow C = 1$. So, the particular solution is $y = e^{-5x}$.

21. When $C = 1$, the graph is the exponential curve $y = e^{3/x}$.

When $C = 2$, the graph is the exponential curve $y = 2e^{3/x}$.

When $C = 4$, the graph is the exponential curve $y = 4e^{3/x}$.

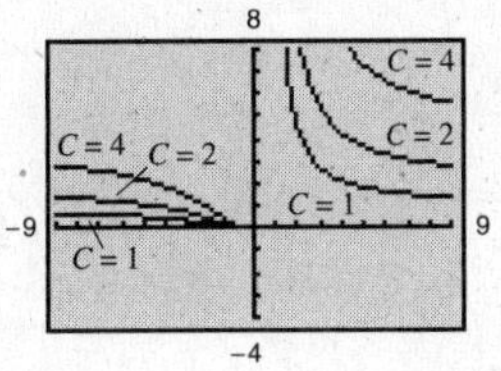

23. Because $y = 1$ when $x = 1$, you have $1 = C(1^2 + 1)$, which implies that $C = \frac{1}{2}$, and the particular solution is $y = \frac{1}{2}(x^2 + 1)$.

25. Yes, $\dfrac{dy}{dx} = \dfrac{y}{x + 3}$

$\dfrac{1}{y}\,dy = \dfrac{1}{x + 3}\,dx$

27. No, the variables cannot be separated.

29. $\dfrac{dy}{dx} = 4x$

$\int dy = \int 4x\,dx$

$y = 2x^2 + C$

31. $4y^3\dfrac{dy}{dx} = 5$

$\int 4y^3\,dy = \int 5\,dx$

$y^4 = 5x + C$

33. $y' + 2xy^2 = 0$

$\dfrac{dy}{dx} = -2xy^2$

$\int \dfrac{1}{y^2}\,dy = \int -2x\,dx$

$-\dfrac{1}{y} = -x^2 + C_1$

$\dfrac{1}{y} = x^2 + C$

$y = \dfrac{1}{x^2 + C}$

35. $\dfrac{dy}{dx} = (x + 1)(y + 1)$

$\int \dfrac{1}{y + 1}\,dy = \int (x + 1)\,dx$

$\ln|y + 1| = \dfrac{1}{2}x^2 + x + C_1$

$y + 1 = e^{1/2x^2 + x + C_1}$

$y = Ce^{x^2/2 + x} - 1$

37. $\dfrac{dy}{dx} = -\dfrac{y + 2}{2x^3}$

$\int \dfrac{1}{y + 2}\,dy = \int -\dfrac{1}{2x^3}\,dx$

$\ln|y + 2| = \dfrac{1}{4x^2} + C_1$

$y + 2 = e^{1/(4x^2) + C_1}$

$y = Ce^{1/(4x^2)} - 2$

39. $\dfrac{dy}{dx} = \dfrac{\cos x}{y}$

$\int y\,dy = \int \cos x\,dx$

$\dfrac{1}{2}y^2 = \sin x + C_1$

$y^2 = 2\sin x + C$

41. (a) $\dfrac{dy}{dx} = 3x^2$

$\int dy = \int 3x^2\,dx$

$y = x^3 + C$

(b)

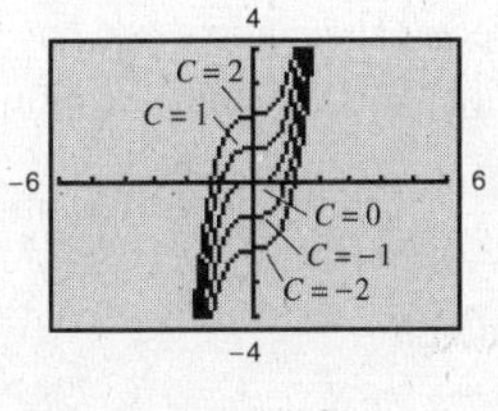

43. $yy' + e^x = 0$

$$\int y\,dy = \int -e^x\,dx$$

$$\tfrac{1}{2}y^2 = -e^x + C_1$$

$$y^2 = -2e^x + C$$

When $x = 0$, $y = 2$. So, $C = 6$ and the particular solution is $y^2 = -2e^x + 6$.

45. $2xy' - \ln x^2 = 0$

$$2x\frac{dy}{dx} = \ln x^2$$

$$\int dy = \int \frac{\ln x^2}{2x}\,dx$$

$$y = \frac{1}{2}(\ln x)^2 + C$$

When $x = 1$, $y = 2$. So, $C = 2$ and the particular solution is $y = \frac{1}{2}(\ln x)^2 + 2$.

47. $\dfrac{dy}{dx} = -\dfrac{9x}{16y}$

$$\int 16y\,dy = \int -9x\,dx$$

$$8y^2 = -\frac{9}{2}x^2 + C$$

When $x = 1$, $y = 1$. So, $C = \frac{25}{2}$ and the equation is

$$8y^2 = -\frac{9}{2}x^2 + \frac{25}{2}$$

$$9x^2 + 16y^2 = 25.$$

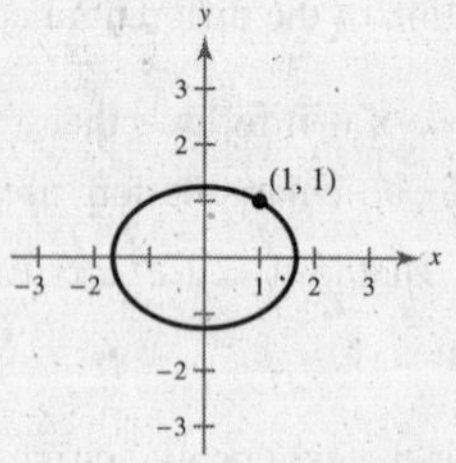

49. (a) $\dfrac{dy}{ds} = -0.012y$

$$\int -\frac{1}{0.012y}\,dy = \int ds$$

$$-\frac{1}{0.012}\ln|y| = s + C_1$$

$$\ln|y| = -0.012s + C_2$$

$$y = e^{-0.012s + C_2} = Ce^{-0.012s}$$

When $s = 50$, $y = 28$. So, $C = 28e^{0.6}$ and the particular solution is $y = 28e^{0.6-0.012s}$.

(b)

Speed	50	55	60	65	70
Miles per gallon	28	26.37	24.83	23.39	22.03

51. $x^4 + 4x^2y' - 4y = 0$

$$4x^2y' - 4y = -x^4$$

$$y' - \frac{1}{x^2}y = -\frac{1}{4}x^2$$

53. $x = 2x^3(y' - y)$

$$2x^3y' - 2x^3y = x$$

$$y' - y = \frac{1}{2x^2}$$

55. For this linear differential equation, you have $P(x) = -4$ and $Q(x) = 8$. So, the integrating factor is $u(x) = e^{\int -4\,dx} = e^{-4x}$ and the general solution is

$$y = \frac{1}{u(x)}\int Q(x)u(x)\,dx$$

$$= e^{4x}\int 8e^{-4x}\,dx = e^{4x}\left(-2e^{-4x} + C\right) = -2 + Ce^{4x}.$$

57. For this linear differential equation, you have $P(x) = -\frac{1}{x}$ and $Q(x) = 2x - 3$. So, the integrating factor is $u(x) = e^{\int -1/x\,dx} = e^{-\ln x} = x^{-1} = \frac{1}{x}$ and the general solution is

$$y = \frac{1}{u(x)}\int Q(x)u(x)\,dx = x\int (2x - 3)\left(\frac{1}{x}\right)dx$$

$$= x\int\left(2 - \frac{3}{x}\right)dx$$

$$= x(2x - 3\ln x + C)$$

$$= 2x^2 - 3x\ln|x| + Cx$$

59. For this linear differential equation, you have $P(x) = -1$ and $Q(x) = 9$. So, the integrating factor is

$u(x) = e^{\int -dx} = e^{-x}$ and the general solution is

$$\begin{aligned} y &= \frac{1}{u(x)} \int Q(x)u(x)\,dx \\ &= e^x \int 9e^{-x}\,dx \\ &= e^x\left(-9e^{-x} + C\right) \\ &= -9 + Ce^x. \end{aligned}$$

61. For this linear differential equation, you have $P(x) = \dfrac{1}{x}$ and $Q(x) = 3x + 4$. So, the integrating factor is

$u(x) = e^{\int 1/x\,dx} = e^{\ln x} = x$ and the general solution is

$$\begin{aligned} y = \frac{1}{u(x)} \int Q(x)u(x)\,dx &= \frac{1}{x}\int (3x + 4)x\,dx \\ &= \frac{1}{x}\int \left(3x^2 + 4x\right)dx \\ &= \frac{1}{x}\left(x^3 + 2x^2 + C\right) \\ &= x^2 + 2x + \frac{C}{x}. \end{aligned}$$

63. Separation of variables:

$$\begin{aligned} \frac{dy}{dx} &= 6 - y \\ \int \frac{dy}{6 - y} &= \int dx \\ -\ln|6 - y| &= x + C_1 \\ 6 - y &= e^{-(x + C_1)} \\ y &= 6 + Ce^{-x} \end{aligned}$$

First-order linear: $P(x) = 1$, $Q(x) = 6$

$$\begin{aligned} u(x) &= e^{\int dx} = e^x \\ y &= e^{-x}\int 6e^x\,dx \\ &= e^{-x}\left(6e^x + C\right) \\ &= 6 + Ce^{-x} \end{aligned}$$

65. Separation of variables:

$$\begin{aligned} \frac{dy}{dx} &= x - 2xy \\ \int \frac{1}{1 - 2y}\,dy &= \int x\,dx \\ -\frac{1}{2}\ln|1 - 2y| &= \frac{1}{2}x^2 + C_1 \\ \ln|1 - 2y| &= -x^2 + C_2 \\ 1 - 2y &= e^{-x^2 + C_2} \\ -2y &= C_3 e^{-x^2} - 1 \\ y &= \frac{1}{2}\left(1 - C_3 e^{-x^2}\right) \\ &= \frac{1}{2} + Ce^{-x^2} \end{aligned}$$

First-order linear:

$P(x) = 2x$, $Q(x) = x$

$$\begin{aligned} u(x) &= e^{\int 2x\,dx} = e^{x^2} \\ y &= e^{-x^2}\int xe^{x^2}\,dx \\ &= e^{-x^2}\left(\frac{1}{2}e^{x^2} + C\right) \\ &= \frac{1}{2} + Ce^{-x^2} \end{aligned}$$

67.

$$\begin{aligned} y' &= k(y - 90) \\ \left(\frac{1}{y - 90}\right)dy &= k\,dt \\ \int \frac{1}{y - 90}\,dy &= \int k\,dt \\ \ln|y - 90| &= kt + C_1 \\ y - 90 &= e^{kt + C_1} \\ y &= 90 + Ce^{kt} \end{aligned}$$

Because $y = 1500$ when $t = 0$, it follows that $C = 1410$. So, $y = 90 + 1410e^{kt}$. Because $y = 1120$ when $t = 1$, it follows that

$$\begin{aligned} 1120 &= 90 + 1410e^k \\ 1030 &= 1410e^k \\ \frac{103}{141} &= e^k \\ \ln\left(\frac{103}{141}\right) &= k \\ -0.3140 &\approx k \end{aligned}$$

So, the particular solution is $y = 90 + 1410e^{-0.3140t}$.

When $t = 5$: $y = 90 + 1410e^{-0.3140(5)} \approx 383.3°\text{F}$

69. $\frac{dy}{dx} = -k\frac{y}{x}$

$\int \frac{1}{y}\,dy = \int -\frac{k}{x}\,dx$

$\ln|y| = -k\ln|x| + C_1$

$y = e^{-k\ln|x|+C_1} = Cx^{-k}$

71. $\frac{dy}{dx} = -0.2y$

$\int -\frac{5}{y}\,dy = \int dx$

$-5\ln|y| = x + C_1$

$\ln|y| = -\frac{1}{5}x + C_2$

$y = e^{-x/5+C_2}$

$y = Ce^{-x/5}$

When $x = 0$, $y = 29.92$. This implies that $C = 29.92$ and the particular solution is $y = 29.92e^{-x/5}$.

(a) Because there are 5280 feet in one mile, you know that 7310 feet $\approx$ 1.3845 miles. So, when $x = 1.3845$, you have $y = 29.92e^{-1.3845/5} \approx 22.68$ inches of mercury.

(b) Because there are 5280 feet in one mile, you know that 19,340 feet $\approx$ 3.6629 miles. So, when $x = 3.6629$, you have $y = 29.92e^{-3.6629/5} \approx 14.38$ inches of mercury.

73. Let y represent the mass of the carbon and let t represent the time (in years).

$\frac{dy}{dt} = ky$

$\int \frac{1}{y}\,dy = \int k\,dt$

$\ln y = kt + C_1$

$y = e^{kt+C_1}$

$y = Ce^{kt}$

The half-life is 5715 years.

$\frac{1}{2}C = Ce^{k(5715)}$

$\frac{1}{2} = e^{5715k}$

$\ln\frac{1}{2} = 5715k$

$\frac{1}{5715}\ln\frac{1}{2} = k$

$-0.000121 \approx k$

The equation is $y = Ce^{-0.000121t}$. After $t = 1000$ years, $y \approx 0.886C$. So, about 88.6% of a present amount will remain.

75. $\frac{dy}{dt} = k\sqrt[3]{y}$

$\int \frac{1}{y^{1/3}}\,dy = \int k\,dt$

$\frac{3}{2}y^{2/3} = kt + C_1$

$y^{2/3} = \frac{2}{3}kt + C$

Because $y = 27$ when $t = 0$, you know that $C = 9$. Because $y = 8$ when $t = 1$, it follows that

$(8^{2/3}) = \frac{2}{3}k + 9$

$4 = \frac{2}{3}k + 9$

$-5 = \frac{2}{3}k$

$-7.5 = k.$

So, the particular solution is $y^{2/3} = -5t + 9$ or $y = (-5t + 9)^{3/2}$.

77. Let y be the number of gallons of alcohol in the tank at any time t.

$\frac{dy}{dt} = -6\left(\frac{y}{30}\right) + 3$

$y' + \frac{1}{5}y = 3$

$P(t) = \frac{1}{5},\ Q(t) = 3$

$u(t) = e^{\int 1/5\,dt} = e^{t/5}$

$y = e^{-t/5}\int 3e^{t/5}\,dt = e^{-t/5}\left(15e^{t/5} + C\right) = 15 + Ce^{-t/5}$

Because $y = 6$ when $t = 0$, it follows that $C = -9$. So, the particular solution is $y = 15 - 9e^{-t/5}$.

When $t = 10$: $y = 15 - 9e^{-10/5} \approx 13.8$ gallons

79. Let y be the number of pelicans at any time t.

$\frac{dy}{dt} = ky\ln\frac{64}{y}$

$y = 64e^{-Ce^{-kt}}$

Because $y = 12$ when $t = 0$,

$12 = 64e^{-C} \Rightarrow C = \ln\left(\frac{16}{3}\right) \approx 1.6740.$

Because $y = 28$ when $t = 3$,

$28 = 64e^{-1.6740e^{-3k}}$

$\ln\left(\frac{7}{16}\right) = -1.6740e^{-3k}$

$k = -\frac{1}{3}\ln\left(\frac{\ln(7/16)}{-1.6740}\right) = 0.2352$

So, $y = 64e^{-1.6740e^{-0.2352t}}$.

When $t = 8$, $y \approx 50$ pelicans.

Chapter Test Solutions

1. Because $y' = -2e^{-2x}$, you have

$$3y' + 2y = 3\left(-2e^{-2x}\right) + 2\left(e^{-2x}\right) = -4e^{-2x}.$$

2. Because $y' = -\dfrac{1}{(x+1)^2}$ and $y'' = \dfrac{2}{(x+1)^3}$, you have

$$y'' - \frac{2y}{(x+1)^2} = \frac{2}{(x+1)^3} - \frac{2\left(\dfrac{1}{x+1}\right)}{(x+1)^2} = 0.$$

3.

$$\begin{aligned} yy' &= x \\ \int y\,dy &= \int x\,dx \\ \frac{1}{2}y^2 &= \frac{1}{2}x^2 + C_1 \\ y^2 &= x^2 + C \end{aligned}$$

4.

$$\begin{aligned} \frac{dy}{dx} &= \frac{2x}{y} \\ \int y\,dy &= \int 2x\,dx \\ \frac{1}{2}y^2 &= x^2 + C_1 \\ y^2 &= 2x^2 + C \end{aligned}$$

5.

$$\begin{aligned} \frac{dy}{dx} &= \frac{\cos \pi x}{3y^2} \\ \int 3y^2\,dy &= \int \cos \pi x\,dx \\ y^3 &= \frac{1}{\pi}\sin \pi x + C \end{aligned}$$

6.

$$\begin{aligned} y' &= 2(xy - x) \\ \frac{dy}{dx} &= 2x(y-1) \\ \int \frac{1}{y-1}\,dy &= \int 2x\,dx \\ \ln|y-1| &= x^2 + C_1 \\ y - 1 &= e^{x^2 + C_1} \\ y &= 1 + Ce^{x^2} \end{aligned}$$

7. For this linear differential equation, you have $P(x) = -2$ and $Q(x) = e^{2x}$. So, the integrating factor is $u(x) = e^{\int -2\,dx} = e^{-2x}$ and the general solution is

$$\begin{aligned} y &= \frac{1}{u(x)}\int Q(x)u(x)\,dx \\ &= e^{2x}\int e^{2x}\left(e^{-2x}\right)dx \\ &= e^{2x}\int dx \\ &= e^{2x}(x + C) \\ &= xe^{2x} + Ce^{2x}. \end{aligned}$$

8. For this linear differential equation, you have $P(x) = -1$ and $Q(x) = x$. So, the integrating factor is $u(x) = e^{\int -dx} = e^{-x}$ and the general solution is

$$\begin{aligned} y &= \frac{1}{u(x)}\int Q(x)u(x)\,dx \\ &= e^x\int xe^{-x}\,dx \\ &= e^x\left(-xe^{-x} - e^{-x} + C\right) \\ &= -x - 1 + Ce^x. \end{aligned}$$

9. For this linear differential equation, you have $P(x) = \dfrac{1}{x}$ and $Q(x) = x$. So, the integrating factor is $u(x) = e^{\int 1/x\,dx} = e^{\ln x} = x$ and the general solution is

$$\begin{aligned} y &= \frac{1}{u(x)}\int Q(x)u(x)\,dx \\ &= \frac{1}{x}\int x^2\,dx \\ &= \frac{1}{x}\left(\frac{1}{3}x^3 + C\right) \\ &= \frac{1}{3}x^2 + \frac{C}{x}. \end{aligned}$$

10. For this linear differential equation, you have $P(x) = -x^2$ and $Q(x) = x^2$. So, the integrating factor is $u(x) = e^{\int -x^2\,dx} = e^{-x^3/3}$ and the general solution is

$$\begin{aligned} y &= \frac{1}{u(x)}\int Q(x)u(x)\,dx \\ &= e^{x^3/3}\int x^2 e^{-x^3/3}\,dx \\ &= e^{x^3/3}\left(-e^{-x^3/3} + C\right) \\ &= -1 + Ce^{x^3/3}. \end{aligned}$$

11. $y' + x^2y - x^2 = 0$

$$\frac{dy}{dx} = x^2(1 - y)$$

$$\int \frac{1}{1-y}\,dy = \int x^2\,dx$$

$$-\ln|1 - y| = \frac{1}{3}x^3 + C_1$$

$$\ln|1 - y| = -\frac{1}{3}x^3 + C_2$$

$$1 - y = e^{-x^3/3 + C_2}$$

$$y = 1 - Ce^{-x^3/3}$$

When $x = 0$, $y = 1$. So, $C = 1$ and the particular solution is $y = 1 - e^{-x^3/3}$.

12. $y'e^{-x^2} = 2xy$

$$\frac{dy}{dx} = 2xe^{x^2}y$$

$$\int \frac{1}{y}\,dy = \int 2xe^{x^2}\,dx$$

$$\ln|y| = e^{x^2} + C_1$$

$$y = e^{e^{x^2} + C_1} = Ce^{e^{x^2}}$$

When $x = 0$, $y = e$. So, $C = 1$ and the particular solution is $y = e^{e^{x^2}}$.

13. $x\dfrac{dy}{dx} = \dfrac{1}{7}\ln x$

$$\int 7\,dy = \int \frac{1}{x}\ln x\,dx$$

$$7y = \frac{1}{2}(\ln x)^2 + C_1$$

$$y = \frac{1}{14}(\ln x)^2 + C$$

When $x = 1$, $y = -2$. So, $C = -2$ and the particular solution is $y = \frac{1}{14}(\ln x)^2 - 2$.

14. $\dfrac{dw}{dt} = k(200 - w)$

$$\int \frac{1}{200 - w}\,dw = \int k\,dt$$

$$-\ln|200 - w| = kt + C_1$$

$$\ln|200 - w| = -kt + C_2$$

$$200 - w = e^{-kt + C_2}$$

$$w = 200 - Ce^{-kt}$$

When $t = 0$, $w = 10$. So, $C = 190$ and the particular solution is $w = 200 - 190e^{-kt}$.

(a)

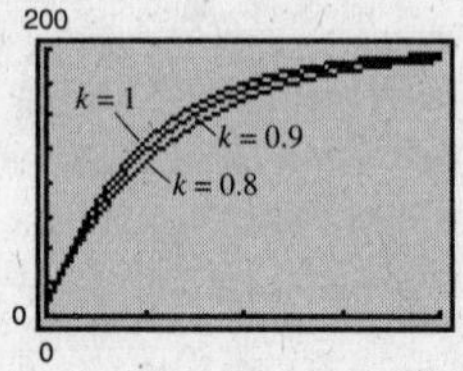

(b) For $k = 0.8$ and $w = 150$:

$$150 = 200 - 190e^{-0.8t}$$

$$-50 = -190e^{-0.8t}$$

$$\frac{5}{19} = e^{-0.8t}$$

$$\ln\left(\frac{5}{19}\right) = -0.8t$$

$$-1.25\ln\left(\frac{5}{19}\right) = t$$

$$1.7 \text{ years} \approx t$$

For $k = 0.9$ and $w = 150$:

$$150 = 200 - 190e^{-0.9t}$$

$$-50 = -190e^{-0.9t}$$

$$\frac{5}{19} = e^{-0.9t}$$

$$\ln\left(\frac{5}{19}\right) = -0.9t$$

$$-\frac{10}{9}\ln\left(\frac{5}{19}\right) = t$$

$$1.5 \text{ years} \approx t$$

For $k = 1$ $w = 150$:

$$150 = 200 - 190e^{-t}$$

$$-50 = -190e^{-t}$$

$$\frac{5}{19} = e^{-t}$$

$$\ln\left(\frac{5}{19}\right) = -t$$

$$-\ln\left(\frac{5}{19}\right) = t$$

$$1.3 \text{ years} \approx t$$

(c) The weight of the lamb approaches 200 pounds as t increases, for each of the models.

15.

$$y' = k(y - 72)$$

$$\int \frac{1}{y - 72}\, dy = \int k\, dt$$

$$\ln|y - 72| = kt + C_1$$

$$y - 72 = e^{kt + C_1}$$

$$y = 72 + Ce^{kt}$$

Because $y = 400$ when $t = 0$, it follows that $C = 328$. So, $y = 72 + 328e^{kt}$. Because $y = 160$ when $t = 40$, it follows that

$$160 = 72 + 328e^{40k}$$

$$88 = 328e^{40k}$$

$$\frac{11}{41} = e^{40k}$$

$$\ln\left(\frac{11}{41}\right) = 40k$$

$$\frac{1}{40}\ln\left(\frac{11}{41}\right) = k$$

$$-0.0329 \approx k$$

So, the particular solution is $y = 72 + 328e^{-0.0329t}$.

When $y = 100$:

$$100 = 72 + 328e^{-0.0329t}$$

$$28 = 328e^{-0.0329t}$$

$$\frac{7}{82} = e^{-0.0329t}$$

$$\ln\left(\frac{7}{82}\right) = -0.0329t$$

$$-\frac{1}{0.0329}\ln\left(\frac{7}{82}\right) = t$$

$$75 \text{ minutes} \approx t$$

So, it will require about 35 more minutes for the object to cool to a temperature of 100°F.

Practice Test for Chapter 10

1. Verify that $y = x^2 \ln x$ is a solution of the differential equation $xy'' - y' = 2x$.

2. Verify that $y = C_1e^{-x} + C_2xe^{-x}$ is a solution of $y'' + 2y' + y = 0$.

3. For problem #2, find the particular solution that satisfies $y'(0) = 2$ and $y(0) = -3$.

4. Find the general solution to the differential equation $y' = y^2/(x + 1)$ using separation of variables.

5. Find the particular solution to the differential equation in #4 that satisfies $y(0) = 1$.

6. Solve the first-order linear differential equation $y' + 3x^2y = e^{x^3}$.

7. The rate of increase in sales (in thousands) is given by $dS/dt = 0.05(100 - S) + 0.2t$ for t in years. If $S = 60$ when $t = 0$, find sales when $t = 10$ years.

8. In a town of 10,000 people the rate at which a new product is introduced is proportional to the number of people who haven't heard of the product yet.

$$\frac{dN}{dt} = k(10{,}000 - N)$$

where N is the number of people who have heard of the product. If $N(0) = 0$ and $N(2) = 2000$, find $N(t)$.

9. A container of hot liquid is placed in a freezer that is kept at a constant temperature of 20°F. The initial temperature of the liquid is 160°F. After 5 minutes, the liquid's temperature is 60°F. How much longer will it take for the temperature to decrease to 30°F?

10. Suppose the net rate of growth of the population is $dP/dt = 0.2P + 1000$. If $P(0) = 30{,}000$, find $P(t)$.

CHAPTER 11
Probability and Calculus

CHAPTER 11
Probability and Calculus

Section 11.1 Discrete Probability

Skills Review

1. $\frac{1}{9} + \frac{2}{3} + \frac{2}{9} = x$

$\frac{1}{9} + \frac{6}{9} + \frac{2}{9} = x$

$\frac{9}{9} = x$

$1 = x$

2. $\frac{1}{3} + \frac{5}{12} + \frac{1}{8} + \frac{1}{12} + \frac{x}{24} = 1$

$\frac{8}{24} + \frac{10}{24} + \frac{3}{24} + \frac{2}{24} + \frac{x}{24} = \frac{24}{24}$

$\frac{23 + x}{24} = \frac{24}{24}$

$23 + x = 24$

$x = 1$

3. $0\left(\frac{1}{16}\right) + 1\left(\frac{3}{16}\right) + 2\left(\frac{8}{16}\right) + 3\left(\frac{3}{16}\right) + 4\left(\frac{1}{16}\right) = \frac{32}{16} = 2$

4. $0\left(\frac{1}{12}\right) + 1\left(\frac{2}{12}\right) + 2\left(\frac{6}{12}\right) + 3\left(\frac{2}{12}\right) + 4\left(\frac{1}{12}\right) = \frac{24}{12} = 2$

5. $(0 - 1)^2\left(\frac{1}{4}\right) + (1 - 1)^2\left(\frac{1}{2}\right) + (2 - 1)^2\left(\frac{1}{4}\right) = \frac{1}{4} + \frac{1}{4} = \frac{1}{2}$

6. $(0 - 2)^2\left(\frac{1}{12}\right) + (1 - 2)^2\left(\frac{2}{12}\right) + (2 - 2)^2\left(\frac{6}{12}\right) + (3 - 2)^2\left(\frac{2}{12}\right) + (4 - 2)^2\left(\frac{1}{12}\right) = \frac{4}{12} + \frac{2}{12} + \frac{2}{12} + \frac{4}{12} = 1$

7. $\frac{3}{8} = 0.375 = 37.5\%$

8. $\frac{9}{11} \approx 0.8182 = 81.82\%$

9. $\frac{13}{24} \approx 0.5417 = 54.17\%$

10. $\frac{112}{256} = 0.4375 = 43.75\%$

1. (a) $S = \{HHH, HHT, HTH, THH, HTT, THT, TTH, TTT\}$

(b) $A = \{HHH, HHT, HTH, THH\}$

(c) $B = \{HTT, THT, TTH, TTT\}$

3. (a) $S = \{III, IIO, IIU, IOI, IUI, OII, UII, IOO, IOU, IUO, IUU, OIO, OIU, UIO, UIU, OOI,$
$OUI, UOI, UUI, OOO, OOU, OUO, UOO, OUU, UOU, UUO, UUU\}$

(b) $A = \{III, IIO, IIU, IOI, IUI, OII, UII\}$

(c) $B = \{III, IIO, IIU, IOI, IUI, OII, UII, IOU, IUO, IUU, OIU, UIO, UIU, OUI, UOI, UUI,$
$OUU, UOU, UUO, UUU\}$

5.

Random variable	0	1	2
Frequency	1	2	1

7.

Random variable	0	1	2	3
Frequency	1	3	3	1

9. $P(\text{third candidate}) = 1 - 0.29 - 0.47 = 0.24$

11. $P(\text{both failing}) = 1 - 0.9855 = 0.0145$

13. $P(3) = 1 - (0.20 + 0.35 + 0.15 + 0.05) = 0.25$

15. The table represents a probability distribution because each distinct value of x corresponds to a probability $P(x)$ such that $0 \le P(x) \le 1$, and the sum of all $P(x)$ is 1.

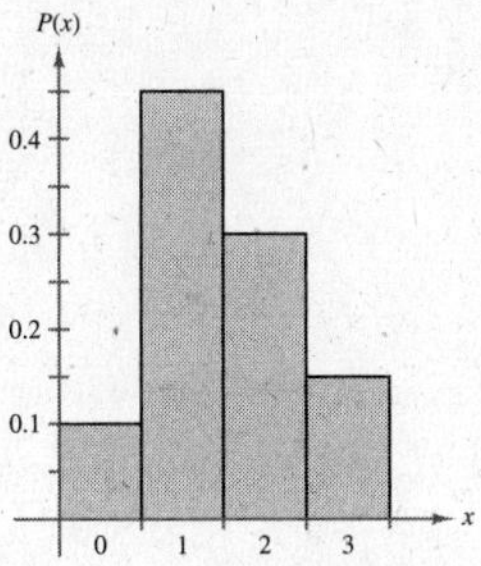

17. The table does not represent a probability distribution because $P(4)$ is not a probability satisfying the inequality $0 \le P(x) \le 1$, and because the sum of all values of $P(x)$ is not equal to 1.

19. (a) $P(1 \le x \le 3) = P(1) + P(2) + P(3)$

$= \frac{3}{20} + \frac{6}{20} + \frac{6}{20}$

$= \frac{15}{20} = \frac{3}{4}$

(b) $P(x \ge 2) = P(2) + P(3) + P(4)$

$= \frac{6}{20} + \frac{6}{20} + \frac{4}{20}$

$= \frac{16}{20}$

$= \frac{4}{5}$

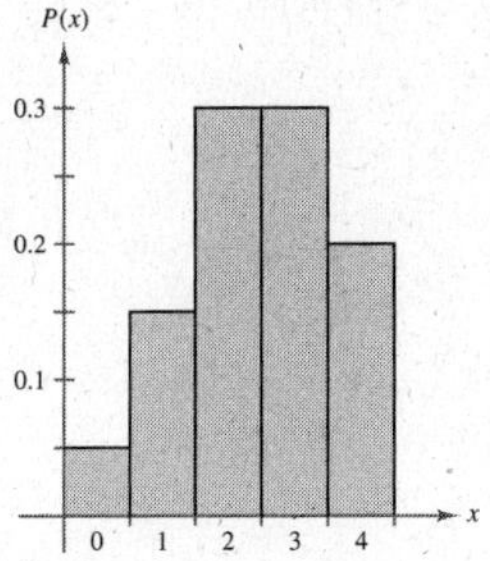

21. (a) $P(x \le 3) = P(0) + P(1) + P(2) + P(3)$

$= 0.041 + 0.189 + 0.247 + 0.326$

$= 0.803$

(b) $P(x > 3) = P(4) + P(5)$

$= 0.159 + 0.038$

$= 0.197$

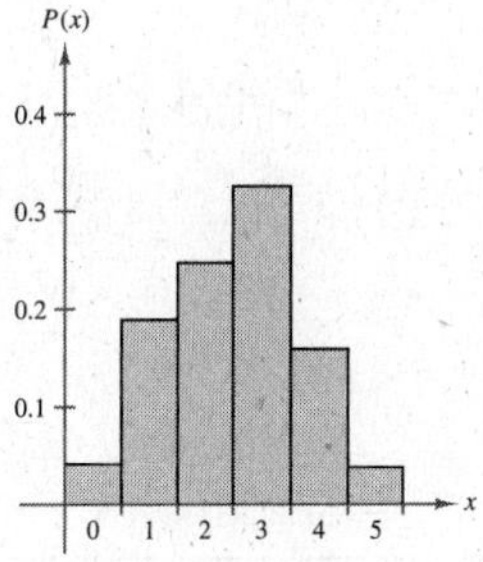

23. (a) $S = \{gggg, gggb, ggbg, gbgg, bggg, ggbb, gbbg, gbgb, bgbg, bbgg, bggb, gbbb, bgbb, bbgb, bbbg, bbbb\}$

(b)

x	0	1	2	3	4
$P(x)$	$\frac{1}{16}$	$\frac{4}{16}$	$\frac{6}{16}$	$\frac{4}{16}$	$\frac{1}{16}$

(c)

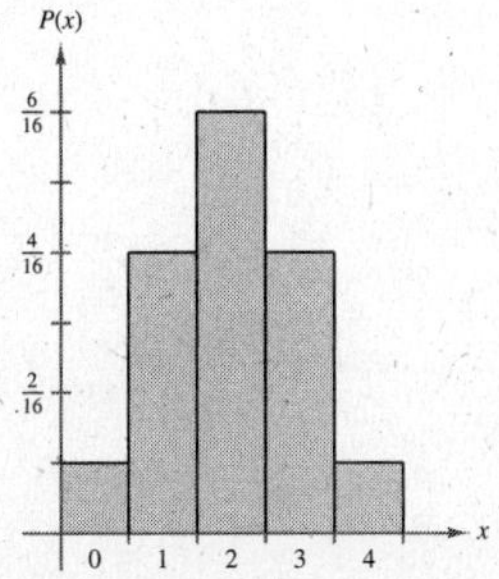

(d) Probability of at least one boy = 1 − probability of all girls

$P = 1 - \frac{1}{16} = \frac{15}{16}$

25. $E(x) = 1\left(\frac{1}{16}\right) + 2\left(\frac{3}{16}\right) + 3\left(\frac{8}{16}\right) + 4\left(\frac{3}{16}\right) + 5\left(\frac{1}{16}\right) = \frac{48}{16} = 3$

$V(x) = (1-3)^2\left(\frac{1}{16}\right) + (2-3)^2\left(\frac{3}{16}\right) + (3-3)^2\left(\frac{8}{16}\right) + (4-3)^2\left(\frac{3}{16}\right) + (5-3)^2\left(\frac{1}{16}\right) = 14\left(\frac{1}{16}\right) = \frac{7}{8} = 0.875$

$\sigma = \sqrt{V(x)} = 0.9354$

27. (a)

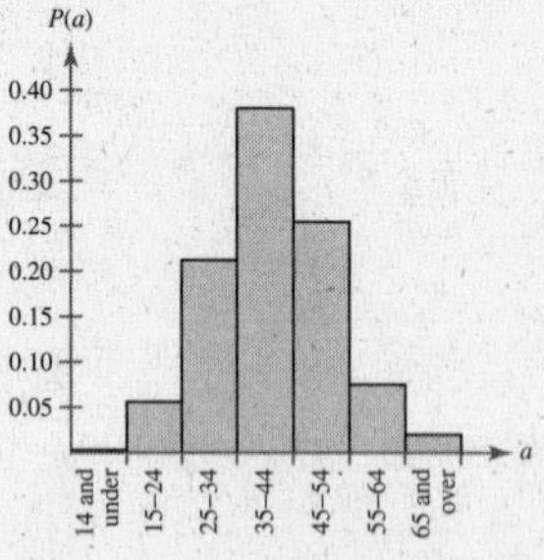

(b) $P(15\text{–}44) = P(15\text{–}24) + P(25\text{–}34) + P(35\text{–}44)$

$= 0.056 + 0.212 + 0.380 = 0.648$

(c) $P(a \geq 35) = P(35\text{–}44) + P(45\text{–}54) + P(55\text{–}64) + P(65 \text{ and over})$

$= 0.380 + 0.254 + 0.075 + 0.020 = 0.729$

29. (a) $E(x) = 1\left(\frac{1}{4}\right) + 2\left(\frac{1}{4}\right) + 3\left(\frac{1}{4}\right) + 4\left(\frac{1}{4}\right) = \frac{10}{4} = 2.5$

$V(x) = (1-2.5)^2\left(\frac{1}{4}\right) + (2-2.5)^2\left(\frac{1}{4}\right) + (3-2.5)^2\left(\frac{1}{4}\right) + (4-2.5)^2\left(\frac{1}{4}\right) = 1.25$

(b) $E(x) = 2\left(\frac{1}{16}\right) + 3\left(\frac{2}{16}\right) + 4\left(\frac{3}{16}\right) + 5\left(\frac{4}{16}\right) + 6\left(\frac{3}{16}\right) + 7\left(\frac{2}{16}\right) + 8\left(\frac{1}{16}\right) = \frac{80}{16} = 5$

$V(x) = (2-5)^2\left(\frac{1}{16}\right) + (3-5)^2\left(\frac{2}{16}\right) + (4-5)^2\left(\frac{3}{16}\right) + (5-5)^2\left(\frac{4}{16}\right) + (6-5)^2\left(\frac{3}{16}\right) + (7-5)^2\left(\frac{2}{16}\right) + (8-5)^2\left(\frac{1}{16}\right)$

$= \frac{5}{2}$

$= 2.5$

31. (a) Total number $= 14 + 26 + 7 + 2 + 1 = 50$

x	0	1	2	3	4
$P(x)$	$\frac{14}{50}$	$\frac{26}{50}$	$\frac{7}{50}$	$\frac{2}{50}$	$\frac{1}{50}$

(b)

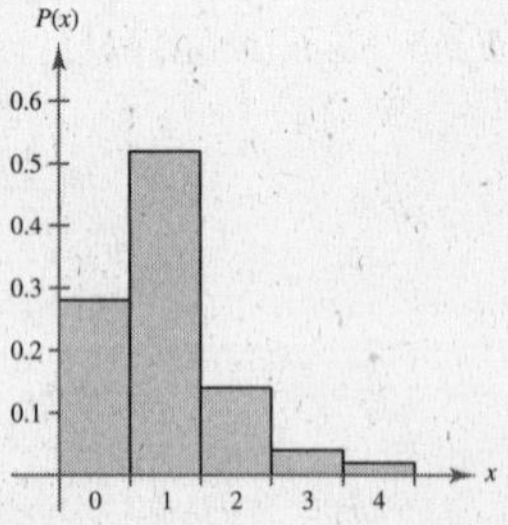

(c) $P(1 \leq x \leq 3) = \frac{26}{50} + \frac{7}{50} + \frac{2}{50} = \frac{35}{50}$

(d) $E(x) = 0 \cdot \frac{14}{50} + 1 \cdot \frac{26}{50} + 2 \cdot \frac{7}{50} + 3 \cdot \frac{2}{50} + 4 \cdot \frac{1}{50} = \frac{50}{50} = 1$

On the average, you can expect the player to get 1 hit per game.

$V(x) = (0-1)^2\frac{14}{50} + (1-1)^2\frac{26}{50} + (2-1)^2\frac{7}{50} + (3-1)^2\frac{2}{50} + (4-1)^2\frac{1}{50} \approx 0.76$

$\sigma = \sqrt{V(x)} \approx 0.87$

$V(x)$ is the variance and σ is the standard deviation. Both are measures of how spread out the data are.

Section 11.2 Continuous Random Variables

Skills Review

1. The domain of the rational function $f(x) = \frac{1}{x}$ consists of all real numbers except $x = 0$. The value of $\frac{1}{x}$ is positive for all positive real numbers x. So, $f(x)$ is continuous and nonnegative on $[0, 1]$.

2. The polynomial function $f(x) = x^2 - 1$ is continuous at every real number. Because $f(0) = 0^2 - 1 = -1$, however, $f(x)$ is continuous but *not* nonnegative on $[0, 1]$.

3. The polynomial function $f(x) = 3 - x$ is continuous at every real number, but $f(5) = 3 - 5 = -2$. So, $f(x)$ is continuous but *not* nonnegative on $[1, 5]$.

4. The function $f(x) = e^{-x} = \frac{1}{e^x}$ is continuous at every real number and its value is positive on $[0, 1]$. So, $f(x)$ is continuous and nonnegative on $[0, 1]$.

5. $\int_0^4 \frac{1}{4}\,dx = \frac{1}{4}x\Big]_0^4 = 1$

6. $\int_1^3 \frac{1}{4}\,dx = \frac{1}{4}x\Big]_1^3 = \frac{3}{4} - \frac{1}{4} = \frac{1}{2}$

7. $$\int_0^2 \frac{2 - x}{2}\,dx = \int_0^2 \left(1 - \frac{x}{2}\right) dx = \left[x - \frac{1}{4}x^2\right]_0^2 = 2 - \frac{4}{4} = 1$$

8. $$\int_1^2 \frac{2 - x}{2}\,dx = \int_1^2 \left(1 - \frac{x}{2}\right) dx = \left[x - \frac{1}{4}x^2\right]_1^2 = 1 - \frac{3}{4} = \frac{1}{4}$$

9. $$\int_0^\infty 0.4e^{-0.4t}\,dt = \lim_{b\to\infty} \int_0^b 0.4e^{-0.4t}\,dt = \lim_{b\to\infty}\left[-e^{-0.4t}\right]_0^b = \lim_{b\to\infty}\left(-e^{-0.4b} + 1\right) = 1$$

10. $$\int_0^\infty 3e^{-3t}\,dt = \lim_{b\to\infty} \int_0^b 3e^{-3t}\,dt = \lim_{b\to\infty}\left[-e^{-3t}\right]_0^b = \lim_{b\to\infty}\left(-e^{-3b} + 1\right) = 1$$

1. 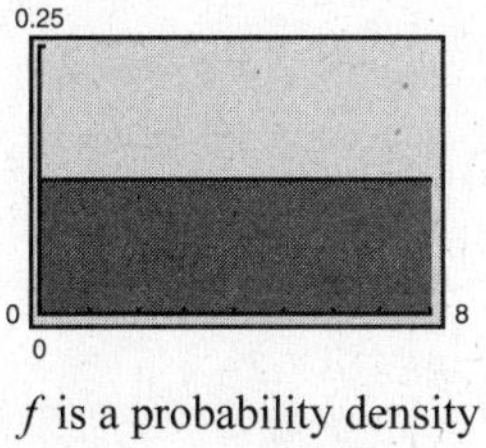

f is a probability density function because

$\int_0^8 \frac{1}{8}\,dx = \frac{1}{8}x\Big]_0^8 = 1$ and $f(x) = \frac{1}{8} \geq 0$ on $[0, 8]$.

3. 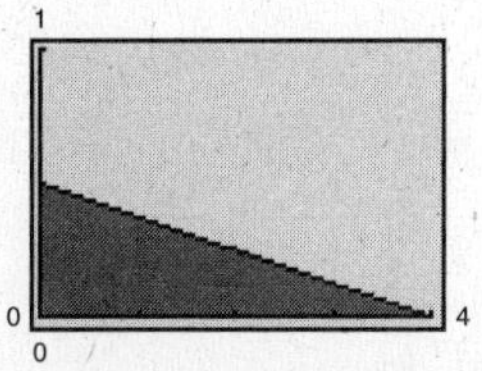

f is a probability density function because

$$\int_0^4 \frac{4 - x}{8}\,dx = \frac{1}{8}\left(4x - \frac{x^2}{2}\right)\Bigg]_0^4 = 1 \text{ and}$$

$$f(x) = \frac{4 - x}{8} \geq 0 \text{ on } [0, 4].$$

5. 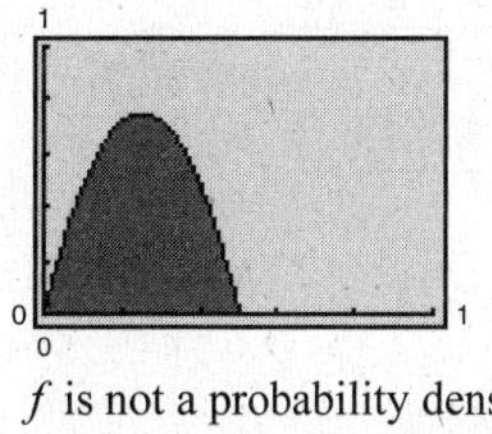

f is not a probability density function because

$$\int_0^1 6x(1 - 2x)\,dx = \int_0^1 \left(6x - 12x^2\right) dx = \left[3x^2 - 4x^3\right]_0^1 = -1 \neq 1$$

and because $f(x) = 6x(1 - 2x)$ is negative on $\left(\frac{1}{2}, 1\right)$.

7. 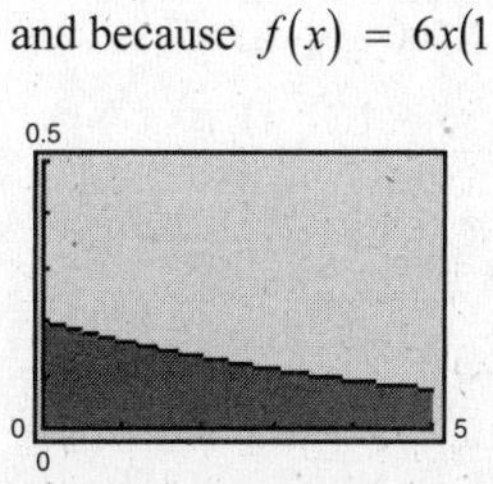

f is not a probability density function because

$$\int_0^5 \frac{1}{5}e^{-x/5}\,dx = \left[-e^{-x/5}\right]_0^5 = -e^{-1} + 1 \approx 0.632 \neq 1.$$

9.

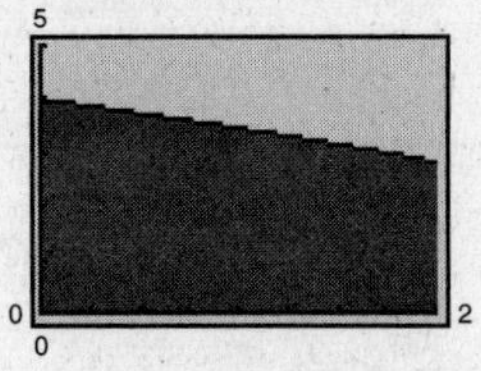

f is not a probability density function because

$$\int_0^2 2\sqrt{4-x}\,dx = -\tfrac{4}{3}\Big[(4-x)^{3/2}\Big]_0^2 = -\tfrac{4}{3}\sqrt{8} + \tfrac{32}{3}$$

$$\approx 6.895 \neq 1.$$

11.

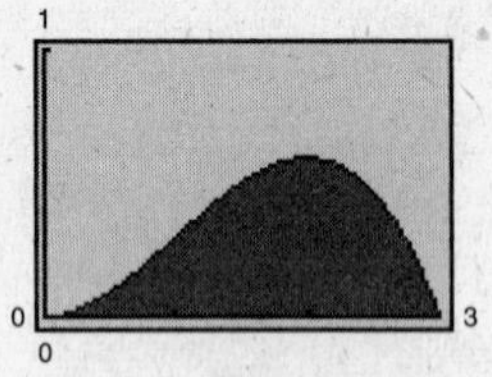

f is a probability density function because

$$\int_0^3 \frac{4}{27}x^2(3-x)\,dx = \frac{4}{27}\int_0^3 \left(3x^2 - x^3\right) dx$$

$$= \frac{4}{27}\left[x^3 - \frac{x^4}{4}\right]_0^3$$

$$= \frac{4}{27}\left(27 - \frac{81}{4}\right) = 1$$

and $f(x) = \dfrac{4}{27}x^2(3-x) \geq 0$ on $[0, 3]$.

13.

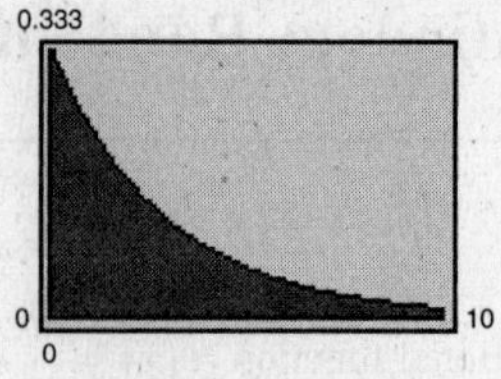

f is a probability density function because

$$\int_0^\infty \tfrac{1}{3}e^{-x/3}\,dx = \lim_{b\to\infty}\left[-e^{-x/3}\right]_0^b = 0 - (-1) = 1$$

and $f(x) = \tfrac{1}{3}e^{-x/3} \geq 0$ on $[0, \infty)$.

15. $\displaystyle\int_a^b \frac{1}{5}\,dx = \frac{x}{5}\Big]_a^b = \frac{b-a}{5}$

(a) $P(0 < x < 3) = \dfrac{3-0}{5} = \dfrac{3}{5}$

(b) $P(1 < x < 3) = \dfrac{3-1}{5} = \dfrac{2}{5}$

(c) $P(3 < x < 5) = \dfrac{5-3}{5} = \dfrac{2}{5}$

(d) $P(x \geq 1) = P(1 < x < 5) = \dfrac{5-1}{5} = \dfrac{4}{5}$

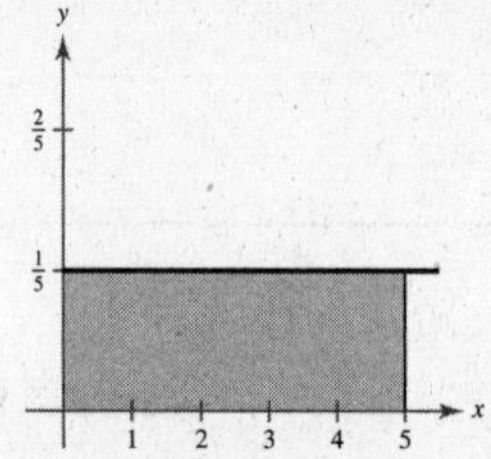

17. $\displaystyle\int_a^b \frac{x}{50}\,dx = \left[\frac{x^2}{100}\right]_a^b = \frac{b^2-a^2}{100}$

(a) $P(0 < x < 6) = \dfrac{36-0}{100} = \dfrac{9}{25}$

(b) $P(4 < x < 6) = \dfrac{36-16}{100} = \dfrac{1}{5}$

(c) $P(8 < x < 10) = \dfrac{100-64}{100} = \dfrac{9}{25}$

(d) $P(x \geq 2) = P(2 \leq x \leq 10) = \dfrac{100-4}{100} = \dfrac{24}{25}$

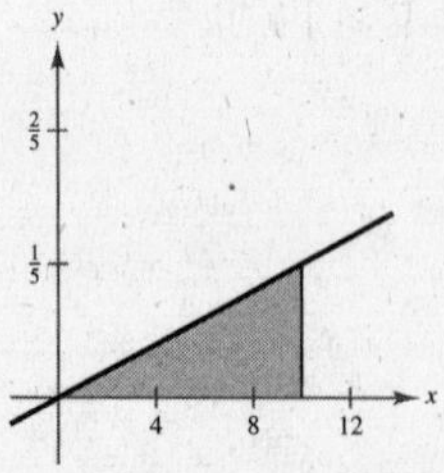

19. $\int_a^b \frac{3}{16}\sqrt{x}\,dx = \left(\frac{3}{16}\right)\frac{2}{3}x^{3/2}\Big]_a^b = \frac{1}{8}\left[b\sqrt{b} - a\sqrt{a}\right]$

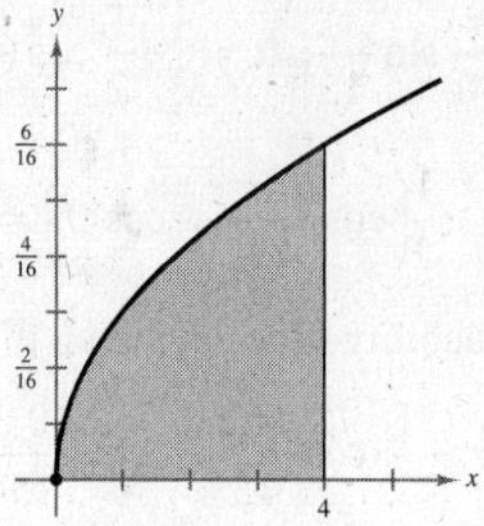

(a) $P(0 < x < 2) = \frac{\sqrt{2}}{4} \approx 0.354$

(b) $P(2 < x < 4) = 1 - \frac{\sqrt{2}}{4} \approx 0.646$

(c) $P(1 < x < 3) = \frac{1}{8}\left(3\sqrt{3} - 1\right) \approx 0.525$

(d) $P(x \le 3) = \frac{3\sqrt{3}}{8} \approx 0.650$

21. $\int_a^b \frac{1}{3}e^{-t/3}\,dt = e^{-t/3}\Big]_a^b = e^{-a/3} - e^{-b/3}$

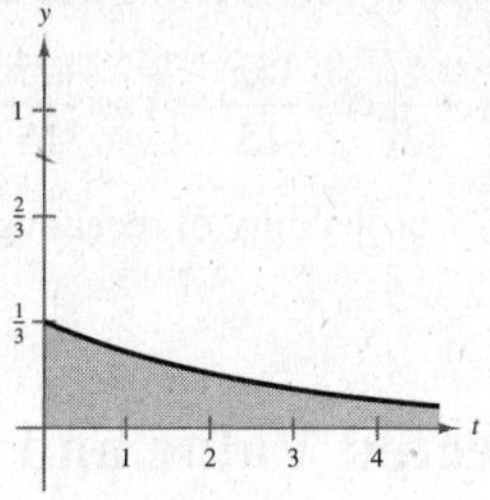

(a) $P(t < 2) = e^{-0/3} - e^{-2/3} \approx 0.4866$

(b) $P(t \ge 2) = e^{-2/3} - 0 \approx 0.5134$

(c) $P(1 < t < 4) = e^{-1/3} - e^{-4/3} \approx 0.4529$

(d) $P(t = 3) = 0$

23. $\int_1^4 kx\,dx = \left[\frac{kx^2}{2}\right]_1^4 = \frac{15}{2}k = 1 \Rightarrow k = \frac{2}{15}$

25. $\int_a^b \frac{k}{b-a}\,dx = \frac{kx}{b-a}\Big]_a^b = k = 1 \Rightarrow k = 1$

27. $\int_0^1 k\sqrt{x}(1-x)\,dx = k\int_0^1 \left(x^{1/2} - x^{3/2}\right)dx$

$= k\left[\frac{2}{3}x^{3/2} - \frac{2}{5}x^{5/2}\right]_0^1$

$= \frac{4k}{15} = 1 \Rightarrow k = \frac{15}{4}$

29. $P(a < x < b) = \int_a^b (0.41 - 0.08x)\,dx = \left[0.41x - 0.04x^2\right]_a^b = \left(0.41b - 0.04b^2\right) - \left(0.41a - 0.04a^2\right)$

(a) $P(0 \le x \le 3) = \left[0.41(3) - 0.04(3)^2\right] - \left[0.41(0) - 0.04(0)^2\right] = 0.87$

(b) $P(2 \le x \le 4) = \left[0.41(4) - 0.04(4)^2\right] - \left[0.41(2) - 0.04(2)^2\right] = 0.34$

31. $\int_a^b \frac{1}{3}e^{-t/3}\,dt = -e^{-t/3}\Big]_a^b = e^{-a/3} - e^{-b/3}$

(a) $P(0 < t < 2) = e^{-0/3} - e^{-2/3} = 1 - e^{-2/3} \approx 0.487$

(b) $P(2 < t < 4) = e^{-2/3} - e^{-4/3} \approx 0.250$

(c) $P(t > 2) = 1 - P(0 < t < 2) = e^{-2/3} \approx 0.513$

33. $\int_a^b \frac{1}{5}e^{-t/5}\,dt = -e^{-t/5}\Big]_a^b = e^{-a/5} - e^{-b/5}$

(a) $P(0 < t < 6) = 1 - e^{-6/5} \approx 0.699$

(b) $P(2 < t < 6) = e^{-2/5} - e^{-6/5} \approx 0.369$

(c) $P(t > 8) = e^{-8/5} - \lim_{b\to\infty} e^{-b/5}$

$= e^{-8/5} \approx 0.202$

35. $P(a \le x \le b) = \int_a^b \frac{\pi}{30} \sin \frac{\pi x}{15}\, dx = \left[-\frac{1}{2}\cos\frac{\pi x}{15}\right]_a^b = -\frac{1}{2}\left(\cos\frac{\pi b}{15} - \cos\frac{\pi a}{15}\right)$

(a) $P(0 \le x \le 10) = -\frac{1}{2}\left(\cos\frac{10\pi}{15} - \cos 0\right) = -\frac{1}{2}\left(-\frac{3}{2}\right) = 0.75$

There is a 75% probability of receiving up to 10 inches of rain.

(b) $P(10 \le x \le 15) = -\frac{1}{2}\left(\cos\frac{15\pi}{15} - \cos\frac{10\pi}{15}\right) = -\frac{1}{2}\left(-\frac{1}{2}\right) = 0.25$

There is a 25% probability of receiving between 10 and 15 inches of rain.

(c) $P(0 \le x < 5) = -\frac{1}{2}\left(\cos\frac{5\pi}{15} - \cos 0\right) = -\frac{1}{2}\left(-\frac{1}{2}\right) = 0.25$

There is a 25% probability of receiving less than 5 inches of rain.

(d) $P(12 \le x \le 15) = -\frac{1}{2}\left(\cos\frac{15\pi}{15} - \cos\frac{12\pi}{15}\right) \approx 0.095$

There is about a 9.5% probability of receiving between 12 and 15 inches of rain.

37. Answers will vary.

Section 11.3 Expected Value and Variance

Skills Review

1. $\int_0^m \frac{1}{10}\, dx = 0.5$

$\frac{1}{10}x\Big]_0^m = 0.5$

$\frac{1}{10}m = 0.5$

$m = 5$

2. $\int_0^m \frac{1}{16}\, dx = 0.5$

$\frac{1}{16}x\Big]_0^m = 0.5$

$\frac{1}{16}m = 0.5$

$m = 8$

3. $\int_0^m \frac{1}{3}e^{-t/3}\, dt = 0.5$

$-e^{-t/3}\Big]_0^m = 0.5$

$-\left(e^{-m/3} - e^0\right) = 0.5$

$-e^{-m/3} + 1 = 0.5$

$-e^{-m/3} = -0.5$

$e^{-m/3} = 0.5$

$-\frac{m}{3} = \ln 0.5$

$m = -3 \ln 0.5 = 3 \ln 2$

4. $\int_0^m \frac{1}{9}e^{-t/9}\, dt = 0.5$

$-e^{-t/9}\Big]_0^m = 0.5$

$-\left(e^{-m/9} - e^0\right) = 0.5$

$-e^{-m/9} + 1 = 0.5$

$-e^{-m/9} = -0.5$

$e^{-m/9} = 0.5$

$-\frac{m}{9} = \ln 0.5$

$m = -9 \ln 0.5 = 9 \ln 2$

5. $\int_0^2 \frac{x^2}{2}\, dx = \frac{1}{6}x^3\Big]_0^2 = \frac{1}{6}\left(2^3 - 0^3\right) = \frac{8}{6} = \frac{4}{3}$

6. $\int_1^2 x(4 - 2x)\, dx = \int_1^2 \left(4x - 2x^2\right) dx$

$= \left[2x^2 - \frac{2}{3}x^3\right]_1^2 = \frac{8}{3} - \frac{4}{3} = \frac{4}{3}$

7. $\int_2^5 x^2\left(\frac{1}{3}\right) dx - \left(\frac{7}{2}\right)^2 = \frac{1}{9}x^3\Big]_2^5 - \frac{49}{4}$

$= \frac{125}{9} - \frac{8}{9} - \frac{49}{4} = \frac{3}{4}$

Skills Review —*continued*—

8. $\int_2^4 x^2\left(\frac{4-x}{2}\right) dx - \left(\frac{8}{3}\right)^2 = \frac{1}{2}\int_2^4 \left(4x^2 - x^3\right) dx - \frac{64}{9}$

$= \frac{1}{2}\left[\frac{4}{3}x^3 - \frac{1}{4}x^4\right]_2^4 - \frac{64}{9}$

$= \frac{1}{2}\left(\frac{64}{3} - \frac{20}{3}\right) - \frac{64}{9}$

$= \frac{2}{9}$

9. $P(a \le x \le b) = \int_a^b \frac{1}{8}\, dx = \frac{1}{8}x\Big]_a^b = \frac{b-a}{8}$

(a) $P(x \le 2) = P(0 \le x \le 2) = \frac{2-0}{8} = \frac{1}{4}$

(b) $P(3 < x < 7) = \frac{7-3}{8} = \frac{1}{2}$

10. $P(a \le x \le b) = \int_a^b \left(6x - 6x^2\right) dx = \left[3x^2 - 2x^3\right]_a^b = \left(3b^2 - 2b^3\right) - \left(3a^2 - 2a^3\right)$

(a) $P\left(x \le \frac{1}{2}\right) = P\left(0 \le x \le \frac{1}{2}\right) = \left[3\left(\frac{1}{2}\right)^2 - 2\left(\frac{1}{2}\right)^3\right] - \left[3(0)^2 - 2\left(0^3\right)\right] = \frac{1}{2}$

(b) $P\left(\frac{1}{4} \le x \le \frac{3}{4}\right) = \left[3\left(\frac{3}{4}\right)^2 - 2\left(\frac{3}{4}\right)^3\right] - \left[3\left(\frac{1}{4}\right)^2 - 2\left(\frac{1}{4}\right)^3\right] = \frac{27}{32} - \frac{5}{32} = \frac{11}{16}$

1. (a) $\mu = \int_a^b x f(x)\, dx = \int_0^3 x\left(\frac{1}{3}\right) dx = \frac{x^2}{6}\Big]_0^3 = \frac{3}{2}$

(b) $\sigma^2 = \int_a^b x^2 f(x)\, dx - \mu^2 = \int_0^3 x^2\left(\frac{1}{3}\right) dx - \left(\frac{3}{2}\right)^2 = \frac{x^3}{9}\Big]_0^3 - \frac{9}{4} = 3 - \frac{9}{4} = \frac{3}{4}$

(c) $\sigma = \sqrt{\frac{3}{4}} = \frac{\sqrt{3}}{2}$

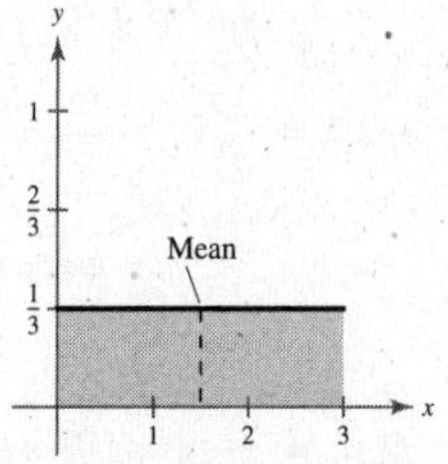

3. (a) $\mu = \int_a^b t f(t)\, dt = \int_0^6 t\left(\frac{t}{18}\right) dt = \frac{t^3}{54}\Big]_0^6 = 4$

(b) $\sigma^2 = \int_a^b t^2 f(t)\, dt - \mu^2 = \int_0^6 t^2\left(\frac{t}{18}\right) dt - 4^2 = \frac{t^4}{72}\Big]_0^6 - 4^2 = 18 - 16 = 2$

(c) $\sigma = \sqrt{2}$

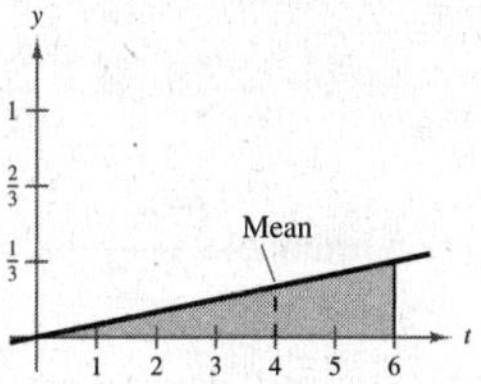

5. (a) $\mu = \int_a^b x f(x)\, dx = \int_0^1 x\left(\frac{5}{2}x^{3/2}\right) dx = \frac{5}{2}\int_0^1 x^{5/2}\, dx = \left(\frac{5}{2}\right)\frac{2}{7}x^{7/2}\Big]_0^1 = \frac{5}{7}$

(b) $\sigma^2 = \int_a^b x^2 f(x)\, dx - \mu^2 = \int_0^1 x^2\left(\frac{5}{2}x^{3/2}\right) dx - \left(\frac{5}{7}\right)^2 = \frac{5}{2}\int_0^1 x^{7/2}\, dx - \frac{25}{49} = \left(\frac{5}{2}\right)\frac{2}{9}x^{9/2}\Big]_0^1 - \frac{25}{49} = \frac{5}{9} - \frac{25}{49} = \frac{20}{441}$

(c) $\sigma = \frac{2\sqrt{5}}{21}$

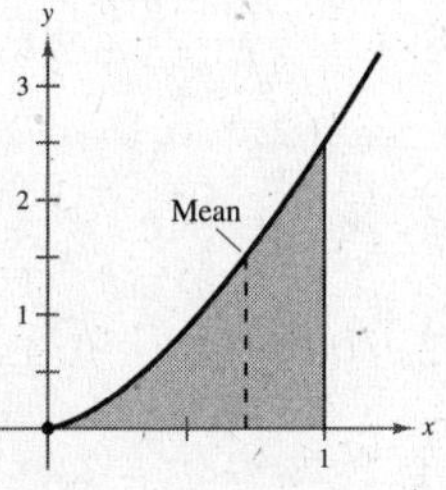

7. $\mu = \int_a^b xf(x)\,dx = \int_0^1 (x)6x(1-x)\,dx = \frac{1}{2}$

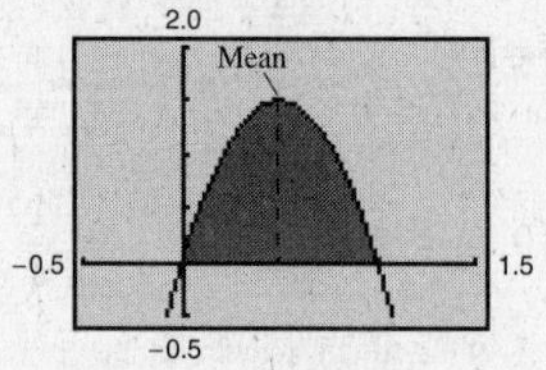

9. $\mu = \int_a^b xf(x)\,dx = \int_0^3 x\frac{4}{3(x+1)^2}\,dx = \frac{4}{3}\ln 4 - 1$

≈ 0.848

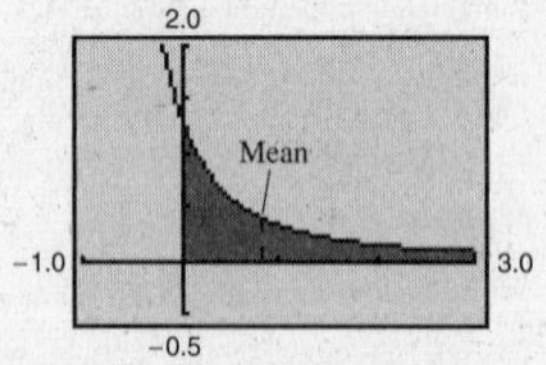

11. $\text{Median} = \int_0^m \frac{1}{9}e^{-t/9}\,dt = -e^{t/9}\Big]_0^m = 1 - e^{m/9} = \frac{1}{2}$

$e^{-m/9} = \frac{1}{2} \Rightarrow -\frac{m}{9} = \ln\frac{1}{2}$

$m = -9\ln\frac{1}{2} = 9\ln 2 \approx 6.238$

13. $f(x) = \frac{1}{10}$, $[0, 10]$ is a uniform probability density function.

Mean: $\frac{a+b}{2} = \frac{0+10}{2} = 5$

Variance: $\frac{(b-a)^2}{12} = \frac{(10-0)^2}{12} = \frac{100}{12} = \frac{25}{3}$

Standard deviation: $\frac{b-a}{\sqrt{12}} = \frac{10-0}{\sqrt{12}} \approx 2.887$

15. $f(x) = \frac{1}{8}e^{-x/8}$, $[0, \infty)$ is an exponential probability density function with $a = \frac{1}{8}$.

Mean: $\frac{1}{a} = 8$

Variance: $\frac{1}{a^2} = 64$

Standard deviation: $\frac{1}{a} = 8$

17. $f(x) = \frac{1}{11\sqrt{2\pi}}e^{-(x-100)^2/242}$, $(-\infty, \infty)$ is a normal probability density function with $\mu = 100$ and $\sigma = 11$.

Mean: $\mu = 100$

Variance: $\sigma^2 = 121$

Standard deviation: $\sigma = 11$

19. Mean $= 0$

Standard deviation $= 1$

$P(0 \le x \le 0.85) \approx 0.3023$

21. Mean $= 6$

Standard deviation $= 6$

$P(x \ge 2.23) \approx 0.6896$

23. Mean $= 8$

Standard deviation $= 2$

$P(3 \le x \le 13) \approx 0.9876$

25. $\mu = 50$, $\sigma = 10$

(a) $P(x > 55) = \int_{55}^{\infty} \frac{1}{10\sqrt{2\pi}}e^{-(x-50)^2/2(10^2)}\,dx$

≈ 0.3085

(b) $P(x > 60) = \int_{60}^{\infty} \frac{1}{10\sqrt{2\pi}}e^{-(x-50)^2/2(10^2)}\,dx$

≈ 0.1587

(c) $P(x < 60) = \int_{-\infty}^{60} \frac{1}{10\sqrt{2\pi}}e^{-(x-50)^2/2(10^2)}\,dx$

≈ 0.8413

(d) $P(30 < x < 55) = \int_{30}^{55} \frac{1}{10\sqrt{2\pi}}e^{-(x-50)^2/2(10^2)}\,dx$

≈ 0.6687

27. $f(t) = \frac{1}{10}$, $[0, 10]$ where $t = 0$ corresponds to 10:00 A.M.

(a) Mean $= \frac{10}{2} = 5$

The mean is 10:05 A.M.

Standard deviation $= \frac{10}{\sqrt{12}} \approx 2.887$ minutes

(b) $1 - \int_3^{10} \frac{1}{10}\,dx = 1 - \frac{7}{10} = \frac{3}{10} = 0.30$

29. (a) Because $\mu = 2$, $f(t) = \frac{1}{2}e^{-t/2}$, $0 \le t < \infty$.

(b) $P(0 < t < 1) = \int_0^1 \frac{1}{2}e^{-t/2}\,dt = -e^{-t/2}\Big]_0^1$

$= 1 - e^{-1/2} \approx 0.3935$

31. (a) Because $\mu = 5$, $f(t) = \frac{1}{5}e^{-t/5}$, $0 \le t < \infty$.

(b) $P(\mu - \sigma < t < \mu + \sigma) = P(0 < t < 10) = \int_0^{10} \frac{1}{5}e^{-t/5}\,dt = -e^{-t/5}\Big]_0^{10} = 1 - e^{-2} \approx 0.865 = 86.5\%$

33. (a) $\dfrac{174 - 150}{16} = \dfrac{3}{2} = 1.5$

Your score exceeded the national mean by 1.5 standard deviations.

(b) $P(x < 174) = 0.9332$

Thus, $0.9332 = 93.32\%$ of those who took the exam had scores lower than yours.

35. $\mu = \int_0^1 \dfrac{1155}{32}x^4(1 - x)^{3/2}\,dx = \dfrac{1155}{32}\int_0^1 x^4(1 - x)^{3/2}\,dx$

Let $u = \sqrt{1 - x}$. Then $x = 1 - u^2$ and $dx = -2u\,du$.

$$\frac{1155}{32}\int_0^1 x^4(1 - x)^{3/2}\,dx = \frac{1155}{32}\int_1^0 \left(1 - u^2\right)^4 u^3(-2u)\,du = -\frac{1155}{16}\int_1^0 \left(u^4 - 4u^6 + 6u^8 - 4u^{10} + u^{12}\right)du$$

$$= -\frac{1155}{16}\left[\frac{u^5}{5} - \frac{4u^7}{7} + \frac{2u^9}{3} - \frac{4u^{11}}{11} + \frac{u^{13}}{13}\right]_1^0 \approx 0.615 = 61.5\%$$

37. Mean $= \dfrac{1}{2}(0 + 11) = 5.5$

Median $= m$: $\int_0^m \dfrac{1}{11}\,dx = 0.5$

$\dfrac{m}{11} = 0.5$

$m = 5.5$

39. Mean $= \int_0^{1/2} x(4)(1 - 2x)\,dx = \frac{1}{6}$

Median $= m$: $\int_0^m 4(1 - 2x)\,dx = \frac{1}{2}$

$4x - 4x^2\Big]_0^m = \frac{1}{2}$

$4m - 4m^2 = \frac{1}{2} \Rightarrow m \approx 0.1465$

$\left(m \approx 0.8536 \text{ is not in the interval } \left[0, \tfrac{1}{2}\right]\right)$

41. Mean $= 5$

Median $= 5 \ln 2 \approx 3.4657$

43. $\int_0^m f(x)\,dx = \int_0^m 0.28e^{-0.28x}\,dx = 0.5$

$-e^{-0.28m} + 1 = 0.5$

$e^{-0.28m} = 0.5$

$m = \dfrac{1}{-0.28}\ln 0.5$

≈ 2.4755

45. $\mu = 50{,}000$, $\sigma = 3000$

$P(0 < x < m) = \int_0^m \dfrac{1}{3000\sqrt{2\pi}}e^{-(x-50{,}000)^2/2(3000)^2}\,dx = 0.10$

By trial and error and using a graphing utility, $m \approx 46{,}156$ miles.

47. $u = 4.5$, $\sigma = 0.5$

(a) $P(4 < x < 5) = \int_4^5 \dfrac{1}{0.5\sqrt{2\pi}}e^{-(x-4.5)^2/2(0.5)^2}\,dx = \dfrac{2}{\sqrt{2\pi}}\int_4^5 e^{-2(x-4.5)^2}\,dx \approx 0.6827$

(b) $P(x < 3) = \int_0^3 \dfrac{1}{0.5\sqrt{2\pi}}e^{-(x-4.5)^2/2(0.5)^2}\,dx = \dfrac{2}{\sqrt{2\pi}}\int_0^3 e^{-2(x-4.5)^2}\,dx \approx 0.0013 = 0.13\%$

No, only about 0.13% of the batteries will last less than 3 years.

49. $f(x) = \dfrac{1}{16\sqrt{2\pi}} e^{-(x-266)^2/2(16)^2} = \dfrac{1}{40.106} e^{-(x-266)^2/512}$

(a)

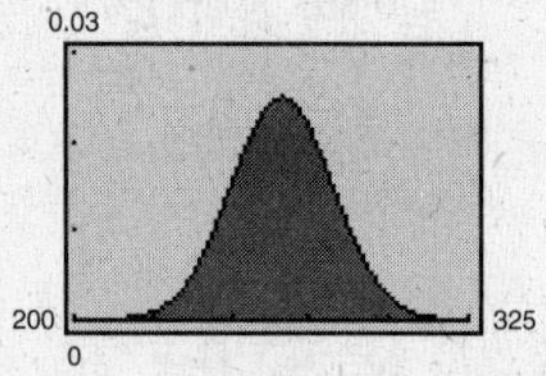

(b) $P(240 \le x \le 280) = \int_{240}^{280} f(x)\, dx$

≈ 0.757 or 76%

(c) $P(x > 280) = \int_{280}^{\infty} f(x)\, dx \approx 0.191$ or 19.1%

51. $f(x) = \dfrac{1}{4.9\sqrt{2\pi}} e^{-(x-20.6)^2/2(4.9)^2} = \dfrac{1}{4.9\sqrt{2\pi}} e^{-(x-20.6)^2/48.02}$

(a)

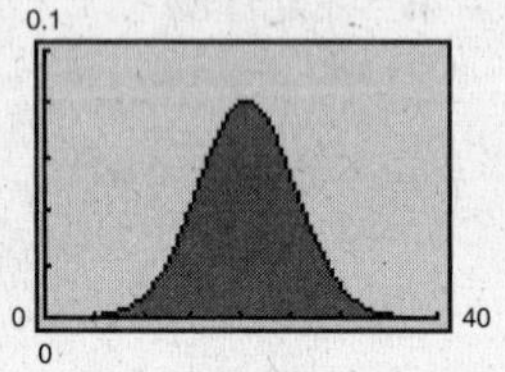

(b) $P(25 \le x \le 30) = \int_{25}^{30} f(x)\, dx \approx 0.157$ or 15.7%

(c) $P(x < 18) = \int_{0}^{18} f(x)\, dx \approx 0.298$ or 29.8%

Review Exercises for Chapter 11

1. The sample space consists of the twelve months of the year.

$S = \{$January, February, March, April, May, June, July, August, September, October, November, December$\}$

3. If the questions are numbered 1, 2, 3, and 4,

$S = \{123, 124, 134, 234\}$

5. $S = \{0, 1, 2, 3\}$

7.

x	0	1	2	3
$n(x)$	1	3	3	1

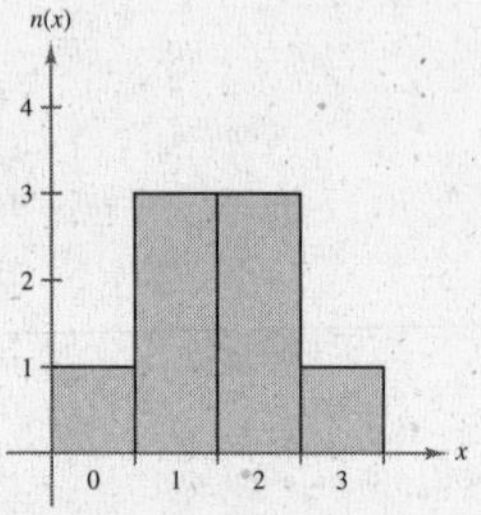

9. (a) $P(2 \le x \le 4) = P(2) + P(3) + P(4)$

$= \frac{7}{18} + \frac{5}{18} + \frac{3}{18} = \frac{15}{18} = \frac{5}{6}$

(b) $P(x \ge 3) = P(3) + P(4) + P(5)$

$= \frac{5}{18} + \frac{3}{18} + \frac{2}{18} = \frac{10}{18} = \frac{5}{9}$

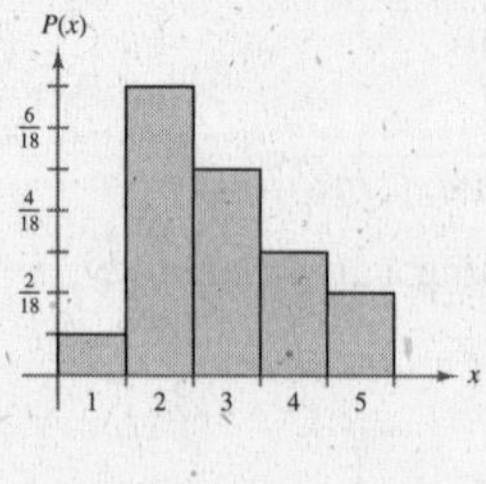

11.

x	2	3	4	5	6	7	8	9	10	11	12
$n(x)$	1	2	3	4	5	6	5	4	3	2	1

$n(S) = 36$

(a) $P(x = 8) = \frac{5}{36}$

(b) $P(x > 4) = 1 - P(x \le 4) = 1 - \frac{6}{36} = \frac{5}{6}$

(c) $P(\text{doubles}) = \frac{6}{36} = \frac{1}{6}$

(d) $P(\text{double sixes}) = \frac{1}{36}$

13. $$\text{Mean} = \frac{9 + 11 + 3(15) + 4(16) + 7(17) + 3(18) + 9(20) + 11(21) + 6(22) + 3(23) + 4(25)}{52} = \frac{1014}{52} = 19.5$$

15.

x	\$2995	\$995	\$15	−\$5
$P(x)$	$\frac{1}{2000}$	$\frac{1}{2000}$	$\frac{50}{2000}$	$\frac{1948}{2000}$

$E(x) = 2995\left(\frac{1}{2000}\right) + 995\left(\frac{1}{2000}\right) + 15\left(\frac{50}{2000}\right) + (-5)\left(\frac{1948}{2000}\right) = -\$2.50$

Expected net gain: −\$2.50

17. $$E(x) = \frac{30(5) + 25(12) + 20(30) + 18(49) + 12(65)}{105} \approx \$25.83$$

$$V(x) = (5 - 25.83)^2\left(\frac{30}{105}\right) + (12 - 25.83)^2\left(\frac{25}{105}\right) + (30 - 25.83)^2\left(\frac{20}{105}\right) + (49 - 25.83)^2\left(\frac{18}{105}\right) + (65 - 25.83)^2\left(\frac{12}{105}\right)$$

≈ 440.1992

$\sigma = \sqrt{V(x)} \approx 20.9809$

19. $E(x) = 0(0.12) + 1(0.31) + 2(0.43) + 3(0.12) + 4(0.02) = 1.61$

$V(x) = (0 - 1.61)^2(0.12) + (1 - 1.61)^2(0.31) + (2 - 1.61)^2(0.43) + (3 - 1.61)^2(0.12) + (4 - 1.61)^2(0.02) = 0.8379$

$\sigma = \sqrt{V(x)} \approx 0.9154$

21.

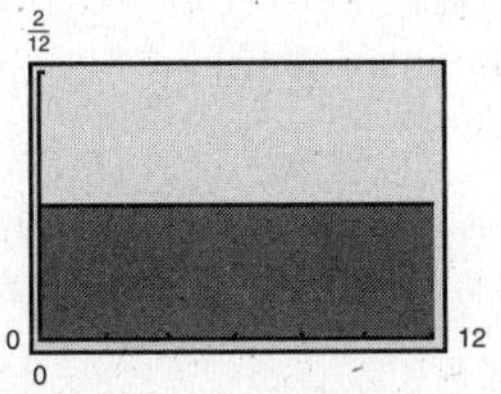

f is a probability density function because

$\int_0^{12} \frac{1}{12}\,dx = \left[\frac{1}{12}x\right]_0^{12} = 1$ and $f(x) = \frac{1}{12} \ge 0$ on $[0, 12]$.

23.

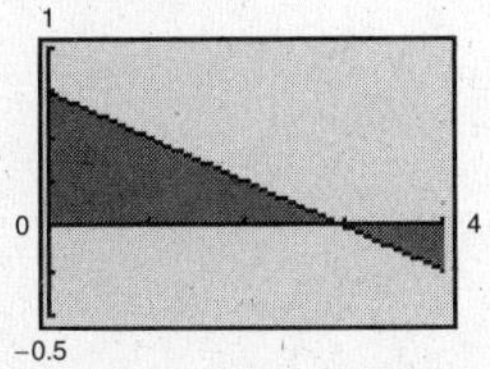

f is not a probability density function because $f(x) = \frac{1}{4}(3 - x)$ is negative on $(3, 4]$.

25.

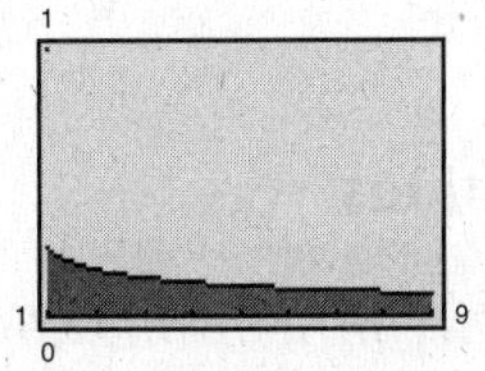

f is a probability density function because

$\int_1^9 \frac{1}{4\sqrt{x}}\,dx = \frac{1}{4}\left[2\sqrt{x}\right]_1^9 = 1$ and $f(x) = \frac{1}{4\sqrt{x}} \ge 0$ on $[1, 9]$.

27. $$P(0 < x < 2) = \int_0^2 \frac{1}{50}(10 - x)\,dx = \frac{1}{50}\left[10x - \frac{x^2}{2}\right]_0^2 = \frac{9}{25}$$

29. $P\left(0 < x < \frac{1}{2}\right) = \int_0^{1/2} \frac{2}{(x+1)^2}\,dx$

$= -\frac{2}{x+1}\Big]_0^{1/2} = -\frac{4}{3} + 2 = \frac{2}{3}$

31. (a) $P(t \le 10) = \int_0^{10} \frac{1}{20}\,dt = \frac{1}{20}t\Big]_0^{10} = \frac{1}{2}$

(b) $P(t \ge 15) = \int_{15}^{20} \frac{1}{20}\,dt = \frac{1}{20}t\Big]_{15}^{20} = 1 - \frac{3}{4} = \frac{1}{4}$

33. $\mu = \frac{1}{2}(a+b) = \frac{1}{2}(0+7) = \frac{7}{2}$

35. $\mu = \frac{1}{a} = \frac{1}{1/6} = 6$

37. $\mu = \int_0^3 x\frac{2}{9}x(3-x)\,dx = \frac{2}{9}\left[x^3 - \frac{x^4}{4}\right]_0^3 = \frac{3}{2}$

$V(x) = \int_0^3 \left(x - \frac{3}{2}\right)^2 \frac{2}{9}x(3-x)\,dx = \frac{9}{20}$

$\sigma = \sqrt{V(x)} = \frac{3}{2\sqrt{5}}$

39. $\mu = \int_0^\infty x\left(\frac{1}{2}e^{-x/2}\right)dx = \frac{1}{a} = \frac{1}{1/2} = 2$

$V(x) = \frac{1}{a^2} = \frac{1}{(1/2)^2} = 4$

$\sigma = \sqrt{V(x)} = 2$

41. $\int_0^m 6x(1-x)\,dx = \frac{1}{2}$

$3x^2 - 2x^3\Big]_0^m = \frac{1}{2}$

$3m^2 - 2m^3 = \frac{1}{2}$

$m = \frac{1}{2}$

43. $\int_0^m 0.25e^{-x/4}\,dx = \frac{1}{2}$

$1 - e^{-m/4} = \frac{1}{2}$

$m \approx 2.7726$

45. $f(t) = \frac{1}{15}e^{-t/15}\,dt, \quad 0 \le t < \infty$

(a) $P(t < 10) = \int_0^{10} f(t)\,dt \approx 0.4866$

(b) $P(10 < t < 20) = \int_{10}^{20} f(t)\,dt \approx 0.2498$

47. $f(x) = \frac{1}{\sigma\sqrt{2\pi}}e^{-(x-\mu)^2/2\sigma^2}$

$P(x \ge 50) = \int_{50}^{\infty} \frac{1}{3\sqrt{2\pi}}e^{-(x-42)^2/[2(3)^2]}\,dx \approx 0.00383$

49. $f(x) = \frac{1}{25\sqrt{2\pi}}e^{-(x-130)^2/2(25)^2} = \frac{1}{25\sqrt{2\pi}}e^{-(x-130)^2/1250}$

(a)

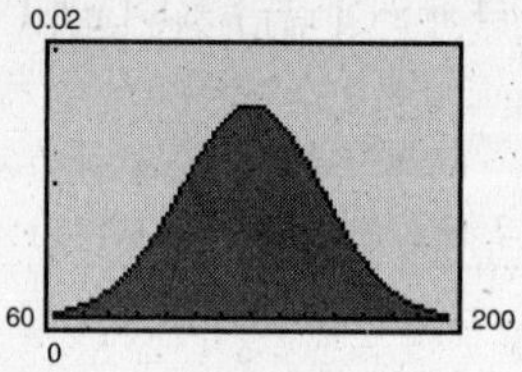

(b) $P(70 < x < 105) = \int_{70}^{108} f(x)\,dx \approx 0.150$

(c) $P(x > 120) = \int_{120}^{\infty} f(x)\,dx \approx 0.655$

51. $\mu = 3.75,\ \sigma = 0.5$

$P(3.5 < x < 4) = \int_{3.5}^{4} \frac{1}{0.5\sqrt{2\pi}}e^{-(x-3.75)^2/2(0.5)^2}\,dx$

$= \frac{2}{\sqrt{2\pi}}\int_{3.5}^{4} e^{-2(x-3.75)^2}\,dx \approx 0.3829$

(Using Simpson's Rule with $n = 12$)

Chapter Test Solutions

1. (a) $S = \{HHHH, HHHT, HHTH, HTHH, THHH, HHTT, HTHT, HTTH,$
$THTH, TTHH, HTTT, THTT, TTHT, TTTH, THHT, TTTT\}$

Let x represent the number of heads in each outcome.

x	0	1	2	3	4
$n(x)$	1	4	6	4	1

(b) P (at least 2 heads) $= P(x \ge 2) = \frac{6+4+1}{1+4+6+4+1} = \frac{11}{16}$

2. P (red and not a face card) $= \frac{20}{52} = \frac{5}{13}$

3. (a) $P(x < 3) = P(1) + P(2) = \frac{3}{16} + \frac{7}{16} = \frac{5}{8}$

(b) $P(x \ge 3) = P(3) + P(4) = \frac{1}{16} + \frac{5}{16} = \frac{3}{8}$

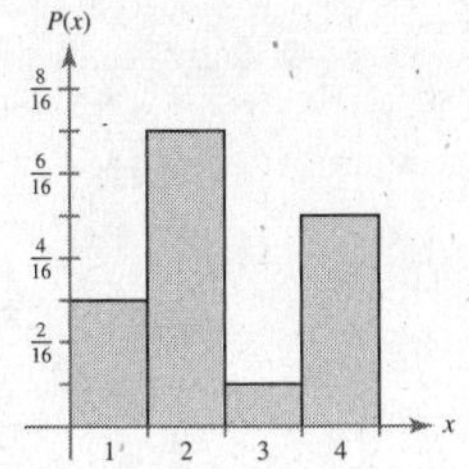

4. (a) $P(7 \le x \le 10) = P(7) + P(8) + P(9) + P(10)$

$= 0.21 + 0.13 + 0.19 + 0.42$

$= 0.95$

(b) $P(x > 8) = P(9) + P(10) + P(11)$

$= 0.19 + 0.42 + 0.05 = 0.66$

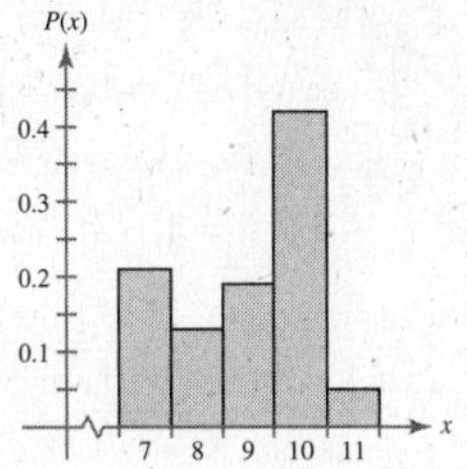

5. $E(x) = 0\left(\frac{2}{10}\right) + 1\left(\frac{1}{10}\right) + 2\left(\frac{4}{10}\right) + 3\left(\frac{3}{10}\right) = \frac{9}{5} = 1.8$

$V(x) = (0 - 1.8)^2\left(\frac{2}{10}\right) + (1 - 1.8)^2\left(\frac{1}{10}\right) + (2 - 1.8)^2\left(\frac{4}{10}\right) + (3 - 1.8)^2\left(\frac{3}{10}\right) = 1.16$

$\sigma = \sqrt{V(x)} \approx 1.077$

6. $E(x) = -2(0.141) + (-1)(0.305) + 0(0.257) + 1(0.063) + 2(0.234) = -0.056$

$V(x) = (-2 + 0.056)^2(0.141) + (-1 + 0.056)^2(0.305) + (0 + 0.056)^2(0.257) + (1 + 0.056)^2 0.063 + (2 + 0.056)^2(0.234)$

$= 1.864864$

$\sigma = \sqrt{V(x)} \approx 1.366$

7.

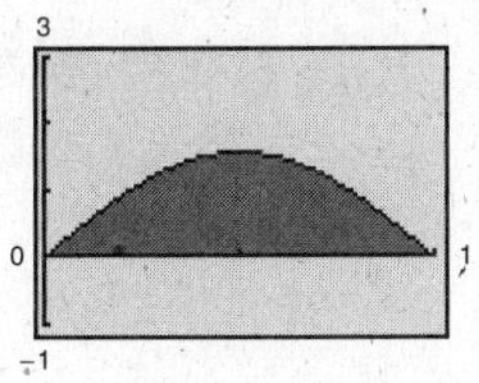

f is a probability density function because

$$\int_0^1 \frac{\pi}{2} \sin \pi x \, dx = \left[-\frac{1}{2} \cos \pi x\right]_0^1 = 1 \text{ and}$$

$f(x) = \frac{\pi}{2} \sin \pi x \ge 0$ on $[0, 1]$.

8.

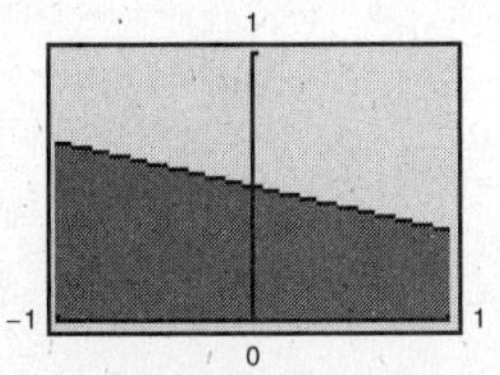

f is a probability density function because

$$\int_{-1}^1 \frac{3 - x}{6} \, dx = \left[\frac{1}{2}x - \frac{x^2}{12}\right]_{-1}^1 = 1.$$

9.

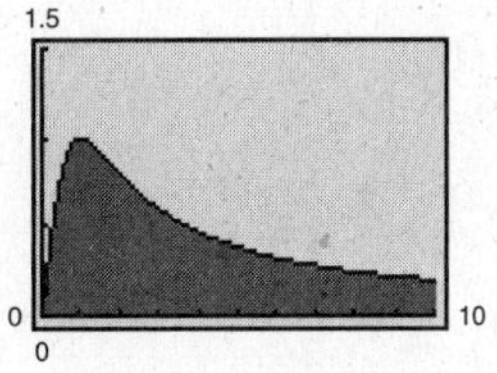

f is not a probability density function because

$$\int_0^\infty \frac{2x}{x^2 + 1} \, dx = \lim_{b \to \infty} \left[\ln(x^2 + 1)\right]_0^b = \infty \ne 1.$$

10. $\int_a^b \frac{x}{32} \, dx = \left[\frac{x^2}{64}\right]_a^b = \frac{b^2 - a^2}{64}$

(a) $P(1 \le x \le 4) = \frac{16 - 1}{64} = \frac{15}{64}$

(b) $P(3 \le x \le 6) = \frac{36 - 9}{64} = \frac{27}{64}$

11. $\int_a^b 4(x - x^3)\,dx = \int_a^b (4x - 4x^3)\,dx = \Big[2x^2 - x^4\Big]_a^b = (2b^2 - b^4) - (2a^2 - a^4)$

(a) $P(0 < x < 0.5) = \Big[2(0.5)^2 - 0.5^4\Big] - \Big[2(0)^2 - 0^4\Big] = \frac{7}{16} = 0.4375$

(b) $P(0.25 \le x < 1) = \Big[2(1)^2 - 1^4\Big] - \Big[2(0.25)^2 - 0.25^4\Big] = \frac{225}{256} \approx 0.879$

12. $\int_a^b 2xe^{-x^2}\,dx = -e^{-x^2}\Big]_a^b = -\left(e^{-b^2} - e^{-a^2}\right)$

(a) $P(x < 1) = -\left(e^{-1^2} - e^{0^2}\right) = 1 - e^{-1} \approx 0.632$

(b) $P(x \ge 1) = \int_1^\infty 2xe^{-x^2}\,dx = \lim_{b\to\infty}\Big[-e^{-x^2}\Big]_1^b = \lim_{b\to\infty}\left(-e^{-b^2} + e^{-1}\right) = e^{-1} \approx 0.368$

13. $f(x) = \frac{1}{14}, [0, 14]$

$\mu = \frac{a + b}{2} = \frac{0 + 14}{2} = 7$

$V(x) = \frac{(b - a)^2}{12} = \frac{(14 - 0)^2}{12} = \frac{49}{3}$

$\sigma = \frac{b - a}{\sqrt{12}} = \frac{14}{\sqrt{12}} \approx 4.041$

14. $f(x) = 3x - \frac{3}{2}x^2, [0, 1]$

$\mu = \int_0^1 x\left(3x - \frac{3}{2}x^2\right)dx = \int_0^1 \left(3x^2 - \frac{3}{2}x^3\right)dx$

$= \Big[x^3 - \frac{3}{8}x^4\Big]_0^1 = \frac{5}{8}$

$V(x) = \int_0^1 x^2\left(3x - \frac{3}{2}x^2\right)dx - \left(\frac{5}{8}\right)^2$

$= \int_0^1 \left(3x^3 - \frac{3}{2}x^4\right)dx - \frac{25}{64}$

$= \Big[\frac{3}{4}x^4 - \frac{3}{10}x^5\Big]_0^1 - \frac{25}{64}$

$= \frac{9}{20} - \frac{25}{64}$

$= \frac{19}{320} \approx 0.0594$

$\sigma = \sqrt{V(x)} \approx 0.244$

15. $f(x) = e^{-x}, [0, \infty)$; f is an exponential probability density function with $a = 1$.

$\mu = \frac{1}{a} = \frac{1}{1} = 1$

$V(x) = \frac{1}{a^2} = \frac{1}{1^2} = 1$

$\sigma = \sqrt{V(x)} = 1$

16. $f(x) = \frac{1}{10\sqrt{2\pi}}e^{-(x-110)^2/2(10)^2}$; $\mu = 110, \sigma = 10$

$P(\mu - \sigma < x < \mu + \sigma) = P(100 < x < 120)$

$= \int_{100}^{120} f(x)\,dx \approx 0.683$

Practice Test for Chapter 11

1. A coin is tossed four times. What is the probability that at least two head occur?

2. A card is chosen at random from a standard 52-card deck of playing cards. What is the probability that the card will be red and not a face card?

3. Find $E(x)$, $V(x)$, and σ for the given probability distribution.

x	-2	-1	0	3	4
$P(x)$	$\frac{2}{10}$	$\frac{1}{10}$	$\frac{4}{10}$	$\frac{2}{10}$	$\frac{1}{10}$

4. Find the constant k so that $f(x) = ke^{-x/4}$ is a probability density function over the interval $[0, \infty)$.

5. Find (a) $P(0 < x < 5)$ and (b) $P(x > 1)$ for the probability density function

   $f(x) = \frac{x}{32}, [0, 8]$.

6. Find (a) the mean, (b) the standard deviation, and (c) the median for the probability density function

   $f(x) = \frac{3}{256}x(8 - x), [0, 8]$.

7. Find (a) the mean, (b) the standard deviation, and (c) the median for the probability density function

   $f(x) = \frac{6}{x^2}, [2, 3]$.

8. Find the expected value, median, and standard deviation of the exponential density function

   $f(x) = 7e^{-7x}, [0, \infty)$.

Technology Required

9. Find $P(1.67 < x < 3.24)$ using the standard normal probability density function.

10. The monthly revenue x (in thousands of dollars) of a given shop is normally distributed with $\mu = 20$ and $\sigma = 4$. Approximate $P(19 < x < 24)$.

APPENDIX B
Additional Topics in Differentiation

APPENDIX B
Additional Topics in Differentiation

Appendix B.1 Implicit Differentiation

Skills Review

1. $x - \frac{y}{x} = 2$

$x^2 - y = 2x$

$-y = 2x - x^2$

$y = x^2 - 2x$

2. $\frac{4}{x-3} = \frac{1}{y}$

$4y = x - 3$

$y = \frac{x-3}{4}$

3. $xy - x + 6y = 6$

$xy + 6y = 6 + x$

$y(x + 6) = 6 + x$

$y = \frac{6+x}{x+6}$

$y = 1, x \neq -6$

4. $12 + 3y = 4x^2 + x^2y$

$3y - x^2y = 4x^2 - 12$

$y(3 - x^2) = 4x^2 - 12$

$y = \frac{4x^2 - 12}{3 - x^2} = -4, x \neq \pm\sqrt{3}$

5. $x^2 + y^2 = 5$

$y^2 = 5 - x^2$

$y = \pm\sqrt{5 - x^2}$

6. $x = \pm\sqrt{6 - y^2}$

$x^2 = 6 - y^2$

$x^2 - 6 = -y^2$

$6 - x^2 = y^2$

$\pm\sqrt{6 - x^2} = y$

7. $\frac{3x^2 - 4}{3y^2}, (2, 1)$

$\frac{3(2^2) - 4}{3(1^2)} = \frac{3(4) - 4}{3}$

$= \frac{8}{3}$

8. $\frac{x^2 - 2}{1 - y}, (0, -3)$

$\frac{0^2 - 2}{1 - (-3)} = \frac{-2}{4}$

$= -\frac{1}{2}$

9. $\frac{5x}{3y^2 - 12y + 5}, (-1, 2)$

$\frac{5(-1)}{3(2^2) - 12(2) + 5} = \frac{-5}{3(4) - 24 + 5}$

$= \frac{-5}{-7}$

$= \frac{5}{7}$

10. $\frac{1}{y^2 - 2xy + x^2}, (4, 3)$

$\frac{1}{3^2 - 2(4)(3) + 4^2} = \frac{1}{9 - 24 + 16}$

$= \frac{1}{1}$

$= 1$

1. $xy = 4$

$x\frac{dy}{dx} + y = 0$

$x\frac{dy}{dx} = -y$

$\frac{dy}{dx} = -\frac{y}{x}$

3. $y^2 = 1 - x^2$

$2y\frac{dy}{dx} = -2x$

$\frac{dy}{dx} = -\frac{x}{y}$

5.
$$x^2y^2 - 2x = 3$$
$$2x^2\frac{dy}{dx}y + 2xy^2 - 2 = 0$$
$$2x^2y\frac{dy}{dx} = 2 - 2xy^2$$
$$\frac{dy}{dx} = \frac{2 - 2xy^2}{2x^2y}$$
$$= \frac{1 - xy^2}{x^2y}$$

7.
$$4y^2 - xy = 2$$
$$8y\frac{dy}{dx} - x\frac{dy}{dx} - y = 0$$
$$(8y - x)\frac{dy}{dx} = y$$
$$\frac{dy}{dx} = \frac{y}{8y - x}$$

9.
$$\frac{2y - x}{y^2 - 3} = 5$$
$$2y - x = 5y^2 - 15$$
$$2\frac{dy}{dx} - 1 = 10y\frac{dy}{dx}$$
$$-1 = (10y - 2)\frac{dy}{dx}$$
$$\frac{dy}{dx} = -\frac{1}{2(5y - 1)}$$

11.
$$\frac{x + y}{2x - y} = 1$$
$$x + y = 2x - y$$
$$2y = x$$
$$y = \frac{1}{2}x$$
$$\frac{dy}{dx} = \frac{1}{2}$$

13.
$$x^2 + y^2 = 16$$
$$2x + 2y\frac{dy}{dx} = 0$$
$$2y\frac{dy}{dx} = -2x$$
$$\frac{dy}{dx} = -\frac{x}{y}$$

At $(0, 4)$, $\frac{dy}{dx} = -\frac{0}{4} = 0$.

15.
$$y + xy = 4$$
$$\frac{dy}{dx} + x\frac{dy}{dx} + y = 0$$
$$\frac{dy}{dx}(1 + x) = -y$$
$$\frac{dy}{dx} = -\frac{y}{x + 1}$$

At $(-5, -1)$, $\frac{dy}{dx} = -\frac{1}{4}$.

17.
$$x^3 - xy + y^2 = 4$$
$$3x^2 - x\frac{dy}{dx} - y + 2y\frac{dy}{dx} = 0$$
$$\frac{dy}{dx}(2y - x) = y - 3x^2$$
$$\frac{dy}{dx} = \frac{y - 3x^2}{2y - x}$$

At $(0, -2)$, $\frac{dy}{dx} = \frac{1}{2}$.

19.
$$x^3y^3 - y = x$$
$$3x^3y^2\frac{dy}{dx} + 3x^2y^3 - \frac{dy}{dx} = 1$$
$$\frac{dy}{dx}(3x^3y^2 - 1) = 1 - 3x^2y^3$$
$$\frac{dy}{dx} = \frac{1 - 3x^2y^3}{3x^3y^2 - 1}$$

At $(0, 0)$, $\frac{dy}{dx} = -1$.

21.
$$x^{1/2} + y^{1/2} = 9$$
$$\frac{1}{2}x^{-1/2} + \frac{1}{2}y^{-1/2}\frac{dy}{dx} = 0$$
$$x^{-1/2} + y^{-1/2}\frac{dy}{dx} = 0$$
$$\frac{dy}{dx} = \frac{-x^{-1/2}}{y^{-1/2}} = -\sqrt{\frac{y}{x}}$$

At $(16, 25)$, $\frac{dy}{dx} = -\frac{5}{4}$.

23.
$$x^{2/3} + y^{2/3} = 5$$
$$\frac{2}{3}x^{-1/3} + \frac{2}{3}y^{-1/3}\frac{dy}{dx} = 0$$
$$\frac{dy}{dx} = \frac{-x^{-1/3}}{y^{-1/3}} = -\frac{y^{1/3}}{x^{1/3}} = -\sqrt[3]{\frac{y}{x}}$$

At $(8, 1)$, $\frac{dy}{dx} = -\frac{1}{2}$.

25. $3x^2 - 2y + 5 = 0$

$$6x - 2\frac{dy}{dx} = 0$$

$$\frac{dy}{dx} = 3x$$

At (1, 4), $\frac{dy}{dx} = 3$.

27. $x^2 + y^2 = 4$

$$2x + 2y\frac{dy}{dx} = 0$$

$$\frac{dy}{dx} = -\frac{x}{y}$$

At (0, 2), $\frac{dy}{dx} = 0$.

29. $4x^2 + 9y^2 = 36$

$$8x + 18y\frac{dy}{dx} = 0$$

$$\frac{dy}{dx} = -\frac{4x}{9y}$$

At $\left(\sqrt{5}, \frac{4}{3}\right)$, $\frac{dy}{dx} = -\frac{4\sqrt{5}}{9(4/3)} = -\frac{\sqrt{5}}{3}$.

31. Implicitly: $2x + 2y\frac{dy}{dx} = 0$

$$\frac{dy}{dx} = -\frac{x}{y}$$

Explicitly: $y = \pm\sqrt{25 - x^2}$

$$\frac{dy}{dx} = \pm\left(\frac{1}{2}\right)\left(25 - x^2\right)^{-1/2}(-2x)$$

$$= \pm\frac{-x}{\sqrt{25 - x^2}} = -\frac{x}{\pm\sqrt{25 - x^2}} = -\frac{x}{y}$$

At $(-4, 3)$, $\frac{dy}{dx} = \frac{4}{3}$.

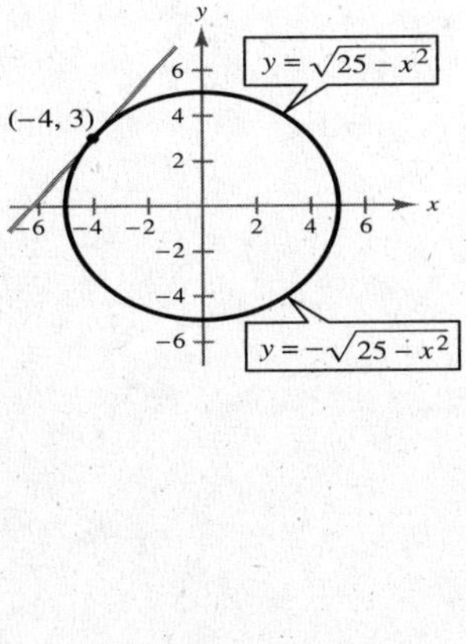

33. Implicitly: $1 - 2y\frac{dy}{dx} = 0$

$$\frac{dy}{dx} = \frac{1}{2y}$$

Explicitly: $y = \pm\sqrt{x - 1}$

$$= \pm(x - 1)^{1/2}$$

$$\frac{dy}{dx} = \pm\frac{1}{2}(x - 1)^{-1/2}(1)$$

$$= \pm\frac{1}{2\sqrt{x - 1}}$$

$$= \frac{1}{2\left(\pm\sqrt{x - 1}\right)}$$

$$= \frac{1}{2y}$$

At $(2, -1)$, $\frac{dy}{dx} = -\frac{1}{2}$.

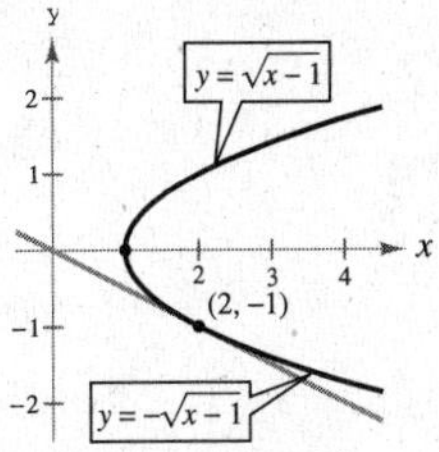

35. $x^2 + y^2 = 100$

$$2x + 2y\frac{dy}{dx} = 0$$

$$\frac{dy}{dx} = -\frac{x}{y}$$

At (8, 6):

$$m = -\frac{4}{3}$$

$$y - 6 = -\frac{4}{3}(x - 8)$$

$$y = -\frac{4}{3}x + \frac{50}{3}$$

At $(-6, 8)$:

$$m = \frac{3}{4}$$

$$y - 8 = \frac{3}{4}(x + 6)$$

$$y = \frac{3}{4}x + \frac{25}{2}$$

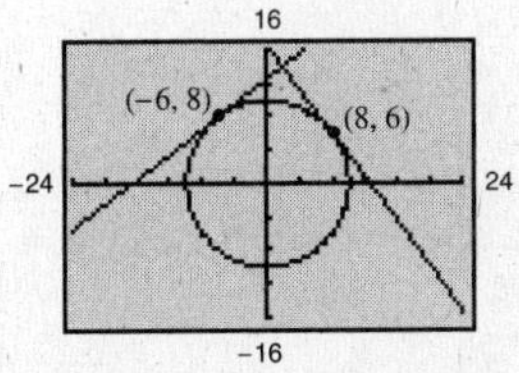

37. $y^2 = 5x^3$

$$2y\frac{dy}{dx} = 15x^2$$

$$\frac{dy}{dx} = \frac{15x^2}{2y}$$

At $(1, \sqrt{5})$:

$$m = \frac{15}{2\sqrt{5}} = \frac{3\sqrt{5}}{2}$$

$$y - \sqrt{5} = \frac{3\sqrt{5}}{2}(x - 1)$$

$$y = \frac{3\sqrt{5}}{2}x - \frac{\sqrt{5}}{2}$$

At $(1, -\sqrt{5})$:

$$m = \frac{-15}{2\sqrt{5}} = -\frac{3\sqrt{5}}{2}$$

$$y + \sqrt{5} = -\frac{3\sqrt{5}}{2}(x - 1)$$

$$y = -\frac{3\sqrt{5}}{2}x + \frac{\sqrt{5}}{2}$$

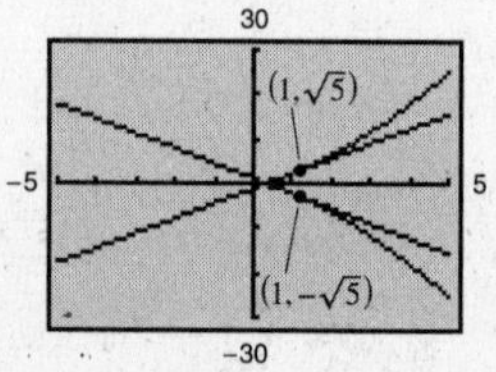

39. $x^3 + y^3 = 8$

$$3x^2 + 3y^2\frac{dy}{dx} = 0$$

$$3y^2\frac{dy}{dx} = -3x^2$$

$$\frac{dy}{dx} = -\frac{x^2}{y^2}$$

At $(0, 2)$:

$$m = \frac{dy}{dx} = 0$$

$$y - 2 = 0(x - 0)$$

$$y = 2$$

At $(2, 0)$:

$m = \frac{dy}{dx}$ is undefined.

The tangent line is $x = 2$.

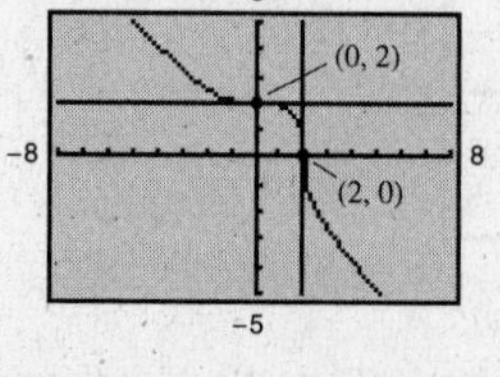

41. $y = \dfrac{2}{0.00001x^3 + 0.1x}, \ x > 0$

$$0.00001x^3 + 0.1x = \frac{2}{y}$$

$$0.00003x^2\frac{dx}{dy} + 0.1\frac{dx}{dy} = -\frac{2}{y^2}$$

$$\left(0.00003x^2 + 0.1\right)\frac{dx}{dy} = -\frac{2}{y^2}$$

$$\frac{dx}{dy} = -\frac{2}{y^2\left(0.00003x^2 + 0.1\right)}$$

43. $y = \sqrt{\dfrac{200 - x}{2x}}, \ 0 < x \le 200$

$$2xy^2 = 200 - x$$

$$2x(2y) + y^2\left(2\frac{dx}{dy}\right) = -\frac{dx}{dy}$$

$$\left(2y^2 + 1\right)\frac{dx}{dy} = -4xy$$

$$\frac{dx}{dy} = -\frac{4xy}{2y^2 + 1}$$

45. $100x^{0.75}y^{0.25} = 135{,}540$

$$100x^{0.75}\left(0.25y^{-0.75}\frac{dy}{dx}\right) + y^{0.25}\left(75x^{-0.25}\right) = 0$$

$$\frac{25x^{0.75}}{y^{0.75}} \cdot \frac{dy}{dx} = -\frac{75y^{0.25}}{x^{0.25}}$$

$$\frac{dy}{dx} = -\frac{3y}{x}$$

When $x = 1500$ and $y = 1000$, $\dfrac{dy}{dx} = -2$.

Appendix B.2 Related Rates

Skills Review

1. $A = \pi r^2$

2. $V = \frac{4}{3}\pi r^3$

3. $SA = 6s^2$

4. $V = s^3$

5. $V = \frac{1}{3}\pi r^2 h$

6. $A = \frac{1}{2}bh$

7.
$$\begin{aligned} x^2 + y^2 &= 9 \\ \frac{d}{dx}\left[x^2 + y^2\right] &= \frac{d}{dx}[9] \\ 2x + 2y\frac{dy}{dx} &= 0 \\ 2y\frac{dy}{dx} &= -2x \\ \frac{dy}{dx} &= \frac{-2x}{2y} \\ &= \frac{-x}{y} \end{aligned}$$

8.
$$\begin{aligned} 3xy - x^2 &= 6 \\ \frac{d}{dx}\left[3xy - x^2\right] &= \frac{d}{dx}[6] \\ 3y + 3x\frac{dy}{dx} - 2x &= 0 \\ 3x\frac{dy}{dx} &= 2x - 3y \\ \frac{dy}{dx} &= \frac{2x - 3y}{3x} \end{aligned}$$

9.
$$\begin{aligned} x^2 + 2y + xy &= 12 \\ \frac{d}{dx}\left[x^2 + 2y + xy\right] &= \frac{d}{dx}(12) \\ 2x + 2\frac{dy}{dx} + y + x\frac{dy}{dx} &= 0 \\ 2\frac{dy}{dx} + x\frac{dy}{dx} &= -y - 2x \\ \frac{dy}{dx}(2 + x) &= -y - 2x \\ \frac{dy}{dx} &= \frac{-y - 2x}{2 + x} \end{aligned}$$

10.
$$\begin{aligned} x + xy^2 - y^2 &= xy \\ \frac{d}{dx}\left[x + xy^2 - y^2\right] &= \frac{d}{dx}[xy] \\ 1 + y^2 + 2xy\frac{dy}{dx} - 2y\frac{dy}{dx} &= y + x\frac{dy}{dx} \\ 2xy\frac{dy}{dx} - 2y\frac{dy}{dx} - x\frac{dy}{dx} &= y - y^2 - 1 \\ \frac{dy}{dx}(2xy - 2y - x) &= y - y^2 - 1 \\ \frac{dy}{dx} &= \frac{y - y^2 - 1}{2xy - 2y - x} \end{aligned}$$

1. $y = \sqrt{x}$, $\frac{dy}{dt} = \frac{1}{2}x^{-1/2}\frac{dx}{dt} = \frac{1}{2\sqrt{x}}\frac{dx}{dt}$,

$\frac{dx}{dt} = 2\sqrt{x}\,\frac{dy}{dt}$

(a) When $x = 4$ and $\frac{dx}{dt} = 3$, $\frac{dy}{dt} = \left(\frac{1}{2\sqrt{4}}\right)(3) = \frac{3}{4}$.

(b) When $x = 25$ and $\frac{dy}{dt} = 2$, $\frac{dx}{dt} = 2\sqrt{25}(2) = 20$.

3. $xy = 4$, $x\frac{dy}{dt} + y\frac{dx}{dt} = 0$, $\frac{dy}{dt} = \left(-\frac{y}{x}\right)\frac{dx}{dt}$,

$\frac{dx}{dt} = \left(-\frac{x}{y}\right)\frac{dy}{dt}$

(a) When $x = 8$, $y = \frac{1}{2}$, and

$\frac{dx}{dt} = 10$, $\frac{dy}{dt} = -\frac{1/2}{8}(10) = -\frac{5}{8}$.

(b) When $x = 1$, $y = 4$, and

$\frac{dy}{dt} = -6$, $\frac{dx}{dt} = -\frac{1}{4}(-6) = \frac{3}{2}$.

5. $A = \pi r^2, \dfrac{dA}{dt} = 2\pi r\dfrac{dr}{dt}$

If $\dfrac{dr}{dt}$ is constant, then $\dfrac{dA}{dt}$ is not constant; $\dfrac{dA}{dt}$ is proportional to r.

7. $V = \dfrac{4}{3}\pi r^3, \dfrac{dV}{dt} = 10, \dfrac{dV}{dt} = 4\pi r^2\dfrac{dr}{dt},$

$\dfrac{dr}{dt} = \left(\dfrac{1}{4\pi r^2}\right)\dfrac{dV}{dt}$

(a) When $r = 1$, $\dfrac{dr}{dt} = \dfrac{1}{4\pi(1)^2}(10) = \dfrac{5}{2\pi}$ ft/min.

(b) When $r = 2$, $\dfrac{dr}{dt} = \dfrac{1}{4\pi(2)^2}(10) = \dfrac{5}{8\pi}$ ft/min.

9. $V = x^3, \dfrac{dx}{dt} = 3, \dfrac{dV}{dt} = 3x^2\dfrac{dx}{dt}$

(a) When $x = 1$, $\dfrac{dV}{dt} = 3(1)^2(3) = 9$ cm³/sec.

(b) When $x = 10$, $\dfrac{dV}{dt} = 3(10)^2(3) = 900$ cm³/sec.

11. Let y be the distance from the ground to the top of the ladder and let x be the distance from the house to the base of the ladder.

$x^2 + y^2 = 25^2, 2x\dfrac{dx}{dt} + 2y\dfrac{dy}{dt} = 0,$

$\dfrac{dy}{dt} = \dfrac{-x}{y}\dfrac{dx}{dt} = -\dfrac{2x}{y}$ because $\dfrac{dx}{dt} = 2.$

(a) When $x = 7$, $y = \sqrt{576} = 24$,

$\dfrac{dy}{dt} = -\dfrac{2(7)}{24} = -\dfrac{7}{12}$ ft/sec.

(b) When $x = 15$, $y = \sqrt{400} = 20$,

$\dfrac{dy}{dt} = -\dfrac{2(15)}{20} = -\dfrac{3}{2}$ ft/sec.

(c) When $x = 24$, $y = \sqrt{576} = 7$,

$\dfrac{dy}{dt} = -\dfrac{2(24)}{7} = -\dfrac{48}{7}$ ft/sec.

13. $V = \pi r^2 h, h = 0.08, V = 0.08\pi r^2, \dfrac{dV}{dt} = 0.16\pi r\dfrac{dr}{dt}$

When $r = 150$ and $\dfrac{dr}{dt} = \dfrac{1}{2}$,

$$\begin{aligned}\frac{dV}{dt} &= 0.16\pi(150)\left(\frac{1}{2}\right)\\ &= 12\pi\\ &= 37.70 \text{ ft}^3/\text{min}.\end{aligned}$$

15. $x^2 + 6^2 = s^2$

$$\begin{aligned}2x\frac{dx}{dt} &= 2s\frac{ds}{dt}\\ \frac{dx}{dt} &= \frac{s}{x}\frac{ds}{dt}\end{aligned}$$

When $s = 10$, $x = 8$ and $\dfrac{ds}{dt} = 240$:

$\dfrac{dx}{dt} = \dfrac{10}{8}(-240) = 300$ mi/hr.

17. $rg \tan\theta = v^2$, $g = 32$, r is a constant.

$$\begin{aligned}rg\sec^2\theta\frac{d\theta}{dt} &= 2v\frac{dv}{dt}\\ 32r\sec^2\theta\frac{d\theta}{dt} &= 2v\frac{dv}{dt}\\ \frac{16r}{v}\sec^2\theta\frac{d\theta}{dt} &= \frac{dv}{dt}\end{aligned}$$

Likewise, $\dfrac{d\theta}{dt} = \dfrac{v}{16r}\cos^2\theta\dfrac{dv}{dt}$.

19. (a) Total volume of pool $= \frac{1}{2}(2)(12)(6) + (1)(6)(12) = 144 \text{ m}^3$

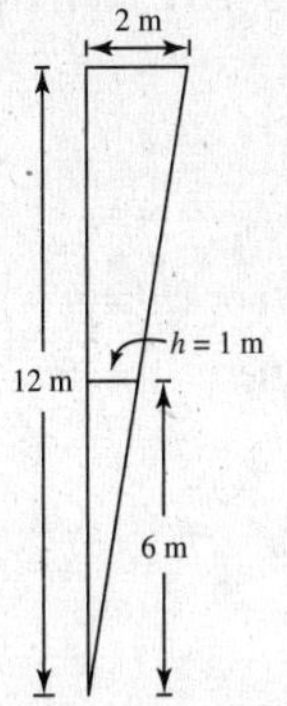

Volume of 1 m of water $= \frac{1}{2}(1)(6)(6) = 18 \text{ m}^3$.

% pool filled $= \frac{18}{144} = 0.125$, or 12.5%

Because for $0 \le h \le 2$, $b = 6h$, you have

$$V = \frac{1}{2}bh(6) = 3bh = 3(6h)(h) = 18h^2.$$

$$\frac{dV}{dt} = 36h\frac{dh}{dt}$$

$$\frac{1}{4} = 36h\frac{dh}{dt}$$

$$\frac{1}{144h} = \frac{dh}{dt}$$

$$\frac{1}{144(1)} = \frac{dh}{dt}$$

$$\frac{1}{144} \text{ m/min} = \frac{dh}{dt}$$

21.

$$pV^{1.3} = k$$

$$1.3pV^{0.3}\frac{dV}{dt} + V^{1.3}\frac{dp}{dt} = 0$$

$$V^{0.3}\left(1.3p\frac{dV}{dt} + V\frac{dp}{dt}\right) = 0$$

$$1.3p\frac{dV}{dt} = -V\frac{dp}{dt}$$

Practice Test Solutions for Chapter 0

1. Rational (Sec. 0.1)

2. (Sec. 0.1)
(a) Satisfies (b) Does not satisfy
(c) Satisfies (d) Satisfies

3. $x \geq 3$ (Sec. 0.1)

4. $-1 < x < 7$ (Sec. 0.1)

5. $\sqrt{19} > \frac{13}{3}$ (Sec. 0.1)

6. (Sec. 0.2)
(a) $d = 10$ (b) Midpoint: 2

7. $-\frac{11}{3} \leq x \leq 3$ (Sec. 0.2)

8. $x < -5$ or $x > \frac{33}{5}$ (Sec. 0.2)

9. $-\frac{25}{2} < x < \frac{55}{2}$ (Sec. 0.2)

10. $|x - 1| \leq 4$ (Sec. 0.2)

11. $3x^5$ (Sec. 0.3)

12. 1 (Sec. 0.3)

13. $2xy\sqrt[3]{4x}$ (Sec. 0.3)

14. $\frac{1}{4}(x + 1)^{-1/3}(x + 7)$ (Sec. 0.3)

15. $x < 5$ (Sec. 0.3)

16. $(3x + 2)(x - 7)$ (Sec. 0.4)

17. $(5x + 9)(5x - 9)$ (Sec. 0.4)

18. $(x + 2)(x^2 - 2x + 4)$ (Sec. 0.4)

19. $-3 \pm \sqrt{11}$ (Sec. 0.4)

20. $-1, 2, 3$ (Sec. 0.4)

21. $\dfrac{-3}{(x - 1)(x + 3)}$ (Sec. 0.5)

22. $\dfrac{x + 13}{2\sqrt{x + 5}}$ (Sec. 0.5)

23. $\dfrac{1}{\sqrt{x}(x + 2)^{3/2}}$ (Sec. 0.5)

24. $\dfrac{3y\sqrt{y^2 + 9}}{y^2 + 9}$ (Sec. 0.5)

25. $-\dfrac{1}{2\left(\sqrt{x} - \sqrt{x + 7}\right)}$ (Sec. 0.5)

26. $-1, 2, 4$ (Sec. 0.4)

Practice Test Solutions for Chapter 1

1. $d = \sqrt{82}$ (Sec. 1.1)

2. Midpoint: $(1, 3)$ (Sec. 1.1)

3. Collinear (Sec. 1.1)

4. $x = \pm 3\sqrt{5}$ (Sec. 1.1)

5. x-intercepts: $(\pm 2, 0)$ (Sec. 1.2)
y-intercept: $(0, 4)$

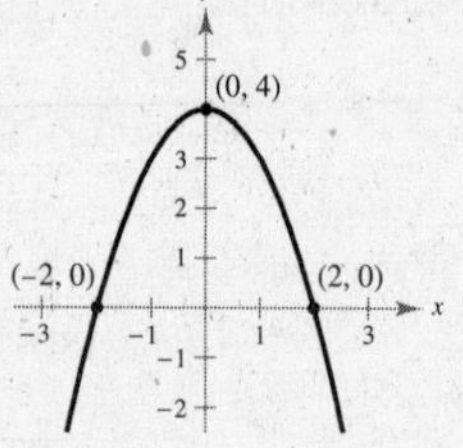

6. x-intercepts: $(2, 0)$ (Sec. 1.2)
No y-intercept

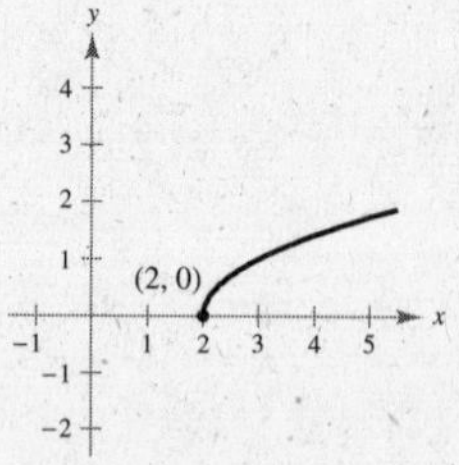

7. x-intercepts: $(3, 0)$ (Sec. 1.2)
y-intercept: $(0, 3)$

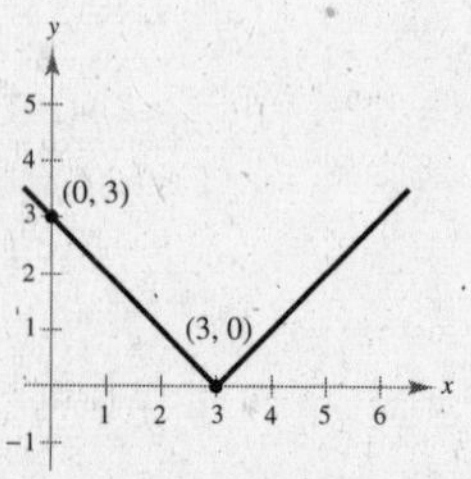

8. $(x-4)^2 + (y+1)^2 = 9$ (Sec. 1.2)

Center: $(4, -1)$

Radius: 3

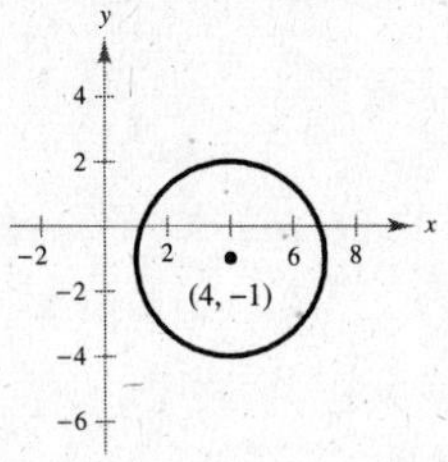

9. $(0, -5)$ and $(4, -3)$ (Sec. 1.2)

10. $6x - y - 38 = 0$ (Sec. 1.3)

11. $2x - 3y + 1 = 0$ (Sec. 1.3)

12. $x - 6 = 0$ (Sec. 1.3)

13. $5x + 2y - 6 = 0$ (Sec. 1.3)

14. (a) 4 (Sec. 1.4)

(b) 31

(c) $x^2 - 10x + 20$

(d) $x^2 + 2x(\Delta x) + (\Delta x)^2 - 5$

15. Domain: $(-\infty, 3]$ (Sec. 1.4)

Range: $[0, \infty)$

16. (a) $2x^2 + 1$ (Sec. 1.4)

(b) $4(x+1)(x+2)$

17. $f^{-1}(x) = \sqrt[3]{x-6}$ (Sec. 1.4)

18. 22 (Sec. 1.5)

19. 12 (Sec. 1.5)

20. Does not exist (Sec. 1.5)

21. $\dfrac{\sqrt{5}}{10}$ (Sec. 1.5)

22. 5 (Sec. 1.5)

23. Discontinuities: $x = \pm 8$ (Sec. 1.6)

$x = 8$ is removable.

24. $x = 3$ is a nonremovable discontinuity. (Sec. 1.6)

25. (Sec. 1.6)

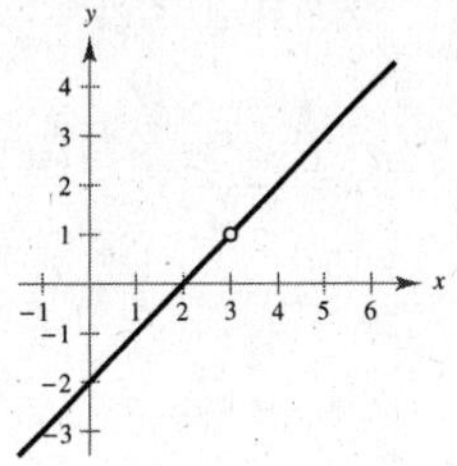

26. $y = \pm\sqrt{-x^2 - 6x - 5}$ (Sec. 1.2)

Domain: $[-5, -1]$

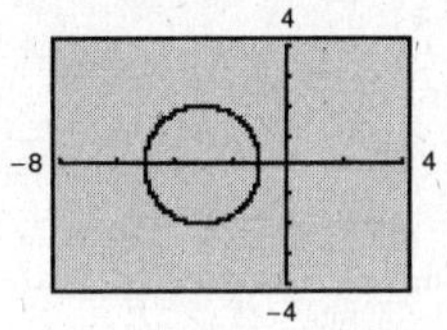

27. The graph does **not** show that the function does not exist at $x = 3$ on many graphing utilities.

$\lim_{x\to 3} f(x) = 6$ (Sec. 1.5)

Practice Test Solutions for Chapter 2

1. $$\lim_{\Delta x\to 0} \frac{f(x+\Delta x) - f(x)}{\Delta x} = \lim_{\Delta x\to 0} (4x + 2\Delta x + 3) = 4x + 3$$

(Sec. 2.1)

2. $$\lim_{\Delta x\to 0} \frac{f(x+\Delta x) - f(x)}{\Delta x} = \lim_{\Delta x\to 0} \frac{-1}{(x+\Delta x-4)(x-4)} = -\frac{1}{(x-4)^2}$$

(Sec. 2.1)

3. $x - 4y + 2 = 0$ (Sec. 2.1)

4. $15x^2 - 12x + 15$ (Sec. 2.2)

5. $\dfrac{4x-2}{x^3}$ (Sec. 2.2)

6. $\dfrac{2}{3\sqrt[3]{x}} + \dfrac{3}{5\sqrt[5]{x^2}}$ (Sec. 2.2)

7. (Sec. 2.3)

Average rate of change: 4

Instantaneous rates of change: $f'(0) = 0$, $f'(2) = 12$

8. (Sec. 2.3)

Marginal cost: $4.31 - 0.0002x$

9. $5x^4 + 28x^3 - 39x^2 - 56x + 36$ (Sec. 2.4)

10. $-\dfrac{x^2 + 14x + 8}{\left(x^2 - 8\right)^2}$ (Sec. 2.4)

11. $\dfrac{3x^4 + 14x^3 - 45x^2}{(x + 5)^2}$ (Sec. 2.4)

12. $-\dfrac{3x^2 + 4x + 1}{2\sqrt{x}\left(x^2 + 4x - 1\right)^2}$ (Sec. 2.4)

13. $72(6x - 5)^{11}$ (Sec. 2.5)

14. $-\dfrac{12}{\sqrt{4 - 3x}}$ (Sec. 2.5)

15. $\dfrac{18x}{\left(x^2 + 1\right)^4}$ (Sec. 2.5)

16. $\dfrac{\sqrt{10x}}{x(x + 2)^{3/2}}$ (Sec. 2.5)

17. $24x - 54$ (Sec. 2.6)

18. $-\dfrac{15}{16(3 - x)^{7/2}}$ (Sec. 2.6)

19. (Sec. 2.4)

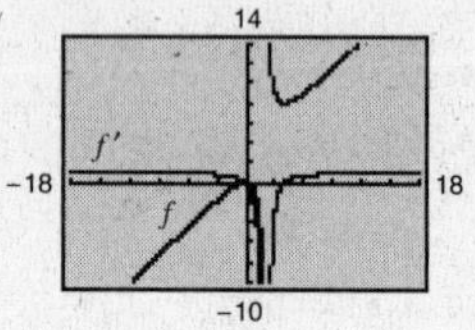

Horizontal Tangents at $x = 0$ and $x = 4$.

$f'(0) = f'(4) = 0$

Practice Test Solutions for Chapter 3

1. Increasing: $(-\infty, 0), (4, \infty)$

Decreasing: $(0, 4)$

(Sec. 3.1)

2. Increasing: $\left(-\infty, \frac{2}{3}\right)$

Decreasing: $\left(\frac{2}{3}, 1\right)$

(Sec. 3.1)

3. Relative minimum: $(2, -45)$ (Sec. 3.2)

4. Relative minimum: $(-3, 0)$ (Sec. 3.2)

5. Maximum: $(5, 0)$ (Sec. 3.2)

Minimum: $(2, -9)$

6. No inflection points (Sec. 3.3)

7. Points of inflection: (Sec. 3.3)

$\left(-\dfrac{1}{\sqrt{3}}, \dfrac{1}{4}\right), \left(\dfrac{1}{\sqrt{3}}, \dfrac{1}{4}\right)$

8. $S = x + \dfrac{600}{x}$ (Sec. 3.4)

First number: $10\sqrt{6}$

Second number: $\dfrac{10\sqrt{6}}{3}$

9. $A = 3xy = 3x\left(\dfrac{3000 - 6x}{4}\right)$

$3x = 750$ feet, $y = 375$ feet

(Sec. 3.4)

10. $-\infty$ (Sec. 3.5)

11. -2 (Sec. 3.5)

12. (Sec. 3.6)

Intercept: $(0\ 0)$

Vertical asymptote: $x = \pm 3$

Horizontal asymptote: $y = 1$

Relative maximum: $(0, 0)$

No inflection points

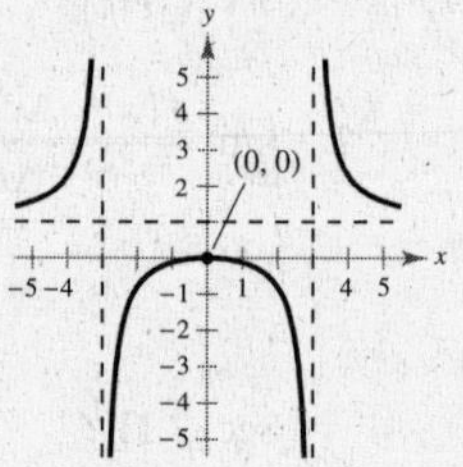

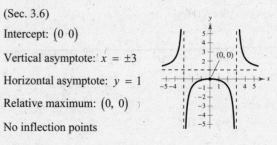

13. (Sec. 3.6)

Intercepts: $(-2, 0)$, $\left(0, \frac{2}{5}\right)$

Horizontal asymptote: $y = 0$

Relative maximum: $\left(1, \frac{1}{2}\right)$

Relative minimum: $\left(-5, -\frac{1}{10}\right)$

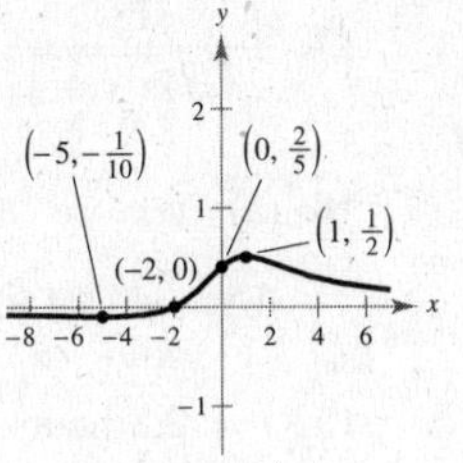

14. (Sec. 3.6)

Intercept: $(0, -1)$

No relative extrema

Inflection point: $(-1, -2)$

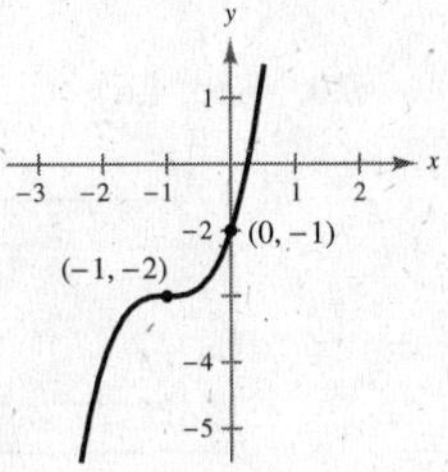

15. (Sec. 3.6)

Intercepts: $(0, 4)$, $(2, 0)$

Relative minimum: $(2, 0)$

No inflection points

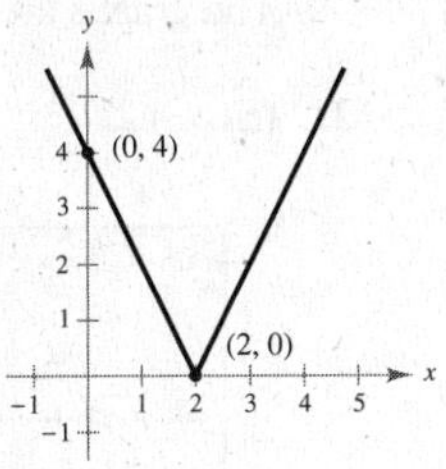

16. (Sec. 3.6)

Intercepts: $(2, 0)$, $\left(0, \sqrt[3]{4}\right)$

Relative minimum: $(2, 0)$

No inflection points

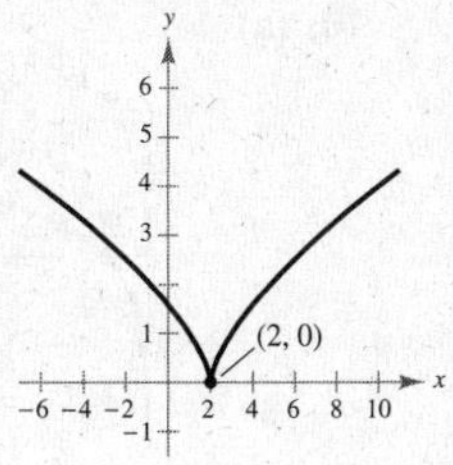

17. $\sqrt[3]{65} \approx 4.0208$ (Sec. 3.7)

18. (Sec. 3.3)

Point of inflection: (7.79, 296)

During the first eight years, the pancreas transplant rate was increasing, but near the end of the seventh year the rate began to decrease.

19. (Sec. 3.6)

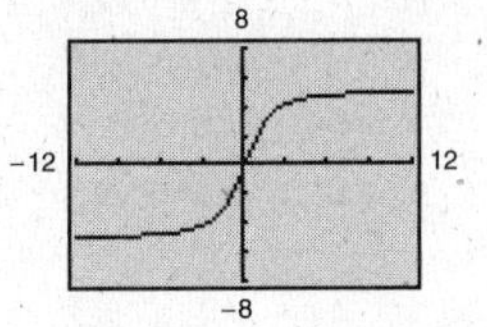

Horizontal asymptotes at $y = \pm 5$.

No relative extrema

20. (Sec. 3.6)

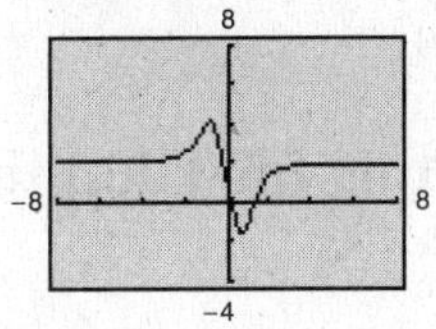

Yes, the graph crosses the horizontal asymptote $y = 2$.

Practice Test Solutions for Chapter 4

1. (a) 81 (Sec. 4.1)

(b) $\frac{1}{32}$

(c) 1

2. (a) $x = 2$ (Sec. 4.1)

(b) $x = 32$

(c) $x = 5$

3. (a)

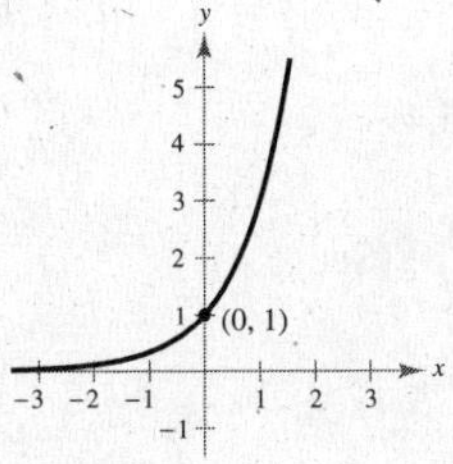

(b)

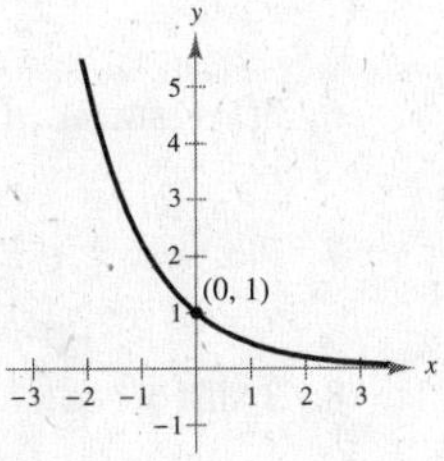

(Sec. 4.1)

4. $6xe^{3x^2}$ (Sec. 4.3)

5. $\dfrac{e^{\sqrt[3]{x}}}{3\sqrt[3]{x^2}}$ (Sec. 4.3)

6. $\dfrac{e^x - e^{-x}}{2\sqrt{e^x + e^{-x}}}$ (Sec. 4.3)

7. $x^2e^{2x}(2x + 3)$ (Sec. 4.3)

8. $\dfrac{xe^x - e^x - 3}{4x^2}$ (Sec. 4.3)

9. $e^{1.6094...} = 5$ (Sec. 4.4)

10. (a)

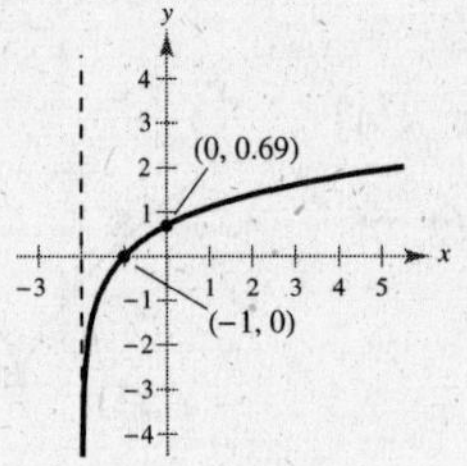

(b)

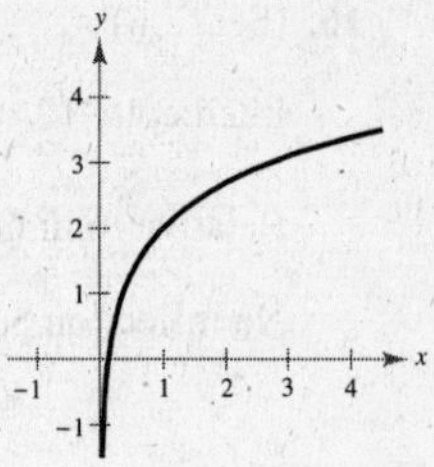

(Sec. 4.4)

11. (a) $\ln\left(\dfrac{3x+1}{2x-5}\right)$ (Sec. 4.4)

(b) $\ln\left(\dfrac{x^4}{y^3\sqrt{z}}\right)$

12. (a) $x = e^{17}$ (Sec. 4.4)

(b) $x = \dfrac{\ln 2}{3 \ln 5}$

13. $\dfrac{6}{6x-7}$ (Sec. 4.5)

14. $\dfrac{4x+15}{x(2x+5)}$ (Sec. 4.5)

15. $\dfrac{1}{x(x+3)}$ (Sec. 4.5)

16. $x^3(1 + 4 \ln x)$ (Sec. 4.5)

17. $\dfrac{1}{2x\sqrt{\ln x + 1}}$ (Sec. 4.5)

18. (a) $y = 7e^{-0.7611t}$ (Sec. 4.6)

(b) $y = 0.1501e^{0.4970t}$

19. $t \approx 2.8$ hours (Sec. 4.2)

$t \approx 4.4$ hours

20. (Sec. 4.6)

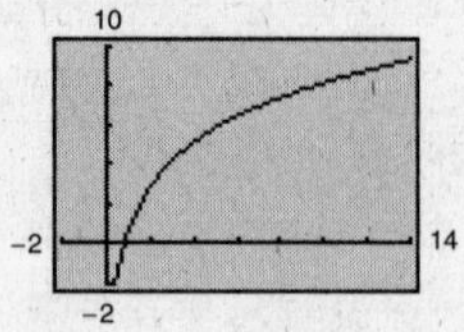

The graphs are the same.

21. (Sec. 4.2)

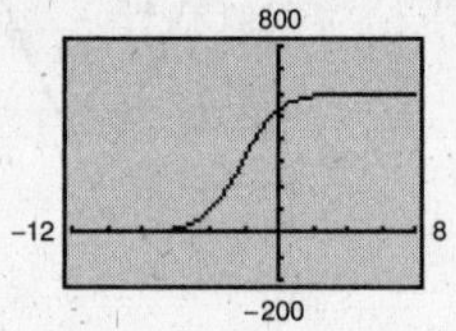

$\lim_{t\to\infty} f(t) = 600$

$\lim_{x\to-\infty} f(t) = 0$

Practice Test Solutions for Chapter 5

1. (a) 93.913° (Sec. 5.1)

(b) $\dfrac{7\pi}{12}$

2. (a) 140°, −580° (Sec. 5.1)

(b) $\dfrac{25\pi}{9}, -\dfrac{11\pi}{9}$

3. $\sin\theta = \dfrac{y}{r} = -\dfrac{5}{13}$ $\quad\csc\theta = \dfrac{r}{y} = -\dfrac{13}{5}$

$\cos\theta = \dfrac{x}{r} = \dfrac{12}{13}$ $\quad\sec\theta = \dfrac{r}{x} = \dfrac{13}{12}$

$\tan\theta = \dfrac{y}{x} = -\dfrac{5}{12}$ $\quad\cot\theta = \dfrac{x}{y} = -\dfrac{12}{5}$

(Sec. 5.2)

4. $\theta = 0, \dfrac{\pi}{2}, \dfrac{3\pi}{2}$ (Sec. 5.2)

5. (Sec. 5.3)

(a) Period: 8π

Amplitude: 3

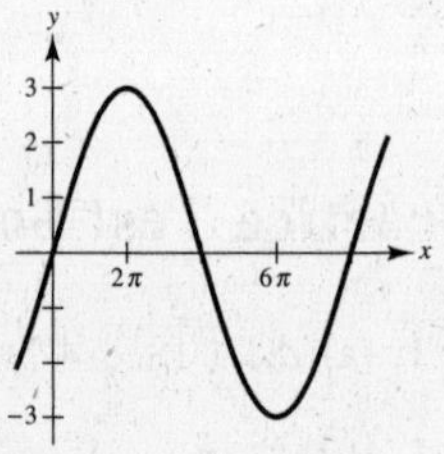

(b) Period: $\frac{1}{2}$

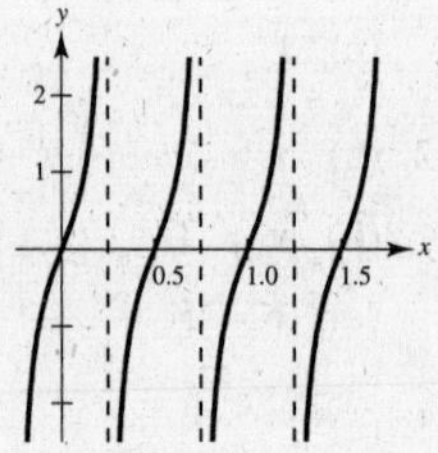

6. $3(1 + \sin x)$ (Sec. 5.4)

7. $x(x \sec^2 x + 2 \tan x)$ (Sec. 5.4)

8. $3 \sin^2 x \cos x$ (Sec. 5.4)

9. $\dfrac{\sec x(x \tan x - 2)}{x^3}$ (Sec. 5.4)

10. $5 \cos 10x$ (Sec. 5.4)

11. $-\frac{1}{2}\sqrt{\csc x}\cot x$ (Sec. 5.4)

12. $\sec x$ (Sec. 5.4)

13. $-2e^{2x} \csc^2 e^{2x}$ (Sec. 5.4)

14.

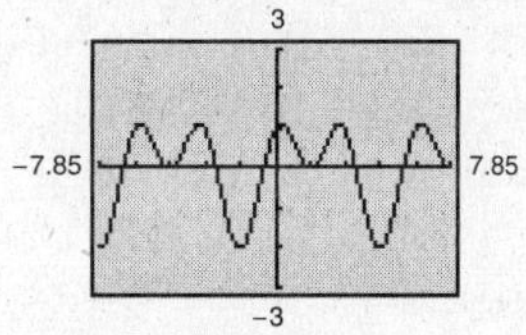

Minimum: -2

Maximum: 1.125

(Sec. 5.3)

Practice Test Solutions for Chapter 6

1. $x^3 - 4x^2 + 5x + C$ (Sec. 6.1)

2. $\dfrac{x^4}{4} + \dfrac{7x^3}{3} - 2x^2 - 28x + C$ (Sec. 6.1)

3. $\dfrac{x^2}{2} - 9x - \dfrac{1}{x} + C$ (Sec. 6.1)

4. $-\frac{1}{5}(1 - x^4)^{5/4} + C$ (Sec. 6.2)

5. $\frac{9}{14}(7x)^{2/3} + C$ (Sec. 6.2)

6. $-\frac{2}{33}(6 - 11x)^{3/2} + C$ (Sec. 6.2)

7. $\frac{4}{5}x^{5/4} + \frac{6}{7}x^{7/6} + C$ (Sec. 6.1)

8. $-\dfrac{1}{3x^3} + \dfrac{1}{4x^4} + C$ (Sec. 6.1)

9. $x - x^3 + \frac{3}{5}x^5 - \frac{1}{7}x^7 + C$ (Sec. 6.1)

10. $-\dfrac{5}{12(1 + 3x^2)^2} + C$ (Sec. 6.2)

11. $\left(\frac{1}{7}\right)e^{7x} + C$ (Sec. 6.3)

12. $\left(\frac{1}{8}\right)e^{4x^2} + C$ (Sec. 6.3)

13. $\left(\frac{1}{16}\right)(1 + 4e^x)^4 + C$ (Sec. 6.3)

14. $\left(\frac{1}{2}\right)e^{2x} + 4e^x + 4x + C$ (Sec. 6.3)

15. $\left(\frac{1}{2}\right)e^{2x} - 4x - e^{-x} + C$ (Sec. 6.3)

16. $\ln|x + 6| + C$ (Sec. 6.3)

17. $-\left(\frac{1}{3}\right) \ln|8 - x^3| + C$ (Sec. 6.3)

18. $\frac{1}{3} \ln(1 + 3e^x) + C$ (Sec. 6.3)

19. $\dfrac{(\ln x)^7}{7} + C$ (Sec. 6.3)

20. $\dfrac{x^2}{2} + x + 6 \ln|x - 1| + C$ (Sec. 6.3)

(Use long division first)

21. -3 (Sec. 6.4)

22. $\frac{381}{7}$ (Sec. 6.4)

23. 2 (Sec. 6.4)

24. $A = 36$ (Sec. 6.5)

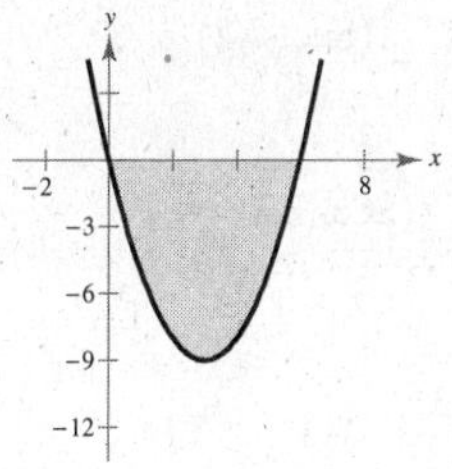

25. $A = \frac{1}{2}$ (Sec. 6.5)

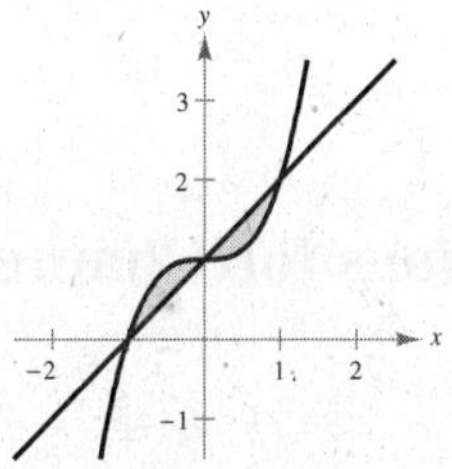

26. $A = \frac{2}{3}$ (Sec. 6.5)

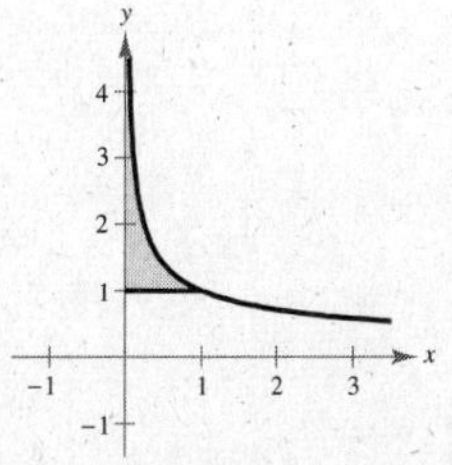

27. $V = 3\pi$ (Sec 6.6)

28. $V = \dfrac{5000\pi}{3}$ (Sec. 6.6)

29. (c) Actual area is 4.5.

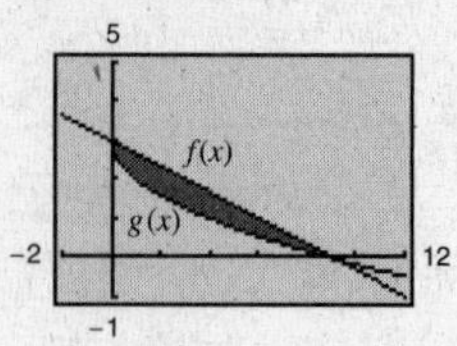

(Sec. 6.5)

Practice Test Solutions for Chapter 7

1. $\frac{1}{4}e^{2x}(2x - 1) + C$ (Sec. 7.1)

2. $\dfrac{x^4}{16}\left[4(\ln x) - 1\right] + C$ (Sec. 7.1)

3. $\frac{2}{35}(x - 6)^{3/2}(5x^2 + 24x + 96) + C$ (Sec. 7.1)

4. $\frac{1}{32}e^{4x}(8x^2 - 4x + 1) + C$ (Sec. 7.1)

5. $\ln\left|\dfrac{x + 3}{x - 2}\right| + C$ (Sec. 7.2)

6. $\ln\left|\dfrac{x^3}{(x + 4)^2}\right| + C$ (Sec. 7.2)

7. $5 \ln|x + 2| + \dfrac{7}{x + 2} + C$ (Sec. 7.2)

8. $\dfrac{3}{2}x^2 + \ln\dfrac{|x|}{(x + 3)^2} + C$ (Sec. 7.2)

9. $-8 \cot \dfrac{x}{8} + C$ (Sec. 7.3)

10. $\dfrac{\sin^6 x}{6} + C$ (Sec. 7.3)

11. $\dfrac{\pi^2 - 8\sqrt{2}}{16}$ (Sec. 7.3)

12. 1.4949 (Sec. 7.4)

13. 0.1472 (Sec. 7.4)

14. (a) 15.567 (Sec. 7.5)
    (b) 15.505

15. (a) 1.191 (Sec. 7.5)
    (b) 1.196

16. Convergent; 6 (Sec. 7.6)

17. Divergent (Sec. 7.6)

18. Divergent (Sec. 7.6)

19. $n = 100$: 8.935335 (Sec. 7.5 and 7.6)
    $n = 1000$: 2.288003
    $n = 10{,}000$: 1.636421
    Converges $\left(\text{Actual answer is } \dfrac{\pi}{2}.\right)$

20. $n = 50$: 22.442278 (Sec. 7.4)
    $n = 100$: 22.443875

21. $n = 50$: 1.652674 (Sec. 7.5)
    $n = 100$: 1.652674

Practice Test Solutions for Chapter 8

1. $(4, -3)$ (Sec. 8.1)

2. $\left(-1, \frac{2}{3}\right)$ (Sec. 8.1)

3. 40, 70 (Sec. 8.1)

4. 60 ft by 25 ft (Sec. 8.1)

5. $(17, -2)$ (Sec. 8.1)

6. $\left(\frac{15}{19}, \frac{23}{19}\right)$ (Sec. 8.1)

7. $(0.03, 0.2)$ (Sec. 8.1)

8. $(-2, 5, 3)$ (Sec. 8.2)

9. \$6500 at 11% (Sec. 8.1)
   \$10,500 at 13%

10. $(3, -5, 0)$ (Sec. 8.2)

11. $\left(-\frac{3}{5}a + \frac{3}{5}, \frac{7}{5}a + \frac{8}{5}, a\right)$ (Sec. 8.2)

12. $y = 2x^2 + 3x - 1$ (Sec. 8.2)

13. $\begin{bmatrix} 1 & 0 & -2 \\ 0 & 1 & -3 \end{bmatrix}$ (Sec. 8.3)

14. $(-4, 3)$ (Sec. 8.3)

15. $(6, -5)$ (Sec. 8.3)

16. $(1, -2, -2)$ (Sec. 8.3)

17. $\begin{bmatrix} -4 & -12 \\ 5 & 6 \end{bmatrix}$ (Sec. 8.4)

18. $\begin{bmatrix} -3 & 13 \\ -27 & -1 \end{bmatrix}$ (Sec. 8.4)

19. False; Matrix multiplication is not commutative. Specifically, $BA + 3AB \neq 4AB$. (Sec. 8.4)

20. $\begin{bmatrix} -5 & 2 \\ 3 & -1 \end{bmatrix}$ (Sec. 8.5)

21. $\begin{bmatrix} 1 & -1 & \frac{1}{2} \\ -3 & -1 & 1 \\ 3 & 2 & -\frac{3}{2} \end{bmatrix}$ (Sec. 8.5)

22. (Sec. 8.5)

(a) $(-18, 11)$

(b) $(-19, 11)$

Practice Test Solutions for Chapter 9

1. (a) $d = 14\sqrt{2}$ (Sec. 9.1)

(b) Midpoint: $(4, 2, -2)$

2. $(x - 1)^2 + (y + 3)^2 + z^2 = 5$ (Sec. 9.1)

3. Center: $(2, -1, -4)$ (Sec. 9.1)

Radius: $\sqrt{21}$

4. (Sec. 9.2)

(a) x-intercept: $(8, 0, 0)$

y-intercept: $(0, 3, 0)$

z-intercept: $(0, 0, 4)$

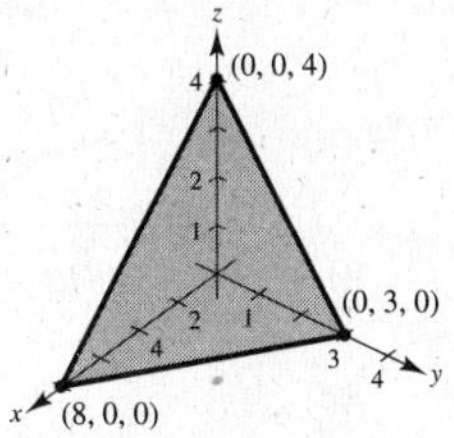

(b) $y = 2$

Parallel to xz-plane

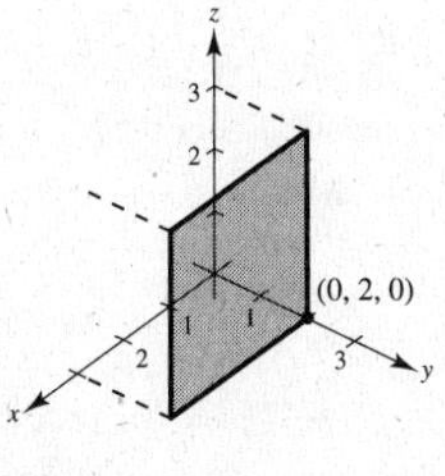

5. (Sec. 9.2)

(a) Hyperboloid of one sheet

(b) Elliptic paraboloid

6. (a) Domain: $x + y < 3$ (Sec. 9.3)

(b) Domain: all points in the xy-plane except the origin

7. $f_x(x, y) = 6x + 9y^2 - 3$ (Sec. 9.4)

$f_y(x, y) = 18xy + 12y^2 - 6$

8. $f_x(x, y) = \dfrac{2x}{x^2 + y^2 + 5}$ (Sec. 9.4)

$f_y(x, y) = \dfrac{2y}{x^2 + y^2 + 5}$

9. $\dfrac{\partial w}{\partial x} = 2xy^3\sqrt{z}$ (Sec. 9.4)

$\dfrac{\partial w}{dy} = 3x^2y^2\sqrt{z}$

$\dfrac{\partial w}{dz} = \dfrac{x^2y^3}{2\sqrt{z}}$

10. $\dfrac{\partial^2 z}{\partial x^2} = 2x\left(\dfrac{x^2 - 3y^2}{(x^2 + y^2)^3}\right)$ (Sec. 9.4)

$\dfrac{\partial^2 z}{\partial y \partial x} = 2y\left(\dfrac{3x^2 - y^2}{(x^2 + y^2)^3}\right)$

$\dfrac{\partial^2 z}{\partial x \partial y} = 2y\left(\dfrac{3x^2 - y^2}{(x^2 + y^2)^3}\right)$

$\dfrac{\partial^2 z}{\partial y^2} = 2x\left(\dfrac{3y^2 - x^2}{(x^2 + y^2)^3}\right)$

11. Relative minimum: $(1, -2, -23)$ (Sec. 9.5)

12. Saddle point: $(0, 0, 0)$ (Sec. 9.5)

Relative maxima: $(1, 1, 2), (-1, -1, 2)$

13. $y = \frac{1}{65}(-51x + 355)$ (Sec. 9.6)

14. $y = \frac{1}{6}x^2 - \frac{7}{26}x + \frac{7}{3}$ (Sec. 9.6)

15. $\frac{81}{16}$ (Sec. 9.7)

16. $-\frac{135}{4}$ (Sec. 9.7)

17. (Sec. 9.7)

(a) $A = \int_{-2}^{2} \int_{3}^{7-x^2} dy\,dx = \int_{3}^{7} \int_{-\sqrt{7-y}}^{\sqrt{7-y}} dx\,dy$

(b) $A = \int_{0}^{1} \int_{x^2+2}^{x+2} dy\,dx = \int_{2}^{3} \int_{y-2}^{\sqrt{y-2}} dx\,dy$

18. $V = \int_{0}^{4} \int_{0}^{4-x} (4 - x - y)\,dy\,dx = \frac{32}{3}$ (Sec. 9.8)

19. $y \approx 0.832t + 20.432$ (Sec. 9.6)

$r \approx 0.983$

20. 1.028531×10^{17} (Sec. 9.7)

Practice Test Solutions for Chapter 10

1. $y' = x + 2x \ln x$

$y'' = 3 + 2 \ln x$

$xy'' - y' = 2x$ (Sec. 10.1)

2. $y' = -C_1e^{-x} - C_2xe^{-x} + C_2e^{-x}$

$y'' = C_1e^{-x} + C_2xe^{-x} - 2C_2e^{-x}$

$y'' + 2y' + y = 0$ (Sec. 10.1)

3. $y = -3e^{-x} - xe^{-x}$ (Sec. 10.1)

4. $y = -\dfrac{1}{\ln|x + 1| + C}$ (Sec. 10.2)

5. $y = -\dfrac{1}{\ln|x + 1| - 1}$ (Sec. 10.2)

6. $y = (x + C)e^{x^3}$ (Sec. 10.3)

7. $S = 60 + 40e^{-0.5} \Rightarrow \$84,261$ (Sec. 10.3)

8. $N = 10,000\left(1 - e^{[(\ln 0.8)/2]t}\right)$ (Sec. 10.3)

9. about 5.53 minutes (Sec. 10.4)

10. $P = 35,000e^{0.2t} - 5000$ (Sec. 10.4)

Practice Test Solutions for Chapter 11

1. $\frac{11}{16}$ (Sec. 11.1)

2. $\frac{5}{13}$ (Sec. 11.1)

3. $E(x) = \frac{1}{2}$ (Sec. 11.1)

$V(x) = 4.05$

$\sigma \approx 2.012$

4. $k = \frac{1}{4}$ (Sec. 11.2)

5. (a) $\frac{25}{64}$ (Sec. 11.2)

(b) $\frac{63}{64}$

6. (a) 4 (Sec. 11.3)

(b) $\dfrac{4\sqrt{5}}{5}$

(c) 4

7. (a) $6 \ln\left(\frac{3}{2}\right) \approx 2.433$ (Sec. 11.3)

(b) $\sqrt{6 - 36\left(\ln \frac{3}{2}\right)^2} \approx 0.286$

(c) $\frac{12}{5}$

8. $\mu = \dfrac{1}{7}$ (Sec. 11.3)

Median: $\dfrac{\ln 2}{7}$

$\sigma = \dfrac{1}{7}$

9. 0.0469 (Sec. 11.3)

10. $P(19 < x < 24) = 0.4401$ (Sec. 11.3)